MW00837249

Nonlinear Programming

SECOND EDITION

Dimitri P. Bertsekas

Massachusetts Institute of Technology

WWW site for book information and orders

http://world.std.com/~athenasc/index.html

Athena Scientific, Belmont, Massachusetts

Athena Scientific
Post Office Box 391
Belmont, Mass. 02178-9998
U.S.A.

Email: athenasc@world.std.com
WWW: http://world.std.com/~athenasc/index.html

Cover Design: *Ann Gallager*

Publisher's Cataloging-in-Publication Data

Bertsekas, Dimitri P.
Nonlinear Programming: Second Edition
Includes bibliographical references and index
1. Nonlinear Programming. 2. Mathematical Optimization. I. Title.
T57.8.B47 1999 519.703 99-73208

ISBN 1-886529-00-0

ABOUT THE AUTHOR

Dimitri Bertsekas studied Mechanical and Electrical Engineering at the National Technical University of Athens, Greece, and obtained his Ph.D. in system science from the Massachusetts Institute of Technology. He has held faculty positions with the Engineering-Economic Systems Dept., Stanford University and the Electrical Engineering Dept. of the University of Illinois, Urbana. He is currently Professor of Electrical Engineering and Computer Science at the Massachusetts Institute of Technology. He consults regularly with private industry and has held editorial positions in several journals. He has been elected Fellow of the IEEE.

Professor Bertsekas has done research in a broad variety of subjects from optimization theory, control theory, parallel and distributed computation, data communication networks, and systems analysis. He has written numerous papers in each of these areas.

Other books by the author:

1) *Dynamic Programming and Stochastic Control*, Academic Press, 1976.
2) *Stochastic Optimal Control: The Discrete-Time Case*, Academic Press, 1978; republished by Athena Scientific, 1997 (with S. E. Shreve; translated in Russian).
3) *Constrained Optimization and Lagrange Multiplier Methods*, Academic Press, 1982; republished by Athena Scientific, 1996 (translated in Russian).
4) *Dynamic Programming: Deterministic and Stochastic Models*, Prentice-Hall, 1987.
5) *Data Networks*, Prentice-Hall, 1987; 2nd Edition 1992 (with R. G. Gallager; translated in Russian and Japanese).
6) *Parallel and Distributed Computation: Numerical Methods*, Prentice-Hall, 1989; republished by Athena Scientific, 1997 (with J. N. Tsitsiklis).
7) *Linear Network Optimization: Algorithms and Codes*, M.I.T. Press, 1991.
8) *Dynamic Programming and Optimal Control*, (2 Vols.), Athena Scientific, 1995.
9) *Neuro-Dynamic Programming*, Athena Scientific, 1996 (with J. N. Tsitsiklis).
10) *Network Optimization: Continuous and Discrete Models*, Athena Scientific, 1998.

ATHENA SCIENTIFIC

OPTIMIZATION AND COMPUTATION SERIES

1. Dynamic Programming and Optimal Control, Vols. I and II, by Dimitri P. Bertsekas, 1995, ISBN 1-886529-11-6, 704 pages
2. Nonlinear Programming, Second Edition, by Dimitri P. Bertsekas, 1999, ISBN 1-886529-00-0, 791 pages
3. Neuro-Dynamic Programming, by Dimitri P. Bertsekas and John N. Tsitsiklis, 1996, ISBN 1-886529-10-8, 512 pages
4. Constrained Optimization and Lagrange Multiplier Methods, by Dimitri P. Bertsekas, 1996, ISBN 1-886529-04-3, 410 pages
5. Stochastic Optimal Control: The Discrete-Time Case by Dimitri P. Bertsekas and Steven E. Shreve, 1996, ISBN 1-886529-03-5, 330 pages
6. Introduction to Linear Optimization by Dimitris Bertsimas and John N. Tsitsiklis, 1997, ISBN 1-886529-19-1, 608 pages
7. Parallel and Distributed Computation: Numerical Methods by Dimitri P. Bertsekas and John N. Tsitsiklis, 1997, ISBN 1-886529-01-9, 718 pages
8. Network Flows and Monotropic Optimization by R. Tyrrell Rockafellar, 1998, ISBN 1-886529-06-X, 634 pages
9. Network Optimization: Continuous and Discrete Models by Dimitri P. Bertsekas, 1998, ISBN 1-886529-02-7, 608 pages

Contents

Preface

Nonlinear programming is a mature field that has experienced major developments in the last ten years. The first such development is the merging of linear and nonlinear programming algorithms through the use of interior point methods. This has resulted in a profound rethinking of how we solve linear programming problems, and in a major reassessment of how we treat constraints in nonlinear programming. A second development, less visible but still important, is the increased emphasis on large-scale problems, and the associated algorithms that take advantage of problem structure as well as parallel hardware. A third development has been the extensive use of iterative unconstrained optimization to solve the difficult least squares problems arising in the training of neural networks. As a result, simple gradient-like methods and stepsize rules have attained increased importance.

The purpose of this book is to provide an up-to-date, comprehensive, and rigorous account of nonlinear programming at the beginning graduate student level. In addition to the classical topics, such as descent algorithms, Lagrange multiplier theory, and duality, some of the important recent developments are covered: interior point methods for linear and nonlinear programs, major aspects of large-scale optimization, and least squares problems and neural network training.

A further noteworthy feature of the book is that it treats Lagrange multipliers and duality using two different and complementary approaches: a variational approach based on the implicit function theorem, and a convex analysis approach based on geometrical arguments. The former approach applies to a broader class of problems, while the latter is more elegant and more powerful for the convex programs to which it applies.

The chapter-by-chapter description of the book follows:

Chapter 1: This chapter covers unconstrained optimization: main concepts, optimality conditions, and algorithms. The material is classic, but there are discussions of topics frequently left untreated, such as the behavior of algorithms for singular problems, neural network training, and discrete-time optimal control.

Chapter 2: This chapter treats constrained optimization over a convex set without the use of Lagrange multipliers. I prefer to cover this material before dealing with the complex machinery of Lagrange multipliers because I have found that students absorb easily algorithms such as conditional gradient, gradient projection, and coordinate descent, which can be viewed as natural extensions of unconstrained descent algorithms. This chapter contains also a treatment of the affine scaling method for linear programming.

Chapter 3: This chapter gives a detailed treatment of Lagrange multipliers, the associated necessary and sufficient conditions, and sensitivity analysis. The first three sections deal with nonlinear equality and inequality constraints. The last section deals with linear constraints and develops a simple form of duality theory for linearly constrained problems with differentiable cost, including linear and quadratic programming.

Chapter 4: This chapter treats constrained optimization algorithms that use penalties and Lagrange multipliers, including barrier, augmented Lagrangian, sequential quadratic programming, and primal-dual interior point methods for linear programming. The treatment is extensive, and borrows from my 1982 research monograph on Lagrange multiplier methods.

Chapter 5: This chapter provides an in-depth coverage of duality theory (Lagrange and Fenchel). The treatment is totally geometric, and everything is explained in terms of intuitive figures.

Chapter 6: This chapter deals with large-scale optimization methods based on duality. Some material is borrowed from my Parallel and Distributed Algorithms book (coauthored by John Tsitsiklis), but there is also an extensive treatment of nondifferentiable optimization, including subgradient, ϵ-subgradient, and cutting plane methods. Decomposition methods such as Dantzig-Wolfe and Benders are also discussed.

Appendixes: Four appendixes are given. The first gives a summary of calculus, analysis, and linear algebra results used in the text. The second is a fairly extensive account of convexity theory, including proofs of the basic polyhedral convexity results on extreme points and Farkas' lemma, as well the basic facts about subgradients. The third appendix covers one-dimensional minimization methods. The last appendix discusses an implementation of Newton's method for unconstrained optimization.

Inevitably, some coverage compromises had to be made. The subject of nonlinear optimization has grown so much that leaving out a number of important topics could not be avoided. For example, a discussion of variational inequalities, a deeper treatment of optimality conditions, and a more detailed development of Quasi-Newton methods are not provided. Also, a larger number of sample applications would have been desirable. I hope that instructors will supplement the book with the type of practical examples that their students are most familiar with.

The book was developed through a first-year graduate course that I taught at the Univ. of Illinois and at M.I.T. over a period of 20 years. The mathematical prerequisites are matrix-vector algebra and advanced calculus, including a good understanding of convergence concepts. A course in analysis and/or linear algebra should also be very helpful, and would provide the mathematical maturity needed to follow and to appreciate the mathematical reasoning used in the book. Some of the sections in the book may be ommited at first reading without loss of continuity. These sections have been marked by a star. The rule followed here is that the material discussed in a starred section is not used in a non-starred section.

The book can be used to teach several different types of courses.

(a) A two-quarter course that covers most sections of every chapter.

(b) A one-semester course that covers Chapter 1 except for Section 1.9, Chapter 2 except for Sections 2.4 and 2.5, Chapter 3 except for Section 3.4, Chapter 4 except for parts of Sections 4.2 and 4.3, the first three sections of Chapter 5, and a selection from Section 5.4 and Chapter 6. This is the course I usually teach at MIT.

(c) A one-semester course that covers most of Chapters 1, 2, and 3, and selected algorithms from Chapter 4. I have taught this type of course several times. It is less demanding of the students because it does not require the machinery of convex analysis, yet it still provides a fairly powerful version of duality theory (Section 3.4).

(d) A one-quarter course that covers selected parts of Chapters 1, 2, 3, and 4. This is a less comprehensive version of (c) above.

(e) A one-quarter course on convex analysis and optimization that starts with Appendix B and covers Sections 1.1, 2.1, 3.4, and Chapter 5.

There is a very extensive literature on nonlinear programming and to give a complete bibliography and a historical account of the research that led to the present form of the subject would have been impossible. I thus have not attempted to compile a comprehensive list of original contributions to the field. I have cited sources that I have used extensively, that provide important extensions to the material of the book, that survey important topics, or that are particularly well suited for further reading. I have also cited selectively a few sources that are historically significant, but the reference list is far from exhaustive in this respect. Generally, to aid researchers in the field, I have preferred to cite surveys and textbooks for subjects that are relatively mature, and to give a larger number of references for relatively recent developments.

Finally, I would like to express my thanks to a number of individuals for their contributions to the book. My conceptual understanding of the subject was formed at Stanford University while I interacted with David

Luenberger and I taught using his books. This experience had a lasting influence on my thinking. My research collaboration with several colleagues, particularly Joe Dunn, Eli Gafni, Paul Tseng, and John Tsitsiklis, were very useful and are reflected in the book. I appreciate the suggestions and insights of a number of people, particularly David Castanon, Joe Dunn, Terry Rockafellar, Paul Tseng, and John Tsitsiklis. I am thankful to the many students and collaborators whose comments led to corrections and clarifications. Steve Patek, Serap Savari, and Cynara Wu were particularly helpful in this respect. David Logan, Steve Patek, and Lakis Polymenakos helped me to generate the graph of the cover, which depicts the cost function of a simple neural network training problem. My wife Joanna cheered me up with her presence and humor during the long hours of writing, as she has with her companionship of over 30 years. I dedicate this book to her with my love.

Dimitri P. Bertsekas
November, 1995

Preface to the Second Edition

This second edition has expanded by about 130 pages the coverage of the original. Nearly 40% of the new material represents miscellaneous additions scattered throughout the text. The remainder deals with three new topics. These are:

(a) A new section in Chapter 3 that focuses on a simple but far-reaching treatment of Fritz John necessary conditions and constraint qualifications, and also includes semi-infinite programming.

(b) A new section in Chapter 5 on the use of duality and Lagrangian relaxation for solving discrete optimization problems. This section describes several motivating applications, and provides a connecting link between continuous and discrete optimization.

(c) A new section in Chapter 6 on approximate and incremental subgradient methods. This material is the subject of ongoing joint research with Angelia Geary, but it was thought sufficiently significant to be included in summary here.

One of the aims of the revision was to highlight the connections of nonlinear programming with other branches of optimization, such as linear programming, network optimization, and discrete/integer optimization. This should provide some additional flexibility for using the book in the classroom. In addition, the presentation was improved, the mathematical background material of the appendixes has been expanded, the exercises were reorganized, and a substantial number of new exercises were added.

A new internet-based feature was added to the book, which significantly extends its scope and coverage. Many of the theoretical exercises, quite a few of them new, have been solved in detail and their solutions have been posted in the book's www page

http://world.std.com/~ athenasc/nonlinbook.html

These exercises have been marked with the symbol 

The book's www page also contains links to additional resources, such as computer codes and my lecture transparencies from my MIT Nonlinear Programming class.

I would like to express my thanks to the many colleagues who contributed suggestions for improvement of the second edition. I would like to thank particularly Angelia Geary for her extensive help with the internet-posted solutions of the theoretical exercises.

Dimitri P. Bertsekas
bertsekas@lids.mit.edu
June, 1999

1

Unconstrained Optimization

Contents

Mathematical models of optimization can be generally represented by a *constraint set* X and a *cost function* f that maps elements of X into real numbers. The set X consists of the available decisions x and the cost $f(x)$ is a scalar measure of undesirability of choosing decision x. We want to find an optimal decision, i.e., an $x^* \in X$ such that

$$f(x^*) \leq f(x), \qquad \forall\, x \in X.$$

In this book we focus on the case where each decision x is an n-dimensional vector; that is, x is an n-tuple of real numbers $(x_1, \ldots, x_n)$. Thus the constraint set X is a subset of $\Re^n$, the n-dimensional Euclidean space.

The optimization problem just stated is very broad and contains as special cases several important classes of problems that have widely differing structures. Our focus will be on nonlinear programming problems, so let us provide some orientation about the character of these problems and their relations with other types of optimization problems.

Perhaps the most important characteristic of an optimization problem is whether it is *continuous* or *discrete*. Continuous problems are those where the constraint set X is infinite and has a "continuous" character. Typical examples of continuous problems are those where there are no constraints, i.e., where $X = \Re^n$, or where X is specified by some equations and inequalities. Generally, continuous problems are analyzed using the mathematics of calculus and convexity.

Discrete problems are basically those that are not continuous, usually because of finiteness of the constraint set X. Typical examples are combinatorial problems, arising for example in scheduling, route planning, and matching. Another important type of discrete problems is *integer programming*, where there is a constraint that the optimization variables must take only integer values from some range (such as 0 or 1). Discrete problems are addressed with combinatorial mathematics, and other special methodology, some of which relates to continuous problems.

Nonlinear programming, the case where either the cost function f is nonlinear or the constraint set X is specified by nonlinear equations and inequalities, lies squarely within the continuous problem category. Several other important types of optimization problems have more of a hybrid character, but are strongly connected with nonlinear programming.

In particular, *linear programming* problems, the case where f is linear and X is a polyhedron specified by linear inequality constraints, have many of the characteristics of continuous problems. However, they also have in part a combinatorial structure: according to a fundamental theorem [Prop. B.21(d) in Appendix B], optimal solutions of a linear program can be found by searching among the (finite) set of extreme points of the polyhedron X. Thus the search for an optimum can be confined within this finite set, and indeed one of the most popular methods for linear programming, the simplex method, is based on this idea. We note, however, that other important linear programming methods, such as the interior point methods to be discussed in Sections 2.6, 4.1, and 4.4, and some of the duality methods in Chapters 5 and 6, rely on the continuous structure of linear programs and are based on nonlinear programming ideas.

Another major class of problems with a strongly hybrid character is *network optimization*. Here the constraint set X is a polyhedron in $\Re^n$ that is defined in terms of a graph consisting of nodes and directed arcs. The salient feature of this constraint set is that its extreme points have *integer components*, something that is not true for general polyhedra. As a result, important combinatorial or integer programming problems, such as for example some matching and shortest path problems, can be embedded and solved within a continuous network optimization framework.

Our objective in this book is to focus on nonlinear programming problems, their continuous character, and the associated mathematical analysis. However, we will maintain a view to other broad classes of problems that have in part a discrete character. In particular, we will consider extensively those aspects of linear programming that bear a close relation to nonlinear programming methodology, such as interior point methods and polyhedral convexity (see Sections 2.5, 2.6, 4.1, 4.4, B.3, and B.4). We will discuss various aspects of network optimization problems that relate to both their continuous and their discrete character in Sections 2.1 and 5.5 (a far more extensive treatment, which straddles the boundary between continuous and discrete optimization, can be found in the author's network optimization textbook [Ber98]). Finally, we will discuss some of the major methods for integer programming and combinatorial optimization, such as branch-and-bound and Lagrangian relaxation, which rely on duality and the solution of continuous optimization subproblems (see Sections 5.5 and 6.3).

In this chapter we consider unconstrained nonlinear programming problems where $X = \Re^n$:

$$\begin{aligned} &\text{minimize} \quad f(x) \\ &\text{subject to} \quad x \in \Re^n. \end{aligned} \tag{UP}$$

In subsequent chapters, we focus on problems where X is a subset of $\Re^n$ and may be specified by equality and inequality constraints. For the most part, we assume that f is a continuously differentiable function, and we often also assume that f is twice continuously differentiable. The first and second derivatives of f play an important role in the characterization of optimal solutions via necessary and sufficient conditions, which are the main subject of Section 1.1. The first and second derivatives are also central in numerical algorithms for computing approximately optimal solutions. There is a broad range of such algorithms, with a rich theory, which is discussed in Sections 1.2-1.8. Section 1.9 specializes the methodology of the earlier sections to the important class of optimal control problems that involves a discrete-time dynamic system. While the focus is on unconstrained optimization, many of the ideas discussed in this first chapter are fundamental to the material in the remainder of the book.

1.1 OPTIMALITY CONDITIONS

1.1.1 Variational Ideas

The main ideas underlying optimality conditions in nonlinear programming usually admit simple explanations although their detailed proofs are sometimes tedious. For this reason, we have chosen to first discuss informally these ideas in the present subsection, and to leave detailed statements of results and proofs for the next subsection.

Local and Global Minima

A vector x^* is an *unconstrained local minimum* of f if it is no worse than its neighbors; that is, if there exists an $\epsilon > 0$ such that

$$f(x^*) \le f(x), \qquad \forall\, x \ \text{ with } \|x - x^*\| < \epsilon.$$

(Unless stated otherwise, we use the standard Euclidean norm $\|x\| = \sqrt{x'x}$. Appendix A describes in detail our mathematical notation and terminology.)

A vector x^* is an *unconstrained global minimum* of f if it is no worse than all other vectors; that is,

$$f(x^*) \le f(x), \qquad \forall\, x \in \Re^n.$$

The unconstrained local or global minimum x^* is said to be *strict* if the corresponding inequality above is strict for $x \neq x^*$. Figure 1.1.1 illustrates these definitions.

The definitions of local and global minima can be extended to the case where f is a function defined over a subset X of $\Re^n$. In particular, we say that x^* is a local minimum of f over X if $x^* \in X$ and there is an $\epsilon > 0$ such that $f(x^*) \le f(x)$ for all $x \in X$ with $\|x - x^*\| < \epsilon$. The definitions of a global and a strict minimum of f over X are analogous.

Local and global *maxima* are similarly defined. In particular, x^* is an unconstrained local (global) maximum of f, if x^* is an unconstrained local (global) minimum of the function $-f$.

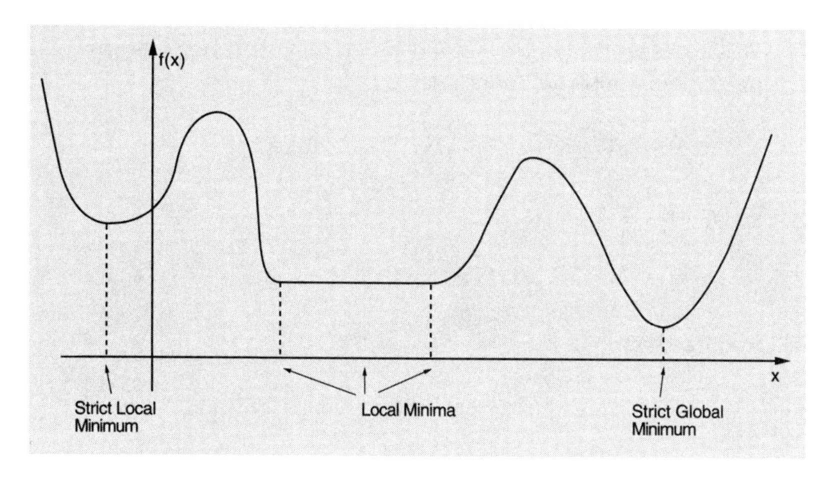

Figure 1.1.1. Unconstrained local and global minima in one dimension.

Necessary Conditions for Optimality

If the cost function is differentiable, we can use gradients and Taylor series expansions to compare the cost of a vector with the cost of its close neighbors. In particular, we consider small variations Δx from a given vector x^*, which approximately, up to first order, yield a cost variation

$$f(x^* + \Delta x) - f(x^*) \approx \nabla f(x^*)'\Delta x$$

and, up to second order, yield a cost variation

$$f(x^* + \Delta x) - f(x^*) \approx \nabla f(x^*)'\Delta x + \tfrac{1}{2}\Delta x'\nabla^2 f(x^*)\Delta x.$$

We expect that if x^* is an unconstrained local minimum, the first order cost variation due to a small variation Δx is nonnegative:

$$\nabla f(x^*)'\Delta x = \sum_{i=1}^{n} \frac{\partial f(x^*)}{\partial x_i}\Delta x_i \geq 0.$$

In particular, by taking Δx to be positive and negative multiples of the unit coordinate vectors (all coordinates equal to zero except for one which is equal to unity), we obtain $\partial f(x^*)/\partial x_i \geq 0$ and $\partial f(x^*)/\partial x_i \leq 0$, respectively, for all coordinates $i = 1, \ldots, n$. Equivalently, we have the necessary condition

$$\nabla f(x^*) = 0,$$

[originally formulated by Fermat in 1637 in the short treatise "Methodus as Disquirendam Maximam et Minimam" without proof (of course!)]. This condition is proved in Prop. 1.1.1, given in the next subsection.

We also expect that the second order cost variation due to a small variation Δx must also be nonnegative:

$$\nabla f(x^*)'\Delta x + \tfrac{1}{2}\Delta x'\nabla^2 f(x^*)\Delta x \geq 0.$$

Since $\nabla f(x^*)'\Delta x = 0$, we obtain

$$\Delta x'\nabla^2 f(x^*)\Delta x \geq 0,$$

which implies that

$$\nabla^2 f(x^*) : \text{positive semidefinite.}$$

We prove this necessary condition in the subsequent Prop. 1.1.1. Appendix A reviews the definition and properties of positive definite and positive semidefinite matrices.

In what follows, we refer to a vector x^* satisfying the condition $\nabla f(x^*) = 0$ as a *stationary point.*

The Case of a Convex Cost Function

Convexity notions, reviewed in Appendix B, play a very important role in nonlinear programming. One reason is that when the cost function f is convex, there is no distinction between local and global minima; every local minimum is also global. The idea is illustrated in Fig. 1.1.2 and the formal proof is given in Prop. B.10 of Appendix B.

Another important fact is that the first order condition $\nabla f(x^*) = 0$ is also sufficient for optimality if f is convex. This is established in Prop. 1.1.2 in the next subsection. The proof is based on a basic property of a convex function f: the linear approximation at a point x^* based on the gradient, that is, $f(x^*) + \nabla f(x^*)'(x - x^*)$, underestimates $f(x)$, so if $\nabla f(x^*) = 0$, then $f(x^*) \leq f(x)$ for all x (see Prop. B.3 in Appendix B).

Figure 1.1.3 shows how the first and second order necessary conditions can fail to guarantee local optimality of x^* if f is not convex.

Sufficient Conditions for Optimality

Suppose we have a vector x^* that satisfies the first order necessary optimality condition

$$\nabla f(x^*) = 0, \tag{1.1}$$

and also satisfies the following strengthened form of the second order necessary optimality condition

$$\nabla^2 f(x^*) : \text{ positive definite,} \tag{1.2}$$

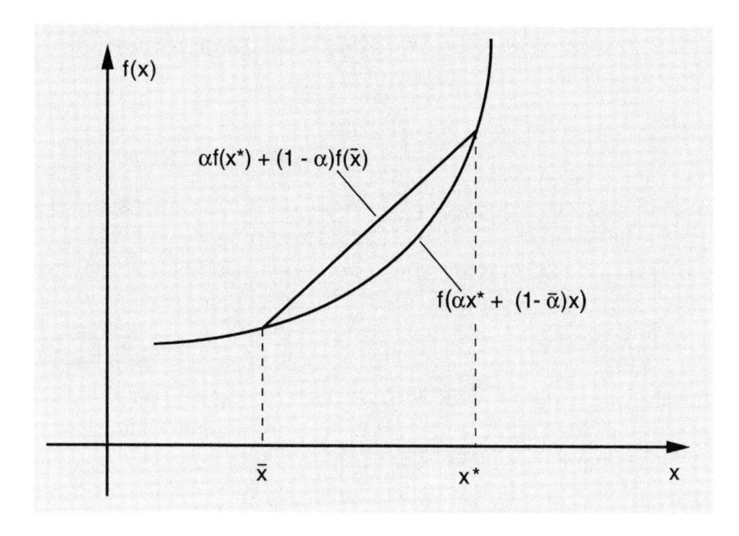

Figure 1.1.2. Illustration of why local minima of convex functions are also global. Suppose that f is convex, and let x^* and $\bar{x}$ be such that $f(\bar{x}) < f(x^*)$. By convexity, for all $\alpha \in (0,1)$,

$$f\big(\alpha x^* + (1-\alpha)\bar{x}\big) \leq \alpha f(x^*) + (1-\alpha) f(\bar{x}) < f(x^*).$$

Thus, f has strictly lower value than $f(x^*)$ at every point on the line segment connecting x^* with $\bar{x}$, except x^*. Hence, if x^* is not a global minimum [so that there exists $\bar{x}$ with $f(\bar{x}) < f(x^*)$], then x^* cannot be a local minimum.

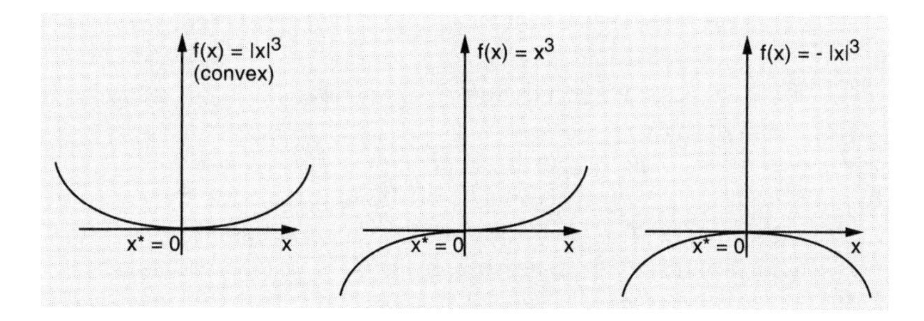

Figure 1.1.3. Illustration of the first order necessary optimality condition of zero slope $[\nabla f(x^*) = 0]$ and the second order necessary optimality condition of nonnegative curvature $[\nabla^2 f(x^*) \geq 0]$ for functions of one variable. The first order condition is satisfied not only by local minima, but also by local maxima and "inflection" points, such as the one on the middle figure above. In some cases [e.g. $f(x) = x^3$ and $f(x) = -|x|^3$] the second order condition is also satisfied by local maxima and inflection points. If the function f is convex, the condition $\nabla f(x^*) = 0$ is necessary and sufficient for global optimality of x^*.

(that is, the Hessian is positive definite rather than semidefinite). Then, for all $\Delta x \neq 0$ we have

$$\Delta x' \nabla^2 f(x^*) \Delta x > 0,$$

implying that at x^* the second order variation of f due to a small nonzero variation Δx is positive. Thus, f tends to increase strictly with small excursions from x^*, suggesting that the above conditions (1.1) and (1.2) are sufficient for local optimality of x^*. This is established in Prop. 1.1.3.

Local minima that don't satisfy the sufficiency conditions (1.1) and (1.2) are called *singular*; otherwise they are called *nonsingular*. Singular local minima are harder to deal with for two reasons. First, in the absence of convexity of f, their optimality cannot be ascertained using easily verifiable sufficiency conditions. Second, in their neighborhood, the behavior of the most commonly used optimization algorithms tends to be slow and/or erratic, as we will see in the subsequent sections.

Quadratic Cost Functions

Consider the quadratic function

$$f(x) = \tfrac{1}{2} x' Q x - b' x,$$

where Q is a symmetric $n \times n$ matrix and b is a vector in $\Re^n$. If x^* is a local minimum of f, we must have, by the necessary optimality conditions,

$$\nabla f(x^*) = Q x^* - b = 0, \qquad \nabla^2 f(x^*) = Q : \text{ positive semidefinite.}$$

Thus, if Q is not positive semidefinite, f can have no local minima. If Q is positive semidefinite, f is convex [Prop. B.4(d) of Appendix B], so any vector x^* satisfying the first order condition $\nabla f(x^*) = Qx^* - b = 0$ is a global minimum of f. On the other hand there might not exist a solution of the equation $\nabla f(x^*) = Qx^* - b = 0$ if Q is singular. If, however, Q is positive definite (and hence invertible, by Prop. A.20 of Appendix A), the equation $Qx^* - b = 0$ can be solved uniquely and the vector $x^* = Q^{-1}b$ is the unique global minimum. This is consistent with Prop. 1.1.2(a) to be given shortly, which asserts that strictly convex functions can have at most one global minimum [f is strictly convex if and only if Q is positive definite; Prop. B.4(d) of Appendix B]. Figure 1.1.4 illustrates the various special cases considered.

Quadratic cost functions are important in nonlinear programming because they arise frequently in applications, but they are also important for another reason. From the Taylor expansion

$$f(x) = f(x^*) + \tfrac{1}{2}(x - x^*)' \nabla^2 f(x^*)(x - x^*) + o\big(\|x - x^*\|^2\big),$$

it is seen that a nonquadratic cost function can be approximated well by a quadratic function near a nonsingular local minimum x^* [$\nabla^2 f(x^*)$: positive definite]. This means that we can carry out much of our analysis and experimentation with algorithms using positive definite quadratic functions and expect that the conclusions will largely carry over to more general cost functions near convergence to such local minima. However, for local minima near which the Hessian matrix either does not exist or is singular, the higher than second order terms in the Taylor expansion are not negligible and an algorithmic analysis based on quadratic cost functions will likely be seriously flawed.

Existence of Optimal Solutions

In many cases it is useful to know that there exists at least one global minimum of a function f over a set X. Generally, such a minimum need not exist. For example, the scalar functions $f(x) = x$ and $f(x) = e^x$ have no global minima over the set of real numbers. The first function decreases without bound to $-\infty$ as x tends toward $-\infty$, while the second decreases toward 0 as x tends toward $-\infty$ but always takes positive values. Given the range of values that $f(x)$ takes as x ranges over X, that is, the set of real numbers

$$\big\{f(x) \mid x \in X\big\},$$

there are two possibilities:

1. The set $\big\{f(x) \mid x \in X\big\}$ is bounded below; that is, there exists a scalar M such that $M \leq f(x)$ for all $x \in X$. In this case, the greatest lower bound of $\big\{f(x) \mid x \in X\big\}$ is a real number, which is denoted by $\inf_{x \in X} f(x)$. For example, $\inf_{x \in \Re} e^x = 0$ and $\inf_{x < 0} e^x = 0$.

2. The set $\big\{f(x) \mid x \in X\big\}$ is unbounded below (i.e., contains arbitrarily small real numbers). In this case we write

$$\inf_{x \in X} f(x) = -\infty.$$

Existence of at least one global minimum is guaranteed if f is a continuous function and X is a compact subset of $\Re^n$. This is the *Weierstrass theorem*, (see Prop. A.8 in Appendix A). By a related result, also shown in Prop. A.8 of Appendix A, existence of an optimal solution is guaranteed if $f : \Re^n \mapsto \Re$ is a continuous function, X is closed, and f is coercive, that is, $f(x) \to \infty$ when $\|x\| \to \infty$.

Why do we Need Optimality Conditions?

Hardly anyone would doubt that optimality conditions are fundamental to the analysis of an optimization problem. In practice, however, optimality

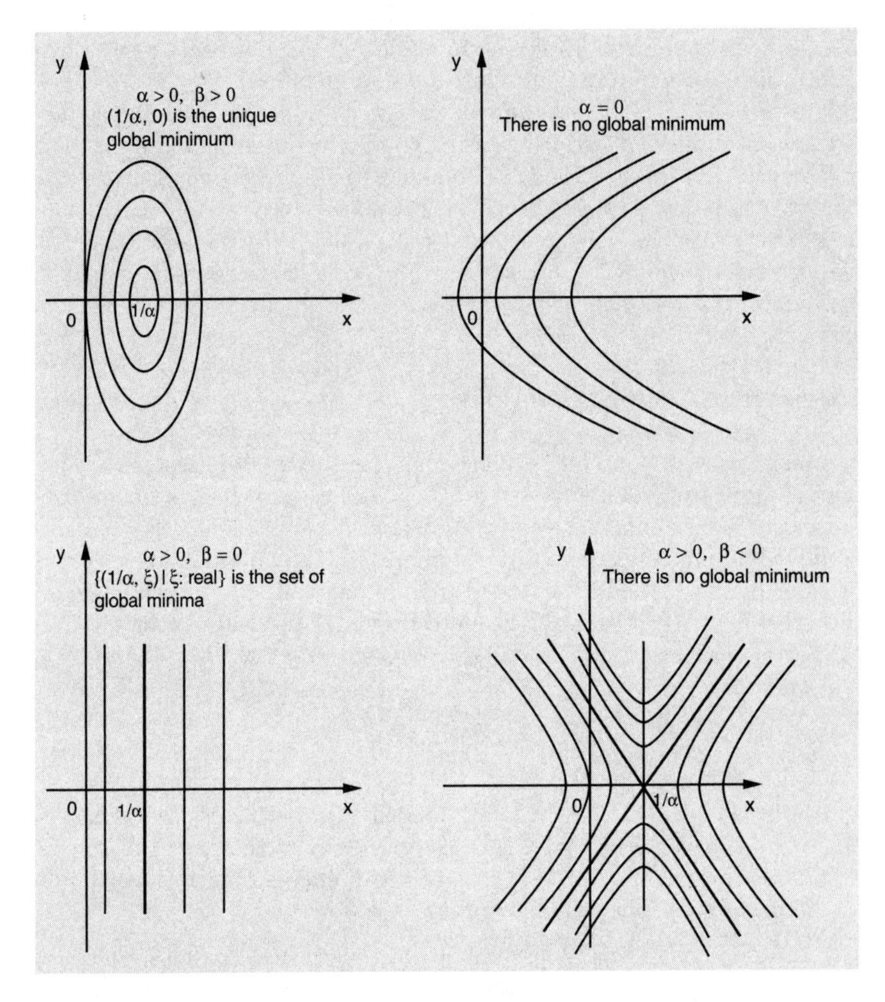

Figure 1.1.4. Illustration of the isocost surfaces of the quadratic cost function $f : \Re^2 \mapsto \Re$ given by

$$f(x, y) = \frac{1}{2}\left(\alpha x^2 + \beta y^2\right) - x$$

for various values of α and β.

conditions play an important role in a variety of contexts, some of which may not be readily apparent.

The most straightforward method to use optimality conditions to solve an optimization problem, is as follows: First, find all points satisfying the first order necessary condition $\nabla f(x) = 0$; then (if f is not known to be convex), check the second order necessary condition ($\nabla^2 f$ is positive semidefinite) for each of these points, filtering out those that do not satisfy it; finally for the remaining candidates, check if $\nabla^2 f$ is positive definite, in which case we are sure that they are strict local minima.

A slightly different alternative is to find all points satisfying the necessary conditions, and to declare as global minimum the one with smallest cost value. However, here it is essential to know that a global minimum exists. As an example, for the one-dimensional function

$$f(x) = x^2 - x^4,$$

the points satisfying the necessary condition

$$\nabla f(x) = 2x - 4x^3 = 0$$

are 0, $1/\sqrt{2}$, and $-1/\sqrt{2}$, and of these, 0 gives the smallest cost value. However, we cannot declare 0 as the global minimum, because we don't know if a global minimum exists. Indeed, in this example none of the points 0, $1/\sqrt{2}$, and $-1/\sqrt{2}$ is a global minimum, because f decreases to $-\infty$ as $|x| \to \infty$, and has no global minimum. Here is another example:

Example 1.1.1 (Arithmetic-Geometric Mean Inequality)

We want to show the following classical inequality [due to Cauchy (1821)]

$$(x_1 x_2 \cdots x_n)^{1/n} \leq \frac{\sum_{i=1}^n x_i}{n}$$

for any set of positive numbers x_i, $i = 1, \ldots, n$. By making the change of variables

$$y_i = \ln(x_i), \qquad i = 1, \ldots, n,$$

we have $x_i = e^{y_i}$, so that this inequality is equivalently written as

$$e^{\frac{y_1 + \cdots + y_n}{n}} \leq \frac{e^{y_1} + \cdots + e^{y_n}}{n},$$

which must be shown for all scalars $y_1, \ldots, y_n$. Note that with this transformation, the nonnegativity requirements on the variables have been eliminated.

One approach to proving the above inequality is to minimize the function

$$\frac{e^{y_1} + \cdots + e^{y_n}}{n} - e^{\frac{y_1 + \cdots + y_n}{n}},$$

and to show that its minimal value is 0. An alternative, which works better if we use optimality conditions, is to minimize instead

$$e^{y_1} + \cdots + e^{y_n},$$

over all $y = (y_1, \ldots, y_n)$ such that $y_1 + \cdots + y_n = s$ for an arbitrary scalar s, and to show that the optimal value is no less than $ne^{s/n}$. To this end, we use the constraint $y_1 + \cdots + y_n = s$ to eliminate the variable y_n, thereby obtaining the equivalent *unconstrained* problem of minimizing

$$g(y_1, \ldots, y_{n-1}) = e^{y_1} + \cdots + e^{y_{n-1}} + e^{s - y_1 - \cdots - y_{n-1}},$$

over $y_1, \ldots, y_{n-1}$. The necessary conditions $\partial g/\partial y_i = 0$ yield the system of equations

$$e^{y_i} = e^{s-y_1-\cdots-y_{n-1}}, \qquad i = 1, \ldots, n-1,$$

or

$$y_i = s - y_1 - \cdots - y_{n-1}, \qquad i = 1, \ldots, n-1.$$

This system has only one solution: $y_i^* = s/n$ for all i. This solution must be the unique global minimum if we can show that that there exists a global minimum. Indeed, it can be seen that the function $g(y_1, \ldots, y_{n-1})$ is coercive, so it has an unconstrained global minimum. Therefore, $(s/n, \ldots, s/n)$ is this minimum. Thus the optimal value of $e^{y_1} + \cdots + e^{y_n}$ is $ne^{s/n}$, which as argued above, is sufficient to show the arithmetic-geometric mean inequality.

It is important to realize, however, that except under very favorable circumstances, using optimality conditions to obtain a solution as described above does *not* work. The reason is that solving for x the system of equations $\nabla f(x) = 0$ is usually nontrivial; algorithmically, it is typically as difficult as solving the original optimization problem.

The principal context in which optimality conditions become useful will not become apparent until we consider iterative optimization algorithms in subsequent sections. We will see that optimality conditions often provide the basis for the development and the analysis of algorithms. In particular, algorithms recognize solutions by checking whether they satisfy various optimality conditions and terminate when such conditions hold approximately. Furthermore, the behavior of various algorithms in the neighborhood of a local minimum often depends on whether various optimality conditions are satisfied at that minimum. Thus, for example, sufficiency conditions play a key role in assertions regarding the speed of convergence of various algorithms.

There is one other important context, prominently arising in microeconomic theory, where optimality conditions provide the basis for analysis. Here one is interested primarily not in finding an optimal solution, but rather in how the optimal solution is affected by changes in the problem data. For example, an economist may be interested in how the prices of some raw materials will affect the availability of certain goods that are produced by using these raw materials; the assumption here is that the amounts produced are the variables of a profit optimization problem, which is solved by the corresponding producers. This type of analysis is known as *sensitivity analysis*, and is discussed next.

Sensitivity*

Suppose that we want to quantify the variation of the optimal solution as a vector of parameters changes. In particular, consider the optimization problem

$$\begin{aligned}&\text{minimize} \;\; f(x,a)\\&\text{subject to} \;\; x \in \Re^n,\end{aligned}$$

where $f : \Re^{m+n} \mapsto \Re$ is a twice continuously differentiable function involving the m-dimensional parameter vector a. Let $x(a)$ denote the global minimum corresponding to a, assuming for the moment that it exists and is unique. By the first order necessary condition we have

$$\nabla_x f\big(x(a),a\big) = 0, \qquad \forall\ a \in \Re^m,$$

and by differentiating this relation with respect to a, we obtain

$$\nabla x(a) \nabla^2_{xx} f\big(x(a),a\big) + \nabla^2_{xa} f\big(x(a),a\big) = 0,$$

where the elements of the $m \times n$ gradient matrix $\nabla x(a)$ are the first partial derivatives of the components of $x(a)$ with respect to the different components of a. Assuming that the inverse below exists, we have

$$\nabla x(a) = -\nabla^2_{xa} f\big(x(a),a\big) \left(\nabla^2_{xx} f\big(x(a),a\big)\right)^{-1}, \tag{1.3}$$

which gives the first order variation of the components of the optimal x with respect to the components of a.

For the preceding analysis to be precise, we must be sure that $x(a)$ exists and is differentiable as a function of a. The principal analytical framework for this is the implicit function theorem (Prop. A.25 in Appendix A). With the aid of this theorem, we can define $x(a)$ in some sphere around a minimum $\bar{x} = x(\bar{a})$ corresponding to a nominal parameter value $\bar{a}$, assuming that the Hessian matrix $\nabla^2_{xx} f\big(\bar{x},\bar{a}\big)$ is positive definite. Thus, the preceding development and the formula (1.3) for the matrix $\nabla x(a)$ can be justified provided the nominal local minimum $\bar{x}$ is nonsingular.

We will postpone further discussion of sensitivity analysis for Section 3.2, where we will provide a constrained version of the expression (1.3) for $\nabla x(a)$.

1.1.2 Main Optimality Conditions

We now provide formal statements and proofs of the optimality conditions discussed in the preceding section.

Proposition 1.1.1: (Necessary Optimality Conditions) Let x^* be an unconstrained local minimum of $f : \Re^n \mapsto \Re$, and assume that f is continuously differentiable in an open set S containing x^*. Then

$$\nabla f(x^*) = 0. \qquad \text{(First Order Necessary Condition)}$$

If in addition f is twice continuously differentiable within S, then

$$\nabla^2 f(x^*) : \text{positive semidefinite.} \qquad \text{(Second Order Necessary Condition)}$$

Proof: Fix some $d \in \Re^n$. Then, using the chain rule to differentiate the function $g(\alpha) = f(x^* + \alpha d)$ of the scalar α, we have

$$0 \le \lim_{\alpha \downarrow 0} \frac{f(x^* + \alpha d) - f(x^*)}{\alpha} = \frac{dg(0)}{d\alpha} = d'\nabla f(x^*),$$

where the inequality follows from the assumption that x^* is a local minimum. Since d is arbitrary, the same inequality holds with d replaced by $-d$. Therefore, $d'\nabla f(x^*) = 0$ for all $d \in \Re^n$, which shows that $\nabla f(x^*) = 0$.

Assume that f is twice continuously differentiable, and let d be any vector in $\Re^n$. For all $\alpha \in \Re$, the second order Taylor series expansion yields

$$f(x^* + \alpha d) - f(x^*) = \alpha \nabla f(x^*)'d + \frac{\alpha^2}{2} d'\nabla^2 f(x^*)d + o(\alpha^2).$$

Using the condition $\nabla f(x^*) = 0$ and the local optimality of x^*, we see that there is a sufficiently small $\epsilon > 0$ such that for all α with $\alpha \in (0, \epsilon)$,

$$0 \le \frac{f(x^* + \alpha d) - f(x^*)}{\alpha^2} = \tfrac{1}{2} d'\nabla^2 f(x^*)d + \frac{o(\alpha^2)}{\alpha^2}.$$

Taking the limit as $\alpha \to 0$ and using the fact

$$\lim_{\alpha \to 0} \frac{o(\alpha^2)}{\alpha^2} = 0,$$

we obtain $d'\nabla^2 f(x^*)d \ge 0$, showing that $\nabla^2 f(x^*)$ is positive semidefinite. **Q.E.D.**

The following proposition handles the case of a convex cost.

Proposition 1.1.2: (Convex Cost Function) Let $f : X \mapsto \Re$ be a convex function over the convex set X.

(a) A local minimum of f over X is also a global minimum over X. If in addition f is strictly convex, then there exists at most one global minimum of f.

(b) If f is convex and the set X is open, then $\nabla f(x^*) = 0$ is a necessary and sufficient condition for a vector $x^* \in X$ to be a global minimum of f over X.

Proof: Part (a) is proved in Prop. B.10 of Appendix B. To show part (b), note that by Prop. B.3 of Appendix B, we have

$$f(x) \geq f(x^*) + \nabla f(x^*)'(x - x^*), \qquad \forall\, x \in X.$$

If $\nabla f(x^*) = 0$, we obtain $f(x) \geq f(x^*)$ for all $x \in X$, so x^* is a global minimum. **Q.E.D.**

In the absence of convexity, we have the following sufficiency conditions for local optimality.

Proposition 1.1.3: (Second Order Sufficient Optimality Conditions) Let $f : \Re^n \mapsto \Re$ be twice continuously differentiable in an open set S. Suppose that a vector $x^* \in S$ satisfies the conditions

$$\nabla f(x^*) = 0, \qquad \nabla^2 f(x^*) : \text{positive definite.}$$

Then, x^* is a strict unconstrained local minimum of f. In particular, there exist scalars $\gamma > 0$ and $\epsilon > 0$ such that

$$f(x) \geq f(x^*) + \frac{\gamma}{2}\|x - x^*\|^2, \qquad \forall\, x \text{ with } \|x - x^*\| < \epsilon. \tag{1.4}$$

Proof: Let λ be the smallest eigenvalue of $\nabla^2 f(x^*)$. By Prop. A.20(b) of Appendix A, λ is positive since $\nabla^2 f(x^*)$ is positive definite. Furthermore, by Prop. A.18(b) of Appendix A, $d'\nabla^2 f(x^*)d \geq \lambda\|d\|^2$ for all $d \in \Re^n$. Using this relation, the hypothesis $\nabla f(x^*) = 0$, and the second order Taylor series expansion, we have for all d

$$\begin{aligned} f(x^* + d) - f(x^*) &= \nabla f(x^*)'d + \tfrac{1}{2}d'\nabla^2 f(x^*)d + o(\|d\|^2) \\ &\geq \frac{\lambda}{2}\|d\|^2 + o(\|d\|^2) \end{aligned}$$

$$= \left(\frac{\lambda}{2} + \frac{o(\|d\|^2)}{\|d\|^2} \right) \|d\|^2.$$

It is seen that Eq. (1.4) is satisfied for any $\epsilon > 0$ and $\gamma > 0$ such that

$$\frac{\lambda}{2} + \frac{o(\|d\|^2)}{\|d\|^2} \geq \frac{\gamma}{2}, \qquad \forall\, d \text{ with } \|d\| < \epsilon.$$

Q.E.D.

EXERCISES

1.1.1

For each value of the scalar β, find the set of all stationary points of the following function of the two variables x and y

$$f(x, y) = x^2 + y^2 + \beta xy + x + 2y.$$

Which of these stationary points are global minima?

1.1.2

In each of the following problems fully justify your answer using optimality conditions.

(a) Show that the 2-dimensional function $f(x, y) = (x^2 - 4)^2 + y^2$ has two global minima and one stationary point, which is neither a local maximum nor a local minimum.

(b) Find all local minima of the 2-dimensional function $f(x, y) = \frac{1}{2}x^2 + x \cos y$.

(c) Find all local minima and all local maxima of the 2-dimensional function $f(x, y) = \sin x + \sin y + \sin(x + y)$ within the set $\{(x, y) \mid 0 < x < 2\pi,\ 0 < x < 2\pi\}$.

(d) Show that the 2-dimensional function $f(x, y) = (y - x^2)^2 - x^2$ has only one stationary point, which is neither a local maximum nor a local minimum.

(e) Consider the minimization of the function f in part (d) subject to no constraint on x and the constraint $-1 \leq y \leq 1$ on y. Show that there exists at least one global minimum and find all global minima.

1.1.3 [Hes75]

Let $f : \Re^n \mapsto \Re$ be a differentiable function. Suppose that a point x^* is a local minimum of f along every line that passes through x^*; that is, the function

$$g(\alpha) = f(x^* + \alpha d)$$

is minimized at $\alpha = 0$ for all $d \in \Re^n$.

(a) Show that $\nabla f(x^*) = 0$.

(b) Show by example that x^* need not be a local minimum of f. *Hint*: Consider the function of two variables $f(y, z) = (z - py^2)(z - qy^2)$, where $0 < p < q$; see Fig. 1.1.5. Show that $(0, 0)$ is a local minimum of f along every line that passes through $(0, 0)$. Furthermore, if $p < m < q$, then $f(y, my^2) < 0$ if $y \neq 0$ while $f(0, 0) = 0$.

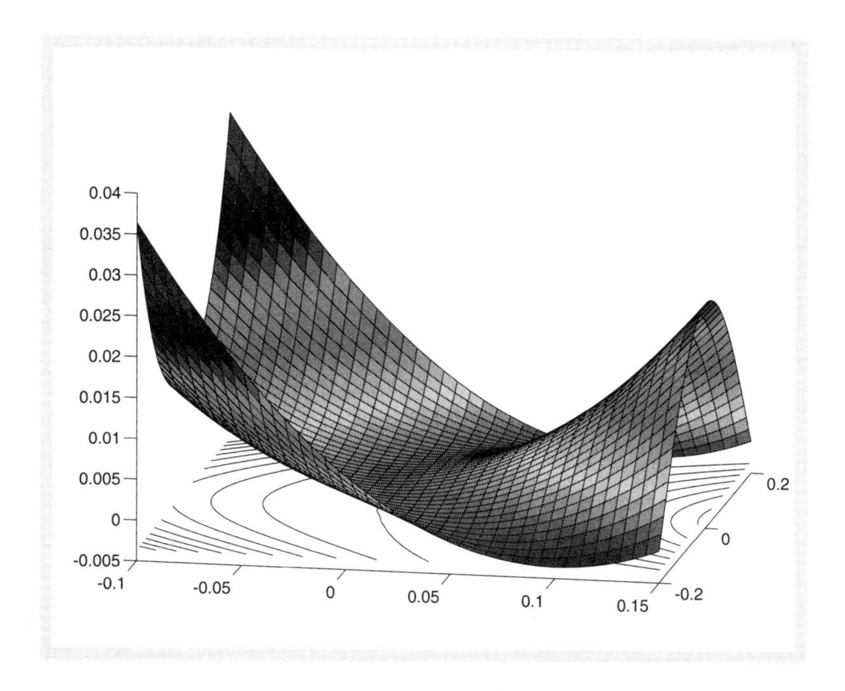

Figure 1.1.5. Three-dimensional graph of the function $f(y, z) = (z - py^2)(z - qy^2)$ for $p = 1$ and $q = 4$ (cf. Exercise 1.1.3). The origin is a local minimum with respect to every line that passes through it, but is not a local minimum of f.

1.1.4

Use optimality conditions to show that for all $x > 0$ we have

$$\frac{1}{x} + x \geq 2.$$

1.1.5

Find the rectangular parallelepiped of unit volume that has the minimum surface area. *Hint*: By eliminating one of the dimensions, show that the problem is equivalent to the minimization over $x > 0$ and $y > 0$ of

$$f(x, y) = xy + \frac{1}{x} + \frac{1}{y}.$$

Show that the sets $\{(x, y) \mid f(x, y) \leq \gamma,\ x > 0,\ y > 0\}$ are compact for all scalars γ.

1.1.6 (The Weber Point of a Set of Points)

We want to find a point x in the plane whose sum of weighted distances from a given set of points $y_1, \ldots, y_m$ is minimized. Mathematically, the problem is

$$\begin{aligned} &\text{minimize} \quad \sum_{i=1}^{m} w_i \|x - y_i\| \\ &\text{subject to} \quad x \in \Re^n, \end{aligned}$$

where $w_1, \ldots, w_m$ are given positive scalars.

(a) Show that there exists a global minimum for this problem and that it can be realized by means of the mechanical model shown in Fig. 1.1.6.

(b) Is the optimal solution always unique?

(c) Show that an optimal solution minimizes the potential energy of the mechanical model of Fig. 1.1.6, defined as $\sum_{i=1}^{m} w_i h_i$, where h_i is the height of the ith weight, measured from some reference level.

Note: This problem stems from Weber's work [Web29], which is generally viewed as the starting point of location theory.

1.1.7 (Fermat-Torricelli-Viviani Problem)

Given a triangle in the plane, consider the problem of finding a point whose sum of distances from the vertices of the triangle is minimal. Show that such a point is either a vertex, or else it is such that each side of the triangle is seen from that point at a 120 degree angle (this is known as the Torricelli point). *Note*: This problem, whose detailed history is traced in [BMS99], was suggested by Fermat to Torricelli who solved it. Viviani also solved the problem a little later and proved the following generalization: Suppose that x_i, $i = 1, \ldots, m$, are points in the plane, and x is a point in their convex hull such that $x \neq x_i$ for all i, and the angles $x_i \widehat{x x_{i+1}}$, $i < m$, and $x_m \widehat{x x_1}$ are all equal to $2\pi/m$. Then x minimizes $\sum_{i=1}^{m} \|z - x_i\|$ over all z in the plane (show this as an exercise by using sufficient optimality conditions; compare with the preceding exercise). Fermat is credited with being the first to study systematically optimization problems in geometry.

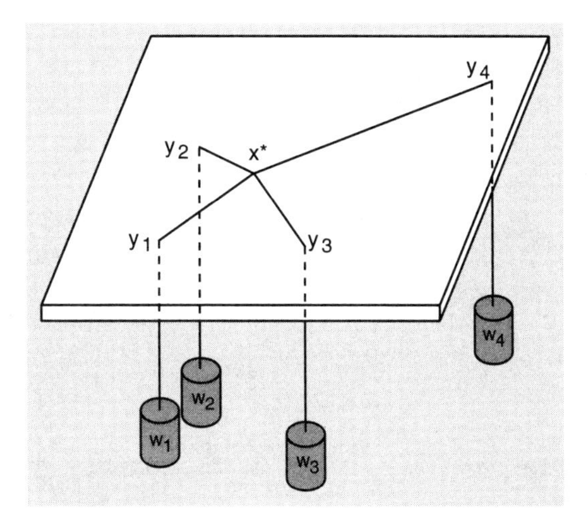

Figure 1.1.6. Mechanical model (known as the Varignon frame) associated with the Weber problem (Exercise 1.1.6). It consists of a board with a hole drilled at each of the given points y_i. Through each hole, a string is passed with the corresponding weight w_i attached. The other ends of the strings are tied with a knot as shown. In the absence of friction or tangled strings, the forces at the knot reach equilibrium when the knot is located at an optimal solution x^*.

1.1.8 (Diffraction Law in Optics)

Let p and q be two points on the plane that lie on opposite sides of a horizontal axis. Assume that the speed of light from p and from q to the horizontal axis is v and w, respectively, and that light reaches a point from other points along paths of minimum travel time. Find the path that a ray of light would follow from p to q.

1.1.9 (www)

Let $f : \Re^n \mapsto \Re$ be a twice continuously differentiable function that satisfies

$$m\|y\|^2 \le y'\nabla^2 f(x)y \le M\|y\|^2, \qquad \forall\ x,\ y \in \Re^n,$$

where m and M are some positive scalars. Show that f has a unique global minimum x^*, which satisfies

$$\frac{1}{2M}\|\nabla f(x)\|^2 \le f(x) - f(x^*) \le \frac{1}{2m}\|\nabla f(x)\|^2, \qquad \forall\ x \in \Re^n,$$

and

$$\frac{m}{2}\|x - x^*\|^2 \le f(x) - f(x^*) \le \frac{M}{2}\|x - x^*\|^2, \qquad \forall\ x \in \Re^n.$$

Hint: Use a second order expansion and the relation

$$\min_{y\in\Re^n}\left\{\nabla f(x)'(y-x) + \frac{\alpha}{2}\|y-x\|^2\right\} = -\frac{1}{2\alpha}\|\nabla f(x)\|^2, \qquad \forall\ \alpha > 0.$$

1.1.10 (Nonconvex Level Sets [Dun87])

Let $f : \Re^2 \mapsto \Re$ be the function

$$f(x) = x_2^2 - ax_2\|x\|^2 + \|x\|^4,$$

where $0 < a < 2$ (see Fig. 1.1.7). Show that $f(x) > 0$ for all $x \neq 0$, so that the origin is the unique global minimum. Show also that there exists a $\bar{\gamma} > 0$ such that for all $\gamma \in (0, \bar{\gamma}]$, the level set $L_\gamma = \{x \mid f(x) \leq \gamma\}$ is not convex. *Hint*: Show that for $\gamma \in (0, \bar{\gamma}]$, there is a $p > 0$ and a $q > 0$ such that the vectors $(-p, q)$ and (p, q) belong to L_γ, but $(0, q)$ does not belong to L_γ.

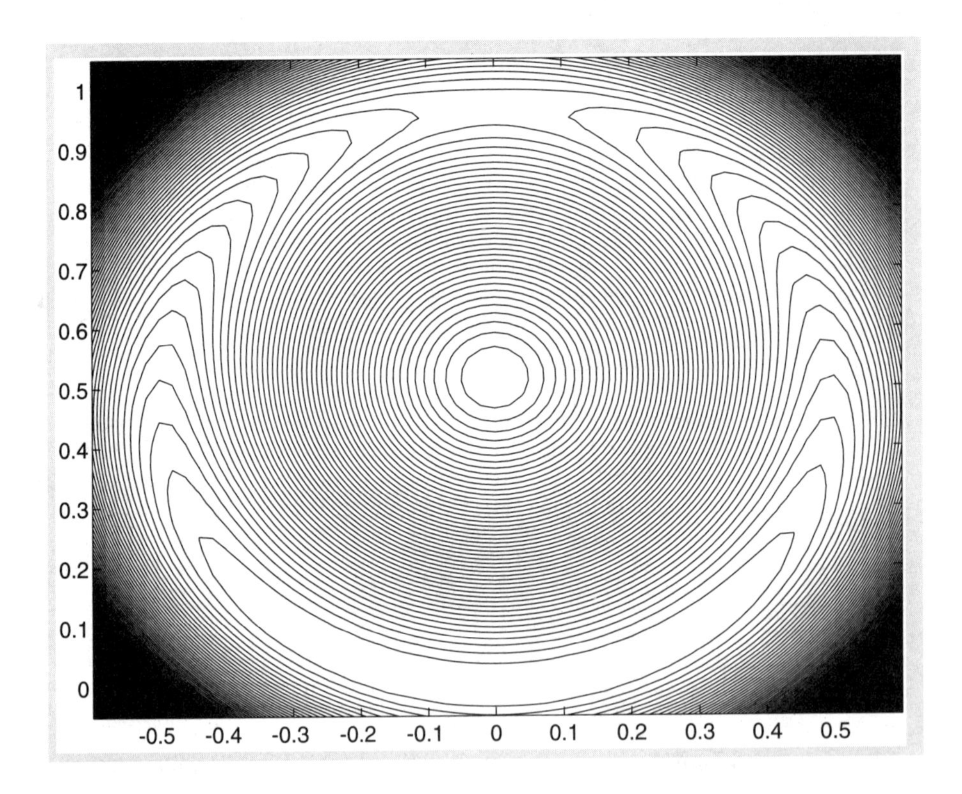

Figure 1.1.7. Level sets of the function f of Exercise 1.1.10 for the case where $a = 1.98$. The unique global minimum is the origin, but the level sets of f are nonconvex.

1.1.11 (Singular Strict Local Minima [Dun87]) (www)

Show that if x^* is a nonsingular strict local minimum of a twice continuously differentiable function $f : \Re^n \mapsto \Re$, then x^* is an isolated stationary point; that is, there is a sphere centered at x^* such that x^* is the only stationary point of f within that sphere. Use the following example function $f : \Re \mapsto \Re$ to show that this need not be true if x^* is a singular strict local minimum:

$$f(x) = \begin{cases} x^2 \left(\sqrt{2} - \sin\left(\frac{4\pi}{3} - \sqrt{3}\ln(x^2)\right)\right) & \text{if } x \neq 0 \\ 0 & \text{if } x = 0. \end{cases}$$

In particular, show that $x^* = 0$ is the unique (singular) global minimum, while the sequence $\{x^k\}$ of nonsingular local minima, where

$$x^k = e^{\frac{(1-8k)\pi}{8\sqrt{3}}},$$

converges to x^* (cf. Fig. 1.1.8). Verify also that there is a sequence of nonsingular local maxima that converges to x^*.

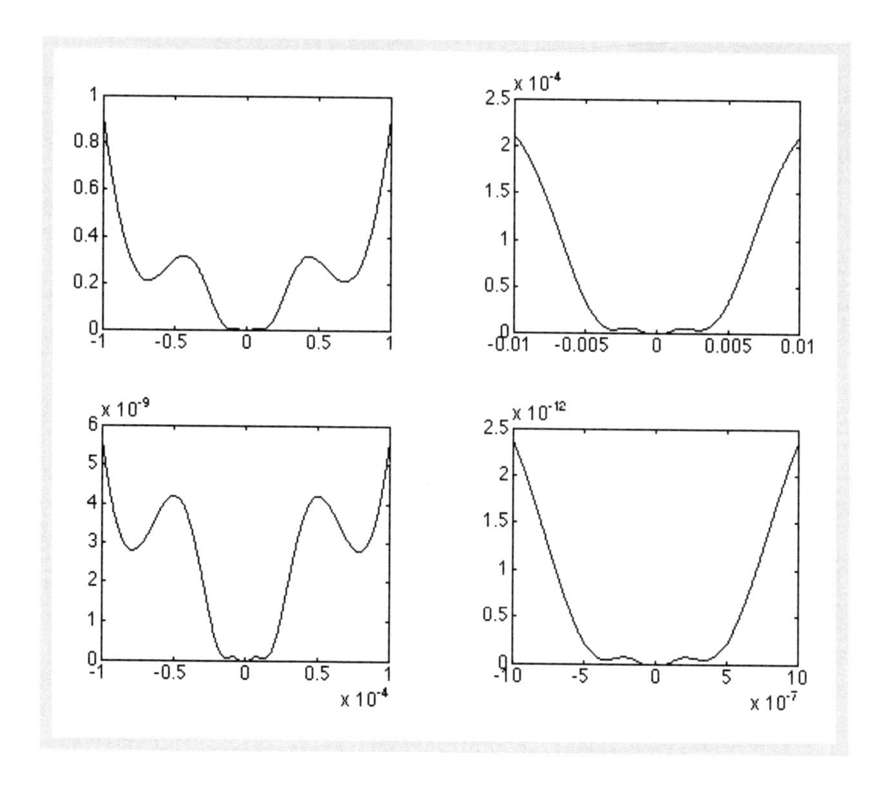

Figure 1.1.8. Illustration of the function f of Exercise 1.1.11 in progressively finer scale. $x^* = 0$ is the unique (singular) global minimum, but there are sequences of nonsingular local minima and local maxima that converge to x^*.

1.1.12 (Stability) (www)

We are often interested in whether optimal solutions change radically when the problem data are slightly perturbed. This issue is addressed by *stability analysis*, to be contrasted with sensitivity analysis, which deals with *how much* optimal solutions change when problem data change. An unconstrained local minimum x^* of a function f is said to be *locally stable* if there exists a $\delta > 0$ such that all sequences $\{x^k\}$ with $f(x^k) \to f(x^*)$ and $||x^k - x^*|| < \delta$, for all k, converge to x^*. Suppose that f is a continuous function and let x^* be a local minimum of f.

(a) Show that x^* is locally stable if and only if x^* is a strict local minimum.

(b) Let g be a continuous function. Show that if x^* is locally stable, there exists a $\delta > 0$ such that for all sufficiently small $\epsilon > 0$, the function $f(x) + \epsilon g(x)$ has an unconstrained local minimum x_ϵ that lies within the sphere centered at x^* with radius δ. Furthermore, $x_\epsilon \to x^*$ as $\epsilon \to 0$.

1.1.13 (Sensitivity) (www)

Let $f : \Re^n \mapsto \Re$ and $g : \Re^n \mapsto \Re$ be twice continuously differentiable functions, and let x^* be a nonsingular local minimum of f. Show that there exists an $\bar{\epsilon} > 0$ and a $\delta > 0$ such that for all $\epsilon \in [0, \bar{\epsilon})$ the function $f(x) + \epsilon g(x)$ has a unique local minimum x_ϵ within the sphere $\{x \mid \|x - x^*\| < \delta\}$, and we have

$$x_\epsilon = x^* - \epsilon\big(\nabla^2 f(x^*)\big)^{-1}\nabla g(x^*) + o(\epsilon).$$

Hint: Use the implicit function theorem (Prop. A.25 in Appendix A).

1.2 GRADIENT METHODS – CONVERGENCE

We now start our development of computational methods for unconstrained optimization. The conceptual framework of this section is fundamental in nonlinear programming and applies to constrained optimization methods as well.

1.2.1 Descent Directions and Stepsize Rules

As in the case of optimality conditions, the main ideas of unconstrained optimization methods have simple geometrical explanations, but the corresponding convergence analysis is often complex. For this reason we first discuss informally the methods and their behavior in the present subsection, and we substantiate our conclusions with rigorous analysis in Section 1.2.2.

Consider the problem of unconstrained minimization of a continuously differentiable function $f : \Re^n \mapsto \Re$. Most of the interesting algorithms for this problem rely on an important idea, called *iterative descent* that works as follows: We start at some point x^0 (an initial guess) and successively generate vectors $x^1, x^2, \ldots$, such that f is decreased at each iteration, that is

$$f(x^{k+1}) < f(x^k), \qquad k = 0, 1, \ldots,$$

(cf. Fig. 1.2.1). In doing so, we successively improve our current solution estimate and we hope to decrease f all the way to its minimum. In this section, we introduce a general class algorithms based on iterative descent, and we analyze their convergence to local minima. In Section 1.3 we examine their rate of convergence properties.

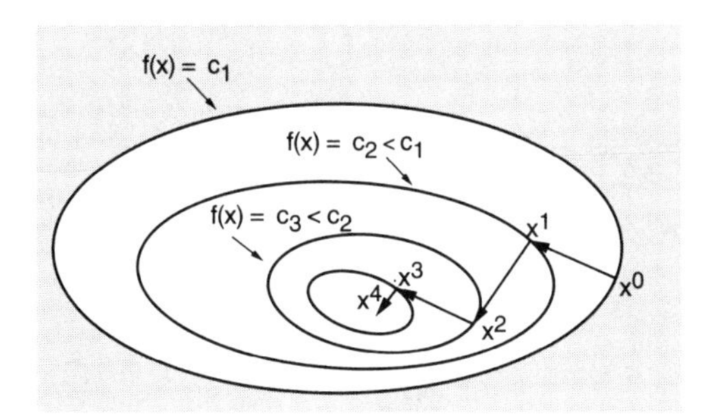

Figure 1.2.1. Iterative descent for minimizing a function f. Each vector in the generated sequence has a lower cost than its predecessor.

Gradient Methods

Given a vector $x \in \Re^n$ with $\nabla f(x) \neq 0$, consider the half line of vectors

$$x_\alpha = x - \alpha \nabla f(x), \qquad \forall\, \alpha \geq 0.$$

From the first order Taylor series expansion around x we have

$$\begin{aligned} f(x_\alpha) &= f(x) + \nabla f(x)'(x_\alpha - x) + o\big(\|x_\alpha - x\|\big) \\ &= f(x) - \alpha \|\nabla f(x)\|^2 + o\big(\alpha \|\nabla f(x)\|\big), \end{aligned}$$

so we can write

$$f(x_\alpha) = f(x) - \alpha \|\nabla f(x)\|^2 + o(\alpha).$$

The term $\alpha \|\nabla f(x)\|^2$ dominates $o(\alpha)$ for α near zero, so for positive but sufficiently small α, $f(x_\alpha)$ is smaller than $f(x)$ as illustrated in Fig. 1.2.2.

Carrying this idea one step further, consider the half line of vectors

$$x_\alpha = x + \alpha d, \qquad \forall\, \alpha \geq 0,$$

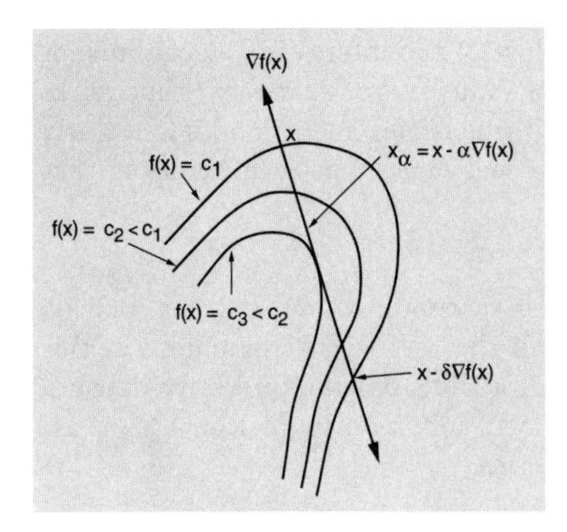

Figure 1.2.2. If $\nabla f(x) \neq 0$, there is an interval $(0, \delta)$ of stepsizes such that $f(x - \alpha\nabla f(x)) < f(x)$ for all $\alpha \in (0, \delta)$.

where the direction vector $d \in \Re^n$ makes an angle with $\nabla f(x)$ that is greater than 90 degrees, that is,

$$\nabla f(x)'d < 0.$$

Again by Taylor's theorem we have

$$f(x_\alpha) = f(x) + \alpha\nabla f(x)'d + o(\alpha).$$

For α near zero, the term $\alpha\nabla f(x)'d$ dominates $o(\alpha)$ and as a result, for positive but sufficiently small α, $f(x+\alpha d)$ is smaller than $f(x)$ as illustrated in Fig. 1.2.3.

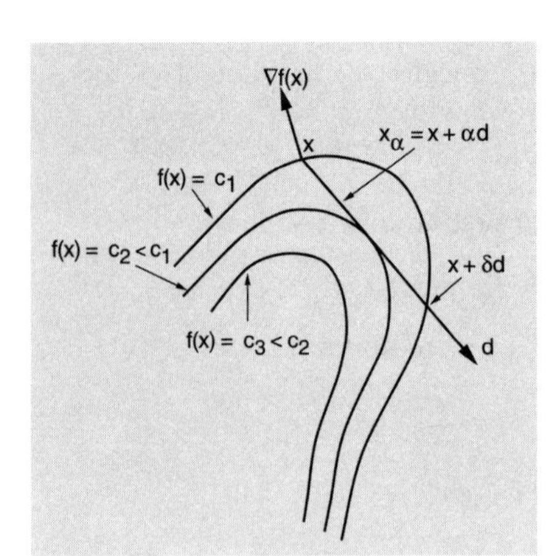

Figure 1.2.3. If the direction d makes an angle with $\nabla f(x)$ that is greater than 90 degrees, that is, $\nabla f(x)'d < 0$, there is an interval $(0, \delta)$ of stepsizes such that $f(x + \alpha d) < f(x)$ for all $\alpha \in (0, \delta)$.

The preceding observations form the basis for the broad and important class of algorithms

$$x^{k+1} = x^k + \alpha^k d^k, \qquad k = 0, 1, \ldots \tag{1.5}$$

where, if $\nabla f(x^k) \neq 0$, the direction d^k is chosen so that

$$\nabla f(x^k)'d^k < 0, \tag{1.6}$$

and the stepsize α^k is chosen to be positive. If $\nabla f(x^k) = 0$, the method stops, i.e., $x^{k+1} = x^k$ (equivalently we choose $d^k = 0$). In view of the relation (1.6) of the direction d^k and the gradient $\nabla f(x^k)$, we call algorithms of this type *gradient methods*. [There is no universally accepted name for these algorithms; some authors reserve the name "gradient method" for the special case where $d^k = -\nabla f(x^k)$.] The majority of the gradient methods that we will consider are also descent algorithms; that is, the stepsize α^k is selected so that

$$f(x^k + \alpha^k d^k) < f(x^k), \qquad k = 0, 1, \ldots \tag{1.7}$$

However, there are some exceptions.

There is a large variety of possibilities for choosing the direction d^k and the stepsize α^k in a gradient method. Indeed there is no single gradient method that can be recommended for all or even most problems. Otherwise said, given any one of the numerous methods and variations thereof that we will discuss, there are interesting types of problems for which this method is well-suited. Our principal analytical aim is to develop a few guiding principles for understanding the performance of broad classes of methods and for appreciating the practical contexts in which their use is most appropriate.

Selecting the Descent Direction

Many gradient methods are specified in the form

$$x^{k+1} = x^k - \alpha^k D^k \nabla f(x^k), \tag{1.8}$$

where D^k is a positive definite symmetric matrix. Since $d^k = -D^k \nabla f(x^k)$, the descent condition $\nabla f(x^k)'d^k < 0$ is written as

$$\nabla f(x^k)' D^k \nabla f(x^k) > 0,$$

and holds thanks to the positive definiteness of D^k.

Here are some examples of choices of the matrix D^k, resulting in methods that are widely used:

Steepest Descent

$$D^k = I, \qquad k = 0, 1, \ldots,$$

where I is the $n \times n$ identity matrix. This is the simplest choice but it often leads to slow convergence, as we will see in Section 1.3. The difficulty is illustrated in Fig. 1.2.4 and motivates the methods of the subsequent examples.

The name "steepest descent" is derived from an interesting property of the (normalized) negative gradient direction $d^k = -\nabla f(x^k)/\|\nabla f(x^k)\|$; among all directions $d \in \Re^n$ that are normalized so that $\|d\| = 1$, it is the one that minimizes the slope $\nabla f(x^k)'d$ of the cost $f(x^k + \alpha d)$ along the direction d at $\alpha = 0$. Indeed, by the Schwartz inequality (Prop. A.2 in Appendix A), we have for all d with $\|d\| = 1$,

$$\nabla f(x^k)'d \geq -\|\nabla f(x^k)\| \cdot \|d\| = -\|\nabla f(x^k)\|,$$

and it is seen that equality is attained above for d equal to $-\nabla f(x^k)/\|\nabla f(x^k)\|$.

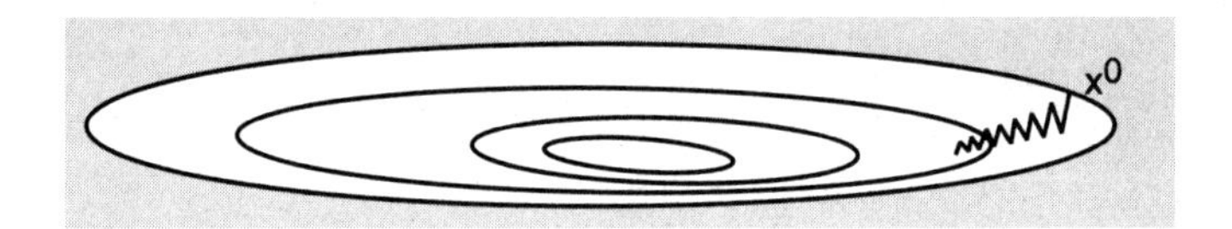

Figure 1.2.4. Slow convergence of the steepest descent method

$$x^{k+1} = x^k - \alpha^k \nabla f(x^k)$$

when the equal cost surfaces of f are "elongated." The difficulty is that the gradient direction is almost orthogonal to the direction that leads to the minimum. As a result the method is zig-zagging without making fast progress.

Newton's Method

$$D^k = \left(\nabla^2 f(x^k)\right)^{-1}, \qquad k = 0, 1, \ldots,$$

provided $\nabla^2 f(x^k)$ is positive definite. If $\nabla^2 f(x^k)$ is not positive definite, some modification is necessary as will be explained in Section 1.4. The idea in Newton's method is to minimize at each iteration the quadratic approximation of f around the current point x^k given by

$$f^k(x) = f(x^k) + \nabla f(x^k)'(x - x^k) + \tfrac{1}{2}(x - x^k)'\nabla^2 f(x^k)(x - x^k),$$

(see Fig. 1.2.5). By setting the derivative of $f^k(x)$ to zero,

$$\nabla f(x^k) + \nabla^2 f(x^k)(x - x^k) = 0,$$

we obtain the next iterate x^{k+1} as the minimum of $f^k(x)$:

$$x^{k+1} = x^k - \left(\nabla^2 f(x^k)\right)^{-1} \nabla f(x^k).$$

This is the pure Newton iteration. It corresponds to the more general iteration

$$x^{k+1} = x^k - \alpha^k \left(\nabla^2 f(x^k)\right)^{-1} \nabla f(x^k),$$

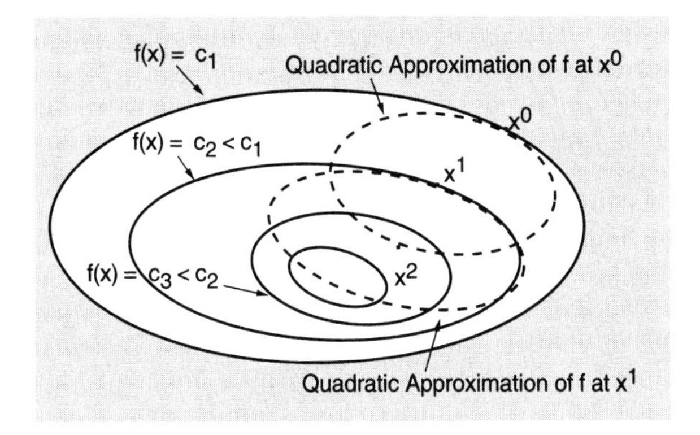

Figure 1.2.5. Illustration of the fast convergence rate of Newton's method with a stepsize equal to one. Given x^k, the method obtains x^{k+1} as the minimum of a quadratic approximation of f based on a second order Taylor expansion around x^k.

where the stepsize $\alpha^k = 1$. Note that Newton's method finds the global minimum of a positive definite quadratic function in a single iteration (assuming $\alpha^k = 1$). More generally, Newton's method typically converges very fast asymptotically and does not exhibit the zig-zagging behavior of steepest descent, as we will show in Section 1.4. For this reason many other methods try to emulate Newton's method. Some examples are given below.

Diagonally Scaled Steepest Descent

$$D^k = \begin{pmatrix} d_1^k & 0 & 0 & \cdots & 0 & 0 & 0 \\ 0 & d_2^k & 0 & \cdots & 0 & 0 & 0 \\ \vdots & \vdots & \vdots & \ddots & \vdots & \vdots & \\ 0 & 0 & 0 & \cdots & 0 & d_{n-1}^k & 0 \\ 0 & 0 & 0 & \cdots & 0 & 0 & d_n^k \end{pmatrix}, \qquad k = 0, 1, \ldots,$$

where d_i^k are positive scalars, thus ensuring that D^k is positive definite. A popular choice, resulting in a method known as a *diagonal approximation to Newton's method*, is to take d_i^k to be an approximation to the inverted second partial derivative of f with respect to x_i, that is,

$$d_i^k \approx \left(\frac{\partial^2 f(x^k)}{(\partial x_i)^2} \right)^{-1}.$$

Modified Newton's Method

$$D^k = \left(\nabla^2 f(x^0)\right)^{-1}, \qquad k = 0, 1, \ldots,$$

provided $\nabla^2 f(x^0)$ is positive definite. This method is the same as Newton's method except that to economize on overhead, the Hessian matrix is not recalculated at each iteration. A related method is obtained when the Hessian is recomputed every $p > 1$ iterations.

Discretized Newton's Method

$$D^k = \big(H(x^k)\big)^{-1}, \qquad k = 0, 1, \ldots,$$

where $H(x^k)$ is a positive definite symmetric approximation of $\nabla^2 f(x^k)$, formed by using finite difference approximations of the second derivatives, based on first derivatives or values of f.

Gauss-Newton Method

This method is applicable to the problem of minimizing the sum of squares of real valued functions $g_1, \ldots, g_m$, a problem often encountered in statistical data analysis and in the context of neural network training (see Section 1.5). By denoting $g = (g_1, \ldots, g_m)$, the problem is written as

$$\begin{aligned} &\text{minimize} \quad f(x) = \tfrac{1}{2}\|g(x)\|^2 = \tfrac{1}{2}\sum_{i=1}^{m}\big(g_i(x)\big)^2 \\ &\text{subject to} \quad x \in \Re^n. \end{aligned}$$

We choose

$$D^k = \big(\nabla g(x^k)\nabla g(x^k)'\big)^{-1}, \qquad k = 0, 1, \ldots,$$

assuming the matrix $\nabla g(x^k)\nabla g(x^k)'$ is invertible. The latter matrix is always positive semidefinite, and it is positive definite and hence invertible if and only if the matrix $\nabla g(x^k)$ has rank n (Prop. A.20 in Appendix A). Since

$$\nabla f(x^k) = \nabla g(x^k) g(x^k),$$

the Gauss-Newton method takes the form

$$x^{k+1} = x^k - \alpha^k \big(\nabla g(x^k)\nabla g(x^k)'\big)^{-1}\nabla g(x^k) g(x^k). \tag{1.9}$$

We will see in Section 1.5 that the Gauss-Newton method may be viewed as an approximation to Newton's method, particularly when the optimal value of $\|g(x)\|^2$ is small.

Other choices of D^k yield the class of *Quasi-Newton methods* discussed in Section 1.7. There are also some interesting methods where the direction d^k is not usually expressed as $d^k = -D^k\nabla f(x^k)$. Important examples are the *conjugate gradient method* and the *coordinate descent methods* discussed in Sections 1.6 and 1.8, respectively.

Stepsize Selection

There are a number of rules for choosing the stepsize α^k in a gradient method. We list some that are used widely in practice:

Minimization Rule

Here α^k is such that the cost function is minimized along the direction d^k, that is, α^k satisfies

$$f(x^k + \alpha^k d^k) = \min_{\alpha \geq 0} f(x^k + \alpha d^k). \tag{1.10}$$

Limited Minimization Rule

This is a version of the minimization rule, which is more easily implemented in many cases. A fixed scalar $s > 0$ is selected and α^k is chosen to yield the greatest cost reduction over all stepsizes in the interval $[0, s]$, i.e.,

$$f(x^k + \alpha^k d^k) = \min_{\alpha \in [0,\, s]} f(x^k + \alpha d^k).$$

The minimization and limited minimization rules must typically be implemented with the aid of one-dimensional line search algorithms (see Appendix C). In general, the minimizing stepsize cannot be computed exactly, and in practice, the line search is stopped once a stepsize α^k satisfying some termination criterion is obtained. Some stopping criteria are discussed in Exercise 1.2.16.

Successive Stepsize Reduction – Armijo Rule

To avoid the often considerable computation associated with the line minimization rules, it is natural to consider rules based on successive stepsize reduction. In the simplest rule of this type an initial stepsize s is chosen, and if the corresponding vector $x^k + sd^k$ does not yield an improved value of f, that is, $f(x^k + sd^k) \geq f(x^k)$, the stepsize is reduced, perhaps repeatedly, by a certain factor, until the value of f is improved. While this method often works in practice, it is theoretically unsound because the cost improvement obtained at each iteration may not be substantial enough to guarantee convergence to a minimum. This is illustrated in Fig. 1.2.6.

The Armijo rule is essentially the successive reduction rule just described, suitably modified to eliminate the theoretical convergence difficulty shown in Fig. 1.2.6. Here, fixed scalars s, β, and σ, with $0 < \beta < 1$, and $0 < \sigma < 1$ are chosen, and we set $\alpha^k = \beta^{m_k} s$, where m_k is the first nonnegative integer m for which

$$f(x^k) - f(x^k + \beta^m s d^k) \geq -\sigma \beta^m s \nabla f(x^k)' d^k. \tag{1.11}$$

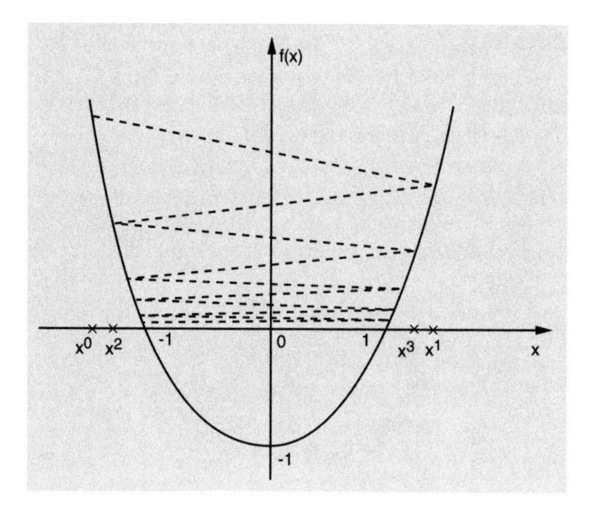

Figure 1.2.6. Example of failure of the successive stepsize reduction rule for the one-dimensional function

$$f(x) = \begin{cases} \dfrac{3(1-x)^2}{4} - 2(1-x) & \text{if } x > 1, \\ \dfrac{3(1+x)^2}{4} - 2(1+x) & \text{if } x < -1, \\ x^2 - 1, & \text{if } -1 \le x \le 1. \end{cases}$$

The gradient of f is given by

$$\nabla f(x) = \begin{cases} \dfrac{3x}{2} + \frac{1}{2} & \text{if } x > 1, \\ \dfrac{3x}{2} - \frac{1}{2} & \text{if } x < -1, \\ 2x, & \text{if } -1 \le x \le 1. \end{cases}$$

It is seen that f is strictly convex, continuously differentiable, and is minimized at $x^* = 0$. Furthermore, for any two scalars x, and $\tilde{x}$ we have

$$f(x) < f(\tilde{x}) \qquad \text{if and only if} \qquad |x| < |\tilde{x}|.$$

We have for $x > 1$

$$x - \nabla f(x) = x - \frac{3x}{2} - \tfrac{1}{2} = -\left(1 + \frac{x-1}{2}\right),$$

from which it can be verified that $|x - \nabla f(x)| < |x|$, so that $f(x - \nabla f(x)) < f(x)$ and $x - \nabla f(x) < -1$. Similarly, for $x < -1$, we have $f(x - \nabla f(x)) < f(x)$ and $x - \nabla f(x) > 1$. Consider now the steepest descent iteration where the stepsize is successively reduced from an initial stepsize $s = 1$ until descent is obtained. Let the starting point satisfy $|x^0| > 1$. From the preceding equations, it follows that $f(x^0 - \nabla f(x^0)) < f(x^0)$ and the stepsize $s = 1$ will be accepted by the method. Thus, the next point is $x^1 = x^0 - \nabla f(x^0)$, which satisfies $|x^1| > 1$. By repeating the preceding argument, we see that the generated sequence $\{x^k\}$ satisfies $|x^k| > 1$ for all k, and cannot converge to the unique stationary point $x^* = 0$. In fact, it can be shown that $\{x^k\}$ will have two limit points, $\bar{x} = 1$ and $\bar{x} = -1$, for every x^0 with $|x^0| > 1$.

In other words, the stepsizes $\beta^m s$, $m = 0, 1, \ldots$, are tried successively until the above inequality is satisfied for $m = m_k$. Thus, the cost improvement must not be just positive; it must be sufficiently large as per the test (1.11). Figure 1.2.7 illustrates the rule.

Usually σ is chosen close to zero, for example, $\sigma \in [10^{-5}, 10^{-1}]$. The reduction factor β is usually chosen from 1/2 to 1/10 depending on the confidence we have on the quality of the initial stepsize s. We can always take $s = 1$ and multiply the direction d^k by a scaling factor. Many methods, such as Newton-like methods, incorporate some type of implicit scaling of the direction d^k, which makes $s = 1$ a good stepsize choice (see the discussion on rate of convergence in Section 1.3). If a suitable scaling factor for d^k is not known, one may use various ad hoc schemes to determine one. For example, a simple possibility is based on quadratic interpolation of the function

$$g(\alpha) = f(x^k + \alpha d^k),$$

which is the cost along the direction d^k, viewed as a function of the stepsize α. In this scheme, we select some stepsize $\bar{\alpha}$, evaluate $g(\bar{\alpha})$, and perform the quadratic interpolation of g on the basis of $g(0) = f(x^k)$, $dg(0)/d\alpha = \nabla f(x^k)'d^k$, and $g(\bar{\alpha})$. If $\tilde{\alpha}$ minimizes the quadratic interpolation, we replace d^k by $\tilde{d}^k = \tilde{\alpha} d^k$, and we use an initial stepsize $s = 1$.

Goldstein Rule

Here, a fixed scalar $\sigma \in (0, 1/2)$ is selected, and α^k is chosen to satisfy

$$\sigma \leq \frac{f(x^k + \alpha^k d^k) - f(x^k)}{\alpha^k \nabla f(x^k)'d^k} \leq 1 - \sigma,$$

(cf. Fig. 1.2.8). It is possible to show that if f is bounded below, there exists an interval of stepsizes α^k for which the relation above is satisfied. There are fairly simple algorithms for finding such a stepsize but we will not go into the details, since in practice the simpler Armijo rule seems to be universally preferred. The Goldstein rule is included here primarily because of its historical significance: it was the first sound proposal for a general-purpose stepsize rule that did not rely on line minimization, and it embodies the fundamental idea on which the subsequently proposed Armijo rule was based.

Constant Stepsize

Here a fixed stepsize $s > 0$ is selected and

$$\alpha^k = s, \qquad k = 0, 1, \ldots$$

The constant stepsize rule is very simple. However, if the stepsize is too large, divergence will occur, while if the stepsize is too small, the rate of

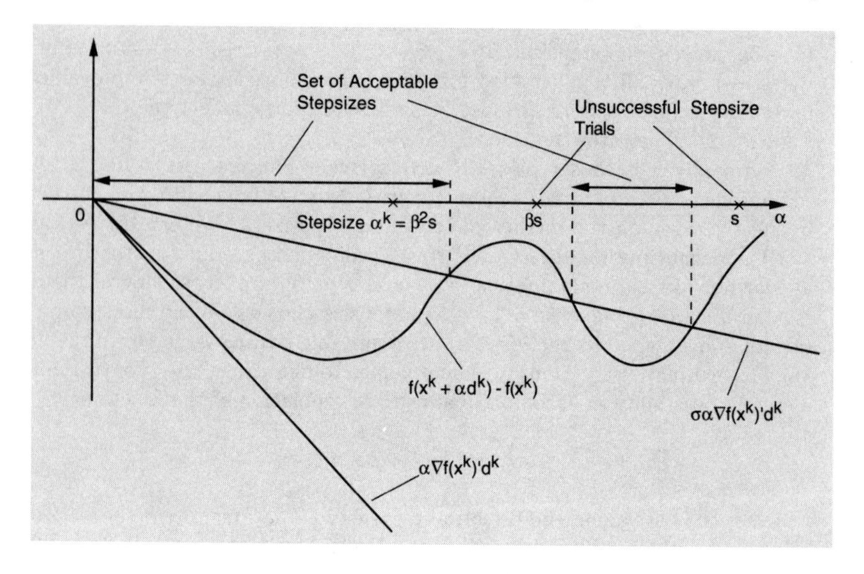

Figure 1.2.7. Line search by the Armijo rule. We start with the trial stepsize s and continue with $\beta s, \beta^2 s, \ldots$, until the first time that $\beta^m s$ falls within the set of stepsizes α satisfying the inequality

$$f(x^k) - f(x^k + \alpha d^k) \geq -\sigma\alpha\nabla f(x^k)'d^k.$$

While this set need not be an interval, it will always contain an interval of the form $[0, \delta]$ with $\delta > 0$, provided $\nabla f(x^k)'d^k < 0$. For this reason the stepsize α^k chosen by the Armijo rule is well defined and will be found after a finite number of trial evaluations of f at the points $(x^k + sd^k), (x^k + \beta s d^k), \ldots$

convergence may be very slow. Thus, the constant stepsize rule is useful only for problems where an appropriate constant stepsize value is known or can be determined fairly easily. (A method that attempts to determine automatically an appropriate value of stepsize is given in Exercise 1.2.20.)

Diminishing Stepsize

Here the stepsize converges to zero,

$$\alpha^k \to 0.$$

This stepsize rule is different than the preceding ones in that it does not guarantee descent at each iteration, although descent becomes more likely as the stepsize diminishes. One difficulty with a diminishing stepsize is that it may become so small that substantial progress cannot be maintained, even when far from a stationary point. For this reason, we require that

$$\sum_{k=0}^{\infty} \alpha^k = \infty.$$

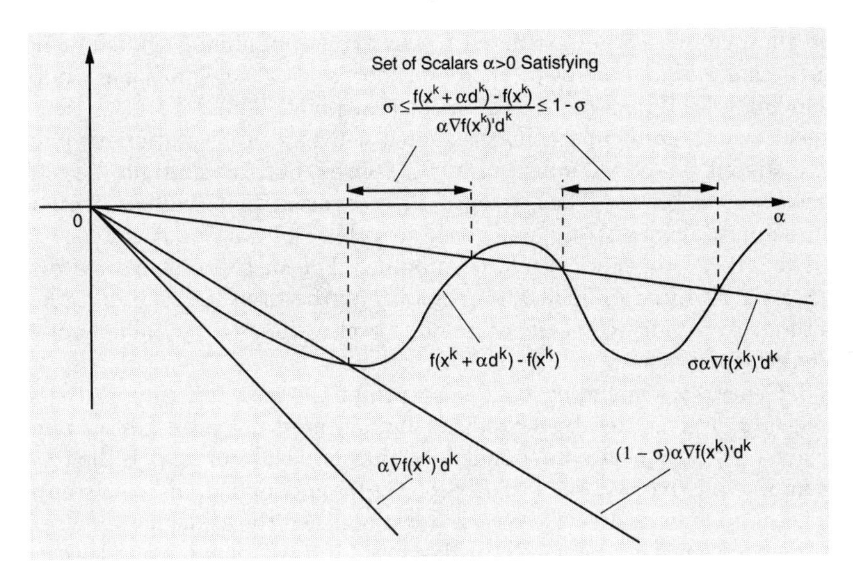

Figure 1.2.8. Illustration of the set of stepsizes that are acceptable in the Goldstein rule.

The last condition guarantees that $\{x^k\}$ does not converge to a nonstationary point. Indeed, if $x^k \to \bar{x}$, then for any large indexes m and n $(m > n)$ we have

$$x^m \approx x^n \approx \bar{x}, \qquad x^m \approx x^n - \left(\sum_{k=n}^{m-1} \alpha^k\right) \nabla f(\bar{x}),$$

which is a contradiction when $\bar{x}$ is nonstationary and $\sum_{k=n}^{m-1} \alpha^k$ can be made arbitrarily large. Generally, the diminishing stepsize rule has good theoretical convergence properties (see Prop. 1.2.4, and Exercises 1.2.13 and 1.2.14). The associated convergence rate tends to be slow, so this stepsize rule is used primarily in situations where slow convergence is inevitable; for example, in singular problems or when the gradient is calculated with error (see the discussion later in this section).

Convergence Issues

Let us now delineate the type of convergence issues that we would like to clarify. We will first discuss informally these issues and we will state and prove the associated convergence results in Section 1.2.2. Given a gradient method, ideally we would like the generated sequence $\{x^k\}$ to converge to a global minimum. Unfortunately, however, this is too much to expect, at least when f is not convex, because of the presence of local minima that are not global. Indeed a gradient method is guided downhill by the form of f near the current iterate, while being oblivious to the global structure of f, and thus, can easily get attracted to any type of minimum, global or not. Furthermore, if a gradient method starts or lands at any stationary point, including a local maximum, it stops at that point. Thus, the most we can expect from a gradient method is that it converges to a stationary point. Such a point is a global minimum if f is convex, but this need not be so for nonconvex problems. Thus, it must be recognized that gradient methods can be quite inadequate, particularly if little is known about the location and/or other properties of global minima. For such problems one must either try an often difficult and frustrating process of running a gradient method from multiple starting points, or else resort to a fundamentally different approach.

Generally, depending on the nature of the cost function f, the sequence $\{x^k\}$ generated by a gradient method need not have a limit point; in fact $\{x^k\}$ is typically unbounded if f has no local minima. If, however, we know that the level set $\{x \mid f(x) \leq f(x^0)\}$ is bounded, and the stepsize is chosen to enforce descent at each iteration, then the sequence $\{x^k\}$ must be bounded since it belongs to this level set. It must then have at least one limit point; this is because every bounded sequence has at least one limit point (see Prop. A.5 of Appendix A).

Even if $\{x^k\}$ is bounded, convergence to a single limit point may not be easy to guarantee. However, it can be shown that local minima, which are isolated stationary points (unique stationary points within some open sphere), tend to attract most types of gradient methods, that is, once a gradient method gets sufficiently close to such a local minimum, it converges to it. This is the subject of a simple and remarkably powerful result, the *capture theorem*, which is given in the next subsection (Prop. 1.2.5). Exercise 1.2.13 develops another type of convergence result for the steepest descent method, which applies to the case where there are multiple nonisolated local minima. Generally, if there is a connected set of multiple global minima, it is theoretically possible for $\{x^k\}$ to have multiple limit points (see Exercise 1.2.18), but the occurance of such a phenomenon has never been documented in practice.

Limit Points of Gradient Methods

We now address the question of whether each limit point of a sequence $\{x^k\}$ generated by a gradient method is a stationary point. From the first order Taylor expansion

$$f(x^{k+1}) = f(x^k) + \alpha^k \nabla f(x^k)' d^k + o(\alpha^k),$$

we see that if the slope of f at x^k along the direction d^k, which is $\nabla f(x^k)' d^k$, has "substantial" magnitude, the rate of progress of the method will also tend to be substantial. If on the other hand, the directions d^k tend to become asymptotically orthogonal to the gradient direction,

$$\frac{\nabla f(x^k)'d^k}{\|\nabla f(x^k)\|\|d^k\|} \to 0,$$

as x^k approaches a nonstationary point, there is a chance that the method will get "stuck" near that point. To ensure that this does not happen, we consider rather technical conditions on the directions d^k, which are either naturally satisfied or can be easily enforced in most algorithms of interest.

One such condition for the case where

$$d^k = -D^k \nabla f(x^k),$$

is to assume that the eigenvalues of the positive definite symmetric matrix D^k are bounded above and bounded away from zero, that is, for some positive scalars c_1 and c_2, we have

$$c_1\|z\|^2 \le z'D^k z \le c_2\|z\|^2, \qquad \forall\ z \in \Re^n,\ k = 0, 1, \ldots \tag{1.12}$$

It can be seen then that

$$|\nabla f(x^k)'d^k| = |\nabla f(x^k)'D^k\nabla f(x^k)| \ge c_1\|\nabla f(x^k)\|^2$$

and

$$\|d^k\|^2 = |\nabla f(x^k)'(D^k)^2\nabla f(x^k)| \le c_2^2\|\nabla f(x^k)\|^2,$$

where we have used the fact that, from Eq. (1.12), c_2 is no less than the largest eigenvalue of D^k, and that the eigenvalues of $(D^k)^2$ are equal to the squares of the corresponding eigenvalues of D^k (Props. A.18 and A.13 in Appendix A). Thus, as long as $\nabla f(x^k)$ does not tend to zero, $\nabla f(x^k)$ and d^k cannot become asymptotically orthogonal.

We now introduce another "nonorthogonality" type of condition, which is more general than the "bounded eigenvalues" condition (1.12). Let us assume that the direction d^k is uniquely determined by the corresponding iterate x^k; that is, d^k is obtained as a given function of x^k. We say that the direction sequence $\{d^k\}$ is *gradient related* to $\{x^k\}$ if the following property can be shown:

For any subsequence $\{x^k\}_{k\in\mathcal{K}}$ that converges to a nonstationary point, the corresponding subsequence $\{d^k\}_{k\in\mathcal{K}}$ is bounded and satisfies

$$\limsup_{k\to\infty,\, k\in\mathcal{K}} \nabla f(x^k)'d^k < 0. \tag{1.13}$$

In particular, if $\{d^k\}$ is gradient related, it follows that if a subsequence $\{\nabla f(x^k)\}_{k\in\mathcal{K}}$ tends to a nonzero vector, the corresponding subsequence of directions d^k is bounded and does not tend to be orthogonal to $\nabla f(x^k)$.

Roughly, this means that d^k does not become "too small" or "too large" relative to $\nabla f(x^k)$, and that the angle between d^k and $\nabla f(x^k)$ does not get "too close" to 90 degrees.

We can often guarantee *a priori* that $\{d^k\}$ is gradient related. In particular, if $d^k = -D^k \nabla f(x^k)$ and the eigenvalues of D^k are bounded as in the "bounded eigenvalues" condition (1.12), it can be seen that $\{d^k\}$ is gradient related, provided x^k is nonstationary for all k (if x^k is stationary for some k, the issue of convergence in effect does not arise). Two other examples of conditions that, if satisfied for some scalars $c_1 > 0$, $c_2 > 0$, $p_1 \geq 0$, $p_2 \geq 0$, and all k, guarantee that $\{d^k\}$ is gradient related are

(a)

$$c_1 \|\nabla f(x^k)\|^{p_1} \leq -\nabla f(x^k)' d^k, \qquad \|d^k\| \leq c_2 \|\nabla f(x^k)\|^{p_2}.$$

(b)

$$d^k = -D^k \nabla f(x^k),$$

with D^k a positive definite symmetric matrix satisfying

$$c_1 \|\nabla f(x^k)\|^{p_1} \, \|z\|^2 \leq z' D^k z \leq c_2 \|\nabla f(x^k)\|^{p_2} \, \|z\|^2, \qquad \forall \, z \in \Re^n.$$

This condition generalizes the "bounded eigenvalues" condition (1.12), which is obtained for $p_1 = p_2 = 0$.

An important convergence result is that if $\{d^k\}$ is gradient related and the minimization rule, or the limited minimization rule, or the Armijo rule is used, then all limit points of $\{x^k\}$ are stationary. This is shown in Prop. 1.2.1, given in the next subsection. Proposition 1.2.2 provides a similar result for the Goldstein rule.

When a constant stepsize is used, convergence can be proved assuming that the stepsize is sufficiently small and that f satisfies some further conditions (cf. Prop. 1.2.3). Under the same conditions, convergence can also be proved for a diminishing stepsize.

There is a common line of proof for these convergence results. The main idea is that the cost function is improved at each iteration and that, based on our assumptions, the improvement is "substantial" near a nonstationary point, i.e., it is bounded away from zero. We then argue that the algorithm cannot approach a nonstationary point, since in this case the total cost improvement would accumulate to infinity.

Termination of Gradient Methods

Generally, gradient methods are not finitely convergent, so it is necessary to have criteria for terminating the iterations with some assurance that we are reasonably close to at least a local minimum. A typical approach is to stop the computation when the norm of the gradient becomes sufficiently small, that is, when a point x^k is obtained with

$$\|\nabla f(x^k)\| \leq \epsilon,$$

where ϵ is a small positive scalar. Unfortunately, it is not known a priori how small one should take ϵ in order to guarantee that the final point x^k is a "good" approximation to a stationary point. The appropriate value of ϵ depends on how the problem is scaled. In particular, if f is multiplied by some scalar, the appropriate value of ϵ is also multiplied by the same scalar. It is possible to correct this difficulty by replacing the criterion $\|\nabla f(x^k)\| \leq \epsilon$ with

$$\frac{\|\nabla f(x^k)\|}{\|\nabla f(x^0)\|} \leq \epsilon.$$

Still, however, the gradient norm $\|\nabla f(x^k)\|$ depends on all the components of the gradient, and depending on how the optimization variables are scaled, the preceding termination criterion may not work well. In particular, some components of the gradient may be naturally much smaller than others, thus requiring a smaller value of ϵ than the other components.

Assuming that the direction d^k captures the relative scaling of the optimization variables, it may be appropriate to terminate computation when the norm of the direction d^k becomes sufficiently small, that is,

$$\|d^k\| \leq \epsilon.$$

Still the appropriate value of ϵ may not be easy to guess, and it may be necessary to experiment prior to settling on a reasonable termination criterion for a given problem. Sometimes, other problem-dependent criteria are used, in addition to or in place of $\|\nabla f(x^k)\| \leq \epsilon$ and $\|d^k\| \leq \epsilon$.

When $\nabla^2 f(x)$ is positive definite, the condition $\|\nabla f(x^k)\| \leq \epsilon$ yields bounds on the distance from local minima. In particular, if x^* is a local minimum of f and there exists $m > 0$ such that for all x in a sphere S centered at x^* we have

$$m\|z\|^2 \leq z'\nabla^2 f(x)z, \qquad \forall\ z \in \Re^n,$$

then every $x \in S$ satisfying $\|\nabla f(x)\| \leq \epsilon$ also satisfies

$$\|x - x^*\| \leq \frac{\epsilon}{m}, \qquad f(x) - f(x^*) \leq \frac{\epsilon^2}{m},$$

(see Exercise 1.2.10).

In the absence of positive definiteness conditions on $\nabla^2 f(x)$, it may be very difficult to infer the proximity of the current iterate to the optimal solution set by just using the gradient norm. We will return to this point when we will discuss singular local minima in the next section.

Spacer Steps

Often, optimization problems are solved with complex descent algorithms in which the rule used to determine the next point may depend on several previous points or on the iteration index k. Some of the conjugate direction algorithms discussed in Section 1.6 are of this type. Other algorithms consist of a combination of different methods and switch from one method to the other in a manner that may either be prespecified or may depend on the progress of the algorithm. Such combinations are usually introduced in order to improve speed of convergence or reliability. However, their convergence analysis can become extremely complicated. It is thus often valuable to know that if in such algorithms one inserts, perhaps irregularly but infinitely often, an iteration of a convergent algorithm such as the gradient methods of this section, then the theoretical convergence properties of the overall algorithm are quite satisfactory. Such an iteration is known as a *spacer step*. The related convergence result is given in Prop. 1.2.6. The only requirement imposed on the iterations of the algorithm other than the spacer steps is that they do not increase the cost; these iterations, however, need not strictly decrease the cost.

Gradient Methods with Random and Nonrandom Errors*

Frequently in optimization problems the gradient $\nabla f(x^k)$ is not computed exactly. Instead, one has available

$$g^k = \nabla f(x^k) + e^k,$$

where e^k is an uncontrollable error vector. There are several potential sources of error; roundoff error, and discretization error due to finite difference approximations to the gradient are two possibilities, but there are others that will be discussed in more detail in Section 1.5. Let us for concreteness focus on the steepest descent method with errors,

$$x^{k+1} = x^k - \alpha^k g^k,$$

and let us consider several qualitatively different cases:

(a) e^k **is small relative to the gradient**, that is,

$$\|e^k\| < \|\nabla f(x^k)\|, \qquad \forall\ k.$$

Then, assuming $\nabla f(x^k) \neq 0$, $-g^k$ is a direction of cost improvement, that is, $\nabla f(x^k)'g^k > 0$. This is illustrated in Fig. 1.2.9, and is verified by the calculation

$$\begin{aligned} \nabla f(x^k)'g^k &= \|\nabla f(x^k)\|^2 + \nabla f(x^k)'e^k \\ &\geq \|\nabla f(x^k)\|^2 - \|\nabla f(x^k)\|\,\|e^k\| \\ &= \|\nabla f(x^k)\|\big(\|\nabla f(x^k)\| - \|e^k\|\big) \\ &> 0. \end{aligned} \tag{1.14}$$

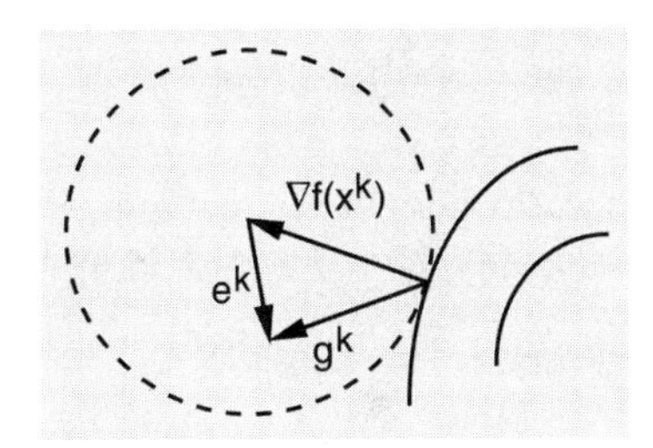

Figure 1.2.9. Illustration of the descent property of the direction $g^k = \nabla f(x^k) + e^k$. If the error e^k has smaller norm than the gradient $\nabla f(x^k)$, then g^k lies strictly within the sphere centered at $\nabla f(x^k)$ with radius $\|\nabla f(x^k)\|$, and thus makes an angle less than 90 degrees with $\nabla f(x^k)$.

In this case convergence results that are analogous to Props. 1.2.3 and 1.2.4 can be shown.

(b) $\{e^k\}$ **is bounded**, that is,

$$\|e^k\| \leq \delta, \qquad \forall\ k,$$

where δ is some scalar. Then by the preceding calculation (1.14), the method operates like a descent method within the region

$$\big\{x \mid \|\nabla f(x)\| > \delta\big\}.$$

In the complementary region where $\|\nabla f(x)\| \leq \delta$, the method can behave quite unpredictably. For example, if the errors e^k are constant, say $e^k \equiv e$, then since $g^k = \nabla f(x^k) + e$, the method will essentially be trying to minimize $f(x) + e'x$ and will typically converge to a point $\bar{x}$ with $\nabla f(\bar{x}) = -e$. If the errors e^k vary substantially, the method will tend to oscillate within the region where $\|\nabla f(x)\| \leq \delta$ (see Exercise 1.2.17 and also Exercise 1.3.4 in the next section). The precise behavior will depend on the precise nature of the errors, and on whether a constant or a diminishing stepsize is used (see also the following cases).

(c) $\{e^k\}$ **is proportional to the stepsize**, that is,

$$\|e^k\| \leq \alpha^k q, \qquad \forall\ k,$$

where q is some scalar. If the stepsize is constant, we come under case (b), while if the stepsize is diminishing, the behavior described in case (b) applies, but with $\delta \to 0$, so the method will tend to converge to a stationary point of f. Important situations where the condition $\|e^k\| \leq \alpha^k q$ holds will be encountered in Section 1.5 (see Prop. 1.5.1 of that section). A more general condition under which similar behavior occurs is

$$\|e^k\| \leq \alpha^k\big(q + p\|\nabla f(x^k)\|\big), \qquad \forall\ k,$$

where q and p are some scalars. Generally, under this condition and with a diminishing stepsize, the convergence behavior is similar to the case where there are no errors (see the following Prop. 1.2.4 and Exercise 1.2.21).

(d) $\{e^k\}$ **are independent zero mean random vectors with finite variance.** An important special case where such errors arise is when f is of the form

$$f(x) = E_w\{F(x, w)\}, \tag{1.15}$$

where $F : \Re^{m+n} \to \Re$ is some function, w is a random vector in $\Re^m$, and $E_w\{\cdot\}$ denotes expected value. Under very mild assumptions it can be shown that if F is continuously differentiable, the same is true of f and furthermore,

$$\nabla f(x) = E_w\{\nabla_x F(x, w)\}.$$

Often an approximation g^k to $\nabla f(x^k)$ is computed by simulation or by using a limited number of samples of $\nabla F(x, w)$, with potentially substantial error resulting. In an extreme case, we have

$$g^k = \nabla_x F(x^k, w^k),$$

where w^k is a single sample value corresponding to x^k. Then the error

$$e^k = \nabla_x F(x^k, w^k) - \nabla f(x^k) = \nabla_x F(x^k, w^k) - E_w\{\nabla_x F(x^k, w)\}$$

need not diminish with $\|\nabla f(x^k)\|$, but has zero mean, and under appropriate conditions, its effects are "averaged out." What is happening here is that the descent condition $\nabla f(x^k)'g^k > 0$ holds *on the average* at nonstationary points x^k. It is still possible that for some sample values of e^k, the direction g^k is "bad", but with a diminishing stepsize, the occasional use of a bad direction cannot deteriorate the cost enough for the method to oscillate, given that on the average the method uses "good" directions. The detailed analysis of gradient methods with random errors is beyond the scope of this text. We refer to the literature (see e.g. [BeT89], [KuC78], [KuY97], [LPW92], [Pfl96], [PoT73a], [Pol87], [TBA86], [BeT96]). Let us mention one representative convergence result, due to [BeT99], which parallels the following Prop. 1.2.4 that deals with a gradient method without errors: if in the iteration

$$x^{k+1} = x^k - \alpha^k \nabla_x F(x^k, w^k)$$

the random variables $w^0, w^1, \ldots$ are independent with finite variance, the stepsize is diminishing and satisfies

$$\alpha^k \to 0, \qquad \sum_{k=0}^{\infty} \alpha^k = \infty, \qquad \sum_{k=0}^{\infty} (\alpha^k)^2 < \infty,$$

and some additional technical conditions hold, then with probability one, we either have $f(x^k) \to -\infty$ or else $\nabla f(x^k) \to 0$.

The Role of Convergence Analysis

The following subsection gives a number of mathematical propositions relating to the convergence properties of gradient methods. The meaning of these propositions is usually quite intuitive but their statement often requires complicated mathematical assumptions. Furthermore, their proof often involves tedious ϵ-δ arguments, so at first sight students may wonder whether "we really have to go through all this."

When Euclid was faced with a similar question from king Ptolemy of Alexandria, he replied that "there is no royal road to geometry." In our case, however, the answer is not so simple because we are not dealing with a pure subject such as geometry that may be developed without regard for its practical application. In the eyes of most people, the value of an analysis or algorithm in nonlinear programming is judged primarily by its practical impact in solving various types of problems. It is therefore important to give some thought to the interface between convergence analysis and its practical application. To this end it is useful to consider two extreme viewpoints; most workers in the field find themselves somewhere between the two.

In the first viewpoint, convergence analysis is considered primarily a mathematical subject. The properties of an algorithm are quantified to the extent possible through mathematical statements. General and broadly applicable assertions, and simple and elegant proofs are at a premium here. The rationale is that simple statements and proofs are more readily understood, and general statements apply not only to the problems at hand but also to other problems that are likely to appear in the future. On the negative side, one may remark that simplicity is not always compatible with relevance, and broad applicability is often achieved through assumptions that are hard to verify or appreciate.

The second viewpoint largely rejects the role of mathematical analysis. The rationale here is that the validity and the properties of an algorithm for a given class of problems must be verified through practical experimentation anyway, so if an algorithm looks promising on intuitive grounds, why bother with a convergence analysis. Furthermore, there are a number of important practical questions that are hard to address analytically, such as roundoff error, multiple local minima, and a variety of finite termination and approximation issues. The main criticism of this viewpoint is that mathematical analysis often reveals (and explains) fundamental flaws of algorithms that experimentation may miss. These flaws often point the way to better algorithms or modified algorithms that are tailored to the type of practical problem at hand. Similarly, analysis may be more effective than experimentation in delineating the types of problems for which particular algorithms are well-suited.

Our own mathematical approach is tempered by practical concerns, but we note that the balance between theory and practice in nonlinear programming is particularly delicate, subjective, and problem dependent. Aside from the fact that the mathematical proofs themselves often provide valuable insight into algorithms, here are some of our reasons for insisting on a rigorous convergence analysis:

(a) We want to delineate the range of applicability of various methods. In particular, we want to know for what type of cost function (once or twice differentiable, convex or nonconvex, with singular or nonsingular minima) each algorithm is best suited. If the cost function violates the assumptions under which a given algorithm can be proved to converge, it is reasonable to suspect that the algorithm is unsuitable for this cost function.

(b) We want to provide information about the qualitative behavior of various methods. For example, we want to know whether convergence of the method depends on the availability of a good starting point, whether the iterates x^k or just the function values $f(x^k)$ are guaranteed to converge, etc. This information may supplement and/or guide the computational experimentation.

(c) We want to provide guidelines for choosing a few algorithms for further experimentation out of the often bewildering array of candidate algorithms that are applicable for the solution of a given type of problem. One of the principal means for this is the rate of convergence analysis to be given in Section 1.3. Note here that while an algorithm may provably converge, in practice it may be entirely inappropriate for a given problem because it converges very slowly. Experience has shown that without a good understanding of the rate of convergence properties of algorithms it may be difficult to exclude bad candidates from consideration without costly experimentation.

At the same time one should be aware of some of the limitations of the mathematical results that we will provide. For example, some of the assumptions under which an algorithm will be proved convergent may be hard to verify for a given type of problem. Furthermore, our convergence rate analysis of Section 1.3 is largely asymptotic; that is, it applies near the eventual limit of the generated sequence. It is possible, that an algorithm has a good asymptotic rate of convergence but it works poorly in practice for a given type of problem because it is very slow in its initial phase.

There is still another viewpoint, which is worth addressing because it is often adopted by the casual user of nonlinear programming algorithms. This user is really interested in a particular application of nonlinear programming in his/her special field, and is counting on an existing code or package to solve the problem (several such packages are commercially or publicly available). Since the package will do most of the work, the user may hope that a superficial acquaintance with the properties of the algorithms underlying the package will suffice. This hope is sometimes realized but unfortunately in the majority of the cases it is not. There are a number of reasons for this. First, there are many packages implementing a lot of different methods, and to choose the right package, one needs to have insight into the suitability of different methods for the special features of the application at hand. Second, to use a package one must often know how to suitably formulate the problem, how to set various parameters (e.g. termination criteria, stepsize parameters, etc.), and how to interpret the results of the computation (particularly when things don't work out as hoped initially, which is often the case). For this, one needs considerable insight into the inner workings of the algorithm underlying the package. Finally, for a challenging practical optimization problem (e.g. one of large dimension), it may be essential to exploit its special structure, and packages often do not have this capability. As a result the user may have to modify the package or write an altogether new code that is tailored to the application at hand. Both of these require an intimate understanding of the convergence properties and other characteristics of the relevant nonlinear programming algorithms.

1.2.2 Convergence Results

We now provide an analysis of the convergence behavior of gradient methods. The following proposition is the main convergence result.

Proposition 1.2.1: (Stationarity of Limit Points for Gradient Methods) Let $\{x^k\}$ be a sequence generated by a gradient method $x^{k+1} = x^k + \alpha^k d^k$, and assume that $\{d^k\}$ is gradient related [cf. Eq. (1.13)] and α^k is chosen by the minimization rule, or the limited minimization rule, or the Armijo rule. Then every limit point of $\{x^k\}$ is a stationary point.

Proof: Consider the Armijo rule, and to arrive at a contradiction, assume that $\bar{x}$ is a limit point of $\{x^k\}$ with $\nabla f(\bar{x}) \neq 0$. Note that since $\{f(x^k)\}$ is monotonically nonincreasing, $\{f(x^k)\}$ either converges to a finite value or diverges to $-\infty$. Since f is continuous, $f(\bar{x})$ is a limit point of $\{f(x^k)\}$, so it follows that the entire sequence $\{f(x^k)\}$ converges to $f(\bar{x})$. Hence,

$$f(x^k) - f(x^{k+1}) \to 0.$$

By the definition of the Armijo rule, we have

$$f(x^k) - f(x^{k+1}) \geq -\sigma \alpha^k \nabla f(x^k)' d^k. \tag{1.16}$$

Hence, $\alpha^k \nabla f(x^k)' d^k \to 0$. Let $\{x^k\}_{\mathcal{K}}$ be a subsequence converging to $\bar{x}$. Since $\{d^k\}$ is gradient related, we have

$$\limsup_{\substack{k \to \infty \\ k \in \mathcal{K}}} \nabla f(x^k)' d^k < 0,$$

and therefore

$$\{\alpha^k\}_{\mathcal{K}} \to 0.$$

Hence, by the definition of the Armijo rule, we must have for some index $\bar{k} \geq 0$

$$f(x^k) - f\big(x^k + (\alpha^k/\beta)d^k\big) < -\sigma(\alpha^k/\beta)\nabla f(x^k)'d^k, \qquad \forall\, k \in \mathcal{K},\ k \geq \bar{k}, \tag{1.17}$$

that is, the initial stepsize s will be reduced at least once for all $k \in \mathcal{K}$, $k \geq \bar{k}$. Denote

$$p^k = \frac{d^k}{\|d^k\|}, \qquad \bar{\alpha}^k = \frac{\alpha^k \|d^k\|}{\beta}.$$

Since $\{d^k\}$ is gradient related, $\{\|d^k\|\}_{\mathcal{K}}$ is bounded, and it follows that

$$\{\bar{\alpha}^k\}_{\mathcal{K}} \to 0.$$

Since $\|p^k\| = 1$ for all $k \in \mathcal{K}$, there exists a subsequence $\{p^k\}_{\bar{\mathcal{K}}}$ of $\{p^k\}_{\mathcal{K}}$ such that

$$\{p^k\}_{\bar{\mathcal{K}}} \to \bar{p},$$

where $\bar{p}$ is some vector with $\|\bar{p}\| = 1$ [Prop. A.5(c) in Appendix A]. From Eq. (1.17), we have

$$\frac{f(x^k) - f(x^k + \bar{\alpha}^k p^k)}{\bar{\alpha}^k} < -\sigma \nabla f(x^k)'p^k, \qquad \forall\, k \in \bar{\mathcal{K}},\ k \geq \bar{k}. \tag{1.18}$$

By using the mean value theorem, this relation is written as

$$-\nabla f(x^k + \tilde{\alpha}^k p^k)'p^k < -\sigma \nabla f(x^k)'p^k, \qquad \forall\, k \in \bar{\mathcal{K}},\ k \geq \bar{k},$$

where $\tilde{\alpha}^k$ is a scalar in the interval $[0, \bar{\alpha}^k]$. Taking limits in the above equation we obtain

$$-\nabla f(\bar{x})'\bar{p} \leq -\sigma \nabla f(\bar{x})'\bar{p}$$

or

$$0 \leq (1-\sigma)\nabla f(\bar{x})'\bar{p}.$$

Since $\sigma < 1$, it follows that

$$0 \leq \nabla f(\bar{x})'\bar{p}. \tag{1.19}$$

On the other hand, we have

$$\nabla f(x^k)'p^k = \frac{\nabla f(x^k)'d^k}{\|d^k\|}.$$

By taking the limit as $k \in \bar{\mathcal{K}}$, $k \to \infty$,

$$\nabla f(\bar{x})'\bar{p} \leq \frac{\limsup_{k\to\infty,\, k\in\bar{\mathcal{K}}} \nabla f(x^k)'d^k}{\limsup_{k\to\infty,\, k\in\bar{\mathcal{K}}} \|d^k\|} < 0,$$

which contradicts Eq. (1.19). This proves the result for the Armijo rule.

Consider now the minimization rule, and let $\{x^k\}_\mathcal{K}$ converge to $\bar{x}$ with $\nabla f(\bar{x}) \neq 0$. Again we have that $\{f(x^k)\}$ decreases monotonically to $f(\bar{x})$. Let $\tilde{x}^{k+1}$ be the point generated from x^k via the Armijo rule, and let $\tilde{\alpha}^k$ be the corresponding stepsize. We have

$$f(x^k) - f(x^{k+1}) \geq f(x^k) - f(\tilde{x}^{k+1}) \geq -\sigma\tilde{\alpha}^k \nabla f(x^k)'d^k.$$

By repeating the arguments of the earlier proof following Eq. (1.16), replacing α^k by $\tilde{\alpha}^k$, we can obtain a contradiction. In particular, we have

$$\{\tilde{\alpha}^k\}_\mathcal{K} \to 0,$$

and by the definition of the Armijo rule, we have for some index $\bar{k} \geq 0$

$$f(x^k) - f\big(x^k + (\tilde{\alpha}^k/\beta)d^k\big) < -\sigma(\tilde{\alpha}^k/\beta)\nabla f(x^k)'d^k, \qquad \forall\, k \in \mathcal{K},\ k \geq \bar{k},$$

[cf. Eq. (1.17)]. Proceeding as earlier, we obtain Eqs. (1.18) and (1.19) (with $\bar{\alpha}^k = \tilde{\alpha}^k\|d^k\|/\beta$), and a contradiction of Eq. (1.19).

The line of argument just used establishes that any stepsize rule that gives a larger reduction in cost at each iteration than the Armijo rule inherits its convergence properties. This also proves the proposition for the limited minimization rule. **Q.E.D.**

The following proposition can be shown similar to Prop. 1.2.1. Its proof is left for the reader.

Proposition 1.2.2: The conclusions of Prop. 1.2.1 hold if $\{d^k\}$ is gradient related and α^k is chosen by the Goldstein rule.

The next proposition establishes, among other things, convergence for the case of a constant stepsize. The idea is that if the rate of growth of the gradient of f is limited (i.e., the curvature of f is limited), then one can construct a quadratic function f^k that majorizes f; see Fig. 1.2.10. Given x^k and d^k, an appropriate constant stepsize α^k can then be obtained within an interval around the scalar $\bar{\alpha}^k$ that minimizes f^k along the direction d^k.

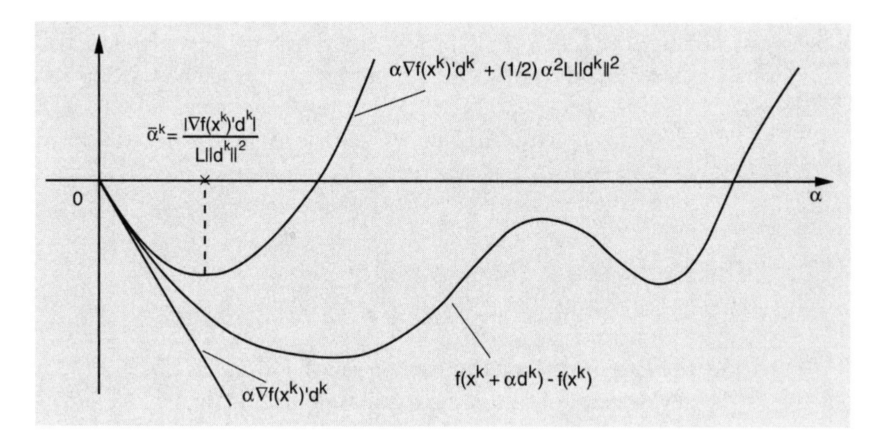

Figure 1.2.10. The idea of the proof of Prop. 1.2.3. Given x^k and the descent direction d^k, the cost difference $f(x^k+\alpha d^k)-f(x^k)$ is majorized by $\alpha\nabla f(x^k)'d^k+\frac{1}{2}\alpha^2L\|d^k\|^2$ (see the proof of Prop. 1.2.3). Minimization of this function over α yields the stepsize

$$\bar{\alpha}^k=\frac{|\nabla f(x^k)'d^k|}{L\|d^k\|^2},$$

which reduces the cost function f as well (see the proof of Prop. 1.2.3).

Proposition 1.2.3: (Convergence for a Constant Stepsize) Let $\{x^k\}$ be a sequence generated by a gradient method $x^{k+1}=x^k+\alpha^k d^k$, where $\{d^k\}$ is gradient related. Assume that for some constant $L>0$, we have

$$\|\nabla f(x)-\nabla f(y)\|\le L\|x-y\|,\qquad \forall\ x,y\in\Re^n, \tag{1.20}$$

and that for all k we have $d^k\neq 0$ and

$$\epsilon\le\alpha^k\le(2-\epsilon)\bar{\alpha}^k, \tag{1.21}$$

where

$$\bar{\alpha}^k=\frac{|\nabla f(x^k)'d^k|}{L\|d^k\|^2},$$

and ϵ is a fixed positive scalar. Then every limit point of $\{x^k\}$ is a stationary point of f.

Note: The gradient related assumption on $\{d^k\}$ does not quite guarantee that an ϵ satisfying Eq. (1.21) exists (since this assumption relates to subsequences only). However, standard conditions that imply the gradient related assumption also guarantee that an ϵ satisfying Eq. (1.21) exist. In particular, if $\{d^k\}$ is such that there exist positive scalars c_1, c_2 such that for all k we have

$$c_1\|\nabla f(x^k)\|^2 \le -\nabla f(x^k)'d^k, \qquad \|d^k\|^2 \le c_2\|\nabla f(x^k)\|^2, \tag{1.22}$$

then Eq. (1.21) is satisfied if for all k we have

$$\epsilon \le \alpha^k \le \frac{c_1(2-\epsilon)}{Lc_2}. \tag{1.23}$$

In the case of steepest descent $[d^k = -\nabla f(x^k)]$, we can take $c_1 = c_2 = 1$, and the condition on the stepsize becomes

$$\epsilon \le \alpha^k \le \frac{2-\epsilon}{L}.$$

Furthermore, if for all k, we have $d^k = -D^k\nabla f(x^k)$ with D^k positive definite symmetric and having eigenvalues in an interval $[\gamma, \Gamma]$, the condition (1.22) can be seen to hold with

$$c_1 = \gamma, \qquad c_2 = \Gamma^2.$$

Proof: By using the descent lemma (Prop. A.24 of Appendix A), we obtain

$$\begin{aligned} f(x^k + \alpha^k d^k) - f(x^k) &\le \alpha^k\nabla f(x^k)'d^k + \tfrac{1}{2}(\alpha^k)^2L\|d^k\|^2 \\ &= \alpha^k\big(\tfrac{1}{2}\alpha^kL\|d^k\|^2 - |\nabla f(x^k)'d^k|\big). \end{aligned} \tag{1.24}$$

The right-hand side of Eq. (1.21) yields

$$\tfrac{1}{2}\alpha^kL\|d^k\|^2 - |\nabla f(x^k)'d^k| \le -\tfrac{1}{2}\epsilon|\nabla f(x^k)'d^k|.$$

Using this relation together with the condition $\alpha^k \ge \epsilon$ in the inequality (1.24), we obtain the following bound on the cost improvement obtained at iteration k:

$$f(x^k) - f(x^k + \alpha^k d^k) \ge \tfrac{1}{2}\epsilon^2|\nabla f(x^k)'d^k|. \tag{1.25}$$

Now if a subsequence $\{x^k\}_K$ converges to a nonstationary point $\bar{x}$, we must have, as in the proof of Prop. 1.2.1, $f(x^k) - f(x^{k+1}) \to 0$, and Eq. (1.25) implies that $|\nabla f(x^k)'d^k| \to 0$. This contradicts the hypothesis that $\{d^k\}$ is gradient related. Hence, every limit point of $\{x^k\}$ is stationary. **Q.E.D.**

A condition of the form

$$\|\nabla f(x) - \nabla f(y)\| \le L\|x - y\|, \qquad \forall\, x, y \in \Re^n,$$

[cf. Eq. (1.20)] is called a *Lipschitz continuity* condition on ∇f. Exercise 1.2.3 provides an example showing that this condition is essential for the validity of Prop. 1.2.3. The Lipschitz condition requires roughly that the "curvature" of f is no more than L in all directions. In particular, it is possible to show that this condition is satisfied for some $L > 0$, if f is twice differentiable and the Hessian $\nabla^2 f$ is bounded over $\Re^n$. Unfortunately, however, it is generally difficult to obtain an estimate of L, so in most cases the range of stepsizes that guarantee convergence [cf. Eq. (1.21) or (1.23)] is unknown, and experimentation may be necessary to obtain an appropriate range of stepsizes (a method that attempts to determine an appropriate value of stepsize automatically is given in Exercise 1.2.20). Even worse, many types of cost function f, though twice differentiable have Hessian $\nabla^2 f$ that is unbounded over $\Re^n$ [this is so for any function $f(x)$ that grows faster than a quadratic as $x \to \infty$, such as $f(x) = \|x\|^3$].

Fortunately, the Lipschitz condition can be significantly weakened, as shown in Exercise 1.2.5. In particular, it is sufficient that it holds for all x, y in the set $\{z \mid f(z) \le f(x^0)\}$, in which case, however, the range of stepsizes that guarantee convergence depends on the starting point x^0.

The Lipschitz continuity condition also essentially guarantees convergence for a diminishing stepsize, as shown by the following proposition.

Proposition 1.2.4: (Convergence for a Diminishing Stepsize) Let $\{x^k\}$ be a sequence generated by a gradient method $x^{k+1} = x^k + \alpha^k d^k$. Assume that for some constant $L > 0$, we have

$$\|\nabla f(x) - \nabla f(y)\| \le L\|x - y\|, \qquad \forall\, x, y \in \Re^n, \tag{1.26}$$

and that there exist positive scalars c_1, c_2 such that for all k we have

$$c_1\|\nabla f(x^k)\|^2 \le -\nabla f(x^k)'d^k, \qquad \|d^k\|^2 \le c_2\|\nabla f(x^k)\|^2. \tag{1.27}$$

Suppose also that

$$\alpha^k \to 0, \qquad \sum_{k=0}^{\infty} \alpha^k = \infty.$$

Then either $f(x^k) \to -\infty$ or else $\{f(x^k)\}$ converges to a finite value and $\nabla f(x^k) \to 0$. Furthermore, every limit point of $\{x^k\}$ is a stationary point of f.

Proof: Combining Eqs. (1.24) and (1.27), we have

$$\begin{aligned} f(x^{k+1}) &\le f(x^k) + \alpha^k \left(\tfrac{1}{2}\alpha^k L\|d^k\|^2 - |\nabla f(x^k)'d^k|\right) \\ &\le f(x^k) - \alpha^k \left(c_1 - \tfrac{1}{2}\alpha^k c_2 L\right) \|\nabla f(x^k)\|^2. \end{aligned}$$

Since the linear term in α^k dominates the quadratic term in α^k for sufficiently small α^k, and $\alpha^k \to 0$, we have for some positive constant c and all k greater than some index $\bar{k}$,

$$f(x^{k+1}) \le f(x^k) - \alpha^k c\|\nabla f(x^k)\|^2. \tag{1.28}$$

From this relation, we see that for $k \ge \bar{k}$, $\{f(x^k)\}$ is monotonically decreasing, so either $f(x^k) \to -\infty$ or $\{f(x^k)\}$ converges to a finite value. In the latter case, by adding Eq. (1.28) over all $k \ge \bar{k}$, we obtain

$$c\sum_{k=\bar{k}}^{\infty} \alpha^k \|\nabla f(x^k)\|^2 \le f(x^{\bar{k}}) - \lim_{k\to\infty} f(x^k) < \infty.$$

We see that there cannot exist an $\epsilon > 0$ such that $\|\nabla f(x^k)\|^2 > \epsilon$ for all k greater than some $\hat{k}$, since this would contradict the assumption $\sum_{k=0}^{\infty} \alpha^k = \infty$. Therefore, we must have $\liminf_{k\to\infty} \|\nabla f(x^k)\| = 0$.

To show that $\nabla f(x^k) \to 0$, assume the contrary; that is,

$$\limsup_{k\to\infty} \|\nabla f(x^k)\| \ge \epsilon > 0. \tag{1.29}$$

Let $\{m_j\}$ and $\{n_j\}$ be sequences of indexes such that

$$m_j < n_j < m_{j+1},$$

$$\frac{\epsilon}{3} < \|f(x^k)\| \qquad \text{for } m_j \le k < n_j, \tag{1.30}$$

$$\|f(x^k)\| \le \frac{\epsilon}{3} \qquad \text{for } n_j \le k < m_{j+1}. \tag{1.31}$$

Let also $\bar{j}$ be sufficiently large so that

$$\sum_{k=m_{\bar{j}}}^{\infty} \alpha^k \|\nabla f(x^k)\|^2 < \frac{\epsilon^2}{9L\sqrt{c_2}}. \tag{1.32}$$

For any $j \geq \bar{j}$ and any m with $m_j \leq m \leq n_j - 1$, we have

$$\begin{aligned}
\|\nabla f(x^{n_j}) - \nabla f(x^m)\| &\leq \sum_{k=m}^{n_j-1} \|\nabla f(x^{k+1}) - \nabla f(x^k)\| \\
&\leq L \sum_{k=m}^{n_j-1} \|x^{k+1} - x^k\| \\
&= L \sum_{k=m}^{n_j-1} \alpha^k \|d^k\| \\
&\leq L\sqrt{c_2} \sum_{k=m}^{n_j-1} \alpha^k \|\nabla f(x^k)\| \\
&\leq \frac{3L\sqrt{c_2}}{\epsilon} \sum_{k=m}^{n_j-1} \alpha^k \|\nabla f(x^k)\|^2 \\
&\leq \frac{3L\sqrt{c_2}}{\epsilon} \frac{\epsilon^2}{9L\sqrt{c_2}} \\
&= \frac{\epsilon}{3},
\end{aligned}$$

where the last two inequalities follow using Eqs. (1.30) and (1.32). Thus

$$\|\nabla f(x^m)\| \leq \|\nabla f(x^{n_j})\| + \frac{\epsilon}{3} \leq \frac{2\epsilon}{3}, \qquad \forall\, j \geq \bar{j},\ m_j \leq m \leq n_j - 1.$$

Thus, using also Eq. (1.31), we have for all $m \geq m_{\bar{j}}$

$$\|\nabla f(x^m)\| \leq \frac{2\epsilon}{3}.$$

This contradicts Eq. (1.29), implying that $\lim_{k\to\infty} \nabla f(x^k) = 0$.

Finally, if $\bar{x}$ is a limit point of x^k, then $f(x^k)$ converges to the finite value $f(\bar{x})$. Thus we have $\nabla f(x^k) \to 0$, implying that $\nabla f(\bar{x}) = 0$. **Q.E.D.**

Under the assumptions of the preceding proposition, descent is not guaranteed in the initial iterations. However, if the stepsizes are all sufficiently small [e.g., they satisfy the right-hand side inequality of Eq. (1.23)], descent is guaranteed at all iterations. In this case, it is sufficient that the Lipschitz condition $\|\nabla f(x) - \nabla f(y)\| \leq L\|x - y\|$ holds for all x, y in the set $\{z \mid f(z) \leq f(x^0)\}$ (see Exercise 1.2.5); otherwise the Lipschitz condition must hold over a set larger than $\{z \mid f(z) \leq f(x^0)\}$ to guarantee convergence (see Exercise 1.2.15).

The following proposition explains to some extent why sequences generated by gradient methods tend in practice to have unique limit points.

It states that local minima which happen to be isolated stationary points tend to attract gradient methods; once the method gets close enough to such a minimum it remains close and converges to it.

Proposition 1.2.5: (Capture Theorem) Let f be continuously differentiable and let $\{x^k\}$ be a sequence satisfying $f(x^{k+1}) \le f(x^k)$ for all k and generated by a gradient method $x^{k+1} = x^k + \alpha^k d^k$, which is convergent in the sense that every limit point of sequences that it generates is a stationary point of f. Assume that there exist scalars $s > 0$ and $c > 0$ such that for all k there holds

$$\alpha^k \le s, \qquad \|d^k\| \le c\|\nabla f(x^k)\|.$$

Let x^* be a local minimum of f, which is the only stationary point of f within some open set. Then there exists an open set S containing x^* such that if $x^{\bar{k}} \in S$ for some $\bar{k} \ge 0$, then $x^k \in S$ for all $k \ge \bar{k}$ and $\{x^k\} \to x^*$. Furthermore, given any scalar $\bar{\epsilon} > 0$, the set S can be chosen so that $\|x - x^*\| < \bar{\epsilon}$ for all $x \in S$.

Note: The conditions $f(x^{k+1}) \le f(x^k)$ and $\alpha^k \le s$ are satisfied for the Armijo rule and the limited minimization rule. They are also satisfied for a constant and a diminishing stepsize under conditions that guarantee descent at each iteration (see the proofs of Props. 1.2.3 and 1.2.4). The condition $\|d^k\| \le c\|\nabla f(x^k)\|$ is satisfied if $d^k = -D^k \nabla f(x^k)$ with the eigenvalues of D^k bounded from above.

Proof: Suppose that $\rho > 0$ is such that

$$f(x^*) < f(x), \qquad \forall\, x \text{ with } \|x - x^*\| \le \rho.$$

Define for $t \in [0, \rho]$

$$\phi(t) = \min_{\{x \mid t \le \|x - x^*\| \le \rho\}} f(x) - f(x^*),$$

and note that ϕ is a monotonically nondecreasing function of t, and that $\phi(t) > 0$ for all $t \in (0, \rho]$. Given any $\epsilon \in (0, \rho]$, let $r \in (0, \epsilon]$ be such that

$$\|x - x^*\| < r \quad \Rightarrow \quad \|x - x^*\| + sc\|\nabla f(x)\| < \epsilon. \tag{1.33}$$

Consider the open set

$$S = \big\{x \mid \|x - x^*\| < \epsilon,\ f(x) < f(x^*) + \phi(r)\big\}.$$

We claim that if $x^k \in S$ for some k, then $x^{k+1} \in S$.

Indeed if $x^k \in S$, from the definition of ϕ and S we have

$$\phi(\|x^k - x^*\|) \leq f(x^k) - f(x^*) < \phi(r).$$

Since ϕ is monotonically nondecreasing, the above relation implies that $\|x^k - x^*\| < r$, so that by Eq. (1.33),

$$\|x^k - x^*\| + sc\|\nabla f(x^k)\| < \epsilon.$$

We also have by using the hypotheses $\alpha^k \leq s$ and $\|d^k\| \leq c\|\nabla f(x^k)\|$

$$\|x^{k+1} - x^*\| \leq \|x^k - x^*\| + \|\alpha^k d^k\| \leq \|x^k - x^*\| + sc\|\nabla f(x^k)\|,$$

so from the last two relations it follows that $\|x^{k+1} - x^*\| < \epsilon$. Since $f(x^{k+1}) < f(x^k)$, we also obtain $f(x^{k+1}) - f(x^*) < \phi(r)$, so we conclude that $x^{k+1} \in S$.

By using induction it follows that if $x^{\bar{k}} \in S$ for some $\bar{k}$, we have $x^k \in S$ for all $k \geq \bar{k}$. Let $\bar{S}$ be the closure of S. Since $\bar{S}$ is compact, the sequence $\{x^k\}$ will have at least one limit point, which by assumption must be a stationary point of f. Now the only stationary point of f within $\bar{S}$ is the point x^* (since we have $\|x - x^*\| \leq \rho$ for all $x \in \bar{S}$). Hence $x^k \to x^*$. Finally given any $\bar{\epsilon} > 0$, we can choose $\epsilon \leq \bar{\epsilon}$ in which case we have $\|x - x^*\| < \bar{\epsilon}$ for all $x \in S$. **Q.E.D.**

Finally, we state a result that deals with the convergence of algorithms involving a combination of different methods. It shows that for convergence it is enough to insert, perhaps irregularly but infinitely often, an iteration of a convergent gradient algorithm, provided that the other iterations do not degrade the value of the cost function. The proof is similar to the one of Prop. 1.2.1, and is left for the reader.

Proposition 1.2.6: (Convergence for Spacer Steps) Consider a sequence $\{x^k\}$ such that

$$f(x^{k+1}) \leq f(x^k), \qquad k = 0, 1, \ldots$$

Assume that there exists an infinite set $\mathcal{K}$ of integers for which

$$x^{k+1} = x^k + \alpha^k d^k, \qquad \forall\, k \in \mathcal{K},$$

where $\{d^k\}_{\mathcal{K}}$ is gradient related and α^k is chosen by the minimization rule, or the limited minimization rule, or the Armijo rule. Then every limit point of the subsequence $\{x^k\}_{\mathcal{K}}$ is a stationary point.

EXERCISES

1.2.1

Consider the problem of minimizing the function of two variables $f(x, y) = 3x^2 + y^4$.

(a) Apply one iteration of the steepest descent method with $(1, -2)$ as the starting point and with the stepsize chosen by the Armijo rule with $s = 1$, $\sigma = 0.1$, and $\beta = 0.5$.

(b) Repeat (a) using $s = 1$, $\sigma = 0.1$, $\beta = 0.1$ instead. How does the cost of the new iterate compare to that obtained in (a)? Comment on the tradeoffs involved in the choice of β.

(c) Apply one iteration of Newton's method with the same starting point and stepsize rule as in (a). How does the cost of the new iterate compare to that obtained in (a)? How about the amount of work involved in finding the new iterate?

1.2.2

Describe the behavior of the steepest descent method with constant stepsize s for the function $f(x) = \|x\|^{2+\beta}$, where $\beta \geq 0$. For which values of s and x^0 does the method converge to $x^* = 0$. Relate your answer to the assumptions of Prop. 1.2.3.

1.2.3

Consider the function $f : \Re^n \mapsto \Re$ given by

$$f(x) = \|x\|^{3/2},$$

and the method of steepest descent with a constant stepsize. Show that for this function, the Lipschitz condition $\|\nabla f(x) - \nabla f(y)\| \leq L\|x - y\|$ for all x and y is not satisfied for any L. Furthermore, for any value of constant stepsize, the method either converges in a *finite* number of iterations to the minimizing point $x^* = 0$ or else it does not converge to x^*.

1.2.4

Apply the steepest descent method with constant stepsize α to the function f of Exercise 1.1.11. Show that the gradient ∇f satisfies the Lipschitz condition

$$\|\nabla f(x) - \nabla f(y)\| \leq L\|x - y\|, \qquad \forall\ x, y \in \Re,$$

for some constant L. Write a computer program to verify that the method is a descent method for $\alpha \in (0, 2/L)$. Do you expect to get in the limit the global minimum $x^* = 0$?

1.2.5 (www)

Suppose that the Lipschitz condition

$$\|\nabla f(x) - \nabla f(y)\| \leq L\|x - y\|, \qquad \forall\, x, y \in \Re^n,$$

[cf. Eq. (1.20)] is replaced by the condition that for every bounded set $A \subset \Re^n$, there exists some constant L such that

$$\|\nabla f(x) - \nabla f(y)\| \leq L\|x - y\|, \qquad \forall\, x, y \in A.$$

Show that:

(a) Condition (i) is always satisfied if the level sets $\{x \mid f(x) \leq c\}$ are bounded for all $c \in \Re$, and f is twice continuously differentiable.

(b) The convergence result of Prop. 1.2.3 remains valid provided that the level set

$$A = \{x \mid f(x) \leq f(x^0)\}$$

is bounded and the stepsize is allowed to depend on the choice of the initial vector x^0. *Hint*: The key idea is to show that x^k stays in the set A, and to use a stepsize α^k that depends on the constant L corresponding to this set. Let

$$R = \max\{\|x\| \mid x \in A\},$$

$$G = \max\{\|\nabla f(x)\| \mid x \in A\},$$

and

$$B = \{x \mid \|x\| \leq R + 2G\}.$$

Using condition (i), there exists some constant L such that $\|\nabla f(x) - \nabla f(y)\| \leq L\|x - y\|$, for all $x, y \in B$. Suppose the stepsize α^k satisfies

$$0 < \epsilon \leq \alpha^k \leq (2 - \epsilon)\gamma^k \min\{1, 1/L\},$$

where

$$\gamma^k = \frac{|\nabla f(x^k)'d^k|}{\|d^k\|^2}.$$

Let $\beta^k = \alpha^k(\gamma^k - L\alpha^k/2)$, which can be seen to satisfy $\beta^k \geq \epsilon^2\gamma^k/2$ by our choice of α^k. Show by induction on k that with such a choice of stepsize, we have $x^k \in A$ and

$$f(x^{k+1}) \leq f(x^k) - \beta^k\|d^k\|^2, \qquad \forall\, k \geq 0.$$

1.2.6

Suppose that f is quadratic and of the form $f(x) = \frac{1}{2}x'Qx - b'x$, where Q is positive definite and symmetric.

(a) Show that the Lipschitz condition $\|\nabla f(x) - \nabla f(y)\| \leq L\|x - y\|$ is satisfied with L equal to the maximal eigenvalue of Q.

(b) Consider the gradient method $x^{k+1} = x^k - sD\nabla f(x^k)$, where D is positive definite and symmetric. Show that the method converges to $x^* = Q^{-1}b$ for every starting point x^0 if and only if $s \in (0, 2/\bar{L})$, where $\bar{L}$ is the maximum eigenvalue of $D^{1/2}QD^{1/2}$.

1.2.7

An electrical engineer wants to maximize the current I between two points A and B of a complex network by adjusting the values x_1 and x_2 of two variable resistors, where $0 \leq x_1 \leq R_1$, $0 \leq x_2 \leq R_2$, and R_1, R_2 are given. The engineer does not have an adequate mathematical model of the network and decides to adopt the following procedure. She keeps the value x_2 of the second resistor fixed and adjusts the value of the first resistor until the current I is maximized. She then keeps the value x_1 of the first resistor fixed and adjusts the value of the second resistor until the current I is maximized. She then repeats the procedure until no further progress can be made. She knows *a priori* that during this procedure, the values x_1 and x_2 can never reach their extreme values 0, R_1, and R_2. Explain whether there is a sound theoretical basis for the engineer's procedure. *Hint*: Consider how the steepest descent method works for two-dimensional problems.

1.2.8

Consider the gradient method $x^{k+1} = x^k + \alpha^k d^k$, where α^k is chosen by the Armijo rule or the line minimization rule and

$$d^k = -\begin{bmatrix} 0 \\ \vdots \\ 0 \\ \frac{\partial f(x^k)}{\partial x_i} \\ 0 \\ \vdots \\ 0 \end{bmatrix},$$

where i is the index for which $\left|\partial f(x^k)/\partial x_j\right|$ is maximized over $j = 1, \ldots, n$. Show that every limit point of $\{x^k\}$ is stationary.

1.2.9

Consider the gradient method $x^{k+1} = x^k + \alpha^k d^k$ for the case where f is positive definite quadratic, and let $\bar{\alpha}^k$ be the stepsize corresponding to the line minimization rule. Show that a stepsize α^k satisfies the inequalities of the Goldstein rule if and only if

$$2\sigma\bar{\alpha}^k \leq \alpha^k \leq 2(1-\sigma)\bar{\alpha}^k.$$

1.2.10 (www)

Let f be twice continuously differentiable. Suppose that x^* is a local minimum such that for all x in an open sphere S centered at x^*, we have, for some $m > 0$,

$$m\|d\|^2 \leq d'\nabla^2 f(x)d, \qquad \forall\, d \in \Re^n.$$

Show that for every $x \in S$, we have

$$\|x - x^*\| \leq \frac{\|\nabla f(x)\|}{m}, \qquad f(x) - f(x^*) \leq \frac{\|\nabla f(x)\|^2}{2m}.$$

Hint: Use the relation

$$\nabla f(y) = \nabla f(x) + \int_0^1 \nabla^2 f\big(x + t(y-x)\big)(y-x)dt.$$

See also Exercise 1.1.9.

1.2.11 (Alternative Assumptions for Convergence) (www)

Consider the gradient method $x^{k+1} = x^k + \alpha^k d^k$. Instead of $\{d^k\}$ being gradient related, assume *one* of the following two conditions:

(i) It can be shown that for any subsequence $\{x^k\}_{k\in\mathcal{K}}$ that converges to a nonstationary point, the corresponding subsequence $\{d^k\}_{k\in\mathcal{K}}$ is bounded and satisfies

$$\liminf_{k\to\infty,\, k\in\mathcal{K}} \nabla f(x^k)'d^k < 0.$$

(ii) α^k is chosen by the minimization rule, and for some $c > 0$ and all k, we have

$$|\nabla f(x^k)'d^k| \geq c\|\nabla f(x^k)\|\,\|d^k\|.$$

Show that the result of Prop. 1.2.1 holds.

1.2.12 (Behavior of Steepest Descent Near a Saddle Point)

Let $f(x) = (1/2)x'Qx$, where Q is symmetric, invertible, and has at least one negative eigenvalue. Consider the steepest descent method with constant stepsize and show that unless the starting point x^0 belongs to the subspace spanned by the eigenvectors of Q corresponding to the nonnegative eigenvalues, the generated sequence $\{x^k\}$ diverges.

1.2.13 (Convergence to a Single Limit) (www)

Consider the steepest descent method $x^{k+1} = x^k - \alpha^k \nabla f(x^k)$ and assume that for all x, y, we have

$$\frac{\|\nabla f(x) - \nabla f(y)\|^2}{L} \leq \big(\nabla f(x) - \nabla f(y)\big)'(x - y).$$

(It can be shown that this condition holds if f is convex, twice continuously differentiable and its Hessian matrix has eigenvalues that are less than or equal to L.) Assume also that f has at least one stationary point. Show that $\{x^k\}$ converges to a stationary point of f under *one* of the following two conditions:

(i) For some $\epsilon > 0$, we have

$$\epsilon \leq \alpha^k \leq \frac{2 - \epsilon}{L}, \qquad \forall\, k.$$

(ii) $\alpha^k \to 0$ and $\sum_{k=0}^{\infty} \alpha^k = \infty$.

Hint: Show that for any stationary point $\bar{x}$ we have

$$\|x^{k+1} - \bar{x}\|^2 \leq \|x^k - \bar{x}\|^2 - \alpha^k \left(\frac{2}{L} - \alpha^k\right) \|\nabla f(x^k)\|^2.$$

1.2.14 (Steepest Descent with Diminishing Stepsize [CoL94]) (www)

Consider the steepest descent method

$$x^{k+1} = x^k - \alpha^k \nabla f(x^k),$$

assuming that the function f is convex.

(a) Use the convexity of f to show that for any $y \in \Re^n$, we have

$$\|x^{k+1} - y\|^2 \leq \|x^k - y\|^2 - 2\alpha^k \big(f(x^k) - f(y)\big) + \big(\alpha^k \|\nabla f(x^k)\|\big)^2.$$

(b) Assume that

$$\sum_{k=0}^{\infty} \alpha^k = \infty, \qquad \alpha^k \|\nabla f(x^k)\|^2 \to 0.$$

Show that $\liminf_{k\to\infty} f(x^k) = \inf_{x\in\Re^n} f(x)$. *Hint*: Argue by contradiction. Assume that for some $\delta > 0$, there exists y with $f(y) < f(x^k) - \delta$ for all k sufficiently large. Use part (a).

(c) Assume that

$$\alpha^k = \frac{s^k}{\|\nabla f(x^k)\|},$$

where

$$\sum_{k=0}^{\infty} s^k = \infty, \qquad \sum_{k=0}^{\infty} (s^k)^2 < \infty.$$

Show that $\liminf_{k\to\infty} f(x^k) = \inf_{x\in\Re^n} f(x)$, and that if f has at least one global minimum, then $\{x^k\}$ converges to some global minimum. *Hint*: In part (a), set y to some x^* such that $f(x^*) < f(x^k)$ for all k (if no such x^* exists, we are done). Show that the relation

$$\|x^{k+1} - x^*\|^2 \leq \|x^k - x^*\|^2 + (s^k)^2$$

implies that $\{x^k\}$ is bounded and hence also that $\{\nabla f(x^k)\}$ is bounded. Use part (b).

1.2.15 (Divergence with Diminishing Stepsize)

Consider the one-dimensional function

$$f(x) = \frac{2}{3}|x|^3 + \frac{1}{2}x^2$$

and the method of steepest descent with stepsize $\alpha^k = \gamma/(k+1)$, where γ is a positive scalar.

(a) Show that for $\gamma = 1$ and $|x^0| \geq 1$ the method diverges. In particular, show that $|x^k| \geq k+1$ for all k.

(b) Characterize as best as you can the set of pairs (γ, x^0) for which the method converges to $x^* = 0$.

(c) How do you reconcile the results of (a) and (b) with Prop. 1.2.4.

1.2.16

There are several criteria for implementing approximately the minimization rule in a gradient method. An example of such a criterion is that α^k satisfies simultaneously

$$f(x^k) - f(x^k + \alpha^k d^k) \geq -\sigma\alpha^k \nabla f(x^k)' d^k, \tag{1.34}$$

$$\nabla f(x^k + \alpha^k d^k)' d^k \geq \beta \nabla f(x^k)' d^k, \tag{1.35}$$

where α and β are some scalars with $\sigma \in (0, 1/2)$ and $\beta \in (\sigma, 1)$. If α^k is indeed a minimizing stepsize, then $\nabla f(x^k + \alpha^k d^k)' d^k = dg(\alpha^k)/d\alpha = 0$, where g is the function $g(\alpha) = f(x^k + \alpha d^k)$, so Eq. (1.35) is in effect a test on the accuracy of the minimization (see Fig. 1.2.11).

(a) Show that if conditions (1.34) and (1.35) are satisfied by a gradient method at each iteration and the direction sequence is gradient related, then all limit points of the generated sequence $\{x^k\}$ are stationary points of f.

(b) Assume that there is a scalar M such that $f(x) \geq M$. Show that there exists an interval $[c_1, c_2]$ with $0 < c_1 < c_2$, such that every $\alpha \in [c_1, c_2]$ satisfies Eqs. (1.34) and (1.35).

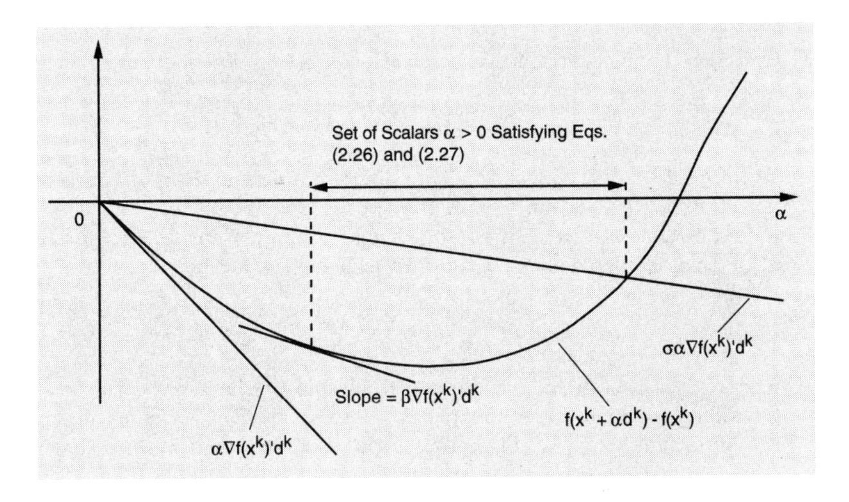

Figure 1.2.11. Illustration of the stepsize selection criterion based on Eqs. (1.34) and (1.35).

1.2.17 (Steepest Descent with Errors) (www)

Consider the steepest descent method $x^{k+1} = x^k - \alpha^k\big(\nabla f(x^k) + e^k\big)$, where e^k is an error satisfying $\|e^k\| \leq \delta$ for all k. Assume that ∇f is Lipschitz continuous. Show that for any $\delta' > \delta$, there exists a range of positive stepsizes $[\underline{\alpha}, \bar{\alpha}]$ such that if $\alpha^k \in [\underline{\alpha}, \bar{\alpha}]$ for all sufficiently large k, then either $f(x^k) \to -\infty$ or $\|\nabla f(x^k)\| < \delta'$ for infinitely many values of k. *Hint*: Use the reasoning of Prop. 1.2.3.

1.2.18 (A Continuum of Limit Points for Steepest Descent [Zou76])

Consider the two-dimensional function

$$f(x) = \begin{cases} (r-1)^2 - \frac{1}{2}(r-1)^2 \cos\left(\dfrac{1}{r-1} - \phi\right) & \text{if } r \neq 1, \\ 0 & \text{if } r = 1, \end{cases}$$

where

$$r = \sqrt{x_1^2 + x_2^2}, \qquad \phi = \arctan(x_1/x_2).$$

This function is minimized at each point of the circle where $r = 1$. Consider a nonoptimal starting point and the method of steepest descent where x^{k+1} is set equal to the first local minimum along the line $\{x^k - \alpha \nabla f(x^k) \mid \alpha \geq 0\}$. Show that this method follows a spiral path that comes arbitrarily close to every point of the circle of optimal points.

1.2.19 (Simplified Steepest Descent) (www)

(a) Consider the unconstrained minimization of a function f of the form

$$f(x) = F\big(x, g(x)\big),$$

where $g : \Re^m \to \Re^n$ is continuously differentiable and $F(x, y)$ is a continuously differentiable function of the two arguments $x \in \Re^n$ and $y \in \Re^m$. It is sometimes convenient to approximate the gradient of $F\big(x, g(x)\big)$ by neglecting the dependence on g. This leads to the method

$$x^{k+1} = x^k - \alpha^k \nabla_x F\big(x^k, g(x^k)\big),$$

where α^k is chosen by the minimization rule or the Armijo rule on the function f. (Such a method makes sense when $\nabla_x F$ is much easier to compute than $\nabla g \nabla_y F$.) Show that if there exists $\gamma \in (0, 1)$ such that

$$\Big\|\nabla g(x) \nabla_y F\big(x, g(x)\big)\Big\| \leq \gamma \Big\|\nabla_x F\big(x, g(x)\big)\Big\|, \qquad \forall\, x \in \Re^n,$$

then the method is convergent in the sense that all limit points of the sequences that it generates are stationary points of f.

(b) Consider the constrained minimization problem

$$\begin{aligned} &\text{minimize} \quad f(x, y) \\ &\text{subject to} \quad h(x, y) = 0 \end{aligned}$$

where $f : \Re^{n+m} \to \Re$ and $h : \Re^{n+m} \to \Re^m$ are continuously differentiable functions of the two arguments $x \in \Re^n$ and $y \in \Re^m$. Consider also a method of the form

$$x^{k+1} = x^k - \alpha^k \nabla_x f\big(x^k, y^k\big),$$

where y^k is a solution of $h(x^k, y) = 0$, viewed as a system of m equations in the unknown vector y, and α^k is chosen by the minimization rule or the Armijo rule. Formulate conditions that guarantee that this method is convergent.

1.2.20 (A Stepsize Reduction Rule) (www)

Suppose that the cost function f is convex, and consider a gradient method $x^{k+1} = x^k + \alpha^k d^k$ where the assumptions of Prop. 1.2.4 are satisfied, except that the stepsize α^k is determined by the following rule:

$$\alpha^{k+1} = \begin{cases} \alpha^k & \text{if } \nabla f(x^{k+1})' d^k \leq 0, \\ \beta\alpha^k & \text{otherwise,} \end{cases}$$

where $\beta \in (0,1)$ is a fixed scalar and α^0 is any positive scalar.

(a) Show that the stepsize is reduced after iteration k if and only if the interval I^k connecting x^k and x^{k+1} contains in its interior all the vectors $\bar{x}^k$ that minimize $f(x)$ over $x \in I^k$.

(b) Show that the stepsize will be constant after a finite number of iterations.

(c) Show that either $f(x^k) \to -\infty$ or else $\{f(x^k)\}$ converges to a finite value and $\nabla f(x^k) \to 0$. Furthermore, every limit point of $\{x^k\}$ is a global minimum of f.

1.2.21 (Convergence of Gradient Method with Errors [BeT96], [BeT99]) (www)

Let $\{x^k\}$ be a sequence generated by the gradient method with errors

$$x^{k+1} = x^k + \alpha^k (d^k + e^k),$$

where ∇f satisfies the Lipschitz assumption of Prop. 1.2.4, d^k satisfies

$$c_1 \|\nabla f(x^k)\|^2 \leq -\nabla f(x^k)' d^k, \quad \|d^k\| \leq c_2 \big(1 + \|\nabla f(x^k)\|\big), \qquad \forall\, k,$$

where c_1 and c_2 are some scalars, the stepsizes α^k satisfy

$$\sum_{k=0}^{\infty} \alpha^k = \infty, \qquad \sum_{k=0}^{\infty} (\alpha^k)^2 < \infty,$$

and the errors e^k satisfy

$$\|e^k\| \leq \alpha^k \big(q + p\|\nabla f(x^k)\|\big), \qquad \forall\, k,$$

where q and p are some scalars. Show that either $f(x^k) \to -\infty$ or else $\{f(x^k)\}$ converges to a finite value and $\nabla f(x^k) \to 0$. Furthermore, every limit point of $\{x^k\}$ is a stationary point of f. *Hint*: Show that for sufficiently large k, we have

$$f(x^{k+1}) \leq f(x^k) - \alpha^k b_1 \|\nabla f(x^k)\|^2 + (\alpha^k)^2 b_2$$

for some constants b_1 and b_2. Use the line of argument of Prop. 1.2.4.

1.3 GRADIENT METHODS – RATE OF CONVERGENCE

The second major issue regarding gradient methods relates to the rate (or speed) of convergence of the generated sequences $\{x^k\}$. The mere fact that $\{x^k\}$ converges to a stationary point x^* will be of little practical value unless the points x^k are reasonably close to x^* after relatively few iterations. Thus, the study of the rate of convergence provides what are often the dominant criteria for selecting one algorithm in favor of others for solving a particular problem.

Approaches for Rate of Convergence Analysis

There are several approaches towards quantifying the rate of convergence of nonlinear programming algorithms. We will discuss briefly three possibilities and then concentrate on the third.

(a) *Computational complexity approach*: Here we try to estimate the number of elementary operations needed by a given method to find an optimal solution exactly or within an ϵ-tolerance. Usually, this approach provides worst-case estimates, that is, upper bounds on the number of required operations over a class of problems of given dimension and type (e.g. linear, convex, etc.). These estimates may also involve parameters such as the distance of the starting point from the optimal solution set, etc.

(b) *Informational complexity approach*: One difficulty with the computational complexity approach is that for a diverse class of problems, it is often difficult or meaningless to quantify the amount of computation needed for a single function or gradient evaluation. For example, in estimating the computational complexity of the gradient method applied to the entire class of differentiable convex functions, how are we to compare the overhead for finding the stepsize and for updating the x vector with the work needed to compute the cost function value and its gradient? The informational complexity approach, which is discussed in detail by Nemirovsky and Yudin [NeY83] (see also [TrW80]), bypasses this difficulty by estimating the number of function (and possibly gradient) evaluations needed to find an exact or approximately optimal solution (as opposed to the number of necessary computational operations). In other respects, the informational and computational complexity approaches are similar.

(c) *Local analysis*: In this approach we focus on the local behavior of the method in a neighborhood of an optimal solution. Local analysis can describe quite accurately the behavior of a method near the solution by using Taylor series approximations, but ignores entirely the behavior of the method when far from the solution.

The main potential advantage of the computational and informational complexity approaches is that they provide information about the method's progress when far from the eventual limit. Unfortunately, however, this information is usually pessimistic as it accounts for the worst possible problem instance within the class considered. This has resulted in some striking discrepancies between the theoretical model predictions and practical real-world observations. For example, the most widely used linear programming method, the simplex method, is categorized as a "bad" method by worst-case complexity analysis, because it performs very poorly on some specially constructed examples, which, however, are highly unlikely in practice. On the other hand, the ellipsoid method of Khachiyan [Kha79] (see [BGT81] or [BeT97] for a survey and discussion of this method), which was the first linear programming method with a polynomial complexity bound, is categorized as much better than the simplex method by worst-case complexity analysis, even though it performs very poorly on most practical linear programs.

The computational complexity approach has received considerable attention in the context of interior point methods. These methods, discussed in Sections 2.6, 4.2, and 4.4, were primarily motivated by Karmarkar's development of a linear programming algorithm with a polynomial complexity bound that was more favorable than the one of the ellipsoid method [Kar84]. It turned out, however, that the worst-case predictions for the required number of iterations of these methods were off by many orders of magnitude from the practically observed number of iterations. Furthermore, the interior point methods that perform best in practice have poor worst-case complexity, while the ones with the best complexity bounds are very slow in practice.

The local analysis approach, which will be adopted exclusively in this text, has enjoyed considerable success in predicting the behavior of various methods near nonsingular local minima where the cost function can be well approximated by a quadratic. However, the local analysis approach also has some important drawbacks, the most important of which is that it does not account for the rate of progress in the initial iterations. Nonetheless, in many practical situations this is not a serious omission because progress is fast in the initial iterations and slows down only in the limit (the reasons for this seem hard to understand; they are problem-dependent). Furthermore, often in practice, starting points that are near a solution are easily obtainable by a combination of heuristics and experience, in which case local analysis becomes more meaningful.

Local analysis is not very helpful for problems which either involve singular local minima or which are difficult in the sense that the principal methods take many iterations to get near their solution where local analysis applies. It may be said that at present there is little theory and experience to help a practitioner who is faced with such a problem.

1.3.1 The Local Analysis Approach

We now formalize the basic ingredients of our local rate of convergence analysis approach. These are:

(a) We restrict attention to sequences $\{x^k\}$ that converge to a unique limit point x^*.

(b) Rate of convergence is evaluated using an *error function* $e : \Re^n \mapsto \Re$ satisfying $e(x) \geq 0$ for all $x \in \Re^n$ and $e(x^*) = 0$. Typical choices are the Euclidean distance

$$e(x) = \|x - x^*\|$$

and the cost difference

$$e(x) = |f(x) - f(x^*)|.$$

(c) Our analysis is asymptotic, that is, we look at the rate of convergence of the tail of the error sequence $\{e(x^k)\}$.

(d) The generated error sequence $\{e(x^k)\}$ is compared with some "standard" sequences. In our case, we compare $\{e(x^k)\}$ with the geometric progression

$$\beta^k, \qquad k = 0, 1, \ldots,$$

where $\beta \in (0, 1)$ is some scalar. In particular, we say that $\{e(x^k)\}$ *converges linearly or geometrically*, if there exist $q > 0$ and $\beta \in (0, 1)$ such that for all k

$$e(x^k) \leq q\beta^k.$$

It is possible to show that linear convergence is obtained if for some $\beta \in (0, 1)$ we have

$$\limsup_{k \to \infty} \frac{e(x^{k+1})}{e(x^k)} \leq \beta,$$

that is, asymptotically, the error is dropping by a factor of at least β at each iteration (see Exercise 1.3.6, which gives several additional convergence rate characterizations). If for every $\beta \in (0, 1)$, there exists q such that the condition $e(x^k) \leq q\beta^k$ holds for all k, we say that $\{e(x^k)\}$ converges *superlinearly*. This is true in particular, if

$$\limsup_{k \to \infty} \frac{e(x^{k+1})}{e(x^k)} = 0.$$

To quantify further the notion of superlinear convergence, we may compare $\{e(x^k)\}$ with the sequence

$$(\beta)^{p^k}, \qquad k = 0, 1, \ldots,$$

(this is β raised to the power p raised to the power k) where $\beta \in (0,1)$, and $p > 1$ are some scalars. This sequence converges much faster than a geometric progression. We say that $\{e(x^k)\}$ *converges at least superlinearly with order* p, if there exist $q > 0$, $\beta \in (0,1)$, and $p > 1$ such that for all k

$$e(x^k) \leq q(\beta)^{p^k}.$$

The case where $p = 2$ is referred to as *quadratic convergence*. It is possible to show that superlinear convergence with order p is obtained if

$$\limsup_{k \to \infty} \frac{e(x^{k+1})}{e(x^k)^p} < \infty,$$

or equivalently, $e(x^{k+1}) = O\big(e(x^k)^p\big)$; see Exercise 1.3.7.

Most optimization algorithms that are of interest in practice produce sequences converging either linearly or superlinearly, at least when they converge to nonsingular local minima. Linear convergence is a fairly satisfactory rate of convergence for nonlinear programming algorithms, provided the factor β of the associated geometric progression is not too close to unity. Several nonlinear programming algorithms converge superlinearly for particular classes of problems. Newton's method is an important example, as we will see in the present section and also in Section 1.4. For convergence to singular local minima, slower than linear convergence rate is quite common.

1.3.2 The Role of the Condition Number

Many of the important convergence rate characteristics of gradient methods reveal themselves when the cost function is quadratic. To see why, assume that a gradient method is applied to minimization of a twice continuously differentiable function function $f : \Re^n \mapsto \Re$, and it generates a sequence $\{x^k\}$ converging to a nonsingular local minimum x^*. By Taylor's theorem we have

$$f(x) = f(x^*) + \tfrac{1}{2}(x - x^*)'\nabla^2 f(x^*)(x - x^*) + o\big(\|x - x^*\|^2\big).$$

Therefore, since $\nabla^2 f(x^*)$ is positive definite, f can be accurately approximated near x^* by the quadratic function

$$f(x^*) + \tfrac{1}{2}(x - x^*)'\nabla^2 f(x^*)(x - x^*).$$

We thus expect that asymptotic convergence rate results obtained for the quadratic cost case have direct analogs for the general case. This conjecture can indeed be established by rigorous analysis and has been substantiated by extensive numerical experimentation. For this reason, we take the positive definite quadratic case as our point of departure. We subsequently discuss what happens when $\nabla^2 f(x^*)$ is not positive definite, in which case an analysis based on a quadratic model is inadequate.

Convergence Rate of Steepest Descent for Quadratic Functions

Suppose that the cost function f is quadratic with positive definite Hessian Q. We may assume without loss of generality that f is minimized at $x^* = 0$ and that $f(x^*) = 0$ [otherwise we can use the change of variables $y = x - x^*$ and subtract the constant $f(x^*)$ from $f(x)$]. Thus we have

$$f(x) = \tfrac{1}{2}x'Qx, \qquad \nabla f(x) = Qx, \qquad \nabla^2 f(x) = Q.$$

The steepest descent method takes the form

$$x^{k+1} = x^k - \alpha^k \nabla f(x^k) = (I - \alpha^k Q)x^k.$$

Therefore, we have

$$\|x^{k+1}\|^2 = x^{k'}(I - \alpha^k Q)^2 x^k.$$

Since by Prop. A.18(b) of Appendix A, we have for all $x \in \Re^n$

$$x'(I - \alpha^k Q)^2 x \leq \big(\text{maximum eigenvalue of } (I - \alpha^k Q)^2\big)\|x\|^2,$$

we obtain

$$\|x^{k+1}\|^2 \leq \big(\text{maximum eigenvalue of } (I - \alpha^k Q)^2\big)\|x^k\|^2.$$

Using Prop. A.13 of Appendix A, it can be seen that the eigenvalues of $(I - \alpha^k Q)^2$ are equal to $(1 - \alpha^k \lambda_i)^2$, where λ_i are the eigenvalues of Q. Therefore, we have

$$\text{maximum eigenvalue of } (I - \alpha^k Q)^2 = \max\big\{(1 - \alpha^k m)^2, (1 - \alpha^k M)^2\big\},$$

where

$$m: \text{ smallest eigenvalue of } Q,$$

$$M: \text{ largest eigenvalue of } Q.$$

It follows that for $x^k \neq 0$, we have

$$\frac{\|x^{k+1}\|}{\|x^k\|} \leq \max\big\{|1 - \alpha^k m|, |1 - \alpha^k M|\big\}. \tag{1.36}$$

It can be seen that if $|1 - \alpha^k m| \geq |1 - \alpha^k M|$, this inequality holds as an equation if x^k is proportional to an eigenvector corresponding to m. Otherwise, the inequality holds as an equation if x^k is proportional to an eigenvector corresponding to M.

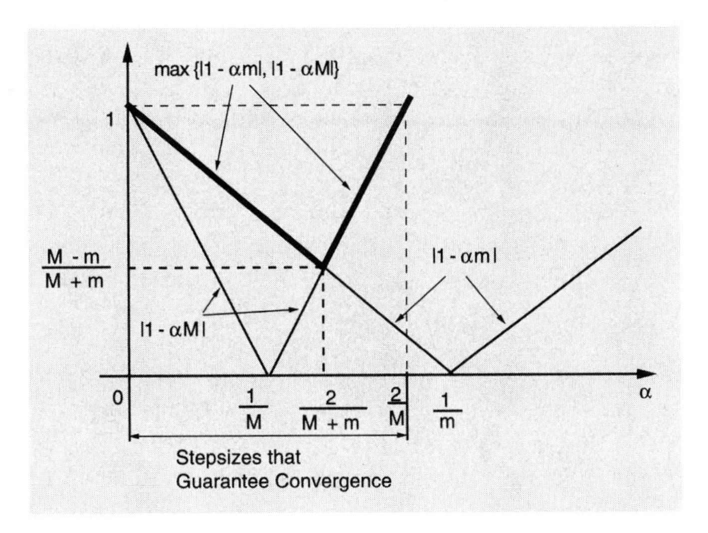

Figure 1.3.1. Illustration of the convergence rate bound $\|x^{k+1}\|/\|x^k\| \leq \max\big\{|1-\alpha m|, |1-\alpha M|\big\}$ for steepest descent. The bound is minimized when α is such that $1-\alpha m = \alpha M - 1$, i.e., for $\alpha = 2/(M+m)$.

Figure 1.3.1 illustrates the convergence rate bound of Eq. (1.36) as a function of the stepsize α^k. It can be seen that the value of α^k that minimizes the bound is

$$\alpha^* = \frac{2}{M+m},$$

in which case

$$\frac{\|x^{k+1}\|}{\|x^k\|} \leq \frac{M-m}{M+m}.$$

This is the best convergence rate bound for steepest descent with constant stepsize.

There is another interesting convergence rate result, which holds when α^k is chosen by the line minimization rule. This result quantifies the rate at which the cost decreases and has the form

$$\frac{f(x^{k+1})}{f(x^k)} \leq \left(\frac{M-m}{M+m}\right)^2. \tag{1.37}$$

The above inequality is verified in Prop. 1.3.1, given in the next subsection, where we collect and prove the more formal results of this section. It can be shown that the inequality is sharp in the sense that given any Q, there is a starting point x^0 such that this inequality holds as an equation for all k (see Fig. 1.3.2).

The ratio M/m is called the *condition number* of Q, and problems where M/m is large are referred as *ill-conditioned*. Such problems are characterized by very elongated elliptical level sets. The steepest descent method converges slowly for these problems as indicated by the convergence rate bounds of Eqs. (1.36) and (1.37), and as illustrated in Fig. 1.3.2.

Scaling and Steepest Descent

Consider now the more general method

$$x^{k+1} = x^k - \alpha^k D^k \nabla f(x^k), \tag{1.38}$$

where D^k is positive definite and symmetric; most of the gradient methods of interest have this form as discussed in Section 1.2. It turns out that we may view this iteration as a *scaled version of steepest descent*. In particular, this iteration is just steepest descent applied in a different coordinate system, which depends on D^k.

Indeed, let

$$S = (D^k)^{1/2}$$

denote the positive definite square root of D^k (cf. Prop. A.21 in Appendix A), and consider a transformation of variables defined by

$$x = Sy.$$

Then, in the space of y, the problem is written as

$$\begin{aligned} &\text{minimize} \quad h(y) \equiv f(Sy) \\ &\text{subject to} \quad y \in \Re^n. \end{aligned} \tag{1.39}$$

The steepest descent method for this problem takes the form

$$y^{k+1} = y^k - \alpha^k \nabla h(y^k). \tag{1.40}$$

Multiplying with S, we obtain

$$Sy^{k+1} = Sy^k - \alpha^k S \nabla h(y^k).$$

By passing back to the space of x, using the relations

$$x^k = Sy^k, \qquad \nabla h(y^k) = S\nabla f(x^k), \qquad S^2 = D^k, \tag{1.41}$$

we obtain

$$x^{k+1} = x^k - \alpha^k D^k \nabla f(x^k).$$

Thus the above gradient iteration is nothing but the steepest descent method (1.40) in the space of y.

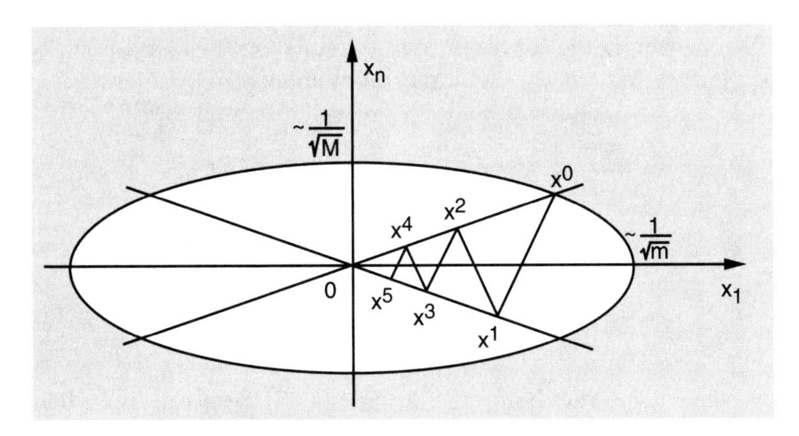

Figure 1.3.2. Example showing that the convergence rate bound

$$\frac{f(x^{k+1})}{f(x^k)} \leq \left(\frac{M-m}{M+m}\right)^2$$

is sharp for the steepest descent method with the line minimization rule. Consider the quadratic function

$$f(x) = \tfrac{1}{2}\sum_{i=1}^{n} \lambda_i x_i^2,$$

where $0 < m = \lambda_1 \leq \lambda_2 \leq \cdots \leq \lambda_n = M$. Any positive definite quadratic function can be put into this form by transformation of variables. Consider the starting point

$$x^0 = \left(m^{-1}, 0, \ldots, 0, M^{-1}\right)'$$

and apply the steepest descent method $x^{k+1} = x^k - \alpha^k \nabla f(x^k)$ with α^k chosen by the line minimization rule. We have $\nabla f(x^0) = (1, 0, \ldots, 0, 1)'$ and it can be verified that the minimizing stepsize is $\alpha^0 = 2/(M+m)$. Thus we obtain $x_1^1 = 1/m - 2/(M+m)$, $x_n^1 = 1/M - 2/(M+m)$, $x_i^1 = 0$ for $i = 2, \ldots, n-1$. Therefore,

$$x^1 = \left(\frac{M-m}{M+m}\right)\left(m^{-1}, 0, \ldots, 0, -M^{-1}\right)'$$

and, we can verify by induction that for all k,

$$x^{2k} = \left(\frac{M-m}{M+m}\right)^{2k} x^0, \qquad x^{2k+1} = \left(\frac{M-m}{M+m}\right)^{2k} x^1.$$

Thus, there exist starting points on the plane of points x of the form $x = (\xi_1, 0, \ldots, 0, \xi_n)'$, $\xi_1 \in \Re$, $\xi_n \in \Re$, in fact two lines shown in the figure, for which steepest descent converges in a way that the inequality

$$\frac{f(x^{k+1})}{f(x^k)} \leq \left(\frac{M-m}{M+m}\right)^2$$

is satisfied as an equation at each iteration.

We now apply the convergence rate results for steepest descent to the scaled iteration $y^{k+1} = y^k - \alpha^k \nabla h(y^k)$, obtaining

$$\frac{\|y^{k+1}\|}{\|y^k\|} \le \max\{|1 - \alpha^k m^k|, |1 - \alpha^k M^k|\}$$

and

$$\frac{h(y^{k+1})}{h(y^k)} \le \left(\frac{M^k - m^k}{M^k + m^k}\right)^2,$$

[cf. the convergence rate bounds (1.36) and (1.37), respectively], where m^k and M^k are the smallest and largest eigenvalues of the Hessian $\nabla^2 h(y)$, which is equal to $S\nabla^2 f(x)S = (D^k)^{1/2}Q(D^k)^{1/2}$. Using the equations

$$y^k = (D^k)^{-1/2}x^k, \qquad y^{k+1} = (D^k)^{-1/2}x^{k+1}$$

to pass back to the space of x, we obtain the convergence rate bounds

$$\frac{x^{k+1'}(D^k)^{-1}x^{k+1}}{x^{k'}(D^k)^{-1}x^k} \le \max\{(1 - \alpha^k m^k)^2, (1 - \alpha^k M^k)^2\} \tag{1.42}$$

and

$$\frac{f(x^{k+1})}{f(x^k)} \le \left(\frac{M^k - m^k}{M^k + m^k}\right)^2, \tag{1.43}$$

where

m^k : smallest eigenvalue of $(D^k)^{1/2}Q(D^k)^{1/2}$,

M^k : largest eigenvalue of $(D^k)^{1/2}Q(D^k)^{1/2}$.

The stepsize that minimizes the right-hand side bound of Eq. (1.42) is

$$\frac{2}{M^k + m^k}. \tag{1.44}$$

The important point is that if M^k/m^k is much larger than unity, the convergence rate can be very slow, even if an optimal stepsize is used. Furthermore, we see that it is desirable to choose D^k as close as possible to Q^{-1}, so that $(D^k)^{1/2}$ is close to $Q^{-1/2}$ (cf. Prop. A.21 in Appendix A) and $M^k \approx m^k \approx 1$. Note that if D^k is so chosen, Eq. (1.44) shows that the stepsize $\alpha = 1$ is near optimal.

Diagonal Scaling

Many practical problems are ill-conditioned because of poor relative scaling of the optimization variables. By this we mean that the units in which the variables are expressed are incongruent in the sense that single unit changes of different variables have disproportionate effects on the cost.

As an example, consider a financial problem with two variables, *investment* denoted x_1 and expressed in dollars, and *interest rate* denoted x_2 and expressed in percentage points. If the effect on the cost function f due to a million dollar increment of investment is comparable to the effect due to a percentage point increment of interest rate, then the condition number will be of the order of 10^{12}!! [This rough calculation is based on estimating the condition number by the ratio

$$\frac{\partial^2 f(x_1, x_2)}{(\partial x_2^2)} \Bigg/ \frac{\partial^2 f(x_1, x_2)}{(\partial x_1)^2},$$

approximating the second partial derivatives by the finite difference formulas

$$\frac{\partial^2 f(x_1, x_2)}{(\partial x_1)^2} \approx \frac{f(x_1 + h_1, x_2) + f(x_1 - h_1, x_2) - 2f(x_1, x_2)}{h_1^2},$$

$$\frac{\partial^2 f(x_1, x_2)}{(\partial x_2)^2} \approx \frac{f(x_1, x_2 + h_2) + f(x_1, x_2 - h_2) - 2f(x_1, x_2)}{h_2^2},$$

and using the relations $f(x_1 + h_1, x_2) \approx f(x_1, x_2 + h_2)$, $f(x_1 - h_1, x_2) \approx f(x_1, x_2 - h_2)$, and $h_1 = 10^6$, $h_2 = 1$, which express the comparability of the effects of a million dollar investment increment and an interest rate percentage point increment.]

The ill-conditioning in such problems can be significantly alleviated by changing the units in which the optimization variables are expressed, which amounts to diagonal scaling of the variables. By this, we mean working in a new coordinate system of a vector y related to x by a transformation,

$$x = Sy,$$

where S is a diagonal matrix. In the absence of further information, a reasonable choice of S is one that makes all the diagonal elements of the Hessian of the cost

$$S\nabla^2 f(x)S$$

in the y-coordinate system approximately equal to unity. For this, we must have

$$s_i \approx \left(\frac{\partial^2 f(x)}{(\partial x_i)^2}\right)^{-1/2},$$

where s_i is the ith diagonal element of S. As discussed earlier, we may express any gradient algorithm in the space of variables y as a gradient algorithm in the space of variables x. In particular, steepest descent in the y-coordinate system, when translated in the x-coordinate system, yields the *diagonally scaled steepest descent method*

$$x^{k+1} = x^k - \alpha^k D^k \nabla f(x^k),$$

where

$$D^k = \begin{pmatrix} d_1^k & 0 & 0 & \cdots & 0 & 0 & 0 \\ 0 & d_2^k & 0 & \cdots & 0 & 0 & 0 \\ \vdots & \vdots & \vdots & \ddots & \vdots & \vdots & \\ 0 & 0 & 0 & \cdots & 0 & d_{n-1}^k & 0 \\ 0 & 0 & 0 & \cdots & 0 & 0 & d_n^k \end{pmatrix},$$

and

$$d_i^k \approx \left(\frac{\partial^2 f(x^k)}{(\partial x_i)^2} \right)^{-1}.$$

This method is also valid for nonquadratic problems as long as d_i^k are chosen to be positive. It is not guaranteed to improve the convergence rate of steepest descent, but it is simple and often surprisingly effective in practice.

Nonquadratic Cost Functions

It is possible to show that our main conclusions on rate of convergence carry over to the nonquadratic case for sequences converging to nonsingular local minima.

Let f be twice continuously differentiable and consider the gradient method

$$x^{k+1} = x^k - \alpha^k D^k \nabla f(x^k), \tag{1.45}$$

where D^k is positive definite and symmetric. Consider a generated sequence $\{x^k\}$, and assume that

$$x^k \to x^*, \qquad \nabla f(x^*) = 0, \qquad \nabla^2 f(x^*) : \text{ positive definite}, \tag{1.46}$$

and that $x^k \neq x^*$ for all k. Then, denoting

$$m^k : \text{ smallest eigenvalue of } (D^k)^{1/2} \nabla^2 f(x^k) (D^k)^{1/2},$$

$$M^k : \text{ largest eigenvalue of } (D^k)^{1/2} \nabla^2 f(x^k) (D^k)^{1/2},$$

it is possible to show the following:

(a) There holds

$$\limsup_{k \to \infty} \frac{(x^{k+1} - x^*)' (D^k)^{-1} (x^{k+1} - x^*)}{(x^k - x^*)' (D^k)^{-1} (x^k - x^*)} = \limsup_{k \to \infty} \max\{|1 - \alpha^k m^k|^2, |1 - \alpha^k M^k|^2\}.$$

(b) If α^k is chosen by the line minimization rule, there holds

$$\limsup_{k\to\infty} \frac{f(x^{k+1}) - f(x^*)}{f(x^k) - f(x^*)} \le \limsup_{k\to\infty} \left(\frac{M^k - m^k}{M^k + m^k}\right)^2. \tag{1.47}$$

The proof of these facts essentially involves a repetition of the proofs for the quadratic case. However, the details are complicated and tedious and will not be given. Exercise 1.3.11 provides an alternative result for the case of the Armijo rule.

From Eq. (1.47), we see that if D^k converges to some positive definite matrix as $x^k \to x^*$, the sequence $\{f(x^k)\}$ converges to $f(x^*)$ linearly. When

$$D^k \to \nabla^2 f(x^*)^{-1},$$

we have $\lim_{k\to\infty} M^k = \lim_{k\to\infty} m^k = 1$ and Eq. (1.47) shows that the convergence rate of $\{f(x^k)\}$ is superlinear. A somewhat more general version of this result for the case of the Armijo rule is given in the Prop. 1.3.2, which is given in the next subsection. In particular, it is shown that if the direction

$$d^k = -D^k \nabla f(x^k)$$

approaches asymptotically the Newton direction $-\big(\nabla^2 f(x^k)\big)^{-1} \nabla f(x^k)$ and the Armijo rule is used with initial stepsize equal to one, the rate of convergence is superlinear.

There is a consistent theme that emerges from our analysis, namely that to achieve asymptotically fast convergence of the gradient method

$$x^{k+1} = x^k - \alpha^k D^k \nabla f(x^k),$$

one should try to choose the matrices D^k as close as possible to $\big(\nabla^2 f(x^*)\big)^{-1}$ so that the maximum and minimum eigenvalues of $(D^k)^{1/2} \nabla^2 f(x^*) (D^k)^{1/2}$ satisfy $M^k \approx 1$ and $m^k \approx 1$. Furthermore, when D^k is so chosen, the initial stepsize $s = 1$ is a good choice for the Armijo rule and other related rules, or as a starting point for one-dimensional minimization procedures used in minimization stepsize rules. This finding has been supported by extensive numerical experience and is one of the most reliable guidelines for selecting and designing optimization algorithms for unconstrained problems. Note, however, that this guideline is valid only for problems where the cost function is twice differentiable and has positive definite Hessian near the points of interest. We discuss next problems where this condition is not satisfied.

Singular and Difficult Problems

We now consider problems where the Hessian matrix either does not exist or is not positive definite at or near local minima of interest. Expressed mathematically, there are local minima x^* and directions d such that the slope of f along d, which is $\nabla f(x^* + \alpha d)'d$, changes very slowly or very rapidly with α, that is, either

$$\lim_{\alpha \to 0} \frac{\nabla f(x^* + \alpha d)'d - \nabla f(x^*)'d}{\alpha} = 0, \tag{1.48}$$

or

$$\lim_{\alpha \to 0} \frac{\nabla f(x^* + \alpha d)'d - \nabla f(x^*)'d}{\alpha} = \infty. \tag{1.49}$$

The case of Eq. (1.48) is characterized by flatness of the cost along the direction d; large excursions from x^* along d produce small changes in cost. In the case of Eq. (1.49) the reverse is true; the cost rises steeply along d. An example is the function

$$f(x_1, x_2) = |x_1|^4 + |x_2|^{3/2},$$

where for the minimum $x^* = (0,0)$, Eq. (1.48) holds along the direction $d = (1,0)$ and Eq. (1.49) holds along the direction $d = (0,1)$. Gradient methods that use directions that are comparable in size to the gradient may require very large stepsizes in the case of Eq. (1.48) and very small stepsizes in the case of Eq. (1.49). This suggests potential difficulties in the implementation of a good stepsize rule; certainly a constant stepsize does not look like an attractive possibility. Furthermore, in the Armijo rule, the initial stepsize should not be taken constant; it should be adjusted according to a suitable scheme, although designing such a scheme may not be easy.

From the point of view of speed of convergence one may view the cases of Eqs. (1.48) and (1.49) as corresponding to an "infinite condition number," thereby suggesting slower than linear convergence rate for the method of steepest descent. Proposition 1.3.3 of the next subsection quantifies the rate of convergence of gradient methods for the case of a convex function whose gradient satisfies the Lipschitz condition

$$\|\nabla f(x) - \nabla f(y)\| \le L\|x - y\|, \tag{1.50}$$

for some L, and all x and y in a neighborhood of x^* [this assumption is consistent with the "flat" cost case of Eq. (1.48), but not with the "steep" cost case of Eq. (1.49)]. It is shown in particular that for a gradient method with the minimization rule, we have

$$f(x^k) - f(x^*) = o(1/k).$$

This type of estimate suggests that for many practical singular problems one may be unable to obtain a highly accurate approximation of an optimal solution. In the "steep" cost case where Eq. (1.49) holds for some directions d, computational examples suggest that the rate of convergence can be slower than linear for the method of steepest descent, although a formal analysis of this conjecture does not seem to have been published.

It should be noted that problems with singular local minima are not the only ones for which gradient methods may converge slowly. There are problems where a given method may have excellent asymptotic rate of convergence, but its progress when far from the eventual limit can be very slow. A prominent example is when the cost function is continuously differentiable but its Hessian matrix is discontinuous and possibly singular in some regions that are outside a small neighborhood of the solution; such functions arise for example in augmented Lagrangian methods for inequality constrained problems (see Section 4.2). Then the powerful Newton-like methods may require a very large number of iterations to get to the small neighborhood of the eventual limit where their convergence rate is favorable. What happens here is that these methods use second derivative information in sophisticated ways, but this information may be misleading due to the Hessian discontinuities.

Generally, there is a tendency to think that difficult problems should be addressed with sophisticated methods, such as Newton-like methods. This is often true, particularly for problems with nonsingular local minima that are poorly conditioned. However, it is important to realize that *often the reverse is true*, namely that for problems with "difficult" cost functions and singular local minima, it is best to use simple methods such as (perhaps diagonally scaled) steepest descent with simple stepsize rules such as a constant or a diminishing stepsize. The reason is that methods that use sophisticated descent directions and stepsize rules often rely on assumptions that are likely to be violated in difficult problems. We also note that for difficult problems, it may be helpful to supplement the steepest descent method with features that allow it to deal better with multiple local minima and peculiarities of the cost function. An often useful modification is the *heavy ball method*, discussed in Exercise 1.3.9.

1.3.3 Convergence Rate Results

We first derive the convergence rate of steepest descent with the minimization stepsize rule when the cost is quadratic.

Proposition 1.3.1: Consider the quadratic function

$$f(x) = \tfrac{1}{2}x'Qx, \tag{1.51}$$

where Q is positive definite and symmetric, and the method of steepest descent

$$x^{k+1} = x^k - \alpha^k \nabla f(x^k), \tag{1.52}$$

where the stepsize α^k is chosen according to the minimization rule

$$f\big(x^k - \alpha^k \nabla f(x^k)\big) = \min_{\alpha \geq 0} f\big(x^k - \alpha \nabla f(x^k)\big).$$

Then, for all k,

$$f(x^{k+1}) \leq \left(\frac{M-m}{M+m}\right)^2 f(x^k),$$

where M and m are the largest and smallest eigenvalues of Q, respectively.

Proof: Let us denote

$$g^k = \nabla f(x^k) = Qx^k. \tag{1.53}$$

The result clearly holds if $g^k = 0$, so we assume $g^k \neq 0$. We first compute the minimizing stepsize α^k. We have

$$\frac{d}{d\alpha} f(x^k - \alpha g^k) = -g^{k\prime} Q(x^k - \alpha g^k) = -g^{k\prime} g^k + \alpha g^{k\prime} Q g^k.$$

By setting this derivative equal to zero, we obtain

$$\alpha^k = \frac{g^{k\prime} g^k}{g^{k\prime} Q g^k}. \tag{1.54}$$

We have, using Eqs. (1.51)-(1.53),

$$\begin{aligned} f(x^{k+1}) &= \tfrac{1}{2}(x^k - \alpha^k g^k)' Q (x^k - \alpha^k g^k) \\ &= \tfrac{1}{2}\left(x^{k\prime} Q x^k - 2\alpha^k g^{k\prime} Q x^k + (\alpha^k)^2 g^{k\prime} Q g^k\right) \\ &= \tfrac{1}{2}\left(x^{k\prime} Q x^k - 2\alpha^k g^{k\prime} g^k + (\alpha^k)^2 g^{k\prime} Q g^k\right) \end{aligned}$$

and using Eq. (1.54),

$$f(x^{k+1}) = \tfrac{1}{2}\left(x^{k\prime} Q x^k - \frac{(g^{k\prime} g^k)^2}{g^{k\prime} Q g^k}\right).$$

Thus, using the fact $f(x^k) = \frac{1}{2} x^{k\prime} Q x^k = \frac{1}{2} g^{k\prime} Q^{-1} g^k$, we obtain

$$f(x^{k+1}) = \left(1 - \frac{(g^{k\prime} g^k)^2}{(g^{k\prime} Q g^k)(g^{k\prime} Q^{-1} g^k)}\right) f(x^k). \tag{1.55}$$

At this point we need the following lemma.

Lemma 3.1: (Kantorovich Inequality) Let Q be a positive definite and symmetric $n \times n$ matrix. Then for any vector $y \in \Re^n, y \neq 0$, there holds

$$\frac{(y'y)^2}{(y'Qy)(y'Q^{-1}y)} \geq \frac{4Mm}{(M+m)^2},$$

where M and m are the largest and smallest eigenvalues of Q, respectively.

Proof: Let $\lambda_1, \ldots, \lambda_n$ denote the eigenvalues of Q and assume that

$$0 < m = \lambda_1 \leq \lambda_2 \leq \cdots \leq \lambda_n = M.$$

Let S be the matrix consisting of the n orthogonal eigenvectors of Q, normalized so that they have unit norm (cf. Prop. A.17 in Appendix A). Then, it can be seen that $S'QS$ is diagonal with diagonal elements $\lambda_1, \ldots, \lambda_n$. By using if necessary a transformation of the coordinate system that replaces y by Sx, we may assume that Q is diagonal and that its diagonal elements are $\lambda_1, \ldots, \lambda_n$. We have for $y = (y_1, \ldots, y_n)' \neq 0$

$$\frac{(y'y)^2}{(y'Qy)(y'Q^{-1}y)} = \frac{\left(\sum_{i=1}^n y_i^2\right)^2}{\left(\sum_{i=1}^n \lambda_i y_i^2\right)\left(\sum_{i=1}^n \frac{y_i^2}{\lambda_i}\right)}.$$

By letting

$$\xi_j = \frac{y_j^2}{\sum_{i=1}^n y_i^2}$$

and by defining

$$\phi(\xi) = \frac{1}{\sum_{i=1}^n \xi_i \lambda_i}, \qquad \psi(\xi) = \sum_{i=1}^n \frac{\xi_i}{\lambda_i},$$

we obtain

$$\frac{(y'y)^2}{(y'Qy)(y'Q^{-1}y)} = \frac{\phi(\xi)}{\psi(\xi)}.$$

Figure 1.3.3 shows that we have

$$\frac{\phi(\xi)}{\psi(\xi)} \geq \frac{4\lambda_1\lambda_n}{(\lambda_1+\lambda_n)^2},$$

which proves the desired inequality. **Q.E.D.**

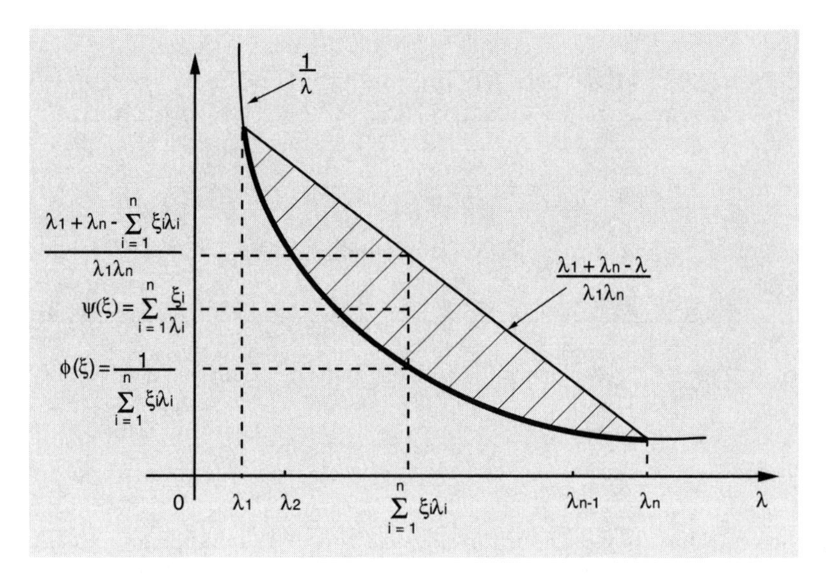

Figure 1.3.3. Proof of the Kantorovich inequality. Consider the function $1/\lambda$. The scalar $\sum_{i=1}^n \xi_i\lambda_i$ represents, for any $\xi = (\xi_1, \ldots, \xi_n)$ with $\xi_i \geq 0$, $\sum_{i=1}^n \xi_i = 1$, a point in the line segment $[\lambda_1, \lambda_n]$. Thus, the values $\phi(\xi) = 1/\sum_{i=1}^n \xi_i\lambda_i$ correspond to the thick part of the curve $1/\lambda$. On the other hand, the value $\psi(\xi) = \sum_{i=1}^n (\xi_i/\lambda_i)$ is a convex combination of $1/\lambda_1, \ldots, 1/\lambda_n$ and hence corresponds to a point in the shaded area in the figure. For the same vector ξ, both $\phi(\xi)$ and $\psi(\xi)$ are represented by points on the same vertical line. Hence,

$$\frac{\phi(\xi)}{\psi(\xi)} \geq \min_{\lambda_1 \leq \lambda \leq \lambda_n} \frac{\frac{1}{\lambda}}{\frac{\lambda_1 + \lambda_n - \lambda}{\lambda_1\lambda_n}}.$$

The minimum is attained for $\lambda = (\lambda_1 + \lambda_n)/2$ and we obtain

$$\frac{\phi(\xi)}{\psi(\xi)} \geq \frac{4\lambda_1\lambda_n}{(\lambda_1 + \lambda_n)^2},$$

which is used to show the result.

Returning to the proof of Prop. 1.3.1, we have by using the Kantorovich inequality in Eq. (1.55)

$$f(x^{k+1}) \leq \left(1 - \frac{4Mm}{(M+m)^2}\right) f(x^k) = \left(\frac{M-m}{M+m}\right)^2 f(x^k).$$

Q.E.D.

The following proposition shows superlinear convergence for methods where d^k approaches the Newton direction $-\big(\nabla^2 f(x^*)\big)^{-1}\nabla f(x^k)$ and the Armijo rule is used.

Proposition 1.3.2: (Superlinear Convergence of Newton-Like Methods) Let f be twice continuously differentiable. Consider a sequence $\{x^k\}$ generated by the gradient method $x^{k+1} = x^k + \alpha^k d^k$ and suppose that

$$x^k \to x^*, \qquad \nabla f(x^*) = 0, \qquad \nabla^2 f(x^*): \text{ positive definite.} \tag{1.56}$$

Assume further that $\nabla f(x^k) \neq 0$ for all k and

$$\lim_{k \to \infty} \frac{\left\| d^k + \left(\nabla^2 f(x^*)\right)^{-1} \nabla f(x^k) \right\|}{\|\nabla f(x^k)\|} = 0. \tag{1.57}$$

Then, if α^k is chosen by means of the Armijo rule with initial stepsize $s = 1$ and $\sigma < 1/2$, we have

$$\lim_{k \to \infty} \frac{\|x^{k+1} - x^*\|}{\|x^k - x^*\|} = 0. \tag{1.58}$$

Furthermore, there exists an integer $\bar{k} \geq 0$ such that $\alpha^k = 1$ for all $k \geq \bar{k}$ (i.e., eventually no reduction of the initial stepsize will be taking place).

Proof: We first prove that there exists a $\bar{k} \geq 0$ such that for all $k \geq \bar{k}$,

$$f(x^k + d^k) - f(x^k) \leq \sigma \nabla f(x^k)' d^k,$$

that is, the unity initial stepsize passes the test of the Armijo rule. By the mean value theorem, we have

$$f(x^k + d^k) - f(x^k) = \nabla f(x^k)' d^k + \tfrac{1}{2} d^{k\prime} \nabla^2 f(\tilde{x}^k) d^k,$$

where $\tilde{x}^k$ is a point on the line segment joining x^k and $x^k + d^k$. Thus, it will be sufficient to show that for k sufficiently large, we have

$$\nabla f(x^k)' d^k + \tfrac{1}{2} d^k \nabla^2 f(\tilde{x}^k) d^k \leq \sigma \nabla f(x^k)' d^k.$$

By defining

$$p^k = \frac{\nabla f(x^k)}{\|\nabla f(x^k)\|}, \qquad q^k = \frac{d^k}{\|\nabla f(x^k)\|},$$

this condition is written

$$(1 - \sigma) p^{k\prime} q^k + \tfrac{1}{2} q^{k\prime} \nabla^2 f(\tilde{x}^k) q^k \leq 0. \tag{1.59}$$

From Eq. (1.57), we have $q^k + (\nabla^2 f(x^*))^{-1} p^k \to 0$. Since $\nabla^2 f(x^*)$ is positive definite and $\|p^k\| = 1$, it follows that $\{q^k\}$ is a bounded sequence, and in view of $q^k = d^k / \|\nabla f(x^k)\|$ and $\nabla f(x^k) \to 0$, we obtain $d^k \to 0$. Hence, $x^k + d^k \to x^*$, and it follows that $\bar{x}^k \to x^*$ and $\nabla^2 f(\bar{x}^k) \to \nabla^2 f(x^*)$. We now write Eq. (1.57) as

$$q^k = -\big(\nabla^2 f(x^*)\big)^{-1} p^k + \beta^k,$$

where $\{\beta^k\}$ denotes a vector sequence with $\beta^k \to 0$. By using the above relation and the fact $\nabla^2 f(\bar{x}^k) \to \nabla^2 f(x^*)$, we may write Eq. (1.59) as

$$(1-\sigma) p^{k\prime} \big(\nabla^2 f(x^*)\big)^{-1} p^k - \tfrac{1}{2} p^{k\prime} \big(\nabla^2 f(x^*)\big)^{-1} p^k \geq \gamma^k,$$

where $\{\gamma^k\}$ is some scalar sequence with $\gamma^k \to 0$. Thus Eq. (1.59) is equivalent to

$$\left(\tfrac{1}{2} - \sigma\right) p^{k\prime} \big(\nabla^2 f(x^*)\big)^{-1} p^k \geq \gamma^k.$$

Since $1/2 > \sigma$, $\|p^k\| = 1$, and $\nabla^2 f(x^*)$ is positive definite, the above relation holds for sufficiently large k. Thus, the unity initial stepsize is acceptable for sufficiently large k, as desired.

To complete the proof, we note that from Eq. (1.57), we have

$$d^k + \big(\nabla^2 f(x^*)\big)^{-1} \nabla f(x^k) = \|\nabla f(x^k)\| \delta^k, \tag{1.60}$$

where δ^k is some vector sequence with $\delta^k \to 0$. From Taylor's theorem we obtain

$$\nabla f(x^k) = \nabla^2 f(x^*)(x^k - x^*) + o(\|x^k - x^*\|),$$

from which

$$\big(\nabla^2 f(x^*)\big)^{-1} \nabla f(x^k) = x^k - x^* + o(\|x^k - x^*\|),$$

$$\|\nabla f(x^k)\| = O(\|x^k - x^*\|).$$

Using the above two relations in Eq. (1.60), we obtain

$$d^k + x^k - x^* = o(\|x^k - x^*\|). \tag{1.61}$$

Since for sufficiently large k we have $d^k + x^k = x^{k+1}$, Eq. (1.61) yields

$$x^{k+1} - x^* = o(\|x^k - x^*\|),$$

from which

$$\lim_{k \to \infty} \frac{\|x^{k+1} - x^*\|}{\|x^k - x^*\|} = \lim_{k \to \infty} \frac{o(\|x^k - x^*\|)}{\|x^k - x^*\|} = 0.$$

Q.E.D.

Note that the equation

$$\lim_{k\to\infty} \frac{\|x^{k+1} - x^*\|}{\|x^k - x^*\|} = 0$$

implies that $\{\|x^k - x^*\|\}$ converges superlinearly (see Exercise 1.3.6). In particular, we see that Newton's method, combined with the Armijo rule with unity initial stepsize, has the property that when it converges to a local minimum x^* such that $\nabla^2 f(x^*)$ is positive definite, its rate of convergence is superlinear. The capture theorem (Prop. 1.2.5) together with the preceding proposition suggest that Newton-like methods with the Armijo rule and a unity initial stepsize converge to a local minimum x^* such that $\nabla^2 f(x^*)$ is positive definite, whenever they are started sufficiently close to such a local minimum. The proof of this is left as Exercise 1.3.2 for the reader.

We finally consider the convergence rate of gradient methods for singular problems whose cost is sufficiently flat for a Lipschitz condition on the gradient to hold.

Proposition 1.3.3: (Convergence Rate of Gradient Methods for Singular Problems) Suppose that the cost function f is convex and its gradient satisfies for some L the Lipschitz condition

$$\|\nabla f(x) - \nabla f(y)\| \le L\|x - y\|, \qquad \forall\, x, y \in \Re^n. \tag{1.62}$$

Consider a gradient method $x^{k+1} = x^k + \alpha^k d^k$ where α^k is chosen by the minimization rule, and for some $c > 0$ and all k we have

$$\nabla f(x^k)' d^k \le -c\|\nabla f(x^k)\|\,\|d^k\|. \tag{1.63}$$

Suppose that the set of global minima of f is nonempty and bounded. Then

$$f(x^k) - f^* = o(1/k),$$

where $f^* = \min_x f(x)$ is the optimal value.

Proof: We assume that $\nabla f(x^k) \neq 0$ and therefore also $d^k \neq 0$ for all k; otherwise the method terminates finitely at a global minimum and the result holds trivially. Let

$$\tilde{\alpha}^k = \frac{|\nabla f(x^k)' d^k|}{L\|d^k\|^2}, \tag{1.64}$$

$$\tilde{x}^k = x^k + \tilde{\alpha}^k d^k. \tag{1.65}$$

By using the descent lemma (Prop. A.24 in Appendix A), and Eqs. (1.63) and (1.64), we have

$$\begin{aligned} f(\tilde{x}^k) - f(x^k) &\le -\tilde{\alpha}^k |\nabla f(x^k)'d^k| + \tfrac{1}{2}(\tilde{\alpha}^k)^2 L \|d^k\|^2 \\ &= \tilde{\alpha}^k \big(-|\nabla f(x^k)'d^k| + \tfrac{1}{2}|\nabla f(x^k)'d^k|\big) \\ &= -\frac{\tilde{\alpha}^k}{2} |\nabla f(x^k)'d^k| \\ &= -\frac{|\nabla f(x^k)'d^k|^2}{2L\|d^k\|^2} \\ &\le -\frac{c^2 \|\nabla f(x^k)\|^2}{2L}. \end{aligned}$$

Using this relation together with the fact $f(x^{k+1}) \le f(\tilde{x}^k)$, we obtain

$$f(x^{k+1}) \le f(x^k) - \frac{c^2 \|\nabla f(x^k)\|^2}{2L}. \tag{1.66}$$

Let X^* be the set of global minima of f. Since X^* is nonempty and compact, all the level sets of f are compact (Prop. B.9 in Appendix B). Thus, $\{x^k\}$ is bounded, and by Prop. 1.2.1, all limit points of $\{x^k\}$ belong to X^*, and the distance of x^k from X^*, defined by

$$d(x^k, X^*) = \min_{x^* \in X^*} \|x^k - x^*\|,$$

converges to 0. Using the convexity of f, we also have for every global minimum x^*

$$f(x^k) - f(x^*) \le \nabla f(x^k)'(x^k - x^*) \le \|\nabla f(x^k)\| \cdot \|x^k - x^*\|,$$

from which, by minimizing over $x^* \in X^*$,

$$f(x^k) - f^* \le \|\nabla f(x^k)\| \, d(x^k, X^*). \tag{1.67}$$

Let us denote for all k

$$e^k = f(x^k) - f^*.$$

Combining Eqs. (1.66) and (1.67), we obtain

$$e^{k+1} \le e^k - \frac{c^2 (e^k)^2}{2L d(x^k, X^*)^2}, \qquad \forall\, k, \tag{1.68}$$

where we assume without loss of generality that $d(x^k, X^*) \neq 0$.

We will show that Eq. (1.68) implies that $e^k = o(1/k)$. Indeed we have

$$0 < e^{k+1} \le e^k \left(1 - \frac{c^2 e^k}{2L d(x^k, X^*)^2}\right),$$

$$0 < 1 - \frac{c^2 e^k}{2Ld(x^k, X^*)^2},$$

from which

$$\begin{aligned}(e^{k+1})^{-1} &\geq (e^k)^{-1}\left(1 - \frac{c^2 e^k}{2Ld(x^k, X^*)^2}\right)^{-1} \geq (e^k)^{-1}\left(1 + \frac{c^2 e^k}{2Ld(x^k, X^*)^2}\right) \\ &= (e^k)^{-1} + \frac{c^2}{2Ld(x^k, X^*)^2}.\end{aligned}$$

Summing this inequality over all k, we obtain

$$e^k \leq \left((e^0)^{-1} + \frac{c^2}{2L}\sum_{i=0}^{k-1} d(x^i, X^*)^{-2}\right)^{-1},$$

or

$$ke^k \leq \left(\frac{1}{ke^0} + \frac{c^2}{2Lk}\sum_{i=0}^{k-1} d(x^i, X^*)^{-2}\right)^{-1}. \tag{1.69}$$

Since $d(x^i, X^*) \to 0$, we have $d(x^i, X^*)^{-2} \to \infty$ and

$$\frac{c^2}{2Lk}\sum_{i=0}^{k-1} d(x^i, X^*)^{-2} \to \infty.$$

Therefore the right-hand side of Eq. (1.69) tends to 0, implying that $e^k = o(1/k)$. **Q.E.D.**

Note that the preceding proof can be modified to cover the case where the Lipschitz condition (1.62) holds within the set $\{x \mid f(x) \leq f(x^0)\}$. Furthermore, the proof goes through for any stepsize rule for which a relation of the form $f(x^{k+1}) \leq f(x^k) - \gamma\|\nabla f(x^k)\|^2$ can be established for some $\gamma > 0$ [cf. Eq. (1.66)]; see Exercise 1.3.8.

With additional assumptions on the structure of the function f some more precise convergence rate results can be obtained. In particular, if f is convex, has a unique minimum x^*, and satisfies the following growth condition

$$f(x) - f(x^*) \geq q\|x - x^*\|^\beta, \qquad \forall\ x \text{ such that } f(x) \leq f(x^0),$$

for some scalars $q > 0$ and $\beta > 2$, it can be shown (see [Dun81]) that for the method of steepest descent with the Armijo rule we have

$$f(x^k) - f(x^*) = O\left(\frac{1}{k^{\frac{\beta}{\beta-2}}}\right).$$

EXERCISES

1.3.1

Estimate the rate of convergence of steepest descent with the line minimization rule when applied to the function of two variables $f(x, y) = x^2 + 1.999xy + y^2$. Find a starting point for which this estimate is sharp (cf. Fig. 1.3.2).

1.3.2 (www)

Let f be twice continuously differentiable. Consider a sequence $\{x^k\}$ generated by the gradient method $x^{k+1} = x^k + \alpha^k d^k$ and suppose that x^* is a nonsingular local minimum. Assume that, for all k, $\nabla f(x^k) \neq 0$ and $d^k = d(x^k)$, where $d(\cdot)$ is a continuous function of x with

$$\lim_{x \to x^*,\, \nabla f(x) \neq 0} \frac{\left\| d(x) + \left(\nabla^2 f(x)\right)^{-1} \nabla f(x) \right\|}{\left\| \nabla f(x) \right\|} = 0.$$

Furthermore, α^k is chosen by means of the Armijo rule with initial stepsize $s = 1$ and $\sigma < 1/2$. Show that there exists an $\epsilon > 0$ such that if $\|x^0 - x^*\| < \epsilon$, then:

(a) $\{x^k\}$ converges to x^*.

(b) $\alpha^k = 1$ for all k.

(c) $\lim_{k \to \infty} \left(\|x^{k+1} - x^*\| / \|x^k - x^*\| \right) = 0$.

Hint: Use the line of argument of Prop. 1.3.2 together with the capture theorem (Prop. 1.2.5). Alternatively, instead of using the capture theorem, consult the proof of the subsequent Prop. 1.4.1.

1.3.3

Consider a positive definite quadratic problem with Hessian matrix Q. Suppose we use scaling with the diagonal matrix whose ith diagonal element is q_{ii}^{-1}, where q_{ii} is the ith diagonal element of Q. Show that if Q is 2×2, this diagonal scaling improves the condition number of the problem and the convergence rate of steepest descent. (*Note*: This need not be true for dimensions higher than 2.)

1.3.4 (Steepest Descent with Errors)

Consider the steepest descent method

$$x^{k+1} = x^k - s\left(\nabla f(x^k) + e^k\right),$$

where s is a constant stepsize, e^k is an error satisfying $\|e^k\| \leq \delta$ for all k, and f is the positive definite quadratic function

$$f(x) = \tfrac{1}{2}(x - x^*)'Q(x - x^*).$$

Let

$$q = \max\big\{|1 - sm|, |1 - sM|\big\},$$

where

$$m : \text{ smallest eigenvalue of } Q, \qquad M : \text{ largest eigenvalue of } Q,$$

and assume that $q < 1$. Show that for all k, we have

$$\|x^k - x^*\| \leq \frac{s\delta}{1 - q} + q^k \|x^0 - x^*\|.$$

1.3.5

Consider the positive definite quadratic function $f(x) = \frac{1}{2}x'Qx$ and the steepest descent method with the stepsize α^k chosen by the Goldstein rule. Show that for all k,

$$f(x^{k+1}) \leq \left(1 - \frac{16\sigma(1-\sigma)Mm}{(M+m)^2}\right) f(x^k).$$

Explain why when $\sigma = 1/2$ this relation yields the result of Prop. 1.3.1. *Hint*: Use the result of Exercise 1.2.9.

1.3.6 [Ber82a], p. 14

Consider a scalar sequence $\{e^k\}$ with $e^k \geq 0$ for all k, and $e^k \to 0$. We say that $\{e^k\}$ converges *faster than linearly with convergence ratio* β, where $0 < \beta < 1$, if for every $\bar{\beta} \in (\beta, 1)$ and $q > 0$, there exists $\bar{k}$ such that

$$e^k \leq q\bar{\beta}^k, \qquad \forall\, k \geq \bar{k}.$$

We say that $\{e^k\}$ converges *slower than linearly with convergence ratio* β, where $0 < \beta < 1$, if for every $\bar{\beta} \in (\beta, 1)$ and $q > 0$, there exists $\bar{k}$ such that

$$q\bar{\beta}^k \leq e^k, \qquad \forall\, k \geq \bar{k}.$$

We say that $\{e^k\}$ converges *linearly with convergence ratio* β if it converges both faster and slower than linearly with convergence ratio β. Show that:

(a) $\{e^k\}$ converges faster than linearly with convergence ratio β if and only if

$$\limsup_{k\to\infty} (e^k)^{1/k} \leq \beta.$$

$\{e^k\}$ converges slower than linearly with convergence ratio β if and only if

$$\liminf_{k\to\infty}(e^k)^{1/k} \geq \beta.$$

$\{e^k\}$ converges linearly with convergence ratio β if and only if

$$\lim_{k\to\infty}(e^k)^{1/k} = \beta.$$

(b) Assume that $e^k \neq 0$ for all k, and denote

$$\beta_1 = \liminf_{k\to\infty} \frac{e^{k+1}}{e^k}, \qquad \beta_2 = \limsup_{k\to\infty} \frac{e^{k+1}}{e^k}.$$

Show that if $0 < \beta_1 \leq \beta_2 < 1$, then $\{e^k\}$ converges faster than linearly with convergence ratio β_2 and slower than linearly with convergence ratio β_1. Furthermore, if $\beta_1 = \beta_2 = 0$, then $\{e^k\}$ converges superlinearly.

1.3.7

Consider a scalar sequence $\{e^k\}$ with $e^k > 0$ for all k, and $e^k \to 0$. Show that $\{e^k\}$ converges superlinearly with order p if

$$\limsup_{k\to\infty} \frac{e^{k+1}}{(e^k)^p} < \infty.$$

1.3.8 (www)

Prove the result of Prop. 1.3.3 for the steepest descent case $[d^k = -\nabla f(x^k)]$, and assuming that the stepsize is not chosen by the line minimization rule but is instead a sufficiently small constant.

1.3.9 (The Heavy Ball Method [Pol64]) (www)

Consider the following variant of the steepest descent method:

$$x^{k+1} = x^k - \alpha\nabla f(x^k) + \beta(x^k - x^{k-1}), \qquad k = 1, 2, \ldots,$$

where α is a constant positive stepsize and β is a scalar with $0 < \beta < 1$. (For another similar variant, see Exercise 1.6.5 in Section 1.6.)

(a) Let f be the quadratic function $f(x) = (1/2)x'Qx + c'x$, where Q is positive definite and symmetric, and let m and M be the minimum and the maximum eigenvalues of Q, respectively. Show that the method converges linearly to the unique solution if $0 < \alpha < 2(1+\beta)/M$. Show that with optimal choices of α and β, the ratio of linear convergence is

$$\frac{\sqrt{M}-\sqrt{m}}{\sqrt{M}+\sqrt{m}},$$

which, if $m < M$, is faster than the corresponding ratio of the steepest descent method where $\beta = 0$ and α is chosen optimally [cf. Eq. (1.36)]. *Hint*: Consider the iteration

$$\begin{pmatrix} x^{k+1} \\ x^k \end{pmatrix} = \begin{pmatrix} (1+\beta)I - \alpha Q & -\beta I \\ I & 0 \end{pmatrix} \begin{pmatrix} x^k \\ x^{k-1} \end{pmatrix}$$

and show that v is an eigenvalue of the matrix in the above equation if and only if $v + \beta/v$ is equal to $1 + \beta - \alpha\lambda$ where λ is an eigenvalue of Q. (This is a challenging exercise.)

(b) It is generally conjectured that in comparison to steepest descent, the method is less prone to getting trapped at "shallow" local minima, and tends to behave better for difficult problems where the cost function is alternatively very flat and very steep. Argue for or against this conjecture.

(c) In support of your answer in (b), write a computer program to test the method with $\beta = 0$ and $\beta > 0$ with one-dimensional cost functions of the form

$$f(x) = \tfrac{1}{2}x^2\big(1 + \gamma\cos(x)\big),$$

where $\gamma \in (0,1)$, and

$$f(x) = \tfrac{1}{2}\sum_{i=1}^{m} |z_i - \tanh(xy_i)|^2,$$

where z_i and y_i are given scalars.

1.3.10 (www)

Suppose that a vector sequence $\{e^k\}$ satisfies

$$\|e^{k+1} - e^k\| \le \beta\|e^k - e^{k-1}\|, \qquad \forall\, k \ge \bar{k},$$

where $\bar{k}$ is a positive integer and $\beta \in (0,1)$ is a scalar. Show that $\{e^k\}$ converges to some vector e^* linearly, and in fact we have

$$\|e^k - e^*\| \le q\beta^k$$

for some scalar q and all k. *Hint*: Show that $\{e^k\}$ is a Cauchy sequence.

1.3.11 (Convergence Rate of Steepest Descent with the Armijo Rule) (www)

Let $f : \Re^n \mapsto \Re$ be a twice continuously differentiable function that satisfies

$$m\|y\|^2 \leq y'\nabla^2 f(x)y \leq M\|y\|^2, \qquad \forall\, x,\, y \in \Re^n,$$

where m and M are some positive scalars. Consider the steepest descent method $x^{k+1} = x^k - \alpha^k \nabla f(x^k)$ with α^k determined by the Armijo rule. Let x^* be the unique unconstrained minimum of f and let

$$r = 1 - \frac{4m\beta\sigma(1-\sigma)}{M}.$$

Show that for all k, we have

$$f(x^{k+1}) - f(x^*) \leq r\big(f(x^k) - f(x^*)\big),$$

and

$$\|x^k - x^*\|^2 \leq qr^k,$$

where q is some constant.

1.4 NEWTON'S METHOD AND VARIATIONS

In the last two sections we emphasized a basic tradeoff in gradient methods: implementation simplicity versus fast convergence. We have already discussed steepest descent, one of the simplest but also one of the slowest methods. We now consider its opposite extreme, Newton's method, which is arguably the most complex and also the fastest of the gradient methods (under appropriate conditions).†

---

† Newton's method is often referred to as the "Newton-Raphson method." We all know of Newton, one of the greatest mathematicians of all time, but who was Raphson? According to a popular joke, Raphson was "Newton's programmer," but this is unfair, for Raphson was an eminent scientist in his time and a personal friend of Newton. In fact Raphson was made a member of the Royal Society in 1691 on the strength of his book "Analysis Aequationum Universalis," which was published in 1690 and contained the Newton method for approximating the roots of an equation. The same method is described by Newton in "Method of Fluxions," which although written in 1671, was not published until 1736. Thus Raphson published the method nearly 50 years before Newton. While the origins of the discovery of the method are not entirely clear, it appears that Raphson has as much claim to it as Newton.

Newton's method consists of the iteration

$$x^{k+1} = x^k - \alpha^k \big(\nabla^2 f(x^k)\big)^{-1} \nabla f(x^k), \tag{1.70}$$

assuming that the Newton direction

$$d^k = -\big(\nabla^2 f(x^k)\big)^{-1} \nabla f(x^k) \tag{1.71}$$

is defined and is a direction of descent [i.e., $d_k{}'\nabla f(x^k) < 0$]. As explained in the preceding section, one may view this iteration as a scaled version of steepest descent where the "optimal" scaling matrix $\big(\nabla^2 f(x^k)\big)^{-1}$ is used. It is worth mentioning in this connection that *Newton's method is "scale-free,"* in the sense that it cannot be affected by a change in coordinate system as is true for steepest descent (see Exercise 1.4.1).

When the Armijo rule is used with initial stepsize $s = 1$, then no reduction of the stepsize will be necessary near a nonsingular minimum (positive definite Hessian), as shown in Prop. 1.3.2. Thus, near convergence the method takes the form

$$x^{k+1} = x^k - \big(\nabla^2 f(x^k)\big)^{-1} \nabla f(x^k), \tag{1.72}$$

which will be referred to as the *pure form of Newton's method*. On the other hand, far from such a local minimum, the Hessian matrix may be singular or the Newton direction of Eq. (1.71) may not be a direction of descent because the Hessian $\nabla^2 f(x^k)$ is not positive definite. Thus the analysis of Newton's method has two principal aspects:

(a) Local convergence, dealing with the behavior of the pure form of the method near a nonsingular local minimum.

(b) Global convergence, addressing the modifications that are necessary to ensure that the method is valid and is likely to converge to a local minimum when started far from all local minima.

We consider these issues in this section and we also discuss some variations of Newton's method, which are aimed at reducing the overhead for computing the Newton direction.

Local Convergence

The local convergence result for gradient methods (Prop. 1.2.5) together with the superlinear convergence result for Newton-like methods (Prop. 1.3.2) suggest that the pure form of Newton's method converges superlinearly when started close enough to a nonsingular local minimum. Results of this type hold for a more general form of Newton's method, that can be used to solve the system of n equations with n unknowns

$$g(x) = 0, \tag{1.73}$$

where $g : \Re^n \mapsto \Re^n$ is a continuously differentiable function. This method has the form

$$x^{k+1} = x^k - \big(\nabla g(x^k)'\big)^{-1} g(x^k), \tag{1.74}$$

and for the special case where $g(x)$ is equal to the gradient $\nabla f(x)$, it yields the pure form of Eq. (1.72). [A continuously differentiable function $g : \Re^n \mapsto \Re^n$ need not be equal to the gradient of some function. In particular, $g(x) = \nabla f(x)$ for some $f : \Re^n \mapsto \Re$, if and only if the $n \times n$ matrix $\nabla g(x)$ is symmetric for all x ([OrR70], p. 95). Thus, the equation version of Newton's method (1.74) is more broadly applicable than the optimization version of Eq. (1.72).]

There is a simple argument that shows the fast convergence of Newton's method (1.74). Suppose that the method generates a sequence $\{x^k\}$ that converges to a vector x^* such that $g(x^*) = 0$ and $\nabla g(x^*)$ is invertible. Let us use Taylor's theorem to write

$$0 = g(x^*) = g(x^k) + \nabla g(x^k)'(x^* - x^k) + o\big(\|x^k - x^*\|\big).$$

By multiplying this relation with $\big(\nabla g(x^k)'\big)^{-1}$ we have

$$x^k - x^* - \big(\nabla g(x^k)'\big)^{-1} g(x^k) = o\big(\|x^k - x^*\|\big),$$

so for the pure Newton iteration $x^{k+1} = x^k - \big(\nabla g(x^k)'\big)^{-1} g(x^k)$ we obtain

$$x^{k+1} - x^* = o\big(\|x^k - x^*\|\big),$$

or, for $x^k \neq x^*$,

$$\lim_{k\to\infty} \frac{\|x^{k+1} - x^*\|}{\|x^k - x^*\|} = \lim_{k\to\infty} \frac{o\big(\|x^k - x^*\|\big)}{\|x^k - x^*\|} = 0,$$

implying superlinear convergence. This argument can also be used to show convergence to x^* if the initial vector x^0 is sufficiently close to x^*. The following proposition proves a more detailed result.

Proposition 1.4.1: Consider a function $g : \Re^n \mapsto \Re^n$, and a vector x^* such that $g(x^*) = 0$. For $\delta > 0$, let S_δ denote the sphere $\{x \mid \|x - x^*\| \leq \delta\}$. Assume that g is continuously differentiable within some sphere $S_{\bar{\delta}}$ and that $\nabla g(x^*)$ is invertible.

(a) There exists $\delta > 0$ such that if $x^0 \in S_\delta$, the sequence $\{x^k\}$ generated by the iteration

$$x^{k+1} = x^k - \big(\nabla g(x^k)'\big)^{-1} g(x^k)$$

is defined, belongs to S_δ, and converges to x^*. Furthermore, $\big\{\|x^k - x^*\|\big\}$ converges superlinearly.

(b) If for some $L > 0$, $M > 0$, $\delta > 0$, and for all x and y in S_δ,

$$\|\nabla g(x) - \nabla g(y)\| \leq L\|x - y\|, \qquad \left\|\big(\nabla g(x)'\big)^{-1}\right\| \leq M, \quad (1.75)$$

then, if $x^0 \in S_\delta$, we have

$$\|x^{k+1} - x^*\| \leq \frac{LM}{2}\|x^k - x^*\|^2, \qquad \forall\, k = 0, 1, \ldots,$$

so if $LM\delta/2 < 1$ and $x^0 \in S_\delta$, $\{\|x^k - x^*\|\}$ converges superlinearly with order at least two.

Proof: (a) Let $\delta > 0$ be such that $\big(\nabla g(x)'\big)^{-1}$ exists within S_δ and let $M > 0$ be such that

$$\left\|\big(\nabla g(x)'\big)^{-1}\right\| \leq M, \qquad \forall\, x \in S_\delta.$$

Assuming $x \in S_\delta$, and using the relation

$$g(x^k) = \int_0^1 \nabla g\big(x^* + t(x^k - x^*)\big)' dt (x^k - x^*),$$

we estimate $\|x^{k+1} - x^*\|$ as

$$\begin{aligned}
\|x^{k+1} - x^*\| &= \left\|x^k - x^* - \big(\nabla g(x^k)'\big)^{-1} g(x^k)\right\| \\
&= \left\|\big(\nabla g(x^k)'\big)^{-1}\big(\nabla g(x^k)'(x^k - x^*) - g(x^k)\big)\right\| \\
&= \left\|\big(\nabla g(x^k)'\big)^{-1}\left(\nabla g(x^k)' - \int_0^1 \nabla g\big(x^* + t(x^k - x^*)\big)' dt\right)(x^k - x^*)\right\| \\
&= \left\|\big(\nabla g(x^k)'\big)^{-1}\left(\int_0^1 \left[\nabla g(x^k)' - \nabla g\big(x^* + t(x^k - x^*)\big)'\right] dt\right)(x^k - x^*)\right\| \\
&\leq M\left(\int_0^1 \left\|\nabla g(x^k) - \nabla g\big(x^* + t(x^k - x^*)\big)\right\| dt\right)\|x^k - x^*\|.
\end{aligned}$$

By continuity of ∇g, we can take δ sufficiently small to ensure that the term under the integral sign is arbitrarily small. The convergence $x^k \to x^*$ and the superlinear convergence of $\|x^k - x^*\|$ follow.

(b) If the condition (1.75) holds, the preceding relation yields

$$\|x^{k+1} - x^*\| \leq M\left(\int_0^1 Lt\|x^k - x^*\| dt\right)\|x^k - x^*\| = \frac{LM}{2}\|x^k - x^*\|^2.$$

Q.E.D.

A related result is the following. Its proof is left for the reader.

Proposition 1.4.2: Under the assumptions of Prop. 1.4.1(a), given any $r > 0$, there exists a $\delta > 0$ such that if $\|x^k - x^*\| < \delta$, then

$$\|x^{k+1} - x^*\| \leq r\|x^k - x^*\|, \qquad \|g(x^{k+1})\| \leq r\|g(x^k)\|.$$

Thus, once it gets "near" a solution x^* where $\nabla g(x^*)$ is invertible, the pure form of Newton's method converges extremely fast, typically taking a handful of iterations to achieve very high solution accuracy; see Fig. 1.4.1. Unfortunately, it is usually difficult to predict whether a given starting point is sufficiently near to a solution for the fast convergence rate of Newton's method to become effective right away. Thus, in practice one can only expect that *eventually* the fast convergence rate of Newton's method will become effective. Figure 1.4.2 illustrates how the method can fail to converge when started far from a solution.

Global Convergence

Newton's method in its pure form for unconstrained minimization of f has several serious drawbacks.

(a) The inverse $\big(\nabla^2 f(x^k)\big)^{-1}$ may fail to exist, in which case the method breaks down. This will happen, for example, in regions where f is linear ($\nabla^2 f = 0$).

(b) The pure form is not a descent method; it may happen that $f(x^{k+1}) > f(x^k)$.

(c) The pure form tends to be attracted by local maxima just as much as it is attracted by local minima. It just tries to solve the system of equations $\nabla f(x) = 0$.

For these reasons, it is necessary to modify the pure form of Newton's method to turn it into a reliable minimization algorithm. There are several schemes that accomplish this by converting the pure form into a gradient method with a gradient related direction sequence. Simultaneously the modifications are such that, near a nonsingular local minimum, the algorithm assumes the pure form of Newton's method (1.72) and achieves the attendant fast convergence rate.

A simple possibility is to replace the Newton direction by the steepest descent direction (possibly after diagonal scaling), whenever the Newton

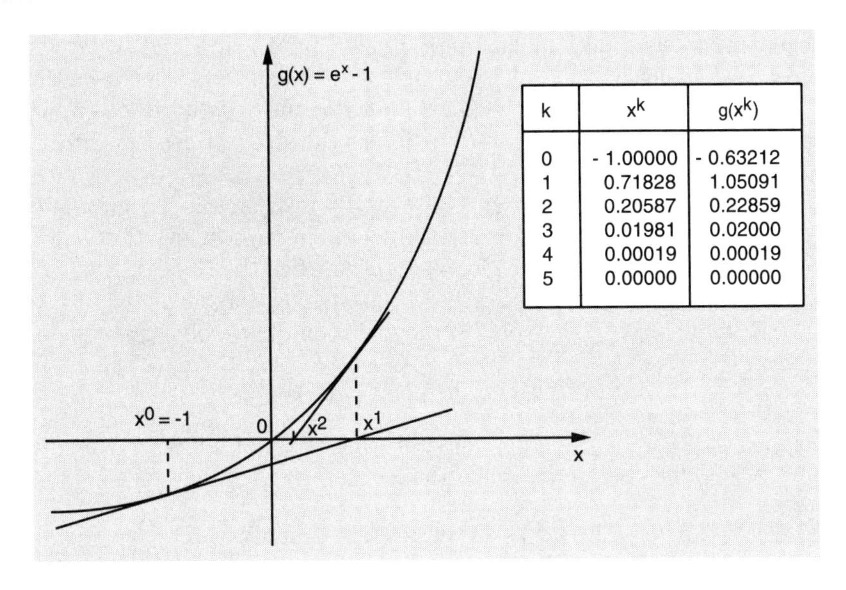

k	x^k	$g(x^k)$
0	- 1.00000	- 0.63212
1	0.71828	1.05091
2	0.20587	0.22859
3	0.01981	0.02000
4	0.00019	0.00019
5	0.00000	0.00000

Figure 1.4.1. Fast convergence of Newton's method for solving the equation $e^x - 1 = 0$.

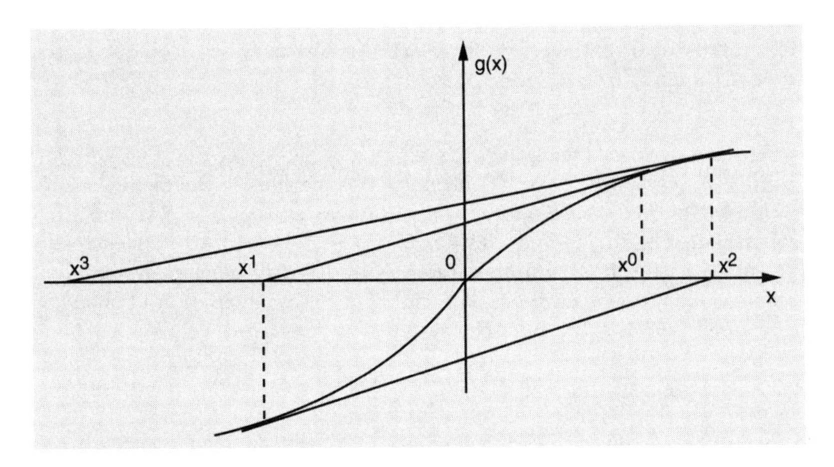

Figure 1.4.2. Divergence of Newton's method for solving an equation $g(x) = 0$ of a single variable x, when the starting point is far from the solution. This phenomenon typically happens when $\|\nabla g(x)\|$ tends to decrease as $\|x\| \to \infty$.

direction is either not defined or is not a descent direction.† With proper

---

† Interestingly, this motivated the development of steepest descent by M. Augustin Cauchy. In his original paper [Cau47], Cauchy states as motivation for the steepest descent method its capability to obtain a close approximation to the solution, in which case "... one can obtain new approximations very rapidly with

safeguards, such a method has appropriate convergence and asymptotic rate of convergence properties (see Exercise 1.4.3 and for a related method, see Exercise 1.4.4). However, its performance at the early iterations may be quite slow, whether the Newton direction or the steepest descent direction is used in these iterations.

Generally, no modification of Newton's method can be guaranteed to converge fast in the early iterations, but there are schemes that can use second derivative information effectively, even when the Hessian is not positive definite. These schemes are based on making diagonal modifications to the Hessian; that is, they obtain the direction d^k by solving a system of the form

$$\big(\nabla^2 f(x^k) + \Delta^k\big) d^k = -\nabla f(x^k),$$

whenever the Newton direction does not exist or is not a descent direction. Here Δ^k is a diagonal matrix such that

$$\nabla^2 f(x^k) + \Delta^k : \ \text{positive definite.}$$

We outline some possibilities.

*Modified Cholesky Factorization**

It can be shown that every positive definite matrix Q has a unique factorization of the form

$$Q = LL',$$

where L is a lower triangular matrix; this is known as the *Cholesky factorization of* Q (see Appendix D). Systems of equations of the form $Qx = b$ can be solved by first solving for y the triangular system $Ly = b$, and then by solving for x the triangular system $L'x = y$. These triangular systems can be solved easily [in $O(n^2)$ operations as opposed to general systems, which require $O(n^3)$ operations; see Appendix D]. Since calculation of the Newton direction involves solution of the system

$$\nabla^2 f(x^k) d^k = -\nabla f(x^k),$$

it is natural to compute d^k by attempting to form the Cholesky factorization of $\nabla^2 f(x^k)$. During this process, one can detect whether $\nabla^2 f(x^k)$ is either nonpositive definite or nearly singular, in which case some of the diagonal elements of $\nabla^2 f(x^k)$ are suitably increased to ensure that the resulting matrix is positive definite. This is done sequentially during the factorization process, so in the end we obtain

$$L^k L^{k'} = \nabla^2 f(x^k) + \Delta^k,$$

where L^k is lower triangular and nonsingular, and Δ^k is diagonal.

---

the aid of the linear or Newton's method ..." (Note the attribution to Newton by Cauchy.)

As an illustration, consider the 2-dimensional case (for the general case, see Appendix D). Let

$$\nabla^2 f(x^k) = \begin{pmatrix} h_{11} & h_{12} \\ h_{21} & h_{22} \end{pmatrix}$$

and let the desired factorization be of the form

$$LL' = \begin{pmatrix} \alpha & 0 \\ \gamma & \beta \end{pmatrix} \cdot \begin{pmatrix} \alpha & \gamma \\ 0 & \beta \end{pmatrix}.$$

We choose α, β, and γ, so that $\nabla^2 f(x^k) = LL'$ if $\nabla^2 f(x^k)$ is positive definite, and we appropriately modify h_{11} and h_{22} otherwise. This determines the first diagonal element α according to the relation

$$\alpha = \begin{cases} \sqrt{h_{11}} & \text{if } h_{11} > 0 \\ \sqrt{h_{11} + \delta_1} & \text{otherwise} \end{cases}$$

where δ_1 is such that $h_{11} + \delta_1 > 0$. Given α, we can calculate γ by equating the corresponding elements of $\nabla^2 f(x^k)$ and LL'. We obtain $\gamma\alpha = h_{12}$ or

$$\gamma = \frac{h_{12}}{\alpha}.$$

We can now calculate the second diagonal element β by equating the corresponding elements of $\nabla^2 f(x^k)$ and LL', after appropriately modifying h_{22} if necessary,

$$\beta = \begin{cases} \sqrt{h_{22} - \gamma^2} & \text{if } h_{22} > \gamma^2, \\ \sqrt{h_{22} - \gamma^2 + \delta_2} & \text{otherwise,} \end{cases}$$

where δ_2 is such that $h_{22} - \gamma^2 + \delta_2 > 0$. The method for choosing the increments δ_1 and δ_2 is largely heuristic. One possibility is discussed in Appendix D, which also describes more sophisticated versions of the above procedure where a positive increment is added to the diagonal elements of the Hessian even when the corresponding diagonal elements of the factorization are positive but very close to zero.

Given the $L^k L^{k'}$ factorization, the direction d^k is obtained by solving the system

$$L^k L^{k'} d^k = -\nabla f(x^k).$$

The next iterate is

$$x^{k+1} = x^k + \alpha^k d^k,$$

where α^k is chosen according to the Armijo rule or one of the other stepsize rules we have discussed.

To guarantee convergence, the increments added to the diagonal elements of the Hessian can be chosen so that $\{d^k\}$ is gradient related (cf. Prop. 1.2.1). Also, these increments can be chosen to be zero near a nonsingular local minimum. In particular, with proper safeguards, near such a point, the method becomes identical to the pure form of Newton's method and achieves the corresponding superlinear convergence rate (see Appendix D).

*Trust Region Methods**

As explained in Section 1.2, the pure Newton step is obtained by minimizing over d the second order Taylor series approximation of f around x^k, given by

$$f^k(d) = f(x^k) + \nabla f(x^k)'d + \tfrac{1}{2}d'\nabla^2 f(x^k)d.$$

We know that $f^k(d)$ is a good approximation of $f(x^k + d)$ when d is in a small neighborhood of zero, but the difficulty is that with unconstrained minimization of $f^k(d)$ one may obtain a step that lies outside this neighborhood. It therefore makes sense to consider a *restricted Newton step* d^k obtained by minimizing $f^k(d)$ over a suitably small neighborhood of zero, called the *trust region*:

$$d^k = \arg \min_{||d|| \leq \gamma^k} f^k(d), \tag{1.76}$$

where γ^k is some positive scalar. [It can be shown that the restricted Newton step d^k also solves a system of the form $\big(\nabla^2 f(x^k) + \delta^k I\big)d = -\nabla f(x^k)$, where I is the identity matrix and δ^k is a nonnegative scalar (a Lagrange multiplier in the terminology of Chapter 3), so the preceding method of determining d^k fits the general framework of using a correction of the Hessian matrix by a positive semidefinite matrix.] An approximate solution of the constrained minimization problem of Eq. (1.76) can be obtained quickly using the fact that it has only one constraint (see [MoS83]).

An important observation here is that even if $\nabla^2 f(x^k)$ is not positive definite or, more generally, even if the pure Newton direction is not a descent direction, the restricted Newton step d^k improves the cost, provided $\nabla f(x^k) \neq 0$ and γ^k is sufficiently small. To see this, note that we have for all d with $||d|| \leq \gamma^k$

$$f(x^k + d) = f^k(d) + o\big((\gamma^k)^2\big),$$

so that

$$\begin{aligned} f(x^k + d^k) &= f^k(d^k) + o\big((\gamma^k)^2\big) \\ &= f(x^k) + \min_{||d|| \leq \gamma^k} \big\{\nabla f(x^k)'d + \tfrac{1}{2}d'\nabla^2 f(x^k)d\big\} + o\big((\gamma^k)^2\big). \end{aligned}$$

Therefore, denoting

$$\tilde{d}^k = -\frac{\nabla f(x^k)}{||\nabla f(x^k)||}\gamma^k,$$

we have

$$\begin{aligned} f(x^k + d^k) \leq & f(x^k) + \nabla f(x^k)'\tilde{d}^k + \tfrac{1}{2}\tilde{d}^{k\prime}\nabla^2 f(x^k)\tilde{d}^k + o\big((\gamma^k)^2\big) \\ = & f(x^k) - \gamma^k||\nabla f(x^k)|| + \frac{(\gamma^k)^2}{2||\nabla f(x^k)||^2}\nabla f(x^k)'\nabla^2 f(x^k)\nabla f(x^k) \\ & + o\big((\gamma^k)^2\big). \end{aligned}$$

For γ^k sufficiently small, the negative term $-\gamma^k||\nabla f(x^k)||$ dominates the last two terms on the right-hand side above, showing that

$$f(x^{k+1}) < f(x^k).$$

It can be seen in fact from the preceding relations that a cost improvement is possible even when $\nabla f(x^k) = 0$, provided γ^k is sufficiently small and f has a direction of negative curvature at x^k, that is, $\nabla^2 f(x^k)$ is not positive semidefinite. Thus the preceding procedure will fail to improve the cost only if $\nabla f(x^k) = 0$ and $\nabla^2 f(x^k)$ is positive semidefinite, that is, x^k satisfies the first and second order necessary conditions. In particular, one can typically make progress even if x^k is a stationary point that is not a local minimum.

We are thus motivated to consider a method of the form

$$x^{k+1} = x^k + d^k,$$

where d^k is the restricted Newton step corresponding to a suitably chosen scalar γ^k as per Eq. (1.76). Here, for a given x^k, γ^k should be small enough so that there is cost improvement; one possibility is to start from an initial trial γ^k and successively reduce γ^k by a certain factor as many times as necessary until a cost reduction occurs $[f(x^{k+1}) < f(x^k)]$. The choice of the initial trial value for γ^k is crucial here; if it is chosen too large, a large number of reductions may be necessary before a cost improvement occurs; if it is chosen too small the convergence rate may be poor. In particular, to maintain the superlinear convergence rate of Newton's method, as x^k approaches a nonsingular local minimum, one should select the initial trial value of γ^k sufficiently large so that the restricted Newton step and the pure Newton step coincide.

A reasonable way to adjust the initial trial value for γ^k is to increase this value when the method appears to be progressing well and to decrease this value otherwise. One can measure progress by using the ratio of actual over predicted cost improvement [based on the approximation $f^k(d)$]

$$r^k = \frac{f(x^k) - f(x^{k+1})}{f(x^k) - f^k(d^k)}.$$

In particular, it makes sense to increase the initial trial value for γ ($\gamma^{k+1} > \gamma^k$) if this ratio is close to or above unity, and decrease γ otherwise. The following algorithm is a typical example of such a method. Given x^k and an initial trial value γ^k, it determines x^{k+1} and an initial trial value γ^{k+1} by using two threshold values σ_1, σ_2 with $0 < \sigma_1 \leq \sigma_2 \leq 1$ and two factors β_1, β_2 with $0 < \beta_1 < 1 < \beta_2$ (typical values are $\sigma_1 = 0.2$, $\sigma_2 = 0.8$, $\beta_1 = 0.25$, $\beta_2 = 2$).

Step 1: Find

$$d^k = \arg \min_{||d|| \leq \gamma^k} f^k(d), \tag{1.77}$$

If $f^k(d^k) = f(x^k)$ stop (x^k satisfies the first and second order necessary conditions for a local minimum); else go to Step 2.

Step 2: If $f(x^k + d^k) < f(x^k)$ set

$$x^{k+1} = x^k + d^k \tag{1.78}$$

calculate

$$r^k = \frac{f(x^k) - f(x^{k+1})}{f(x^k) - f^k(d^k)} \tag{1.79}$$

and go to Step 3; else set $\gamma^k := \beta_1 ||d^k||$ and go to Step 1.
Step 3: Set

$$\gamma^{k+1} = \begin{cases} \beta_1 ||d^k|| & \text{if } r^k < \sigma_1, \\ \beta_2 \gamma^k & \text{if } \sigma_2 \le r_k \text{ and } ||d^k|| = \gamma^k, \\ \gamma^k & \text{otherwise.} \end{cases} \tag{1.80}$$

Go to the next iteration.

Assuming that f is twice continuously differentiable, it is possible to show that the above algorithm is convergent in the sense that if $\{x^k\}$ is a bounded sequence, there exists a limit point of $\{x^k\}$ that satisfies the first and the second order necessary conditions for optimality. Furthermore, if $\{x^k\}$ converges to a nonsingular local minimum x^*, then asymptotically, the method is identical to the pure form of Newton's method, thereby attaining a superlinear convergence rate; see the references given at the end of the chapter for proofs of these and other related results for trust region methods.

Newton's Method with Periodic Reevaluation of the Hessian

A variation of Newton's method is obtained if the Hessian matrix $\nabla^2 f$ is recomputed every $p > 1$ iterations rather than at every iteration. In particular, this method, in unmodified form, is given by

$$x^{k+1} = x^k - \alpha^k D^k \nabla f(x^k),$$

where

$$D^{ip+j} = \left(\nabla^2 f(x^{ip})\right)^{-1}, \qquad j = 0, 1, \ldots, p-1, \ i = 0, 1, \ldots$$

The idea here is to save the computation and the inversion (or factorization) of the Hessian for the iterations where $j \neq 0$. This reduction in overhead is achieved at the expense of what is usually a small degradation in speed of convergence.

Truncated Newton Methods*

We have so far implicitly assumed that the system $\nabla^2 f(x^k) d^k = -\nabla f(x^k)$ will be solved for the direction d^k by Cholesky factorization or Gaussian elimination, which require a finite number of arithmetic operations $[O(n^3)]$. When the dimension n is large, the calculation required for exact solution of the system may be prohibitive, and one may have to be satisfied with only an approximate solution. Such an approximation may be obtained by using an iterative method. This approach is often used for solving very large linear systems of equations, arising in the solution of partial differential equations, where an adequate approximation to the solution can often be obtained by iterative methods quite fast, while the computation to find the exact solution can be overwhelming.

Generally, solving for d any system of the form $H^k d = -\nabla f(x^k)$, where H^k is a positive definite symmetric $n \times n$ matrix, can be done by solving the quadratic optimization problem

$$\begin{aligned} &\text{minimize} \quad \tfrac{1}{2} d' H^k d + \nabla f(x^k)' d \\ &\text{subject to} \quad d \in \Re^n, \end{aligned}$$

whose cost function gradient is zero at d if and only if $H^k d = -\nabla f(x^k)$. Suppose that an iterative descent method is used for solution and the starting point is $d^0 = 0$. Since the quadratic cost is reduced at each iteration and its value at the starting point is zero, we obtain after each iteration a vector d^k satisfying $\frac{1}{2} d^{k\prime} H^k d^k + \nabla f(x^k)' d^k < 0$, from which, using the positive definiteness of H^k,

$$\nabla f(x^k)' d^k < 0.$$

Thus the approximate solution d^k of the system $H^k d = -\nabla f(x^k)$, obtained after any positive number of iterations, is a descent direction. Possible iterative descent methods include the conjugate gradient method to be presented in Section 1.6 and the coordinate descent methods to be discussed in Section 1.8.

Conditions on the accuracy of the approximate solution d^k that ensure linear or superlinear rate of convergence are given in Exercise 1.4.5. Generally, the superlinear convergence rate property of the method to a nonsingular local minimum is maintained if the approximate Newton directions d^k satisfy

$$\lim_{k \to \infty} \frac{\left\| \nabla^2 f(x^k) d^k + \nabla f(x^k) \right\|}{\left\| \nabla f(x^k) \right\|} = 0,$$

(compare with Prop. 1.3.2). Thus, for superlinear convergence rate, the norm of the residual error in solving the Newton system must become negligible relative to the gradient norm in the limit.

EXERCISES

1.4.1

The purpose of this exercise is to show that Newton's method is unaffected by linear scaling of the variables. Consider a linear invertible transformation of variables $x = Sy$. Write Newton's method in the space of the variables y and show that it generates the sequence $y^k = S^{-1} x^k$, where $\{x^k\}$ is the sequence generated by Newton's method in the space of the variables x.

1.4.2 (www)

Show Prop. 1.4.2. *Hint*: For the second relation, let $M(x) = \int_0^1 \nabla g\big(x^* + t(x - x^*)\big)' dt$, so that $g(x) = M(x)(x - x^*)$. Argue that for some $\delta > 0$ the eigenvalues of $M(x)'M(x)$ lie between some positive scalars γ and Γ for all x with $\|x - x^*\| \le \delta$. Show that

$$\gamma\|x - x^*\|^2 \le \|g(x)\|^2 \le \Gamma\|x - x^*\|^2, \qquad \forall\, x \text{ with } \|x - x^*\| \le \delta.$$

1.4.3 (1st Combination with Steepest Descent)

This exercise and the next provide alternative modifications of Newton's method to improve its global convergence properties, while retaining its fast convergence rate. Consider the iteration $x^{k+1} = x^k + \alpha^k d^k$ where α^k is chosen by the Armijo rule with initial stepsize $s = 1$, $\sigma \in (0, 1/2)$, and d^k is equal to

$$d_N^k = -\big(\nabla^2 f(x^k)\big)^{-1} \nabla f(x^k)$$

if $\nabla^2 f(x^k)$ is nonsingular and the following two inequalities hold:

$$c_1\big\|\nabla f(x^k)\big\|^{p_1} \le -\nabla f(x^k)' d_N^k,$$

$$\|d_N^k\|^{p_2} \le c_2\big\|\nabla f(x^k)\|;$$

otherwise

$$d^k = -D\nabla f(x^k),$$

where D is a fixed positive definite symmetric matrix. The scalars c_1, c_2, p_1, and p_2 satisfy $c_1 > 0$, $c_2 > 0$, $p_1 > 2$, $p_2 > 1$. Show that the sequence $\{d^k\}$ is gradient related. Furthermore, every limit point of $\{x^k\}$ is stationary, and if $\{x^k\}$ converges to a nonsingular local minimum x^*, the rate of convergence of $\{\|x^k - x^*\|\}$ is superlinear.

1.4.4 (2nd Combination with Steepest Descent)

Assume that $f : \Re^n \mapsto \Re$ is twice continuously differentiable. At a point x^k, let $d_S^k = -D\nabla f(x^k)$ be a scaled steepest descent direction, where D is a fixed positive definite symmetric matrix. Let also d_N^k be equal to the Newton direction $-\big(\nabla^2 f(x^k)\big)^{-1}\nabla f(x^k)$ if $\nabla^2 f(x^k)$ is nonsingular, and be equal to d_S^k otherwise. Consider the method

$$x^{k+1} = x^k + \alpha^k\big((1 - \alpha^k)d_S^k + \alpha^k d_N^k\big),$$

where $\alpha^k = \beta^{m^k}$ and m^k is the first nonnegative integer m such that the following three inequalities hold:

$$f(x^k) - f\big(x^k + \beta^m\big((1-\beta^m)d_S^k + \beta^m d_N^k\big)\big) \ge -\sigma\beta^m \nabla f(x^k)'\big((1-\beta^m)d_S^k + \beta^m d_N^k\big),$$

$$c_1 \nabla f(x^k)' D \nabla f(x^k) \leq -\nabla f(x^k)' \big((1-\beta^m) d_S^k + \beta^m d_N^k \big),$$

$$\|(1-\beta^m) d_S^k + \beta^m d_N^k\| \leq c_2 \|D \nabla f(x^k)\|,$$

where β and σ are scalars satisfying $0 < \beta < 1$ and $0 < \sigma < 1/2$, and c_1 and c_2 are scalars satisfying $c_1 < 1$ and $c_2 > 1$. Show that the method is well defined in the sense that the stepsize α^k will be obtained after a finite number of trials. Furthermore, every limit point of $\{x^k\}$ is stationary and if $\{x^k\}$ converges to a nonsingular local minimum x^*, the rate of convergence of $\{\|x^k - x^*\|\}$ is superlinear. *Hint*: For a given x^k, each of the three inequalities is satisfied for m sufficiently large, so α^k is obtained after a finite number of trials. The directions $d^k = (1-\alpha^k)d_S^k + \alpha^k d_N^k$ are gradient related by construction (cf. the last two inequalities). Use the line of proof of Prop. 1.2.1 to show stationarity of the limit points of $\{x^k\}$. Use the line of proof of Prop. 1.3.2 to show that if $\{x^k\}$ converges to a nonsingular local minimum x^*, the convergence is superlinear (including the fact that $\alpha^k = 1$ for all sufficiently large k).

1.4.5 (www)

Consider a truncated Newton method with the stepsize chosen by the Armijo rule with initial stepsize $s = 1$ and $\sigma < 1/2$, and assume that $\{x^k\}$ converges to a nonsingular local minimum x^*. Assume that the matrices H^k and the directions d^k satisfy

$$\lim_{k\to\infty} \big\|H^k - \nabla^2 f(x^k)\big\| = 0, \qquad \lim_{k\to\infty} \frac{\big\|H^k d^k + \nabla f(x^k)\big\|}{\big\|\nabla f(x^k)\big\|} = 0.$$

Show that $\big\{\|x^k - x^*\|\big\}$ converges superlinearly.

1.4.6 (www)

Apply Newton's method with a constant stepsize to minimization of the function $f(x) = \|x\|^3$. Identify the range of stepsizes for which convergence is obtained, and show that it includes the unit stepsize. Show that for any stepsize within this range, the method converges linearly to $x^* = 0$. Explain this fact in light of Prop. 1.4.1.

1.4.7

Apply Newton's method with the trust region implementation to a positive definite quadratic function. Show that the method terminates in a finite number of iterations.

1.4.8

(a) Apply the pure form of Newton's method to minimization of the function $f(x) = \|x\|^\beta$, where $\beta > 1$. For what starting points and values of β does the method converge to the optimal solution? What happens when $\beta \leq 1$?

(b) Repeat part (a) for the case where Newton's method with the Armijo rule is used.

1.5 LEAST SQUARES PROBLEMS

In this section we consider methods for solving least squares problems of the form

$$\begin{aligned} &\text{minimize} \quad f(x) = \tfrac{1}{2}\|g(x)\|^2 = \tfrac{1}{2}\sum_{i=1}^{m}\|g_i(x)\|^2 \\ &\text{subject to} \quad x \in \Re^n, \end{aligned} \tag{1.81}$$

where g is a continuously differentiable function with component functions $g_1, \ldots, g_m$, where $g_i : \Re^n \to \Re^{r_i}$. Usually $r_i = 1$, but it is sometimes notationally convenient to consider the more general case.

Least squares problems are very common in practice. A principal case arises when g consists of n scalar-valued functions and we want to solve the system of n equations with n unknowns $g(x) = 0$. We can formulate this as the least squares optimization problem (1.81) [x^* solves the system $g(x) = 0$ if and only if it minimizes $\frac{1}{2}\|g(x)\|^2$ and the optimal value is zero]. Here are some other examples:

Example 1.5.1 (Model Construction – Curve Fitting)

Suppose that we want to estimate n parameters of a mathematical model so that it fits well a physical system, based on a set of measurements. In particular, we hypothesize an approximate relation of the form

$$z = h(x, y),$$

where h is a known function representing the model and

$x \in \Re^n$ is a vector of unknown parameters,
$z \in \Re^r$ is the model's output,
$y \in \Re^p$ is the model's input.

Given a set of m input-output data pairs $(y_1, z_1), \ldots, (y_m, z_m)$ from measurements of the physical system that we try to model, we want to find the vector of parameters x that matches best the data in the sense that it minimizes the sum of squared errors

$$\frac{1}{2}\sum_{i=1}^{m}\|z_i - h(x, y_i)\|^2.$$

For example, to fit the data pairs by a cubic polynomial approximation, we would choose

$$h(x, y) = x_3 y^3 + x_2 y^2 + x_1 y + x_0,$$

where $x = (x_0, x_1, x_2, x_3)$ is the vector of unknown coefficients of the cubic polynomial.

The next two examples are really special cases of the preceding one.

Example 1.5.2 (Dynamic System Identification)

A common model for a single input-single output dynamic system is to relate the input sequence $\{y_k\}$ to the output sequence $\{z_k\}$ by a linear equation of the form

$$\sum_{j=0}^{n} \alpha_j z_{k-j} = \sum_{j=0}^{n} \beta_j y_{k-j}.$$

Given a record of inputs and outputs $y_1, z_1, \ldots, y_m, z_m$ from the true system, we would like to find a set of parameters $\{\alpha_j, \beta_j \mid j = 0, 1, \ldots, n\}$ that matches this record best in the sense that it minimizes

$$\sum_{k=n}^{m}\left(\sum_{j=0}^{n} \alpha_j z_{k-j} - \sum_{j=0}^{n} \beta_j y_{k-j}\right)^2.$$

This is a least-squares problem.

Example 1.5.3 (Neural Networks)

A least squares modeling problem that has received a lot of attention is provided by *neural networks*. Here the model is specified by a multistage system, also called a *multilayer perceptron*. The kth stage consists of n_k *activation units*, each of which is a single input-single output mapping of a given form $\phi : \Re \mapsto \Re$ to be described shortly. The output of the jth activation unit of the $(k+1)$st stage is denoted by x_{k+1}^j and the input is a linear function of the output vector $x_k = (x_k^1, \ldots, x_k^{n_k})$ of the kth stage. Thus

$$x_{k+1}^j = \phi\left(u_k^{0j} + \sum_{s=1}^{n_k} x_k^s u_k^{sj}\right), \qquad j = 1, \ldots, n_{k+1}, \tag{1.82}$$

where the coefficients u_k^{sj} (also called *weights*) are to be determined.

Suppose that the multilayer perceptron has N stages, and let u denote the vector of the weights of all the stages:

$$u = \{u_k^{sj} \mid k = 0, \ldots, N-1,\ s = 0, \ldots, n_k,\ j = 1, \ldots, n_{k+1}\}.$$

Then, for a given vector u of weights, an input vector x_0 to the first stage produces a unique output vector x_N from the Nth stage via Eq. (1.82). Thus, we may view the multilayer perceptron as a mapping h that is parameterized by u and transforms the input vector x_0 into an output vector of the form $x_N = h(u, x_0)$. Suppose that we have m sample input-output pairs $(y_1, z_1), \ldots, (y_m, z_m)$ from a physical system that we are trying to model. Then, by selecting u appropriately, we can try to match the mapping of the multilayer perceptron with the mapping of the physical system. A common way to do this is to minimize over u the sum of squared errors

$$\tfrac{1}{2} \sum_{i=1}^{m} \|z_i - h(u, y_i)\|^2.$$

In the terminology of neural network theory, the process of finding the optimal weights is known as *training the network*.

Common examples of activation units are functions such as

$$\phi(\xi) = \frac{1}{1 + e^{-\xi}}, \qquad \text{(sigmoidal function)},$$

$$\phi(\xi) = \frac{e^{\xi} - e^{-\xi}}{e^{\xi} + e^{-\xi}}, \qquad \text{(hyperbolic tangent function)},$$

whose gradients are zero as the argument ξ approaches $-\infty$ and ∞. For these functions ϕ, it is possible to show that with a sufficient number of activation units and a number of stages $N \geq 2$, a multilayer perceptron can approximate arbitrarily closely very complex input-output maps; see [Cyb89].

Neural network training problems can be quite challenging. Their cost function is typically nonconvex and involves multiple local minima. For large values of the weights u_k^{ij}, the cost becomes "flat." In fact, as illustrated in Fig. 1.5.1, the cost function tends to a constant as u is changed along rays of the form $r\bar{u}$, where $r > 0$ and $\bar{u}$ is a fixed vector. For u near the origin, the cost function can be quite complicated alternately involving flat and steep regions.

The next example deals with an important context where neural networks are often used:

Example 1.5.4 (Pattern Classification)

Consider the problem of classifying objects based on the values of their characteristics. (We use the term "object" generically; in some contexts, the classification may relate to persons or situations.) Each object is presented to us with a vector y of features, and we wish to classify it in one of s categories

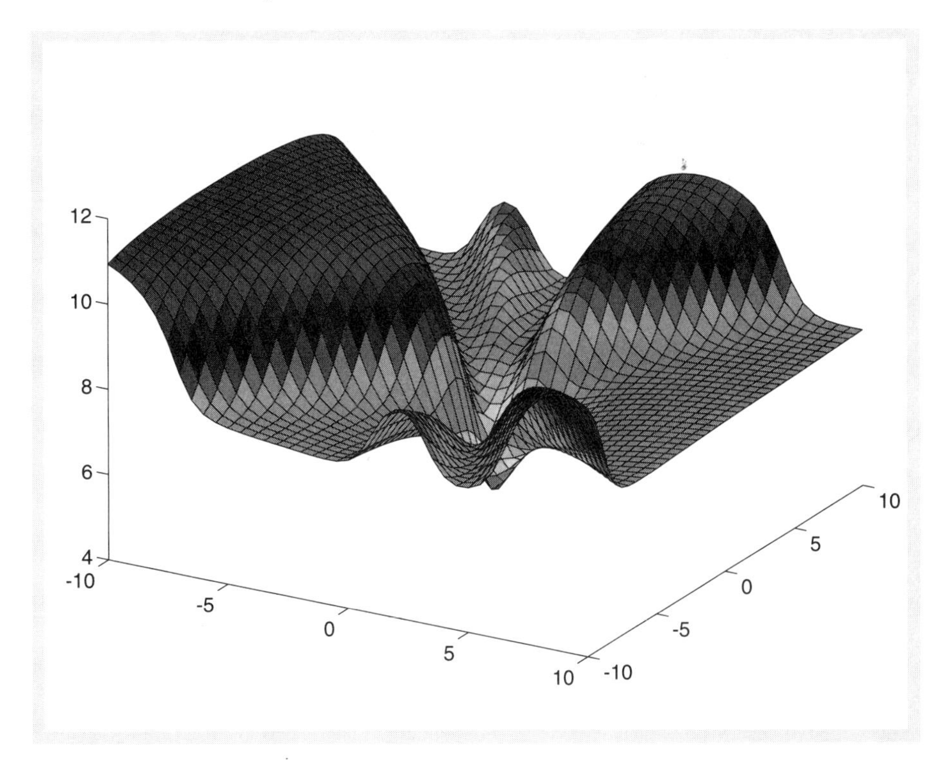

Figure 1.5.1. Three-dimensional plot of a least squares cost function

$$\frac{1}{2}\sum_{i=1}^{5}\big(z_i - \phi(u_1 y_i + u_0)\big)^2,$$

for a neural network training problem where there are only two weights u_0 and u_1, five data pairs, and ϕ is the hyperbolic tangent function. The data of the problem are given in Exercise 1.5.3. The cost function tends to a constant as u is changed along rays of the form $r\bar{u}$, where $r > 0$ and $\bar{u}$ is a fixed vector.

$1, \ldots, s$. For example, the vector y may represent the results of a collection of tests on a medical patient, and we may wish to classify the patient as being healthy or as having one of several types of illnesses.

A classical pattern classification approach is to assume that for each category $j = 1, \ldots, s$, we know the probability $p(j|y)$ that an object with feature vector y is of category j. Then we may associate an object with feature vector y with the category $j^*(y)$ having maximum posterior probability, that is,

$$j^*(y) = \arg \max_{j=1,\ldots,s} p(j|y). \tag{1.83}$$

Suppose now that the probabilities $p(j|y)$ are unknown, but instead we have a sample consisting of m object-category pairs. Then we may try to estimate $p(j|y)$ based on the following simple fact: out of all functions $f_j(y)$ of y, $p(j|y)$ is the one that minimizes the expected value of $(z_j - f_j(y))^2$, where

$$z_j = \begin{cases} 1 & \text{if } y \text{ is of category } j, \\ 0 & \text{otherwise.} \end{cases}$$

In particular, let y_i denote the feature vector of the ith object. For each category $j = 1, \ldots, s$, we estimate the probability $p(j|y)$ with a function $h_j(x_j, y)$ that is parameterized by a vector x_j. The function h_j may be provided for example by a neural network (cf. Example 1.5.3). Then, we can obtain x_j by minimizing the least squares function

$$\tfrac{1}{2} \sum_{i=1}^{m} \big(z_j^i - h_j(x_j, y_i)\big)^2,$$

where

$$z_j^i = \begin{cases} 1 & \text{if } y_i \text{ is of category } j, \\ 0 & \text{otherwise.} \end{cases}$$

This minimization approximates the minimization of the expected value of $(z_j - f_j(y))^2$. Once the optimal parameter vectors x_j^*, $j = 1, \ldots, s$, have been obtained, we may use them to classify a new object with feature vector y according to the rule

$$\text{Estimated Object Category} = \arg \max_{j=1,\ldots,s} h_j(x_j^*, y),$$

which approximates the maximum posterior probability rule (1.83).

For the simpler case where there are just two categories, say A and B, a similar formulation is to hypothesize a relation of the following form between feature vector y and category of an object:

$$\text{Object Category} = \begin{cases} A & \text{if } h(x, y) = 1, \\ B & \text{if } h(x, y) = -1, \end{cases}$$

where h is a given function and x is an unknown vector of parameters. Given a set of m data pairs $(z_1, y_1), \ldots, (z_m, y_m)$ of representative objects of known category, where y_i is the feature vector of the ith object, and

$$z_i = \begin{cases} 1 & \text{if } y \text{ is of category } A, \\ -1 & \text{if } y \text{ is of category } B, \end{cases}$$

we obtain x by minimizing the least squares function

$$\tfrac{1}{2} \sum_{i=1}^{m} \big(z_i - h(x, y_i)\big)^2.$$

The optimal parameter vector x^* is used to classify a new object with feature vector y according to the rule

$$\text{Estimated Object Category} = \begin{cases} A & \text{if } h(x^*, y) > 0, \\ B & \text{if } h(x^*, y) < 0. \end{cases}$$

There are several other variations on the above theme, for which we refer to the specialized literature. Furthermore, there are several alternative optimization-based methods for pattern classification, some of which are based on linear programming (see Mangasarian [Man93] for a survey).

1.5.1 The Gauss-Newton Method

Let us consider now specialized methods for minimizing the least squares cost $(1/2)\|g(x)\|^2$, starting with the most commonly used method, the *Gauss-Newton method.* Given a point x^k, the pure form of the Gauss-Newton iteration is based on linearizing g to obtain

$$\tilde{g}(x, x^k) = g(x^k) + \nabla g(x^k)'(x - x^k)$$

and then minimizing the norm of the linearized function $\tilde{g}$:

$$\begin{aligned} x^{k+1} &= \arg \min_{x \in \Re^n} \tfrac{1}{2}\|\tilde{g}(x, x^k)\|^2 \\ &= \arg \min_{x \in \Re^n} \tfrac{1}{2}\big\{\|g(x^k)\|^2 + 2(x - x^k)'\nabla g(x^k) g(x^k) \\ &\qquad\qquad + (x - x^k)'\nabla g(x^k)\nabla g(x^k)'(x - x^k)\big\}. \end{aligned}$$

Assuming that the $n \times n$ matrix $\nabla g(x^k)\nabla g(x^k)'$ is invertible, the above quadratic minimization yields

$$x^{k+1} = x^k - \big(\nabla g(x^k)\nabla g(x^k)'\big)^{-1}\nabla g(x^k) g(x^k). \tag{1.84}$$

Note that if g is already a linear function, we have $\|g(x)\|^2 = \|\tilde{g}(x, x^k)\|^2$, and the method converges in a single iteration. Note also that the direction

$$-\big(\nabla g(x^k)\nabla g(x^k)'\big)^{-1}\nabla g(x^k) g(x^k)$$

used in the above iteration is a descent direction since $\nabla g(x^k) g(x^k)$ is the gradient at x^k of the least squares cost function $(1/2)\|g(x)\|^2$ and $\big(\nabla g(x^k)\nabla g(x^k)'\big)^{-1}$ is a positive definite matrix.

To ensure descent, and also to deal with the case where the matrix $\nabla g(x^k)\nabla g(x^k)'$ is singular (as well as enhance convergence when this matrix is nearly singular), the method is often implemented in the modified form

$$x^{k+1} = x^k - \alpha^k\big(\nabla g(x^k)\nabla g(x^k)' + \Delta^k\big)^{-1}\nabla g(x^k) g(x^k), \tag{1.85}$$

where α^k is a stepsize chosen by one of the stepsize rules that we have discussed, and Δ^k is a diagonal matrix such that

$$\nabla g(x^k)\nabla g(x^k)' + \Delta^k : \text{positive definite.}$$

For example, Δ^k may be chosen in accordance with the Cholesky factorization scheme outlined in Section 1.4. An early proposal, known as the *Levenberg-Marquardt method*, is to choose Δ^k to be a positive multiple of the identity matrix. With these choices of Δ^k, it can be seen that the directions used by the method are gradient related, and the convergence results of Section 1.2 apply.

Relation to Newton's Method

The Gauss-Newton method bears a close relation to Newton's method. In particular, assuming each g_i is a scalar function, the Hessian of the cost $(1/2)\|g(x)\|^2$ is

$$\nabla g(x^k)\nabla g(x^k)' + \sum_{i=1}^{m} \nabla^2 g_i(x^k) g_i(x^k), \tag{1.86}$$

so it is seen that the Gauss-Newton iterations (1.84) and (1.85) are approximate versions of their Newton counterparts, where the second order term

$$\sum_{i=1}^{m} \nabla^2 g_i(x^k) g_i(x^k) \tag{1.87}$$

is neglected. Thus, in the Gauss-Newton method, we save the computation of this term at the expense of some deterioration in the convergence rate. If, however, the neglected term (1.87) is relatively small near a solution, the convergence rate of the Gauss-Newton method is satisfactory. This is often true in many applications such as for example when g is nearly linear, and also when the components $g_i(x)$ are small near the solution. In the case where $m = n$ and the problem is to solve the system $g(x) = 0$, the neglected term is zero at a solution. In this case, assuming $\nabla g(x^k)$ is invertible, we have

$$\big(\nabla g(x^k)\nabla g(x^k)'\big)^{-1}\nabla g(x^k)g(x^k) = \big(\nabla g(x^k)'\big)^{-1} g(x^k),$$

and the pure form of the Gauss-Newton method (1.84) takes the form

$$x^{k+1} = x^k - \big(\nabla g(x^k)'\big)^{-1} g(x^k),$$

which is identical to Newton's method for solving the system $g(x) = 0$ [rather than Newton's method for minimizing $\|g(x)\|^2$]. Thus, the convergence rate is typically superlinear in this case, as discussed in Section 1.4.

1.5.2 Incremental Gradient Methods*

Let us return to the model construction Example 1.5.1, where we want to find a vector $x \in \Re^n$ of model parameters based on data obtained from a physical system. Each component g_i in the least squares formulation is referred to as a *data block*, and the entire function $g = (g_1, \ldots, g_m)$ is referred to as the *data set*.

In many problems of interest where there are many data blocks, the Gauss-Newton method may be ineffective, because the size of the data set makes each iteration very costly. For such problems it may be more attractive to use an incremental method that does not wait to process the entire data set before updating x; instead, the method cycles through the data blocks in sequence and updates the estimate of x after each data block is processed.

An example of such a method operates as follows: given x^k, we obtain x^{k+1} at the last step of a cycle through the data blocks, which starts with

$$\psi_0 = x^k$$

and consists of the following m steps

$$\psi_i = \psi_{i-1} - \alpha^k \nabla g_i(\psi_{i-1}) g_i(\psi_{i-1}), \qquad i = 1, \ldots, m. \tag{1.88}$$

Here $\alpha^k > 0$ is a stepsize, and the direction used is the gradient of the ith data block,

$$\nabla\left(\tfrac{1}{2}\|g_i(x)\|^2\right)\Big|_{x=\psi_{i-1}} = \nabla g_i(\psi_{i-1}) g_i(\psi_{i-1}). \tag{1.89}$$

This method can be written as

$$x^{k+1} = x^k - \alpha^k \sum_{i=1}^{m} \nabla g_i(\psi_{i-1}) g_i(\psi_{i-1}), \tag{1.90}$$

and it differs from the steepest descent method in that it uses the direction

$$\sum_{i=1}^{m} \nabla g_i(\psi_{i-1}) g_i(\psi_{i-1})$$

in place of the gradient

$$\nabla f(x^k) = \sum_{i=1}^{m} \nabla g_i(x^k) g_i(x^k).$$

We consequently call the method (1.90) *incremental gradient method*, and we note that it can also be viewed as *a steepest descent method with errors*. In particular, we have

$$x^{k+1} = x^k - \alpha^k \big(\nabla f(x_k) + e_k\big),$$

where the errors e_k are given by

$$e_k = \sum_{i=1}^{m} \big(\nabla g_i(\psi_{i-1}) g_i(\psi_{i-1}) - \nabla g_i(x^k) g_i(x^k)\big).$$

There are two significant potential advantages to an incremental approach:

(a) Estimates of x become available as data is accumulated, making the approach suitable for real-time operation.

(b) If the data set is very large, one hopes to obtain a good estimate of x after a *single* cycle through the data set. This is particularly so if the data blocks are *statistically homogeneous*, in the sense that the minimizers of all the terms $\|g_j(x)\|^2$ are not too different. More generally, for a large data set one hopes to attain a faster rate of convergence with an incremental approach than with an ordinary gradient method, particularly when far from the eventual limit. This type of behavior is most vividly illustrated in the case where the data blocks are linear and the vector x is one-dimensional, as in the following example.

Example 1.5.5

Assume that x is a scalar, and that the least squares problem has the form

$$\begin{aligned} &\text{minimize} \quad f(x) = \tfrac{1}{2}\sum_{i=1}^{m}(a_i x - b_i)^2 \\ &\text{subject to} \quad x \in \Re, \end{aligned}$$

where a_i and b_i are given scalars with $a_i \neq 0$ for all i. The minimum of each of the squared data blocks

$$f_i(x) = \tfrac{1}{2}(a_i x - b_i)^2$$

is

$$x_i^* = \frac{b_i}{a_i},$$

while the minimum of the least squares cost function f is

$$x^* = \frac{\sum_{i=1}^{m} a_i b_i}{\sum_{i=1}^{m} a_i^2}.$$

It can be seen that x^* lies within the range of the data block minima

$$R = \left[\min_i x_i^*,\ \max_i x_i^*\right],$$

and that for all x *outside* the range R, the gradient

$$\nabla f_i(x) = a_i(a_i x - b_i)$$

has the same sign as $\nabla f(x)$ (see Fig. 1.5.2). As a result, when outside the region R, the incremental gradient method

$$\psi_i = \psi_{i-1} - \alpha^k a_i(a_i\psi_{i-1} - b_i)$$

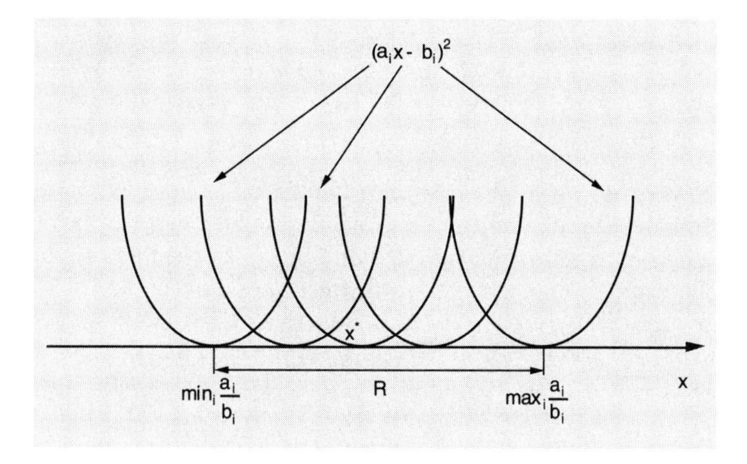

Figure 1.5.2. Illustrating the advantage of incrementalism when far from the optimal solution. The ith step in an incremental gradient cycle is a gradient step for minimizing $(a_i x - b_i)^2$, so if x lies outside the region of data block minima

$$R = \left[\min_i x_i^*,\ \max_i x_i^*\right],$$

and the stepsize is small enough, progress towards the solution x^* is made.

approaches x^* at each step [cf. Eq. (1.88)], provided the stepsize α^k is small enough. In fact it is sufficient that

$$\alpha^k \leq \min_i \frac{1}{a_i^2}.$$

However, for x *inside* the region R, the ith step of a cycle of the incremental gradient method need not make progress. It will approach x^* (for small enough stepsize α^k) only if the current point ψ_{i-1} does not lie in the interval connecting x_i^* and x^*. This induces an oscillatory behavior within the region R, and as a result, the incremental gradient method will typically not converge to x^* unless $\alpha^k \to 0$. By contrast, it can be shown that the steepest descent method, which takes the form

$$x^{k+1} = x^k - \alpha^k \sum_{i=1}^{m} a_i(a_i x^k - b_i),$$

converges to x^* for any constant stepsize satisfying

$$\alpha^k \leq \frac{1}{\sum_{i=1}^{m} a_i^2}.$$

However, unless the stepsize choice is particularly favorable, for x outside the region R, a full iteration of steepest descent need not make more progress towards the solution than a single step of the incremental gradient method.

In other words, with comparably intelligent stepsize choices, *far from the solution (outside R), a single pass through the entire data set by incremental gradient is roughly as effective as m passes through the data set by steepest descent.*

The preceding example relies on x being one-dimensional, but in many multidimensional problems the same qualitative behavior can be observed. In particular, a pass through the ith data block g_i by the incremental gradient method can make progress towards the solution in the region where the data block gradient $\nabla g_i(\psi_{i-1})g_i(\psi_{i-1})$ makes an angle less than 90 degrees with the cost function gradient $\nabla f(\psi_{i-1})$. If the data blocks g_i are not "too dissimilar", this is likely to happen in a region of points that are not too close to the optimal solution set; see also Exercise 1.5.5. This behavior has been verified in many practical contexts, including the training of neural networks (cf. Example 1.5.3), where incremental gradient methods (1.90) have been used extensively, frequently under the name *back-propagation methods* (see e.g., [BeT96], [Bis95], and [Hay98]).

The choice of the stepsize α^k plays an important role in the performance of incremental gradient methods. On close examination, it turns out that the direction used by the method differs from the gradient direction by an error that is proportional to the stepsize, and for this reason a diminishing stepsize is essential for convergence to a stationary point of f. However, it turns out that a peculiar form of convergence also typically occurs for a constant but sufficiently small stepsize. In this case, the iterates converge to a "limit cycle", whereby the ith iterates ψ_i within the cycles converge to a different limit than the jth iterates ψ_j for $i \neq j$. The sequence $\{x^k\}$ that consists of the iterates obtained at the end of cycles converges, except that the limit obtained *need not* be a stationary point of f (see Exercise 1.5.2 and the following analysis). The limit tends to be close to a stationary point when the constant stepsize is small (see the following Prop. 1.5.1). In practice, it is common to use a constant stepsize for a (possibly prespecified) number of iterations, then decrease the stepsize by a certain factor, and repeat, up to the point where the stepsize reaches a prespecified minimum. An alternative possibility is to use a stepsize rule of the form

$$\alpha^k = \min\left\{\gamma, \frac{\gamma_1}{k + \gamma_2}\right\},$$

where γ, γ_1, and γ_2 are some positive scalars. There are also variants of the incremental gradient method that use a constant stepsize throughout, and generically converge to a stationary point of f at a linear rate. The degree of incrementalism in these methods gradually diminishes as the method progresses (see Exercise 1.5.4 and [Ber95b]).

We note a popular modification of the incremental gradient method, which uses the update

$$\psi_i = \begin{cases} \psi_0 - \alpha^k \nabla g_1(\psi_0) g_1(\psi_0) & \text{if } i = 1, \\ \psi_{i-1} - \alpha^k \nabla g_i(\psi_{i-1}) g_i(\psi_{i-1}) + \beta(\psi_{i-1} - \psi_{i-2}) & \text{if } i = 2, \ldots, m, \end{cases} \tag{1.91}$$

where $\beta \in [0, 1)$, in place of the update

$$\psi_i = \psi_{i-1} - \alpha^k \nabla g_i(\psi_{i-1}) g_i(\psi_{i-1}).$$

This method may be viewed as an incremental version of the heavy ball method given in Exercise 1.3.9 of Section 1.3. The term $(\psi_{i-1} - \psi_{i-2})$ is known as a *momentum term*, and is often helpful in dealing with the peculiar features of the cost functions of neural network training problems.

Another popular technique for incremental methods is to reshuffle randomly the order of the data blocks after each cycle through the data set. A related alternative is to select randomly the data block from the data set at each iteration. If we take the view that an incremental method is basically a gradient method with errors, we see that randomization of the order of the data blocks tends to randomize the size of the errors, and under certain circumstances, this tends to improve the convergence properties of the method. The following example illustrates some of the issues involved.

Example 1.5.6

Assume that x is a scalar, and that the least squares problem has the form

$$\begin{aligned} &\text{minimize} \quad f(x) = \tfrac{1}{2} \left(\sum_{i=1}^{m} (x-1)^2 + \sum_{i=1}^{m} (x+1)^2 \right) \\ &\text{subject to} \quad x \in \Re. \end{aligned}$$

[This example is somewhat peculiar because the $2m$ terms in the least squares cost function consist of m copies of just two functions, $(x-1)^2$ and $(x+1)^2$, but is suitable for our purposes.] Let us assume that the stepsize is a constant α. Then, if $\alpha < 2$, it can be shown that the method will converge to a limit cycle, the size of which depends on the order for processing the data blocks. Let us determine the size of the limit cycle under various rules for selecting the data block to be iterated on.

Consider the case where the m terms $(x-1)^2$ are processed first, the m terms $(x+1)^2$ are processed next, and the process is repeated (it can be seen that this is the order that produces the largest size of limit cycle). Then, if ψ_0 is the iterate in the beginning of a cycle, the first m iterates within the cycle are given by

$$\psi_i = \psi_{i-1} - \alpha(\psi_{i-1} - 1) = (1 - \alpha)\psi_{i-1} + \alpha, \quad i = 1, \ldots, m,$$

leading to the mid-cycle iterate

$$\begin{aligned} \psi_m &= (1-\alpha)^m \psi_0 + \alpha\big(1 + (1-\alpha) + \cdots + (1-\alpha)^{m-1}\big) \\ &= (1-\alpha)^m \psi_0 + \big(1 - (1-\alpha)^m\big). \end{aligned} \tag{1.92}$$

The final m iterates within the cycle are given by

$$\psi_{m+i} = \psi_{m+i-1} - \alpha(\psi_{m+i-1} + 1) = (1 - \alpha)\psi_{m+i-1} - \alpha, \quad i = 1, \ldots, m,$$

leading similarly to the final iterate in the cycle

$$\psi_{2m} = (1 - \alpha)^m \psi_m - \big(1 - (1 - \alpha)^m\big). \tag{1.93}$$

Thus, by combining Eqs. (1.92) and (1.93), we have

$$\psi_{2m} = (1 - \alpha)^{2m} \psi_0 - \big(1 - (1 - \alpha)^m\big)^2.$$

In the limit, we must have $\psi_{2m} = \psi_0$, from which we obtain the limit values of the start (and end) iterate of the limit cycle

$$\bar{\psi}_0 = -\frac{1 - (1 - \alpha)^m}{1 + (1 - \alpha)^m}.$$

Using this equation and Eq. (1.92), we see that the limit value of the iterate ψ_m is

$$\bar{\psi}_m = \frac{1 - (1 - \alpha)^m}{1 + (1 - \alpha)^m}.$$

Thus the width of the limit cycle is

$$\bar{\psi}_m - \bar{\psi}_0 = \frac{2\big(1 - (1 - \alpha)^m\big)}{1 + (1 - \alpha)^m}, \tag{1.94}$$

and for small values of α it is of order $O(m\alpha)$.

Consider now the variant of the incremental gradient method that selects randomly (with equal probability $1/2m$) the data block from the data set at each iteration. This method has the form

$$x^{k+1} = x^k - \alpha(x^k - w^k) = (1 - \alpha)x^k + \alpha w^k,$$

where w^k takes the value 1 with probability 1/2 [corresponding to the data block $(x - 1)$], and the value -1 with probability 1/2 [corresponding to the data block $(x + 1)$]. The second moment of x^k obeys the recursion

$$E\big[(x^{k+1})^2\big] = (1 - \alpha)^2 E\big[(x^k)^2\big] + \alpha^2 E\big[(w^k)^2\big].$$

The steady-state value of the second moment of x^k (which is also the steady-state value of the variance of x^k, since the expected value $E[x^k]$ converges to 0) is given by

$$\lim_{k \to \infty} E\big[(x^k)^2\big] = \frac{\alpha^2}{1 - (1 - \alpha)^2} = \frac{\alpha}{2 - \alpha}.$$

Thus, the standard deviation of x^k is of order $O(\sqrt{\alpha})$. Compared with the estimate (1.94) of the width of the limit cycle of the (deterministic) incremental gradient method, we see that for small values of α, the deterministic method is superior in that it has a smaller limit cycle width. For moderate values of α and large enough values of m, we see that the stochastic method is superior.

Let us also note that diagonal scaling of the gradients of the squared norm terms $\|g_i(x)\|^2$ is possible and is helpful in many problems. Such scaling can be implemented by replacing the incremental gradient iteration $\psi_i = \psi_{i-1} - \alpha^k h_i$, where $h_i = \nabla g_i(\psi_{i-1}) g_i(\psi_{i-1})$, with the iteration

$$\psi_i = \psi_{i-1} - \alpha^k D h_i, \qquad i = 1, \ldots, m,$$

where D is a diagonal positive definite matrix, which may change from cycle to cycle.

Generally, it may be said that while incremental methods are used widely in practice, their effective use often requires skill, insight into the problem's structure, and trial and error.

Convergence Analysis of Incremental Gradient Methods

We now analyze the convergence of the incremental gradient method of Eq. (1.90) for the case where the data blocks are linear. The main idea is that the method can be viewed as a steepest descent iteration where the gradient is perturbed by an error term that is proportional to the stepsize (see the subsequent analysis). The qualitative behavior of the method then is as discussed in Section 1.2. The following lemma will be used in the proof.

Lemma 1.5.1: Let $\{e^k\}$ be a nonnegative sequence satisfying

$$e^{k+1} \le (1 - \gamma^k) e^k + \beta^k, \qquad \forall\, k = 0, 1, \ldots,$$

where $\{\beta^k\}$ and $\{\gamma^k\}$ are nonnegative sequences such that

$$\gamma^k \to 0, \qquad \sum_{k=0}^{\infty} \gamma^k = \infty, \qquad \frac{\beta^k}{\gamma^k} \to 0.$$

Then $e^k \to 0$.

Proof: We first show that given any $\epsilon > 0$, we have $e^k < \epsilon$ for infinitely many k. Indeed, assuming this were not so and letting $\bar{k}$ be such that $e^k \ge \epsilon$ and $\beta^k/\gamma^k \le \epsilon/2$ for all $k \ge \bar{k}$, we would have for all $k \ge \bar{k}$

$$e^{k+1} \le e^k - \gamma^k e^k + \beta^k \le e^k - \gamma^k \epsilon + \gamma^k \epsilon/2 = e^k - \gamma^k \epsilon/2.$$

Therefore, for all $m \geq \bar{k}$ we would have

$$e^{m+1} \leq e^{\bar{k}} - (\epsilon/2) \sum_{k=\bar{k}}^{m} \gamma^k.$$

This contradicts the nonnegativity of $\{e^k\}$ and the assumption $\sum_{k=0}^{\infty} \gamma^k = \infty$.

Thus, given any $\epsilon > 0$, there exists $\bar{k}$ such that $\beta^k/\gamma^k < \epsilon$ for all $k \geq \bar{k}$ and $e^{\bar{k}} < \epsilon$. We then have

$$e^{\bar{k}+1} \leq (1 - \gamma^k)e^{\bar{k}} + \beta^k < (1 - \gamma^k)\epsilon + \gamma^k \epsilon = \epsilon.$$

By repeating this argument, we obtain $e^k < \epsilon$ for all $k \geq \bar{k}$. Since ϵ can be arbitrarily small, it follows that $e^k \to 0$. **Q.E.D.**

Proposition 1.5.1: Consider the case of linear data blocks,

$$g_i(x) = z_i - C_i x, \qquad i = 1, \ldots, m, \tag{1.95}$$

and the incremental gradient method

$$x^{k+1} = x^k + \alpha^k \sum_{i=1}^{m} C_i'(z_i - C_i \psi_{i-1}), \tag{1.96}$$

where $\psi_0 = x^k$ and

$$\psi_i = \psi_{i-1} + \alpha^k C_i'(z_i - C_i \psi_{i-1}), \qquad i = 1, \ldots, m. \tag{1.97}$$

Assume that $\sum_{i=1}^{m} C_i' C_i$ is a positive definite matrix and let x^* be the optimal solution of the corresponding least squares problem. Then:

(a) There exists $\bar{\alpha} > 0$ such that if α^k is equal to some constant $\alpha \in (0, \bar{\alpha}]$ for all k, $\{x^k\}$ converges to some vector $x(\alpha)$. Furthermore, the error $\|x^k - x(\alpha)\|$ converges to 0 linearly. In addition, we have $\lim_{\alpha \to 0} x(\alpha) = x^*$.

(b) If $\alpha^k > 0$ for all k, and

$$\alpha^k \to 0, \qquad \sum_{k=0}^{\infty} \alpha^k = \infty,$$

then $\{x^k\}$ converges to x^*.

Proof: (a) We first show by induction that for all $i = 1, \ldots, m$, the vectors ψ_i of Eq. (1.97) have the form

$$\psi_i = x^k + \alpha \sum_{j=1}^{i} C_j'(z_j - C_j x^k) + \sum_{j=1}^{i-1} \alpha^{j+1}(\Phi_{ij} x^k + \phi_{ij}), \tag{1.98}$$

where Φ_{ij} and ϕ_{ij} are some matrices and vectors, respectively, which are independent of k. Indeed, this relation holds by definition for $i = 1$ [the last term in Eq. (1.98) is zero when $i = 1$]. Suppose that it holds for $i = p$. We have, using the induction hypothesis,

$$\begin{aligned}
\psi_{p+1} &= \psi_p + \alpha C_{p+1}'(z_{p+1} - C_{p+1}\psi_p) \\
&= x^k + \alpha \sum_{j=1}^{p} C_j'(z_j - C_j x^k) + \sum_{j=1}^{p-1} \alpha^{j+1}(\Phi_{pj} x^k + \phi_{pj}) \\
&\quad + \alpha C_{p+1}'(z_{p+1} - C_{p+1} x^k) - \alpha C_{p+1}' C_{p+1}(\psi_p - x^k) \\
&= x^k + \alpha \sum_{j=1}^{p+1} C_j'(z_j - C_j x^k) + \sum_{j=1}^{p-1} \alpha^{j+1}(\Phi_{pj} x^k + \phi_{pj}) \\
&\quad - \alpha C_{p+1}' C_{p+1} \left(\alpha \sum_{j=1}^{p} C_j'(z_j - C_j x^k) + \sum_{j=1}^{p-1} \alpha^{j+1}(\Phi_{pj} x^k + \phi_{pj}) \right),
\end{aligned}$$

which is of the form (1.98), thereby completing the induction.

For $i = m$, Eq. (1.98) yields

$$x^{k+1} = A(\alpha) x^k + b(\alpha), \tag{1.99}$$

where

$$A(\alpha) = I - \alpha \sum_{j=1}^{m} C_j' C_j + \sum_{j=1}^{m-1} \alpha^{j+1} \Phi_{mj}, \tag{1.100}$$

$$b(\alpha) = \alpha \sum_{j=1}^{m} C_j' z_j + \sum_{j=1}^{m-1} \alpha^{j+1} \phi_{mj}. \tag{1.101}$$

Let us choose α small enough so that the eigenvalues of $A(\alpha)$ are all strictly within the unit circle; this is possible since $\sum_{j=1}^{m} C_j' C_j$ is assumed positive definite and the last term in Eq. (1.100) involves powers of α that are greater or equal to 2. Define

$$x(\alpha) = \big(I - A(\alpha)\big)^{-1} b(\alpha). \tag{1.102}$$

Then $b(\alpha) = \big(I - A(\alpha)\big) x(\alpha)$, and by substituting this expression in Eq. (1.99), it can be seen that

$$x^{k+1} - x(\alpha) = A(\alpha)\big(x^k - x(\alpha)\big),$$

from which

$$x^k - x(\alpha) = A(\alpha)^k \big(x^0 - x(\alpha)\big), \qquad \forall\, k. \tag{1.103}$$

Since all the eigenvalues of $A(\alpha)$ are strictly within the unit circle, we have $A(\alpha)^k \to 0$ (Prop. A.16 in Appendix A), so $\|x^k - x(\alpha)\| \to 0$ linearly.

To prove that $\lim_{\alpha\to 0} x(\alpha) = x^*$, we first calculate x^*. We set the gradient of $\|g(x)\|^2$ to zero, to obtain

$$\sum_{i=1}^m C_i'(C_i x^* - z_i) = 0, \tag{1.104}$$

so that

$$x^* = \left(\sum_{i=1}^m C_i' C_i\right)^{-1} \sum_{i=1}^m C_i' z_i. \tag{1.105}$$

Then, we use Eq. (1.102) to write $x(\alpha) = \big(I/\alpha - A(\alpha)/\alpha\big)^{-1}\big(b(\alpha)/\alpha\big)$, and we see from Eqs. (1.100) and (1.101) that

$$\lim_{\alpha\to 0} x(\alpha) = \left(\sum_{i=1}^m C_i' C_i\right)^{-1} \sum_{i=1}^m C_i' z_i = x^*.$$

(b) For $i = m$ and $\alpha = \alpha^k$, Eq. (1.98) yields

$$x^{k+1} = x^k + \alpha^k \sum_{j=1}^m C_j'(z_j - C_j x^k) + (\alpha^k)^2 H^k (x^k - x^*) + (\alpha^k)^2 h^k, \tag{1.106}$$

where

$$H^k = \sum_{j=1}^{m-1} (\alpha^k)^{j-1} \Phi_{mj}, \tag{1.107}$$

$$h^k = \sum_{j=1}^{m-1} (\alpha^k)^{j-1} (\Phi_{mj} x^* + \phi_{mj}). \tag{1.108}$$

Using also the expression (1.105) for x^*, we can write Eq. (1.106) as

$$x^{k+1} - x^* = \left(I - \alpha^k \sum_{j=1}^m C_j' C_j + (\alpha^k)^2 H^k\right)(x^k - x^*) + (\alpha^k)^2 h^k. \tag{1.109}$$

For large enough k, the eigenvalues of $\alpha^k \sum_{j=1}^m C_j' C_j$ are bounded from above by 1, and hence the matrix $I - \alpha^k \sum_{j=1}^m C_j' C_j$ is positive definite. Without loss of generality, we assume that this is so for all k. Then we have

$$\left\| \left(I - \alpha^k \sum_{j=1}^m C_j' C_j\right)(x^k - x^*) \right\| \le (1 - \alpha^k A)\|x^k - x^*\|, \tag{1.110}$$

where A is the smallest eigenvalue of $\sum_{j=1}^{m} C_j'C_j$. Let also B and δ be positive scalars such that for all k we have

$$\|H^k(x^k - x^*)\| \leq B\|x^k - x^*\|, \qquad \|h^k\| \leq \delta. \tag{1.111}$$

Combining Eqs. (1.109)-(1.111), we have

$$\begin{aligned} \|x^{k+1} - x^*\| &\leq \left\| \left(I - \alpha^k \sum_{j=1}^{m} C_j'C_j \right) (x^k - x^*) \right\| \\ &\quad + (\alpha^k)^2 \|H^k(x^k - x^*)\| + (\alpha^k)^2 \|h^k\| \\ &\leq \big(1 - \alpha^k A + (\alpha^k)^2 B\big)\|x^k - x^*\| + (\alpha^k)^2 \delta. \end{aligned} \tag{1.112}$$

Let $\bar{k}$ be such that $\alpha^k B \leq A/2$ for all $k \geq \bar{k}$. Then from Eq. (1.112) we obtain

$$\|x^{k+1} - x^*\| \leq (1 - \alpha^k A/2)\|x^k - x^*\| + (\alpha^k)^2 \delta, \qquad \forall\ k \geq \bar{k},$$

and Lemma 1.5.1 can be used to show that $\|x^k - x^*\| \to 0$. **Q.E.D.**

In the case where the data blocks are nonlinear, stationarity of the limit points of sequences $\{x^k\}$ generated by the incremental gradient method has been shown under certain assumptions (including Lipschitz continuity of the data block gradients) for the case of the stepsize $\alpha^k = \gamma/(k+1)$, where γ is a positive scalar [see [BeT96], Prop. 3.8, [BeT99], and [MaS94], which in addition analyzes the heavy ball-like method (1.91)]. The convergence proof is similar to the one of the preceding proposition, but it is technically more involved.

In the case of a constant stepsize and nonlinear data blocks, it is also possible to show a result analogous to Prop. 1.5.1(a), but again the proof is technically complex.

1.5.3 Incremental Forms of the Gauss-Newton Method*

We now consider incremental versions of the Gauss-Newton method. An example of such a method starts with some x^0, then updates x via a Gauss-Newton-like iteration aimed at minimizing

$$\|g_1(x)\|^2,$$

then updates x via a Gauss-Newton-like iteration aimed at minimizing

$$\lambda\|g_1(x)\|^2 + \|g_2(x)\|^2,$$

where λ is a scalar with

$$0 \leq \lambda \leq 1,$$

and similarly continues, with the ith step consisting of a Gauss-Newton-like iteration aimed at minimizing the weighted partial sum

$$\sum_{j=1}^{i} \lambda^{i-j} \|g_j(x)\|^2. \tag{1.113}$$

Once the entire data set is processed, the cycle is restarted.

The parameter λ determines the influence of old data blocks on new estimates. Generally, with smaller values of λ, the effect of old data blocks is discounted faster, and successive estimates produced by the method tend to change more rapidly. Thus one may obtain a faster rate of progress of the method when $\lambda < 1$, and this is often desirable, particularly when the data blocks are obtained in real time from a model whose parameters are slowly changing.

The Kalman Filter for Linear Least Squares

When the data blocks are linear functions, it takes a single pure Gauss-Newton iteration to find the least squares estimate, and it turns out that this iteration can be implemented with an incremental algorithm known as the *Kalman filter*. This algorithm has many important applications in control and communication theory, and has been studied extensively. We first develop this algorithm for linear data blocks, and we then extend it to the case of nonlinear data blocks.

Suppose that the functions g_i are linear of the form

$$g_i(x) = z_i - C_i x, \tag{1.114}$$

where $z_i \in \Re^{r_i}$ are given vectors and C_i are given $r_i \times n$ matrices. In other words, we are trying to fit a linear model to the set of measurements $z_1, \ldots, z_m$. Let us consider the incremental method that sequentially generates the vectors

$$\psi_i = \arg \min_{x \in \Re^n} \sum_{j=1}^{i} \lambda^{i-j} \|z_j - C_j x\|^2, \qquad i = 1, \ldots, m. \tag{1.115}$$

Then, for $\lambda = 1$, the least squares solution is obtained at the last step as

$$x^* = \psi_m.$$

Furthermore, the method can be conveniently implemented as shown by the following proposition:

Proposition 1.5.2: (Kalman Filter) Assuming that the matrix $C_1'C_1$ is positive definite, the least squares estimates

$$\psi_i = \arg\min_{x\in\Re^n} \sum_{j=1}^{i} \lambda^{i-j} \|z_j - C_j x\|^2, \qquad i = 1, \ldots, m,$$

can be generated by the algorithm

$$\psi_i = \psi_{i-1} + H_i^{-1} C_i'(z_i - C_i \psi_{i-1}), \qquad i = 1, \ldots, m, \tag{1.116}$$

where ψ_0 is an arbitrary vector, and the positive definite matrices H_i are generated by

$$H_i = \lambda H_{i-1} + C_i' C_i, \qquad i = 1, \ldots, m, \tag{1.117}$$

with

$$H_0 = 0.$$

More generally, for all integers i and $\bar{i}$ with $1 \le \bar{i} < i \le m$ we have

$$\psi_i = \psi_{\bar{i}} + H_i^{-1} \sum_{j=\bar{i}+1}^{i} \lambda^{i-j} C_j'(z_j - C_j \psi_{\bar{i}}). \tag{1.118}$$

Proof: We first establish the result for the case of two data blocks in the following lemma:

Lemma 1.5.2: Let ζ_1, ζ_2 be given vectors, and Γ_1, Γ_2 be given matrices such that $\Gamma_1'\Gamma_1$ is positive definite. Then the vectors

$$\psi_1 = \arg\min_{x\in\Re^n} \|\zeta_1 - \Gamma_1 x\|^2, \tag{1.119}$$

and

$$\psi_2 = \arg\min_{x\in\Re^n} \left\{ \|\zeta_1 - \Gamma_1 x\|^2 + \|\zeta_2 - \Gamma_2 x\|^2 \right\}, \tag{1.120}$$

are also given by

$$\psi_1 = \psi_0 + (\Gamma_1'\Gamma_1)^{-1} \Gamma_1'(\zeta_1 - \Gamma_1 \psi_0), \tag{1.121}$$

and

$$\psi_2 = \psi_1 + (\Gamma_1'\Gamma_1 + \Gamma_2'\Gamma_2)^{-1}\Gamma_2'(\zeta_2 - \Gamma_2\psi_1), \qquad (1.122)$$

where ψ_0 is an arbitrary vector.

Proof: By carrying out the minimization in Eq. (1.119), we obtain

$$\psi_1 = \left(\Gamma_1'\Gamma_1\right)^{-1}\Gamma_1'\zeta_1, \qquad (1.123)$$

yielding for any ψ_0,

$$\psi_1 = \psi_0 - \left(\Gamma_1'\Gamma_1\right)^{-1}\Gamma_1'\Gamma_1\psi_0 + \left(\Gamma_1'\Gamma_1\right)^{-1}\Gamma_1'\zeta_1,$$

from which the desired Eq. (1.121) follows.

Also, by carrying out the minimization in Eq. (1.120), we obtain

$$\psi_2 = \left(\Gamma_1'\Gamma_1 + \Gamma_2'\Gamma_2\right)^{-1}\left(\Gamma_1'\zeta_1 + \Gamma_2'\zeta_2\right),$$

or equivalently, using also Eq. (1.123),

$$\begin{aligned}(\Gamma_1'\Gamma_1 + \Gamma_2'\Gamma_2)\psi_2 &= \Gamma_1'\zeta_1 + \Gamma_2'\zeta_2 \\ &= \Gamma_1'\Gamma_1\psi_1 + \Gamma_2'\zeta_2 \\ &= (\Gamma_1'\Gamma_1 + \Gamma_2'\Gamma_2)\psi_1 - \Gamma_2'\Gamma_2\psi_1 + \Gamma_2'\zeta_2,\end{aligned}$$

from which, by multiplying both sides with $(\Gamma_1'\Gamma_1 + \Gamma_2'\Gamma_2)^{-1}$, the desired Eq. (1.122) follows. **Q.E.D.**

Proof of Prop. 1.5.2: Equation (1.118) follows by applying Lemma 1.5.2 with the correspondences $\psi_0 \sim \psi_0$, $\psi_1 \sim \psi_{\bar{i}}$, $\psi_2 \sim \psi_i$, and

$$\zeta_1 \sim \begin{pmatrix} \sqrt{\lambda^{i-1}}z_1 \\ \vdots \\ \sqrt{\lambda^{i-\bar{i}}}z_{\bar{i}} \end{pmatrix}, \qquad \Gamma_1 \sim \begin{pmatrix} \sqrt{\lambda^{i-1}}C_1 \\ \vdots \\ \sqrt{\lambda^{i-\bar{i}}}C_{\bar{i}} \end{pmatrix},$$

$$\zeta_2 \sim \begin{pmatrix} \sqrt{\lambda^{i-\bar{i}-1}}z_{\bar{i}+1} \\ \vdots \\ z_i \end{pmatrix}, \qquad \Gamma_2 \sim \begin{pmatrix} \sqrt{\lambda^{i-\bar{i}-1}}C_{\bar{i}+1} \\ \vdots \\ C_i \end{pmatrix},$$

and by carrying out the straightforward algebra. Equation (1.116) is the special case of Eq. (1.118) corresponding to $\bar{i} = i - 1$. **Q.E.D.**

Note that the positive definiteness assumption on $C_1'C_1$ in Prop. 1.5.2 is needed to guarantee that the first matrix H_1 is positive definite and hence invertible; then the positive definiteness of the subsequent matrices

$H_2, \ldots, H_m$ follows from Eq. (1.117). As a practical matter, it is possible to guarantee the positive definiteness of $C_1'C_1$ by lumping a sufficient number of measurements into the first data block (C_1 should contain n linearly independent columns). An alternative is to redefine ψ_i as

$$\psi_i = \arg\min_{x\in\Re^n} \left\{ \delta\lambda^i \|x - \psi_0\|^2 + \sum_{j=1}^{i} \lambda^{i-j} \|z_j - C_j x\|^2 \right\}, \qquad i = 1, \ldots, m,$$

where δ is a small positive scalar. Then it can be seen from the proof of Prop. 1.5.2 that ψ_i is generated by the same equations (1.116) and (1.117), except that the initial condition $H_0 = 0$ is replaced by

$$H_0 = \delta I,$$

so that $H_1 = \lambda\delta I + C_1'C_1$ is positive definite even if $C_1'C_1$ is not. Note, however, that in this case, the last estimate ψ_m is only approximately equal to the least squares estimate x^*, even if $\lambda = 1$ (the approximation error depends on the size of δ).

The Extended Kalman Filter

Consider now the general case where the data blocks g_i are nonlinear. Then a generalization of the Kalman filter, known as the *extended Kalman filter* (EKF for short), can be used. Like the Gauss-Newton method, it involves linearization of the data blocks and solution of linear least squares problems. However, these problems are solved incrementally using the Kalman filtering algorithm, and *the linearization of each data block is done at the latest iterate that is available when this data block is processed*. In particular, a cycle through the data set of the algorithm sequentially generates the vectors

$$\psi_i = \arg\min_{x\in\Re^n} \sum_{j=1}^{i} \lambda^{i-j} \|\tilde{g}_j(x, \psi_{j-1})\|^2, \qquad i = 1, \ldots, m, \tag{1.124}$$

where $\tilde{g}_j(x, \psi_{j-1})$ are the linearized functions

$$\tilde{g}_j(x, \psi_{j-1}) = g_j(\psi_{j-1}) + \nabla g_j(\psi_{j-1})'(x - \psi_{j-1}), \tag{1.125}$$

and ψ_0 is an initial estimate of x. Using the formulas (1.116) and (1.117) of Prop. 1.5.2 with the identifications

$$z_i = g_i(\psi_{i-1}) - \nabla g_i(\psi_{i-1})'\psi_{i-1}, \qquad C_i = -\nabla g_i(\psi_{i-1})',$$

this algorithm can be written in the incremental form

$$\psi_i = \psi_{i-1} - H_i^{-1} \nabla g_i(\psi_{i-1}) g_i(\psi_{i-1}), \qquad i = 1, \ldots, m, \tag{1.126}$$

where the matrices H_i are generated by

$$H_i = \lambda H_{i-1} + \nabla g_i(\psi_{i-1}) \nabla g_i(\psi_{i-1})', \qquad i = 1, \ldots, m, \tag{1.127}$$

with

$$H_0 = 0. \tag{1.128}$$

To contrast the EKF with the pure form of the Gauss-Newton method (unit stepsize), note that a single iteration of the latter can be written as

$$x^{k+1} = \arg \min_{x \in \Re^n} \sum_{i=1}^{m} \|\tilde{g}_i(x, x^k)\|^2. \tag{1.129}$$

Using the formulas of Prop. 1.5.2 with the identifications

$$z_i = g_i(x^k) - \nabla g_i(x^k)' x^k, \qquad C_i = -\nabla g_i(x^k)',$$

we can generate x^{k+1} by an incremental algorithm as

$$x^{k+1} = \bar{\psi}_m,$$

where

$$\bar{\psi}_i = \bar{\psi}_{i-1} - \bar{H}_i^{-1} \nabla g_i(x^k) \big(g_i(x^k) + \nabla g_i(x^k)'(\bar{\psi}_{i-1} - x^k)\big), \qquad i = 1, \ldots, m, \tag{1.130}$$

$\bar{\psi}_0 = x^k$, and the matrices $\bar{H}_i$ are generated by

$$\bar{H}_i = \bar{H}_{i-1} + \nabla g_i(x^k) \nabla g_i(x^k)', \qquad i = 1, \ldots, m, \tag{1.131}$$

with

$$\bar{H}_0 = 0. \tag{1.132}$$

Thus, by comparing Eqs. (1.126)-(1.128) with Eqs. (1.130)-(1.132), we see that, if $\lambda = 1$, a cycle of the EKF through the data set differs from a pure Gauss-Newton iteration only in that the linearization of the data blocks g_i is done at the corresponding current estimates ψ_{i-1} rather than at the estimate x^k available at the start of the cycle.

Convergence Issues for the Extended Kalman Filter

We have considered so far a single cycle of the EKF. To obtain an algorithm that cycles through the data set multiple times, we can simply create a larger data set by concatenating multiple copies of the original data set, that is, by forming what we refer to as *the extended data set*

$$(g_1, g_2, \ldots, g_m, g_1, g_2, \ldots, g_m, g_1, g_2, \ldots). \tag{1.133}$$

The EKF when applied to the extended data set often works well in practice, although its convergence properties as the number of cycles increases to infinity have not been fully investigated, particularly when $\lambda < 1$.

The basic reason why the algorithm works is that asymptotically it resembles a gradient method with diminishing stepsize of the type described in Section 1.2. To get a sense of this, assume that the EKF is applied to the extended data set (1.133) with $\lambda = 1$. Let us denote by x^k the iterate at the end of the kth cycle through the data set, that is,

$$x^k = \psi_{km}, \qquad k = 1, 2, \ldots$$

Then by using Eq. (1.128) with $i = (k+1)m$ and $\bar{i} = km$, we obtain

$$x^{k+1} = x^k - H_{(k+1)m}^{-1} \left(\sum_{i=1}^{m} \nabla g_i(\psi_{km+i-1}) g_i(\psi_{km+i-1}) \right). \tag{1.134}$$

Now $H_{(k+1)m}$ grows roughly in proportion to $k+1$ because, by Eq. (1.117), we have

$$H_{(k+1)m} = \sum_{j=0}^{k} \sum_{i=1}^{m} \nabla g_i(\psi_{jm+i-1}) \nabla g_i(\psi_{jm+i-1})'. \tag{1.135}$$

It is therefore reasonable to expect that the method tends to make slow progress when k is large, which means that the vectors ψ_{km+i-1} in Eq. (1.134) are roughly equal to x^k. Thus for large k, the sum in the right-hand side of Eq. (1.134) is roughly equal to the gradient $\nabla g(x^k) g(x^k)$, while from Eq. (1.135), $H_{(k+1)m}$ is roughly equal to $(k+1)\nabla g(x^k)\nabla g(x^k)'$, where $g = (g_1, g_2, \ldots, g_m)$ is the original data set. It follows that for large k, the EKF iteration (1.134) can be written approximately as

$$x^{k+1} \approx x^k - \frac{1}{k+1} \big(\nabla g(x^k) \nabla g(x^k)'\big)^{-1} \nabla g(x^k) g(x^k), \tag{1.136}$$

that is, as an approximate Gauss-Newton iteration with diminishing stepsize.

When $\lambda < 1$, the matrix H_i^{-1} generated by the EKF recursion (1.127) will typically not diminish to zero, and $\{x^k\}$ may not converge to a stationary point of $\sum_{i=1}^{m} \lambda^{m-i} \|g_i(x)\|^2$. Furthermore, as the following example shows, the sequences $\{\psi_{km+i}\}$ produced by the EKF using Eq. (1.126), may converge to different limits for different i:

Example 1.5.7

Consider the case where there are two data blocks, $g_1(x) = x - c_1$ and $g_2(x) = x - c_2$, where c_1 and c_2 are given scalars. Each cycle of the EKF consists of two steps. At the second step of the kth cycle, we minimize

$$\sum_{i=1}^{k} \left(\lambda^{2i-1}(x - c_1)^2 + \lambda^{2i-2}(x - c_2)^2 \right),$$

which is equal to the following scalar multiple of $\lambda(x-c_1)^2 + (x-c_2)^2$,

$$(1+\lambda^2+\cdots+\lambda^{2k-2})\big(\lambda(x-c_1)^2+(x-c_2)^2\big).$$

Thus at the second step, we obtain the minimizer of $\lambda(x-c_1)^2 + (x-c_2)^2$,

$$\psi_{2k} = \frac{\lambda c_1 + c_2}{\lambda + 1}.$$

At the first step of the kth cycle, we minimize

$$(x-c_1)^2 + \lambda\sum_{i=1}^{k-1}\left(\lambda^{2i-1}(x-c_1)^2 + \lambda^{2i-2}(x-c_2)^2\right),$$

which is equal to the following scalar multiple of $(x-c_1)^2 + \lambda(x-c_2)^2$

$$(1+\lambda^2+\cdots+\lambda^{2k-4})\big((x-c_1)^2+\lambda(x-c_2)^2\big),$$

plus the diminishing term $\lambda^{2k-2}(x-c_1)^2$. Thus at the first step, we obtain approximately (for large k) the minimizer of $(x-c_1)^2 + \lambda(x-c_2)^2$,

$$\psi_{2k-1} \approx \frac{c_1 + \lambda c_2}{1+\lambda}.$$

We see therefore that within each cycle, there is an oscillation around the minimizer $(c_1+c_2)/2$ of $(x-c_1)^2 + (x-c_2)^2$. The size of the oscillation diminishes as λ approaches 1.

Generally, for a nonlinear least squares problem, the convergence rate tends to be faster when $\lambda < 1$ than when $\lambda = 1$, essentially because the implicit stepsize does not diminish as in the case $\lambda = 1$. For this reason, a hybrid method that uses a different value of λ within each cycle may work best in practice. One may start with a relatively small λ to attain a fast initial rate of convergence, and then progressively increase λ towards 1 in order to attain high solution accuracy. An example that the reader may try is given in Exercise 1.5.7. The following proposition shows convergence for the case where λ tends to 1 at a sufficiently fast rate.

Proposition 1.5.3: Assume that $\nabla g_i(x)$ has full rank for all x and $i = 1,\ldots,m$, and that for some $L > 0$, we have

$$\|\nabla g_i(x)g_i(x) - \nabla g_i(y)g_i(y)\| \le L\|x-y\|, \qquad \forall\, x,y \in \Re^n,\ i=1,\ldots,m.$$

Assume also that there is a constant $c > 0$ such that the scalar λ used in the updating formula (1.127) within the kth cycle, call it $\lambda(k)$, satisfies

$$0 \le 1 - \big(\lambda(k)\big)^m \le \frac{c}{k}, \qquad \forall\, k = 1,2,\ldots.$$

Then if the EKF applied to the extended data set (1.133) generates a bounded sequence of vectors ψ_i, each of the limit points of $\{x^k\}$ is a stationary point of the least squares problem.

Proof: *(Abbreviated; for a complete proof see [Ber94])* We have using the Kalman filter recursion (1.118) that x^k satisfies

$$x^{k+1} = x^k - H_{(k+1)m}^{-1} \left(\sum_{i=1}^{m} \big(\lambda(k)\big)^{m-i} \nabla g_i(\psi_{km+i-1}) g_i(\psi_{km+i-1}) \right).$$

It can be shown using the rank assumption on $\nabla g_i(x)$, the growth assumption on λ^k, the boundedness of $\{x^k\}$, and the preceding analysis that the eigenvalues of the matrices H_{km} are within an interval $[c_1 k, c_2 k]$, where c_1 and c_2 are some positive constants (see Exercise 1.5.9). The proof then follows the line of argument of the convergence proof of gradient methods with diminishing stepsize (Prop. 1.2.4 in Section 1.2). **Q.E.D.**

E X E R C I S E S

1.5.1

Consider the least squares problem (1.81) for the case where $m < n$ and $r_i = 1$ for all i.

(a) Show that the Hessian matrix is singular at any optimal solution x^* for which $g(x^*) = 0$.

(b) Consider the case where g is linear and of the form $g(x) = z - Ax$, where A is an $m \times n$ matrix. Show that there are infinitely many optimal solutions. Show also that if A has linearly independent rows, $x^* = A'(AA')^{-1}z$ is one of these solutions.

1.5.2 [Luo91]

Consider the least squares problem

$$\begin{aligned} &\text{minimize} \ \ \tfrac{1}{2}\big\{(z_1 - x)^2 + (z_2 - x)^2\big\} \\ &\text{subject to} \ \ x \in \Re \end{aligned}$$

and the incremental gradient algorithm that generates x^{k+1} from x^k according to

$$x^{k+1} = y^k - \alpha(y^k - z_2),$$

where

$$y^k = x^k - \alpha(x^k - z_1),$$

and α is positive stepsize. Assuming that $\alpha < 1$, show that $\{x^k\}$ and $\{y^k\}$ converge to limits $x(\alpha)$ and $y(\alpha)$, respectively. However, unless $z_1 = z_2$, $x(\alpha)$ and $y(\alpha)$ are neither equal to each other, nor equal to the least squares solution $x^* = (z_1 + z_2)/2$. Consistently with Prop. 1.5.1, verify that

$$\lim_{\alpha \to 0} x(\alpha) = \lim_{\alpha \to 0} y(\alpha) = x^*.$$

1.5.3 (Computational Problem)

Consider the least squares cost function of Fig. 1.5.1 for the case where the five data pairs (y_s, z_s) are $(1.165, 1)$, $(0.626, -1)$, $(0.075, -1)$, $(0.351, 1)$, $(-0.696, 1)$ (these correspond to the three-dimensional plot of Fig. 1.5.1). Minimize this cost function using appropriate versions of the Gauss-Newton method, the Extended Kalman Filter, the incremental gradient method, and its heavy ball version (1.91).

1.5.4 (A Generalized Incremental Gradient Method [Ber95b])

The purpose of this exercise is to embed the incremental gradient method and the method of steepest descent within a one-parameter family of methods for the least squares problem. In contrast with the incremental gradient method, some of the methods in this family have a generically linear convergence rate. For a fixed $\mu \geq 0$, define

$$\xi_i(\mu) = \frac{1}{1 + \mu + \cdots + \mu^{m-i}}, \qquad i = 1, \ldots, m.$$

Consider a method which given x^k, generates x^{k+1} according to $x^{k+1} = \psi_m$, where

$$\psi_i = x^k - \alpha^k h_i,$$

$$h_i = \mu h_{i-1} + \sum_{j=1}^{i} \xi_j(\mu) \nabla g_j(\psi_{j-1}) g_j(\psi_{j-1}),$$

with the initial condition $h_0 = 0$.

(a) Show that when $\mu = 0$, the method coincides with the incremental gradient method, and when $\mu \to \infty$, the method approaches the steepest descent method.

(b) Show that parts (a) and (b) of Prop. 1.5.1, which were proved for the case $\mu = 0$, hold as stated when $\mu > 0$ as well.

(c) Suppose that in the kth iteration, a k-dependent value of μ, say $\mu(k)$, and a constant stepsize $\alpha^k = \alpha$ are used. Under the assumptions of Prop. 1.5.1, show that if for some $q > 1$ and all k greater than some index $\bar{k}$, we have $\mu(k) \geq q^k$, then there exists $\bar{\alpha} > 0$ such that for all $\alpha \in (0, \bar{\alpha}]$, the iteration converges linearly to the optimal solution x^*.

1.5.5 (Gradient Method for Infinitely Many Data Blocks)

Consider the gradient method

$$x^{k+1} = x^k - \alpha \nabla f_k(x^k) \qquad k = 0, 1, \ldots,$$

where $f_0, f_1, \ldots,$ are quadratic functions with eigenvalues lying within some interval $[\gamma, \Gamma]$, where $\gamma > 0$. Suppose that for a given $\epsilon > 0$, there is a vector x^* such that

$$\|\nabla f_k(x^*)\| \leq \epsilon, \qquad \forall\ k = 0, 1, \ldots.$$

Show that for all α with $0 < \alpha \leq 2/(\gamma + \Gamma)$, we have

$$\limsup_{k \to \infty} \|x^k - x^*\| \leq \frac{2\epsilon}{\gamma}.$$

Hint: Let Q_k be the positive definite symmetric matrix corresponding to f_k, and write

$$x^{k+1} - x^* = (I - \alpha Q_k)(x^k - x^*) - \alpha \nabla f_k(x^*).$$

Use this relation to show that

$$\|x^k - x^*\| > \frac{2\epsilon}{\gamma} \qquad \Rightarrow \qquad \|x^{k+1} - x^*\| < \left(1 - \frac{\alpha\gamma}{2}\right) \|x^k - x^*\|,$$

while

$$\|x^k - x^*\| \leq \frac{2\epsilon}{\gamma} \qquad \Rightarrow \qquad \|x^{k+1} - x^*\| \leq \frac{2\epsilon}{\gamma}.$$

1.5.6

Show that if the sequence $\{x^k\}$ produced by the EKF with $\lambda = 1$ converges to some $\bar{x}$, then $\bar{x}$ must be a stationary point of $\sum_{i=1}^m \|g_i(x)\|^2$.

1.5.7

Consider the EKF of Eqs. (1.126)-(1.128) for the data set $g_1(x) = x^2 - 2$ and $g_2(x) = x^2$, where $x \in \Re$.

(a) Verify that each cycle of the EKF consists of executing sequentially the iterations

$$H_y := \lambda H_x + (2x)^2,$$
$$y := x - H_y^{-1}(2x)(x^2 - 2),$$
$$H_x := \lambda H_y + (2y)^2,$$
$$x := y - H_x^{-1}(2y)(y^2).$$

(b) Write a computer program that implements the EKF, and compare the limit of $(x + y)/2$ with the least squares solutions, which are 1 and -1, for different values of λ. Try also a version of the method where λ is progressively increased towards 1.

1.5.8 (Incremental Gauss-Newton Method with Restart)

Consider a version of the EKF where the matrix H is reset to 0 at the end of each cycle. In particular, this method, given x^k, generates x^{k+1} according to $x^{k+1} = \psi_m$, where

$$\psi_i = \psi_{i-1} - \alpha^k H_i^{-1} \nabla g_i(\psi_{i-1}) g_i(\psi_{i-1}), \qquad i = 1, \ldots, m,$$

$$H_i = H_{i-1} + \nabla g_i(\psi_{i-1}) \nabla g_i(\psi_{i-1})', \qquad i = 1, \ldots, m,$$

with the initial conditions

$$\psi_0 = x^k, \qquad H_0 = 0.$$

(a) Show that if the data blocks g_i are linear and α^k is for all k equal to a constant α with $0 < \alpha < 2$, then $\{x^k\}$ converges to the optimal solution.

(b) Show that if $\sum_{k=0}^{\infty} \alpha^k = \infty$, $\sum_{k=0}^{\infty} (\alpha^k)^2 < \infty$, and the generated sequence $\{x^k\}$ is bounded, then every limit point of $\{x^k\}$ is a stationary point of the least squares problem.

1.5.9

Under the assumptions of Prop. 1.5.3, show that the eigenvalues of the matrices H_{km} lie within an interval $[c_1 k, c_2 k]$, where c_1 and c_2 are some positive constants. *Hint*: Let X be a compact set containing all vectors ψ_i generated by the algorithm, and let B and b be an upper bound and a lower bound, respectively, on the eigenvalues of $\nabla g_i(x) \nabla g_i(x)'$ as x ranges over X. Show that all eigenvalues of H_{km} are less or equal to kmB. If v_k is the smallest eigenvalue of H_{km}, show that $v_{k+1} \geq (1 - c/k) v_k + m(1 - c/k) b$, and use this relation to show that $v_k \geq k\gamma$ for a sufficiently small but positive value of γ.

1.6 CONJUGATE DIRECTION METHODS

Conjugate direction methods are motivated by a desire to accelerate the convergence rate of steepest descent, while avoiding the overhead associated with Newton's method. They were originally developed for solving the quadratic problem

$$\begin{aligned} &\text{minimize} \quad f(x) = \tfrac{1}{2} x' Q x - b' x \\ &\text{subject to} \quad x \in \Re^n, \end{aligned} \tag{1.137}$$

where Q is positive definite, or equivalently, for solving the linear system

$$Qx = b. \tag{1.138}$$

They can also be used for solution of the more general system $Ax = b$, where A is invertible but not symmetric or positive definite, after conversion to the positive definite system $A'Ax = A'b$.

Conjugate direction methods can solve these problems after at most n iterations but they are best viewed as iterative methods, since usually fewer than n iterations are required to attain a sufficiently accurate solution, particularly when n is large. They can also be used to solve nonquadratic optimization problems. For such problems, they do not in general terminate after a finite number of iterations, but still, when properly implemented, they have attractive convergence and rate of convergence properties. We will first develop the methods for quadratic problems and then discuss their application to more general problems.

Given a positive definite $n \times n$ matrix Q, we say that a set of nonzero vectors $d^1, \ldots, d^k$ are *Q-conjugate*, if

$$d^{i\prime} Q d^j = 0, \qquad \text{for all } i \text{ and } j \text{ such that } i \neq j. \tag{1.139}$$

If $d^1, \ldots, d^k$ are Q-conjugate, then they are *linearly independent*, since if one of these vectors, say d^k, were expressed as a linear combination of the others,

$$d^k = \alpha^1 d^1 + \cdots + \alpha^{k-1} d^{k-1},$$

then by multiplication with $d^{k\prime} Q$ we would obtain using the Q-conjugacy of d^k and d^j, $j = 1, \ldots, k-1$,

$$d^{k\prime} Q d^k = \alpha^1 d^{k\prime} Q d^1 + \cdots + \alpha^{k-1} d^{k\prime} Q d^{k-1} = 0,$$

which is impossible since $d^k \neq 0$ and Q is positive definite.

For a given set of n Q-conjugate directions $d^0, \ldots, d^{n-1}$, the corresponding *conjugate direction method* for unconstrained minimization of the quadratic function

$$f(x) = \tfrac{1}{2} x' Q x - b' x, \tag{1.140}$$

is given by

$$x^{k+1} = x^k + \alpha^k d^k, \qquad k = 0, \ldots, n-1, \tag{1.141}$$

where x^0 is an arbitrary starting vector and α^k is obtained by the line minimization rule

$$f(x^k + \alpha^k d^k) = \min_{\alpha} f(x^k + \alpha d^k). \tag{1.142}$$

In particular, by setting to zero the derivative of $f(x^k + \alpha d^k)$ with respect to α, we obtain

$$0 = \frac{\partial f(x^k + \alpha^k d^k)}{\partial \alpha} = d^{k\prime} \nabla f(x^k + \alpha^k d^k) = d^{k\prime} \big(Q(x^k + \alpha^k d^k) - b \big),$$

or

$$\alpha^k = \frac{d^{k\prime}(b - Qx^k)}{d^{k\prime} Q d^k}.$$

The principal result about conjugate direction methods is that *successive iterates minimize f over a progressively expanding linear manifold that eventually includes the global minimum of f*. In particular, for each k, x^{k+1} minimizes f over the linear manifold passing through x^0 and spanned by the conjugate directions $d^0, \ldots, d^k$, that is,

$$x^{k+1} = \arg\min_{x \in M^k} f(x), \tag{1.143}$$

where

$$\begin{aligned} M^k &= \{x \mid x = x^0 + v,\ v \in (\text{subspace spanned by } d^0, ..., d^k)\} \\ &= x^0 + (\text{subspace spanned by } d^0, ..., d^k). \end{aligned} \tag{1.144}$$

In particular, x^n minimizes f over $\Re^n$.

To show this, note that by Eq. (1.142), we have for all i

$$\left. \frac{\partial f(x^i + \alpha d^i)}{\partial \alpha} \right|_{\alpha = \alpha^i} = \nabla f(x^{i+1})' d^i = 0$$

and, for $i = 0, \ldots, k-1$,

$$\begin{aligned} \nabla f(x^{k+1})' d^i &= (Qx^{k+1} - b)' d^i \\ &= \left(x^{i+1} + \sum_{j=i+1}^{k} \alpha^j d^j \right)' Q d^i - b' d^i \\ &= x^{i+1\prime} Q d^i - b' d^i \\ &= \nabla f(x^{i+1})' d^i, \end{aligned}$$

where we have used the conjugacy of d^i and d^j, $j = i+1, \ldots, k$. Combining the last two equations we obtain

$$\nabla f(x^{k+1})' d^i = 0, \qquad i = 0, \ldots, k, \tag{1.145}$$

so that

$$\left. \frac{\partial f(x^0 + \gamma^0 d^0 + \cdots + \gamma^k d^k)}{\partial \gamma^i} \right|_{\substack{\gamma^j = \alpha^j \\ j = 0, \ldots, k}} = 0, \qquad i = 0, \ldots, k,$$

which verifies Eq. (1.143).

It is easy to visualize the expanding manifold minimization property of Eq. (1.143) when $b = 0$ and $Q = I$ (the identity matrix). In this case, the equal cost surfaces of f are concentric spheres, and the notion of Q-conjugacy reduces to usual orthogonality. By a simple algebraic argument, we see that minimization along n orthogonal directions yields the global minimum of f, that is, the center of the spheres. (This becomes evident once we rotate the coordinate system so that the given n orthogonal directions coincide with the coordinate directions.)

The case of a general positive definite Q can be reduced to the case where $Q = I$ by means of a scaling transformation. By setting $y = Q^{1/2}x$, minimizing $\frac{1}{2}x'Qx$ is equivalent to minimizing $\frac{1}{2}\|y\|^2$. If $w^0, \ldots, w^{n-1}$ are any set of orthogonal nonzero vectors in $\Re^n$, the algorithm

$$y^{k+1} = y^k + \alpha^k w^k, \qquad k = 0, \ldots, n-1, \tag{1.146}$$

where

$$\alpha^k = \arg\min_{\alpha} \tfrac{1}{2}\|y^k + \alpha w^k\|^2,$$

terminates in at most n steps with $y^n = 0$. To pass back to the x-coordinate system, we multiply Eq. (1.146) by $Q^{-1/2}$ and obtain

$$x^{k+1} = x^k + \alpha^k d^k, \qquad k = 0, \ldots, n-1,$$

where $d^k = Q^{-1/2}w^k$. The orthogonality property of $w^0, \ldots, w^{n-1}$, that is, $w^{i\prime}w^j = 0$ for $i \neq j$, is equivalent to the requirement that the directions $d^0, \ldots, d^{n-1}$ be Q-conjugate, that is, $d^{i\prime}Qd^j = 0$ for $i \neq j$.

Thus, using the transformation $y = Q^{1/2}x$, we can think of any conjugate direction method for minimizing $\frac{1}{2}x'Qx$ as a method that minimizes $\frac{1}{2}\|y\|^2$ by successive minimization along n orthogonal directions (see Fig. 1.6.1).

Generating Q-Conjugate Directions

Given any set of linearly independent vectors $\xi^0, \ldots, \xi^k$, we can construct a set of mutually Q-conjugate directions $d^0, \ldots, d^k$ such that for all $i = 0, \ldots, k$, we have

$$(\text{subspace spanned by } d^0, \ldots, d^i) = (\text{subspace spanned by } \xi^0, \ldots, \xi^i), \tag{1.147}$$

using the so called *Gram-Schmidt procedure*. Indeed, let us do this recursively, starting with

$$d^0 = \xi^0. \tag{1.148}$$

Suppose that, for some $i < k$, we have selected Q-conjugate $d^0, \ldots, d^i$ so that the above property holds. We then take d^{i+1} to be of the form

$$d^{i+1} = \xi^{i+1} + \sum_{m=0}^{i} c^{(i+1)m} d^m \tag{1.149}$$

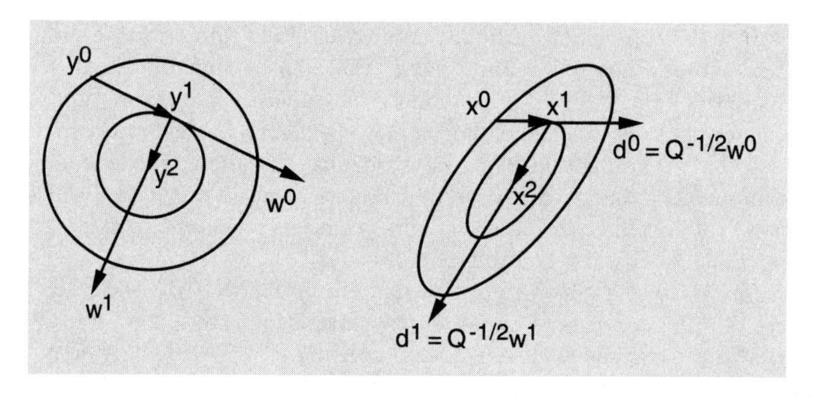

Figure 1.6.1. Geometric interpretation of conjugate direction methods in terms of successive minimization along n orthogonal directions. In (a) the function $\|y\|^2$ is minimized successively along the directions $w^0, \ldots, w^{n-1}$, which are orthogonal in the usual sense ($w^{i\prime} w^j = 0$ for $i \neq j$). When this process is viewed in the coordinate system of variables $x = Q^{-1/2}y$, it yields the conjugate direction method that uses the Q-conjugate directions $d^0, \ldots, d^{n-1}$ with $d^i = Q^{-1/2}w^i$, as shown in (b).

and choose the coefficients $c^{(i+1)m}$ so that d^{i+1} is Q-conjugate to $d^0, \ldots, d^i$. This will be so if for each $j = 0, \ldots, i$,

$$d^{i+1\prime} Q d^j = \xi^{i+1\prime} Q d^j + \left(\sum_{m=0}^{i} c^{(i+1)m} d^m \right)' Q d^j = 0. \tag{1.150}$$

Since $d^0, \ldots, d^i$ are Q-conjugate, we have $d^{m\prime} Q d^j = 0$ if $m \neq j$, and Eq. (1.150) yields

$$c^{(i+1)j} = -\frac{\xi^{i+1\prime} Q d^j}{d^{j\prime} Q d^j}, \qquad j = 0, \ldots, i. \tag{1.151}$$

Note that the denominator $d^{j\prime} Q d^j$ in the above equation is nonzero, since $d^0, \ldots, d^i$ are assumed Q-conjugate and are therefore nonzero. Note also that $d^{i+1} \neq 0$, since otherwise from Eqs. (1.147) and (1.149), ξ^{i+1} would be a linear combination of $\xi^0, \ldots, \xi^i$, contradicting the linear independence of $\xi^0, \ldots, \xi^k$. Finally, note from Eq. (1.149) that ξ^{i+1} lies in the subspace spanned by $d^0, \ldots, d^{i+1}$, while d^{i+1} lies in the subspace spanned by $\xi^0, \ldots, \xi^{i+1}$, since $d^0, \ldots, d^i$ and $\xi^0, \ldots, \xi^i$ span the same space [cf. Eq. (1.147)]. Thus, Eq. (1.147) is satisfied when i is increased to $i+1$ and the Gram-Schmidt procedure defined by Eqs. (1.148), (1.149), and (1.151), has the property claimed. Figure 1.6.2 illustrates the procedure.

It is also worth noting what will happen if the vectors $\xi^0, \ldots, \xi^i$ are linearly independent, but the next vector ξ^{i+1} is linearly dependent on these vectors. In this case it can be seen (compare also with Fig. 1.6.2) that the equations (1.149) and (1.151) are still valid, but the new vector d^{i+1} as

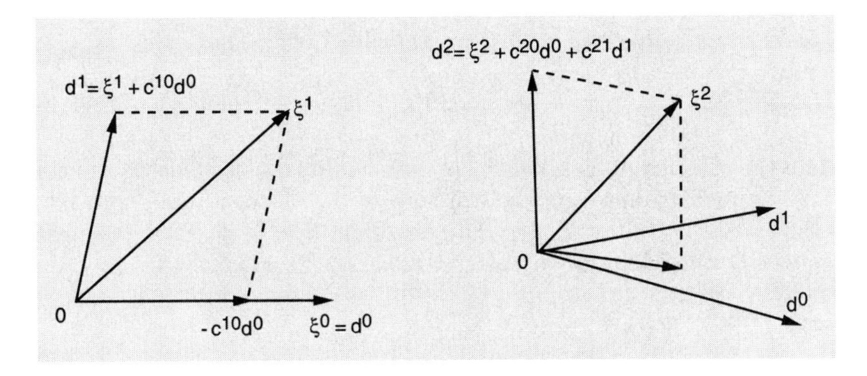

Figure 1.6.2. Illustration of the Gram-Schmidt procedure for generating Q-conjugate directions $d^0, \ldots, d^k$ from a set of linearly independent vectors $\xi^0, \ldots, \xi^k$, so that

$$(\text{subspace spanned by } d^0, \ldots, d^k) = (\text{subspace spanned by } \xi^0, \ldots, \xi^k).$$

Given $d^0, \ldots, d^{i-1}$, the ith direction is obtained as $d^i = \xi^i - \hat{\xi}^i$, where $\hat{\xi}^i$ is a vector on the subspace spanned by $d^0, \ldots, d^{i-1}$ (or $\xi^0, \ldots, \xi^{i-1}$) chosen so that d^i is Q-conjugate to $d^0, \ldots, d^{i-1}$. (It can be shown that the vector $\hat{\xi}^i$ is the projection of ξ^i on this subspace with respect to the norm $\|x\|_Q = \sqrt{x'Qx}$, i.e., minimizes $\|\xi^i - x\|_Q$ over all x in this subspace; see Exercise 1.6.1.)

given by Eq. (1.149) will be zero.† We can use this property to construct a set of Q-conjugate directions that span the same space as a set of vectors $\xi^0, \ldots, \xi^k$ that are not a priori known to be linearly independent. This construction can be accomplished with an extended version of the Gram-Schmidt procedure that generates directions via Eqs. (1.149) and (1.151), but each time the new direction d^{i+1} as given by Eq. (1.149) turns out to be zero, it is simply discarded rather than added to the set of preceding directions.

The Conjugate Gradient Method

The most important conjugate direction method, the conjugate gradient method, is obtained by applying the Gram-Schmidt procedure to the gradient vectors $\xi^0 = -g^0, \ldots, \xi^{n-1} = -g^{n-1}$, where we use the notation

$$g^k = \nabla f(x^k) = Qx^k - b. \tag{1.152}$$

† From Eq. (1.149) and the linear independence of $\xi^0, \ldots, \xi^i$, it is seen that d^{i+1} can be uniquely expressed as $d^{i+1} = \sum_{m=0}^{i} \gamma^m d^m$, where γ^m are some scalars. By multiplying this equation with $d^{j'}Q$ and by using the Q-conjugacy of the directions $d^0, \ldots, d^i$ and Eq. (1.150), we see that $\gamma^m = 0$ for all $m = 0, \ldots, i$.

Thus the conjugate gradient method is defined by

$$x^{k+1} = x^k + \alpha^k d^k, \tag{1.153}$$

where the stepsize α^k is obtained by line minimization, and the direction d^k is obtained by applying the kth step of the Gram-Schmidt procedure to the vector $-g^k$ and the preceding directions $d^0, \ldots, d^{k-1}$. In particular, from the Gram-Schmidt equations (1.149) and (1.151), we have

$$d^k = -g^k + \sum_{j=0}^{k-1} \frac{g^{k\prime} Q d^j}{d^{j\prime} Q d^j} d^j. \tag{1.154}$$

Note here that

$$d^0 = -g^0$$

and that the method terminates with an optimal solution if $g^k = 0$. The method also effectively stops if $d^k = 0$, but we will show that this can only happen if $g^k = 0$.

The key property of the conjugate gradient method is that the direction formula (1.154) can be greatly simplified. In particular, all but one of the coefficients in the sum of Eq. (1.154) turn out to be zero because, in view of the expanding manifold minimization property, the gradient g^k is orthogonal to the subspace spanned by $d^0, \ldots, d^{k-1}$ [cf. Eq. (1.145)]. This simplification of the direction formula will be shown as part of the following proposition.

Proposition 1.6.1: The directions of the conjugate gradient method are generated by

$$d^0 = -g^0,$$

$$d^k = -g^k + \beta^k d^{k-1}, \qquad k = 1, \ldots, n-1,$$

where β^k is given by

$$\beta^k = \frac{g^{k\prime} g^k}{g^{k-1\prime} g^{k-1}}.$$

Furthermore, the method terminates with an optimal solution after at most n steps.

Proof: We first use induction to show that all the gradients g^k generated up to termination are linearly independent. We have that g^0 by itself is linearly independent, unless $g^0 = 0$, in which case the method terminates. Suppose that the method has not terminated after k steps, and that $g^0, \ldots, g^{k-1}$ are linearly independent. Then, since the method is by definition a conjugate direction method, we have

$$(\text{subspace spanned by } d^0, \ldots, d^{k-1}) = (\text{subspace spanned by } g^0, \ldots, g^{k-1}) \tag{1.155}$$

[cf. Eq. (1.147)]. There are two possibilities:

(a) $g^k = 0$, in which case the method terminates.

(b) $g^k \neq 0$, in which case the expanding manifold minimization property of the conjugate direction method implies that

$$g^k \text{ is orthogonal to } d^0, \ldots, d^{k-1} \tag{1.156}$$

[cf. Eq. (1.145)]. Since the subspaces spanned by $(d^0, \ldots, d^{k-1})$ and by $(g^0, \ldots, g^{k-1})$ are the same [cf. Eq. (1.155)], we see that

$$g^k \text{ is orthogonal to } g^0, \ldots, g^{k-1}. \tag{1.157}$$

Therefore, g^k is linearly independent of $g^0, \ldots, g^{k-1}$, thus completing the induction.

Since at most n linearly independent gradients can be generated, it follows that the gradient will be zero after at most n iterations and the method will terminate with the minimum of f.

To conclude the proof, we use the orthogonality properties (1.156) and (1.157) to verify that the calculation of the coefficients multiplying d^j in the Gram-Schmidt formula (1.154) can be simplified as stated in the proposition. We have for all j such that $g^j \neq 0$,

$$g^{j+1} - g^j = Q(x^{j+1} - x^j) = \alpha^j Q d^j, \tag{1.158}$$

[cf. Eqs. (1.152) and (1.153)]. We note that $\alpha^j \neq 0$, since if $\alpha^j = 0$ we would have $g^{j+1} = g^j$ implying, in view of Eq. (1.157), that $g^j = 0$. Therefore, we have using Eqs. (1.157) and (1.158)

$$g^{i\prime} Q d^j = \frac{1}{\alpha^j} g^{i\prime}(g^{j+1} - g^j) = \begin{cases} 0 & \text{if } j = 0, \ldots, i-2, \\ \dfrac{1}{\alpha^j} g^{i\prime} g^i & \text{if } j = i-1, \end{cases}$$

and also that

$$d^{j\prime} Q d^j = \frac{1}{\alpha^j} d^{j\prime}(g^{j+1} - g^j).$$

Substituting the last two relations in the Gram-Schmidt formula (1.154), we obtain

$$d^k = -g^k + \beta^k d^{k-1}, \tag{1.159}$$

where

$$\beta^k = \frac{g^{k\prime} g^k}{d^{k-1\prime}(g^k - g^{k-1})}. \tag{1.160}$$

From Eq. (1.159) we have $d^{k-1} = -g^{k-1} + \beta^{k-1} d^{k-2}$. Using this equation, and the orthogonality of g^k and g^{k-1}, and of d^{k-2} and $g^k - g^{k-1}$ [cf. Eqs. (1.156) and (1.157)], the denominator in Eq. (1.160) is written as $g^{k-1'} g^{k-1}$, and the desired formula for β^k follows. **Q.E.D.**

Note that by using the orthogonality of g^k and g^{k-1} the formula

$$\beta^k = \frac{g^{k'} g^k}{g^{k-1'} g^{k-1}} \tag{1.161}$$

of Prop. 1.6.1 can also be written as

$$\beta^k = \frac{g^{k'} (g^k - g^{k-1})}{g^{k-1'} g^{k-1}}. \tag{1.162}$$

While the alternative formulas (1.161) and (1.162) produce the same results for quadratic problems, their differences become significant when the conjugate gradient method is extended to nonquadratic problems, as we will discuss shortly.

Preconditioned Conjugate Gradient Method

This method is really the conjugate gradient method implemented in a new coordinate system. Suppose we make a change of variables, $x = Sy$, where S is an invertible symmetric $n \times n$ matrix, and we apply the conjugate gradient method to the equivalent problem

$$\begin{aligned} &\text{minimize} \quad h(y) = f(Sy) = \tfrac{1}{2} y' SQSy - b' Sy \\ &\text{subject to} \quad y \in \Re^n. \end{aligned}$$

The method is described by

$$y^{k+1} = y^k + \alpha^k \tilde{d}^k, \tag{1.163}$$

where α^k is obtained by line minimization and $\tilde{d}^k$ is generated by [cf. Eqs. (1.159) and (1.161)]

$$\tilde{d}^0 = -\nabla h(y^0), \qquad \tilde{d}^k = -\nabla h(y^k) + \beta^k \tilde{d}^{k-1}, \qquad k = 1, \ldots, n-1, \tag{1.164}$$

where

$$\beta^k = \frac{\nabla h(y^k)' \nabla h(y^k)}{\nabla h(y^{k-1})' \nabla h(y^{k-1})}. \tag{1.165}$$

Setting $x^k = Sy^k$, $\nabla h(y^k) = Sg^k$, $d^k = S\tilde{d}^k$, and $H = S^2$, we obtain from Eqs. (1.163)-(1.165) the equivalent method

$$x^{k+1} = x^k + \alpha^k d^k, \tag{1.166}$$

$$d^0 = -Hg^0, \qquad d^k = -Hg^k + \beta^k d^{k-1}, \qquad k = 1, \ldots, n-1, \tag{1.167}$$

where

$$\beta^k = \frac{g^{k\prime} H g^k}{g^{k-1\prime} H g^{k-1}}, \tag{1.168}$$

and α^k is obtained by line minimization.

The method described by the above equations is called the *preconditioned conjugate gradient method with scaling matrix H*. To see that this method is a conjugate direction method, note that since $\nabla^2 h(y) = SQS$, the vectors $\tilde{d}^0, \ldots, \tilde{d}^{n-1}$ are (SQS)-conjugate. Since $d^k = S\tilde{d}^k$, we obtain that $d^0, \ldots, d^{n-1}$ are Q-conjugate. Therefore, the scaled method terminates with the minimum of f after at most n iterations, just as the ordinary conjugate gradient method. The motivation for scaling is to improve the rate of convergence within an n-iteration cycle (see the following analysis). This is important for a nonquadratic problem, but it may be important even for a quadratic problem if n is large and we want to obtain an approximate solution without waiting for the method to terminate.

Application to Nonquadratic Problems

The conjugate gradient method can be applied to the nonquadratic problem

$$\begin{aligned} &\text{minimize} \quad f(x) \\ &\text{subject to} \quad x \in \Re^n. \end{aligned}$$

It takes the form

$$x^{k+1} = x^k + \alpha^k d^k, \tag{1.169}$$

where α^k is obtained by line minimization

$$f(x^k + \alpha^k d^k) = \min_{\alpha} f(x^k + \alpha d^k), \tag{1.170}$$

and d^k is generated by

$$d^k = -\nabla f(x^k) + \beta^k d^{k-1}. \tag{1.171}$$

The most common way to compute β^k is [cf. Eq. (1.162)]

$$\beta^k = \frac{\nabla f(x^k)'\big(\nabla f(x^k) - \nabla f(x^{k-1})\big)}{\nabla f(x^{k-1})'\nabla f(x^{k-1})}. \tag{1.172}$$

The direction d^k generated by the formula $d^k = -\nabla f(x^k) + \beta^k d^{k-1}$ is a direction of descent, since from Eq. (1.170) we obtain $\nabla f(x^k)' d^{k-1} = 0$, so that

$$\nabla f(x^k)' d^k = -\big\|\nabla f(x^k)\big\|^2 + \beta^k \nabla f(x^k)' d^{k-1} = -\big\|\nabla f(x^k)\big\|^2.$$

For nonquadratic problems, the formula (1.172) is typically superior to alternative formulas such as

$$\beta^k = \frac{\nabla f(x^k)'\nabla f(x^k)}{\nabla f(x^{k-1})'\nabla f(x^{k-1})}, \tag{1.173}$$

[cf. Eq. (1.161)]. A heuristic explanation is that due to nonquadratic terms in the objective function and possibly inaccurate line searches, conjugacy of the generated directions is progressively lost and a situation may arise where the method "jams" in the sense that the generated direction d^k is nearly orthogonal to the gradient $\nabla f(x^k)$. When this occurs, we have $\nabla f(x^{k+1}) \simeq \nabla f(x^k)$. In that case, the scalar β^{k+1}, generated by

$$\beta^{k+1} = \frac{\nabla f(x^{k+1})'\big(\nabla f(x^{k+1}) - \nabla f(x^k)\big)}{\nabla f(x^k)'\nabla f(x^k)},$$

will be nearly zero and the next direction $d^{k+1} = -\nabla f(x^{k+1}) + \beta^{k+1}d^k$ will be close to $-\nabla f(x^{k+1})$ thereby breaking the jam. By contrast, when Eq. (1.173) is used, under the same circumstances the method typically continues to jam.

Regardless of the direction update formula used, one must deal with the loss of conjugacy that results from nonquadratic terms in the cost function. The conjugate gradient method is often employed in problems where the number of variables n is large, and it is not unusual for the method to start generating nonsensical and inefficient directions of search after a few iterations. For this reason it is important to operate the method in cycles of conjugate direction steps, with the first step in the cycle being a steepest descent step. Some possible restarting policies are:

(a) Restart with a steepest descent step n iterations after the preceding restart.

(b) Restart with a steepest descent step k iterations after the preceding restart with $k < n$. This is recommended when the problem has special structure so that the resulting method has good convergence rate (see the following Prop. 1.6.2).

(c) Restart with a steepest descent step if either n iterations have taken place since the preceding restart or if

$$\big|\nabla f(x^k)'\nabla f(x^{k-1})\big| > \gamma\big\|\nabla f(x^{k-1})\big\|^2, \tag{1.174}$$

where γ is a fixed scalar with $0 < \gamma < 1$. The above relation is a test on loss of conjugacy, for if the generated directions were conjugate, then we would have $\nabla f(x^k)'\nabla f(x^{k-1}) = 0$.

Note that in all these restart procedures the steepest descent iteration serves as a spacer step and guarantees global convergence (Prop. 1.2.6 in

Section 1.2). If the scaled version of the conjugate gradient method is used, then a scaled steepest descent iteration is used to restart a cycle. The scaling matrix may change at the beginning of a cycle but should remain unchanged during the cycle.

An important practical issue relates to the line search accuracy that is necessary for efficient computation. On one hand, an accurate line search is needed to limit the loss of direction conjugacy and the attendant deterioration of convergence rate. On the other hand, insisting on a very accurate line search can be computationally expensive. Some trial and error may therefore be required in practice. For a discussion of implementations that are tolerant of line search inaccuracies see Perry [Per78] and Shanno [Sha78] (see also Exercise 1.7.6).

Conjugate Gradient-Like Methods for Linear Systems*

The conjugate gradient method can be used to solve the linear system of equations

$$Ax = b,$$

where A is an invertible $n \times n$ matrix and b is a given vector in $\Re^n$. One way to do this is to apply the conjugate gradient method to the positive definite quadratic optimization problem

$$\begin{aligned} &\text{minimize} \ \ \tfrac{1}{2}x'A'Ax - b'Ax \\ &\text{subject to} \ \ x \in \Re^n, \end{aligned}$$

which corresponds to the equivalent linear system $A'Ax = A'b$. This, however, has several disadvantages, including the need to form the matrix $A'A$, which may have a much less favorable sparsity structure than A.

An alternative possibility is to introduce the vector z defined by

$$x = A'z$$

and to solve the system $AA'z = b$ or equivalently the positive definite quadratic problem

$$\begin{aligned} &\text{minimize} \ \ \tfrac{1}{2}z'AA'z - b'z \\ &\text{subject to} \ \ z \in \Re^n, \end{aligned}$$

whose cost function gradient is zero at z if and only if $AA'z = b$. By streamlining the computations, it is possible to write the conjugate gradient method for the preceding problem directly in terms of the vector x, and without explicitly forming the product AA'. The resulting method is known as *Craig's method*; it is given by the following iteration where H is a positive definite symmetric preconditioning matrix:

$$x^{k+1} = x^k + \alpha^k d^k, \qquad \alpha^k = \frac{r^{k\prime} r^k}{d^{k\prime} d^k},$$

where the vectors r^k and d^k are generated by the recursions

$$r^{k+1} = r^k + \alpha^k H A d^k, \qquad d^{k+1} = -A' H r^{k+1} + \frac{r^{k+1'} r^{k+1}}{r^{k'} r^k} d^k$$

with the initial conditions

$$r^0 = H(Ax^0 - b), \qquad d^0 = -A' H r^0.$$

The verification of these equations is left for the reader.

There are other conjugate gradient-like methods for the system $Ax = b$, which are not really equivalent to the conjugate gradient method for any quadratic optimization problem. One possibility, due to Saad and Schultz [SaS86], and known as the *Generalized Minimum Residual* method (GMRES), is to start with a vector x^0 and obtain x^k as the vector that minimizes $\|Ax - b\|^2$ over the linear manifold $x^0 + S^k$, where

$$S^k = (\text{subspace spanned by the vectors } r, Ar, A^2 r, \ldots, A^{k-1} r),$$

and r is the initial residual

$$r = Ax^0 - b.$$

This successive subspace minimization process can be efficiently implemented, but we will not get into the details further (see [SaS86]). It can be shown that x^k is a solution of the system $Ax = b$ if and only if $A^k r$ belongs to the subspace S^k (write the minimization of $\|Ax - b\|^2$ over $x^0 + S^k$ as the equivalent minimization of $\|\xi - r\|^2$ over all ξ in the subspace AS^k). Thus if none of the vectors $x^0, \ldots, x^{n-2}$ is a solution, the subspace S^{n-1} is equal to $\Re^n$, implying that x^{n-1} is an unconstrained minimum of $\|Ax - b\|^2$, and therefore solves the system $Ax = b$. It follows that the method will terminate after at most n iterations.

GMRES can be viewed as a conjugate gradient method only in the special case where *A is positive definite and symmetric*. In that case it can be shown that the method is equivalent to a preconditioned conjugate gradient method applied to the quadratic cost $\|Ax - b\|^2$. This is based on the expanding subspace minimization property of the conjugate gradient method (see also Exercise 1.6.4). Note, however, that GMRES can be used for any matrix A that is invertible.

Rate of Convergence of the Conjugate Gradient Method*

There are a number of convergence rate results for the conjugate gradient method. Since the method terminates in at most n steps for a quadratic cost, one would expect that when viewed in cycles of n steps, its rate of convergence for a nonquadratic cost would be comparable to the rate of

Newton's method. Indeed there are results which roughly state that if the method is restarted every n iterations and $\{x^k\}$ converges to a nonsingular local minimum x^*, then the error $e^k = \|x^{nk} - x^*\|$ converges superlinearly. (Note that here the error is considered at the end of cycles of n iterations rather than at the end of each iteration.) Such results are reassuring but not terribly interesting because the conjugate gradient method is most useful in problems where n is large (see the discussion at the end of Section 1.7), and for such problems, one hopes that practical convergence will occur after fewer than n iterations. Therefore, the single-step rate of convergence of the method is more interesting than its rate of convergence in terms of n-step cycles. The following analysis gives a result of this type, based on an interpretation of the conjugate gradient method as an optimal process.

Assume that the cost is positive definite quadratic of the form

$$f(x) = \tfrac{1}{2}x'Qx.$$

(To simplify the following exposition, we have assumed that the linear term $b'x$ is zero, but with minor modifications, the following analysis holds also when $b \neq 0$.) Let g^i denote as usual the gradient $\nabla f(x^i)$ and consider an algorithm of the form

$$x^1 = x^0 + \gamma^{00}g^0,$$

$$x^2 = x^0 + \gamma^{10}g^0 + \gamma^{11}g^1,$$

$$\cdots \qquad \cdots$$

$$x^{k+1} = x^0 + \gamma^{k0}g^0 + \cdots + \gamma^{kk}g^k, \tag{1.175}$$

where γ^{ij} are arbitrary scalars. Since $g^i = Qx^i$, we see that for suitable scalars c^{ki}, the above algorithm can be written for all k as

$$x^{k+1} = x^0 + c^{k0}Qx^0 + c^{k1}Q^2x^0 + \cdots + c^{kk}Q^{k+1}x^0 = \big(I + QP^k(Q)\big)x^0,$$

where P^k is a polynomial of degree k.

Among algorithms of the form (1.175), the conjugate gradient method is optimal in the sense that for every k, it minimizes $f(x^{k+1})$ over all sets of coefficients $\gamma^{k0}, ..., \gamma^{kk}$. It follows from the equation above that in the conjugate gradient method we have, for every k,

$$f(x^{k+1}) = \min_{P^k} \tfrac{1}{2}x^{0\prime}Q\big(I + QP^k(Q)\big)^2x^0. \tag{1.176}$$

Let $\lambda_1, \ldots, \lambda_n$ be the eigenvalues of Q, and let $e_1, \ldots, e_n$ be corresponding orthogonal eigenvectors, normalized so that $\|e_i\| = 1$. Since $e_1, \ldots, e_n$ form a basis, any vector $x^0 \in \Re^n$ can be written as

$$x^0 = \sum_{i=1}^{n} \xi_i e_i$$

for some scalars ξ_i. Since

$$Qx^0 = \sum_{i=1}^{n} \xi_i Q e_i = \sum_{i=1}^{n} \xi_i \lambda_i e_i,$$

we have, using the orthogonality of $e_1, \ldots, e_n$ and the fact $\|e_i\| = 1$,

$$f(x^0) = \tfrac{1}{2} x^{0\prime} Q x^0 = \tfrac{1}{2} \left(\sum_{i=1}^{n} \xi_i e_i \right)' \left(\sum_{i=1}^{n} \xi_i \lambda_i e_i \right) = \tfrac{1}{2} \sum_{i=1}^{n} \lambda_i \xi_i^2.$$

Applying the same process to Eq. (1.176), we obtain for *any* polynomial P^k of degree k

$$f(x^{k+1}) \le \tfrac{1}{2} \sum_{i=1}^{n} \big(1 + \lambda_i P^k(\lambda_i)\big)^2 \lambda_i \xi_i^2,$$

and it follows that

$$f(x^{k+1}) \le \max_i \big(1 + \lambda_i P^k(\lambda_i)\big)^2 f(x^0), \qquad \forall\ P^k,\ k. \tag{1.177}$$

One can use this relationship for different choices of polynomials P^k to obtain a number of convergence rate results. We provide one such result, which explains a behavior often observed in practice: after a few relatively ineffective iterations, the conjugate gradient method produces substantial and often spectacular cost improvement. In particular, the following proposition shows that the first k conjugate gradient iterations in an n-iteration cycle eliminate the effect of the k largest eigenvalues of Q.

Proposition 1.6.2: Assume that Q has $n - k$ eigenvalues in an interval $[a, b]$ with $a > 0$, and the remaining k eigenvalues are greater than b. Then for every x^0, the vector x^{k+1} generated after $k + 1$ steps of the conjugate gradient method satisfies

$$f(x^{k+1}) \le \left(\frac{b - a}{b + a} \right)^2 f(x^0). \tag{1.178}$$

This relation also holds for the preconditioned conjugate gradient method (1.166)-(1.168) if the eigenvalues of Q are replaced by those of $H^{1/2} Q H^{1/2}$.

Proof: Let $\lambda_1, \ldots, \lambda_k$ be the eigenvalues of Q that are greater than b and consider the polynomial P^k defined by

$$1 + \lambda P^k(\lambda) = \frac{2}{(a + b)\lambda_1 \cdots \lambda_k} \left(\frac{a + b}{2} - \lambda \right) (\lambda_1 - \lambda) \cdots (\lambda_k - \lambda). \tag{1.179}$$

Since $1 + \lambda_i P^k(\lambda_i) = 0$, we have, using Eqs. (1.177), (1.179), and a simple calculation,

$$f(x^{k+1}) \leq \max_{a \leq \lambda \leq b} \frac{\left(\lambda - \frac{1}{2}(a+b)\right)^2}{\left(\frac{1}{2}(a+b)\right)^2} f(x^0) = \left(\frac{b-a}{b+a}\right)^2 f(x^0).$$

Q.E.D.

One consequence of the above proposition is that if the eigenvalues of Q take only k distinct values, then the conjugate gradient method will find the minimum of the quadratic function f in at most k iterations. (Take $a = b$.) Some other possibilities are explored in the problem section.

It is worth mentioning some additional rate of convergence results regarding the conjugate gradient method, as applied to the positive definite quadratic function

$$f(x) = \tfrac{1}{2}(x - x^*)'Q(x - x^*).$$

Let M and m be the largest and smallest eigenvalues of Q, respectively. Then, for any starting point x^0 and any iteration index k, it can be shown that

$$\|x^k - x^*\| \leq 2\left(\frac{M}{m}\right)^{1/2} \left(\frac{\sqrt{M} - \sqrt{m}}{\sqrt{M} + \sqrt{m}}\right)^k \|x^0 - x^*\|.$$

This result is a consequence of a view of the conjugate gradient method as a two-step iterative method (see Exercise 1.6.5 and compare also with the heavy ball method of Exercise 1.3.9). Another related convergence rate estimate is

$$f(x^k) \leq \frac{M\|x^0 - x^*\|^2}{2(2k+1)^2};$$

see [Pol87]. These estimates suggest a more favorable convergence rate than the one of steepest descent; compare with the results of Section 1.3.

EXERCISES

1.6.1

Show that the Gram-Schmidt procedure has the projection property stated in Fig. 1.6.2.

1.6.2 [Lue84]

Assume that Q has all its eigenvalues concentrated in two intervals of the form

$$[a, b], \qquad [a + \delta, b + \delta],$$

where a, b, δ are some positive scalars. Show that for every x^0, the vector x^2 generated after two steps of the conjugate gradient method satisfies

$$f(x^2) \leq \left(\frac{b-a}{b+a}\right)^2 f(x^0).$$

Hint: Proceed as in the proof of Prop. 1.6.2, using an appropriate polynomial.

1.6.3 (Hessian with Clustered Eigenvalues [Ber82a]) (www)

Assume that Q has all its eigenvalues concentrated at k intervals of the form

$$[z_i - \delta_i, z_i + \delta_i], \qquad i = 1, \ldots, k,$$

where we assume that $\delta_i \geq 0$, $i = 1, \ldots, k$, $0 < z_1 - \delta_1$, and

$$0 < z_1 < z_2 < \cdots < z_k, \qquad z_i + \delta_i \leq z_{i+1} - \delta_{i+1}, \qquad i = 1, \ldots, k-1.$$

Show that the vector x^{k+1} generated after $k+1$ steps of the conjugate gradient method satisfies

$$f(x^{k+1}) \leq R f(x^0),$$

where

$$R = \max\left\{ \frac{\delta_1^2}{z_1^2}, \frac{\delta_2^2 (z_2 + \delta_2 - z_1)^2}{z_1^2 z_2^2}, \ldots, \right.$$
$$\left. \frac{\delta_k^2 (z_k + \delta_k - z_1)^2 (z_k + \delta_k - z_2)^2 \cdots (z_k + \delta_k - z_{k-1})^2}{z_1^2 z_2^2 \cdots z_k^2} \right\}.$$

1.6.4 (www)

Consider the conjugate gradient method applied to the minimization of $f(x) = \frac{1}{2} x' Q x - b' x$, where Q is positive definite and symmetric. Show that the iterate x^k minimizes f over the linear manifold

$$x^0 + (\text{subspace spanned by } g^0, Q g^0, \ldots, Q^{k-1} g^0),$$

where $g^0 = \nabla f(x^0)$.

1.6.5 (www)

Consider a method whereby the first iteration is a steepest descent iteration with the stepsize determined by line minimization, and subsequent iterations have the form

$$x^{k+1} = x^k - \alpha^k g^k + \beta^k (x^k - x^{k-1}), \qquad k = 1, 2, \ldots,$$

where α^k and β^k are chosen to minimize f (compare with the heavy ball method of Exercise 1.3.9). Thus, for $k \geq 1$, the $(k+1)$st iteration finds x^{k+1} that minimizes f over the two-dimensional linear manifold

$$x^k + (\text{subspace spanned by } g^k \text{ and } x^k - x^{k-1}).$$

Show that if $f(x) = \frac{1}{2}x'Qx - b'x$, where Q is positive definite and symmetric, this method is equivalent to the conjugate gradient method, in the sense that the two methods generate identical iterates if they start with the same point.

1.6.6 (PARTAN)

Let the cost function be $f(x) = \frac{1}{2}x'Qx - b'x$, where Q is positive definite and symmetric. Show that in the method of Exercise 1.6.5 the minimization over the two-dimensional manifold

$$x^k + (\text{subspace spanned by } g^k \text{ and } x^k - x^{k-1})$$

can be done using just two line searches:

(1) Find the vector y^k that minimizes f over the line $y_\alpha = x^k - \alpha g^k$, $\alpha \geq 0$.

(2) Find x^{k+1} as the vector that minimizes f over the line connecting x^{k-1} and y^k.

(This is known as the method of *parallel tangents* or PARTAN, due to Shah, Buehler, and Kempthorne [SBK64].) *Hint*: Show that g^k is orthogonal to both g^{k-1} and $\nabla f(y^k)$. Conclude that g^k is orthogonal to the gradient $\nabla f(y)$ of any vector y on the line segment connecting x^{k-1} and y^k.

1.6.7 (www)

Let the cost function be $f(x) = \frac{1}{2}x'Qx - b'x$, where Q is positive semidefinite and symmetric. Apply the conjugate gradient method to this function. Show that if an optimal solution exists, the method will find one such solution in at most m steps, where m is the rank of Q. What happens if the problem has no optimal solution?

1.6.8 [Pow64], [Zan67a] (www)

Let the cost function be $f(x) = \frac{1}{2}x'Qx - b'x$, where Q is positive definite and symmetric. Suppose that x_1 and x_2 minimize f over linear manifolds that are parallel to subspaces S_1 and S_2, respectively. Show that if $x_1 \neq x_2$, then $x_1 - x_2$ is Q-conjugate to all vectors in the intersection of S_1 and S_2. Use this property to suggest a conjugate direction method that does not evaluate gradients and uses only line minimizations.

1.6.9

Suppose that $d^0, \ldots, d^k$ are Q-conjugate directions, let $x^1, \ldots, x^{k+1}$ be the vectors generated by the corresponding conjugate direction method, and assume that $x^{i+1} \neq x^i$ for all $i = 0, \ldots, k$. Show that a vector d^{k+1} is Q-conjugate to $d^0, \ldots, d^k$ if and only if $d^{k+1} \neq 0$ and d^{k+1} is orthogonal to the gradient differences $g^{i+1} - g^i$, $i = 0, \ldots, k$.

1.6.10 [Ber74]

Let the cost function be $f(x) = \frac{1}{2}x'Qx - b'x$, where Q has the form

$$Q = M + \sum_{i=1}^{k} v_i v_i',$$

where M is positive definite and symmetric, and v_i are some vectors in $\Re^n$. Show that the vector x^{k+1} generated after $k+1$ steps of the conjugate gradient method satisfies

$$f(x^{k+1}) \leq \left(\frac{b-a}{b+a}\right)^2 f(x^0),$$

where a and b are the smallest and largest eigenvalues of M, respectively. Show also that the vector x^{k+1} generated by the preconditioned conjugate gradient method with $H = M^{-1}$ minimizes f. *Hint*: Use the interlocking eigenvalues lemma [Prop. A.18(d) in Appendix A].

1.7 QUASI-NEWTON METHODS

Quasi-Newton methods are gradient methods of the form

$$x^{k+1} = x^k + \alpha^k d^k, \tag{1.180}$$

$$d^k = -D^k \nabla f(x^k), \tag{1.181}$$

where D^k is a positive definite matrix, which may be adjusted from one iteration to the next so that the direction d^k tends to approximate the Newton direction. Some of these methods are quite popular because they typically converge fast, while avoiding the second derivative calculations associated with Newton's method. Their main drawback relative to the conjugate gradient method is that they require storage of the matrix D^k as well as the matrix-vector multiplication overhead associated with the calculation of the direction d^k (see the subsequent discussion).

An important idea for many quasi-Newton methods is that two successive iterates x^k, x^{k+1} together with the corresponding gradients $\nabla f(x^k)$, $\nabla f(x^{k+1})$, yield curvature information by means of the approximate relation

$$q^k \approx \nabla^2 f(x^{k+1}) p^k, \tag{1.182}$$

where

$$p^k = x^{k+1} - x^k, \tag{1.183}$$

$$q^k = \nabla f(x^{k+1}) - \nabla f(x^k). \tag{1.184}$$

In particular, given n linearly independent iteration increments $p^0, \ldots, p^{n-1}$ together with the corresponding gradient increments $q^0, \ldots, q^{n-1}$, we can obtain approximately the Hessian as

$$\nabla^2 f(x^n) \approx \begin{bmatrix} q^0 & \cdots & q^{n-1} \end{bmatrix} \begin{bmatrix} p^0 & \cdots & p^{n-1} \end{bmatrix}^{-1},$$

and the inverse Hessian as

$$\nabla^2 f(x^n)^{-1} \approx \begin{bmatrix} p^0 & \cdots & p^{n-1} \end{bmatrix} \begin{bmatrix} q^0 & \cdots & q^{n-1} \end{bmatrix}^{-1}.$$

When the cost is quadratic, this relation is exact. Many interesting quasi-Newton methods use similar but more sophisticated ways to build curvature information into the matrix D^k so that it progressively approaches the inverse Hessian. This is true in particular for the quasi-Newton methods we will focus on in this section, and which we now introduce.

In the most popular class of quasi-Newton methods, the matrix D^{k+1} is obtained from D^k, and the vectors p^k and q^k by means of the equation

$$D^{k+1} = D^k + \frac{p^k p^{k\prime}}{p^{k\prime} q^k} - \frac{D^k q^k q^{k\prime} D^k}{q^{k\prime} D^k q^k} + \xi^k \tau^k v^k v^{k\prime}, \tag{1.185}$$

where

$$v^k = \frac{p^k}{p^{k\prime} q^k} - \frac{D^k q^k}{\tau^k}, \tag{1.186}$$

$$\tau^k = q^{k\prime} D^k q^k, \tag{1.187}$$

the scalars ξ^k satisfy, for all k,

$$0 \le \xi^k \le 1, \tag{1.188}$$

and D^0 is an arbitrary positive definite matrix. The scalars ξ^k parameterize the method. If $\xi^k = 0$ for all k, we obtain the *Davidon-Fletcher-Powell (DFP) method*, which is historically the first quasi-Newton method. If $\xi^k = 1$ for all k, we obtain the *Broyden-Fletcher-Goldfarb-Shanno (BFGS) method*, for which there is substantial evidence that it is the best general purpose quasi-Newton method currently known.

We will show that the matrices D^k generated in the above manner indeed encode curvature information as indicated in Eqs. (1.182)-(1.184). In particular, we will see in the proof of Prop. 1.7.2 that for a quadratic problem and all k we have

$$D^{k+1} q^i = p^i, \qquad i = 0, \ldots, k,$$

from which after $n-1$ iterations, we obtain

$$D^n = \begin{bmatrix} p^0 & \cdots & p^{n-1} \end{bmatrix} \begin{bmatrix} q^0 & \cdots & q^{n-1} \end{bmatrix}^{-1} = \nabla^2 f(x^n)^{-1}.$$

We first show that under a mild assumption, the matrices D^k generated by Eq. (1.185) are positive definite. This is a very important property, since it guarantees that d^k is a descent direction.

Proposition 1.7.1: If D^k is positive definite and the stepsize α^k is chosen so that x^{k+1} satisfies

$$\nabla f(x^k)' d^k < \nabla f(x^{k+1})' d^k, \tag{1.189}$$

then D^{k+1} as given by Eq. (1.185) is positive definite.

Note: If x^k is not a stationary point, we have $\nabla f(x^k)' d^k < 0$, so in order to satisfy condition (1.189), it is sufficient to carry out the line search to a point where

$$|\nabla f(x^{k+1})' d^k| < |\nabla f(x^k)' d^k|.$$

In particular, if α^k is determined by the line minimization rule, then we have $\nabla f(x^{k+1})' d^k = 0$ and Eq. (1.189) is satisfied.

Proof: We first note that Eq. (1.189) implies that $\alpha^k \neq 0$, $q^k \neq 0$, and

$$p^{k\prime} q^k = \alpha^k d^{k\prime} \big(\nabla f(x^{k+1}) - \nabla f(x^k) \big) > 0. \tag{1.190}$$

Thus all denominator terms in Eqs. (1.185) and (1.186) are nonzero, and D^{k+1} is well defined.

Now for any $z \neq 0$ we have

$$z' D^{k+1} z = z' D^k z + \frac{(z' p^k)^2}{p^{k\prime} q^k} - \frac{(q^{k\prime} D^k z)^2}{q^{k\prime} D^k q^k} + \xi^k \tau^k (v^{k\prime} z)^2. \tag{1.191}$$

Using the notation $a = (D^k)^{1/2}z$, $b = (D^k)^{1/2}q^k$, this equation is written as

$$z'D^{k+1}z = \frac{\|a\|^2\|b\|^2 - (a'b)^2}{\|b\|^2} + \frac{(z'p^k)^2}{p^{k'}q^k} + \xi^k\tau^k(v^{k'}z)^2. \tag{1.192}$$

From Eqs. (1.187) and (1.190), and the Schwartz inequality [Eq. (A.2) in Appendix A], we have that all terms on the right-hand side of Eq. (1.192) are nonnegative. In order that $z'D^{k+1}z > 0$, it will suffice to show that we cannot have simultaneously

$$\|a\|^2\|b\|^2 = (a'b)^2 \qquad \text{and} \qquad z'p^k = 0.$$

Indeed if $\|a\|^2\|b\|^2 = (a'b)^2$, we must have $a = \lambda b$ or equivalently, $z = \lambda q^k$. Since $z \neq 0$, it follows that $\lambda \neq 0$, so if $z'p^k = 0$, we must have $q^{k'}p^k = 0$, which is impossible by Eq. (1.190). **Q.E.D.**

An important property of the algorithm is that when applied to the positive definite quadratic function $f(x) = \frac{1}{2}x'Qx - b'x$, with the stepsize α^k determined by line minimization, it generates a Q-conjugate direction sequence, while simultaneously constructing the inverse Hessian Q^{-1} after n iterations. This is the subject of the next proposition.

Proposition 1.7.2: Let $\{x^k\}$, $\{d^k\}$, and $\{D^k\}$ be sequences generated by the quasi-Newton algorithm (1.180)-(1.181), (1.185)-(1.188), applied to minimization of the positive definite quadratic function

$$f(x) = \tfrac{1}{2}x'Qx - b'x,$$

with α^k chosen by

$$f(x^k + \alpha^k d^k) = \min_{\alpha} f(x^k + \alpha d^k). \tag{1.193}$$

Assume that none of the vectors $x^0, \ldots, x^{n-1}$ is optimal. Then:

(a) The vectors $d^0, \ldots, d^{n-1}$ are Q-conjugate.

(b) There holds

$$D^n = Q^{-1}.$$

Proof: We will show that for all k

$$d^{i'}Qd^j = 0, \qquad 0 \leq i < j \leq k, \tag{1.194}$$

$$D^{k+1}Qp^i = p^i, \qquad 0 \leq i \leq k. \tag{1.195}$$

Equation (1.194) proves part (a) and it can be shown that Eq. (1.195) proves part (b). Indeed, since for $i < n$, none of the vectors x^i is optimal and d^i is a descent direction [cf. Eq. (1.181) and Prop. 1.7.1], we have that $p^i \neq 0$. Since $p^i = \alpha^i d^i$ and $d^0, \ldots, d^{n-1}$ are Q-conjugate, it follows that $p^0, \ldots, p^{n-1}$ are linearly independent and therefore, Eq. (1.195) implies that $D^n Q$ is equal to the identity matrix.

We first verify that

$$D^{k+1} Q p^k = p^k, \qquad \forall\, k. \tag{1.196}$$

From the equation $Qp^k = q^k$ and the updating formula (1.185), we have

$$\begin{aligned} D^{k+1} Q p^k &= D^{k+1} q^k \\ &= D^k q^k + \frac{p^k p^{k\prime} q^k}{p^{k\prime} q^k} - \frac{D^k q^k q^{k\prime} D^k q^k}{q^{k\prime} D^k q^k} + \xi^k \tau^k v^k v^{k\prime} q^k \\ &= D^k q^k + p^k - D^k q^k + \xi^k \tau^k v^k v^{k\prime} q^k \\ &= p^k + \xi^k \tau^k v^k v_k{}' q^k. \end{aligned}$$

From Eqs. (1.186) and (1.187), we have that $v^{k\prime} q^k = 0$ and Eq. (1.196) follows.

We now show Eqs. (1.194) and (1.195) simultaneously by induction. For $k = 0$ there is nothing to show for Eq. (1.194), while Eq. (1.195) holds in view of Eq. (1.196). Assuming that Eqs. (1.194) and (1.195) hold for k, we prove them for $k+1$. We have, for $i < k$,

$$\nabla f(x^{k+1}) = \nabla f(x^{i+1}) + Q(p^{i+1} + \cdots + p^k). \tag{1.197}$$

The vector p^i is orthogonal to each vector in the right-hand side of this equation; it is orthogonal to $Qp^{i+1}, \ldots, Qp^k$ because $p^0, \ldots, p^k$ are Q-conjugate (since $p^i = \alpha^i d^i$) and it is orthogonal to $\nabla f(x^{i+1})$ because of the line minimization property of the stepsize [cf. Eq. (1.193)]. Therefore from Eq. (1.197) we obtain

$$p^{i\prime} \nabla f(x^{k+1}) = 0, \qquad 0 \leq i < k.$$

From this equation and Eq. (1.195),

$$p^{i\prime} Q D^{k+1} \nabla f(x^{k+1}) = 0, \qquad 0 \leq i \leq k,$$

and since $p^i = \alpha^i d^i$ and $d^{k+1} = -D^{k+1} \nabla f(x^{k+1})$, we obtain

$$d^{i\prime} Q d^{k+1} = 0, \qquad 0 \leq i \leq k. \tag{1.198}$$

This proves Eq. (1.194) for $k+1$.

From the induction hypothesis (1.195) and Eq. (1.196), we have for all i with $0 \leq i \leq k$,

$$q^{k+1'} D^{k+1} Q p^i = q^{k+1'} p^i = p^{k+1'} Q p^i = \alpha^{k+1} \alpha^i d^{k+1'} Q d^i = 0. \tag{1.199}$$

From Eq. (1.185), we have, for $0 \leq i \leq k$,

$$\begin{aligned} D^{k+2} q^i = D^{k+1} q^i &+ \frac{p^{k+1} p^{k+1'} q^i}{p^{k+1'} q^{k+1}} - \frac{D^{k+1} q^{k+1} q^{k+1'} D^{k+1} q^i}{q^{k+1'} D^{k+1} q^{k+1}} \\ &+ \xi^{k+1} \tau^{k+1} v^{k+1} v^{k+1'} q^i. \end{aligned} \tag{1.200}$$

Since $p^{k+1'} q^i = p^{k+1'} Q p^i = \alpha^{k+1} \alpha^i d^{k+1'} Q d^i = 0$, we see that the second term in the right-hand side of Eq. (1.200) is zero. Similarly, Eq. (1.195) implies that $q^{k+1'} D^{k+1} q^i = q^{k+1'} D^{k+1} Q p^i = q^{k+1'} p^i = p^{k+1'} Q p^i = 0$ and we see that the third term in the right-hand side of Eq. (1.200) is zero. Finally, a similar argument using the definition (1.186) of v^{k+1}, shows that the fourth term in the right-hand side of Eq. (1.200) is zero as well. Therefore, Eqs. (1.200) and (1.195) yield

$$D^{k+2} Q p^i = D^{k+2} q^i = D^{k+1} q^i = D^{k+1} Q p^i = p^i, \qquad 0 \leq i \leq k.$$

Taking into account also Eq. (1.196), this proves Eq. (1.195) for $k+1$. **Q.E.D.**

The following proposition sharpens the preceding one and clarifies further the relation between quasi-Newton and conjugate gradient methods.

Proposition 1.7.3: The sequence of iterates $\{x^k\}$ generated by the quasi-Newton algorithm applied to minimization of a positive definite quadratic function as in Prop. 1.7.2 is identical to the one that would be generated by the preconditioned conjugate gradient method with scaling matrix $H = D^0$.

Proof: It is sufficient to show that for $k = 0, 1, \ldots, n-1$, the vector x^{k+1} minimizes f over the linear manifold

$$M^k = \bigl\{ z \mid z = x^0 + \gamma^0 D^0 \nabla f(x^0) + \cdots + \gamma^k D^0 \nabla f(x^k),\ \gamma^0, \ldots, \gamma^k \in \Re \bigr\}.$$

This can be proved for the case where $D^0 = I$ by verifying through induction that for all k there exist scalars β_{ij}^k such that

$$D^k = I + \sum_{i=0}^{k} \sum_{j=0}^{k} \beta_{ij}^k \nabla f(x^i) \nabla f(x^j)'.$$

Therefore, for some scalars b_i^k and all k, we have

$$d^k = -D^k \nabla f(x^k) = \sum_{i=0}^{k} b_i^k \nabla f(x^i).$$

Since by Prop. 1.7.2, the algorithm is a conjugate direction method, for each k, x^{k+1} minimizes f over the manifold

$$\{z \mid z = x^0 + \gamma^0 d^0 + \cdots + \gamma^k d^k,\ \gamma^0, \ldots, \gamma^k \in \Re\},$$

which by the equation $d^k = \sum_{i=0}^{k} b_i^k \nabla f(x^i)$ just shown and the linear independence of $d^0, \ldots, d^k$, is the same as the manifold

$$M^k = \{z \mid z = x^0 + \gamma^0 \nabla f(x^0) + \cdots + \gamma^k \nabla f(x^k),\ \gamma^0, \ldots, \gamma^k \in \Re\}.$$

Thus, when $D^0 = I$, the algorithm satisfies the defining property of the conjugate gradient method (for all k, x^{k+1} is the unique minimum of f over M^k). For the case where $D^0 \neq I$, the proof follows by making a transformation of variables so that in the transformed space the initial matrix is the identity. **Q.E.D.**

A consequence of the above proposition is that if the minimization stepsize rule is used and the cost is quadratic, the generated iterates by the quasi-Newton algorithm do not depend on the values of the scalar ξ^k. Thus the DFP and the BFGS methods perform identically on a quadratic function if the line minimization is perfect. It turns out that this can be shown even when the cost is nonquadratic (as shown by Dixon [Dix72a], [Dix72b]), which is a rather surprising result. Thus the choice of ξ^k makes a difference only if the line minimization is inaccurate.

We finally note that multiplying the initial matrix D^0 by a positive scaling factor can have a significant beneficial effect on the behavior of the algorithm in the initial iterations of an n-iteration cycle, and also more generally in the case of a nonquadratic problem. A popular choice is to compute

$$\tilde{D}^0 = \frac{p^{0'} q^0}{q^{0'} D^0 q^0} D^0, \tag{1.201}$$

once the vector x^1 (and hence also p^0 and q^0) has been obtained, and use $\tilde{D}^0$ in place of D^0 in computing D^1. Sometimes it is beneficial to scale D^k even after the first iteration by multiplication with $p^{k'} q^k / q^{k'} D^k q^k$. This is the idea behind the self-scaling algorithms introduced by Oren [Ore73], and Oren and Luenberger [OrL74]. The rationale for these algorithms is that if the scaling (1.201) is used, then the condition number M^k/m^k, where

$$M^k = \text{max eigenvalue of } (D^k)^{1/2} Q (D^k)^{1/2},$$

$$m^k = \text{min eigenvalue of } (D^k)^{1/2} Q (D^k)^{1/2},$$

is not increased (and is usually decreased) at each iteration (see Exercise 1.7.5).

Comparison of Quasi-Newton Methods with Other Methods

Let us now consider a nonquadratic problem, and compare the quasi-Newton method of Eqs. (1.180)-(1.181), (1.185)-(1.188) with the conjugate gradient method. One advantage of the quasi-Newton method is that when line search is accurate, the algorithm not only tends to generate conjugate directions but also constructs an approximation to the inverse Hessian matrix. As a result, near convergence to a local minimum with positive definite Hessian, it tends to approximate Newton's method thereby attaining a fast convergence rate. It is significant that this property does not depend on the starting matrix D^0, and as a result it is not usually necessary to periodically restart the method with a steepest descent-type step, which is something that is essential for the conjugate gradient method.

A second advantage is that the quasi-Newton method is not as sensitive to accuracy in the line search as the conjugate gradient method. This has been verified by extensive computational experience and can be substantiated to some extent by analysis. A partial explanation is that, under essentially no restriction on the line search accuracy, the method generates positive definite matrices D^k and hence directions of descent (Prop. 1.7.1).

To compare further the conjugate gradient method and the quasi-Newton method, we consider their computational requirements per iteration when n is large. The kth iteration of the conjugate gradient method requires computation of the cost function and its gradient (perhaps several times in the course of the line minimization) together with $O(n)$ operations to compute the conjugate direction d^k and the next point x^{k+1}. The quasi-Newton method requires roughly the same amount of computation for function and gradient evaluations together with $O(n^2)$ operations to compute the matrix D^k and the next point x^{k+1}. If the computation needed for a function and gradient evaluation is larger or comparable to $O(n^2)$ operations, the quasi-Newton method requires only slightly more computation per iteration than the conjugate gradient method and holds the edge in view of its other advantages mentioned earlier. In problems where a function and gradient evaluation requires computation time much less than $O(n^2)$ operations, the conjugate gradient method is typically preferable. As an example, we will see in Section 1.9, that in optimal control problems where typically n is very large, a function and a gradient evaluation typically requires $O(n)$ operations. For this reason the conjugate gradient method is typically preferable for these problems.

In general, both the conjugate gradient method and the quasi-Newton algorithm require less computation per iteration than Newton's method, which requires a function, gradient, and Hessian evaluation, as well as $O(n^3)$ operations at each step for computing the Newton direction. This is counterbalanced by the faster speed of convergence of Newton's method. Furthermore, in some cases, special structure can be exploited to compute the Newton direction efficiently. For example in optimal control problems,

Newton's method typically requires $O(n)$ operations per iteration versus $O(n^2)$ operations for the quasi-Newton method (see Section 1.9).

EXERCISES

1.7.1 (Rank One Quasi-Newton Methods) (www)

Suppose that D^0 is symmetric and that D^k is updated according to the formula

$$D^{k+1} = D^k + \frac{y^k y^{k'}}{q^{k'} y^k},$$

where $y^k = p^k - D^k q^k$. Show that we have

$$D^{k+1} q^i = p^i, \qquad \text{for all } k \text{ and } i \leq k.$$

Conclude that for a positive definite quadratic problem, after n steps for which n linearly independent increments $q^0, \ldots, q^{n-1}$ are obtained, D^n is equal to the inverse Hessian of the cost function.

1.7.2 (BFGS Update) (www)

Verify the following alternative formula for the BFGS update:

$$D^{k+1} = D^k + \left(1 + \frac{q^{k'} D^k q^k}{p^{k'} q^k}\right) \frac{p^k p^{k'}}{p^{k'} q^k} - \frac{D^k q^k p^{k'} + p^k q^{k'} D^k}{p^{k'} q^k}.$$

1.7.3 (Limited Memory BFGS Method [Noc80]) (www)

A major drawback of quasi-Newton methods for large problems is the large storage requirement. This motivates methods that construct the quasi-Newton direction $d^k = -D^k \nabla f(x^k)$ using only a limited number of the vectors p^i and q^i (for example, the last m). This exercise shows one way to do this.

(a) Use Exercise 1.7.2 to show that the BFGS updating formula can be written as

$$D^{k+1} = V^{k'} D^k V^k + \rho^k p^k p^{k'},$$

where

$$\rho^k = \frac{1}{q^{k'} p^k}, \qquad V^k = I - \rho^k q^k p^{k'}.$$

(b) Show how to calculate the direction $d^k = -D^k \nabla f(x^k)$ using D^0 and the past vectors p^i, q^i, $i = 0, 1, \ldots, k-1$.

1.7.4 (Hessian Approximation Updating Formula) (www)

The matrices D^k generated by the updating formula of this section can be viewed as approximations of the inverse Hessian. An alternative is to generate and update approximations of the Hessian. It turns out that the inverses $H^k = (D^k)^{-1}$ can be updated by formulas similar to the ones for D^k, except that q^k is replaced by p^k and reversely. This can be traced to the fact that the relations

$$D^{k+1}q^i = D^{k+1}Qp^i = p^i, \qquad 0 \leq i \leq k,$$

shown in Prop. 1.7.1 can also be written as

$$H^{k+1}p^i = H^{k+1}Q^{-1}q^i = q^i, \qquad 0 \leq i \leq k.$$

In particular, use the BFGS formula given in Exercise 1.7.2 to show that if D^k (or H^k) is updated by the DFP formula, then H^k (or D^k, respectively) is updated by the BFGS formula.

1.7.5 (Condition Number of a Quasi-Newton Iteration) (www)

Consider a positive definite quadratic problem. If we view the quasi-Newton iteration $x^{k+1} = x^k - \alpha^k D^k \nabla f(x^k)$ as a scaled gradient method, its convergence rate (per step) is governed by the condition number of the matrix $(D^k)^{1/2}Q(D^k)^{1/2}$, or equivalently the condition number of the matrix

$$R^k = Q^{1/2}D^kQ^{1/2}.$$

(a) Suppose that D^k is updated using the DFP formula. Let $\lambda_1 \leq \cdots \leq \lambda_n$ be the eigenvalues of R^k and suppose that $1 \in [\lambda_1, \lambda_n]$. Show that the eigenvalues of R^{k+1} are contained in the range $[\lambda_1, \lambda_n]$, so that the condition number of R^{k+1} is no worse than the condition number of R^k. Furthermore, 1 is an eigenvalue of R^{k+1}. *Hint*: Let

$$r^k = Q^{1/2}p^k.$$

Show that R^k is generated by

$$R^{k+1} = R^k + \frac{r^k r^{k\prime}}{r^{k\prime} r^k} - \frac{R^k r^k r^{k\prime} R^k}{r^{k\prime} R^k r^k}.$$

Use the interlocking eigenvalues lemma [Prop. A.18(d) in Appendix A] to show that the eigenvalues $\mu_1, \ldots, \mu_n$ of the matrix

$$R^k - \frac{R^k r^k r^{k\prime} R^k}{r^{k\prime} R^k r^k}$$

satisfy

$$0 = \mu_1 \leq \lambda_1 \leq \mu_2 \leq \cdots \leq \mu_n \leq \lambda_n.$$

Furthermore, r^k is an eigenvector corresponding to the eigenvalue $\mu_1 = 0$. Use the interlocking eigenvalues lemma again to show that the eigenvalues of R^{k+1} are $\mu_2, \ldots, \mu_n$ and 1.

(b) Show that the property $1 \in [\lambda_1, \lambda_n]$ of part (a) holds if the scaling (1.201) is used.

(c) Show the result of part (a) for the case of the BFGS update. *Hint*: Work with an update formula for $(R^k)^{-1}$; see Exercise 1.7.4.

1.7.6 (Memoryless Quasi-Newton Method [Per78], [Sha78]) 

This exercise provides the basis for an alternative implementation of the conjugate gradient method, which incorporates some of the advantages of quasi-Newton methods. Denote for all k, $g^k = \nabla f(x^k)$, and consider an algorithm of the form

$$x^{k+1} = x^k + \alpha^k d^k,$$

where

$$d^0 = -g^0, \qquad d^{k+1} = -D^{k+1} g^{k+1}, \quad k = 0, 1, \ldots$$

The matrix D^{k+1} is obtained via the BFGS update of Exercise 1.7.2 starting with the identity matrix:

$$D^{k+1} = I + \left(1 + \frac{q^{k'} q^k}{p^{k'} q^k}\right) \frac{p^k p^{k'}}{p^{k'} q^k} - \frac{q^k p^{k'} + p^k q^{k'}}{p^{k'} q^k}.$$

Furthermore, the line search is exact, so that $p^{k'} g^{k+1} = 0$. Assume that the cost function is positive definite quadratic.

(a) Show that

$$d^{k+1} = -g^{k+1} + \frac{(g^{k+1} - g^k)' g^{k+1}}{\|g^k\|^2} d^k,$$

so that the method coincides with the conjugate gradient method.

(b) Let D be a positive definite symmetric matrix and assume that

$$d^0 = -D g^0, \qquad d^{k+1} = -D^{k+1} g^{k+1}, \quad k = 0, 1, \ldots$$

where

$$D^{k+1} = D + \left(1 + \frac{q^{k'} D q^k}{p^{k'} q^k}\right) \frac{p^k p^{k'}}{p^{k'} q^k} - \frac{D q^k p^{k'} + p^k q^{k'} D}{p^{k'} q^k}.$$

Show that the method coincides with a preconditioned version of the conjugate gradient method.

1.8 NONDERIVATIVE METHODS

All the gradient methods examined so far require calculation of at least the gradient $\nabla f(x^k)$ and possibly the Hessian matrix $\nabla^2 f(x^k)$ at each generated point x^k. In many problems, either these derivatives are not available in explicit form or they are given by very complicated expressions. In such cases, it may be preferable to use the same algorithms as earlier with all unavailable derivatives approximated by finite differences.

First derivatives may be approximated by the *forward difference formula*

$$\frac{\partial f(x^k)}{\partial x_i} \approx \frac{1}{h}\big(f(x^k + he_i) - f(x^k)\big) \tag{1.202}$$

or by the *central difference formula*

$$\frac{\partial f(x^k)}{\partial x_i} \approx \frac{1}{2h}\big(f(x^k + he_i) - f(x^k - he_i)\big). \tag{1.203}$$

In these relations, h is a small positive scalar and e_i is the ith unit vector (ith column of the identity matrix). In some cases the same value of h can be used for all partial derivatives, but in other cases, particularly when the problem is poorly scaled, it is essential to use a different value of h for each partial derivative. This is a tricky process that often requires trial and error; see the following discussion.

The central difference formula requires twice as much computation as the forward difference formula. However, it is much more accurate. This can be seen by forming the corresponding Taylor series expansions, and by verifying that (in exact arithmetic) the absolute value of the error between the approximation and the actual derivatives is $O(h)$ for the forward difference formula, while it is $O(h^2)$ for the central difference formula. Note that if the central difference formula is used, one obtains at essentially no extra cost an approximation of each diagonal element of the Hessian using the formula

$$\frac{\partial^2 f(x^k)}{(\partial x_i)^2} \approx \frac{1}{h^2}\big(f(x^k + he_i) + f(x^k - he_i) - 2f(x^k)\big).$$

These approximations can be used in schemes based on diagonal scaling.

To reduce the approximation error, we would like to choose the finite difference interval h as small as possible. Unfortunately, there is a limit on how much h can be reduced due to the roundoff error that occurs when quantities of similar magnitude are subtracted by the computer. In particular, an error δ due to finite precision arithmetic in evaluating the numerator in Eq. (1.202) [or Eq. (1.203)], results in an error of δ/h (or $\delta/2h$, respectively) in the first derivative evaluation. Roundoff error is particularly evident in the approximate formulas (1.202) and (1.203) near a stationary point where ∇f is nearly zero, and the relative error size in the gradient approximation becomes very large.

Practical experience suggests that a good policy is to keep the scalar h for each derivative at a fixed value, which roughly balances the approximation error against the roundoff error. Based on the preceding calculations, this leads to the guideline

$$\frac{\delta}{h} = O(h) \quad \text{or} \quad h = O(\delta^{1/2}), \quad \text{for the forward difference formula (1.202),}$$

$$\frac{\delta}{2h} = O(h^2) \quad \text{or} \quad h = O(\delta^{1/3}), \quad \text{for the central difference formula (1.203),}$$

where δ is the error due to finite precision arithmetic in evaluating the numerator in Eq. (1.202) [or Eq. (1.203)]. Thus, a much larger value of h can be used in conjunction with the central difference formula. A good practical rule is to use the forward formula (1.202) until the absolute value of the corresponding approximate derivative becomes less than a certain tolerance; i.e.,

$$\left| \frac{f(x^k + he_i) - f(x^k)}{h} \right| \leq \epsilon,$$

where $\epsilon > 0$ is some prespecified scalar. At that point, a switch to the central difference formula should be made.

Second derivatives may be approximated by the forward difference formula

$$\frac{\partial^2 f(x^k)}{\partial x_i \partial x_j} \approx \frac{1}{h} \left(\frac{\partial f(x^k + he_j)}{\partial x^i} - \frac{\partial f(x^k)}{\partial x_i} \right) \tag{1.204}$$

or the central difference formula

$$\frac{\partial^2 f(x^k)}{\partial x_i \partial x_j} \approx \frac{1}{2h} \left(\frac{\partial f(x^k + he_j)}{\partial x_i} - \frac{\partial f(x^k - he_j)}{\partial x_i} \right). \tag{1.205}$$

Practical experience suggests that in discretized forms of Newton's method, extreme accuracy in approximating second derivatives is not very important in terms of rate of convergence. For this reason, exclusive use of the forward difference formula (1.204) is adequate in most cases. However, one should certainly check for positive definiteness of the discretized Hessian approximation and introduce corresponding modifications if necessary, as discussed in Section 1.4.

1.8.1 Coordinate Descent

There are several other nonderivative methods for minimizing differentiable functions. A particularly important one is the *coordinate descent method.* Here the cost is minimized along one coordinate direction at each iteration. The order in which coordinates are chosen may vary in the course of the algorithm. In the case where this order is cyclical, given x^k, the ith coordinate of x^{k+1} is determined by

$$x_i^{k+1} = \arg\min_{\xi \in \Re} f(x_1^{k+1}, \ldots, x_{i-1}^{k+1}, \xi, x_{i+1}^k, \ldots, x_n^k); \tag{1.206}$$

see Fig. 1.8.1. The method can also be used for minimization of f subject to upper and lower bounds on the variables x^i (the minimization over $\xi \in \Re$ in the preceding equation is replaced by minimization over the appropriate interval). We will analyze the method within this more general context in the next chapter.

An important advantage of the coordinate descent method is that it is well suited for *parallel computation*. In particular, suppose that there is a subset of coordinates $x_{i_1}, x_{i_2}, \ldots, x_{i_m}$, which are not coupled through the cost function, that is, $f(x)$ can be written as $\sum_{r=1}^{m} f_{i_r}(x)$, where for each r, $f_{i_r}(x)$ does not depend on the coordinates x_{i_s} for all $s \neq r$. Then one can perform the m coordinate descent iterations

$$x_{i_r}^{k+1} = \arg\min_{\xi} f_{i_r}(x^k + \xi e_{i_r}), \qquad r = 1, \ldots, m,$$

independently and in parallel. Thus, in problems with special structure where the set of coordinates can be partitioned into p subsets with the independence property just described, one can perform a full cycle of coordinate descent iterations in p (as opposed to n) parallel steps (assuming of course that a sufficient number of parallel processors is available); see [BeT89] for an extensive discussion of these issues.

The coordinate descent method generally has similar convergence properties to steepest descent. For continuously differentiable cost functions, it can be shown to generate sequences whose limit points are stationary, although the proof of this is sometimes complicated and requires some additional assumptions (see Prop. 2.7.1 in Section 2.7, which deals with a constrained version of coordinate descent and requires that the minimum of the cost function along each coordinate is uniquely attained). There is also a great deal of analysis of coordinate descent in a context where its use is particularly favorable, namely in solving dual problems (see Section 6.2). Within this context, the unique attainment assumption of Prop. 2.7.1 is neither satisfied nor is it essential (see the references given in Chapter 6). The convergence rate of coordinate descent to nonsingular and singular local minima can be shown to be linear and sublinear, respectively, similar to steepest descent. Often, the choice between coordinate descent and steepest descent is dictated by the structure of the cost function. Both methods can be very slow, but for many practical contexts, they can be quite effective.

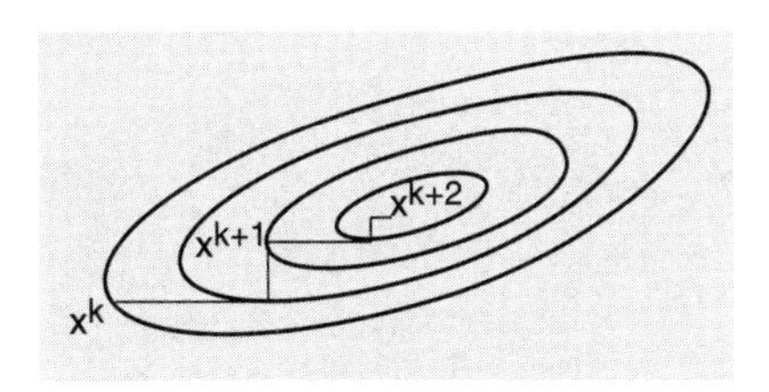

Figure 1.8.1. Illustration of the coordinate descent method.

1.8.2 Direct Search Methods

In the coordinate descent method we search along the fixed set of coordinate directions and we are guaranteed a cost improvement at a nonstationary point because these directions are linearly independent. This idea can be generalized by using a different set of directions and by occasionally changing this set of directions with the aim of accelerating convergence. There are a number of methods of this type: Rosenbrock's method [Ros60a], the pattern search algorithm of Hooke and Jeeves [HoJ61], and the simplex algorithms of Spendley, Hext, and Himsworth [SHH62], and Nelder and Mead [NeM65]. Unfortunately, the rationale of these methods often borders on the heuristic, and their theoretical convergence properties are often unsatisfactory. However, these methods are often fairly simple to implement and like the coordinate descent method, they do not require gradient calculations. We describe the Nelder and Mead simplex method (not to be confused with the simplex method of linear programming), which has enjoyed considerable popularity.

At the typical iteration of this method, we start with a *simplex*, that is, the convex hull of $n+1$ points, $x^0, x^1, \ldots, x^n$, and we end up with another simplex. Let x_{min} and x_{max} denote the "best" and "worst" vertices of the simplex, that is the vertices satisfying

$$f(x_{min}) = \min_{i=0,1,\ldots,n} f(x^i), \tag{1.207}$$

$$f(x_{max}) = \max_{i=0,1,\ldots,n} f(x^i). \tag{1.208}$$

Let also $\hat{x}$ denote the centroid of the face of the simplex formed by the vertices other than x_{max}

$$\hat{x} = \frac{1}{n}\left(-x_{max} + \sum_{i=0}^{n} x^i\right). \tag{1.209}$$

The iteration replaces the worst vertex x_{max} by a "better" one. In particular, the *reflection point* $x_{ref} = 2\hat{x} - x_{max}$ is computed, which lies on the line passing through x_{max} and $\hat{x}$, and is symmetric to x_{max} with respect to $\hat{x}$. Depending on the cost value of x_{ref} relative to the points of the simplex other than x_{max}, a new vertex x_{new} is computed, and a new simplex is formed from the old by replacing the vertex x_{max} by x_{new}, while keeping the other n vertices.

Typical Iteration of the Simplex Method

Step 1: (Reflection Step) Compute

$$x_{ref} = 2\hat{x} - x_{max}. \tag{1.210}$$

Then compute x_{new} according to the following three cases:

(1) **(x_{ref} has min cost)** If $f(x_{min}) > f(x_{ref})$, go to Step 2.

(2) **(x_{ref} has intermediate cost)** If $\max\{f(x^i) \mid x^i \neq x_{max}\} > f(x_{ref}) \geq f(x_{min})$, go to Step 3.

(3) **(x_{ref} has max cost)** If $f(x_{ref}) \geq \max\{f(x^i) \mid x^i \neq x_{max}\}$, go to Step 4.

Step 2: (Attempt Expansion) Compute

$$x_{exp} = 2x_{ref} - \hat{x}. \tag{1.211}$$

Define

$$x_{new} = \begin{cases} x_{exp} & \text{if } f(x_{exp}) < f(x_{ref}), \\ x_{ref} & \text{otherwise,} \end{cases}$$

and form the new simplex by replacing the vertex x_{max} with x_{new}.

Step 3: (Use Reflection) Define $x_{new} = x_{ref}$, and form the new simplex by replacing the vertex x_{max} with x_{new}.

Step 4: (Perform Contraction) Define

$$x_{new} = \begin{cases} \frac{1}{2}(x_{max} + \hat{x}) & \text{if } f(x_{max}) \leq f(x_{ref}), \\ \frac{1}{2}(x_{ref} + \hat{x}) & \text{otherwise,} \end{cases} \tag{1.212}$$

and form the new simplex by replacing the vertex x_{max} with x_{new}.

The reflection step and the subsequent possible steps of the iteration are illustrated in Fig. 1.8.2 (a)-(d). The entire method is illustrated in Fig. 1.8.3. Exercise 1.8.3 shows a cost improvement property of the method in the case where f is strictly convex. However, there are no known convergence results for the method. Furthermore, when the cost function is not convex, it is possible that the new simplex vertex x_{new} has larger cost value than the old vertex x_{max}. In this case a modification that has been suggested is to "shrink" the old simplex towards the best vertex x_{min}, that is, form a new simplex by replacing all the vertices x^i, $i = 0, 1, \ldots, n$, by

$$\bar{x}^i = \tfrac{1}{2}(x^i + x_{min}), \qquad i = 0, 1, \ldots, n.$$

The method as given above seems to work reasonably well in practice, particularly for problems of relatively small dimension (say up to 10). However, it is not guaranteed to have desirable convergence properties, and in fact a convergence counterexample is given in [McK98]. Reference [Tse95a] provides a relatively simple modification with satisfactory convergence properties. There are also a number of related methods, some of which have demonstrable convergence properties; see [DeT91] and [Tor91].

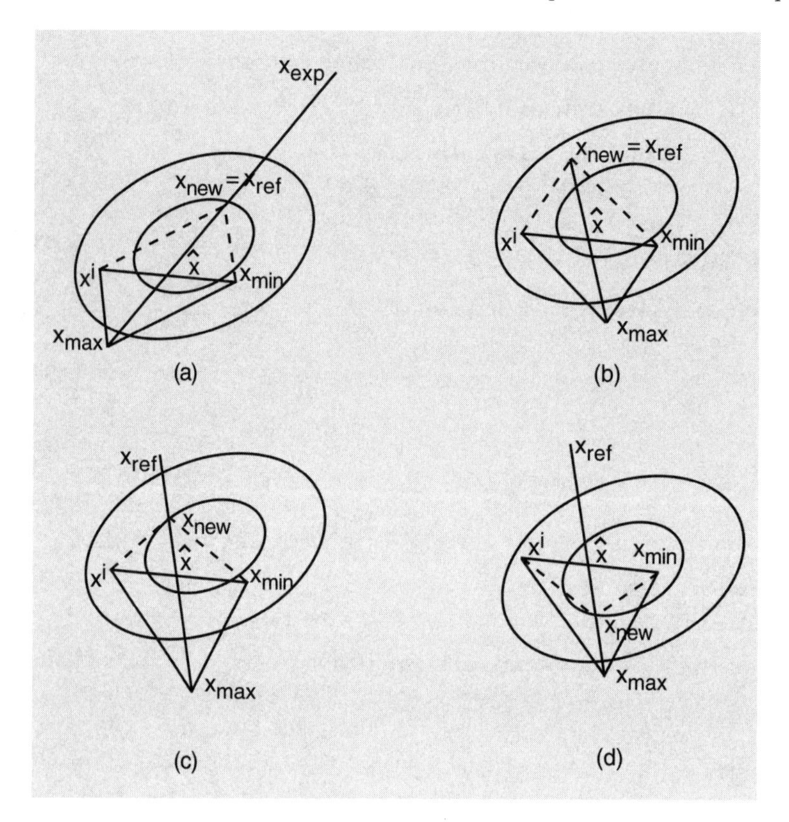

Figure 1.8.2. Illustration of the reflection step and the possible subsequent steps of an iteration of the simplex method. In (a), the new vertex x_{new} is determined via the expansion Step 2. In (b), x_{new} is determined via Step 3, and the reflection step is accepted. In (c) and (d), x_{new} is determined via the contraction Step 4.

Note that the constants used in Eqs. (1.210), (1.211), and (1.212) are somewhat arbitrary, as suggested by the interpretation of the method given in Figs. 1.8.2 and 1.8.3. More general forms of these equations are

$$x_{ref} = \hat{x} + \beta(\hat{x} - x_{max}),$$

$$x_{exp} = x_{ref} + \gamma(x_{ref} - \hat{x}),$$

$$x_{con} = \begin{cases} \theta x_{max} + (1-\theta)\hat{x} & \text{if } f(x_{max}) \leq f(x_{ref}), \\ \theta x_{ref} + (1-\theta)\hat{x} & \text{otherwise,} \end{cases}$$

where $\beta > 0$, $\gamma > 0$, and $\theta \in (0,1)$ are scalars known as the *reflection coefficient*, the *expansion coefficient*, and the *contraction coefficient*, respectively. The formulas of Eqs. (1.210)-(1.212) correspond to $\beta = 1$, $\gamma = 1$, and $\theta = 1/2$, respectively.

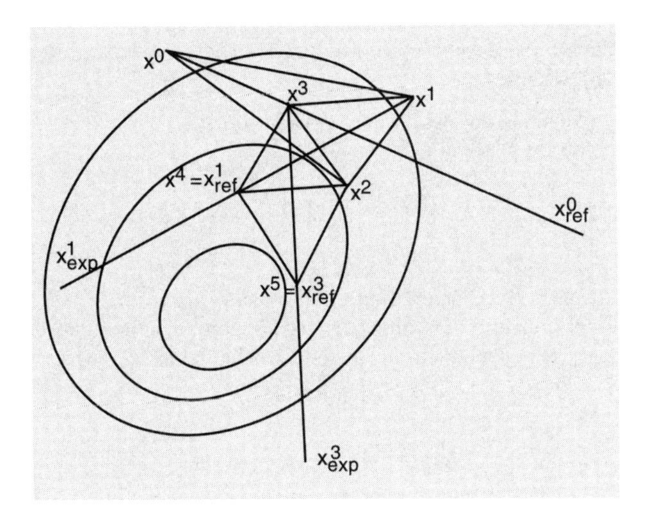

Figure 1.8.3. Illustration of three iterations of the simplex method, which generate the points x^3, x^4, and x^5, starting from the simplex x^0, x^1, x^2. The simplex obtained after these three iterations consists of x^2, x^4, and x^5.

EXERCISES

1.8.1

Let $f : \Re^n \mapsto \Re$ be continuously differentiable, let $p_1, \ldots, p_n$ be linearly independent vectors, and suppose that for some x^*, $\alpha = 0$ is a stationary point of each of the one-dimensional functions $g_i(\alpha) = f(x^* + \alpha p_i)$, $i = 1, \ldots, n$. Show that x^* is a stationary point of f. Suggest a minimization method that does not use derivatives based on this property.

1.8.2 (Stepsize Selection in Jacobi Methods)

Let $f : \Re^n \mapsto \Re$ be a continuously differentiable convex function. For a given $x \in \Re^n$ and all $i = 1, \ldots, n$, define the vector $\bar{x}$ by

$$\bar{x}_i = \arg\min_{\xi \in \Re} f(x_1, \ldots, x_{i-1}, \xi, x_{i+1}, \ldots, x_n).$$

The Jacobi method is defined by the iteration

$$x := x + \alpha(\bar{x} - x),$$

where α is a positive stepsize parameter.

(a) Show that if x does not minimize f, then the Jacobi iteration reduces the value of f when $\alpha = 1/n$.

(b) [Rus95] Consider the case where f has the form

$$f(x) = \sum_{j=1}^{J} f_j\left(\sum_{i=1}^{n} a_{ji}x_i\right),$$

where a_{ji} are given scalars. Let $m = \max_{j=1,\ldots,J} m_j$, where m_j is the number of indices i for which a_{ji} is nonzero. Show that if x does not minimize f, the Jacobi iteration reduces the value of f when $\alpha = 1/m$.

1.8.3

Consider the simplex method applied to a strictly convex function f. Show that at each iteration, either $f(x_{max})$ decreases strictly, or else the number of vertices x^i of the simplex such that $f(x^i) = f(x_{max})$ decreases by at least one.

1.9 DISCRETE-TIME OPTIMAL CONTROL PROBLEMS*

In this section, we consider a class of optimization problems involving a discrete-time dynamic system, i.e., a vector difference equation. Such problems arise often in applications and are frequently challenging because of their large dimension. Continuous-time versions of these problems are the focus of the modern theory of the calculus of variations and the Pontryagin maximum principle, and give rise to some of the most fascinating mathematical problems of optimization. We will focus on some structural aspects of optimal control problems, which can be effectively exploited both in analysis and computation.

Let us consider the problem of finding sequences $u = (u_0, u_1, \ldots, u_{N-1})$ and $x = (x_1, x_2, \ldots, x_N)$, which minimize

$$g_N(x_N) + \sum_{i=0}^{N-1} g_i(x_i, u_i), \tag{1.213}$$

subject to the constraints

$$x_{i+1} = f_i(x_i, u_i), \qquad i = 0, \ldots, N-1, \tag{1.214}$$

$$x_0 : \text{ given},$$

$$x_i \in X_i \subset \Re^n, \qquad i = 1, \ldots, N, \tag{1.215}$$

$$u_i \in U_i \subset \Re^m, \qquad i = 0, \ldots, N-1. \tag{1.216}$$

We refer to u_i as the *control vectors*, and to $u = (u_0, u_1, \ldots, u_{N-1})$ as a *control trajectory*. We refer to x_i as the *state vectors*, and to the sequence $x = (x_0, x_1, \ldots, x_N)$ as a *state trajectory*. The equation $x_{i+1} = f_i(x_i, u_i)$ is called the *system equation*, and for a given initial state x_0, specifies uniquely the state trajectory, which corresponds to a given control trajectory. The functions $f_i : \Re^n \times \Re^m \mapsto \Re^n$ are given and they will be assumed once or twice differentiable for the most part of the following. The sets X_i and U_i specify constraints on the state and control vectors, respectively. They are usually represented by equations or inequalities. However, for the time being we will not specify them further. A more general constraint, that we will discuss on occasion has the form

$$(x_i, u_i) \in \Omega_i \subset \Re^n \times \Re^m, \qquad i = 0, \ldots, N-1.$$

The functions $g_N : \Re^n \mapsto \Re$ and $g_i : \Re^n \times \Re^m \mapsto \Re$, $i = 0, \ldots, N-1$, are given and for the most part, they will be assumed to be once or twice differentiable.

Example 1.9.1 (Reservoir Regulation)

Let x_i denote the volume of water held in a reservoir at the ith of N time periods. The volume x_i evolves according to

$$x_{i+1} = x_i - u_i, \qquad \forall\, i = 0, \ldots, N-1,$$

where u_i is water used for some productive purpose in period i. This is the system equation, with the volume x_i viewed as the state and the outflow u_i viewed as the control. There is a cost $G(x_N)$ for the terminal state being x_N and there is a cost $g_i(u_i)$ for outflow u_i at period i. For example, when u_i is used for electric power generation, $g_i(u_i)$ may be equal to minus the value of power produced from u_i. We want to choose the outflows $u_0, u_1, \ldots, u_{N-1}$ so as to minimize

$$G(x_N) + \sum_{i=0}^{N-1} g_i(u_i),$$

while observing some constraints on the volume x_i (e.g. x_i should lie between some upper and lower bounds) and some constraints on the outflow u_i (e.g. $u_i \geq 0$).

There are also multidimensional versions of the problem, involving several reservoirs that are interconnected in the sense that the outflow u_i from one becomes inflow to another (in addition to serving some other productive purpose). This leads to an optimal control formulation where the state and control vectors are multidimensional with dimension equal to the number of reservoirs.

Let us consider the case where there are no state constraints, i.e.,

$$X_i = \Re^n, \qquad i = 1, \ldots, N.$$

Then, one may reduce problem (1.213)-(1.216) (which is a constrained problem in x and u) to a problem which involves only the control variables $u_0, \ldots, u_{N-1}$. To see this, note that to any given control trajectory

$$u = (u_0, u_1, \ldots, u_{N-1}),$$

there corresponds a unique state trajectory via the system equation $x_{i+1} = f_i(x_i, u_i)$. We may write this correspondence abstractly as

$$x_i = \phi_i(u) = \phi_i(u_0, \ldots, u_{N-1}), \qquad i = 1, \ldots, N, \tag{1.217}$$

where ϕ_i are appropriate functions, determined by the functions f_i. (Actually in each of the functions ϕ_i, only the variables $u_0, \ldots, u_{i-1}$ enter explicitly. However, our notation is technically correct and will prove convenient.)

We may substitute x_i, using the functions ϕ_i of Eq. (1.217), into the cost function (1.213) and write the problem as

$$\begin{aligned} &\text{minimize} \ \ J(u) = g_N\big(\phi_N(u)\big) + \sum_{i=0}^{N-1} g_i\big(\phi_i(u), u_i\big) \\ &\text{subject to} \ \ u_i \in U_i, \qquad i = 0, \ldots, N-1. \end{aligned} \tag{1.218}$$

If the controls are also unconstrained, that is,

$$U_i = \Re^m, \qquad i = 0, \ldots, N-1,$$

problem (1.218) is an unconstrained minimization problem of the type that we have examined in this chapter. Thus, once we calculate the gradient and Hessian matrix of J, we can write explicitly necessary conditions and sufficient conditions for optimality.

We will calculate the derivatives of J in two different ways, both of which are valuable, since they provide complementary insights.

Calculation of $\nabla J(u)$ (1st Method)

Let us first assume that the cost function has the form

$$J(u) = g_N(x_N) = g_N\big(\phi_N(u)\big)$$

i.e., there is only a terminal state cost. As we shall see shortly, the general case can be reduced to this case.

We have for every $i = 0, \ldots, N-1$, using the chain rule,

$$\nabla_{u_i} J(u) = \nabla_{u_i} \phi_N \cdot \nabla g_N, \tag{1.219}$$

where the derivatives are calculated along the current control trajectory u and the corresponding state trajectory x, and where x and u are related by

$$x_i = \phi_i(u), \qquad i = 1, \ldots, N,$$

[cf. Eq. (1.217)]. From Eq. (1.219), we obtain, using again the chain rule,

$$\begin{aligned} \nabla_{u_i} J(u) &= \nabla_{u_i} x_{i+1} \cdot \nabla_{x_{i+1}} x_{i+1} \cdots \nabla_{x_{N-1}} x_N \cdot \nabla g_N \\ &= \nabla_{u_i} f_i \cdot \nabla_{x_{i+1}} f_{i+1} \cdots \nabla_{x_{N-1}} f_{N-1} \cdot \nabla g_N. \end{aligned}$$

By defining the, so called, *costate* vectors $p_i \in \Re^n$, $i = 1, \ldots, N$, via the equations

$$p_i = \nabla_{x_i} f_i \cdots \nabla_{x_{N-1}} f_{N-1} \cdot \nabla g_N, \qquad i = 1, \ldots, N-1, \tag{1.220}$$

$$p_N = \nabla g_N, \tag{1.221}$$

we obtain

$$\nabla_{u_i} J(u) = \nabla_{u_i} f_i \cdot p_{i+1}, \qquad i = 0, \ldots, N-1. \tag{1.222}$$

Note also that $p_1, \ldots, p_N$ are generated (backwards in time) by means of the following equation, known as the *adjoint equation*,

$$p_i = \nabla_{x_i} f_i \cdot p_{i+1}, \qquad i = 1, \ldots, N-1, \tag{1.223}$$

starting from

$$p_N = \nabla g_N. \tag{1.224}$$

Consider now the general case of the cost function

$$J(u) = g_N(x_N) + \sum_{i=0}^{N-1} g_i(x_i, u_i), \tag{1.225}$$

which involves the intermediate cost terms $g_i(x_i, u_i)$. We may reduce the corresponding problem to one which involves a terminal state cost only, by introducing an additional state variable y_i and a corresponding equation by

$$y_{i+1} = y_i + g_i(x_i, u_i),$$

$$y_0 = 0.$$

Defining a new state vector

$$\tilde{x}_i = \begin{pmatrix} x_i \\ y_i \end{pmatrix} \tag{1.226}$$

and a new system equation

$$\tilde{x}_{i+1} = \begin{pmatrix} f_i(x_i, u_i) \\ y_i + g_i(x_i, u_i) \end{pmatrix} \equiv \tilde{f}_i(\tilde{x}_i, u_i), \tag{1.227}$$

the cost function (1.225) takes the form

$$J(u) = g_N(x_N) + y_N \equiv \tilde{g}_N(\tilde{x}_N). \tag{1.228}$$

This cost function involves only a terminal state cost and is of the type examined previously. We have from Eqs. (1.222)-(1.224),

$$\nabla_{u_i} J(u) = \nabla_{u_i} \tilde{f}_i \cdot \tilde{p}_{i+1}, \qquad i = 0, \ldots, N-1, \tag{1.229}$$

where

$$\tilde{p}_i = \nabla_{\tilde{x}_i} \tilde{f}_i \cdot \tilde{p}_{i+1}, \qquad i = 1, \ldots, N-1,$$

$$\tilde{p}_N = \nabla \tilde{g}_N.$$

From Eqs. (1.226)-(1.228), we have by writing

$$\tilde{p}_i = \begin{pmatrix} p_i \\ z_i \end{pmatrix}, \qquad p_i \in \Re^n, \ z_i \in \Re,$$

the following form of the adjoint equation

$$\tilde{p}_N = \begin{pmatrix} p_N \\ z_N \end{pmatrix} = \begin{pmatrix} \nabla g_N \\ 1 \end{pmatrix},$$

$$\tilde{p}_i = \begin{pmatrix} p_i \\ z_i \end{pmatrix} = \begin{pmatrix} \nabla_{x_i} f_i & \nabla_{x_i} g_i \\ 0 & 1 \end{pmatrix} \begin{pmatrix} p_{i+1} \\ z_{i+1} \end{pmatrix}.$$

As a result, we have

$$z_i = 1, \qquad i = 1, \ldots, N,$$

and we obtain the final form of the adjoint equation

$$p_i = \nabla_{x_i} f_i \cdot p_{i+1} + \nabla_{x_i} g_i, \tag{1.230}$$

$$p_N = \nabla g_N. \tag{1.231}$$

Furthermore, from Eq. (1.229),

$$\nabla_{u_i} J(u) = \begin{pmatrix} \nabla_{u_i} f_i & \nabla_{u_i} g_i \end{pmatrix} \begin{pmatrix} p_{i+1} \\ z_{i+1} \end{pmatrix}$$

and finally

$$\nabla_{u_i} J(u) = \nabla_{u_i} f_i \cdot p_{i+1} + \nabla_{u_i} g_i. \tag{1.232}$$

Equation (1.232) yields the cost gradient in terms of the costate vectors obtained from the adjoint equation (1.230) and (1.231), with all derivatives evaluated along the control trajectory u and the corresponding state trajectory under consideration. Thus to compute the gradient $\nabla J(u)$:

(1) We calculate the state trajectory corresponding to u by forwards propagation of the system equation $x_{i+1} = f_i(x_i, u_i)$.

(2) We generate the costate vectors by backwards propagation of the adjoint equation Eqs. (1.230), starting from the terminal condition $p_N = \nabla g_N$. (All partial derivatives are evaluated at the current state and control trajectories.)

(3) We use Eq. (1.232) to compute the components $\nabla_{u_i} J(u)$ of the gradient $\nabla J(u)$.

Calculation of $\nabla J(u)$ and $\nabla^2 J(u)$ (2nd Method)

The alternative method for calculating the derivatives of J is based on writing all the system equations $x_{i+1} = f_i(x_i, u_i)$ compactly as

$$h\big(\phi(u), u\big) = 0. \tag{1.233}$$

Here, $u = (u_0, u_1, \ldots, u_{N-1})$ is the control trajectory and $\phi(u)$ is the corresponding state trajectory

$$x = \phi(u),$$

where the function ϕ maps $\Re^{Nm}$ into $\Re^{Nn}$ and is defined in terms of the functions ϕ_i of Eq. (1.217) as

$$\phi(u) = \begin{pmatrix} \phi_1(u) \\ \vdots \\ \phi_N(u) \end{pmatrix} = \begin{pmatrix} x_1 \\ \vdots \\ x_N \end{pmatrix}. \tag{1.234}$$

The equation $h\big(\phi(u), u\big) = 0$ is a compact representation of the equations

$$\begin{aligned} f_0(x_0, u_0) - x_1 &= 0, \\ f_1(x_1, u_1) - x_2 &= 0, \\ &\vdots \\ f_{N-1}(x_{N-1}, u_{N-1}) - x_N &= 0, \end{aligned} \tag{1.235}$$

and is satisfied by every control trajectory u.

Similarly the cost function may be written abstractly as

$$F\big(\phi(u), u\big) = g_N\big(\phi_N(u)\big) + \sum_{i=0}^{N-1} g_i\big(\phi_i(u), u_i\big), \tag{1.236}$$

and the optimal control problem becomes

$$\begin{aligned} &\text{minimize} \quad J(u) = F\big(\phi(u), u\big) \\ &\text{subject to} \quad u \in \Re^{Nm}. \end{aligned}$$

To obtain the derivatives of $J(u)$, we use the trick of writing for any vector $p \in \Re^{Nn}$,

$$J(u) = F\big(\phi(u), u\big) + h\big(\phi(u), u\big)' p,$$

[cf. Eq. (1.233)] and we select cleverly p so that the derivatives of $J(u)$ are easily calculated. We have

$$\begin{aligned} \nabla J(u) = &\nabla\phi(u)\big(\nabla_x F\big(\phi(u), u\big) + \nabla_x h\big(\phi(u), u\big) \cdot p\big) \\ &+ \nabla_u F\big(\phi(u), u\big) + \nabla_u h\big(\phi(u), u\big) \cdot p, \end{aligned} \tag{1.237}$$

where in the preceding equation, ∇_x denotes gradient with respect to the first argument $[x = \phi(u)]$, ∇_u denotes gradient with respect to the second argument u, and $\nabla \phi(u)$ is the $Nm \times Nn$ gradient matrix of ϕ evaluated at u, i.e.,

$$\nabla \phi(u) = \big[\nabla_u \phi_1(u) \,\vdots\, \nabla_u \phi_2(u) \,\vdots\, \cdots \,\vdots\, \nabla_u \phi_N(u)\big].$$

Equation (1.237) holds for every $p \in \Re^{Nn}$. Suppose that p is selected so that it satisfies the equation

$$\nabla_x F\big(\phi(u), u\big) + \nabla_x h\big(\phi(u), u\big) \cdot p(u) = 0, \tag{1.238}$$

where we denote by $p(u)$ the particular value of p satisfying the equation above. We will see shortly that Eq. (1.238) has a unique solution with respect to p, so our notation is justified. We have then from Eq. (1.236),

$$\nabla J(u) = \nabla_u F\big(\phi(u), u\big) + \nabla_u h\big(\phi(u), u\big) \cdot p(u). \tag{1.239}$$

For notational convenience, we introduce the function $L : \Re^{Nn} \times \Re^{Nm} \times \Re^{Nn} \mapsto \Re$ given by

$$L(x, u, p) = F(x, u) + h(x, u)'p. \tag{1.240}$$

We then have

$$\nabla J(u) = \nabla_u L\big(\phi(u), u, p(u)\big). \tag{1.241}$$

To compute the Hessian matrix of the cost, we write Eq. (1.237) as

$$\nabla J(u) = \nabla \phi(u) \nabla_x L\big(\phi(u), u, p\big) + \nabla_u L\big(\phi(u), u, p\big)$$

and we differentiate this equation with respect to u. By using also the equation

$$\nabla_x L\big(\phi(u), u, p(u)\big) = 0$$

[cf. Eq. (1.238)] we obtain for $p = p(u)$

$$\begin{aligned}\nabla^2 J(u) =& \nabla \phi(u) \nabla^2_{xx} L\big(\phi(u), u, p(u)\big) \nabla \phi(u)' + \nabla \phi(u) \nabla^2_{xu} L\big(\phi(u), u, p(u)\big) \\ &+ \nabla^2_{ux} L\big(\phi(u), u, p(u)\big) \nabla \phi(u)'.\end{aligned} \tag{1.242}$$

In order to calculate the derivatives of J, we must calculate the solution $p(u)$ of Eq. (1.238) and then evaluate the derivatives of L. We have:

$$L(x, u, p) = g_N(x_N) + \sum_{i=0}^{N-1} g_i(x_i, u_i) + \sum_{i=1}^{N} \big(f_{i-1}(x_{i-1}, u_{i-1}) - x_i\big)' p_i. \tag{1.243}$$

Equation (1.238) can be written as $\nabla_x L\big(\phi(u), u, p(u)\big) = 0$, or by using Eq. (1.243),

$$\begin{pmatrix} \nabla g_N \\ \nabla_{x_{N-1}} g_{N-1} \\ \nabla_{x_{N-2}} g_{N-2} \\ \vdots \\ \nabla_{x_1} g_1 \end{pmatrix} + \begin{pmatrix} -I & 0 & 0 & \cdots & 0 \\ \nabla_{x_{N-1}} f_{N-1} & -I & 0 & \cdots & 0 \\ 0 & \nabla_{x_{N-2}} f_{N-2} & -I & \cdots & 0 \\ \vdots & \vdots & \vdots & \cdots & \vdots \\ 0 & 0 & 0 & \cdots & -I \end{pmatrix} \begin{pmatrix} p_N \\ p_{N-1} \\ p_{N-2} \\ \vdots \\ p_1 \end{pmatrix} = 0.$$

By rewriting this equation, we have

$$p_i = \nabla_{x_i} f_i \cdot p_{i+1} + \nabla_{x_i} g_i, \qquad i = 1, \ldots, N-1, \tag{1.244}$$

$$p_N = \nabla g_N, \tag{1.245}$$

which is the same as the adjoint equation derived earlier [Eqs. (1.230) and (1.231)]. Thus, the adjoint equation specifies uniquely $p(u)$.

Also, from Eqs. (1.240), (1.241), and (1.243),

$$\nabla_{u_i} J(u) = \nabla_{u_i} f_i \cdot p_{i+1} + \nabla_{u_i} g_i, \qquad i = 0, \ldots, N-1, \tag{1.246}$$

which is the same equation for the gradient ∇J as the one obtained earlier.

Concerning the Hessian matrix $\nabla^2 J$, it is given by Eqs. (1.242) and (1.243). We only give here the form of the gradient matrix $\nabla \phi(u)$

$$\nabla \phi(u) = \begin{pmatrix} \nabla_{u_0} f_0 & \nabla_{x_1} f_1 \cdot \nabla_{u_0} f_0 & \cdots & \nabla_{x_{N-1}} f_{N-1} \cdots \nabla_{x_1} f_1 \cdot \nabla_{u_0} f_0 \\ 0 & \nabla_{u_1} f_1 & \cdots & \nabla_{x_{N-1}} f_{N-1} \cdots \nabla_{x_2} f_2 \cdot \nabla_{u_1} f_1 \\ \vdots & \vdots & \cdots & \vdots \\ 0 & 0 & \cdots & \nabla_{u_{N-1}} f_{N-1} \end{pmatrix}$$

The expression (1.242) for $\nabla^2 J$ is quite complex, but will be useful when we will discuss Newton's method later in this section.

Example 1.9.2 (Neural Networks)

Recall Example 1.5.3 in Section 5.1, which describes how the neural network training problem can be posed as a least squares problem. It can be seen that this problem can also be viewed as an optimal control problem, where the weight vector at each stage is the control vector for that stage. Mathematically, the problem is to find the control trajectory u that minimizes

$$J(u) = \sum_{j=1}^{m} J_j(u),$$

where

$$J_j(u) = \tfrac{1}{2}\|z_j - h(u, y_j)\|^2,$$

(z_j, y_j) is the jth input-output pair in the training set, and h is an appropriate function defined by the functional relationship between input and output of the neural network. Each gradient $\nabla J_j(u)$ can be calculated as described earlier via the corresponding adjoint equation, and the cost gradient is obtained as

$$\nabla J(u) = \sum_{j=1}^{m} \nabla J_j(u).$$

The process of calculating the gradient by means of the adjoint equation is sometimes called *back-propagation* and corresponding gradient-like methods are sometimes called *back-propagation methods*.

First Order Necessary Condition

The preceding expressions for the cost gradient can be used to write the first order necessary optimality condition

$$\nabla J(u^*) = 0,$$

for

$$u^* = (u_0^*, u_1^*, \ldots, u_{N-1}^*)$$

to be a local minimum of J. It is customary to write this condition in terms of the *Hamiltonian function*, defined for each i by

$$H_i(x_i, u_i, p_{i+1}) = g_i(x_i, u_i) + p_{i+1}' f_i(x_i, u_i). \tag{1.247}$$

By using the expression of $\nabla J(u^*)$ [cf. Eq. (1.244)-(1.246)] we obtain:

Proposition 1.9.1: Let $u^* = (u_0^*, u_1^*, \ldots, u_{N-1}^*)$ be a local minimum control trajectory and let $x^* = (x_1^*, x_2^*, \ldots, x_N^*)$ be the corresponding state trajectory. Then we have

$$\nabla_{u_i} H_i(x_i^*, u_i^*, p_{i+1}^*) = 0, \qquad i = 0, \ldots, N-1, \tag{1.248}$$

where the costate vectors $p_1^*, \ldots, p_N^*$ are obtained from the adjoint equation

$$p_i^* = \nabla_{x_i} H_i(x_i^*, u_i^*, p_{i+1}^*), \qquad i = 1, \ldots, N-1, \tag{1.249}$$

with the terminal condition

$$p_N^* = \nabla g_N(x_N^*). \tag{1.250}$$

Example 1.9.3 (Linear System and Quadratic Cost)

Consider the case where the system is linear

$$x_{i+1} = A_i x_i + B_i u_i, \qquad i = 0, \ldots, N-1 \tag{1.251}$$

and the cost function is quadratic of the form

$$J(u) = \frac{1}{2}\left\{ x_N' Q_N x_N + \sum_{i=0}^{N-1} (x_i' Q_i x_i + u_i' R_i u_i) \right\}. \tag{1.252}$$

The $n \times n$ matrices A_i and the $n \times m$ matrices B_i are given. The matrices Q_i are assumed symmetric and positive semidefinite, and the matrices R_i are assumed symmetric and positive definite. There are no constraints on the state or control vectors.

Let $u^* = (u_0^*, u_1^*, \ldots, u_{N-1}^*)$ be an optimal control trajectory and let $x^* = (x_1^*, x_2^*, \ldots, x_N^*)$ be the corresponding state trajectory. The adjoint equation is given by

$$p_i^* = A_i' p_{i+1}^* + Q_i x_i^*, \qquad i = 1, \ldots, N-1, \tag{1.253}$$

$$p_N^* = Q_N x_N^*. \tag{1.254}$$

It can be seen from Eq. (1.251), that the state trajectory x is linearly related to the control trajectory u, so the cost function $J(u)$ of Eq. (1.252) is a quadratic function of u. Because the matrices Q_i and R_i are assumed positive semidefinite and positive definite, respectively, $J(u)$ is a convex, positive definite, quadratic function. Therefore, the necessary optimality condition $\nabla J(u^*) = 0$ of Prop. 1.8.1 is also sufficient. It can be written as

$$\nabla_{u_i} H_i(x_i^*, u_i^*, p_{i+1}^*) = \nabla_{u_i} \left\{ p_{i+1}^*{}' B_i u_i^* + \tfrac{1}{2} u_i^*{}' R_i u_i^* + \tfrac{1}{2} x_i^*{}' Q_i x_i^* \right\} = 0$$

or equivalently

$$R_i u_i^* + B_i' p_{i+1}^* = 0, \qquad i = 0, \ldots, N-1,$$

from which

$$u_i^* = -R_i^{-1} B_i' p_{i+1}^*, \qquad i = 0, \ldots, N-1. \tag{1.255}$$

We will obtain a more convenient expression for u_i^* by verifying a relation of the form

$$p_{i+1}^* = K_{i+1} x_{i+1}^*, \tag{1.256}$$

where K_{i+1} is a positive semidefinite symmetric matrix, which can be explicitly calculated. We show this relation by induction, first noting that it holds for $i = N-1$ with

$$K_N = Q_N,$$

[cf. Eq. (1.254)]. Assume that it holds for some $i \leq N-1$. Then, by combining Eqs. (1.255) and (1.256), we have

$$u_i^* = -R_1^{-1} B_i' p_{i+1}^* = -R_i^{-1} B_i' K_{i+1} x_{i+1}^*. \tag{1.257}$$

With the substitution $x_{i+1}^* = A_i x_i^* + B_i u_i^*$, this equation yields

$$u_i^* = -R_i^{-1} B_i' K_{i+1} \big(A_i x_i^* + B_i u_i^* \big)$$

or equivalently, by solving for u_i^*,

$$u_i^* = -(R_i + B_i' K_{i+1} B_i)^{-1} B_i' K_{i+1} A_i x_i^*, \qquad i = 0, \ldots, N-1. \tag{1.258}$$

We thus obtain

$$x_{i+1}^* = A_i x_i^* + B_i u_i^* = A_i x_i^* - B_i (R_i + B_i' K_{i+1} B_i)^{-1} B_i' K_{i+1} A_i x_i^*.$$

Multiplying both sides with $A_i' K_{i+1}$ and using Eq. (1.256), we see that

$$A_i' p_{i+1}^* = A_i' K_{i+1} A_i x_i^* - A_i' K_{i+1} B_i (R_i + B_i' K_{i+1})^{-1} B_i' K_{i+1} B_i A_i x_i^*.$$

Adding $Q_i x_i^*$ to both sides, we obtain

$$p_i^* = A_i' p_{i+1}^* + Q_i x_i^* = K_i x_i^*,$$

where

$$K_i = A_i' \big(K_{i+1} - K_{i+1} B_i (R_i + B_i' K_{i+1} B_i)^{-1} B_i' K_{i+1} \big) A_i + Q_i, \tag{1.259}$$

thus completing the induction proof of Eq. (1.256).

Equation (1.259) is known as the *Riccati equation*. It generates the matrices $K_1, K_2, \ldots K_{N-1}$ from the terminal condition

$$K_N = Q_N.$$

Given K_1, we may obtain u_0^* using Eq. (1.258) and the given initial state x_0. Then, we may calculate x_1^* from the system equation (1.251) and the vector u_1^* from Eq. (1.258) using K_2. Similarly we may calculate $x_1^*, u_2^*, \ldots, u_{N-1}^*, x_N^*$.

Gradient Methods for Optimal Control

The application of gradient methods to unconstrained optimal control problems is straightforward in principle. For example the steepest descent method takes the form

$$u_i^{k+1} = u_i^k - \alpha^k \nabla_{u_i} H_i\left(x_i^k, u_i^k, p_{i+1}^k\right), \qquad i = 0, \ldots, N-1, \tag{1.260}$$

where H_i denotes the Hamiltonian function

$$H_i(x_i, u_i, p_{i+1}) = g_i(x_i, u_i) + p_{i+1}' f_i(x_i, u_i),$$

$u^k = \{u_0^k, u_1^k, \ldots, u_{N-1}^k\}$ is the kth control trajectory, $x^k = (x_0^k, x_1^k, \ldots, x_N^k)$ denotes the kth state trajectory, and $p^k = (p_1,^k, \ldots, p_N^k)$ denotes the kth costate trajectory

$$p_i^k = \nabla_{x_i} H_i\big(x_i^k, u_i^k, p_{i+1}^k\big), \qquad i = 1, \ldots, N-1,$$

$$p_N^k = \nabla g_N(x_N^k).$$

Thus, given u^k, one computes x^k by forward propagation of the system equation, and then p^k by backward propagation of the adjoint equation. Subsequently, the steepest descent iteration (1.260) is performed with the stepsize α^k chosen, for example, by the Armijo rule or some line minimization rule. An important point that guides the selection of the stepsize rule is that a gradient evaluation in optimal control problems with many control variables is usually not much more expensive than a function evaluation. Therefore, it is worth considering stepsize rules that aim at efficiency by using intelligently gradient calculations to save substantially on the number of function evaluations.

While steepest descent is simple, its convergence rate in optimal control problems is often very poor. This is particularly so when the underlying dynamic system tends to be unstable (see Exercise 1.9.1). For this reason scaling of the control variables is recommended if suitable scaling factors can be determined with reasonable effort. Otherwise, one should use either the conjugate gradient method or Newton's method.

Conjugate Gradient Method for Optimal Control

The application of the conjugate gradient method to unconstrained optimal control problems is also straightforward. The search direction is a linear combination of the current gradient and the preceding search direction, as discussed in Section 1.6. The performance of the method can often be improved through the use of preconditioning. One possibility is to use a diagonal approximation to the Hessian matrix as preconditioning matrix, but in some cases, far more effective preconditioning schemes are possible; see Exercise 1.9.2 and [Ber74].

Newton's Method for Optimal Control

We first note that the next iterate of the pure form of Newton's method

$$x^{k+1} = x^k - \left(\nabla^2 f(x^k)\right)^{-1} \nabla f(x^k) \tag{1.261}$$

can be obtained by minimizing the second order Taylor series expansion of f around x^k given by

$$f(x^k) + \nabla f(x^k)'(x - x^k) + \tfrac{1}{2}(x - x^k)'\nabla^2 f(x^k)(x - x^k).$$

Carrying the argument one step further, assume that $\tilde{f}^k : \Re^n \mapsto \Re$ is a *quadratic* function such that

$$\nabla \tilde{f}^k(x^k) = \nabla f(x^k), \qquad \nabla^2 \tilde{f}^k(x^k) = \nabla^2 f(x^k).$$

Then $\tilde{f}^k$ has the form

$$\tilde{f}^k(x) = c + \nabla f(x^k)'(x - x^k) + \tfrac{1}{2}(x - x^k)'\nabla^2 f(x^k)(x - x^k),$$

where c is some constant, and hence the next iterate x^{k+1} of Eq. (1.261) is obtained by minimizing the quadratic function $\tilde{f}^k$.

This viewpoint is particularly valuable in unconstrained optimal control problems, where it is impractical to evaluate explicitly the Hessian matrix $\nabla^2 J(u)$ and find the control trajectory u^{k+1} from

$$u^{k+1} = u^k - \left(\nabla^2 J(u^k)\right)^{-1} \nabla J(u^k).$$

Instead, given u^k, we formulate a *linear-quadratic* optimal control problem with cost function $\tilde{J}_k(u)$ such that

$$\nabla \tilde{J}_k(u^k) = \nabla J(u^k), \qquad \nabla^2 \tilde{J}_k(u^k) = \nabla^2 J(u^k). \tag{1.262}$$

The optimal solution u^{k+1} of the linear quadratic problem can be obtained conveniently via a Riccati equation (cf. Example 1.9.3) and according to our previous discussion, it represents the next iterate of the pure form of Newton's method as applied to minimization of $J(u)$.

The linear quadratic problem that corresponds to a control trajectory $u^k = (u_0^k, u_1^k, \ldots, u_{N-1}^k)$ and the associated states $x_0, x_1^k, \ldots, x_N^k$, and costates $p_1^k, p_2^k, \ldots, p_N^k$, involves a linearized version of the original system. It has the form

$$\begin{aligned} \text{minimize} \quad \tilde{J}_k(\delta u) = {} & \tfrac{1}{2}\delta x_N' Q_N \delta x_N + a_N' \delta x_N + \sum_{i=0}^{N-1} \tfrac{1}{2}\delta x_i' Q_i \delta x_i + a_i' \delta x_i \\ & + \sum_{i=0}^{N-1} \tfrac{1}{2}\delta u_i' R_i \delta u_i + b_i' \delta u_i + \sum_{i=1}^{N-1} \delta u_i' M_i \delta x_i \\ \text{subject to} \quad \delta x_0 = 0, \qquad & \delta x_{i+1} = A_i \delta x_i + B_i \delta u_i, \quad i = 1, \ldots, N-1, \end{aligned} \tag{1.263}$$

where

$$A_i = \nabla_{x_i} f_i', \qquad B_i = \nabla_{u_i} f_i', \qquad i = 0, \ldots, N-1,$$

$$Q_N = \nabla^2 g_N, \qquad a_N = \nabla g_N,$$

$$Q_i = \nabla^2_{x_i x_i} H_i, \qquad a_i = \nabla_{x_i} g_i, \qquad i = 1, \ldots, N-1,$$

$$R_i = \nabla^2_{u_i u_i} H_i, \qquad b_i = \nabla_{u_i} g_i, \qquad i = 0, \ldots, N-1,$$

$$M_i = \nabla^2_{u_i x_i} H_i, \qquad i = 0, \ldots, N-1,$$

$$H_i(x_i, u_i, p_{i+1}^k) = g_i(x_i, u_i) + p_{i+1}^k{}' f_i(x_i, u_i). \tag{1.264}$$

The partial derivatives appearing above are evaluated along the current control and state trajectories

$$u^k = (u_0^k, u_1^k, \ldots, u_{N-1}^k), \qquad x^k = (x_0, x_1^k, \ldots, x_N^k), \tag{1.265}$$

and the corresponding costate vectors $p_1^k, p_2^k, \ldots, p_N^k$. The problem (1.263) involves the variations

$$\delta x_i = x_i - \tilde{x}_i^k$$

$$\delta u_i = u_i - u_i^k$$

from the nominal trajectories $u^k, \tilde{x}^k$, where $\tilde{x}^k$ is the trajectory corresponding to the linearized system

$$\tilde{x}_{i+1}^k = A_i \tilde{x}_i^k + B_i u_i^k, \qquad i = 0, \ldots, N-1 \tag{1.266}$$

$$\tilde{x}_0^k = x_0 : \text{given initial state.}$$

It is straightforward to verify that the cost function $J(u)$ for our original problem has a gradient $\nabla J(u^k)$ and Hessian matrix $\nabla^2 J(u^k)$ evaluated at u^k, which are equal to the corresponding gradient and Hessian matrix $\nabla \tilde{J}_k(\delta u^k)$, $\nabla^2 \tilde{J}_k(\delta u^k)$ of the problem (1.263)-(1.264). This can be shown by comparing the expressions for the gradient and Hessian of the two problems given earlier [cf. Eqs. (1.241)-(1.243)]. Hence, according to our earlier discussion, the solution of problem (1.263)-(1.264) yields the next iterate of Newton's method.

We now turn to the linear-quadratic problem (1.263)-(1.264). It is possible to show using the necessary conditions for optimality, similarly as in Example 1.9.3, that the solution of this problem (provided it exists and is unique) may be obtained in feedback form as follows:

$$\delta u_i^k = -(R_i + B_i' K_{i+1} B_i)^{-1} \big((M_i + B_i' K_{i+1} A_i) \delta x_i^k + b_i + B_i' \lambda_{i+1} \big), \qquad i = 0, \ldots, N-1, \tag{1.267}$$

where

$$\delta x_0^k = 0,$$

$$\delta x_{i+1}^k = A_i \delta x_i^k + B_i \delta u_i^k, \qquad i = 0, \ldots, N-1, \tag{1.268}$$

the matrices $K_1, K_2, \ldots, K_N$ are given recursively by the equation

$$K_N = Q_N$$

$$\begin{aligned} K_i = A_i' K_{i+1} A_i + Q_i - (B_i' K_{i+1} A_i + M_i)' (R_i + B_i' K_{i+1} B_i)^{-1} \\ (B_i' K_{i+1} A_i + M_i), \qquad i = 1, \ldots, N-1, \end{aligned} \tag{1.269}$$

and the vectors $\lambda_1, \lambda_2, \ldots, \lambda_N$ are given recursively by

$$\lambda_N = a_N,$$

$$\lambda_i = a_i + A_i' \lambda_{i+1} - A_i' K_{i+1} B_i (R_i + B_i' K_{i+1} B_i)^{-1} (b_i + B_i' \lambda_{i+1}). \tag{1.270}$$

The next iterate of the pure form of Newton's method is obtained from Eq. (1.267) and if a stepsize α^k is introduced, then we obtain the iteration

$$u_i^{k+1} = u_i^k + \alpha^k \delta u_i^k, \qquad i = 0, \ldots, N-1, \tag{1.271}$$

with

$$\delta u_i^k = -(R_i + B_i' K_{i+1} B_i)^{-1} \big((M_i + B_i' K_{i+1} A_i) \delta x_i^k + b_i + B_i' \lambda_{i+1} \big). \tag{1.272}$$

The computations of the typical iteration of Newton's method are carried out in the following sequence.

kth Iteration of Newton's Method

Step 1: Given the current control trajectory u^k the corresponding state trajectory x^k is calculated.

Step 2: The solution $K_1, \ldots, K_N$ of Eq. (1.269) and the solution $\lambda_1, \ldots, \lambda_N$ of Eq. (1.270) are calculated (backwards, from the terminal conditions) together with all the necessary data for calculation of the optimal control in equation (1.267). This is done by calculating simultaneously the costate vectors $p_1^k, \ldots, p_N^k$, which are needed to evaluate the second derivatives of the Hamiltonian (i.e., Q_i, R_i, M_i).

Step 3: The Newton direction δu^k is generated as follows:

1) Calculate δu_0^k using Eq. (1.272),

$$\delta u_0^k = -(R_0 + B_0' K_1 B_0)^{-1} (b_0 + B_0' \lambda_1).$$

2) Calculate δx_1^k using Eq. (1.268) and the fact $\delta x_0^k = 0$.

3) Calculate δu_1^k using δx_1^k and Eq. (1.272).

4) Continue similarly to compute δu_i^k and δx_i^k for all i.

Step 4: A stepsize α^k is obtained using some stepsize rule; the next control trajectory is

$$u_i^{k+1} = u_i^k + \alpha^k \delta u_i^k, \qquad i = 0, \ldots, N-1.$$

In the computation of the solution of the Riccati equation it is necessary that $(R_i + B_i' K_{i+1} B_i)^{-1}$ exist and be positive definite. If any of these inverses is not positive definite, it can be seen that $\nabla^2 J(u^k)$ is not positive definite and hence the Newton iteration must be modified. Under these circumstances one should replace R_i by $(R_i + E_i)$ where E_i is a positive definite diagonal matrix such that $(R_i + E_i + B_i' K_{i+1} B_i)$ is positive definite, as discussed in Section 1.4. In a neighborhood of a local minimum u^* with $\nabla^2 J(u^*) > 0$, the solution of the Riccati equation exists and the Newton iteration can be carried out as described.

It is interesting to note that our analysis yields as a byproduct a sufficient condition for optimality of a control trajectory u^*. Clearly u^* will be optimal if the corresponding linear-quadratic problem (1.263) with all derivatives evaluated at u^* has $\delta u = 0$ as an optimal solution. This can be guaranteed when

$$\nabla_{u_i} H_i = 0, \qquad i = 0, \ldots, N-1, \quad \text{(1st order condition)},$$

$$R^* + B_i^{*\prime} K_{i+1}^* B_i^* > 0, \qquad i = 0, \ldots, N-1, \quad \text{(2nd order condition)},$$

where

$$R_i^* = \nabla^2_{u_i u_i} H_i, \qquad B_i^* = \nabla_{u_i} f_i', \qquad i = 0, \ldots, N-1,$$

K_{i+1}^* is given by the Riccati equation (1.269) corresponding to u^*, and all derivatives above are evaluated along u^* and the corresponding state and costate trajectories.

Finally, we mention a variant of Newton's method that offers some computational advantages. It uses in place of Eqs. (1.271) and (1.272) the following two equations:

$$\begin{aligned} u_i^{k+1} = u_i^k - \alpha^k (R_i + B_i' K_{i+1} B_i)^{-1} \big((M_i + B_i' K_{i-1} A_i)(x_i^{k+1} - x_i^k) \\ + \beta_i + B_i' \lambda_{i+1} \big), \end{aligned} \tag{1.273}$$

where

$$x_i^{k+1} - x_i^k = f_{i-1}(x_{i-1}^{k+1}, u_{i-1}^{k+1}) - f_{i-1}(x_{i-1}^k, u_{i-1}^k). \tag{1.274}$$

In other words, the first order variation δx_i^k, which is obtained from the linearized system equation (1.268), is replaced in Eq. (1.272) by the actual variation $\left(x_i^{k+1} - x_i^k\right)$ of Eq. (1.274). Since $\left(x_i^{k+1} - x_i^k\right)$ and δx_i^k differ by terms which are of second or higher order, the two methods are asymptotically equivalent. The advantage of using Eqs. (1.273) and (1.274) is that the matrices A_i, B_i need not be stored (or recalculated) for use in Eq. (1.268). In addition, the trajectory x^{k+1} corresponding to u^{k+1} may be used in the next iteration, if the pure form of Newton's method is used or the stepsize α^k turns out to be equal to one.

EXERCISES

1.9.1

Calculate the condition number of the optimal control problem involving the quadratic cost function $x_N^2 + \sum_{i=0}^{N-1} u_i^2$ and the scalar system $x_{i+1} = ax_i + u_i$, where $a > 1$. Show that it tends to ∞ as $N \to \infty$.

1.9.2 [Ber74]

Consider the special case of the linear quadratic problem of Eqs. (1.251) and (1.252) where $Q_i = 0$ for all $i = 0, \ldots, N-1$. Derive a preconditioned conjugate gradient method that converges in at most $n+1$ steps. *Hint*: Consider the structure of the Hessian matrix and use the result of Exercise 1.6.10 of Section 1.6.

1.9.3

Consider a discrete-time optimal control problem, which us similar to the one of this section, except that the system equation has the form

$$x_{i+1} = f_i(x_{i+1}, x_i, u_i),$$

and the continuously differentiable function f_i is such that the above equation can be solved uniquely for x_{i+1} in terms of x_i and u_i. Assume also that when linearized along any control trajectory and corresponding state trajectory, the above equation can also be solved uniquely for δx_{i+1} in terms of δx_i and δu_i. Derive analogs of the steepest descent and the conjugate gradient methods.

1.10 SOME PRACTICAL GUIDELINES

Practical nonlinear programming problems can be very challenging. They require an iterative process that can be slow and may in the end lead to just a local minimum. It is thus useful to follow a few basic practical guidelines. Keep in mind, however, that in nonlinear programming, just as there is no foolproof method, there is also no foolproof advice.

Problem Formulation

In the real world, optimization problems seldom come neatly formulated as mathematical problems. Thus, usually the first step is to select the cost function and to delineate the constraints. One potential difficulty here is that there may be multiple and possibly competing objectives; for example in an engineering design problem one may simultaneously want to minimize cost and maximize efficiency. In this case one usually encodes all objectives except one in the constraints and embodies the remaining objective in the cost function. However, the division between objective function and constraints may be arbitrary and one may have to reconsider the formulation after evaluating the results of the optimization. It is important here to properly correlate the problem formulation with the objectives of the investigation. For example if only an approximate answer is required, there is no point in constructing a highly detailed model.

The most critical task in problem formulation is to capture the realism of the practical context while obtaining an analytically or computationally tractable model. This is where experience and insight into both the application and the methodology is very important. In particular, it is important to be able to recognize the models that are easily solvable and those that are essentially impossible to solve. Two rules of thumb here are:

(a) Convex cost and constraint sets are preferable to nonconvex ones.

(b) Linear programming problems are easier than nonlinear programming problems, which are in turn easier than integer/discrete combinatorial problems.

While (a) is generally true because of the lack of local minima and other spurious stationary points in convex problems, there are many exceptions to (b). For example there is a widespread practice of replacing nonlinear functions by piecewise linear ones that lead to linear programming formulations, so that linear programming codes can be used. This may work out fine, but unfortunately it may also lead to vastly increased dimensionality and loss of insight (solutions to continuous problems often have elegant features that are lost in discrete approximations). Another example along similar lines arises when essentially combinatorial problems are formulated as highly nonconvex nonlinear programming problems whose local minima correspond to the feasible solutions of the original problem. For a somewhat absurd example, think of discarding an integrality constraint on a variable x_i and adding to the cost function the (differentiable) term $c\big(x_i - n(x_i)\big)^2$ where $n(x_i)$ is the nearest integer to x_i and c is a very large penalty parameter. Generally, when a convex problem formulation seems impossible and the presence of many local minima is an important concern, one may wish to bring to bear the methods of global optimization, for which we refer to the specialized literature (see e.g., [FlP95], [Flo95], [HPT95], [PaR87]).

We note, however, that depending on the practical situation, there is often merit in replacing integer variables with continuous variables and using a nonlinear programming formulation to obtain a noninteger solution that can then be rounded to integer. For example, consider a variable such as number of telephone lines to establish between two points in a communication network. This number is naturally discrete but if its range is in the hundreds, little will be lost by replacing it by a continuous variable. On the other hand, other variables may naturally be discrete-valued, such as a $\{0, 1\}$-valued variable modeling whether route A or route B is used to send a message in a communication network. A continuous approximation of such a variable may be meaningless.

Another important consideration is whether the cost function is or is not differentiable. The most powerful methods in nonlinear programming (e.g. the ones discussed in this chapter) require a once or twice differentiable cost function. This motivates the smoothing of whatever natural nondifferentiabilities may exist in the cost. An important special case is when nondifferentiable terms of the form $\max\big\{f_1(x), f_2(x)\big\}$ are replaced in the cost function and/or the constraints by a smooth approximation (see Fig. 1.10.1 and Exercise 6.4.6 in Section 6.4). On the other hand there are situations, particularly arising in the context of duality where nondifferentiable cost functions are unavoidable and the use of a nondifferentiable optimization method is naturally suited to the character of the problem (see Chapter 6). Note that the implementation of gradient methods can be enhanced with the use of automatic differentiation programs that compute first and second derivatives from user-supplied programs that compute only function values; see [Gri89], [Gil92].

Scaling

When selecting the optimization variables in a given problem it is important to pay attention to their natural relative magnitudes, and scale them so that their values are neither too large nor too small. This is useful to control roundoff error and it is also often helpful for improving the condition number of the problem, and the natural convergence rate of steepest descent and other algorithms. However, one should also pay attention to the range of the variables as well, and use a translation to somewhere in the middle of that range. In particular, suppose that the optimization variable x_i is expected to take values in the range $[\alpha_i, \beta_i]$. Then it is typically helpful to

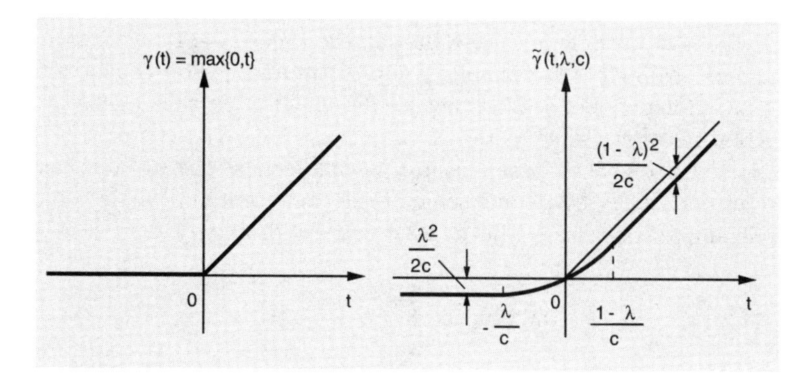

Figure 1.10.1. Smoothing of nondifferentiable terms. The function $f : \Re^n \to \Re$ given by

$$f(x) = \max\big\{f_1(x), f_2(x)\big\} = f_1(x) + \max\big\{0, f_2(x) - f_1(x)\big\}$$

is replaced by

$$f_1(x) + \tilde{\gamma}\big(f_2(x) - f_1(x), \lambda, c\big),$$

where $\tilde{\gamma}(t, \lambda, c)$ is a differentiable function that approximates $\gamma(t) = \max\{0, t\}$. This function depends on a parameter $\lambda \in [0, 1]$ and a parameter $c > 0$. It has the form

$$\tilde{\gamma}(t, \lambda, c) = \begin{cases} t - \dfrac{(1-\lambda)^2}{2c} & \text{if } \frac{1-\lambda}{c} \le t \\ \lambda t + \dfrac{c}{2} t^2 & \text{if } -\frac{\lambda}{c} \le t \le \frac{1-\lambda}{c} \\ -\dfrac{\lambda^2}{2c} & \text{if } t \le -\frac{\lambda}{c} \end{cases}$$

shown in the figure. The accuracy of the approximation increases as c increases. The scalar λ plays the role of a Lagrange multiplier; see Exercise 5.4.6 in Section 5.4. This smoothing method can be generalized for the case

$$f(x) = \max\big\{f_1(x), \ldots, f_m(x)\big\};$$

see [Ber75c], [Ber77], [Geo77], [Ber82a], [Pap81], [Pol79] for further analysis. An alternative smooth approximation of f is the function

$$\frac{1}{c} \ln \left\{ \sum_{i=1}^{m} \lambda_i e^{c f_i(x)} \right\}.$$

where $c > 0$ and λ_i are positive numbers with $\sum_{i=1}^m \lambda_i = 1$ (see [Ber82a], Section 5.1.3).

use the transformation of variables

$$y_i = \frac{x_i - (\alpha_i + \beta_i)/2}{(\beta_i - \alpha_i)/2},$$

so that the new variable y_i takes values in the range $[-1, 1]$. Note that this preliminary scaling of the variables complements but is not a substitute for the iteration-dependent scaling that we discussed in Section 1.3 and which is inherent in Newton-like methods.

Some least squares problems are special because the variables can be scaled automatically by scaling some of the coefficients in the data blocks. As an example consider the linear least squares problem

$$\begin{aligned} &\text{minimize} \quad \sum_{i=1}^{m} (y_i - a_i' x)^2 \\ &\text{subject to} \quad x \in \Re^n \end{aligned}$$

where y_i are given scalars and a_i are given vectors in $\Re^n$ with coordinates a_{ij}, $j = 1, \ldots, n$. The jth diagonal element of the Hessian matrix is

$$d_j = \sum_{i=1}^{m} a_{ij}^2.$$

Thus reasonable scaling of the variables is often obtained by multiplying for each j, all coefficients a_{ij} with a common scalar s_j so that they all lie in the range $[-1, 1]$.

Method Selection

An important consideration here is whether a one-time solution of a given problem is sought or whether the problem (with data variations) will be solved many times on a production basis. In the former case any method that will do the job is sufficient, so one should aim for speed of code development, perhaps through the use of an existing optimization package. In the latter case, efficiency and accuracy of solution become important, so careful experimentation with a variety of methods may be appropriate in order to obtain a method that solves the problem fast and exploits its special structure.

Throughout this chapter we emphasized the fast convergence rate of Newton-like methods when derivatives are available and the problem is nonsingular. It is worth repeating, however, that for singular and difficult problems, simpler steepest descent-like methods with simple scaling may be more effective. Generally, in lack of a clear choice it is best to start with a simple method that can be easily coded and understood, and be prepared to proceed with more sophisticated choices.

Once a type of method is selected, it must be tuned to the problem at hand. In particular, stepsize rule parameters and termination criteria must be chosen. This often requires trial and error, particularly for first order methods, for a diminishing or constant stepsize, and for difficult (e.g., singular) problems. Termination criteria should be scale-free to the extent possible. For example it is preferable to use as termination tests $\|x^k - x^{k-1}\| \leq \epsilon \|x^k\|$ or $\|\nabla f(x^k)\| \leq \epsilon \|\nabla f(x^0)\|$ rather than $\|x^k - x^{k-1}\| \leq \epsilon$ or $\|\nabla f(x^k)\| \leq \epsilon$, respectively.

Validation

Finally one must strive to be convinced that the answers obtained from the computation are reasonable approximations to a (global) minimum. There are no systematic methods for this, but a few heuristics tailored to the problem at hand are adequate in most cases. In particular, suppose that some special case or variation of the given problem has a known optimal solution. Then it is reassuring if the coded method succeeds in finding this solution. Similarly, if a lower bound to the optimal value is known, it is reassuring when the algorithm comes close to achieving this lower bound. For example, in least squares problems, a small cost function value is a strong indication of success.

Another frequently used technique is to start the algorithm from widely varying starting points and see if it will give a better answer. This is particularly recommended in the case of multiple local minima. Finally, one may vary the parameters of the algorithm and also try alternative algorithms to see if substantial improvements can be obtained.

1.11 NOTES AND SOURCES

Section 1.2: The steepest descent method dates to Cauchy [Cau47], who attributes to Newton the unity stepsize version of what we call Newton's method. The modern theory of gradient methods evolved in the 60's, when on one hand, several practical stepsize rules were proposed starting with the work of Goldstein [Gol62], [Gol64], and on the other hand, general methods of convergence analysis were formulated starting with the work of Zangwill [Zan69], which was followed by the works of Ortega and Rheinboldt [OrR70], Daniel [Dan71], and Polak [Pol71]. The Armijo stepsize rule [Arm66] is the most popular of a broad variety of rules that enforce descent without requiring a full line minimization. Stepsize rules that do not enforce strict descent at each iteration, while aiming at faster convergence, are discussed in Grippo, Lampariello, and Lucidi [GLL91], and also in Barzilai and Borwein [BaB88], and Raydan [Ray93]. The capture theorem (Prop. 1.2.5) was formulated and proved by the author for the case of a nonsingular local minimum in [Ber82a], and was extended to the form given here by Dunn [Dun93c]. Gradient methods with errors are discussed in Poljak [Pol87], and Bertsekas and Tsitsiklis [BeT96], [BeT99]. Parallel and asynchronous versions of these methods converge under very weak conditions; see Tsitsiklis, Bertsekas, and Athans [TBA86], and Bertsekas and Tsitsiklis [BeT89].

Section 1.3: For further discussion on various measures of rate of convergence, see Ortega and Rheinboldt [OrR70], Bertsekas [Ber82a], and Barzilai and Dempster [BaD93]. The convergence rate of steepest descent with line minimization was analyzed by Kantorovich [Kan45]. The case of a constant stepsize was analyzed in Goldstein [Gol64], and Levitin and Poljak [LeP65]. For analysis of the convergence rate of steepest descent for singular problems see Dunn [Dun81], [Dun87], and Poljak [Pol87].

Section 1.4: The local convergence of Newton's method was analyzed by Kantorovich [Kan49]. For an analysis of the case where the method converges to a singular point, see Decker and Kelley [DeK80], Decker, Keller, and Kelley [DKK83], and Hughes and Dunn [HuD84]. The modification to extend the region of convergence of Newton's method by modifying the Cholesky factorization is described in Gill and Murray [GiM74]; see also Gill, Murray, and Wright [GMW81]. The use of a trust region has been discussed by Moré and Sorensen [MoS83]. Extensive accounts of various aspects of Newton-like methods are given in Goldstein [Gol67], Gill, Murray, and Wright [GMW81], Dennis and Schnabel [DeS83], and Nazareth [Naz94]. The truncated Newton method is discussed in Dembo, Eisenstadt, and Steihaug [DES82], Nash [Nas85], and Nash and Sofer [NaS89].

Section 1.5: Incremental gradient methods for linear least squares problems are attributed to Widrow and Hoff [WiH59]. For recent analyses, see Gaivoronski [Gai94], Grippo [Gri94], Luo and Tseng [LuT94a], Mangasarian and Solodov [MaS94], Tseng [Tse98], Solodov [Sol98], and Bertsekas and Tsitsiklis [BeT96], [BeT99]. Proposition 1.5.1 is due to Luo [Luo91] and stems from an earlier result of Kohonen [Koh74]. The incremental Gauss-Newton method was proposed in Davidon [Dav76] on an empirical basis. The development given here in terms of the extended Kalman filter and the corresponding convergence analysis is due to Bertsekas [Ber94]. In the case where $\lambda < 1$, and for some x^* we have $g(x^*) = 0$ and $\nabla g(x^*)$: nonsingular, the method has been shown to converge to x^* linearly with convergence ratio λ when started sufficiently close to x^* (Pappas [Pap82]). For a description of methods for neural network training problems, see Bishop [Bis95] and Haykin [Hay98]. The ill-conditioned character of these problems is discussed in Saarinen, Bramley, and Cybenko [SBC93].

Section 1.6: Conjugate direction methods are due to Hestenes and Stiefel [HeS52]. The convergence rate of the conjugate gradient method is discussed in Fadeev and Fadeeva [FaF63], and also in Luenberger [Lue84], which we follow in our discussion. The conjugate gradient method with the formula (1.172) is known as the Poljak-Polak-Ribiere method and its convergence was analyzed by Poljak [Pol69a], and Polak and Ribiere [PoR69]. An extensive discussion of conjugate direction methods can be found in the monograph by Hestenes [Hes80].

Section 1.7: The DFP method is due to Davidon [Dav59]. It was popularized primarily through the work of Fletcher and Powell [FlP63]. The BFGS method is independently due to Broyden [Bro70], Fletcher [Fle70a], Goldfarb [Gol70], and Shanno [Sha70]. For surveys and extensive discussions of quasi-Newton methods, including convergence analysis for a nonquadratic cost function, see Dennis and Moré [DeM77], Dennis and Schnabel [DeS83], Nazareth [Naz94], and Polak [Pol97].

Section 1.8: For further material on direct search methods, see Gill and Murray [GiM74], and Avriel [Avr76].

Section 1.9: An extensive reference on optimality conditions for discrete-time optimal control is Canon, Cullum, and Polak [CCP70]. For computational methods see Polak [Pol71], [Pol97], and also the survey by Polak [Pol73]. The implementation of Newton's method given here is due to Dunn and Bertsekas [DuB89], and was also independently derived by Pantoja and Mayne [PaM89] (see Mitter [Mit66] for a continuous-time version of this algorithm). An alternative approach to discrete-time optimal control is based on dynamic programming; see e.g. Bertsekas [Ber95a]. Discrete-time optimal control problems often arise from discretization of their continuous-time counterparts. The associated issues are discussed in Cullum [Cul71] and Polak [Pol97].

Section 1.10: A great deal of material on the implementation of nonlinear programming methods is given by Gill, Murray, and Wright [GMW81]. A description of available software packages is provided by Moré and Wright [MoW93].

2

Optimization over a Convex Set

Contents

In this chapter we consider the constrained optimization problem

$$\begin{aligned}&\text{minimize } f(x)\\&\text{subject to } x \in X,\end{aligned}$$

where we assume throughout that:

(a) X is a nonempty and convex subset of $\Re^n$. When dealing with algorithms, we assume in addition that X is closed.

(b) The function $f : \Re^n \mapsto \Re$ is continuously differentiable over X.

This problem generalizes the unconstrained optimization problem of the preceding chapter, where $X = \Re^n$. We will see that the main algorithmic ideas for solving the unconstrained and the constrained problems are quite similar.

Usually the set X has structure specified by equations and inequalities. If we take into account this structure, some new ideas, based on Lagrange multipliers and duality theory, come into play. These ideas will not be discussed in the present chapter, but they will be the focus of subsequent chapters.

The problem of this chapter also contains as a special case the general linear programming problem. This is the case where f is linear and X is a polyhedron. This chapter focuses primarily on the general case where f is nonlinear. However, there is quite a bit of algorithmic material that relates to linear programming in this chapter, particularly Sections 2.5 (manifold suboptimization methods) and 2.6 (affine scaling methods). Also, Sections B.3 and B.4 of Appendix B contain a fairly detailed treatment of polyhedral convexity, including a discussion of polyhedral cones, extreme points of polyhedra, and proofs of some of the fundamental theorems of linear programming.

Similar to the preceding chapter, the methods of this chapter are based on iterative descent along suitable directions. However, these directions must have the additional property that they maintain feasibility of the iterates. Such directions are called *feasible*, and as will be seen in Sections 2.2-2.7, they are usually obtained by solving certain optimization subproblems. As a general rule, these direction finding subproblems tend to be simpler when the constraint set X is a polyhedron, and for this reason the methods of this chapter tend to be most suitable for linear constraints.

2.1 OPTIMALITY CONDITIONS

In this section we consider the main necessary and sufficient optimality conditions for our problem, and we provide some examples of their application.

Local and Global Minima

Throughout this book, a vector satisfying the constraints of a given problem will be called a *feasible* vector for that problem. A vector $x^* \in X$ is a *local minimum* of f over the set X if it is no worse than its feasible neighbors; that is, if there exists an $\epsilon > 0$ such that

$$f(x^*) \le f(x), \qquad \forall\, x \in X \text{ with } \|x - x^*\| < \epsilon.$$

A vector $x^* \in X$ is a *global minimum* of f over the set X if it is no worse than all other feasible vectors, that is,

$$f(x^*) \le f(x), \qquad \forall\, x \in X.$$

The local or global minimum x^* is said to be *strict* if the corresponding inequality above is strict for $x \neq x^*$. Figure 2.1.1 illustrates these definitions. Local and global maxima of f over X are similarly defined; they are the local and global minima of the function $-f$ over X.

If in addition to X, the cost function f is also convex, a local minimum is also global. In particular, we have the following proposition, which is proved as Prop. B.10 in Appendix B.

Proposition 2.1.1: If f is a convex function, then a local minimum of f over X is a global minimum. If in addition f is strictly convex over X, then there exists at most one global minimum of f over X.

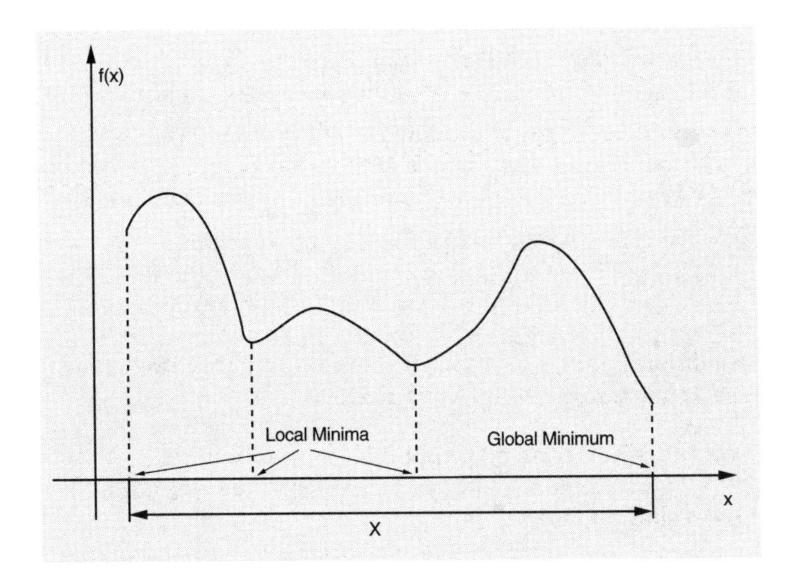

Figure 2.1.1. Local and global minima of f over X.

Necessary and Sufficient Conditions for Optimality

As in unconstrained optimization, we expect that at a local minimum x^*, the first order variation $\nabla f(x^*)'\Delta x$ due to a small feasible variation Δx is nonnegative. Because X is convex, the feasible variations are of the form $\Delta x = x - x^*$, where $x \in X$. Thus the condition $\nabla f(x^*)'\Delta x \geq 0$ translates to the necessary condition $\nabla f(x^*)'(x - x^*) \geq 0$ for all $x \in X$. The following proposition formally proves this condition and also shows that it is a sufficient condition for optimality when f is convex.

Proposition 2.1.2: (Optimality Condition)

(a) If x^* is a local minimum of f over X, then

$$\nabla f(x^*)'(x - x^*) \geq 0, \qquad \forall\, x \in X. \tag{2.1}$$

(b) If f is convex over X, then the condition of part (a) is also sufficient for x^* to minimize f over X.

Proof: (a) Suppose that $\nabla f(x^*)'(x - x^*) < 0$ for some $x \in X$. By the mean value theorem (Prop. A.22 of Appendix A), for every $\epsilon > 0$ there exists an $s \in [0, 1]$ such that

$$f\big(x^* + \epsilon(x - x^*)\big) = f(x^*) + \epsilon \nabla f\big(x^* + s\epsilon(x - x^*)\big)'(x - x^*).$$

[Here the mean value theorem is applied to the function $g(\epsilon) = f\big(x^* + \epsilon(x - x^*)\big)$.] Since ∇f is continuous, we have for all sufficiently small $\epsilon > 0$, $\nabla f\big(x^* + s\epsilon(x - x^*)\big)'(x - x^*) < 0$ and it follows that $f\big(x^* + \epsilon(x - x^*)\big) < f(x^*)$. The vector $x^* + \epsilon(x - x^*)$ is feasible for all $\epsilon \in [0, 1]$ because X is convex, so we obtain a contradiction of the local optimality of x^*.

(b) Using the convexity of f and Prop. B.3 of Appendix B, we have

$$f(x) \geq f(x^*) + \nabla f(x^*)'(x - x^*), \forall\, x \in X.$$

If the condition $\nabla f(x^*)'(x - x^*) \geq 0$ holds for all $x \in X$, we obtain $f(x) \geq f(x^*)$, so x^* minimizes f over X. **Q.E.D.**

Figure 2.1.2 interprets the optimality condition (2.1) geometrically. Figure 2.1.3 shows how the condition can fail in the absence of convexity of the constraint set X.

A vector x^* satisfying the optimality condition (2.1) is referred to as a *stationary point*. In the absence of convexity of f, this condition may also be satisfied by local maxima and other points. To see this, note that

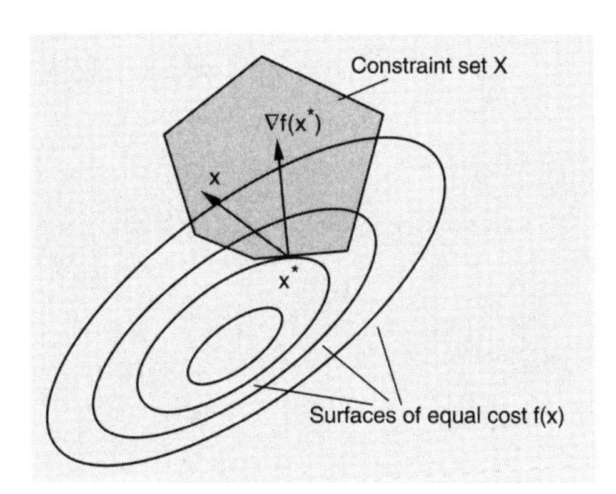

Figure 2.1.2. Geometric interpretation of the optimality condition of Prop. 2.1.2. At a local minimum x^*, the gradient $\nabla f(x^*)$ makes an angle less than or equal to 90 degrees with all feasible variations $x - x^*$, $x \in X$.

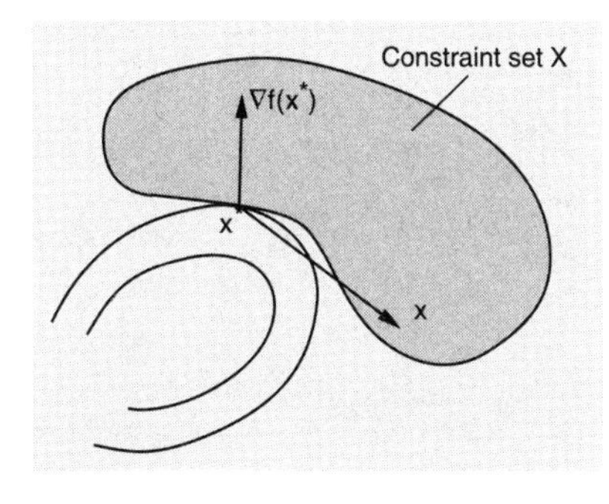

Figure 2.1.3. Illustration of how the necessary optimality condition may fail when X is not convex. Here x^* is a local minimum but we have

$$\nabla f(x^*)'(x - x^*) < 0$$

for the feasible vector x shown.

if $X = \Re^n$ or if x^* is an interior point of X, the condition (2.1) reduces to the stationarity condition $\nabla f(x^*) = 0$ of unconstrained optimization. We now illustrate the condition (2.1) in some examples.

Example 2.1.1 (Optimization Subject to Bounds)

Consider a positive orthant constraint

$$X = \{x \mid x \geq 0\}.$$

Then the necessary condition (2.1) for $x^* = (x_1^*, \ldots, x_n^*)$ to be a local minimum is

$$\sum_{i=1}^{n} \frac{\partial f(x^*)}{\partial x_i}(x_i - x_i^*) \geq 0, \qquad \forall\, x_i \geq 0,\ i = 1, \ldots, n. \tag{2.2}$$

Let us fix i. By letting $x_j = x_j^*$ for $j \neq i$ and $x_i = x_i^* + 1$ in Eq. (2.2), we obtain

$$\frac{\partial f(x^*)}{\partial x_i} \geq 0, \qquad \forall\, i. \tag{2.3}$$

Furthermore, if we have $x_i^* > 0$, by letting $x_j = x_j^*$ for $j \neq i$ and $x_i = \frac{1}{2}x_i^*$ in Eq. (2.2), we obtain $\partial f(x^*)/\partial x_i \leq 0$, which when combined with Eq. (2.3) yields

$$\frac{\partial f(x^*)}{\partial x_i} = 0, \qquad \text{if } x_i^* > 0. \tag{2.4}$$

The necessary conditions of Eqs. (2.3) and (2.4) are illustrated in Fig. 2.1.4.

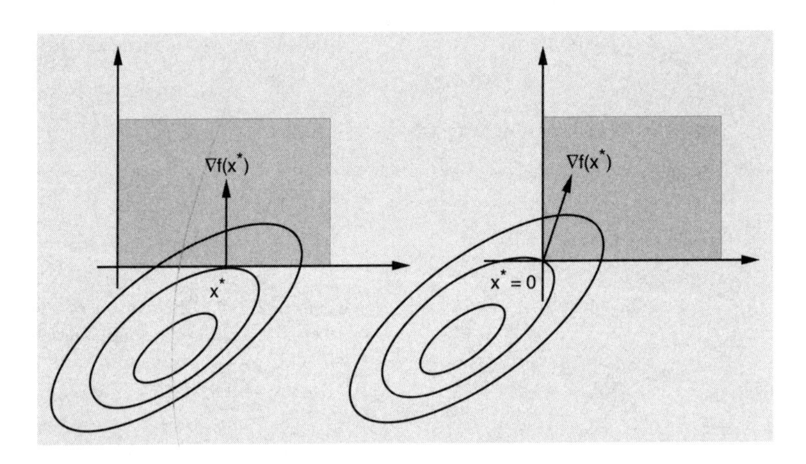

Figure 2.1.4. Illustration of the optimality condition for an orthant constraint. At a minimum x^*, all the partial derivatives $\partial f(x^*)/\partial x_i$ are nonnegative, and they are zero for the inactive constraint indices, i.e. the indices i with $x_i^* > 0$. If all constraints are inactive, the optimality condition reduces to the unconstrained optimization condition $\nabla f(x^*) = 0$.

Consider also the case where the constraints are upper and lower bounds on the variables, that is,

$$X = \{x \mid \alpha_i \leq x_i \leq \beta_i,\ i = 1, \ldots, n\}, \tag{2.5}$$

where α_i and β_i are given scalars. We leave it as an exercise for the reader to verify that if x^* is a local minimum, then

$$\frac{\partial f(x^*)}{\partial x_i} \geq 0, \qquad \text{if } x_i^* = \alpha_i, \tag{2.6}$$

$$\frac{\partial f(x^*)}{\partial x_i} \leq 0, \qquad \text{if } x_i^* = \beta_i, \tag{2.7}$$

$$\frac{\partial f(x^*)}{\partial x_i} = 0, \qquad \text{if } \alpha_i < x_i^* < \beta_i. \tag{2.8}$$

Finally, suppose that f is convex. Then, it can be seen that the conditions (2.3) and (2.4) imply the condition (2.2), which, by Prop. 2.1.2(b), is sufficient for optimality of x^* in the case of nonnegativity constraints. Similarly, the conditions (2.6)-(2.8) are sufficient for optimality in the case of a convex f, and the upper and lower bound constraints of Eq. (2.5).

Example 2.1.2 (Optimization Over a Simplex)

Consider the case where the constraint set is a simplex

$$X = \left\{ x \;\middle|\; x \geq 0,\ \sum_{i=1}^{n} x_i = r \right\},$$

where $r > 0$ is a given scalar. Then the necessary condition (2.1) for $x^* = (x_1^*, \ldots, x_n^*)$ to be a local minimum is

$$\sum_{i=1}^{n} \frac{\partial f(x^*)}{\partial x_i}(x_i - x_i^*) \geq 0, \qquad \forall\ x_i \geq 0 \text{ with } \sum_{i=1}^{n} x_i = r. \tag{2.9}$$

Fix an index i for which $x_i^* > 0$ and let j be any other index. By using the feasible vector $x = (x_1, \ldots, x_n)$ with $x_i = 0$, $x_j = x_j^* + x_i^*$, and $x_m = x_m^*$ for all $m \neq i, j$ in Eq. (2.9), we obtain

$$\left(\frac{\partial f(x^*)}{\partial x_j} - \frac{\partial f(x^*)}{\partial x_i} \right) x_i^* \geq 0,$$

or equivalently

$$x_i^* > 0 \quad \Longrightarrow \quad \frac{\partial f(x^*)}{\partial x_i} \leq \frac{\partial f(x^*)}{\partial x_j}, \qquad \forall\ j. \tag{2.10}$$

Thus, all coordinates which are positive at the optimum must have minimal (and equal) partial cost derivatives.

Assuming that f is convex, we can show that Eq. (2.10) is also sufficient for global optimality of x^*. Indeed, suppose that x^* is feasible and satisfies Eq. (2.10), and let

$$\Delta = \min_{i=1,\ldots,n} \frac{\partial f(x^*)}{\partial x_i}.$$

For every $x \in X$, we have $\sum_{i=1}^{n}(x_i - x_i^*) = 0$, so that

$$0 = \sum_{i=1}^{n} \Delta(x_i - x_i^*) \leq \sum_{\{i \mid x_i > x_i^*\}} \frac{\partial f(x^*)}{\partial x_i}(x_i - x_i^*) + \sum_{\{i \mid x_i < x_i^*\}} \Delta(x_i - x_i^*).$$

If i is such that $x_i < x_i^*$, we must have $x_i^* > 0$ and, by condition (2.10), $\Delta = \partial f(x^*)/\partial x_i$. Thus Δ can be replaced by $\partial f(x^*)/\partial x_i$ in the right-hand side of the preceding inequality, thereby yielding Eq. (2.9), which by Prop. 2.1.2(b), implies that x^* is optimal. Thus, when f is convex, condition (2.10) is necessary and sufficient for optimality of x^*.

There is a straightforward generalization of this example to the case where X is a Cartesian product of several simplices. Then, there is a separate condition of the form (2.10) for each simplex; that is, the condition

$$\frac{\partial f(x^*)}{\partial x_i} \leq \frac{\partial f(x^*)}{\partial x_j}$$

holds for all i with $x_i^* > 0$ and all j for which x_j is constrained by the same simplex as x_i.

Example 2.1.3 (Optimal Routing in a Communication Network)

We are given a directed graph, which is viewed as a model of a data communication network. We are also given a set W of ordered node pairs $w = (i, j)$. The nodes i and j are referred to as the *origin* and the *destination* of w, respectively, and w is referred to as an OD pair. For each w, we are given a scalar r_w referred to as the *input traffic* of w. In the context of routing of data in a communication network, r_w (measured in data units/second) is the arrival rate of traffic entering and exiting the network at the origin and the destination of w, respectively. The routing objective is to divide each r_w among the many paths from origin to destination in a way that the resulting total arc flow pattern minimizes a suitable cost function. We denote:

P_w: A given set of paths that start at the origin and end at the destination of w. All arcs on each of these paths are oriented in the direction from the origin to the destination.

x_p: The portion of r_w assigned to path p, also called the *flow of path* p.

The collection of all path flows $\{x_p \mid p \in P_w,\ w \in W\}$ must satisfy the constraints

$$\sum_{p \in P_w} x_p = r_w, \qquad \forall\ w \in W, \tag{2.11}$$

$$x_p \geq 0, \qquad \forall\ p \in P_w,\ w \in W, \tag{2.12}$$

as shown in Fig. 2.1.5. The total flow F_{ij} of arc (i, j) is the sum of all path flows traversing the arc:

$$F_{ij} = \sum_{\substack{\text{all paths } p \\ \text{containing } (i,j)}} x_p. \tag{2.13}$$

Consider a cost function of the form

$$\sum_{(i,j)} D_{ij}(F_{ij}). \tag{2.14}$$

The problem is to find a set of path flows $\{x_p\}$ that minimize this cost function subject to the constraints of Eqs. (2.11)-(2.13). We assume that D_{ij} is a convex and continuously differentiable function of F_{ij} with first derivative denoted by D'_{ij}. In data routing applications, the form of D_{ij} is often based on a queueing model of average delay (see [BeG92]).

The preceding problem is known as a *multicommodity network flow problem*. The terminology reflects the fact that the arc flows consist of several different commodities; in the present example, the different commodities are the data of the distinct OD pairs.

By expressing the total flows F_{ij} in terms of the path flows in the cost function (2.14) [using Eq. (2.13)], the problem can be formulated in terms of

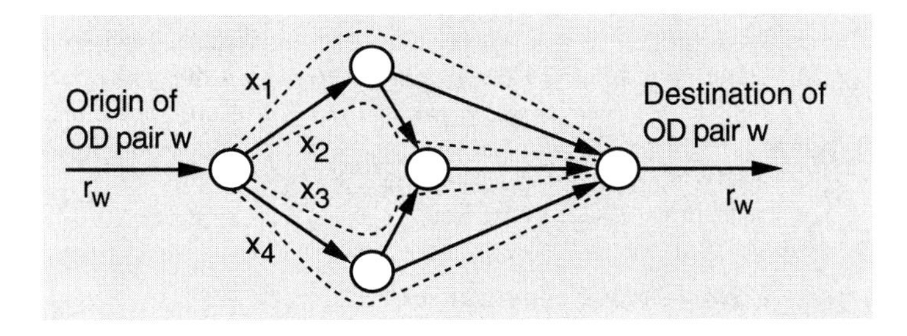

Figure 2.1.5. Constraints for the path flows of an OD pair w. The path flows x_p of the paths $p \in P_w$, should be nonnegative and add up to the given traffic input r_w of the OD pair.

the path flow variables $\{x_p \mid p \in P_w,\ w \in W\}$ as

$$\begin{aligned} &\text{minimize} \quad D(x) \\ &\text{subject to} \quad \sum_{p \in P_w} x_p = r_w, \quad \forall\, w \in W, \\ &\qquad\qquad\quad x_p \geq 0, \quad \forall\, p \in P_w,\ w \in W, \end{aligned}$$

where

$$D(x) = \sum_{(i,j)} D_{ij}\left(\sum_{\substack{\text{all paths } p \\ \text{containing } (i,j)}} x_p \right)$$

and x is the vector of path flows x_p.

This problem, viewed as a problem in the variables $\{x_p\}$, has a convex and differentiable cost function and a constraint set that is a Cartesian product of simplices. Therefore, we can apply the condition (2.10) developed in the preceding example, separately, for the path flows of each OD pair. In particular, we will show that optimal routing directs traffic exclusively along paths that are shortest with respect to arc lengths that depend on the flows carried by the arcs. To this end, we first derive the partial derivative of D with respect to x_p. It is given by

$$\frac{\partial D(x)}{\partial x_p} = \sum_{\substack{\text{all arcs } (i,j) \\ \text{on path } p}} D'_{ij}(F_{ij}), \tag{2.15}$$

where the first derivatives D'_{ij} are evaluated at the arc flows F_{ij} corresponding to x. From this equation, it is seen that $\partial D/\partial x_p$ *is the length of path* p *when the length of each arc* (i,j) *is taken to be the first derivative* D'_{ij} *evaluated at* F_{ij}. Consequently, in what follows, $\partial D/\partial x_p$ is called the *first derivative length of path* p.

By using the necessary and sufficient optimality condition (2.10) of the preceding example, we obtain for all $w \in W$ and $p \in P_w$

$$x_p^* > 0 \quad \Longrightarrow \quad \frac{\partial D(x^*)}{\partial x_p} \leq \frac{\partial D(x^*)}{\partial x_{p'}}, \qquad \forall\, p' \in P_w. \tag{2.16}$$

In words, the above condition says that *a set of path flows is optimal if and only if path flow is positive only on paths with a minimum first derivative length.* The condition also implies that at an optimum, the paths along which the input flow r_w of OD pair w is split must have *equal* length (and less than or equal length to that of all other paths of w).

Example 2.1.4 (Traffic Assignment)

We are given a directed graph, which is viewed as a model of a transportation network. The arcs of the graph represent transportation links such as highways, rail lines, etc. The nodes of the graph represent junction points where traffic can exit from one transportation link and enter another. Similar to the preceding example, we are given a set W of OD pairs. For OD pair $w = (i, j)$, there is a known input $r_w > 0$ representing traffic entering the network at the origin node i of w and exiting the network at the destination node j of w. The input r_w is to be divided among a set P_w of paths starting at the origin node of w and ending at the destination node of w. Let x_p denote the portion of r_w carried by path p and let x be the vector having as coordinates all the path flows x_p, $p \in P_w$, $w \in W$. Thus, the constraint set is the Cartesian product of simplices defined by Eqs. (2.11) and (2.12) of the preceding example.

For each arc (i, j), we are given a function $T_{ij}(F_{ij})$ of the total traffic F_{ij} carried by arc (i, j) [cf. Eq. (2.13)]. This function models the time required for traffic to travel from the start node to the end node of the arc (i, j). An important problem is to find a path flow vector $x^* \in X$ that consists of path flows that are positive only on paths of minimum travel time. That is, for all $w \in W$ and paths $p \in P_w$, we require that

$$x_p^* > 0 \quad \Longrightarrow \quad t_p(x^*) \leq t_{p'}(x^*), \qquad \forall\, p' \in P_w,\ \forall\, w \in W, \tag{2.17}$$

where $t_p(x)$, the travel time of path p, is defined as the sum of the travel times of the arcs of the path,

$$t_p(x) = \sum_{\substack{\text{all arcs } (i,j) \\ \text{on path } p}} T_{ij}(F_{ij}), \qquad \forall\, p \in P_w,\ \forall\, w \in W.$$

The preceding problem draws its validity from a hypothesis, called the *user-optimization principle*, which asserts that traffic equilibrium is established when each user of the network chooses, among all available paths, a path requiring minimum travel time. Thus, assuming that the user-optimization principle holds, a path flow vector x^* that solves the problem also models accurately the distribution of traffic through the network, and can be used for traffic projections when planning modifications to the network.

We now observe that the minimum travel time hypothesis (2.17) is identical with the optimality condition (2.16) if we identify the arc travel time $T_{ij}(F_{ij})$ with the cost derivative $D'_{ij}(F_{ij})$ [cf. Eq. (2.15)]. It follows that we can solve the transportation problem by converting it to the optimal routing problem of the preceding example with the identification

$$D_{ij}(F_{ij}) = \int_0^{F_{ij}} T_{ij}(\xi)d\xi. \tag{2.18}$$

If we assume that T_{ij} is continuous and monotonically nondecreasing, as is natural in a transportation context, it is straightforward to show that the function D_{ij} as defined by Eq. (2.18) is convex with derivative D'_{ij} equal to T_{ij}, so a minimum first derivative length path is a path of minimum travel time.

Projection on a Convex Set

Let z be a fixed vector in $\Re^n$ and consider the problem of finding a vector x^* in a *closed* convex set X, which is at a minimum distance from z; that is,

$$\begin{aligned} &\text{minimize} \quad f(x) = \|z - x\|^2 \\ &\text{subject to} \quad x \in X. \end{aligned}$$

We call this the problem of *projection of z on X*. This problem has important applications in mathematical analysis and plays also an important role in nonlinear programming algorithms. The main facts are summarized in the following proposition.

Proposition 2.1.3: (Projection Theorem) Let X be a nonempty, closed, and convex subset of $\Re^n$.

(a) For every $z \in \Re^n$, there exists a unique $x^* \in X$ that minimizes $\|z - x\|$ over all $x \in X$. This vector is called the *projection* of z on X and is denoted by $[z]^+$.

(b) Given some $z \in \Re^n$, a vector $x^* \in X$ is equal to the projection $[z]^+$ if and only if

$$(z - x^*)'(x - x^*) \leq 0, \qquad \forall\, x \in X.$$

(See Fig. 2.1.6.)

(c) The mapping $f : \Re^n \mapsto X$ defined by $f(x) = [x]^+$ is continuous and nonexpansive, that is,

$$\|[x]^+ - [y]^+\| \leq \|x - y\|, \qquad \forall\, x, y \in \Re^n.$$

(d) In the case where X is a subspace, a vector $x^* \in X$ is equal to the projection $[z]^+$ if and only if $z - x^*$ is orthogonal to X, that is,

$$(z - x^*)'x = 0, \qquad \forall\, x \in X.$$

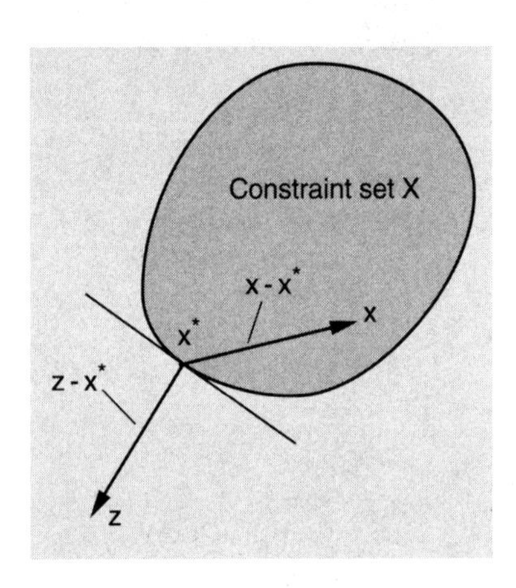

Figure 2.1.6. Illustration of the necessary and sufficient condition for x^* to be the unique projection of z on the closed convex set X. The angle between $z - x^*$ and $x - x^*$ should be greater or equal to 90 degrees for all $x \in X$.

Proof: Parts (a) and (c) are proved in Prop. B.11 of Appendix B. Part (b) is also proved in Prop. B.11 and is equivalent to the necessary and sufficient condition of Prop. 2.1.2, specialized to the projection problem. To prove part (d), note that since X is a subspace, the vectors $x^* + x$ and $x^* - x$ belong to X for all $x \in X$. Applying the condition of part (b) with these vectors replacing x, we obtain $(z - x^*)'x \le 0$ and $(z - x^*)'(-x) \le 0$, which imply the desired result. **Q.E.D.**

Example 2.1.5 (Equality-Constrained Quadratic Programming)

Consider the quadratic programming problem

$$\begin{aligned} &\text{minimize} \quad \tfrac{1}{2}\|x\|^2 + c'x \\ &\text{subject to} \quad Ax = 0, \end{aligned} \tag{2.19}$$

where c is a given vector in $\Re^n$ and A is an $m \times n$ matrix of rank m.

By adding the constant term $\frac{1}{2}\|c\|^2$ to the cost function, we can equivalently write this problem as

$$\begin{aligned} &\text{minimize} \quad \tfrac{1}{2}\|c + x\|^2 \\ &\text{subject to} \quad Ax = 0, \end{aligned}$$

which is the problem of projecting the vector $-c$ on the subspace $X = \{x \mid Ax = 0\}$. By Prop. 2.1.3(d), a vector x^* such that $Ax^* = 0$ is the unique projection if and only if

$$(c + x^*)'x = 0, \qquad \forall\, x \text{ with } Ax = 0.$$

It can be seen that the vector

$$x^* = -\big(I - A'(AA')^{-1}A\big)c \tag{2.20}$$

satisfies this condition and is thus the unique solution of the quadratic programming problem (2.19). [The matrix AA' is invertible because A has rank m (Prop. A.20 in Appendix A).]

Note that x^* is zero if and only if c is orthogonal to the subspace $X = \{x \mid Ax = 0\}$, or equivalently from Eq. (2.20),

$$c = A'(AA')^{-1}Ac.$$

Thus, in this case, c can be expressed in the form $A'\mu$, for some $\mu \in \Re^m$, i.e., as a linear combination of the rows of A. Conversely, it is seen that every vector of the form $A'\mu$, with $\mu \in \Re^m$, is orthogonal to the subspace X.

The conclusion is that *the orthogonal complement $X^\perp$ of X (the set of all vectors orthogonal to all x with $Ax = 0$), is the subspace $\{x \mid x = A'\mu,\ \mu \in \Re^m\}$.* Our proof of this fact assumed that A has rank m but can be easily modified for the case where the rows of A are linearly dependent by eliminating a sufficient number of dependent rows of A. Also a more general version of the preceding analysis, involving cones rather than subspaces, is given in Prop. B.16 of Appendix B.

Consider now the more general quadratic program

$$\begin{aligned} &\text{minimize} \ \ \tfrac{1}{2}(x - \bar{x})'H(x - \bar{x}) + c'(x - \bar{x}) \\ &\text{subject to} \ \ Ax = b, \end{aligned} \tag{2.21}$$

where c and A are as before, H is a positive definite symmetric matrix, b is a given vector in $\Re^m$, and $\bar{x}$ is a given vector in $\Re^n$, which is feasible, that is, satisfies $A\bar{x} = b$. By introducing the vector $y = H^{1/2}(x - \bar{x})$, we can write this problem as

$$\begin{aligned} &\text{minimize} \ \ \tfrac{1}{2}\|y\|^2 + \big(H^{-1/2}c\big)'y \\ &\text{subject to} \ \ AH^{-1/2}y = 0. \end{aligned}$$

Using Eq. (2.20) we see that the solution of this problem is

$$y^* = -\left(I - H^{-1/2}A'\left(AH^{-1}A'\right)^{-1}AH^{-1/2}\right)H^{-1/2}c$$

and by passing to the x-coordinate system through the transformation $x^* - \bar{x} = H^{-1/2}y^*$, we obtain the optimal solution

$$x^* = \bar{x} - H^{-1}(c - A'\lambda), \tag{2.22}$$

where the vector λ is given by

$$\lambda = \left(AH^{-1}A'\right)^{-1} AH^{-1}c. \tag{2.23}$$

The quadratic program (2.21) contains as a special case the program

$$\begin{aligned} &\text{minimize} \ \ \tfrac{1}{2}x'Hx + c'x \\ &\text{subject to} \ \ Ax = b. \end{aligned} \tag{2.24}$$

This special case is obtained when $\bar{x}$ is given by

$$\bar{x} = H^{-1}A'(AH^{-1}A')^{-1}b. \tag{2.25}$$

Indeed $\bar{x}$ as given above satisfies $A\bar{x} = b$ as required, and for all x with $Ax = b$, we have

$$x'H\bar{x} = x'A'(AH^{-1}A')^{-1}b = b'(AH^{-1}A')^{-1}b,$$

which implies that for all x with $Ax = b$

$$\tfrac{1}{2}(x-\bar{x})'H(x-\bar{x})+c'(x-\bar{x}) = \tfrac{1}{2}x'Hx+c'x+\left(\tfrac{1}{2}\bar{x}'H\bar{x}-c'\bar{x}-b'(AH^{-1}A')^{-1}b\right).$$

The last term in parentheses on the right-hand side above is constant, thus establishing that the programs (2.21) and (2.24) have the same optimal solution when $\bar{x}$ is given by Eq. (2.25). By combining Eqs. (2.22) and (2.25), we obtain the optimal solution of program (2.24):

$$x^* = -H^{-1}\left(c - A'\lambda - A'(AH^{-1}A')^{-1}b\right),$$

where λ is given by Eq. (2.23).

EXERCISES

2.1.1 (Heron's Problem)

In three-dimensional space, consider a two-dimensional plane and two points z_1 and z_2 lying outside the plane. Use the optimality condition of Prop. 2.1.2 to characterize the vector x^*, which minimizes $\|z_1 - x\| + \|z_2 - x\|$ over all x in the plane.

2.1.2 (Euclid's Problem)

Among all parallelograms that are contained in a given triangle, find the ones that have maximal area. Show that the maximal area is equal to half the area of the triangle. *Hint*: First argue that an optimal solution can be found among the parallelograms that share one angle with the triangle. Then, reduce the problem to maximizing xy subject to $x + y = \alpha$, $x \geq 0$, $y \geq 0$, where α is a constant.

2.1.3 (Kepler's Planimetric Problem)

Among all rectangles contained in a given circle show that the one that has maximal area is a square.

2.1.4 (Tartaglia's Problem)

Divide the number 8 into two nonnegative parts x and y so as to maximize $xy(x - y)$. *Answer*: $x = 4 + 4/\sqrt{3}$, $y = 4 - 4/\sqrt{3}$.

2.1.5

Among all pyramids whose base is a given triangle and whose height is given, show that the one that has minimal surface area (sum of areas of the triangular sides) has the following property: its vertex that lies outside the triangle projects on the point in the triangle that is at equal distance from all the sides. *Hint*: First argue that the projection point must lie within the triangle. Let x_1, x_2, x_3 be the distances to the sides of the triangle and let a_1, a_2, a_3 be the lengths of the corresponding sides. Express the surface of the pyramid in terms of x_1, x_2, x_3, and show that $\sum_{i=1}^{3} a_i x_i$ is constant. Formulate the problem as a problem of minimization over a simplex.

2.1.6

Consider the problem

$$\begin{aligned} &\text{maximize} \quad x_1^{a_1} x_2^{a_2} \cdots x_n^{a_n} \\ &\text{subject to} \quad \sum_{i=1}^{n} x_i = 1, \qquad x_i \geq 0, \quad i = 1, \ldots, n, \end{aligned}$$

where a_i are given positive scalars. Find a global maximum and show that it is unique.

2.1.7

Let Q be a positive definite symmetric matrix. State and prove a generalization of the projection theorem that involves the cost function $(z - x)'Q(z - x)$ in place of $\|z - x\|^2$. *Hint*: Use a transformation of variables.

2.1.8

Consider a closed convex set X on a plane and a point z outside the plane whose projection on the plane is $\hat{z}$. Characterize the projection of z on X in terms of the position of $\hat{z}$ relative to X. Work out the details of the case where X is a triangle.

2.1.9 (www)

Verify the necessary optimality conditions of Eqs. (2.6)-(2.8). Show that, assuming f is convex, they are also sufficient for optimality of x^*.

2.1.10 (Second Order Necessary Optimality Condition) (www)

Show that if x^* is a local minimum of the twice continuously differentiable function $f : \Re^n \mapsto \Re$ over the convex set X, then

$$(x - x^*)' \nabla^2 f(x^*)(x - x^*) \geq 0$$

for all $x \in X$ such that $\nabla f(x^*)'(x - x^*) = 0$.

2.1.11 (Second Order Sufficiency Conditions) (www)

Show that a vector $x^* \in X$ is a local minimum of a twice continuously differentiable function $f : \Re^n \mapsto \Re$ over the convex set X if

$$\nabla f(x^*)'(x - x^*) \geq 0, \qquad \forall\, x \in X,$$

and one of the following three conditions holds:

(1) X is polyhedral and we have

$$(x - x^*)' \nabla^2 f(x^*)(x - x^*) > 0$$

for all $x \in X$ satisfying $x \neq x^*$ and $\nabla f(x^*)'(x - x^*) = 0$.

(2) We have $p' \nabla^2 f(x^*) p > 0$ for all nonzero p that are in the closure of the set $\{d \mid d = \alpha(x - x^*),\ x \in X,\ \alpha \geq 0\}$ and satisfy $\nabla f(x^*)' p = 0$.

(3) For some $\gamma > 0$, we have

$$(x - x^*)' \nabla^2 f(x^*)(x - x^*) \geq \gamma \|x - x^*\|^2, \qquad \forall\, x \in X.$$

Give an example showing that the polyhedral assumption is needed in (1) above.

2.1.12 (Projection on a Simplex)

(a) Develop an algorithm to find the projection of a vector z on a simplex. This algorithm should be almost as simple as a closed form solution.

(b) Modify the algorithm of part (a) so that it finds the minimum of the cost function

$$f(x) = \sum_{i=1}^{n} \left(\alpha_i x_i + \tfrac{1}{2}\beta_i x_i^2\right)$$

over a simplex, where α_i and β_i are given scalars with $\beta_i > 0$. Consider also the case where β_i may be zero for some indices i.

2.1.13

Let $a_1, \ldots, a_m$ be given vectors in $\Re^n$, and consider the problem of minimizing $\sum_{j=1}^m \|x - a_j\|^2$ over a convex set X. Show that this problem is equivalent to the problem of projecting on X the center of gravity $\frac{1}{m}\sum_{j=1}^m a_j$.

2.1.14 (Optimal Control – Minimum Principle)

Consider the problem of finding sequences $u = (u_0, u_1, \ldots, u_{N-1})$ and $x = (x_1, x_2, \ldots, x_N)$ that minimize

$$g_N(x_N) + \sum_{i=0}^{N-1} g_i(x_i, u_i),$$

subject to the constraints

$$x_{i+1} = f_i(x_i, u_i), \quad i = 0, \ldots, N-1, \qquad x_0 : \text{ given},$$

$$u_i \in U_i \subset \Re^m, \qquad i = 0, \ldots, N-1.$$

All functions g_i and f_i are assumed continuously differentiable, and the sets U_i are assumed convex. Show that if u^* and x^* are optimal, then

$$\nabla_{u_i} H_i\left(x_i^*, u_i^*, p_{i+1}^*\right)'(u_i - u_i^*) \geq 0, \qquad \forall\ u_i \in U_i,\ i = 0, \ldots, N-1,$$

where

$$H_i(x_i, u_i, p_{i+1}) = g_i(x_i, u_i) + p_{i+1}' f_i(x_i, u_i)$$

is the Hamiltonian function, and the vectors $p_1^*, \ldots, p_N^*$ are obtained from the adjoint equation

$$p_i^* = \nabla_{x_i} H_i\left(x_i^*, u_i^*, p_{i+1}^*\right), \qquad i = 1, \ldots, N-1,$$

with the terminal condition

$$p_N^* = \nabla g_N(x_N^*).$$

If, in addition, the Hamiltonian H_i is a convex function of u_i for any fixed x_i and p_{i+1}, we have

$$u_i^* = \arg\min_{u_i \in U_i} H_i\big(x_i^*, u_i, p_{i+1}^*\big), \qquad \forall\, i = 0, \ldots, N-1.$$

(The last condition is known as the *minimum principle.*) *Hint*: Apply Prop. 2.1.2 and use the results of Section 1.9.

2.1.15

A farmer annually producing x_i units of a certain crop stores $(1 - u_i)x_i$ units of his production, where $0 \leq u_i \leq 1$, and invests the remaining $u_i x_i$ units, thus increasing the next year's production to a level x_{i+1} given by

$$x_{i+1} = x_i + w u_i x_i, \qquad i = 0, 1, \ldots, N-1,$$

where w is a given positive scalar. The problem is to find the optimal investment sequence $u_0, \ldots, u_{N-1}$ that maximizes the total product stored over N years

$$x_N + \sum_{i=0}^{N-1} (1 - u_i) x_i.$$

Show that one optimal sequence is given by:

(1) If $w > 1$, $u_0^* = \cdots = u_{N-1}^* = 1$.

(2) If $0 < w < 1/N$, $u_0^* = \cdots = u_{N-1}^* = 0$.

(3) If $1/N \leq w \leq 1$,

$$u_0^* = \cdots = u_{N-\bar{i}-1} = 1,$$
$$u_{N-\bar{i}}^* = \cdots = u_{N-1}^* = 0,$$

where $\bar{i}$ is such that $1/(\bar{i}+1) < w \leq 1/\bar{i}$.

2.1.16

Let x_i denote the number of educators in a certain country at time i and let y_i denote the number of research scientists at time i. New scientists (potential educators or research scientists) are produced during the ith period by educators at a rate γ per educator, while educators and research scientists leave the field due to death, retirement, and transfer at a rate δ, where γ and δ are given scalars with $0 < \gamma$ and $0 < \delta < 1$. By means of incentives a science policy maker can determine the proportion u_i of new scientists produced at time i who become educators. Thus the number of research scientists and educators evolves according to the equations

$$x_{i+1} = (1 - \delta)x_i + u_i \gamma x_i,$$
$$y_{i+1} = (1 - \delta)y_i + (1 - u_i)\gamma x_i.$$

The initial numbers x_0, y_0 are known, and the u_i are constrained by

$$0 < \alpha \leq u_i \leq \beta < 1, \qquad i = 0, 1, \ldots, N-1,$$

where the scalars α and β are given. Find a sequence $\{u_0^*, \ldots, u_{N-1}^*\}$ that maximizes the final number y_N of research scientists.

2.1.17 (Fractional Programming)

Consider the problem

$$\begin{aligned} &\text{minimize} \quad \frac{f(x)}{g(x)} \\ &\text{subject to} \quad x \in X, \end{aligned}$$

where $f : \Re^n \mapsto \Re$ and $g : \Re^n \mapsto \Re$ are given functions, and X is a given subset such that $g(x) > 0$ for all $x \in X$. For $\lambda \in \Re$, define

$$Q(\lambda) = \min_{x \in X} \big\{ f(x) - \lambda g(x) \big\},$$

and suppose that a scalar λ^* and a vector $x^* \in X$ satisfy $Q(\lambda^*) = 0$ and

$$x^* = \arg\min_{x \in X} \big\{ f(x) - \lambda^* g(x) \big\}.$$

Show that x^* is an optimal solution of the original problem. Use this observation to suggest a solution method that does not require dealing with fractions of functions.

2.1.18 (www)

Let $f : \Re^n \mapsto \Re$ be a twice continuously differentiable function that satisfies

$$m\|y\|^2 \le y'\nabla^2 f(x) y \le M\|y\|^2, \qquad \forall\, x,\, y \in \Re^n,$$

where m and M are some positive scalars. Let also X be a closed convex set. Show that f has a unique global minimum x^* over X, which satisfies

$$\theta_M(x) \le f(x) - f(x^*) \le \theta_m(x), \qquad \forall\, x \in \Re^n,$$

where for all $\delta > 0$, we denote

$$\theta_\delta(x) = -\min_{y \in X} \left\{ \nabla f(x)'(y - x) + \frac{\delta}{2}\|y - x\|^2 \right\}.$$

2.2 FEASIBLE DIRECTIONS AND THE CONDITIONAL GRADIENT METHOD

We now turn to computational methods for solving the constrained problem of the preceding section:

$$\begin{aligned} &\text{minimize} \quad f(x) \\ &\text{subject to} \quad x \in X, \end{aligned}$$

where f is continuously differentiable, and X is a nonempty, closed, and convex set. We will see that there is a great variety of algorithms for this problem. However, in this chapter we will restrict ourselves to a limited class of methods that have the following characteristics:

(a) They do not rely on any structure of the constraint set other than its convexity.

(b) They generate sequences of feasible points $\{x^k\}$ by searching along descent directions. In this sense, they can be viewed as constrained versions of the unconstrained descent algorithms of the previous chapter.

Most of the algorithms in this chapter belong to the class of the, so called, feasible direction methods, which we proceed to discuss.

2.2.1 Descent Directions and Stepsize Rules

Given a feasible vector x, a *feasible direction* at x is a vector $d \neq 0$ such that $x + \alpha d$ is feasible for all $\alpha > 0$ that are sufficiently small. Figure 2.2.1 illustrates the set of feasible directions at a point.

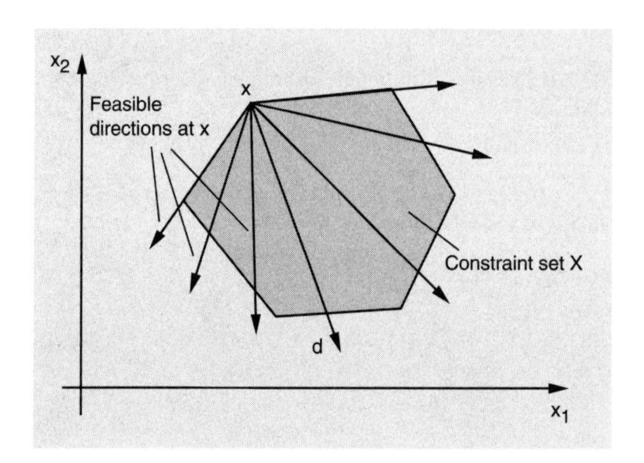

Figure 2.2.1. Feasible directions d at a feasible point x. By definition, d is a feasible direction if changing x by a small amount in the direction d maintains feasibility.

A *feasible direction method* starts with a feasible vector x^0 and generates a sequence of feasible vectors $\{x^k\}$ according to

$$x^{k+1} = x^k + \alpha^k d^k,$$

where, if x^k is not stationary, d^k is a feasible direction at x^k, which is also a descent direction, i.e.,

$$\nabla f(x^k)' d^k < 0,$$

and the stepsize α^k is chosen to be positive and such that

$$x^k + \alpha^k d^k \in X.$$

If x^k is stationary, the method stops, i.e., $x^{k+1} = x^k$ (equivalently, we set $d^k = 0$). We will primarily concentrate on feasible direction methods that are also descent algorithms; i.e., the stepsize α^k is selected so that

$$f(x^k + \alpha^k d^k) < f(x^k), \qquad \forall\, k.$$

Figure 2.2.2 illustrates a feasible direction method.

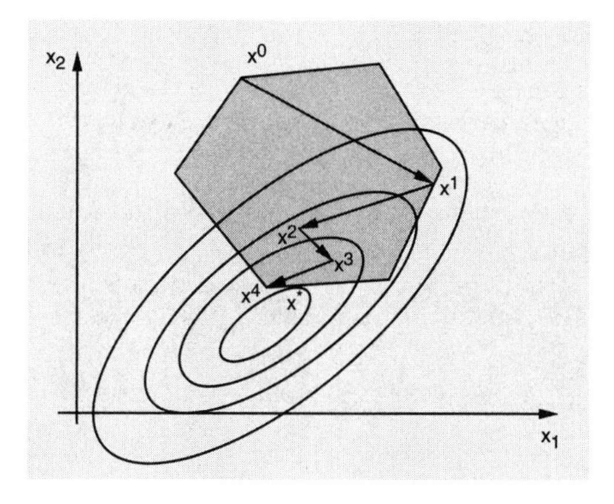

Figure 2.2.2. Sequence generated by a feasible direction method. At the current point x^k, we choose a feasible direction d^k that is also a direction of descent $[\nabla f(x^k)'d^k < 0]$, and we obtain a new feasible point $x^{k+1} = x^k + \alpha^k d^k$ along d^k.

In our case where X is a convex set, there is an alternative and equivalent characterization of feasible direction methods. In particular, it can be seen that the feasible directions at x^k are the vectors of the form

$$d^k = \gamma(\bar{x}^k - x^k), \qquad \gamma > 0,$$

where $\bar{x}^k$ is some feasible vector different from x^k. Thus, taking into account that $x^k + \alpha^k d^k \in X$, a feasible direction method can be written in the form

$$x^{k+1} = x^k + \alpha^k(\bar{x}^k - x^k),$$

where

$$\alpha^k \in (0, 1],$$

and if x^k is nonstationary,

$$\bar{x}^k \in X, \qquad \nabla f(x^k)'(\bar{x}^k - x^k) < 0. \tag{2.26}$$

Note that because the constraint set X is convex, we have $x^k+\alpha^k(\bar{x}^k-x^k) \in X$ for all $\alpha^k \in [0,1]$ when $x^k \in X$, so that the generated sequence of iterates $\{x^k\}$ is feasible. Furthermore, if x^k is nonstationary, there always exists a feasible direction $\bar{x}^k - x^k$ with the descent property (2.26), since otherwise we would have $\nabla f(x^k)'(x - x^k) \geq 0$ for all $x \in X$, implying that x^k is stationary.

We will henceforth assume that the feasible direction d^k in any feasible direction method under consideration is of the form

$$d^k = \bar{x}^k - x^k,$$

where $\bar{x}^k$ satisfies Eq. (2.26).

Stepsize Selection in Feasible Direction Methods

Most of the rules for choosing the stepsize α^k in gradient methods apply also to feasible direction methods. Here are some of the most popular ones, where we assume that the direction d^k is scaled so that the vector $x^k + d^k$ is feasible.

Limited Minimization Rule

Here, α^k is chosen so that

$$f(x^k + \alpha^k d^k) = \min_{\alpha \in [0,1]} f(x^k + \alpha d^k).$$

Note that there is no loss of generality in limiting the stepsize to the interval $[0,1]$, since different stepsize ranges can in effect be used by redefining the direction d^k.

Armijo Rule

Here, fixed scalars β, and $\sigma > 0$, with $\beta \in (0,1)$, and $\sigma \in (0,1)$ are chosen, and we set $\alpha^k = \beta^{m_k} s$, where m_k is the first nonnegative integer m for which

$$f(x^k) - f(x^k + \beta^m d^k) \geq -\sigma \beta^m \nabla f(x^k)' d^k.$$

In other words, the stepsizes $1, \beta, \beta^2, \ldots,$ are tried successively until the above inequality is satisfied for $m = m_k$.

Constant Stepsize

Here, a fixed stepsize

$$\alpha^k = 1, \qquad k = 0, 1, \ldots$$

is used. This choice is not as simple or as restrictive as it may seem. In fact, in any feasible direction method, we can use a constant unity stepsize if we rescale or redefine appropriately the direction d^k. In this case, the burden is placed in effect on the direction finding procedure to yield a direction that guarantees descent and convergence.

Convergence Analysis of Feasible Direction Methods

The convergence analysis of feasible direction methods for the case of the minimization rule and the Armijo rule is very similar to the one for gradient methods. As in Section 1.2, we say that the direction sequence $\{d^k\}$ is *gradient related* to $\{x^k\}$ if the following property can be shown:

For any subsequence $\{x^k\}_{k\in\mathcal{K}}$ that converges to a nonstationary point, the corresponding subsequence $\{d^k\}_{k\in\mathcal{K}}$ is bounded and satisfies

$$\limsup_{k\to\infty,\, k\in\mathcal{K}} \nabla f(x^k)'d^k < 0.$$

A verbatim repetition of the proof of Prop. 1.2.1 of Section 1.2 shows the following result:

Proposition 2.2.1: (Stationarity of Limit Points for Feasible Direction Methods) Let $\{x^k\}$ be a sequence generated by the feasible direction method $x^{k+1} = x^k + \alpha^k d^k$. Assume that $\{d^k\}$ is gradient related and that α^k is chosen by the limited minimization rule or the Armijo rule. Then every limit point of $\{x^k\}$ is a stationary point.

There are also convergence results for feasible direction methods that use a constant stepsize, but these results are best considered in the context of individual methods.

Finding an Initial Feasible Point

To apply a feasible direction method, it is necessary to have an initial feasible point. Finding such a point may be difficult if the constraint set is specified by nonlinear inequality constraints. If, however, the constraint set is a polyhedron (is specified by linear equality and inequality constraints), one can find an initial feasible point by solving the linear programming problem involving the same constraints and an arbitrary linear cost function.

An alternative method is based on introducing an additional artificial variable y, which is constrained to be nonnegative but is heavily penalized in the cost so that it is zero at the optimum. In particular, consider the problem

$$\begin{aligned} &\text{minimize } f(x) \\ &\text{subject to } a_i'x = b_i, \quad i = 1, \ldots, m, \qquad x \geq 0, \end{aligned}$$

where a_i and b_i are given vectors in $\Re^n$ and scalars, respectively. This problem can be replaced by the problem

$$\begin{aligned} &\text{minimize } f(x) + cy \\ &\text{subject to } a_i'x + \left(b_i - \sum_{j=1}^{n} a_{ij} \right) y = b_i, \quad i = 1, \ldots, m, \quad x \geq 0, \ y \geq 0, \end{aligned}$$

where c is a very large cost coefficient, a_{ij} is the jth coordinate of the vector a_i. The vector $(\bar{x}, \bar{y})$ with $\bar{x}_j = 1$, $j = 1, \ldots, n$, $\bar{y} = 1$ is feasible for the modified problem. Note that this vector is interior to the bound constraints $x \geq 0$, $y \geq 0$; this is significant for interior point methods, which will be discussed in Section 2.6.

Similarly, the problem

$$\begin{aligned} &\text{minimize } f(x) \\ &\text{subject to } a_j'x \leq b_j, \quad j = 1, \ldots, r, \end{aligned}$$

can be converted to the problem

$$\begin{aligned} &\text{minimize } f(x) + cy \\ &\text{subject to } a_j'x - y \leq b_j, \quad j = 1, \ldots, r, \quad y \geq 0, \end{aligned}$$

where c is a very large cost coefficient. For any vector $\bar{x}$ that is infeasible for the original problem, the vector $(\bar{x}, \bar{y})$ where

$$\bar{y} = \max_{j=1,\ldots,r} \{a_j'\bar{x} - b_j\}$$

is feasible for the modified problem, while the vector $(\bar{x}, \bar{y} + 1)$ is feasible, as well as interior to the inequality constraints of the modified problem.

2.2.2 The Conditional Gradient Method

The most straightforward way to generate a feasible direction $\bar{x}^k - x^k$ that satisfies the descent condition $\nabla f(x^k)'(\bar{x}^k - x^k) < 0$, and can be used in the method

$$x^{k+1} = x^k + \alpha^k(\bar{x}^k - x^k),$$

is to solve the optimization problem

$$\begin{aligned} &\text{minimize} \quad \nabla f(x^k)'(x - x^k) \\ &\text{subject to} \quad x \in X, \end{aligned} \tag{2.27}$$

and obtain $\bar{x}^k$ as the solution, that is,

$$\bar{x}^k = \arg\min_{x \in X} \nabla f(x^k)'(x - x^k).$$

Here we assume that X is compact, so that the direction generating subproblem (2.27) has a solution. The corresponding feasible direction method is the, so called, *conditional gradient method*, also known as the *Frank-Wolfe method*. The process to obtain $\bar{x}^k$ is illustrated in Fig. 2.2.3. In particular, $\bar{x}^k$ is the remotest point of X along the negative gradient direction.

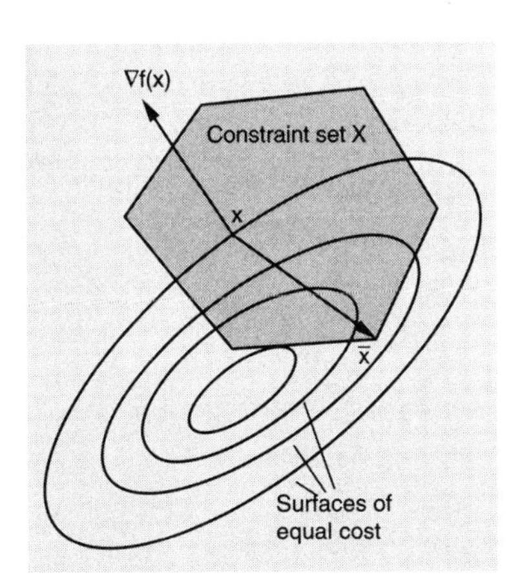

Figure 2.2.3. Finding the feasible direction $\bar{x} - x$ at a point x in the conditional gradient method. $\bar{x}$ is a point of X that lies furthest along the negative gradient direction $-\nabla f(x)$.

Naturally, in order for the method to make practical sense, the subproblem (2.27) must be much simpler than the original. This is typically the case when f is nonlinear, and X is specified by linear equality and inequality constraints. Then, the subproblem (2.27) is a linear program, which in many applications is very easy to solve. A principal example is when X is a simplex or a Cartesian product of simplices, as in the communication and transportation examples of the previous section.

Example 2.2.1 (Simplex Constraint)

Let

$$X = \left\{ x \;\middle|\; x \geq 0, \ \sum_{i=1}^{n} x_i = r \right\},$$

where $r > 0$ is a given scalar (cf. Example 2.1.2 of the previous section). Then, the subproblem (2.27) takes the form

$$\begin{aligned} &\text{minimize} \quad \sum_{i=1}^{n} \frac{\partial f(x^k)}{\partial x_i}(x_i - x_i^k) \\ &\text{subject to} \quad \sum_{i=1}^{n} x_i = r, \qquad x_i \geq 0 \end{aligned}$$

and a solution $\bar{x}^k$ has all coordinates equal to zero except for a single coordinate, say j, which is equal to r; the jth coordinate corresponds to a minimal partial derivative, that is,

$$j = \arg \min_{i=1,\ldots,n} \frac{\partial f(x^k)}{\partial x_i}.$$

This choice, together with the path followed by the conditional gradient method is illustrated in Fig. 2.2.4. At each iteration, the component of x^k that has minimal partial derivative is increased, while the remaining components are decreased by a factor equal to the stepsize α^k.

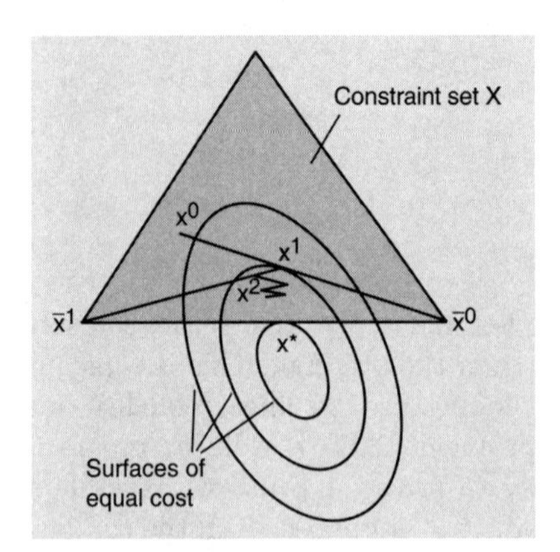

Figure 2.2.4. Successive iterates of the conditional gradient method with the limited minimization stepsize rule, for the case of a simplex constraint.

Convergence of the Conditional Gradient Method

We will show that the direction sequence of the conditional gradient method is gradient related, so Prop. 2.2.1 applies. Indeed, suppose that $\{x^k\}_{k\in K}$ converges to a nonstationary point $\tilde{x}$. We must prove that

$$\limsup_{k\to\infty,\, k\in K} \|\bar{x}^k - x^k\| < \infty, \tag{2.28}$$

$$\limsup_{k\to\infty,\, k\in K} \nabla f(x^k)'(\bar{x}^k - x^k) < 0. \tag{2.29}$$

Equation (2.28) holds because $\bar{x}^k \in X$, $x^k \in X$, and the set X is assumed compact. To show Eq. (2.29), we note that, by the definition of $\bar{x}^k$,

$$\nabla f(x^k)'(\bar{x}^k - x^k) \leq \nabla f(x^k)'(x - x^k), \qquad \forall\, x \in X,$$

and by taking limit we obtain

$$\limsup_{k\to\infty,\, k\in K} \nabla f(x^k)'(\bar{x}^k - x^k) \leq \nabla f(\tilde{x})'(x - \tilde{x}), \qquad \forall\, x \in X.$$

By taking the minimum over $x \in X$ and by using the nonstationarity of $\tilde{x}$, we have

$$\limsup_{k\to\infty,\, k\in K} \nabla f(x^k)'(\bar{x}^k - x^k) \leq \min_{x\in X} \nabla f(\tilde{x})'(x - \tilde{x}) < 0,$$

thereby proving Eq. (2.29).

We have thus shown that every limit point of the conditional gradient method with the limited minimization rule or the Armijo rule is stationary. An alternative line of convergence analysis is given in Exercise 2.2.4.

Rate of Convergence of the Conditional Gradient Method

Unfortunately, the asymptotic rate of convergence of the conditional gradient method is not very fast when the set X is a polyhedron, i.e., it is specified by linear equality and inequality constraints. A partial explanation is that the vectors $\bar{x}^k$ used in the algorithm are typically extreme points (vertices) of X. Thus, the feasible direction used may tend to be orthogonal to the direction leading to the minimum (see Fig. 2.2.4). In fact one may construct simple examples where the error sequences $\{f(x^k) - f(x^*)\}$ and $\{\|x^k - x^*\|\}$ do not converge linearly (see Exercise 2.2.3). Thus, when X is a polyhedron, the method is recommended only for problems where solution accuracy is not very important.

Somewhat peculiarly, the practical performance of the conditional gradient method tends to improve as the number of constraints becomes very large, even if these constraints are linear. An explanation for this is given by Dunn [Dun79], [DuS83], where it is shown that the convergence rate of the method is linear when the constraint set is not polyhedral but rather has a "positive curvature" property (for example it is a sphere). When there are many linear constraints, the constraint set tends to have very many closely spaced extreme points, and has this "positive curvature" property in an approximate sense.

EXERCISES

2.2.1 (Computational Problem)

Consider the three-dimensional problem

$$\begin{aligned} &\text{minimize } f(x) = \tfrac{1}{2}\left(x_1^2 + x_2^2 + 0.1x_3^2\right) + 0.55x_3 \\ &\text{subject to } x_1 + x_2 + x_3 = 1, \qquad 0 \le x_1,\ 0 \le x_2,\ 0 \le x_3. \end{aligned}$$

Show that the global minimum is $x^* = (1/2, 1/2, 0)$. Write a computer program implementing the conditional gradient method with the line minimization stepsize rule. (Here, there is a closed form expression for the minimizing stepsize.) Verify computationally that for a starting point (ξ_1, ξ_2, ξ_3) with $\xi_i > 0$ for all i, and $\xi_1 \neq \xi_2$, the rate of convergence is not linear in the sense that

$$\frac{f(x^{k+1}) - f(x^*)}{f(x^k) - f(x^*)} \to 1.$$

2.2.2

Apply the conditional gradient method to the optimal routing problem of Example 2.1.3 in Section 2.1. Show that the feasible direction is generated by solving certain minimum length path problems.

2.2.3 (Sublinear Rate of Convergence of the Conditional Gradient Method)

Consider the two-dimensional problem

$$\begin{aligned} &\text{minimize } f(x) = x_1^2 + (x_2 + 1)^2 \\ &\text{subject to } -1 \le x_1 \le 1, \qquad 0 \le x_2, \end{aligned}$$

with global minimum $x^* = (0, 0)$. Successive iterates of the conditional gradient method are shown in Fig. 2.2.5. Show that

$$\lim_{k \to \infty} \frac{f(x^{k+1}) - f(x^*)}{f(x^k) - f(x^*)} = 1,$$

and therefore the rate of convergence of $\{f(x^k) - f(x^*)\}$ is not linear. *Hint*: This is a tricky problem. Show that for all $k \geq 1$, we have $f(x^k) - f(x^*) = 2(\cos \beta^k)^2 - 1 = \sin(2\alpha^k)$, where α^k and β^k are the angles shown in Fig. 2.2.5.

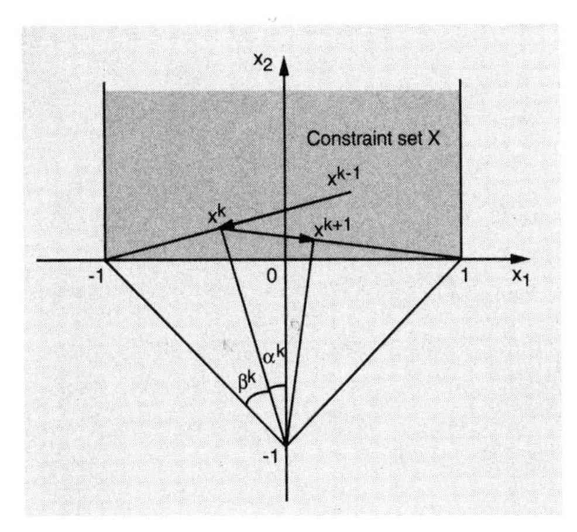

Figure 2.2.5. Successive iterates of the conditional gradient method for the problem of minimizing the distance of the point $(0, -1)$ from the constraint set shown (cf. Exercise 2.2.3). The directions used (asymptotically) become horizontal and the convergence rate is sublinear.

2.2.4 (Another Convergence Proof of the Conditional Gradient Method)

Assume that the gradient of f satisfies

$$\big\|\nabla f(x) - \nabla f(y)\big\| \leq L\|x - y\|, \quad \forall\, x, y \in X,$$

where L is a positive constant.

(a) Show that if d is a descent direction at x, then

$$\min_{\alpha \in [0,1]} f(x + \alpha d) \leq f(x) + \delta,$$

where δ is the negative scalar given by

$$\delta = \begin{cases} \frac{1}{2}\nabla f(x)'d & \text{if } \nabla f(x)'d + L\|d\|^2 < 0, \\ -\dfrac{|\nabla f(x)'d|^2}{2LR^2} & \text{otherwise,} \end{cases}$$

where R is the diameter of X, that is,

$$R = \max_{x, y \in X} \|x - y\|.$$

Hint: Use the descent lemma (Prop. A.24 of Appendix A), which shows that

$$f(x + \alpha d) \leq f(x) + \alpha \nabla f(x)'d + \frac{\alpha^2 L}{2} \|d\|^2.$$

Minimize over $\alpha \in [0, 1]$ both sides of this inequality.

(b) Consider the conditional gradient method

$$x^{k+1} = x^k + \alpha^k d^k$$

with the limited minimization stepsize rule. Show that every limit point of $\{x^k\}$ is stationary. *Hint*: Argue that, if δ^k corresponds to x^k as in part (a), then $\delta^k \to 0$, and therefore we have $\nabla f(x^k)'d^k \to 0$. Take the limit in the relation $\nabla f(x^k)'d^k \leq \nabla f(x^k)'(x - x^k)$ for all $x \in X$.

2.2.5

Consider the conditional gradient method, assuming that the gradient of f satisfies

$$\big\|\nabla f(x) - \nabla f(y)\big\| \leq L\|x - y\|, \quad \forall\, x, y \in X,$$

where L is a positive constant. Show that if the stepsize α^k is given by

$$\alpha^k = \min \left\{ 1, \frac{\nabla f(x^k)'(x^k - \bar{x}^k)}{L\|x^k - \bar{x}^k\|^2} \right\},$$

then every limit point of $\{x^k\}$ is stationary. *Hint*: Use the line of analysis of Exercise 2.2.4.

2.2.6 (Termination Criterion)

Show that if f is convex, then in the conditional gradient method we have the easily computable bounds

$$f(x^k) \geq \min_{x \in X} f(x) \geq f(x^k) + \nabla f(x^k)'(\bar{x}^k - x^k),$$

and the upper and lower bound difference, $\nabla f(x^k)'(\bar{x}^k - x^k)$, tends to zero as $k \to \infty$.

2.2.7 (Simplicial Decomposition Method [Hol74], [Hoh77], [HLV87], [VeH93]) (www)

Assume that X is a bounded polyhedron. Let $\bar{x}^k$ be defined as an extreme point of X satisfying

$$\bar{x}^k = \arg \min_{x \in X} \nabla f(x^k)'(x - x^k),$$

and suppose that x^{k+1} is a stationary point of the optimization problem

$$\begin{aligned} &\text{minimize } f(x) \\ &\text{subject to } x \in X^k, \end{aligned}$$

where X^k is the convex hull of x^0 and the extreme points $\bar{x}^0, \ldots, \bar{x}^k$,

$$X^k = \left\{ x \;\middle|\; x = \beta x^0 + \sum_{i=0}^{k} \gamma_i \bar{x}^i,\ \beta \geq 0, \gamma_i \geq 0,\ \beta + \sum_{i=0}^{k} \gamma_i = 1 \right\}.$$

Using the fact that the number of extreme points of X is finite, show that the method finds a stationary point of f over X in a finite number of iterations.

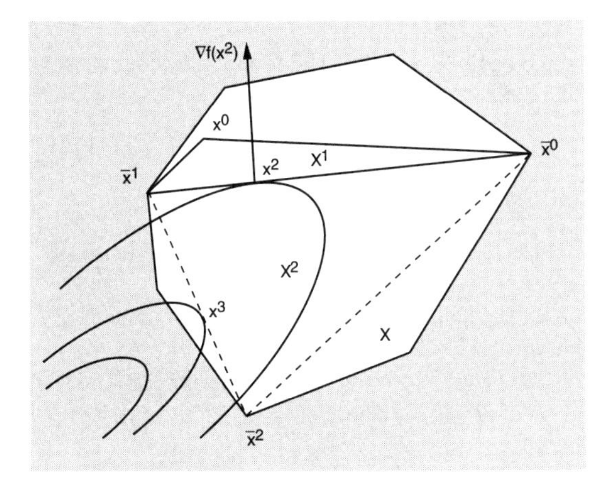

Figure 2.2.6. Successive iterates of the simplicial decomposition method; cf. Exercise 2.2.7. The figure shows how given the initial point x^0, and the calculated extreme points $\bar{x}^0$, $\bar{x}^1$, we determine the next iterate x^2 as a stationary point of f over the convex hull X^1 of x^0, $\bar{x}^0$, and $\bar{x}^1$. Then the new extreme point $\bar{x}^2$ is determined from x^2 as in the conditional gradient method, and the process is repeated to obtain x^3.

2.2.8 (A Variation of the Simplicial Decomposition Method)

Consider the method of Exercise 2.2.7 with the difference that the set X^k is any closed subset of the set

$$\left\{ x \;\middle|\; x = \beta x_0 + \sum_{i=0}^{k} \gamma_i \bar{x}^i,\ \beta \geq 0,\ \gamma_i \geq 0,\ \beta + \sum_{i=0}^{k} \gamma_i = 1 \right\}$$

that contains the segment connecting x^k and $\bar{x}^k$. Prove that every limit point of a sequence $\{x^k\}$ generated by this method is a stationary point. *Hint*: Use the fact that for every k, the cost reduction $f(x^k) - f(x^{k+1})$ is larger than the cost reduction that would be obtained by using an Armijo rule along the segment connecting x^k and $\bar{x}^k$ (cf. Props. 1.2.1 and 2.2.1).

2.2.9 (Zoutendijk's Method of Feasible Directions [Zou60])

Consider the problem

$$\begin{aligned} &\text{minimize} \;\; f(x) \\ &\text{subject to} \;\; g_j(x) \leq 0, \qquad j = 1, \ldots, r, \end{aligned}$$

where $f : \Re^n \mapsto \Re$ is continuously differentiable, and $g_j : \Re^n \mapsto \Re$ are convex and continuously differentiable functions. For any $\epsilon \geq 0$ and feasible vector x, define by $A(x; \epsilon)$ the set of ϵ-active constraints, that is,

$$A(x; \epsilon) = \bigl\{ j \mid -\epsilon \leq g_j(x) \leq 0 \bigr\}$$

and let $\phi(x; \epsilon)$ be the optimal value of the following linear programming problem in the vector $(d, z) \in \Re^{n+1}$

$$\begin{aligned} &\text{minimize} \;\; z \\ &\text{subject to} \;\; \|d\|_\infty \leq 1, \quad \nabla f(x)'d \leq z \\ &\qquad\qquad \nabla g_j(x)'d \leq z, \quad \forall\, j \in A(x; \epsilon) \end{aligned} \qquad (P_{x,\epsilon})$$

where $\|\cdot\|_\infty$ denotes the maximum norm.

Zoutendijk's method uses two scalars $\bar{\epsilon} > 0$ and $\gamma \in (0, 1)$, and determines a feasible descent direction d^k at a nonstationary vector x^k so that $\bigl(d^k, \phi(x^k; \epsilon^k)\bigr)$ is a solution of problem (P_{x^k, ϵ^k}), where $\epsilon^k = \gamma^{m_k} \bar{\epsilon}$ and m_k is the first nonnegative integer m for which

$$\phi\left(x^k, \gamma^m \bar{\epsilon}\right) \leq -\gamma^m \bar{\epsilon}.$$

(a) Show that d^k is obtained after solving a finite number of problems (P_{x^k, ϵ^k}). Furthermore, d^k is a feasible descent direction at x^k.

(b) Prove that $\{d^k\}$ is gradient related, thus establishing stationarity of the limit points of $\{x^k\}$.

2.2.10 (Min-H Method for Optimal Control)

Consider the problem of finding sequences $u = (u_0, u_1, \ldots, u_{N-1})$ and $x = (x_1, x_2, \ldots, x_N)$ that minimize

$$g_N(x_N) + \sum_{i=0}^{N-1} g_i(x_i, u_i),$$

subject to the constraints

$$u_i \in U_i, \qquad i = 0, \ldots, N-1,$$

$$x_{i+1} = \phi_i(x_i) + B_i u_i, \quad i = 0, \ldots, N-1, \qquad x_0 : \text{ given},$$

where B_i are given matrices of appropriate dimension. We assume that the functions g_i and ϕ_i are continuously differentiable, and the sets U_i are convex. We further assume that $g_i(x_i, \cdot)$ is convex as a function of u_i for all x_i. Consider the algorithm that, given $u^k = (u_0^k, \ldots, u_{N-1}^k)$, and the corresponding state and costate trajectories $x^k = (x_1^k, \ldots, x_N^k)$ and $p^k = (p_1^k, \ldots, p_N^k)$, respectively, generates u^{k+1} by

$$u_i^{k+1} = u_i^k + \alpha^k(\bar{u}_i^k - u_i^k), \qquad i = 0, \ldots, N-1,$$

where α^k is generated by the Armijo rule or the limited minimization rule, and $\bar{u}_i^k$ is given by

$$\bar{u}_i^k = \arg\min_{u_i \in U_i} H_i\big(x_i^k, u_i, p_{i+1}^k\big), \qquad i = 0, \ldots, N-1,$$

with

$$H_i(x_i, u_i, p_{i+1}) = g_i(x_i, u_i) + p_{i+1}' f_i(x_i, u_i)$$

denoting the Hamiltonian function. Show that this is a feasible direction method.

2.3 GRADIENT PROJECTION METHODS

The conditional gradient method uses a feasible direction obtained by solving a subproblem with linear cost. Gradient projection methods use instead a subproblem with quadratic cost. While this subproblem may be more complex, the resulting convergence rate is typically better. There are many variations of gradient projection methods, which may be viewed collectively as constrained analogs of the gradient methods of Section 2.2.

2.3.1 Feasible Directions and Stepsize Rules Based on Projection

We first develop informally the main ideas regarding projection methods, leaving the detailed results and proofs for the next subsection.

The simplest gradient projection method is a feasible direction method of the form

$$x^{k+1} = x^k + \alpha^k(\bar{x}^k - x^k), \tag{2.30}$$

where

$$\bar{x}^k = \big[x^k - s^k \nabla f(x^k)\big]^+.$$

Here, $[\cdot]^+$ denotes projection on the set X, $\alpha^k \in (0, 1]$ is a stepsize, and s^k is a positive scalar. Thus, to obtain the vector $\bar{x}^k$, we take a step $-s^k \nabla f(x^k)$ along the negative gradient, as in steepest descent. We then project the result $x^k - s^k \nabla f(x^k)$ on X, thereby obtaining the feasible vector $\bar{x}^k$. Finally, we take a step along the feasible direction $\bar{x}^k - x^k$ using the stepsize α^k.

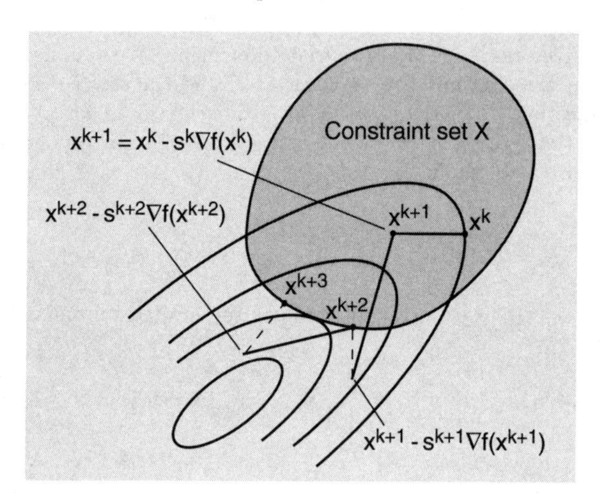

Figure 2.3.1. Illustration of a few iterations of the gradient projection method for the case where $\alpha^k = 1$ for all k. Note that when $x^k - s^k \nabla f(x^k)$ belongs to X, the iteration reduces to an unconstrained steepest descent iteration.

We may also view the scalar s^k as a stepsize. This becomes evident when we select $\alpha^k = 1$ for all k, in which case $x^{k+1} = \bar{x}^k$ and the method becomes

$$x^{k+1} = \left[x^k - s^k \nabla f(x^k)\right]^+ \tag{2.31}$$

(see Fig. 2.3.1). If $x^k - s^k \nabla f(x^k)$ is feasible, the gradient projection iteration becomes an unconstrained steepest descent iteration, in either the form (2.30) or the form (2.31). In this case, Eq. (2.30) takes the form $x^{k+1} = x^k - \alpha^k \nabla f(x^k)$, while Eq. (2.31) takes the form $x^{k+1} = x^k - s^k \nabla f(x^k)$.

Note that we have $x^* = \left[x^* - s\nabla f(x^*)\right]^+$ for all $s > 0$ if and only if x^* is stationary (see Fig. 2.3.2). Thus, the method stops if and only if it encounters a stationary point.

In order for the method to make practical sense, it is necessary that the projection operation is fairly simple. This will be so if X has a relatively simple structure. For example, when the constraints are bounds on the variables,

$$X = \{x \mid \alpha_i \le x_i \le \beta_i,\ i = 1, \ldots, n\},$$

the ith coordinate of the projection of a vector x is given by

$$[x]_i^+ = \begin{cases} \alpha_i & \text{if } x_i \le \alpha_i, \\ \beta_i & \text{if } x_i \ge \beta_i, \\ x_i & \text{otherwise.} \end{cases}$$

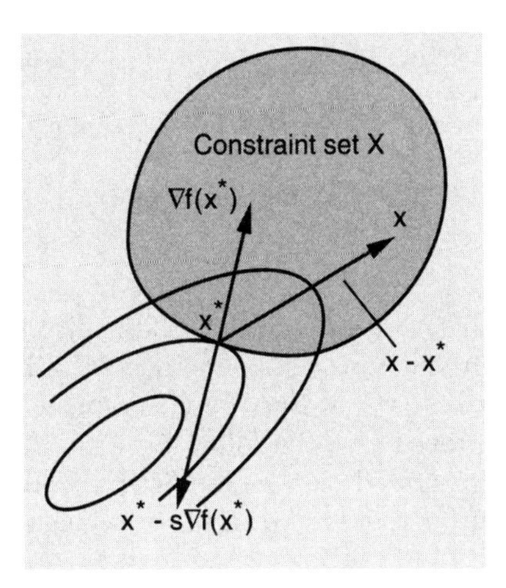

Figure 2.3.2. Illustrating why the gradient projection method stops if and only if it finds a stationary point x^*. By definition, x^* is a stationary point if

$$\nabla f(x^*)'(x - x^*) \geq 0, \qquad \forall\, x \in X,$$

which is equivalent to

$$\Big(\big(x^* - s\nabla f(x^*)\big) - x^*\Big)'(x - x^*) \leq 0, \qquad \forall\, x \in X,\ s > 0.$$

This holds if and only if x^* is the projection of $x^* - s\nabla f(x^*)$ on X (Prop. 2.1.3).

Stepsize Selection and Convergence

There are several stepsize selection procedures in the gradient projection method. The following two are straightforward, although we will later argue that they are often not the best options.

Limited Minimization Rule

Here

$$s^k = s: \text{ constant}, \qquad k = 0, 1, \ldots$$

and α^k is chosen by minimization over $[0,1]$; that is,

$$f\big(x^k + \alpha^k(\bar{x}^k - x^k)\big) = \min_{\alpha \in [0,1]} f\big(x^k + \alpha(\bar{x}^k - x^k)\big).$$

Armijo Rule Along the Feasible Direction

Here

$$s^k = s: \text{ constant}, \qquad k = 0, 1, \ldots$$

and α^k is chosen by the Armijo rule over the interval $[0, 1]$. In particular, fixed scalars β and $\sigma > 0$, with $\beta \in (0, 1)$ and $\sigma \in (0, 1)$ are chosen, and we set $\alpha^k = \beta^{m_k}$, where m_k is the first nonnegative integer m for which

$$f(x^k) - f\big(x^k + \beta^m(\bar{x}^k - x^k)\big) \geq -\sigma\beta^m \nabla f(x^k)'(\bar{x}^k - x^k).$$

Our general convergence result for feasible direction methods (Prop. 2.2.1) applies to the gradient projection method with the above two stepsize rules, provided we can show that the direction sequence is gradient related. Indeed this can be proved (see the following Prop. 2.3.1), so every limit point of a sequence generated by the gradient projection method with the line minimization rule or the Armijo rule is stationary.

Armijo Rule Along the Projection Arc

Here the stepsize α^k is fixed at unity,

$$\alpha^k = 1, \qquad k = 0, 1, \ldots$$

and the stepsize s^k is determined by successive reduction until an Armijo-like inequality is satisfied. This means that x^{k+1} is determined by an Armijo-like search on the "projection arc"

$$\{x^k(s) \mid s > 0\},$$

where, for all $s > 0$, $x^k(s)$ is defined by

$$x^k(s) = \big[x^k - s\nabla f(x^k)\big]^+. \tag{2.32}$$

In particular, fixed scalars $\bar{s}$, β, and $\sigma > 0$, with $\bar{s} > 0$, $\beta \in (0, 1)$, and $\sigma \in (0, 1)$ are chosen, and we set $s^k = \beta^{m_k}\bar{s}$, where m_k is the first nonnegative integer m for which

$$f(x^k) - f\big(x^k(\beta^m\bar{s})\big) \geq \sigma\nabla f(x^k)'\big(x^k - x^k(\beta^m\bar{s})\big). \tag{2.33}$$

Figure 2.3.3 illustrates the Armijo rule on a projection arc.

The convergence properties of the gradient projection method for the Armijo rules along the feasible direction and along the projection arc are essentially the same, although the convergence proofs differ substantially [see the following Prop. 2.3.3, which also shows that the stepsize rules are well defined, i.e., a stepsize will be found after a finite number of trials based on the test (2.33)].

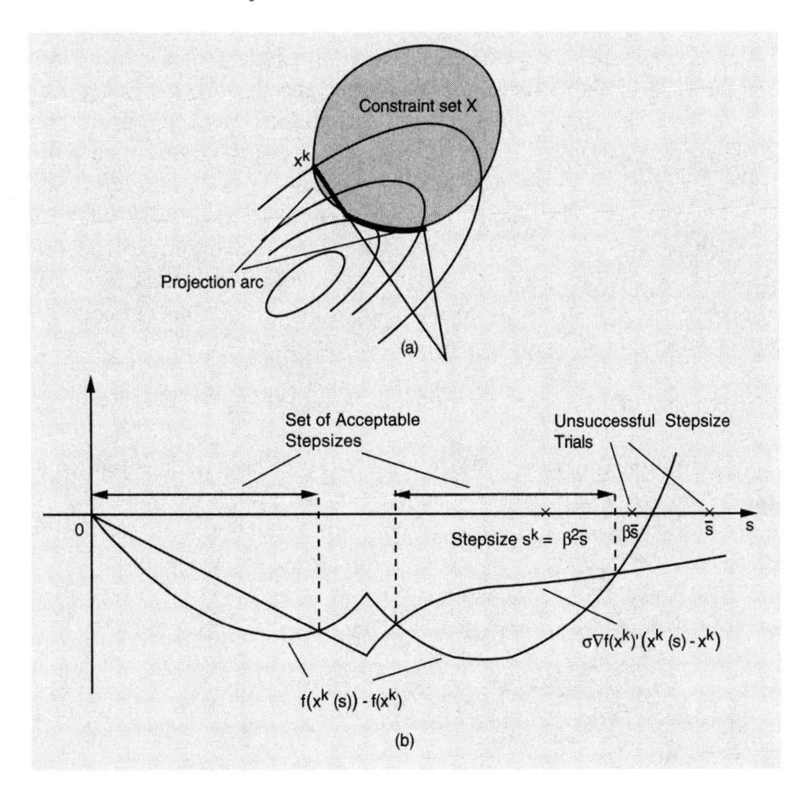

Figure 2.3.3. (a) Illustration of the successive points tested by the Armijo rule along the projection arc. (b) Illustration of the successive tests of the Armijo inequality (2.33).

Constant Stepsize

Here, s^k is also fixed at some constant $s > 0$ and α^k is fixed at unity, that is,

$$s^k = s: \text{ constant}, \qquad \alpha^k = 1, \qquad k = 0, 1, \ldots$$

Similar to unconstrained gradient methods, it is possible to show that the limit points of a sequence generated by the gradient projection method with a constant stepsize are stationary, provided s is sufficiently small and the gradient satisfies a Lipschitz continuity condition (see the following Prop. 2.3.2).

Diminishing Stepsize

Here, α^k is fixed at unity and

$$s^k \to 0, \qquad \sum_{k=0}^{\infty} s^k = \infty.$$

The convergence properties of the diminishing stepsize rule are similar to the ones of their unconstrained counterpart. We cannot guarantee descent at each iteration, but descent is more likely as the stepsize diminishes. The convergence rate tends to be slow, so this stepsize rule is used primarily for singular problems or when there are errors in the gradient calculation. Note that if the constraint set is bounded, the generated sequence $\{x^k\}$ is bounded, and this tends to enhance the convergence properties of the method (see Prop. 1.2.4 in Section 1.2). For this reason, an artificial bounded constraint set is sometimes introduced even for unconstrained problems, when a diminishing stepsize rule is used.

Identification of the Set of Active Constraints*

The main difference between the Armijo rules along the feasible direction and along the projection arc, is that the iterates produced by the latter are more likely to be at the boundary of the constraint set X than the iterates produced by the former. To formalize this idea, let us assume that X is specified by linear inequality constraints

$$X = \{x \mid a_j'x \leq b_j\}.$$

The *set of active constraints at a point* $x \in X$ is the set of indices of the constraints that are satisfied as equations at x, that is, the set

$$A(x) = \{j \mid a_j'x = b_j\}.$$

We say that at the kth iteration the jth constraint is *active*, if $a_j'x^k = b_j$. It can be shown under fairly mild assumptions, that the *set of active constraints at* x^* *are finitely identified* using the Armijo rule along the projection arc (this was first shown in [Ber76c] for the case of bound constraints; for analysis of more general cases, see [GaB84], [Dun87], [BuM88], [Bon89a], [AlK90], [Fla92], [DeT93], [Wri93a]). By this we mean that the set of active constraints at x^k is the same as the set of active constraints at x^* for all sufficiently large k. By contrast an analogous result cannot be shown for the gradient projection method using the Armijo rule along the feasible direction. Generally, the property of finite identification of the active constraints is significant, because it leads to a sharper rate of convergence analysis and because it facilitates the combination of the gradient projection method with other iterations, which are applicable when the set of active constraints stays fixed.

Rate of Convergence

The convergence rate properties of the gradient projection method are essentially the same as those of the unconstrained steepest descent method. As an example, assume that f is quadratic of the form

$$f(x) = \tfrac{1}{2}x'Qx - b'x,$$

where Q is positive definite, and let x^* denote the unique minimum of f over X. Consider the case of a constant stepsize ($a^k = 1$ and $s^k = s$ for all k). Using the nonexpansive property of the projection [Prop. 2.1.3(c)] and the gradient formula $\nabla f(x) = Qx - b$, we have

$$\begin{aligned} \|x^{k+1} - x^*\| &= \left\| \left[x^k - s\nabla f(x^k)\right]^+ - \left[x^* - s\nabla f(x^*)\right]^+ \right\| \\ &\leq \left\| \left(x^k - s\nabla f(x^k)\right) - \left(x^* - s\nabla f(x^*)\right) \right\| \\ &= \left\|(I - sQ)(x^k - x^*)\right\| \\ &\leq \max\left\{|1 - sm|, |1 - sM|\right\}\|x^k - x^*\|, \end{aligned} \tag{2.34}$$

where m and M are the minimum and maximum eigenvalues of Q. This is precisely the rate of convergence estimate obtained for the unconstrained steepest descent method with constant stepsize [cf. Eq. (2.30) and Fig. 1.3.1 in Section 1.3].

We conclude that the gradient projection method suffers from the same type of slow convergence as steepest descent. This motivates us to consider scaling, i.e., application of the gradient projection method in a different coordinate system.

Scaled Gradient Projection

Scaling for gradient projection is derived similar to steepest descent. In particular, at the kth iteration, let H^k be a positive definite matrix and consider a transformation of variables defined by

$$x = (H^k)^{-1/2}y.$$

Then, in the space of y, the problem is written as

$$\begin{aligned} &\text{minimize} \ \ h^k(y) \equiv f\left((H^k)^{-1/2}y\right) \\ &\text{subject to} \ \ y \in Y^k, \end{aligned}$$

where Y^k is the set

$$Y^k = \left\{y \mid (H^k)^{-1/2}y \in X\right\}. \tag{2.35}$$

The gradient projection iteration for this problem takes the form

$$y^{k+1} = y^k + \alpha^k(\bar{y}^k - y^k), \tag{2.36}$$

where

$$\bar{y}^k = \left[y^k - s^k\nabla h^k(y^k)\right]^+.$$

Equivalently, $\bar{y}^k$ can be defined as the vector that minimizes the expression

$$\left\|y - y^k + s^k\nabla h^k(y^k)\right\|^2 = (s^k)^2\left\|\nabla h^k(y^k)\right\|^2 + 2s^k\nabla h^k(y^k)'(y - y^k) + \left\|y - y^k\right\|^2,$$

over $y \in Y^k$. By neglecting the constant term $(s^k)^2\|\nabla h^k(y^k)\|^2$ and by dividing by $2s^k$ in the right-hand side of the above expression, we see that

$$\bar{y}^k = \arg\min_{y \in Y^k} \left\{ \nabla h^k(y^k)'(y - y^k) + \frac{1}{2s^k}\|y - y^k\|^2 \right\}. \tag{2.37}$$

By using the identifications

$$x = (H^k)^{-1/2}y, \qquad x^k = (H^k)^{-1/2}y^k, \qquad \bar{x}^k = (H^k)^{-1/2}\bar{y}^k,$$

$$\nabla h^k(y^k) = (H^k)^{-1/2}\nabla f(x^k)$$

and the definition (2.35) of the set Y^k, we see that the iteration (2.36) can be written as

$$x^{k+1} = x^k + \alpha^k(\bar{x}^k - x^k), \tag{2.38}$$

where $\bar{x}^k$ is given by [cf. Eq. (2.37)]

$$\bar{x}^k = \arg\min_{x \in X} \left\{ \nabla f(x^k)'(x - x^k) + \frac{1}{2s^k}(x - x^k)'H^k(x - x^k) \right\}. \tag{2.39}$$

We refer to the iteration defined by the two preceding equations as the *scaled gradient projection method.*

Note that we can view the quadratic problem in Eq. (2.39) as a generalized projection problem. In particular, it can be verified that $\bar{x}^k$ is the vector of X, which is at minimum distance from the vector $x^k - s^k(H^k)^{-1}\nabla f(x^k)$, but with distance measured in terms of the norm $\|z\|_{H^k} = \sqrt{z'H^kz}$ (see Exercise 2.3.1).

Note also that positive definiteness of the scaling matrix H^k is not strictly necessary for the validity of the method. What is needed is that H^k satisfies

$$(x - x^k)'H^k(x - x^k) > 0, \qquad \forall\, x \in X \text{ with } x \neq x^k.$$

In particular, if the set X is contained in some linear manifold of $\Re^n$, H^k need only be positive definite over the subspace that is parallel to this manifold.

The convergence properties of the scaled gradient projection method are the same as the ones of the unscaled version, provided that the sequence $\{H^k\}$ satisfies a condition guaranteeing that the generated direction sequence is gradient related (see the following Prop. 2.3.4).

The convergence rate of the scaled gradient projection method is governed by m^k and M^k, the smallest and largest eigenvalues of the Hessian $\nabla^2 H^k(y^k)$, which is equal to $(H^k)^{-1/2}\nabla^2 f(x^k)(H^k)^{-1/2}$. As in unconstrained optimization, this suggests that one should try to choose the scaling matrix H^k as close as possible to the Hessian matrix $\nabla^2 f(x^k)$. Using a diagonal approximation to the Hessian is a particularly useful choice because it maintains the simplicity of the quadratic subproblem of Eq. (2.39), when the constraint set X is simple (see Exercise 2.3.3). If $H^k = \nabla^2 f(x^k)$, we obtain a constrained version of Newton's method, which we proceed to describe.

Constrained Newton's Method

Let us assume that f is twice continuously differentiable and that the Hessian matrix $\nabla^2 f(x^k)$ is positive definite for all $x \in X$. Consider the scaled gradient projection method with scaling matrix $H^k = \nabla^2 f(x^k)$. It is given by

$$x^{k+1} = x^k + \alpha^k(\bar{x}^k - x^k), \tag{2.40}$$

where

$$\bar{x}^k = \arg\min_{x \in X} \left\{ \nabla f(x^k)'(x - x^k) + \frac{1}{2s^k}(x - x^k)'\nabla^2 f(x^k)(x - x^k) \right\}. \tag{2.41}$$

Note that if $s^k = 1$, the quadratic cost above is the second order Taylor series expansion of f around x^k (except for a constant term). In particular, when

$$\alpha^k = 1, \qquad s^k = 1,$$

x^{k+1} is the vector that minimizes the second order Taylor series expansion around x^k, just as in the case of unconstrained optimization. We expect, therefore, that for a starting point x^0 sufficiently close to a local minimum x^*, the method with unity stepsizes α^k and s^k converges to x^* superlinearly. Indeed this can be shown using a nearly identical proof to the one for the corresponding unconstrained case (see the following Prop. 2.3.5).

When a good starting point is unknown, one of the Armijo rules with unity initial stepsize may be used to improve the convergence properties of the method. The convergence and rate of convergence results that can be proved for such methods hold no surprises and are very similar to the corresponding unconstrained optimization results of Sections 1.3 and 1.4 (see Exercise 2.3.2).

The main difficulty with Newton's method is that the quadratic direction finding subproblem (2.41) may not be simple, even when the constraint set X has a simple structure. Thus the method typically makes practical sense only for problems of small dimension. This motivates the development of methods that attain the fast convergence of Newton's method, while maintaining the simplicity of the direction finding subproblem when X is a simple set. We will discuss methods of this type in Section 2.4.

Variations of the Scaled Gradient Projection Method*

There are several variations of the scaled gradient projection method. We describe some of the simpler possibilities here and we discuss three major variations in Sections 2.4, 2.5, and 2.6. In all these variations, X is specified by linear inequality constraints

$$X = \{x \mid a_j'x \le b_j,\ j = 1, \ldots, r\}.$$

Projecting on an Expanded Constraint Set

In this variation, we modify the direction finding quadratic subproblem of Eq. (2.39) by considering only the active or nearly active constraints at the current point x^k. In particular, we fix a positive scalar ϵ and obtain a vector $\tilde{x}^k$ as

$$\tilde{x}^k = \arg \min_{x \in X_k} \left\{ \nabla f(x^k)'(x - x^k) + \frac{1}{2s^k}(x - x^k)' H^k (x - x^k) \right\}, \tag{2.42}$$

where

$$X^k = \left\{ x \mid a_j' x \le b_j, \text{ for all } j \text{ with } b_j - \epsilon \le a_j' x^k \le b_j \right\}. \tag{2.43}$$

Note that X^k potentially involves a much smaller number of constraints than X, so there may be an advantage in solving the quadratic subproblem of Eq. (2.42) rather than the subproblem of Eq. (2.39). Note also that $\tilde{x}^k$ need not be a feasible point because X^k may contain X strictly. A feasible descent direction can be obtained, however, as $\bar{x}^k - x^k$, where

$$\bar{x}^k = \bar{\gamma}\tilde{x}^k + (1 - \bar{\gamma})x^k \tag{2.44}$$

and $\bar{\gamma}$ is the largest $\gamma \in [0, 1]$ such that $\gamma\tilde{x}^k + (1 - \gamma)x^k \in X$. The next iterate x^{k+1} is obtained by the usual formula

$$x^{k+1} = x^k + \alpha^k(\bar{x}^k - x^k), \tag{2.45}$$

where α^k is calculated using either the limited minimization or the Armijo rule. The convergence and rate of convergence analysis of this section apply with minor modifications to this method; see Exercise 2.3.4.

Projecting on a Restricted Constraint Set

This variation is similar to the simplicial decomposition method described in Exercise 2.2.7 in Section 2.2.

At the kth iteration, let $\tilde{x}^k$ be defined as an extreme point of X satisfying

$$\tilde{x}^k = \arg \min_{x \in X} \nabla f(x^k)'(x - x^k) \tag{2.46}$$

and define $\bar{x}^k$ as

$$\bar{x}^k = \arg \min_{x \in X^k} \left\{ \nabla f(x^k)'(x - x^k) + \frac{1}{2s^k}(x - x^k)' H^k (x - x^k) \right\}, \tag{2.47}$$

where X^k is the convex hull of x^0 and the extreme points $\tilde{x}^0, \ldots, \tilde{x}^k$,

$$X^k = \left\{ x \;\middle|\; x = \beta x^0 + \sum_{i=0}^{k} \gamma_i \tilde{x}^i,\ \beta \ge 0,\ \gamma_i \ge 0,\ \beta + \sum_{i=0}^{k} \gamma_i = 1 \right\}. \tag{2.48}$$

Note that X^k is the convex hull of the preceding subset X^{k-1} and the extreme point $\tilde{x}^k$ and that it is possible that $\tilde{x}^k \in X^{k-1}$, in which case $X^k = X^{k-1}$. Using the fact that the number of extreme points of X is finite, we can show that the method

$$x^{k+1} = x^k + \alpha^k(\bar{x}^k - x^k), \tag{2.49}$$

with α^k obtained by either the limited minimization or the Armijo rule has the same convergence properties as the regular scaled gradient projection method (Exercise 2.3.5). Note that by using the subset X^k in place of X, the direction finding subproblem may be greatly simplified. For an example where this simplification is important, see Exercise 2.3.6.

Combinations with Unconstrained Optimization Methods

We mentioned earlier that when the Armijo rule along the projection arc is used, the gradient projection method tends to identify the active constraints at a minimum in a finite number of iterations. Thus, for sufficiently large k the method becomes equivalent to steepest descent over a linear manifold

$$\{x \mid a_j'x = b_j,\ j \in A(x^*)\},$$

where $A(x^*)$ is the set of active constraints at the eventual limit x^*. Once this occurs, one may introduce iterations that implement a Newton-like method along this manifold. There is an endless range of possibilities here; any kind of unconstrained minimization method can be adapted fruitfully in this context. The stepsize rule for these methods should be modified so that the generated iterates are feasible, i.e. the stepsize α^k is small enough so that the vector $x^k + \alpha^k(\bar{x}^k - x^k)$ does not violate any of the inactive constraints $a_j'x \le b_j$, $j \notin A(x^k)$.

A difficulty with the type of combined method just described is that one can never really be sure that the active constraints have been identified. Thus a trial and error approach is usually followed, whereby one switches to the unconstrained method after one or more consecutive gradient projection iterations do not change the active set, and one switches to the gradient projection method following a change in the set of active constraints. There are many possibilities along these lines, some of which may be found in [Ber76c], [DeT83], [MoT89], [CGT91].

The Main Limitation of the Gradient Projection Method – Alternatives

The principal drawback of the gradient projection method is the substantial overhead for computing the projection at each iteration. To obtain a good convergence rate, nondiagonal (Newton-like) scaling must often be used, as for example in the iteration (2.40)-(2.41). Unfortunately, in this case, even if the constraints are simple, such as bounds on the variables, the corresponding quadratic program can be very time-consuming.

The next three sections discuss different ways to overcome the difficulty for the case where the constraint set is a polyhedron. This is done by modifying the basic method to simplify the projection problem. One possibility is to solve *equality constrained quadratic programs* in place of inequality constrained quadratic programs. Solving an equality constrained quadratic program is much simpler, because it amounts to projection on a subspace and involves just the solution of an associated system of linear equations (see Example 2.1.5 in Section 2.1).

In Section 2.4, we discuss the *two-metric projection method*, given by

$$x^{k+1} = \left[x^k - \alpha^k D^k \nabla f(x^k)\right]^+,$$

for the case of bound constraints. This is a natural and simple adaptation of unconstrained Newton-like methods. The main difficulty here is that an arbitrary positive definite matrix D^k will not necessarily yield a descent direction. However, it turns out that if some of the off-diagonal terms of D^k are zero, one can obtain descent. Furthermore, one can select D^k as the inverse of a partially diagonalized version of the Hessian matrix $\nabla^2 f(x^k)$ and attain the superlinear rate of Newton's method.

In Section 2.5, we discuss *manifold suboptimization methods*, which are based on scaled gradient projection on the manifold of active constraints. The motivation for these methods is similar to the one for the previously discussed combinations of gradient projection iterations (aimed at identifying the active constraints) with unconstrained iterations on the manifold of active constraints. The main difference is that at the typical iteration of a manifold suboptimization method, at most one constraint can be added or subtracted from the active set.

Finally, in Section 2.6, we discuss some *interior point methods*, which generate sequences of points that are interior to the constraint set. These methods have proved particularly effective for linear programming problems. They can be viewed as scaled projection methods, but the choice of the scaling matrix is crucial for good performance. We discuss the so called *affine scaling method*, which has performed well in practice.

2.3.2 Convergence Analysis*

Generally, the convergence results for (unscaled) gradient projection are similar to those for steepest descent. The details, however, are somewhat more complicated, particularly for the Armijo rule along the projection arc. The following three propositions, each relating to different stepsize rules, are the main results.

Proposition 2.3.1: (Armijo Rule and Limited Minimization Rule Along the Descent Direction) Let $\{x^k\}$ be a sequence generated by the gradient projection method with α^k chosen by the limited minimization rule or by the Armijo rule along the feasible direction. Then every limit point of $\{x^k\}$ is stationary.

Proof: We will show that the direction sequence $\{\bar{x}^k - x^k\}$ is gradient related. Then, application of Prop. 2.2.1 proves the result. Indeed, suppose that $\{x^k\}_{k\in K}$ converges to a nonstationary point $\tilde{x}$. We must prove that

$$\limsup_{k\to\infty,\, k\in K} \|\bar{x}^k - x^k\| < \infty, \tag{2.50}$$

$$\limsup_{k\to\infty,\, k\in K} \nabla f(x^k)'(\bar{x}^k - x^k) < 0. \tag{2.51}$$

By the continuity of the projection [Prop. 2.1.3(c)], we have

$$\lim_{k\to\infty,\, k\in K} \bar{x}^k = [\tilde{x} - s\nabla f(\tilde{x})]^+, \tag{2.52}$$

so Eq. (2.50) holds because $\big\{\|\bar{x}^k - x^k\|\big\}_{k\in K}$ converges to $\|[\tilde{x} - s\nabla f(\tilde{x})]^+ - \tilde{x}\|$. To show Eq. (2.51), we note that, by the characteristic property of the projection [Prop. 2.1.3(b)], we have

$$\big(x^k - s\nabla f(x^k) - \bar{x}^k\big)'(x - \bar{x}^k) \le 0, \qquad \forall\, x \in X.$$

Applying this relation with $x = x^k$, we obtain

$$\nabla f(x^k)'(\bar{x}^k - x^k) \le -\frac{1}{s}\|x^k - \bar{x}^k\|^2. \tag{2.53}$$

By taking limit in the above relation, we obtain

$$\limsup_{k\to\infty,\, k\in K} \nabla f(x^k)'(\bar{x}^k - x^k) \le -\frac{1}{s}\big\|\tilde{x} - \big[\tilde{x} - s\nabla f(\tilde{x})\big]^+\big\|^2.$$

Since $\tilde{x}$ is nonstationary, the right-hand side of the above inequality is negative (cf. Fig. 2.3.2), proving Eq. (2.51). **Q.E.D.**

Proposition 2.3.2: (Constant Stepsize) Let $\{x^k\}$ be a sequence generated by the gradient projection method with $\alpha^k = 1$ and $s^k = s$ for all k. Assume that for some constant $L > 0$, we have

$$\big\|\nabla f(x) - \nabla f(y)\big\| \le L\|x - y\|, \qquad \forall\, x, y \in X. \tag{2.54}$$

Then, if $0 < s < 2/L$, every limit point of $\{x^k\}$ is stationary.

Proof: By using the descent lemma (Prop. A.24 in Appendix A), we have

$$f(x^{k+1}) - f(x^k) = f(\bar{x}^k) - f(x^k) \leq \nabla f(x^k)'(\bar{x}^k - x^k) + \frac{L}{2}\|\bar{x}^k - x^k\|^2.$$

From this equation and Eq. (2.53), we obtain

$$f(x^{k+1}) - f(x^k) \leq \left(\frac{L}{2} - \frac{1}{s}\right)\|\bar{x}^k - x^k\|^2.$$

If $s < 2/L$, the right-hand side of the above relation is nonpositive, so if $\{x^k\}$ has a limit point, the left-hand side tends to 0. Therefore, $\|\bar{x}^k - x^k\| \to 0$, which implies that for every limit point $\tilde{x}$ of $\{x^k\}$ we have $\big[\tilde{x} - s\nabla f(\tilde{x})\big]^+ = \tilde{x}$ [cf. Eq. (2.52))], so $\tilde{x}$ is stationary (cf. Fig. 2.3.2). **Q.E.D.**

An extension of the above proposition that substantially weakens the Lipschitz assumption (2.54), is given in Exercise 2.3.7. The next proposition, originally shown by Gafni and Bertsekas [GaB82], requires a fairly sophisticated proof.

Proposition 2.3.3: (Armijo Rule Along the Projection Arc)

(a) For every $x \in X$ there exists a scalar $s_x > 0$ such that

$$f(x) - f\big(x(s)\big) \geq \sigma \nabla f(x)'\big(x - x(s)\big), \qquad \forall\, s \in [0, s_x], \quad (2.55)$$

where $x(s) = \big[x - s\nabla f(x)\big]^+$ [cf. Eq. (2.32)].

(b) Let $\{x^k\}$ be a sequence generated by the gradient projection method with $\alpha^k = 1$ for all k and with the stepsize s^k chosen by the Armijo rule along the projection arc. Then every limit point of $\{x^k\}$ is stationary.

Proof: We first show the following lemma.

Lemma 2.3.1: For every $x \in X$ and $z \in \Re^n$, the function $g : [0, \infty) \mapsto \Re$ defined by

$$g(s) = \frac{\big\|[x + sz]^+ - x\big\|}{s}, \qquad \forall\, s > 0,$$

is monotonically nonincreasing.

Proof: We take two scalars s_1 and s_2 with $s_1 > 0$ and $s_2 > s_1$, and we show that

$$\frac{\big\|[x + s_2 z]^+ - x\big\|}{s_2} \leq \frac{\big\|[x + s_1 z]^+ - x\big\|}{s_1}.$$

Defining

$$y = s_1 z, \qquad \gamma = \frac{s_2}{s_1},$$

$$a = x + y, \qquad b = x + \gamma y,$$

this inequality is written as

$$\|\bar{b} - x\| \leq \gamma \|\bar{a} - x\|, \tag{2.56}$$

where $\bar{a}$ and $\bar{b}$ are the projections on X of a and b, respectively.

If $\bar{a} = x$ then clearly $\bar{b} = x$, so Eq. (2.56) holds. Also if $a \in X$, then $\bar{a} = a = x+y$, so Eq. (2.56) becomes $||\bar{b}-x|| \leq \gamma||y|| = ||b-x||$, which again holds since the projection is a nonexpansive mapping [cf. Prop. 2.1.3(c)]. Finally, if $\bar{a} = \bar{b}$, then Eq. (2.56) also holds. Therefore, it will suffice to show Eq. (2.56) in the case where $\bar{a} \neq \bar{b}$, $\bar{a} \neq x$, $\bar{b} \neq x$, $a \notin X$, shown in Fig. 2.3.4.

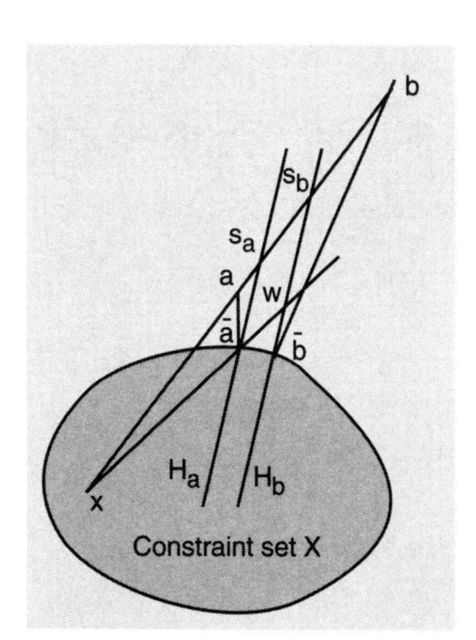

Figure 2.3.4. Construction used in the proof of Lemma 2.3.1.

Let H_a and H_b be the two hyperplanes that are orthogonal to $\bar{b} - \bar{a}$ and pass through $\bar{a}$ and $\bar{b}$, respectively. Since $(\bar{b} - \bar{a})'(b - \bar{b}) \geq 0$ and $(\bar{b} - \bar{a})'(a - \bar{a}) \leq 0$, we have that neither a nor b lie strictly between the two hyperplanes H_a and H_b. Furthermore, x lies on the same side of H_a as a, so $x \notin H_a$. Denote the intersections of the line $\{x + \alpha(b - x) \mid \alpha \in \Re\}$

with H_a and H_b by s_a and s_b, respectively. Denote the intersection of the line $\{x + \alpha(\bar{a} - x) \mid \alpha \in \Re\}$ with H_b by w. We have

$$\begin{aligned}\gamma = \frac{\|b - x\|}{\|a - x\|} &\geq \frac{\|s_b - x\|}{\|s_a - x\|} \\ &= \frac{\|w - x\|}{\|\bar{a} - x\|} = \frac{\|w - \bar{a}\| + \|\bar{a} - x\|}{\|\bar{a} - x\|} \\ &\geq \frac{\|\bar{b} - \bar{a}\| + \|\bar{a} - x\|}{\|\bar{a} - x\|} \geq \frac{\|\bar{b} - x\|}{\|\bar{a} - x\|},\end{aligned} \tag{2.57}$$

where the second equality is by similarity of triangles, the next to last inequality follows from the orthogonality relation $(w - \bar{b})'(\bar{b} - \bar{a}) = 0$, and the last inequality is obtained from the triangle inequality. From Eq. (2.57), we obtain Eq. (2.56), which was to be proved. **Q.E.D.**

We now return to the proof of Prop. 2.3.3:

(a) By the projection theorem, we have

$$\big(x - x(s)\big)'\big(x - s\nabla f(x) - x(s)\big) \leq 0, \qquad \forall\, x \in X,\ s > 0.$$

Hence

$$\nabla f(x)'\big(x - x(s)\big) \geq \frac{\big\|x - x(s)\big\|^2}{s} \qquad \forall\, x \in X,\ s > 0. \tag{2.58}$$

If x is stationary, the conclusion holds with s_x being any positive scalar, so assume that x is nonstationary and therefore $\big\|x - x(s)\big\| \neq 0$ for all $s > 0$. By the mean value theorem, we have for all $x \in X$ and $s \geq 0$,

$$f(x) - f\big(x(s)\big) = \nabla f(x)'\big(x - x(s)\big) + \big(\nabla f(\xi_s) - \nabla f(x)\big)'\big(x - x(s)\big),$$

where ξ_s lies on the line segment joining x and $x(s)$. Therefore, Eq. (2.55) can be written as

$$(1 - \sigma)\nabla f(x)'\big(x - x(s)\big) \geq \big(\nabla f(x) - \nabla f(\xi_s)\big)'\big(x - x(s)\big). \tag{2.59}$$

From Eq. (2.58) and Lemma 2.3.1, we have for all $s \in (0, 1]$,

$$\nabla f(x)'\big(x - x(s)\big) \geq \frac{\big\|x - x(s)\big\|^2}{s} \geq \big\|x - x(1)\big\| \cdot \big\|x - x(s)\big\|.$$

Therefore, Eq. (2.59) is satisfied for all $s \in (0, 1]$ such that

$$(1 - \sigma)\big\|x - x(1)\big\| \geq \big(\nabla f(x) - \nabla f(\xi_s)\big)' \frac{x - x(s)}{\big\|x - x(s)\big\|}.$$

It is seen that there exists $s_x > 0$ such that the above relation, and therefore also Eqs. (2.59) and (2.55), are satisfied for $s \in (0, s_x]$.

(b) Part (a), together with Eq. (2.58) and the definition of the Armijo rule along the projection arc [cf. Eqs. (2.32) and (2.33)], show that s^k is well defined as a positive number for all k. Let $\bar{x}$ be a limit point of $\{x^k\}$ and let $\{x^k\}_K$ be a subsequence converging to $\bar{x}$. Since $\{f(x^k)\}$ is monotonically nonincreasing, we have $f(x^k) \to f(\bar{x})$. Consider two cases:

Case 1: $\liminf_{k\to\infty,\, k\in K} s^k > \hat{s}$ for some $\hat{s} > 0$. Then, from Eq. (2.58) and Lemma 2.3.1, we have for all $k \in K$ that are sufficiently large

$$\begin{aligned} f(x^k) - f(x^{k+1}) &\geq \sigma \nabla f(x^k)'(x^k - x^{k+1}) \\ &\geq \sigma \frac{\|x^k - x^{k+1}\|^2}{s^k} \\ &= \frac{\sigma s^k \|x^k - x^{k+1}\|^2}{(s^k)^2} \\ &\geq \frac{\sigma \hat{s} \|x^k - x^k(\bar{s})\|^2}{\bar{s}^2}, \end{aligned}$$

where $\bar{s}$ is the initial stepsize of the Armijo rule. Taking limit as $k \to \infty$, $k \in K$, we obtain

$$0 \geq \frac{\sigma \hat{s} \|\bar{x} - \bar{x}(\bar{s})\|^2}{\bar{s}^2}.$$

Hence $\bar{x} = \bar{x}(\bar{s})$ and it follows that $\bar{x}$ is stationary (cf. Fig. 2.3.2).

Case 2: $\liminf_{k\to\infty,\, k\in K} s^k = 0$. Then there exists a subsequence $\{s^k\}_{\bar{K}}$, $\bar{K} \subset K$, converging to zero. It follows that for all $k \in \bar{K}$, which are sufficiently large, the Armijo test (2.33) will be failed at least once (i.e., $m^k \geq 1$) and therefore

$$f(x^k) - f\big(x^k(\beta^{-1}s^k)\big) < \sigma \nabla f(x^k)'\big(x^k - x^k(\beta^{-1}s^k)\big). \tag{2.60}$$

Furthermore, for all such $k \in \bar{K}$, x^k cannot be stationary, since for stationary x^k, we have $s^k = \bar{s}$. Therefore,

$$\big\|x^k - x^k(\beta^{-1}s^k)\big\| > 0. \tag{2.61}$$

By the mean value theorem, we have

$$\begin{aligned} f(x^k) - f\big(x^k(\beta^{-1}s^k)\big) = {}& \nabla f(x^k)'\big(x^k - x^k(\beta^{-1}s^k)\big) \\ &+ \big(\nabla f(\xi^k) - \nabla f(x^k)\big)'\big(x^k - x^k(\beta^{-1}s^k)\big), \end{aligned} \tag{2.62}$$

where ξ^k lies in the line segment joining x^k and $x^k(\beta^{-1}s^k)$. Combining Eqs. (2.60) and (2.62), we obtain for all $k \in \bar{K}$ that are sufficiently large

$$(1-\sigma)\nabla f(x^k)'\big(x^k - x^k(\beta^{-1}s^k)\big) < \big(\nabla f(x^k) - \nabla f(\xi^k)\big)'\big(x^k - x^k(\beta^{-1}s^k)\big).$$

Using Eq. (2.58) and Lemma 2.3.1, we obtain

$$\begin{aligned}\nabla f(x^k)'\big(x^k - x^k(\beta^{-1}s^k)\big) &\geq \frac{\big\|x^k - x^k(\beta^{-1}s^k)\big\|^2}{\beta^{-1}s^k} \\ &\geq \frac{1}{\bar{s}}\big\|x^k - x^k(\bar{s})\big\| \cdot \big\|x^k - x^k(\beta^{-1}s^k)\big\|.\end{aligned}$$

Combining the preceding two relations, and using the Schwartz inequality, we obtain for all $k \in \bar{K}$ that are sufficiently large

$$\begin{aligned}\frac{1-\sigma}{\bar{s}}\big\|x^k - x^k(\bar{s})\big\| \cdot \big\|x^k - x^k(\beta^{-1}s^k)\big\| & \\ < \big(\nabla f(x^k) - \nabla f(\xi^k)\big)'\big(x^k - x^k(\beta^{-1}s^k)\big) & \\ \leq \big\|\nabla f(x^k) - \nabla f(\xi^k)\big\| \cdot \big\|x^k - x^k(\beta^{-1}s^k)\big\|. &\end{aligned} \tag{2.63}$$

Using Eqs. (2.61) and (2.63), we obtain

$$\frac{1-\sigma}{\bar{s}}\big\|x^k - x^k(\bar{s})\big\| < \big\|\nabla f(x^k) - \nabla f(\xi^k)\big\|.$$

Since $s^k \to 0$ and $x^k \to \bar{x}$ as $k \to \infty$, $k \in \bar{K}$, it follows that $\xi^k \to \bar{x}$, as $k \to \infty$, $k \in \bar{K}$. Taking the limit in the above relation as $k \to \infty$, $k \in \bar{K}$, we obtain

$$\big\|\bar{x} - \bar{x}(\bar{s})\big\| \leq 0.$$

Hence $\bar{x} = \bar{x}(\bar{s})$ and it follows that $\bar{x}$ is stationary. **Q.E.D.**

Proposition 2.3.4: (Convergence of Scaled Gradient Projection) Let $\{x^k\}$ be a sequence generated by the scaled gradient projection method with α^k chosen by the limited minimization rule or by the Armijo rule along the feasible direction. Assume that, for some positive scalars c_1 and c_2, the scaling matrices H^k satisfy

$$c_1\|z\|^2 \leq z'H^k z \leq c_2\|z\|^2, \qquad \forall\, z \in \Re^n,\ k = 0, 1, \ldots$$

Then every limit point of $\{x^k\}$ is stationary.

Proof: Almost identical to the proof of Prop. 2.3.1; left for the reader. **Q.E.D.**

Proposition 2.3.5: (Convergence of Newton's Method) Let f be twice continuously differentiable with positive definite Hessian for all $x \in X$ and let x^* be a local minimum of f over X. There exists a $\delta > 0$ such that if $\|x_0 - x^*\| < \delta$, then the sequence $\{x^k\}$ generated by the constrained version of Newton's method [Eqs. (2.40) and (2.41)] with $\alpha^k = 1$ and $s^k = 1$ for all k, satisfies $\|x^k - x^*\| < \delta$ for all k and $x^k \to x^*$. Furthermore, $\|x^k - x^*\|$ converges to zero superlinearly.

Proof: (*Abbreviated*) Denote $H^k = \nabla^2 f(x^k)$ and consider the vector norms defined by $\|z\|_{H^k} = \sqrt{z'H^k z}$ for $z \in \Re^n$, and the corresponding norms $\|A\|_{H^k} = \sup_{z \neq 0}\big(\|Az\|_{H^k}/\|z\|_{H^k}\big)$ for $n \times n$ matrices A. Let also $[z]^+$ denote the projection of z on X with respect to the norm $\|\cdot\|_{H^k}$ (cf. Exercise 2.3.1). We have, using the result of Exercise 2.3.1 and the nonexpansive property of projections [Prop. 2.1.3(c) in Section 2.1, which can be generalized for the case of projections with respect to the norm $\|\cdot\|_{H^k}$],

$$\begin{aligned}\|x^{k+1} - x^*\|_{H^k} &= \Big\|\big[x^k - (H^k)^{-1}\nabla f(x^k)\big]^+ - x^*\Big\|_{H^k} \\ &\leq \Big\|x^k - (H^k)^{-1}\nabla f(x^k) - x^*\Big\|_{H^k}.\end{aligned}$$

Using this relation, we obtain as in the proof of Prop. 1.4.1

$$\begin{aligned}\|x^{k+1} - x^*\|_{H^k} \leq M &\left(\int_0^1 \Big\|\nabla^2 f(x^k) - \nabla^2 f\big(x^* + t(x^k - x^*)\big)\Big\|_{H^k} dt\right) \\ &\cdot \big\|x^k - x^*\big\|_{H^k},\end{aligned}$$

for some $M > 0$. By continuity of $\nabla^2 f$, we can take δ sufficiently small to ensure that for $\|x^k - x^*\| < \delta$, the term under the integral sign is arbitrarily small. From this, the superlinear convergence of $\|x^k - x^*\|$ to zero follows. **Q.E.D.**

EXERCISES

2.3.1 (Scaled Gradient Projection Method)

Let H^k be a positive definite matrix and define $\bar{x}^k$ by

$$\bar{x}^k = \arg\min_{x \in X} \left\{\nabla f(x^k)'(x - x^k) + \frac{1}{2s^k}(x - x^k)' H^k (x - x^k)\right\}.$$

Show that $\bar{x}^k$ solves the problem

$$\begin{aligned} &\text{minimize} \ \ \tfrac{1}{2}\left\| x - \left(x^k - s^k (H^k)^{-1}\nabla f(x^k)\right)\right\|_{H^k}^2 \\ &\text{subject to} \ \ x \in X, \end{aligned}$$

where $\|\cdot\|_{H^k}$ is the norm defined by $\|z\|_{H^k} = \sqrt{z'H^k z}$.

2.3.2

Let f be twice continuously differentiable with positive definite Hessian for all $x \in X$. Consider a sequence $\{x^k\}$ generated by the constrained Newton method of Eqs. (2.40) and (2.41), where for each k, either $s^k = 1$ and α^k is chosen by means of the Armijo rule along the feasible direction, or $\alpha^k = 1$ and s^k is chosen by the Armijo rule along the projection arc with initial stepsize $s = 1$. Assume that x^k converges to a local minimum x^* with positive definite Hessian. Show that $\left\{\|x^k - x^*\|\right\}$ converges superlinearly and that there exists an integer $\bar{k} \geq 0$ such that $s^k = 1$ for all $k \geq \bar{k}$.

2.3.3 (Gradient Projection with Diagonal Scaling)

Consider the gradient projection method with a diagonal scaling matrix having the terms H_i^k along the diagonal.

(a) Show that for the case of bound constraints, $X = \{x \mid \alpha_i \leq x_i \leq \beta_i,\ i = 1,\ldots,n\}$, the ith coordinate of the vector $\bar{x}^k$ is given by

$$\bar{x}_i^k = \begin{cases} \alpha_i & \text{if } x_i^k - \frac{s^k}{H_i^k}\frac{\partial f(x^k)}{\partial x_i} \leq \alpha_i \\ \beta_i & \text{if } x_i^k - \frac{s^k}{H_i^k}\frac{\partial f(x^k)}{\partial x_i} \geq \beta_i, \\ x_i^k - \dfrac{s^k}{H_i^k}\dfrac{\partial f(x^k)}{\partial x_i} & \text{otherwise.} \end{cases}$$

(b) Derive a simple algorithm for calculating $\bar{x}^k$ in the case of a simplex constraint.

2.3.4 (www)

Consider the unscaled gradient projection method with projection on an expanded constraint set [Eqs. (2.42)-(2.45) with $s^k = s$]. Show that:

(a) $\bar{x}^k - x^k$ is a feasible descent direction when x^k is nonstationary.

(b) Every limit point of a sequence $\{x^k\}$ generated by the method is stationary.

(c) Consider a variant of the method involving a constant stepsize and show that for this variant, the convergence rate estimate of Eq. (2.34) holds for x^k sufficiently close to x^*.

2.3.5

Consider the unscaled method with projection on a restricted constraint set [Eqs. (2.46)-(2.49)]. Show that the conclusions of parts (a)-(c) of Exercise 2.3.4 hold.

2.3.6 (Gradient Projection for Optimal Routing [BeT89], [BeG92], [LuT94b])

Describe the application of the diagonally scaled gradient projection method with projection on a restricted constraint set to the optimal routing problem of Example 2.1.3 in Section 2.1. Why is this method preferable to the ordinary diagonally scaled gradient projection method? *Hint*: The number of all possible paths may be astronomical, but typically only a few of these paths carry positive flow.

2.3.7 (www)

Suppose that the Lipschitz condition

$$\|\nabla f(x) - \nabla f(y)\| \leq L\|x - y\|, \qquad \forall\, x, y \in X,$$

[cf. Eq. (2.54)] is replaced by the following two conditions:

(i) For every bounded set $A \subset X$, there exists some constant L such that

$$\|\nabla f(x) - \nabla f(y)\| \leq L\|x - y\|, \qquad \forall\, x, y \in A.$$

(ii) The set $\big\{x \in X \mid f(x) \leq c\big\}$ is bounded for every $c \in \Re$.

Show that the convergence result of Prop. 2.3.2 remains valid provided that the constant stepsize s is allowed to depend on the choice of the initial vector x^0. *Hint*: Choose a stepsize that guarantees that x^k stays within the level set $\big\{x \in X \mid f(x) \leq f(x^0)\big\}$.

2.3.8 (The Proximal Minimization Algorithm) (www)

Consider the algorithm

$$x^{k+1} = \arg\min_{x \in X} \left\{ f(x) + \frac{1}{2c^k}\|x - x^k\|^2 \right\}$$

for the case of the quadratic function $f(x) = (1/2)x'Qx + b'x$, and for a sequence of positive scalars $\{c^k\}$ such that $Q + (1/c^k)I$ is positive definite for all k (Q is assumed symmetric but not necessarily positive definite). (This algorithm is reconsidered from a different point of view in Exercise 2.7.1 and in Section 5.4.6.)

(a) Show that this algorithm is equivalent to the scaled gradient projection method (2.38)-(2.39) for the special choice of the scaling matrix $H^k = Q + (1/c^k)I$ and for stepsizes $s^k = 1$, $\alpha^k = 1$. *Hint*: Write f as

$$f(x) = \nabla f(x^k)'(x - x^k) + \tfrac{1}{2}(x - x^k)'Q(x - x^k) + \nabla f(x^k)'x^k - \tfrac{1}{2}x^{k'}Qx^k.$$

(b) Consider the following generalized version of the algorithm:

$$x^{k+1} = x^k + \alpha^k(\bar{x}^k - x^k),$$

where $\alpha^k \in (0, 1]$,

$$\bar{x}^k = \arg\min_{x \in X} \left\{ f(x) + \tfrac{1}{2}(x - x^k)'M^k(x - x^k) \right\},$$

and M^k is such that $Q + M^k$ is positive definite. Show that it can be viewed as a scaled gradient projection method.

(c) Assume that $X = \Re^n$, $M^k = Q$, and that $\alpha^k = 2$. Assuming Q is positive definite, show that the algorithm of part (b) converges in a single step.

2.4 TWO-METRIC PROJECTION METHODS

We mentioned in the preceding section that gradient projection methods make practical sense only when the projection can be carried out fairly easily. A typical example is the case of an unscaled or diagonally scaled gradient projection method and an orthant constraint

$$X = \{x \mid x \geq 0\}. \tag{2.64}$$

On the other hand we have seen that the convergence rate of the diagonally scaled gradient projection method is often unacceptably slow. This motivated the use of more general, nondiagonal scaling. Then, however, the scaled projection becomes a fairly complex quadratic programming problem, even for the case of the simple orthant constraint (2.64).

To resolve the slow convergence/complex implementation dilemma one may think of the method

$$x^{k+1} = \left[x^k - \alpha^k D^k \nabla f(x^k)\right]^+.$$

Here, D^k is a positive definite, not necessarily diagonal, matrix, and $[\cdot]^+$ denotes the usual easy projection on the orthant with respect to the Euclidean norm. Thus, except for the projection operation, this method is identical with the usual scaled gradient iteration for unconstrained minimization. We call this method a *two-metric projection method* because, in contrast with the methods of the preceding section, it embodies two different scaling matrices; one is the matrix D^k, which scales the gradient, and the other is the identity matrix, which is used in the (Euclidean) projection norm. This allows the possibility of incorporating second derivative information within D^k [e.g. $D^k = \big(\nabla^2 f(x^k)\big)^{-1}$, as in unconstrained Newton's method], while maintaining the simplicity of the Euclidean norm projection on the orthant.

In this section we discuss two-metric projection methods for the case of the orthant constraint (2.64), that is, the problem

$$\begin{aligned} &\text{minimize } \ f(x) \\ &\text{subject to } \ x \geq 0. \end{aligned}$$

A similar analysis applies to the case of a box constraint, i.e., upper and lower bounds on the coordinates of x (see Exercise 2.4.1), the case of a Cartesian product of simplices (see Exercise 2.4.2), and more general cases (see the references cited at the end of the chapter).

There is a fundamental difficulty with a two-metric gradient projection method: it is not in general a descent iteration. In particular we may have $f(x^{k+1}) > f(x^k)$ for all positive stepsizes α^k. This situation is illustrated in Fig. 2.4.1, where it can be seen that with an unfavorable choice of D^k, the method does not even recognize (i.e., stop at) a stationary point.

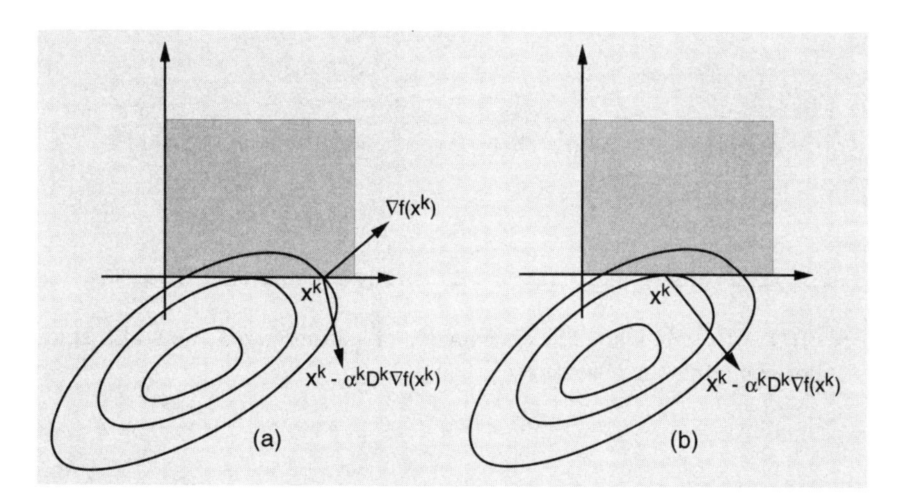

Figure 2.4.1. (a) An example where the iteration

$$x^{k+1} = \big[x^k - \alpha^k D^k \nabla f(x^k)\big]^+$$

fails to make progress at a nonstationary point with a poor choice of the scaling matrix D^k. (b) An example where the iteration fails to stop at a stationary point with a poor choice of the scaling matrix D^k.

It turns out, however, that there is a class of nondiagonal matrices D^k for which descent is guaranteed. This class is sufficiently wide to allow superlinear convergence when D^k properly embodies second derivative information. We introduce this class of matrices and we prove its key properties.

Let us denote for all $x \geq 0$

$$I^+(x) = \left\{ i \;\middle|\; x_i = 0,\ \frac{\partial f(x)}{\partial x_i} > 0 \right\}. \tag{2.65}$$

We say that a symmetric $n \times n$ matrix D with elements d_{ij} is *diagonal with respect to a subset of indices* $I \subset \{1, \ldots, n\}$, if

$$d_{ij} = 0, \qquad \forall\, i \in I,\ j = 1, \ldots, n,\ j \neq i.$$

Proposition 2.4.1: Let $x \geq 0$ and let D be a positive definite symmetric matrix which is diagonal with respect to $I^+(x)$. Denote

$$x(\alpha) = \left[x - \alpha D \nabla f(x)\right]^+, \qquad \forall\, \alpha \geq 0.$$

(a) The vector x is stationary if and only if

$$x = x(\alpha), \qquad \forall\, \alpha \geq 0.$$

(b) If x is not stationary, there exists a scalar $\bar{\alpha} > 0$ such that

$$f\big(x(\alpha)\big) < f(x), \qquad \forall\, \alpha \in (0, \bar{\alpha}]. \tag{2.66}$$

Proof: (a) By relabeling the coordinates of x if necessary, we will assume that for some integer r we have

$$I^+(x) = \{r+1, \ldots, n\}.$$

The following proof also applies with minor modifications when $I^+(x)$ is empty. Since D is diagonal with respect to $I^+(x)$, it has the form

$$D = \begin{pmatrix} \bar{D} & 0 & 0 & \cdots & 0 \\ 0 & d_{r+1} & 0 & \cdots & 0 \\ 0 & 0 & d_{r+2} & \cdots & 0 \\ \vdots & \vdots & \vdots & \vdots & \vdots \\ 0 & 0 & 0 & \cdots & d_n \end{pmatrix}, \tag{2.67}$$

where $\bar{D}$ is positive definite and $d_i > 0$, $i = r+1, \ldots, n$. Denote

$$p = D\nabla f(x). \tag{2.68}$$

Assume that x is a stationary point. Then, from the necessary optimality conditions [Eqs. (2.3) and (2.4) in Section 2.1] and the definition (2.65) of $I^+(x)$, we have

$$\frac{\partial f(x)}{\partial x_i} = 0, \qquad \forall\, i = 1, \ldots, r,$$

$$\frac{\partial f(x)}{\partial x_i} > 0, \qquad \forall\, i = r+1, \ldots, n.$$

These relations and the positivity of d_i imply that

$$p_i = 0, \qquad \forall\, i = 1, \ldots, r,$$

$$p_i > 0, \qquad \forall\, i = r+1, \ldots, n.$$

Since $x_i(\alpha) = [x_i - \alpha p_i]^+$ and $x_i = 0$ for $i = r+1, \ldots, n$, it follows that $x_i(\alpha) = x_i$ for all i, and $\alpha \geq 0$.

Conversely assume that $x = x(\alpha)$ for all $\alpha \geq 0$. Then, we must have

$$p_i = 0, \qquad \forall\, i = 1, \ldots, n \text{ with } x_i > 0,$$

$$p_i \geq 0, \qquad \forall\, i = 1, \ldots, n \text{ with } x_i = 0.$$

Now by the definition of $I^+(x)$, we have that if $x_i = 0$ and $i \notin I^+(x)$, then $\partial f(x)/\partial x_i \leq 0$. This, together with the relations above, imply that

$$\sum_{i=1}^{r} p_i \frac{\partial f(x)}{\partial x_i} \leq 0.$$

Since by Eqs. (2.67) and (2.68),

$$\begin{bmatrix} p_1 \\ \vdots \\ p_r \end{bmatrix} = \bar{D} \begin{bmatrix} \frac{\partial f(x)}{\partial x_1} \\ \vdots \\ \frac{\partial f(x)}{\partial x_r} \end{bmatrix},$$

while $\bar{D}$ is positive definite, we also have

$$\sum_{i=1}^{r} p_i \frac{\partial f(x)}{\partial x_i} \geq 0,$$

and it follows that

$$p_i = \frac{\partial f(x)}{\partial x_i} = 0, \qquad \forall\, i = 1, \ldots, r.$$

Since for $i = r+1, \ldots, n$, we have $\partial f(x)/\partial x_i > 0$ and $x_i = 0$, we see that x is a stationary point.

(b) Let r be as in the proof of part (a). For $i = r+1, \ldots, n$, we have $\partial f(x)/\partial x_i > 0$, $x_i = 0$, and from Eqs. (2.67) and (2.68), we see that

$$x_i = x_i(\alpha) = 0, \qquad \forall\ \alpha \geq 0, \qquad i = r+1, \ldots, n. \tag{2.69}$$

Consider the set of indices

$$I_1 = \{i \mid x_i > 0 \text{ or } (x_i = 0 \text{ and } p_i < 0),\ i = 1, \ldots, r\}, \tag{2.70}$$

$$I_2 = \{i \mid x_i = 0 \text{ and } p_i \geq 0,\ i = 1, \ldots, r\}. \tag{2.71}$$

Let

$$\alpha_1 = \sup\{\alpha \mid x_i - \alpha p_i \geq 0,\ \forall\ i \in I_1\}. \tag{2.72}$$

Note that, in view of the definition of I_1, α_1 is either positive or $+\infty$. Define the vector $\bar{p}$ with coordinates

$$\bar{p}_i = \begin{cases} p_i & \text{if } i \in I_1, \\ 0 & \text{if } i \in I_2 \text{ or } i \in I^+(x). \end{cases} \tag{2.73}$$

Using Eqs. (2.69)-(2.73), we have

$$x(\alpha) = x - \alpha\bar{p}, \qquad \forall\ \alpha \in (0, \alpha_1). \tag{2.74}$$

In view of the definition of I_2 and $I^+(x)$, we obtain

$$\frac{\partial f(x)}{\partial x_i} \leq 0, \qquad \forall\ i \in I_2$$

and hence

$$\sum_{i \in I_2} \frac{\partial f(x)}{\partial x_i} p_i \leq 0. \tag{2.75}$$

Now using Eqs. (2.73) and (2.75), we have

$$\nabla f(x)'\bar{p} = \sum_{i \in I_1} \frac{\partial f(x)}{\partial x_i} p_i \geq \sum_{i=1}^{r} \frac{\partial f(x)}{\partial x_i} p_i. \tag{2.76}$$

Since x is not a stationary point, by part (a) and Eq. (2.74), we must have $x \neq x(\alpha)$ for some $\alpha > 0$ and hence also, in view of Eq. (2.69), $p_i \neq 0$ for some $i \in \{1, \ldots, r\}$. In view of the positive definiteness of $\bar{D}$, and Eqs. (2.67) and (2.68),

$$\sum_{i=1}^{r} \frac{\partial f(x)}{\partial x_i} p_i > 0.$$

It follows from Eq. (2.76) that

$$\nabla f(x)'\bar{p} > 0.$$

Combining this relation with Eq. (2.74) and the fact $\alpha_1 > 0$, yields that $\bar{p}$ is a feasible descent direction at x and there exists a scalar $\bar{\alpha} > 0$ for which the desired relation (2.66) is satisfied. **Q.E.D.**

Based on Prop. 2.4.1 we conclude that to guarantee descent, the matrix D^k in the iteration

$$x^{k+1} = \left[x^k - \alpha^k D^k \nabla f(x^k)\right]^+$$

should be chosen diagonal with respect to a subset of indices that contains

$$I^+(x^k) = \left\{ i \;\middle|\; x_i^k = 0, \frac{\partial f(x^k)}{\partial x_i} > 0 \right\}.$$

However, it turns out that to guarantee convergence one should implement the iteration more carefully. The reason is that the set $I^+(x^k)$ exhibits an undesirable discontinuity at the boundary of the constraint set, whereby given a sequence $\{x^k\}$ of interior points that converges to a boundary point $\bar{x}$ the set $I^+(x^k)$ may be strictly smaller than the set $I^+(\bar{x})$. This causes difficulties in proving convergence of the algorithm and may have an adverse effect on its rate of convergence. To bypass these difficulties one may add to the set $I^+(x^k)$ the indices of those variables x_i^k that satisfy $\partial f(x^k)/\partial x_i > 0$ and are "near" zero (i.e., $0 \leq x_i^k \leq \epsilon$, where ϵ is a small fixed scalar). With such a modification and with a variation of the Armijo rule on the projection arc given in Section 2.3, one can prove a satisfactory convergence result. It is also possible to construct Newton-like algorithms, where the nondiagonal portion of the matrix D^k consists of the inverse of the Hessian submatrix corresponding to the indices not in $I^+(x^k)$, and to show a quadratic rate of convergence result under the appropriate assumptions. This analysis can be found in Bertsekas [Ber82a], [Ber82b].

EXERCISES

2.4.1

Consider the problem

$$\text{minimize } f(x)$$
$$\text{subject to } \alpha \leq x \leq \beta,$$

where α and β are given vectors. Formulate and prove a result that parallels Prop. 2.4.1.

2.4.2 (Simplex Constraints [Ber82b])

Consider the problem

$$\begin{aligned} &\text{minimize } f(x) \\ &\text{subject to } x \geq 0, \quad \sum_{i=1}^{n} x_i = 1. \end{aligned}$$

For a feasible vector x, let $\bar{i}$ be an index such that $x_{\bar{i}} > 0$, and to simplify notation, assume without loss of generality that $\bar{i} = n$. Let $y = (x_1, \ldots, x_{n-1})$, let

$$h(y) = f\left(x_1, \ldots, x_{n-1}, 1 - \sum_{i=1}^{n-1} x_i\right),$$

and let

$$I^+(y) = \left\{ i \;\middle|\; y_i = 0,\ \frac{\partial h(y)}{\partial y_i} > 0 \right\}.$$

Suppose that D is an $(n-1) \times (n-1)$ positive definite matrix, which is diagonal with respect to $I^+(y)$, and define

$$y(\alpha) = \big[y - \alpha D \nabla h(y)\big]^+, \qquad \alpha \geq 0,$$

$$x(\alpha) = \begin{pmatrix} y(\alpha) \\ 1 - \sum_{i=1}^{n-1} y_i(\alpha) \end{pmatrix}$$

Show that:

(a) x is a stationary point if and only if $x = x(\alpha)$ for all $\alpha \geq 0$.

(b) If x is not stationary, there exists a scalar $\bar{\alpha} > 0$ such that

$$f\big(x(\alpha)\big) < f(x), \qquad \forall\, \alpha \in (0, \bar{\alpha}].$$

(c) State a descent method of the Newton type in terms of the first and second derivatives of f.

2.5 MANIFOLD SUBOPTIMIZATION METHODS

The methods of this section are feasible direction methods for the linearly constrained problem

$$\begin{aligned} &\text{minimize } f(x) \\ &\text{subject to } a_j' x \leq b_j, \qquad j = 1, \ldots, r. \end{aligned}$$

They may be viewed as variants of gradient projection methods, the main difference being that, to obtain a feasible descent direction, the gradient is projected on a linear manifold of active constraints rather than on the entire constraint set (see Fig. 2.5.1). This greatly simplifies the projection and constitutes the main advantage of the methods of this section. Furthermore, once the set of active constraints at a solution is identified, the methods behave identically with unconstrained (scaled) gradient methods. The early portion of the computation may be viewed as a systematic effort to identify the set of active constraints at a solution by searching through a sequence of successive manifolds typically differing by a single constraint. Thus the number of iterations required to identify the set of active constraints is typically at least as large as the number of constraints whose active/inactive status is different at the starting point than at the solution. It follows that the methods of this section are well suited only for problems with a relatively small number of constraints.

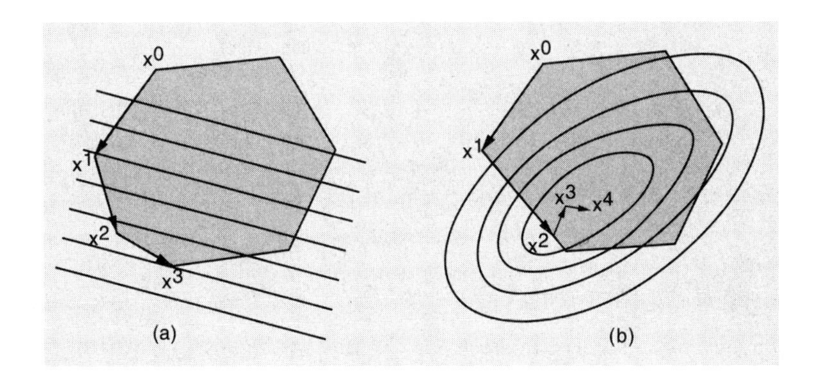

Figure 2.5.1. Illustration of the manifold suboptimization method. The method searches through a sequence of manifolds. Two successive manifolds typically differ by at most one constraint. In (a), the optimal solution is at a vertex, as in the case of linear programs. In (b), the optimal solution is in the interior of the constraint set.

Throughout this section, we assume that at every feasible point x, the set of vectors

$$\{a_j \mid j \in A(x)\}$$

is linearly independent, where $A(x)$ is the set of indices of active constraints at x

$$A(x) = \{j \mid a_j'x = b_j,\ j = 1, \ldots, r\}.$$

This assumption can be relaxed at the expense of some technical complications [basically enough indices have to be dropped from $A(x)$ so that the remaining vectors a_j are linearly independent, but the indices dropped must be chosen carefully so that they define redundant constraints].

The iteration at a vector x^k proceeds roughly as follows:
We try to find a feasible descent direction from the subspace

$$S(x^k) = \{d \mid a_j'd = 0,\ j \in A(x^k)\}. \tag{2.77}$$

This is the subspace which is parallel to the manifold of active constraints, so a small movement along any $d \in S(x^k)$ does not change the set of active constraints.

There are two possibilities:

(a) A feasible descent direction $d^k \in S(x^k)$ is found, in which case the next point x^{k+1} is given by

$$x^{k+1} = x^k + \alpha^k d^k,$$

where α^k is obtained by some stepsize rule within the interval of stepsizes

$$\{\alpha > 0 \mid x^k + \alpha d^k \text{ is feasible}\}.$$

(b) No feasible descent direction $d \in S(x^k)$ can be found because x^k is stationary over the manifold $x^k + S(x^k)$ of active constraints.

In case (b), there are two possibilities: either x^k is stationary over the entire constraint set

$$\{x \mid a_j'x \le b_j,\ j = 1, \ldots, r\},$$

in which case the algorithm stops, or else one of the active constraints, say $\bar{j}$, is relaxed and a feasible descent direction belonging to the subspace

$$\bar{S}(x^k) = \{d \mid a_j'd = 0,\ j \in A(x^k),\ j \neq \bar{j}\}$$

is obtained (see Fig. 2.5.2). We will describe shortly how $\bar{j}$ is selected and how the corresponding direction is computed.

Central to manifold optimization methods are quadratic programming problems of the form

$$\begin{aligned} &\text{minimize} \quad \nabla f(x^k)'d + \tfrac{1}{2}d'H^k d \\ &\text{subject to} \quad d \in S(x^k) \end{aligned} \tag{2.78}$$

where H^k is a symmetric positive definite matrix and $S(x^k)$ is the subspace given by Eq. (2.77). Note that the solution of this problem can be viewed as a scaled projection of the gradient $\nabla f(x^k)$ on the subspace $S(x^k)$. The unique optimal solution is (see Example 2.1.5 in Section 2.1)

$$d^k = -(H^k)^{-1}\big(\nabla f(x^k) + A^{k\prime}\mu\big), \tag{2.79}$$

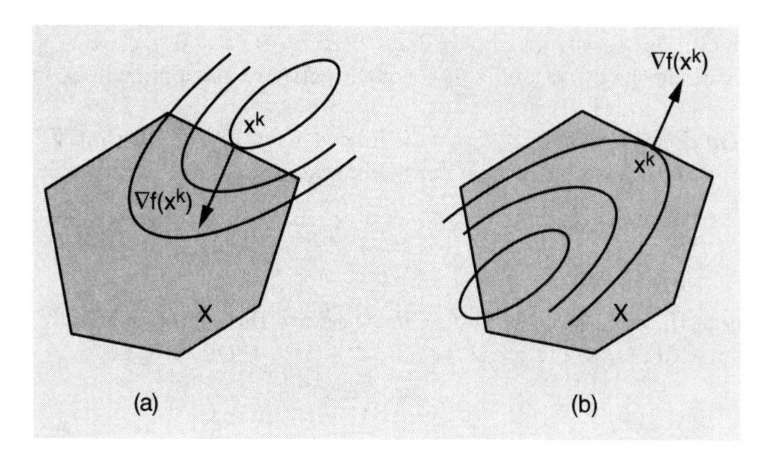

Figure 2.5.2. Illustration of the manifold suboptimization method when x^k is stationary over the current manifold of active constraints. In (a), x^k is stationary over the entire constraint set and the algorithm stops. In (b), one of the active constraints is relaxed and the search continues over a larger manifold.

$$\mu = -\big(A^k (H^k)^{-1} A^{k\prime}\big)^{-1} A^k (H^k)^{-1} \nabla f(x^k), \tag{2.80}$$

where A^k is the matrix that has as rows the vectors a_j, $j \in A(x^k)$. [It is actually sufficient that H^k be positive definite on just the subspace $S(x^k)$; then there will again be a unique solution d^k, but Eqs. (2.79) and (2.80) must be appropriately modified.] Let us describe the typical iteration of the method:

We first note that since d^k is the optimal solution of the quadratic programming problem (2.78), and since the vector $d = 0$ is feasible for this problem, we must have

$$\nabla f(x^k)' d^k + \tfrac{1}{2} d'^k H^k d^k \le 0.$$

If $d^k \neq 0$, we see that

$$\nabla f(x^k)' d^k \le -\tfrac{1}{2} d'^k H^k d^k < 0, \tag{2.81}$$

and d^k is a feasible descent direction at x^k. The iteration is then given by

$$x^{k+1} = x^k + \alpha^k d^k, \tag{2.82}$$

where α^k is obtained by some stepsize rule (e.g., Armijo, limited minimization, etc.) over the interval of stepsizes α for which $x^k + \alpha d^k$ is feasible, that is, the set

$$\big\{\alpha > 0 \mid a_j'(x^k + \alpha d^k) \le b_j,\ j \notin A(x^k)\big\}.$$

[Note that, by construction, we have $a_j'd^k = 0$ for all $j \in A(x^k)$, so by moving along the direction d^k, none of the active constraints will be violated or become inactive.]

On the other hand, if $d^k = 0$, it follows from Eq. (2.79) that $\nabla f(x^k) + A^{k'}\mu = 0$, or equivalently, that for some scalars $\mu_j, j \in A(x^k)$, we have

$$\nabla f(x^k) + \sum_{j \in A(x^k)} \mu_j a_j = 0. \tag{2.83}$$

Now note that the feasible directions at x^k are the vectors d with $a_j'd \leq 0$, for all $j \in A(x^k)$, so if $\mu_j \geq 0$ for all $j \in A(x^k)$, then we have

$$\nabla f(x^k)'d = - \sum_{j \in A(x^k)} \mu_j a_j'd \geq 0$$

for all feasible directions d, so that x^k is stationary. Hence if $d^k = 0$ and x^k is nonstationary, there must exist some index $\bar{j} \in A(x^k)$ such that $\mu_{\bar{j}} < 0$ (see Fig. 2.5.3 for an interpretation). Let $\bar{d}^k$ be the unique solution of the problem

$$\begin{aligned} &\text{minimize} \quad \nabla f(x^k)'d + \tfrac{1}{2} d'\bar{H}^k d \\ &\text{subject to} \quad d \in \bar{S}(x^k) = \big\{ d \mid a_j'd = 0,\ j \in A(x^k),\ j \neq \bar{j} \big\} \end{aligned} \tag{2.84}$$

where $\bar{H}^k$ is a positive definite symmetric matrix. We claim that $\bar{d}^k$ is a feasible descent direction, so that the iteration takes the form

$$x^{k+1} = x^k + \alpha^k \bar{d}^k, \tag{2.85}$$

where α^k is a stepsize chosen as earlier to maintain feasibility of x^{k+1}.

To verify this, we first show that $\bar{d}^k \neq 0$. Indeed, if $\bar{d}^k = 0$, then

$$\nabla f(x^k) + \sum_{j \in A(x^k),\, j \neq \bar{j}} \bar{\mu}_j a_j = 0, \tag{2.86}$$

for some $\bar{\mu}_j$, $j \in A(x^k)$, $j \neq \bar{j}$. By subtracting Eqs. (2.83) and (2.86), we see that

$$\mu_{\bar{j}} a_{\bar{j}} = \sum_{j \in A(x^k),\, j \neq \bar{j}} (\bar{\mu}_j - \mu_j) a_j,$$

which, in view of the choice $\mu_{\bar{j}} < 0$, contradicts the linear independence of $\big\{ a_j \mid j \in A(x^k) \big\}$. Therefore, $\bar{d}^k \neq 0$ and as earlier [cf. Eq. (2.81)], we obtain

$$\nabla f(x^k)'\bar{d}^k \leq -\tfrac{1}{2} \bar{d}'^k \bar{H}^k \bar{d}^k < 0,$$

implying that $\bar{d}^k$ is a descent direction. To show that $\bar{d}^k$ is also a feasible direction, we must show that

$$a_j'\bar{d}^k \leq 0, \qquad \forall\ j \in A(x^k).$$

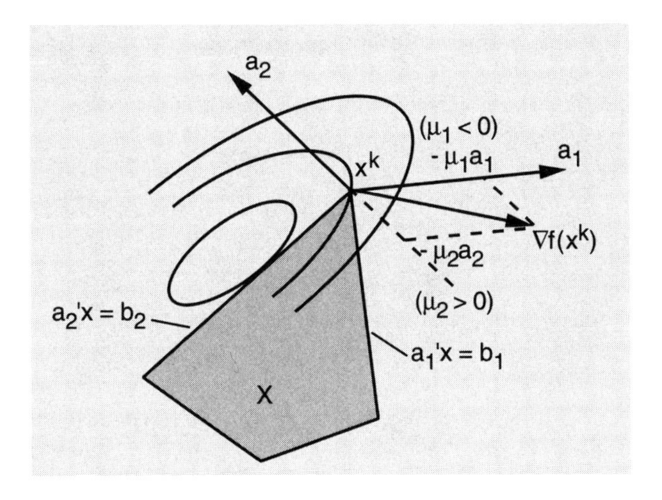

Figure 2.5.3. Dropping a constraint when x^k is a stationary point on the manifold of active constraints. In this example, x^k is a vertex and coincides with the manifold of active constraints. Therefore, x^k is by default stationary on the manifold of active constraints, and one of the two constraints must be dropped. This is the constraint $a_1'x = b_1$ for which the corresponding scalar μ_1 is negative (notice that dropping the other constraint does not allow further descent).

We know that $a_j'\bar{d}^k = 0$ for all $j \in A(x^k)$ except $j = \bar{j}$, and we will show that $a_{\bar{j}}'\bar{d}^k < 0$. Indeed, by taking inner product of Eq. (2.83) with $\bar{d}^k$, we obtain

$$\nabla f(x^k)'\bar{d}^k + \sum_{j \in A(x^k)} \mu_j a_j'\bar{d}^k = 0$$

and since by construction [cf. Eq. (2.84)], we have $a_j'\bar{d}^k = 0$ for all $j \in A(x^k)$, $j \neq \bar{j}$, we obtain

$$a_{\bar{j}}'\bar{d}^k = -\frac{\nabla f(x^k)'\bar{d}^k}{\mu_{\bar{j}}}.$$

Since $\mu_{\bar{j}} < 0$ and $\nabla f(x^k)'\bar{d}^k < 0$, we see that $a_{\bar{j}}'\bar{d}^k < 0$. Thus, $\bar{d}^k$ is a feasible descent direction and a small movement along $\bar{d}^k$ makes the $\bar{j}$th constraint inactive, while it maintains the active status of all other constraints that are active at x^k.

We have thus completed the description of the manifold suboptimization iteration, together with a method for modifying the manifold of active constraints when no more progress is possible on the current active constraint manifold. It is given by iterations (2.82) or (2.85), depending on whether the scaled projection is done on the manifold of active constraints, or on the manifold obtained after one of the constraints is dropped, respectively.

Positive Definite Quadratic Programming

Suppose that f is the positive definite quadratic function

$$f(x) = \tfrac{1}{2}x'Qx + c'x.$$

Consider the algorithm of this section with the matrices H^k and $\bar{H}^k$ in the quadratic subproblems (2.78) and (2.84) chosen to be equal to Q. In this case, the quadratic subproblem (2.78) takes the form

$$\begin{aligned} &\text{minimize} \quad (Qx^k + c)'d + \tfrac{1}{2}d'Qd \\ &\text{subject to} \quad a_j'd = 0, \qquad j \in A(x^k). \end{aligned} \tag{2.87}$$

This problem can be viewed as a restricted version of the original quadratic program. Indeed, since $a_j'x^k = b_j$ for $j \in A(x^k)$, by letting $x = x^k + d$ and by adding the constant term $\frac{1}{2}x^{k\prime}Qx^k + c'x^k$ to its cost function, this problem can be written equivalently as

$$\begin{aligned} &\text{minimize} \quad \tfrac{1}{2}x'Qx + c'x \\ &\text{subject to} \quad a_j'x = b_j, \qquad j \in A(x^k). \end{aligned} \tag{2.88}$$

This is the same problem as the original except that the cost function $f(x)$ is minimized over the currently active manifold $\{x \mid a_j'x = b_j,\ j \in A(x^k)\}$ rather than over the full constraint set $\{x \mid a_j'x \le b_j,\ j = 1, \ldots, r\}$.

Let now d^k be the unique optimal solution of problem (2.87). There are two cases (see Fig. 2.5.4):

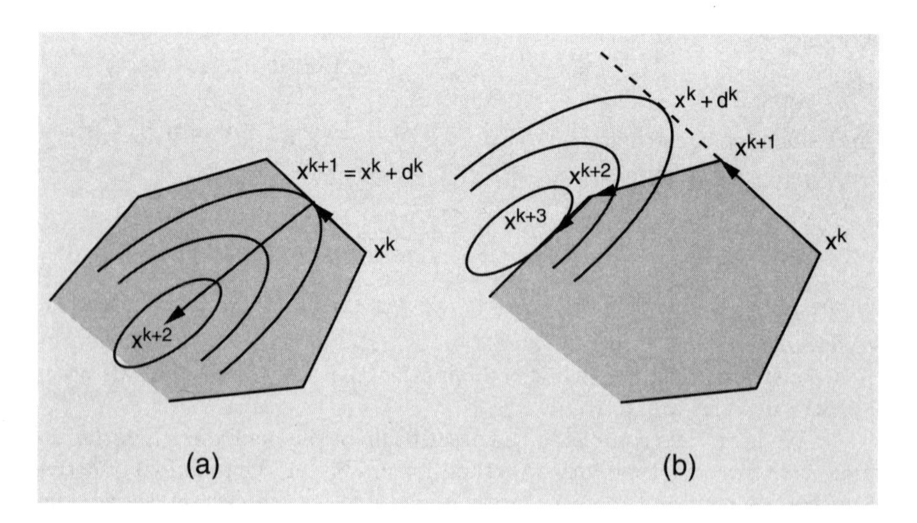

Figure 2.5.4. Illustration of the quadratic programming algorithm. In (a) the vector $x^k + d^k$ is feasible and one of the active constraints must be dropped at the next iteration; in (b) the vector $x^k + d^k$ is not feasible, in which case a new constraint is added to the active set.

(1) $d^k = 0$. Then, one of the active constraints $\bar{j} \in A(x^k)$ involving a negative coefficient $\mu_{\bar{j}}$, must be dropped to form the enlarged manifold of active constraints. Once this is done, we obtain d^k as the unique solution of the problem

$$\begin{aligned} &\text{minimize} \quad (Qx^k + c)'d + \tfrac{1}{2}d'Qd \\ &\text{subject to} \quad a_j'd = 0, \qquad j \in A(x^k),\ j \neq \bar{j} \end{aligned}$$

and proceed as in the following case (2).

(2) $d^k \neq 0$. Then, there are two possibilities:

(a) $x^k + d^k$ *is feasible*, in which case we set

$$x^{k+1} = x^k + d^k.$$

Then, since problems (2.87) and (2.88) are equivalent, x^{k+1} minimizes $f(x)$ over the current manifold $\{x \mid a_j'x = b_j,\ j \in A(x^k)\}$, and in the next iteration, we will drop a constraint as in case (1) above.

(b) $x^k + d^k$ *is not feasible*, in which case we set

$$x^{k+1} = x^k + \alpha^k d^k,$$

where

$$\alpha^k = \max\{\alpha > 0 \mid x^k + \alpha d^k : \text{ feasible}\}.$$

Then, at least one (and typically just one) new constraint will be added to the active set for the next iteration.

In a practical algorithm, the operations of adding and dropping constraints from the active set, together with the solution of the corresponding quadratic subproblems (2.88) can be done using efficient linear algebra operations (see [GiM74] and [GMW81] for a more detailed discussion).

We finally show that the quadratic programming algorithm terminates with the unique optimal solution in a finite number of iterations. To this end, let us classify the manifold of active constraints $x^k + S(x^k)$ as one of two types: if the unique minimizer of f over $x^k + S(x^k)$ is feasible, we say that the manifold is of type F and otherwise we say that the manifold is of type N. We note that a manifold of type F can be visited at most once because if x^k belongs to such a manifold, then x^{k+1} is the unique minimizer of f over the manifold and a descent algorithm cannot generate the same point twice. On the other hand, each time the algorithm visits a manifold of type N, at the next iteration it visits a manifold involving at least one more constraint. Therefore, there can be at most n successive iterations in manifolds of type N. Thus, if the number of manifolds of type F is m, the number of iterations is bounded by mn.

Manifold Suboptimization and Linear Programming

Manifold suboptimization methods relate to one of the oldest and most popular optimization algorithms, the *simplex method* for linear programming. This method applies to problems of the form

$$\begin{aligned} &\text{minimize} \;\; c'x \\ &\text{subject to} \;\; x \in X \end{aligned}$$

where X is a polyhedron in $\Re^n$ that has at least one vertex (extreme point) and c is a given vector. For these problems, the fundamental theorem of linear programming [Prop. B.21(c)] states that if the cost attains a minimum over X, then this minimum is attained at some vertex of X. The idea of the simplex method is to start at some vertex of X and then (assuming the optimal cost is finite) to generate a sequence of vertices of X. These vertices satisfy two properties:

(a) Each vertex, if it is not optimal, it has a lower cost than the preceding vertices.

(b) Two successive vertices are connected by an edge; that is, they both lie on some one-dimensional active constraint manifold.

Assuming the optimal cost is finite, the method is guaranteed to terminate in a finite number of iterations at an optimal vertex. The reason is that X has a finite number of vertices (as shown in Appendix B), and a vertex cannot be repeated because of property (a) above.

The selection of the next vertex starting from a given vertex in the simplex method is guided by iterative descent; that is, out of all the edges that lead to neighboring vertices, one that corresponds to a descent direction is chosen. If we view a vertex as an active constraint manifold of dimension 0 and an edge as an active constraint manifold of dimension 1, it is seen that the movement to a neighboring vertex of lower cost encapsulates the basic operations of the manifold suboptimization method: first dropping a constraint at a vertex, then moving along a descent direction on the corresponding one-dimensional manifold (edge), until a new constraint is encountered and the active manifold becomes a vertex again. The choice of a constraint to drop is identical with the one of the manifold method (except for streamlining the linear algebra). Thus the behavior of the manifold method, when applied to linear programs and is started at a vertex of the constraint polyhedron, is identical to the one of the simplex method. The versions of the simplex method used in practice involve a lot of important implementation technology, which is beyond the scope of this book. This technology is needed to enhance efficiency and also to deal with degeneracy (more than n linear constraints active at a given vertex). Manifold suboptimization methods can benefit from the use of some of this technology. We refer to the literature (particularly, Gill, Murray, and Wright [GiM74], [GMW81], [GMW91]) for more details.

EXERCISES

2.5.1

Use the method of this section to solve the three-dimensional quadratic problem

$$\begin{aligned} &\text{minimize } \; f(x) = x_1^2 + 2x_2^2 + 3x_3^2 \\ &\text{subject to } \; x_1 + x_2 + x_3 \geq 1, \qquad 0 \leq x_1, \; 0 \leq x_2, \; 0 \leq x_3, \end{aligned}$$

starting from the point $x^0 = (0, 0, 1)$.

2.5.2

Show by example that in the method of this section, if several constraints $a_j'x \leq b_j$ with negative values μ_j are simultaneously dropped, then the corresponding vector $\bar{d}^k$ need not be a feasible direction.

2.6 AFFINE SCALING FOR LINEAR PROGRAMMING

The gradient projection methods for linear inequality constraints considered so far obtain the descent direction by solving some quadratic programming problem. We saw two extreme types of methods with several others lying in between. In the first type of method, discussed in Sections 2.3 and 2.4, the projection is done by solving an inequality constrained quadratic problem involving all of the linear constraints; here the set of active constraints may change radically from one iteration to the next. In the second type of method, the manifold suboptimization methods discussed in the preceding section, the projection is done by solving an easier equality constrained quadratic problem; here the set of active constraints typically changes by no more than one constraint per iteration. As a result the method may require a large number of iterations to identify the set of active constraints.

In this section we consider a third type of gradient projection method, which has proved particularly interesting for linear programming problems. It shares with the manifold suboptimization method the advantage of solving equality (rather than inequality) constrained quadratic programming problems; however, the optimal solution is approached through the *interior* of the constraint set, thereby obviating the inefficiency of identifying a large number of active constraints, one-by-one.

We will focus on the linear programming problem

$$\begin{aligned} &\text{minimize} \quad c'x \\ &\text{subject to} \quad Ax = b, \qquad x \geq 0, \end{aligned} \tag{LP1}$$

where $c \in \Re^n$ and $b \in \Re^m$ are given vectors, and A is an $m \times n$ matrix of rank m. We assume that the optimal cost $\min_{Ax=b,\, x\geq 0} c'x$ is finite. Suppose that we have a feasible vector x^k such that $x^k > 0$, that is, $x_i^k > 0$ for $i = 1, \ldots, n$ (an initial vector $x^0 > 0$ can be found as discussed in Section 2.2.1).

The scaled gradient projection iteration considered in Section 2.3.1 [cf. Eqs. (2.38) and (2.39)] is given by

$$x^{k+1} = x^k + \alpha^k(\bar{x}^k - x^k), \tag{2.89}$$

where α^k is a positive stepsize and $\bar{x}^k$ solves a quadratic programming problem of the form

$$\begin{aligned} &\text{minimize} \quad c'(x - x^k) + \frac{1}{2s^k}(x - x^k)'H^k(x - x^k) \\ &\text{subject to} \quad Ax = b, \qquad x \geq 0. \end{aligned} \tag{2.90}$$

Here, H^k is a positive definite matrix and s^k is a positive parameter.

We now observe that for a fixed H^k, the constraints $x \geq 0$ are inactive at the solution $\bar{x}^k$ of the quadratic programming problem ($\bar{x}^k > 0$), provided s^k is small enough. Indeed, based on Example 2.1.5 in Section 2.1, we see that the solution of the *equality* constrained problem

$$\begin{aligned} &\text{minimize} \quad c'(x - x^k) + \frac{1}{2s^k}(x - x^k)'H^k(x - x^k) \\ &\text{subject to} \quad Ax = b \end{aligned}$$

is given by

$$\bar{x}^k = x^k - s^k(H^k)^{-1}(c - A'\lambda^k), \tag{2.91}$$

where

$$\lambda^k = \big(A(H^k)^{-1}A'\big)^{-1}A(H^k)^{-1}c. \tag{2.92}$$

From Eq. (2.91), we see that if s^k is sufficiently small, then since $x^k > 0$, we also have $\bar{x}^k > 0$, implying that the constraints $x \geq 0$ of the quadratic program (2.90) are superfluous, and that $\bar{x}^k$ as given by Eqs. (2.91) and (2.92) solves that program.

By using the expression (2.91) for $\bar{x}^k$ in iteration (2.89), and by lumping the scalar s^k into the stepsize parameter α^k, we obtain the iteration

$$x^{k+1} = x^k - \alpha^k(H^k)^{-1}(c - A'\lambda^k), \tag{2.93}$$

where λ^k is given by Eq. (2.92) and α^k is chosen small enough to ensure that $x^{k+1} > 0$; see Fig. 2.6.1. In particular we must have $\alpha^k < \bar{\alpha}^k$ where $\bar{\alpha}^k$ is the maximum stepsize for which x^{k+1} is feasible, that is,

$$\bar{\alpha}^k = \max\{\alpha \mid x^k - \alpha(H^k)^{-1}(c - A'\lambda^k) \geq 0\}. \tag{2.94}$$

We know that $\bar{x}^k - x^k$ is a descent direction, so the cost decrease is proportional to the size of α^k; this argues for a value of α^k close to $\bar{\alpha}^k$. On the other hand if α^k is very close to $\bar{\alpha}^k$, then x^{k+1} will be very close to the constraint boundary, restricting large cost improvements in subsequent iterations. In practice it is common to choose α^k between $0.9\bar{\alpha}^k$ and $0.999\bar{\alpha}^k$. Note that $\bar{\alpha}^k$ is well-defined as a positive scalar by Eq. (2.94), since if we had $x^k - \alpha(H^k)^{-1}(c - A'\lambda^k) > 0$ for arbitrarily large stepsizes α, then we would be able to make the cost $c'x$ arbitrarily small, contradicting our assumption that the optimal cost is finite.

An important question has to do with the choice of the scaling matrices H^k. Not every type of scaling matrix can lead to convergence through the interior of the constraint set. For example, it can be seen that choosing H^k to be the identity for all k, which corresponds to unscaled gradient projection, does not work because for a linear cost, the projected gradient direction will change only if H^k changes. What is needed is a matrix with the property that it "bends" the direction $\bar{x}^k - x^k$ "away from the boundary" of the constraint set.

A particularly interesting choice is

$$H^k = (X^k)^{-2},$$

where X^k is the diagonal matrix having the (positive) coordinates x_i^k, $i = 1, \ldots, n$, along the diagonal. The iteration (2.93) then takes the form

$$x^{k+1} = x^k - \alpha^k (X^k)^2 (c - A'\lambda^k), \tag{2.95}$$

where

$$\lambda^k = \left(A(X^k)^2 A'\right)^{-1} A(X^k)^2 c, \tag{2.96}$$

and is known as the *affine scaling method*. It was invented in 1967 by Dikin [Dik67], but it received a lot of attention in the West following the proposal by Karmarkar of a different but related method [Kar84]. Karmarkar's method has been shown to have nicer theoretical properties than the affine scaling method (a polynomial worst-case computational complexity), but in practice the affine scaling method is generally thought to be superior.

The use of the scaling matrix $H^k = (X^k)^{-2}$ corresponds to an unscaled gradient projection iteration in the coordinate system $y = (X^k)^{-1}x$. The vector x^k is mapped in this system onto the unit vector $y^k = (1, \ldots, 1)$. Since the coordinates of y^k are uniformly far from the boundary of the constraint $y \geq 0$, this choice of scaling tends to allow large steps and associated cost reductions. Figure 2.6.2 illustrates the iteration in the two coordinate systems.

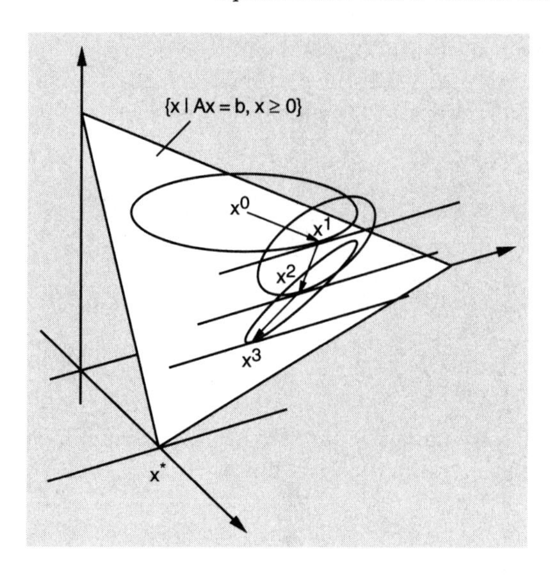

Figure 2.6.1. Illustration of the gradient projection method (2.93) applied to a linear program with three variables and one constraint. The new iterate x^{k+1} is obtained by solving the problem

$$\begin{aligned} &\text{minimize} \quad c'(x - x^k) + \frac{1}{2\alpha^k}(x - x^k)'H^k(x - x^k) \\ &\text{subject to} \quad Ax = b, \end{aligned}$$

where α^k is such that $x^{k+1} > 0$. The scaling matrix H^k determines the shape of the ellipsoid corresponding to the iteration. Clearly, H^k must change at each iteration, so that the ellipsoid "adapts" to the boundary of the feasible region. The affine scaling method uses the choice $H^k = (X^k)^{-2}$, so that if a coordinate x_i is small, the corresponding axis of the ellipsoid is also small.

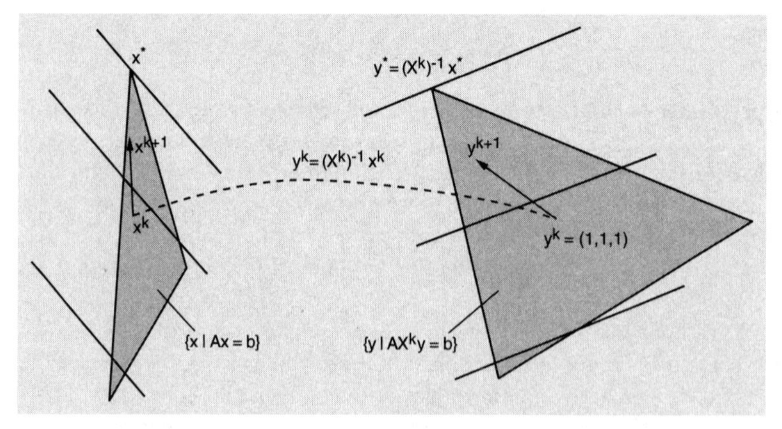

Figure 2.6.2. An iteration of the affine scaling method in the original and in the transformed coordinate system under the transformation $y = (X^k)^{-1}x$.

Inequality Constrained Linear Programs*

Consider now the problem

$$\begin{aligned} &\text{maximize} \quad b'\lambda \\ &\text{subject to} \quad A'\lambda \leq c, \end{aligned} \tag{LP2}$$

where $\lambda \in \Re^m$ is the vector of decision variables, and b, c, and A are as in problem (LP1). It will be shown in Section 3.4 that problems (LP1) and (LP2) are *dual* to each other, in the sense that optimal solutions $\bar{x}$ of (LP1) and $\bar{\lambda}$ of (LP2) are related by $c'\bar{x} = b'\bar{\lambda}$, but this duality relation will not concern us here. We will develop a version of the affine scaling method, applied to problem (LP2).

Let $w \in \Re^n$ be a vector satisfying $Aw = b$. [It can be shown that such a vector exists if problem (LP2) has an optimal solution. A proof is given in Chapter 3; see Section 3.4.2, Prop. 3.4.2, and Example 3.4.2.] Then by making the transformation $x = c - A'\lambda$, we obtain $b'\lambda = w'A'\lambda = w'(c - x)$, so problem (LP2) can equivalently be written as

$$\begin{aligned} &\text{minimize} \quad w'x \\ &\text{subject to} \quad x \geq 0, \text{ and } x = c - A'\lambda \text{ for some } \lambda \in \Re^m. \end{aligned} \tag{2.97}$$

Since the constraint $x = c - A'\lambda$ for some $\lambda \in \Re^m$ defines a linear manifold, this is a linear program of the form (LP1) to which the affine scaling method can be applied. The kth iteration takes the following form:

Given x^k and λ^k with $x^k = c - A'\lambda^k > 0$, we solve the problem

$$\begin{aligned} &\text{minimize} \quad w'(x - x^k) + \frac{1}{2s^k}(x - x^k)'H^k(x - x^k) \\ &\text{subject to} \quad x - x^k = -A'(\lambda - \lambda^k) \text{ for some } \lambda \in \Re^m, \end{aligned} \tag{2.98}$$

where

$$H^k = (X^k)^{-2}$$

and X^k is the diagonal matrix with the coordinates of the vector $x^k = c - A'\lambda^k$ along the diagonal. The affine scaling iteration has the form

$$x^{k+1} = x^k + \alpha^k(\bar{x}^k - x^k), \tag{2.99}$$

where $\bar{x}^k$ solves problem (2.98) and α^k is a positive stepsize that maintains the condition $x^{k+1} > 0$. Equivalently, using the relation $Aw = b$, problem (2.98) can be written in terms of the vector λ as

$$\begin{aligned} &\text{minimize} \quad -b'(\lambda - \lambda^k) + \frac{1}{2s^k}(\lambda - \lambda^k)'AH^kA'(\lambda - \lambda^k) \\ &\text{subject to} \quad x - x^k = -A'(\lambda - \lambda^k) \text{ for some } \lambda \in \Re^m. \end{aligned}$$

The solution is

$$\bar{\lambda}^k = \lambda^k + s^k(AH^kA')^{-1}b$$

and by using the relation $\bar{x}^k - x^k = -A'(\bar{\lambda}^k - \lambda^k)$, the affine scaling iteration (2.99) can be written as

$$x^{k+1} = x^k - \alpha^k A'(AH^kA')^{-1}b.$$

By using the transformation $x^k = c - A'\lambda^k$, we obtain

$$A'\lambda^{k+1} = A'\lambda^k + \alpha^k A'(AH^kA')^{-1}b.$$

By multiplying this iteration by $(AA')^{-1}A$, we can write it as

$$\lambda^{k+1} = \lambda^k + \alpha^k(AH^kA')^{-1}b, \tag{2.100}$$

where α^k is a positive stepsize maintaining the condition $c - A'\lambda^{k+1} > 0$. This is the affine scaling iteration applied to λ^k.

Because of the duality of problems (LP1) and (LP2) mentioned earlier, iteration (2.100) is called the *dual affine scaling method* for problem (LP1). In contrast, the iteration (2.95) is called the *primal affine scaling method* for problem (LP1). However, as we have seen above, the dual method is the same as the primal method applied to the equivalent problem (2.97).

Convergence Analysis of the Affine Scaling Method*

The affine scaling method, in the form described above, typically performs very well in practice, but its theoretical convergence to the optimum can be shown only under certain restrictive nondegeneracy assumptions, to be discussed shortly. In particular, let us write the primal affine scaling method (2.95), (2.96) as

$$x^{k+1} = x^k - \beta\bar{\alpha}^k(X^k)^2z^k, \tag{2.101}$$

where

$$z^k = c - A'\lambda^k,$$

$$\lambda^k = \big(A(X^k)^2A'\big)^{-1}A(X^k)^2c,$$

$\bar{\alpha}^k$ is the maximum stepsize for which the new iterate is feasible [cf. Eq. (2.94)]

$$\begin{aligned}\alpha^k &= \max\big\{\alpha \mid x^k - \alpha(X^k)^2z^k \geq 0\big\} \\ &= \min\big\{\big(x_i^k z_i^k\big)^{-1} \mid z_i^k > 0,\ i = 1,\ldots,n\big\},\end{aligned} \tag{2.102}$$

and β is a scalar in $(0,1)$, which for good performance should be very close to 1 (say $\beta = 0.99$).

We will derive an expression for the amount of cost improvement at the kth iteration. We have, using the fact $A(x^k - x^{k+1}) = 0$,

$$\begin{aligned}\Delta\text{cost} &= c'(x^k - x^{k+1}) \\ &= (c - A'\lambda^k)'(x^k - x^{k+1}) \\ &= z^{k\prime}\big(\beta\bar{\alpha}^k (X^k)^2 z^k\big) \\ &= \beta\bar{\alpha}^k \|X^k z^k\|^2,\end{aligned}$$

and by using the expression (2.102) for $\bar{\alpha}^k$, we have

$$\Delta\text{cost} = \beta\|X^k z^k\|^2 \max\big\{\big(x_i^k z_i^k\big)^{-1} \mid z_i^k > 0,\, i = 1,\ldots,n\big\}.$$

Combining this relation with the fact

$$\begin{aligned}\|X^k z^k\| &= \left(\sum_{i=1}^n \big(x_i^k z_i^k\big)^2\right)^{1/2} \\ &\geq \min\big\{x_i^k z_i^k \mid z_i^k > 0,\, i = 1,\ldots,n\big\} \\ &= \frac{1}{\max\big\{\big(x_i^k z_i^k\big)^{-1} \mid z_i^k > 0,\, i = 1,\ldots,n\big\}},\end{aligned}$$

we finally obtain

$$\begin{aligned}\Delta\text{cost} &\geq \beta\|X^k z^k\| \\ &= \beta\left(\sum_{i=1}^n \big(x_i^k z_i^k\big)^2\right)^{1/2} \\ &\geq \frac{\beta}{2}\sum_{i=1}^n |x_i^k z_i^k|.\end{aligned}$$

Thus the cost improvement at the kth iteration is at least proportional to each of the products $|x_i^k z_i^k|$, implying that

$$x_i^k z_i^k \to 0, \qquad i = 1,\ldots,n. \tag{2.103}$$

Suppose that the sequence $\big\{(x^k, \lambda^k)\big\}$ converges to some $(\bar{x}, \bar{\lambda})$, and let $\bar{z} = c - A'\bar{\lambda}$. Then from Eq. (2.103), we have

$$\bar{x}_i \bar{z}_i = 0, \qquad i = 1,\ldots,n, \tag{2.104}$$

from the feasibility of x^k, we have

$$A\bar{x} = b, \qquad \bar{x} \geq 0, \tag{2.105}$$

and from Eq. (2.101), we have

$$\bar{z} \geq 0 \tag{2.106}$$

[if $\bar{z}_i < 0$ for some i, then from Eq. (2.101) it would follow that $x_i^{k+1} > x_i^k > 0$ for all k, implying $\bar{x}_i > 0$ and contradicting Eq. (2.104)].

It can be shown that Eqs. (2.104)-(2.106) imply that $\bar{x}$ is an optimal solution of the primal problem (LP1), and $\bar{\lambda}$ is an optimal solution of the dual problem (LP2). This is a consequence of the duality theory to be developed in Section 3.4.2 (particularly Example 3.4.2); we refer the reader to that section since the optimality of $\bar{x}$ and $\bar{\lambda}$ will not be used further in this section.

We have thus shown that if the sequence $\{(x^k, \lambda^k)\}$ converges to some $(\bar{x}, \bar{\lambda})$, then $\bar{x}$ and $\bar{\lambda}$ are optimal primal and dual solutions, respectively. Unfortunately, however, it is not possible to guarantee the convergence of $\{(x^k, \lambda^k)\}$ without additional quite restrictive assumptions generally known as *nondegeneracy*. One such assumption is that for each primal feasible solution x we must have $x_i > 0$ for at least m indices i, and that for each dual feasible solution λ and corresponding vector $z = c - A'\lambda$, we must have $z_i > 0$ for at least $n - m$ indices i. Another type of nondegeneracy assumption under which convergence can be proved is developed in Exercise 2.6.2. Such nondegeneracy assumptions are seldom satisfied in practice, but have often been used to simplify various linear programming analyses. If we do not assume nondegeneracy, it is necessary to modify the stepsize procedure of the affine scaling method in order to prove convergence to the optimum (see the references). Unfortunately, while the theoretical convergence properties of the method are then improved, the practical performance becomes substantially slower.

EXERCISES

2.6.1

Consider the linear program

$$\begin{aligned} &\text{minimize} \quad x_1 + 2x_2 + 3x_3 \\ &\text{subject to} \quad x_1 + x_2 + x_3 = 1, \qquad x \geq 0. \end{aligned}$$

Write a computer program to implement the affine scaling method for this problem. Solve the problem for the starting points $x^0 = (.8, .15, .05)$ and $x^0 = (.1, .2, .7)$.

2.6.2 (Convergence of the Affine Scaling Method)

This exercise deals with the convergence of the primal affine scaling method (2.95), (2.96) under a nondegeneracy assumption, and requires some knowledge of the theory of polyhedral convex sets given in Sections B.3 and B.4 of Appendix B. Show that $\{x^k\}$ converges to the optimal primal solution x^* and $\{\lambda^k\}$ converges to an optimal dual solution λ^*, assuming the following:

(1) Every vector z of the form $z = c - A'\lambda$, where $\lambda \in \Re^m$, has at most m zero coordinates.

(2) The extreme points of the polyhedron $X = \{x \mid Ax = b,\, x \geq 0\}$ have distinct costs and the primal optimal solution is unique.

(3) The matrix AX^2A' is nonsingular for all $x \in X$.

Hint: Complete the details in the following argument. Assumption (3) implies that the inverse $(AX^2A')^{-1}$ is continuous over X, and that if $x \in X$, then x must have at most $n-m$ zero coordinates. Assumption (2) implies that for every scalar γ, the level set $\{x \mid c'x \leq \gamma,\, Ax = b,\, x \geq 0\}$ is bounded, so that $\{x^k\}$ is bounded, and therefore also that $\{\lambda^k\}$ and $\{z^k\}$ (where $z^k = c - A'\lambda^k$) are bounded. If (x, z) is any limit point of $\{(x^k, z^k)\}$, we have $Xz = 0$. Then x has exactly $n - m$ zero coordinates, and z has exactly m nonzero coordinates, implying that x is an extreme point of the polyhedron X. Using assumption (2), conclude that $\{x^k\}$ converges, implying that $\{\lambda^k\}$ also converges.

2.7 BLOCK COORDINATE DESCENT METHODS*

We briefly discussed unconstrained coordinate descent methods in Section 1.7. We will now generalize these methods to solve the problem

$$\begin{aligned} &\text{minimize } \ f(x) \\ &\text{subject to } \ x \in X, \end{aligned} \tag{CP}$$

where X is a Cartesian product of closed convex sets $X_1, \ldots, X_m$:

$$X = X_1 \times X_2 \times \cdots \times X_m. \tag{2.107}$$

We assume that X_i is a closed convex subset of $\Re^{n_i}$ and $n = n_1 + \cdots + n_m$. The vector x is partitioned as

$$x = (x_1, x_2, \ldots, x_m),$$

where each x_i belongs to $\Re^{n_i}$, so the constraint $x \in X$ is equivalent to

$$x_i \in X_i, \qquad i = 1, \ldots, m.$$

Let us assume that for every $x \in X$ and every $i = 1, \ldots, m$, the optimization problem

$$\begin{aligned} &\text{minimize } \ f(x_1, \ldots, x_{i-1}, \xi, x_{i+1}, \ldots, x_m) \\ &\text{subject to } \ \xi \in X_i, \end{aligned}$$

has at least one solution. The following algorithm, known as *block coordinate descent* or *nonlinear Gauss-Seidel* method, generates the next iterate $x^{k+1} = (x_1^{k+1}, \ldots, x_m^{k+1})$, given the current iterate $x^k = (x_1^k, \ldots, x_m^k)$, according to the iteration

$$x_i^{k+1} = \arg\min_{\xi \in X_i} f(x_1^{k+1}, \ldots, x_{i-1}^{k+1}, \xi, x_{i+1}^k, \ldots, x_m^k), \qquad i = 1, \ldots, m. \tag{2.108}$$

Thus, at each iteration, the cost is minimized with respect to each of the "block coordinate" vectors x_i^k, taken in cyclic order. Naturally, the method makes practical sense if the minimization in Eq. (2.108) is fairly easy. This is frequently so when each x_i is a scalar, but there are also other cases of interest, where x_i is a multidimensional vector.

The following proposition gives the basic convergence result for the method. It turns out that it is necessary to make an assumption implying that the minimum in Eq. (2.108) is uniquely attained, as first suggested by Zangwill [Zan69]. The need for this assumption is not obvious but has been demonstrated by an example given by Powell [Pow73]. Exercise 2.7.2 provides a modified version of the algorithm, which does not require this assumption.

Proposition 2.7.1: (Convergence of Block Coordinate Descent) Suppose that f is continuously differentiable over the set X of Eq. (2.107). Furthermore, suppose that for each i and $x \in X$, the minimum below

$$\min_{\xi \in X_i} f(x_1, \ldots, x_{i-1}, \xi, x_{i+1}, \ldots, x_m)$$

is uniquely attained. Let $\{x^k\}$ be the sequence generated by the block coordinate descent method (2.108). Then, every limit point of $\{x^k\}$ is a stationary point.

Proof: Let

$$z_i^k = \left(x_1^{k+1}, \ldots, x_i^{k+1}, x_{i+1}^k, \ldots, x_m^k\right).$$

Using the definition (2.108) of the method, we obtain

$$f(x^k) \geq f(z_1^k) \geq f(z_2^k) \geq \cdots \geq f(z_{m-1}^k) \geq f(x^{k+1}), \qquad \forall\, k. \tag{2.109}$$

Let $\bar{x} = \left(\bar{x}_1, \ldots, \bar{x}_m\right)$ be a limit point of the sequence $\{x^k\}$. Notice that $\bar{x} \in X$ because X is closed. Equation (2.109) implies that the sequence $\left\{f(x^k)\right\}$ converges to $f(\bar{x})$. It now remains to show that $\bar{x}$ minimizes f over X.

Let $\{x^{k_j} \mid j = 0, 1, \ldots\}$ be a subsequence of $\{x^k\}$ that converges to $\bar{x}$. We first show that $\{x_1^{k_j+1} - x_1^{k_j}\}$ converges to zero as $j \to \infty$. Assume the contrary or, equivalently, that $\{z_1^{k_j} - x^{k_j}\}$ does not converge to zero. Let $\gamma^{k_j} = \|z_1^{k_j} - x^{k_j}\|$. By possibly restricting to a subsequence of $\{k_j\}$, we may assume that there exists some $\bar{\gamma} > 0$ such that $\gamma^{k_j} \geq \bar{\gamma}$ for all j. Let $s_1^{k_j} = \left(z_1^{k_j} - x^{k_j}\right)/\gamma^{k_j}$. Thus, $z_1^{k_j} = x^{k_j} + \gamma^{k_j} s_1^{k_j}$, $\|s_1^{k_j}\| = 1$, and $s_1^{k_j}$ differs from zero only along the first block-component. Notice that $s_1^{k_j}$ belongs to a compact set and therefore has a limit point $\bar{s}_1$. By restricting to a further subsequence of $\{k_j\}$, we assume that $s_1^{k_j}$ converges to $\bar{s}_1$.

Let us fix some $\epsilon \in [0,1]$. Notice that $0 \leq \epsilon\bar{\gamma} \leq \gamma^{k_j}$. Therefore, $x^{k_j} + \epsilon\bar{\gamma} s_1^{k_j}$ lies on the segment joining x^{k_j} and $x^{k_j} + \gamma^{k_j} s_1^{k_j} = z_1^{k_j}$, and belongs to X because X is convex. Using the fact that $z_1^{k_j}$ minimizes f over all x that differ from x^{k_j} along the first block-component, we obtain

$$f\left(z_1^{k_j}\right) = f\left(x^{k_j} + \gamma^{k_j} s_1^{k_j}\right) \leq f\left(x^{k_j} + \epsilon\bar{\gamma} s_1^{k_j}\right) \leq f\left(x^{k_j}\right).$$

Since $f(x^k)$ converges to $f(\bar{x})$, Eq. (2.109) shows that $f\left(z_1^k\right)$ also converges to $f(\bar{x})$. We now take the limit as j tends to infinity, to obtain $f(\bar{x}) \leq f\left(\bar{x} + \epsilon\bar{\gamma}\bar{s}_1\right) \leq f(\bar{x})$. We conclude that $f(\bar{x}) = f\left(\bar{x} + \epsilon\bar{\gamma}\bar{s}_1\right)$, for every $\epsilon \in [0,1]$. Since $\bar{\gamma}\bar{s}_1 \neq 0$, this contradicts the hypothesis that f is uniquely minimized when viewed as a function of the first block-component. This contradiction establishes that $x_1^{k_j+1} - x_1^{k_j}$ converges to zero. In particular, $z_1^{k_j}$ converges to $\bar{x}$.

From the definition (2.108) of the algorithm, we have

$$f\left(z_1^{k_j}\right) \leq f\left(x_1, x_2^{k_j}, \ldots, x_m^{k_j}\right), \qquad \forall\, x_1 \in X_1.$$

Taking the limit as j tends to infinity, we obtain

$$f(\bar{x}) \leq f(x_1, \bar{x}_2, \ldots, \bar{x}_m), \qquad \forall\, x_1 \in X_1.$$

Using the conditions for optimality over a convex set (Prop. 2.1.2 in Section 2.1), we conclude that

$$\nabla_1 f(\bar{x})'(x_1 - \bar{x}_1) \geq 0, \qquad \forall\, x_1 \in X_1,$$

where $\nabla_i f$ denotes the gradient of f with respect to the component x_i.

Let us now consider the sequence $\{z_1^{k_j}\}$. We have already shown that $z_1^{k_j}$ converges to $\bar{x}$. A verbatim repetition of the preceding argument shows that $x_2^{k_j+1} - x_2^{k_j}$ converges to zero and $\nabla_2 f(\bar{x})'(x_2 - \bar{x}_2) \geq 0$ for every $x_2 \in X_2$. Continuing inductively, we obtain $\nabla_i f(\bar{x})'(x_i - \bar{x}_i) \geq 0$ for every $x_i \in X_i$ and for every i. Adding these inequalities, and using the Cartesian product structure of the set X, we conclude that $\nabla f(\bar{x})'(x - \bar{x}) \geq 0$ for every $x \in X$. **Q.E.D.**

Block coordinate descent methods are often useful in contexts where the cost function and the constraints have a partially decomposable structure with respect to the problem's optimization variables. The following example illustrates the idea.

Example 2.7.1 (Hierarchical Decomposition)

Consider an optimization problem of the form

$$\begin{aligned} &\text{minimize} \quad \sum_{i=1}^{m} f_i(x, y_i) \\ &\text{subject to} \quad x \in X, \qquad y_i \in Y_i, \quad i = 1, \ldots, m, \end{aligned}$$

where X and Y_i, $i = 1, \ldots, m$, are closed, convex subsets of corresponding Euclidean spaces, and the functions f_i are continuously differentiable. This problem is associated with a paradigm of optimization of a system consisting of m subsystems, with the cost function f_i associated with the operations of the ith sybsystem. Here y_i is viewed as vector of local decision variables that influences the cost of the ith subsystem only, and x is viewed as a vector of global or coordinating decision variables that affects the operation of all the subsystems.

The coordinate descent method takes advantage of the decomposable structure and has the form

$$y_i^{k+1} = \arg\min_{y_i \in Y_i} f_i(x^k, y_i), \qquad i = 1, \ldots, m,$$

$$x^{k+1} = \arg\min_{x \in X} \sum_{i=1}^{m} f_i(x, y_i^{k+1}).$$

The method has a natural real-life interpretation: at each iteration, each subsystem optimizes its own cost, taking the global variables as fixed at their current values, and then the coordinator optimizes the overall cost for the current values of the local variables.

EXERCISES

2.7.1 (The Proximal Minimization Algorithm)

Let $f : \Re^n \mapsto \Re$ be a continuously differentiable function, let X be a closed convex set, and let c be a positive scalar.

(a) Show that the algorithm

$$x^{k+1} = \arg\min_{x \in X} \left\{ f(x) + \frac{1}{2c} \|x - x^k\|^2 \right\}$$

(cf. Exercise 2.3.8 in Section 2.3 and Section 5.4.6) is a special case of the block coordinate descent method applied to the problem

$$\begin{aligned}&\text{minimize } \; f(x) + \frac{1}{2c}\|x - y\|^2\\&\text{subject to } \; x \in X, \;\; y \in \Re^n,\end{aligned}$$

which is equivalent to the problem of minimizing f over X.

(b) Derive a convergence result based on Prop. 2.7.1 for the algorithm of part (a).

(c) Assume that f is convex. Show that if f has at least one minimizing point over X, the entire sequence $\{x^k\}$ converges to some such point. *Hint*: We have by definition

$$f(x^{k+1}) + \frac{1}{2c}\|x^{k+1} - x^k\|^2 \le f(x) + \frac{1}{2c}\|x - x^k\|^2, \qquad \forall\, x \in X,$$

so that $\|x^{k+1} - x^k\| \le \|x - x^k\|$ for all x in the set

$$X^k = \big\{x \in X \mid f(x) \le f(x^{k+1})\big\}.$$

Hence x^{k+1} is the unique projection of x^k on X^k, and we have $(x^{k+1} - x^k)'(x - x^k) \ge 0$ for all $x \in X^k$. Conclude that for every x^* that minimizes f over X, we have $\|x^* - x^{k+1}\| \le \|x^* - x^k\|$.

2.7.2 [GrS98]

Consider the following variation of the method (2.108):

$$x_i^{k+1} = \arg\min_{\xi \in X_i} f(x_1^{k+1}, \ldots, x_{i-1}^{k+1}, \xi, x_{i+1}^k, \ldots, x_m^k) + \frac{1}{2c}\|\xi - x_i^k\|^2, \quad i = 1, \ldots, m,$$

where c is a positive scalar. Assuming that f is convex, show that every limit point of the sequence $\{x^k\}$ is a global minimum. *Hint*: Apply the result of Prop. 2.7.1 to the cost function

$$g(x, y) = f(x) + \frac{1}{2c}\|x - y\|^2.$$

For an analysis of this algorithm without the convexity assumption on f, see Grippo and Sciandrone [GrS98].

2.7.3 (Parallel Projections Algorithm)

We are given m closed convex sets $X_1, X_2, \ldots, X_m$ in $\Re^n$, and we want to find a point in their intersection. Consider the equivalent problem

$$\begin{aligned}&\text{minimize } \; \tfrac{1}{2}\sum_{i=1}^m \|y_i - x\|^2\\&\text{subject to } \; x \in \Re^n, \; y_i \in X_i, \; i = 1, \ldots, m,\end{aligned}$$

where the variables of the optimization are $x, y_1, \ldots, y_m$. Derive a block coordinate descent algorithm involving projections on each of the sets X_i that can be carried out independently of each other. State a convergence result for this algorithm.

2.8 NOTES AND SOURCES

Section 2.2: Feasible direction methods were first systematically investigated by Zoutendijk [Zou60]. Convergence analyses and additional methods are given by Topkis and Veinott [ToV67], Polak [Pol71], Pironneau and Polak [PiP73], and Zoutendijk [Zou76]. The conditional gradient method was first proposed for quadratic programs by Frank and Wolfe [FrW56]. Extensive discussions of its application to more general problems are given in Levitin and Poljak [LeP65], and Demjanov and Rubinov [DeR70]. The convergence rate of the method was analyzed in Canon and Cullum [CaC68], Dunn [Dun79], [Dun80a], and Dunn and Sachs [DuS83].

Section 2.3: Gradient projection methods with a constant stepsize were first proposed by Goldstein [Gol64], and Levitin and Poljak [LeP65]. The Armijo rule along the projection arc was proposed by Bertsekas [Ber76c]. For convergence analysis, see Dunn [Dun81], Gafni and Bertsekas [GaB82], [GaB84], Calamai and Moré [CaM87], and Dunn [Dun87], [Dun88a]. For analysis of various aspects of the convergence rate, see Bertsekas [Ber76c], Dunn [Dun81], Bertsekas and Gafni [BeG82], Dunn and Sachs [DuS83], Dunn [Dun87], Tseng [Tse91a], Luo and Tseng [LuT92b], [LuT93a], [LuT93b], [LuT94b], and Gilmore and Kelley [GiK95]. Surveys are given by Dunn [Dun88a], [Dun94]. Variations of gradient projection methods that are based on constraint identification are given by Bertsekas [Ber76c], Dembo and Tulowitzki [DeT83], Moré and Toraldo [MoT89], Burke, Moré, and Toraldo [BMT90], and McKenna, Mesirov, and Zenios [MMZ95].

The gradient projection method is particularly well-suited for large-scale problems with relatively simple constraint structure. Examples of such problems arise in optimal control (see e.g., Polak [Pol73], Bertsekas [Ber76c], [Ber82b], Dunn [Dun88a], [Dun91a], [Dun94]), and in multicommodity flow problems from communications and transportation (see e.g., Bertsekas and Gafni [BeG83], Bertsekas, Gafni, and Gallager [BGG84], Tsitsiklis and Bertsekas [TsB86], Bertsekas and Tsitsiklis [BeT89], Bertsekas and Gallager [BeG92], Luo and Tseng [LuT94b], Florian and Hearn [FlH95]). Newton's method for constrained problems is discussed by Dunn [Dun80b], and Hughes and Dunn [HuD84].

Section 2.4: Two metric-projection methods were proposed by Bertsekas [Ber82b] and were generalized by Gafni and Bertsekas [GaB84]; see also Bertsekas and Gafni [BeG83]. For subsequent work see Dunn [Dun88b], [Dun91a], [Dun91b], [Dun93a], [Dun93b], [Dun94], Gawande and Dunn [GaD88], and Pytlak [Pyt98].

Section 2.5: Manifold suboptimization methods draw their origin from the gradient projection method of Rosen [Ros60b]. Extensive discussions of these methods can be found in Gill and Murray [GiM74], and Gill, Murray, and Wright [GMW81]. An important method of this type, which has been used as the basis for general purpose nonlinear programming software, is the *reduced gradient method*; see Lasdon and Waren [LaW78]. A convergence analysis of Rosen's gradient projection method is given by Du and Zhang [DuZ89]. The application of manifold suboptimization methods to quadratic programming is discussed by Zangwill [Zan69]. Efficient implementations are given by Gill and Murray [GiM74], Gill, Murray, and Wright [GMW81], [GMW91], and Goldfarb and Idnani [GoI82]. Quadratic programming is a special case of the linear complementarity problem, which is discussed in detail by Cottle, Pang, and Stone [CPS92].

Section 2.6: The affine scaling method was first proposed by Dikin [Dik67], and was rediscovered many years later by Barnes [Bar86], and Vanderbei, Meketon, and Freedman [VMF86], following the emergence of interior point methods as serious competitors to the simplex method for linear programming applications. The convergence of the method without restrictive nondegeneracy assumptions is analyzed by Tsuchiya [Tsu91], Tseng and Luo [TsL92], Monteiro, Tsuchiya, and Wang [MTW93], and Tsuchiya and Muramatsu [TsM95]. Note that while we have focused on the linear programming case, the ideas of the affine scaling method apply to problems with a more general cost function.

Section 2.7: The convergence to a unique limit and the rate of convergence of the coordinate descent method are analyzed by Luo and Tseng [LuT91], [LuT92a]. Coordinate descent methods are often well suited for solving dual problems (see Section 6.2), and within this specialized context there has been much convergence analysis (see the references given for Section 6.2).

3

Lagrange Multiplier Theory

Contents

The methods I set forth require neither constructions
nor geometric or mechanical considerations.
They require only algebraic operations
subject to a systematic and uniform course.

Lagrange

The constraint set of an optimization problem is usually specified in terms of equality and inequality constraints. If we take into account this structure, we obtain a sophisticated collection of optimality conditions, involving some auxiliary variables called *Lagrange multipliers*. These variables facilitate the characterization of optimal solutions, but also provide valuable sensitivity information, quantifying up to first order the variation of the optimal cost caused by variations in problem data.

Lagrange multipliers will be central in almost all of the subsequent material in the book. In Chapter 4, we will see how they can be used within computational methods. In Chapter 5, we will see how they can be viewed as the optimization variables of auxiliary optimization problems, called *dual problems*. The computational solution of dual problems will be the subject of Chapter 6. The present chapter also contains an introductory discussion of duality.

The theory of Lagrange multipliers can be developed in a variety of ways, and in this chapter we will follow two basic lines of analysis:

(a) The *penalty viewpoint*, whereby we disregard the constraints, while adding to the cost a high penalty for violating them. By then working with the "penalized" unconstrained problem, and by passing to the limit as the penalty increases, we obtain the desired Lagrange multiplier theorems. This approach is surprisingly simple and powerful, and forms the basis for important algorithms, which will be more fully explored in Chapter 4.

(b) The *feasible direction viewpoint*, which similar to Chapter 2, relies on the fact that at a local minimum there can be no cost improvement when traveling a small distance along a direction that leads to feasible points. This viewpoint needs to be modified somewhat when the constraint set is not convex, resulting in a fair amount of mathematical complication, but in the end works well, and yields powerful results and useful geometric insight.

We will use primarily the simpler penalty viewpoint, but we will develop the feasible direction viewpoint as well, in a parallel and self-contained manner. It is possible to study the present chapter selectively, by focusing initially on just one of these two viewpoints, and the choice may be largely based on taste and background. We note, however, that after gaining some basic familiarity with Lagrange multipliers, the reader will benefit from studying both viewpoints, because they reinforce and complement each other in terms of analysis and algorithmic insight.

3.1 NECESSARY CONDITIONS FOR EQUALITY CONSTRAINTS

In this section we consider problems with equality constraints of the form

$$\begin{aligned} &\text{minimize } f(x) \\ &\text{subject to } h_i(x) = 0, \qquad i = 1, \ldots, m. \end{aligned} \tag{ECP}$$

We assume that $f : \Re^n \mapsto \Re$, $h_i : \Re^n \mapsto \Re$, $i = 1, \ldots, m$, are continuously differentiable functions. *All the necessary and the sufficient conditions of this chapter relating to a local minimum can also be shown to hold if f and h_i are defined and are continuously differentiable within just an open set containing the local minimum.* The proofs are essentially identical to those given here.

For notational convenience, we introduce the constraint function $h : \Re^n \mapsto \Re^m$, where

$$h = (h_1, \ldots, h_m).$$

We can then write the constraints in the more compact form

$$h(x) = 0. \tag{3.1}$$

Our basic Lagrange multiplier theorem states that for a given local minimum x^*, there exist scalars $\lambda_1, \ldots, \lambda_m$, called *Lagrange multipliers*, such that

$$\nabla f(x^*) + \sum_{i=1}^{m} \lambda_i \nabla h_i(x^*) = 0. \tag{3.2}$$

There are two ways to interpret this equation:

(a) The cost gradient $\nabla f(x^*)$ belongs to the subspace spanned by the constraint gradients at x^*. The example of Fig. 3.1.1 illustrates this interpretation.

(b) The cost gradient $\nabla f(x^*)$ is orthogonal to the subspace of *first order feasible variations*

$$V(x^*) = \big\{\Delta x \mid \nabla h_i(x^*)'\Delta x = 0,\ i = 1, \ldots, m\big\}.$$

This is the subspace of variations Δx for which the vector $x = x^* + \Delta x$ satisfies the constraint $h(x) = 0$ up to first order. Thus, according to the Lagrange multiplier condition of Eq. (3.2), at the local minimum x^*, the first order cost variation $\nabla f(x^*)'\Delta x$ is zero for all variations

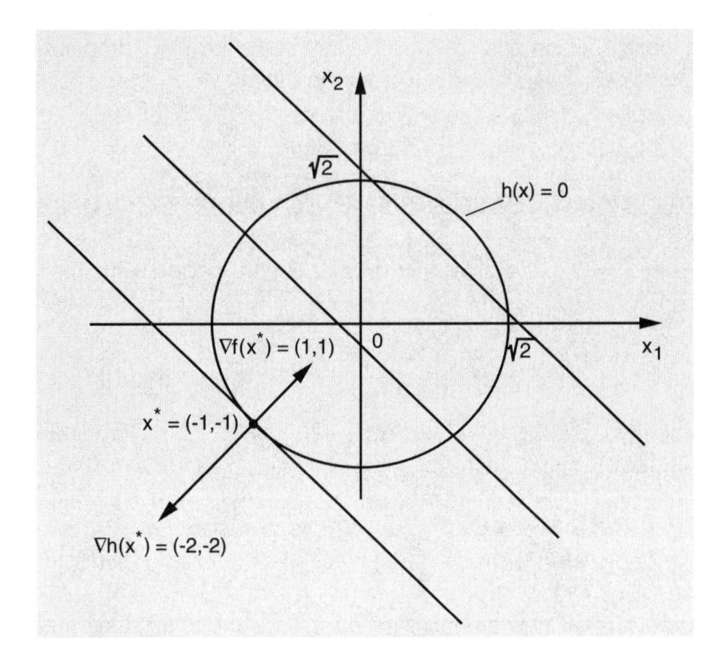

Figure 3.1.1. Illustration of the Lagrange multiplier condition (3.1) for the problem

$$\text{minimize } x_1 + x_2$$
$$\text{subject to } x_1^2 + x_2^2 = 2.$$

At the local minimum $x^* = (-1,-1)$, the cost gradient $\nabla f(x^*)$ is normal to the constraint surface and is therefore, collinear with the constraint gradient $\nabla h(x^*) = (-2,-2)$. The Lagrange multiplier is $\lambda = 1/2$.

Δx in this subspace. This statement is analogous to the "zero gradient condition" $\nabla f(x^*) = 0$ of unconstrained optimization.

Here is a formal statement of the main Lagrange multiplier theorem:

Proposition 3.1.1: (Lagrange Multiplier Theorem – Necessary Conditions) Let x^* be a local minimum of f subject to $h(x) = 0$, and assume that the constraint gradients $\nabla h_1(x^*), \ldots, \nabla h_m(x^*)$ are linearly independent. Then there exists a unique vector $\lambda^* = (\lambda_1^*, \ldots, \lambda_m^*)$, called a *Lagrange multiplier vector*, such that

$$\nabla f(x^*) + \sum_{i=1}^{m} \lambda_i^* \nabla h_i(x^*) = 0. \tag{3.3}$$

If in addition f and h are twice continuously differentiable, we have

$$y'\left(\nabla^2 f(x^*) + \sum_{i=1}^{m} \lambda_i^* \nabla^2 h_i(x^*)\right) y \geq 0, \qquad \text{for all } y \in V(x^*), \quad (3.4)$$

where $V(x^*)$ is the subspace of first order feasible variations

$$V(x^*) = \{y \mid \nabla h_i(x^*)'y = 0,\ i = 1, \ldots, m\}. \quad (3.5)$$

For easy reference, a feasible vector x for which the constraint gradients $\nabla h_1(x), \ldots, \nabla h_m(x)$ are linearly independent will be called *regular*. We will see later (Example 3.1.1) that there may not exist Lagrange multipliers for a local minimum that is not regular.

We will provide two different proofs of the Lagrange multiplier theorem, each providing important insights. These proofs are based on transforming the constrained problem to an unconstrained one, but in different ways. The constrained first and second order necessary conditions are obtained by applying the corresponding unconstrained conditions to the appropriate unconstrained problem. The approaches are:

(a) *The penalty approach*. Here we disregard the constraints, while adding to the cost a high penalty for violating them. By writing the necessary conditions for the "penalized" unconstrained problems, and by passing to the limit as the penalty increases, we obtain the Lagrange multiplier theorem. This approach is simple, and applies to inequality constraints, as will be seen in Section 3.3.

(b) *The elimination approach*. Here we view the constraints as a system of m equations with n unknowns, and we express m of the variables in terms of the remaining $n - m$, thereby reducing the problem to an unconstrained problem. By then applying the corresponding first and second order necessary conditions for unconstrained minima, the Lagrange multiplier theorem follows. This approach requires the use of the implicit function theorem (Prop. A.25 in Appendix A), but is otherwise simple and insightful. Its extension, however, to inequality constraints is complicated.

Before going into the proof of the theorem, let us illustrate how there may not exist Lagrange multipliers for a local minimum that is not regular.

Example 3.1.1 (A Problem with no Lagrange Multipliers)

Consider the problem of minimizing

$$f(x) = x_1 + x_2$$

subject to the two constraints

$$h_1(x) = (x_1 - 1)^2 + x_2^2 - 1 = 0, \qquad h_2(x) = (x_1 - 2)^2 + x_2^2 - 4 = 0.$$

The geometry of this problem is illustrated in Fig. 3.1.2. It can be seen that at the local minimum $x^* = (0,0)$ (the only feasible solution), the cost gradient $\nabla f(x^*) = (1,1)$ cannot be expressed as a linear combination of the constraint gradients $\nabla h_1(x^*) = (-2,0)$ and $\nabla h_2(x^*) = (-4,0)$. Thus the Lagrange multiplier condition

$$\nabla f(x^*) + \lambda_1^* \nabla h_1(x^*) + \lambda_2^* \nabla h_2(x^*) = 0$$

cannot hold for any λ_1^* and λ_2^*.

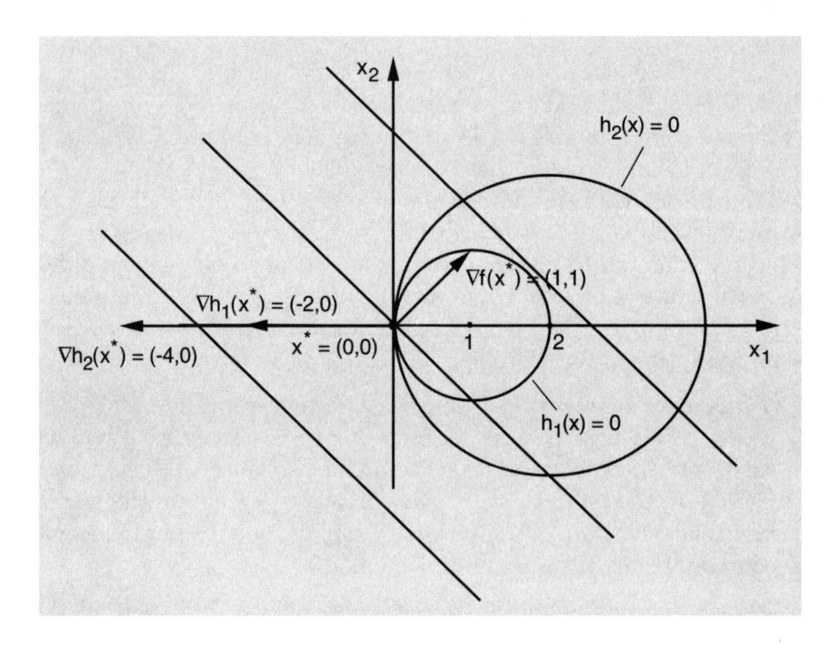

Figure 3.1.2. Illustration of how Lagrange multipliers may not exist (cf. Example 3.1.1). The problem here is

$$\begin{aligned} &\text{minimize} \quad f(x) = x_1 + x_2 \\ &\text{subject to} \quad h_1(x) = (x_1 - 1)^2 + x_2^2 - 1 = 0, \\ &\qquad\qquad\quad h_2(x) = (x_1 - 2)^2 + x_2^2 - 4 = 0, \end{aligned}$$

with a local minimum $x^* = (0,0)$. The cost gradient cannot be expressed as a linear combination of the constraint gradients, so there are no Lagrange multipliers.

The difficulty here is that the subspace of first order feasible variations

$$V(x^*) = \big\{ y \mid \nabla h_1(x^*)' y = 0,\ \nabla h_2(x^*)' y = 0 \big\}$$

[cf. Eq. (3.5)], which is $\{y \mid y_1 = 0,\ y_2 \in \Re\}$, has larger dimension than the true set of feasible variations $\{y \mid y = 0\}$. The optimality of x^* implies that

$\nabla f(x^*)$ is orthogonal to the true set of feasible variations, but for a Lagrange multiplier to exist, $\nabla f(x^*)$ must be orthogonal to the subspace of first order feasible variations. If the constraint gradients $\nabla h_1(x^*)$ and $\nabla h_2(x^*)$ were linearly independent, the set of feasible variations and the subspace of first order feasible variations would have the same dimension and this difficulty would not have occurred. It turns out that the equality of the two sets of feasible variations plays a fundamental role in the existence of Lagrange multipliers, and will be explored further in Section 3.3.6 within a broader context.

3.1.1 The Penalty Approach

Here we approximate the original constrained problem by an unconstrained optimization problem that involves a penalty for violation of the constraints. In particular, for $k = 1, 2, \ldots$, we introduce the cost function

$$F^k(x) = f(x) + \frac{k}{2}||h(x)||^2 + \frac{\alpha}{2}||x - x^*||^2,$$

where x^* is the local minimum of the constrained problem and α is some positive scalar. The term $(k/2)||h(x)||^2$ imposes a penalty for violating the constraint $h(x) = 0$, while the term $(\alpha/2)||x - x^*||^2$ is introduced for technical proof-related reasons [to ensure that x^* is a *strict* local minimum of the function $f(x) + (\alpha/2)||x - x^*||^2$ subject to $h(x) = 0$].

Since x^* is a local minimum, we can select $\epsilon > 0$ such that $f(x^*) \leq f(x)$ for all feasible x in the *closed* sphere

$$S = \{x \mid ||x - x^*|| \leq \epsilon\}.$$

Let x^k be an optimal solution of the problem

$$\begin{aligned} &\text{minimize} \ \ F^k(x) \\ &\text{subject to} \ \ x \in S. \end{aligned} \tag{3.6}$$

[An optimal solution exists because of Weierstrass' theorem (Prop. A.8 in Appendix A), since S is compact.] We will show that the sequence $\{x^k\}$ converges to x^*.

We have for all k

$$F^k(x^k) = f(x^k) + \frac{k}{2}||h(x^k)||^2 + \frac{\alpha}{2}||x^k - x^*||^2 \leq F^k(x^*) = f(x^*) \tag{3.7}$$

and since $f(x^k)$ is bounded over S, we obtain

$$\lim_{k\to\infty} ||h(x^k)|| = 0;$$

otherwise the left-hand side of Eq. (3.7) would become unbounded above as $k \to \infty$. Therefore, every limit point $\bar{x}$ of $\{x^k\}$ satisfies $h(\bar{x}) = 0$. Furthermore, Eq. (3.7) yields $f(x^k) + (\alpha/2)||x^k - x^*||^2 \leq f(x^*)$ for all k, so by taking the limit as $k \to \infty$, we obtain

$$f(\bar{x}) + \frac{\alpha}{2}||\bar{x} - x^*||^2 \leq f(x^*).$$

Since $\bar{x} \in S$ and $\bar{x}$ is feasible, we have $f(x^*) \leq f(\bar{x})$, which when combined with the preceding inequality yields $||\bar{x} - x^*|| = 0$ so that $\bar{x} = x^*$. Thus the sequence $\{x^k\}$ converges to x^*, and it follows that x^k is an interior point of the closed sphere S for sufficiently large k. Therefore x^k is an *unconstrained* local minimum of $F^k(x)$ for sufficiently large k. We will now prove the Lagrange multiplier theorem by working with the corresponding unconstrained necessary optimality conditions.

Proof of the Lagrange multiplier theorem: From the first order necessary condition, we have for sufficiently large k

$$0 = \nabla F^k(x^k) = \nabla f(x^k) + k\nabla h(x^k)h(x^k) + \alpha(x^k - x^*). \tag{3.8}$$

Since $\nabla h(x^*)$ has rank m, the same is true for $\nabla h(x^k)$ if k is sufficiently large. For such k, $\nabla h(x^k)'\nabla h(x^k)$ is invertible (Prop. A.20 in Appendix A), and by premultiplying Eq. (3.8) with $\big(\nabla h(x^k)'\nabla h(x^k)\big)^{-1}\nabla h(x^k)'$, we obtain

$$kh(x^k) = -\big(\nabla h(x^k)'\nabla h(x^k)\big)^{-1}\nabla h(x^k)'\big(\nabla f(x^k) + \alpha(x^k - x^*)\big).$$

By taking the limit as $k \to \infty$ and $x^k \to x^*$, we see that $\{kh(x^k)\}$ converges to the vector

$$\lambda^* = -\big(\nabla h(x^*)'\nabla h(x^*)\big)^{-1}\nabla h(x^*)'\nabla f(x^*).$$

By taking the limit as $k \to \infty$ in Eq. (3.8), we obtain

$$\nabla f(x^*) + \nabla h(x^*)\lambda^* = 0,$$

proving the first order Lagrange multiplier condition (3.3).

By using the second order unconstrained optimality condition for problem (3.6), we see that the matrix

$$\nabla^2 F^k(x^k) = \nabla^2 f(x^k) + k\nabla h(x^k)\nabla h(x^k)' + k\sum_{i=1}^{m} h_i(x^k)\nabla^2 h_i(x^k) + \alpha I$$

is positive semidefinite, for all sufficiently large k and for all $\alpha > 0$. Fix any $y \in V(x^*)$ [that is, $\nabla h(x^*)'y = 0$], and let y^k be the projection of y on the nullspace of $\nabla h(x^k)'$, that is,

$$y^k = y - \nabla h(x^k)\big(\nabla h(x^k)'\nabla h(x^k)\big)^{-1}\nabla h(x^k)'y, \tag{3.9}$$

(cf. Example 2.1.5 in Section 2.1). Since $\nabla h(x^k)'y^k = 0$ and $\nabla^2 F^k(x^k)$ is positive semidefinite, we have

$$0 \leq y^{k\prime}\nabla^2 F^k(x^k)y^k = y^{k\prime}\left(\nabla^2 f(x^k) + k\sum_{i=1}^m h_i(x^k)\nabla^2 h_i(x^k)\right)y^k + \alpha\|y^k\|^2.$$

Since $kh_i(x^k) \to \lambda_i^*$, and from Eq. (3.9), together with the facts $x^k \to x^*$ and $\nabla h(x^*)'y = 0$, we have $y^k \to y$, we obtain

$$0 \leq y'\left(\nabla^2 f(x^*) + \sum_{i=1}^m \lambda_i^*\nabla^2 h_i(x^*)\right)y + \alpha\|y\|^2, \qquad \forall\, y \in V(x^*).$$

Since α can be taken arbitrarily close to 0, we obtain

$$0 \leq y'\left(\nabla^2 f(x^*) + \sum_{i=1}^m \lambda_i^*\nabla^2 h_i(x^*)\right)y, \qquad \forall\, y \in V(x^*),$$

which is the second order Lagrange multiplier condition. The proof of the Lagrange multiplier theorem is thus complete. **Q.E.D.**

3.1.2 The Elimination Approach

We introduce the elimination approach by considering first the easier case where the constraints are linear.

Example 3.1.2 (Lagrange Multipliers for Linear Constraints)

Consider the problem

$$\begin{aligned} &\text{minimize} \ \ f(x) \\ &\text{subject to} \ \ Ax = b, \end{aligned} \tag{3.10}$$

where A is an $m \times n$ matrix with linearly independent rows and $b \in \Re^m$ is a given vector. By rearranging the coordinates of x if necessary, we may assume that the first m columns of A are linearly independent, so that A can be partitioned as

$$A = (\,B \quad R\,),$$

where B is an invertible $m \times m$ matrix and R is an $m \times (n-m)$ matrix. We also partition x as

$$x = \begin{pmatrix} x_B \\ x_R \end{pmatrix},$$

where $x_B \in \Re^m$ and $x_R \in \Re^{n-m}$. We can then write problem (3.10) as

$$\begin{aligned} &\text{minimize} \ \ f(x_B, x_R) \\ &\text{subject to} \ \ Bx_B + Rx_R = b. \end{aligned} \tag{3.11}$$

By using the constraint equation to express x_B in terms of x_R as

$$x_B = B^{-1}(b - Rx_R)$$

and by substitution in the cost function, we can convert problem (3.11) into the unconstrained optimization problem

$$\begin{aligned} &\text{minimize} \;\; F(x_R) \equiv f\big(B^{-1}(b - Rx_R), x_R\big) \\ &\text{subject to} \;\; x_R \in \Re^{n-m}. \end{aligned} \tag{3.12}$$

If (x_B^*, x_R^*) is a local minimum of the constrained problem (3.10), then x_R^* is an unconstrained local minimum of the "reduced" cost function F, so we have

$$0 = \nabla F(x_R^*) = -R'(B')^{-1}\nabla_B f(x^*) + \nabla_R f(x^*), \tag{3.13}$$

where $\nabla_B f$ and $\nabla_R f$ denote the gradients of f with respect to x_B and x_R, respectively. By defining

$$\lambda^* = -(B')^{-1}\nabla_B f(x^*), \tag{3.14}$$

Eq. (3.13) is written as

$$\nabla_R f(x^*) + R'\lambda^* = 0, \tag{3.15}$$

while Eq. (3.14) is written as

$$\nabla_B f(x^*) + B'\lambda^* = 0. \tag{3.16}$$

Equations (3.15) and (3.16) can be combined as

$$\nabla f(x^*) + A'\lambda^* = 0, \tag{3.17}$$

which is the Lagrange multiplier condition of Eq. (3.3) specialized to the case of the linearly constrained problem (3.10). Note that the vector λ^* satisfying this condition is unique because of the linear independence of the columns of A'.

We next show the second order necessary condition (3.4), by showing that it is equivalent to the second order unconstrained necessary condition

$$0 \le d'\nabla^2 F(x_R^*)d, \qquad \forall\, d \in \Re^{n-m}. \tag{3.18}$$

We have using Eqs. (3.12) and (3.13)

$$\begin{aligned} \nabla^2 F(x_R) = \nabla\Big(&- R'(B')^{-1}\nabla_B f\big(B^{-1}(b - Rx_R), x_R\big) \\ &+ \nabla_R f\big(B^{-1}(b - Rx_R), x_R\big)\Big). \end{aligned}$$

By evaluating this expression at $x_R = x_R^*$, and by partitioning the Hessian $\nabla^2 f(x^*)$ as

$$\nabla^2 f(x^*) = \begin{pmatrix} \nabla^2_{BB} f(x^*) & \nabla^2_{BR} f(x^*) \\ \nabla^2_{RB} f(x^*) & \nabla^2_{RR} f(x^*) \end{pmatrix},$$

we obtain

$$\begin{aligned} \nabla^2 F(x_R^*) = {} & R'(B')^{-1}\nabla^2_{BB} f(x^*)B^{-1}R \\ & - R'(B')^{-1}\nabla^2_{BR} f(x^*) - \nabla^2_{RB} f(x^*)B^{-1}R + \nabla^2_{RR} f(x^*). \end{aligned}$$

This expression together with the positive semidefiniteness of $\nabla^2 F(x_R^*)$ [cf. Eq. (3.18)] and the linearity of the constraints [implying that $\nabla^2 h_i(x^*) = 0$], yields for all $d \in \Re^{n-m}$

$$0 \le d'\nabla^2 F(x_R^*)d = y'\nabla^2 f(x^*)y = y'\left(\nabla^2 f(x^*) + \sum_{i=1}^{m} \lambda_i^* \nabla^2 h_i(x^*)\right)y, \quad (3.19)$$

where y is the vector

$$y = \begin{pmatrix} -B^{-1}Rd \\ d \end{pmatrix}.$$

It can be seen that the subspace $V(x^*)$ of feasible variations of Eq. (3.5) is given by

$$\begin{aligned} V(x^*) &= \{(y_B, y_R) \mid By_B + Ry_R = 0\} \\ &= \{(y_B, y_R) \mid y_B = -B^{-1}Rd,\ y_R = d,\ d \in \Re^{n-m}\}, \end{aligned}$$

so Eq. (3.19) is equivalent to the second order Lagrange multiplier condition (3.4).

Note that based on the above proof, the Lagrange multiplier conditions of Prop. 3.1.1 are nothing but the "zero gradient" and "positive semidefinite Hessian" conditions for the unconstrained problem (3.12), which is defined over the reduced space of the vector x_R.

Actually for the case of linear constraints, a stronger version of the Lagrange multiplier theorem can be shown. In particular, the regularity assumption is not needed for the existence of a Lagrange multiplier vector; see Exercise 3.1.4 and Section 3.3.5.

We now prove the Lagrange multiplier theorem by generalizing the analysis of the preceding example.

Proof of the Lagrange multiplier theorem: The proof assumes that $m < n$; if $m = n$, any vector, including $-\nabla f(x^*)$, can be expressed as a linear combination of the linearly independent vectors $\nabla h_1(x^*), \ldots, \nabla h_m(x^*)$, thereby proving the theorem. By reordering the coordinates of x if necessary, we partition the vector x as $x = (x_B, x_R)$, where the square submatrix $\nabla_B h(x^*)$ (the gradient matrix of h with respect to x_B) is invertible. The constraint equation

$$h(x_B, x_R) = 0$$

has the solution (x_B^*, x_R^*), and the implicit function theorem (Prop. A.25 in Appendix A) can be used to express x_B in terms of x_R via a unique continuously differentiable function $\phi : S \mapsto \Re^m$ defined over a sphere S centered at x_R^* (ϕ is twice continuously differentiable if h is). In particular, we have $x_B^* = \phi(x_R^*)$, $h\big(\phi(x_R), x_R\big) = 0$ for all $x_R \in S$, and

$$\nabla\phi(x_R) = -\nabla_R h\big(\phi(x_R), x_R\big)\big(\nabla_B h\big(\phi(x_R), x_R\big)\big)^{-1}, \qquad \forall\, x_R \in S,$$

where $\nabla_R h$ is the gradient matrix of h with respect to x_R.

We now proceed as in the earlier case of linear constraints. We observe that x_R^* is an unconstrained minimum of the "reduced" cost function

$$F(x_R) = f\big(\phi(x_R), x_R\big),$$

and we apply the corresponding unconstrained first and second order necessary conditions. The first order Lagrange multiplier condition (3.3) follows by repeating the calculation of Eqs. (3.14)-(3.17) with the definitions

$$B' = \nabla_B h(x^*), \qquad R' = \nabla_R h(x^*), \qquad A' = \nabla h(x^*),$$

$$\lambda^* = -(B')^{-1}\nabla_B f(x^*).$$

The proof of the second order necessary condition (3.4) requires a lengthy calculation, which resembles the one that yielded Eq. (3.19) in the preceding example. In particular, for $d \in \Re^{n-m}$, denote

$$y = \begin{pmatrix} -B^{-1}Rd \\ d \end{pmatrix}.$$

Denote also

$$H_i(x_R) = h_i\big(\phi(x_R), x_R\big), \qquad i = 1, \ldots, m,$$

and let $\phi_i(x_R)$ be the ith component of ϕ, so that

$$\phi(x_R) = \begin{pmatrix} \phi_1(x_R) \\ \vdots \\ \phi_m(x_R) \end{pmatrix}.$$

Then by differentiating twice the relation

$$F(x_R) = f\big(\phi(x_R), x_R\big),$$

we obtain by a straightforward calculation [compare with the derivation of Eq. (3.19)]

$$d'\nabla^2 F(x_R)d = y'\nabla^2 f(x^*)y + d'\left(\sum_{j=1}^m \nabla^2\phi_j(x_R^*)\frac{\partial f(x^*)}{\partial x_j}\right)d. \qquad (3.20)$$

Similarly by twice differentiating the equation $H_i(x_R) = 0$, we have

$$0 = d'\nabla^2 H_i(x_R)d = y'\nabla^2 h_i(x^*)y + d' \left(\sum_{j=1}^{m} \nabla^2 \phi_j(x_R^*) \frac{\partial h_i(x^*)}{\partial x_j} \right) d.$$

By multiplying the above equation with λ_i^*, and by adding over all i we obtain

$$0 = \sum_{i=1}^{m} \lambda_i^* y'\nabla^2 h_i(x^*)y + d' \left(\sum_{j=1}^{m} \nabla^2 \phi_j(x_R^*) \sum_{i=1}^{m} \lambda_i^* \frac{\partial h_i(x^*)}{\partial x_j} \right) d. \quad (3.21)$$

Adding Eqs. (3.20) and (3.21), and using the relations $d'\nabla^2 F(x_R)d \geq 0$ and

$$\frac{\partial f(x^*)}{\partial x_j} + \sum_{i=1}^{m} \lambda_i^* \frac{\partial h_i(x^*)}{\partial x_j} = 0, \qquad j = 1, \ldots, m,$$

we obtain

$$0 \leq y' \left(\nabla^2 f(x^*) + \sum_{i=1}^{m} \lambda_i^* \nabla^2 h_i(x^*) \right) y,$$

for all y of the form defined above. As shown in Example 3.1.2, y belongs to the subspace $V(x^*)$ if and only if y has this form, thus proving the second order Lagrange multiplier condition (3.4). **Q.E.D.**

3.1.3 The Lagrangian Function

Sometimes it is convenient to write our necessary conditions in terms of the *Lagrangian function* $L : \Re^{n+m} \mapsto \Re$ defined by

$$L(x, \lambda) = f(x) + \sum_{i=1}^{m} \lambda_i h_i(x).$$

Then, if x^* is a local minimum which is regular, the Lagrange multiplier conditions of Prop. 3.1.1 together with the equation $h(x^*) = 0$ are written compactly as

$$\nabla_x L(x^*, \lambda^*) = 0, \qquad \nabla_\lambda L(x^*, \lambda^*) = 0, \quad (3.22)$$

$$y'\nabla_{xx}^2 L(x^*, \lambda^*)y \geq 0, \qquad \text{for all } y \in V(x^*). \quad (3.23)$$

The first order necessary conditions (3.22) represent a system of $n+m$ equations with $n + m$ unknowns – the coordinates of x^* and λ^*. Every local minimum x^* which is regular, together with its associated Lagrange multiplier vector, will be a solution of this system. However, a solution of the system need not correspond to a local minimum, as our experience with unconstrained problems indicates; it could be for example a local maximum.

Example 3.1.3

Consider the problem

$$\begin{aligned} &\text{minimize} \ \ \tfrac{1}{2}\left(x_1^2 + x_2^2 + x_3^2\right) \\ &\text{subject to} \ \ x_1 + x_2 + x_3 = 3. \end{aligned} \tag{3.24}$$

The first order necessary conditions (3.22) yield

$$x_1^* + \lambda^* = 0,$$

$$x_2^* + \lambda^* = 0,$$

$$x_3^* + \lambda^* = 0,$$

$$x_1^* + x_2^* + x_3^* = 3.$$

This is a system of four equations and four unknowns. It has the unique solution

$$x_1^* = x_2^* = x_3^* = 1, \qquad \lambda^* = -1.$$

The constraint gradient here is $(1, 1, 1)$, so all feasible vectors are regular. Therefore, $x^* = (1, 1, 1)$ is the unique candidate for a local minimum. Furthermore, since $\nabla^2_{xx} L(x^*, \lambda^*)$ is the identity matrix for this problem, the second order necessary condition (3.23) is satisfied. We can argue that x^* is a *global* minimum by using the convexity of the cost function and the convexity of the constraint set to verify that x^* satisfies the sufficiency condition of Prop. 2.1.2(b) (alternatively, we can use the sufficiency condition to be given in Section 3.3.4).

If instead we consider the maximization version of problem (3.24), i.e.,

$$\begin{aligned} &\text{minimize} \ \ -\tfrac{1}{2}\left(x_1^2 + x_2^2 + x_3^2\right) \\ &\text{subject to} \ \ x_1 + x_2 + x_3 = 3, \end{aligned} \tag{3.25}$$

then the first order condition (3.22) yields $x^* = (1, 1, 1)$ and $\lambda^* = 1$. However, the second order condition (3.23) is not satisfied and since every feasible vector is also regular, we conclude that the problem has no solution.

Example 3.1.4 (Diffraction Law in Optics)

Consider a smooth curve on the 2-dimensional plane described by the equation

$$h(x) = 0,$$

where $h : \Re^2 \mapsto \Re$ is continuously differentiable. Imagine that the curve separates the plane into two regions and that the velocity of light is different in each region. Let y and z be two points that lie on opposite sides of the curve as shown in part (a) of Fig. 3.1.3. Suppose that light going from y to z follows a ray from y to some point x^* of the curve with velocity v_y, and then follows a ray from x^* to z with velocity v_z. Assuming that light follows a path of minimum travel time, we will show that x^* is characterized by the following diffraction law:

$$\frac{\sin \phi_y}{v_y} = \frac{\sin \phi_z}{v_z}, \tag{3.26}$$

where ϕ_y and ϕ_z are the angles shown in Fig. 3.1.3.

Indeed, the travel time can be expressed as

$$T(x^*) = \frac{\|y - x^*\|}{v_y} + \frac{\|z - x^*\|}{v_z},$$

and the Lagrange multiplier theorem states that the gradient

$$\nabla T(x^*) = \frac{x^* - y}{v_y \|x^* - y\|} + \frac{x^* - z}{v_z \|x^* - z\|} \tag{3.27}$$

is a scalar multiple of $\nabla h(x^*)$. [For this we need that x^* is regular, that is, $\nabla h(x^*) \neq 0$, which we assume.] To interpret this condition, consider the vectors

$$\bar{y} = x^* + \frac{y - x^*}{v_y \|y - x^*\|}, \qquad \bar{z} = x^* + \frac{z - x^*}{v_z \|z - x^*\|},$$

and the parallelogram with sides $\bar{y} - x^*$ and $\bar{z} - x^*$, which by Eq. (3.27), has $-\nabla T(x^*)$ as one of its diagonals. From part (b) of Fig. 3.1.3, we see that the Lagrange multiplier condition states that the diagonal $-\nabla T(x^*)$ of the parallelogram is normal to the curve at x^*. From Euclidean geometry it follows that the distances of $\bar{y}$ and $\bar{z}$ from the vertical diagonal are equal. Since $\|\bar{y} - x^*\| = 1/v_y$ and $\|\bar{z} - x^*\| = 1/v_z$, this is equivalent to the diffraction law (3.26).

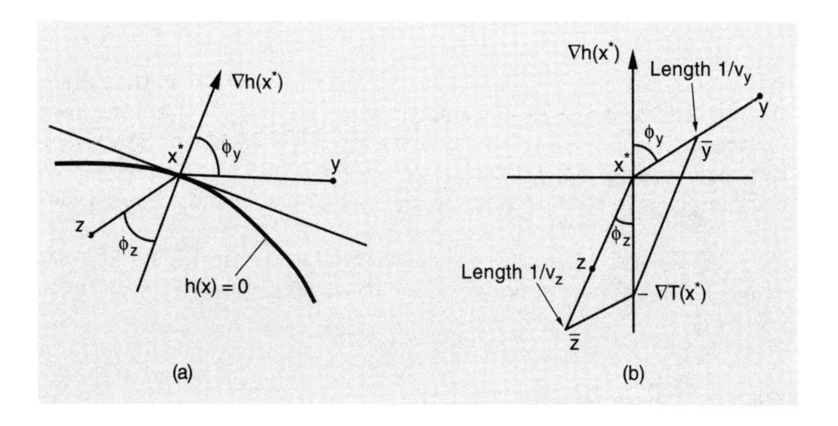

Figure 3.1.3. Diffraction law; cf. Example 3.1.4. The angles ϕ_y and ϕ_z are formed by the normal line to the curve at x^* [which is $\nabla h(x^*)$], and the vectors $y - x^*$ and $z - x^*$, respectively.

Example 3.1.5 (Optimal Portfolio Selection)

Consider an investor who wants to allocate one unit of wealth among n assets offering random rates of return $e_1, \ldots, e_n$, respectively. The means $\bar{e}_i = E\{e_i\}, i = 1, \ldots, n$, and the covariance matrix

$$Q = \begin{pmatrix} E\{(e_1 - \bar{e}_1)^2\} & \ldots & E\{(e_1 - \bar{e}_1)(e_n - \bar{e}_n)\} \\ \vdots & \ldots & \vdots \\ E\{(e_n - \bar{e}_n)(e_1 - \bar{e}_1)\} & \ldots & E\{(e_n - \bar{e}_n)^2\} \end{pmatrix}$$

are known, and we assume that Q is invertible. If x_i is the amount invested in the ith asset, the mean and the variance of the return of the investment $y = \sum_{i=1}^n e_i x_i$ are

$$\bar{y} = E\{y\} = \sum_{i=1}^n \bar{e}_i x_i$$

and

$$\sigma^2 = E\{(y - \bar{y})^2\} = E\left\{\left(\sum_{i=1}^n (e_i - \bar{e}_i)x_i\right)^2\right\}$$

$$= \sum_{i=1}^n \sum_{j=1}^n E\{(e_i - \bar{e}_i)(e_j - \bar{e}_j)\}x_i x_j = x'Qx.$$

The investor's problem is to find the portfolio $x = (x_1, \ldots x_n)$ that minimizes the variance $E\{(y - \bar{y})^2\}$ to achieve a given level of mean return $E\{y\}$, say $E\{y\} = m$. Thus the problem is

$$\begin{aligned} &\text{minimize } x'Qx \\ &\text{subject to } \sum_{i=1}^n x_i = 1, \quad \sum_{i=1}^n \bar{e}_i x_i = m. \end{aligned}$$

We want to see how the solution varies with m. Note that the solution of the problem "scales" with the amount of wealth available for investment. In particular, it can be seen that if w is the amount available for investment and wm is the desired mean, then the optimal portfolio is wx^*, where x^* is the optimal portfolio corresponding to one unit of wealth and a desired mean of m.

Let λ_1 and λ_2 be Lagrange multipliers for the constraints $\sum_{i=1}^n x_i = 1$ and $\sum_{i=1}^n \bar{e}_i x_i = m$, respectively. The first order optimality condition is

$$2Qx^* + \lambda_1 u + \lambda_2 \bar{e} = 0,$$

where $u = (1, \ldots, 1)'$ and $\bar{e} = (\bar{e}_1, \ldots, \bar{e}_n)'$. (Strictly speaking, in order to use the Lagrange multiplier theorem of Prop. 3.1.1, we must assume that the vectors u and $\bar{e}$ are linearly independent. However, even if they are not, the existence of a Lagrange multiplier is guaranteed because the constraints are linear; see Example 3.1.2 and Exercise 3.1.4.) Equivalently,

$$x^* = -\tfrac{1}{2}Q^{-1}u\lambda_1 - \tfrac{1}{2}Q^{-1}\bar{e}\lambda_2, \tag{3.28}$$

and by substitution in the constraints $u'x^* = 1$ and $\overline{e}'x^* = m$, we obtain

$$1 = u'x^* = -\tfrac{1}{2}u'Q^{-1}u\lambda_1 - \tfrac{1}{2}u'Q^{-1}\overline{e}\lambda_2,$$

$$m = \overline{e}'x^* = -\tfrac{1}{2}\overline{e}'Q^{-1}u\lambda_1 - \tfrac{1}{2}\overline{e}'Q^{-1}\overline{e}\lambda_2.$$

By solving these equations for λ_1 and λ_2 we obtain

$$\lambda_1 = \xi_1 + \zeta_1 m,$$

$$\lambda_2 = \xi_m + \zeta_m m,$$

for some scalars $\xi_1, \zeta_1, \xi_m, \zeta_m$. Thus substitution in Eq. (3.28) yields

$$x^* = mv + w \tag{3.29}$$

for some vectors v and w that depend on Q and $\overline{e}$. The corresponding variance of return is

$$\sigma^2 = (mv + w)'Q(mv + w) = (\alpha m + \beta)^2 + \gamma, \tag{3.30}$$

where α, β, and γ are some scalars that depend on Q and $\overline{e}$.

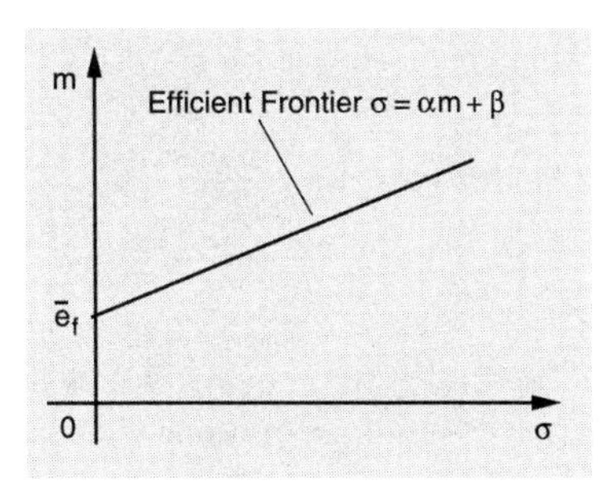

Figure 3.1.4. Efficient frontier for optimal portfolio selection. Every portfolio corresponding to a mean-variance pair on the efficient frontier can be achieved as a linear combination of two portfolios corresponding to two pairs on the efficient frontier.

Equation (3.30) specifies the locus of pairs (σ, m) that are achievable by optimal portfolio selection. Suppose now that one of the assets is riskless, that is, its return is fixed at some known value $\overline{e}_f$ and the variance of its return is zero. Then γ must be zero since when $m = \overline{e}_f$, the variance is minimized (set to zero) by investing exclusively in the riskless asset. In this case, from Eq. (3.30) we obtain

$$\sigma = |\alpha m + \beta|.$$

Thus, assuming $\alpha > 0$, portfolios of interest correspond to mean-variance pairs of the form

$$\sigma = \alpha m + \beta, \qquad m \geq \overline{e}_f,$$

as shown in Fig. 3.1.4. This is known as the *efficient frontier*. To characterize the efficient frontier it is sufficient to determine one more point on it.

Every other point is determined as a linear combination of this point and the point $(0, \bar{e}_f)$. Furthermore, by Eq. (3.29), every pair (σ, m) on the efficient frontier can be achieved by an appropriate linear combination of two portfolios, one consisting of exclusive investment on the riskless asset, and the other corresponding to some pair (σ, m) on the efficient frontier with $\sigma \neq 0$. (Economic arguments can be used to construct a second portfolio on the efficient frontier, leading to the celebrated CAPM theory of finance; see e.g. [HuL88], [Lue98].) Thus, while different investors with different preferences towards risk may prefer different points on the efficient frontier, their preferences can be obtained by appropriate mixtures of just two portfolios.

EXERCISES

3.1.1

Use the Lagrange multiplier theorem to solve the following problems:

(a) $f(x) = ||x||^2$, $h(x) = \sum_{i=1}^n x_i - 1$.

(b) $f(x) = \sum_{i=1}^n x_i$, $h(x) = ||x||^2 - 1$.

(c) $f(x) = ||x||^2$, $h(x) = x'Qx - 1$, where Q is positive definite.

3.1.2

Consider all rectangular parallelepipeds with given length of diagonal; i.e., $x_1^2 + x_2^2 + x_3^2$: fixed where x_1, x_2, x_3 are the lengths of the edges. Find one that has maximal surface area (sum of areas of the rectangular sides) and one that has maximal perimeter (sum of lengths of the edges).

3.1.3 (Fermat's Principle in Optics)

Consider a smooth curve on the plane described by the equation

$$h(x) = 0,$$

where $h : \Re^2 \mapsto \Re$ is continuously differentiable. Let y and z be two points that lie in relation to the curve as shown in Fig. 3.1.5. Show that if a point x^* minimizes the sum of the Euclidean distances

$$||y - x|| + ||z - x||$$

over all points x on the curve, then the angles ϕ_y, ϕ_z shown in Fig. 3.1.6 must be equal.

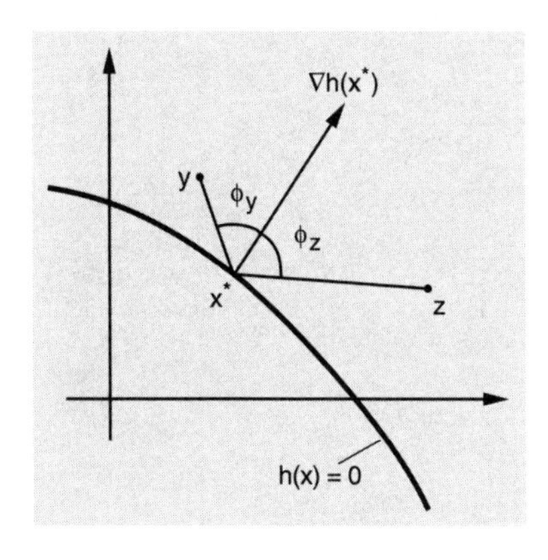

Figure 3.1.5. Fermat's principle; cf. Exercise 3.1.3. The angles ϕ_y, ϕ_z must be equal.

3.1.4

Show that if the constraints are linear, the regularity assumption is not needed in order for the Lagrange multiplier conditions to hold, except that the Lagrange multiplier vector need not be unique. *Hint*: Discard the redundant equality constraints, and assign to them zero Lagrange multipliers.

3.1.5

Consider a symmetric $n \times n$ matrix Q. Define

$$\lambda_1 = \min_{\|x\|^2=1} x'Qx, \qquad e_1 = \arg\min_{\|x\|^2=1} x'Qx,$$

and for $k = 0, \ldots, n-1$,

$$\lambda_{k+1} = \min_{\substack{\|x\|^2=1 \\ e_i'x=0,\ i=1,\ldots,k}} x'Qx, \qquad e_{k+1} = \arg\min_{\substack{\|x\|^2=1 \\ e_i'x=0,\ i=1,\ldots,k}} x'Qx.$$

(a) Show that

$$\lambda_1 \le \lambda_2 \le \cdots \le \lambda_n.$$

(b) Show that the vectors $e_1, \ldots, e_n$ are linearly independent.

(c) Interpret $\lambda_1, \ldots, \lambda_n$ as Lagrange multipliers, and show that $\lambda_1, \ldots, \lambda_n$ are the eigenvalues of Q, while $e_1, \ldots, e_n$ are corresponding eigenvectors.

3.1.6 (Arithmetic-Geometric Mean Inequality)

Let $\alpha_1, \ldots, \alpha_n$ are positive scalars with $\sum_{i=1}^n \alpha_i = 1$. Use a Lagrange multiplier to solve the problem

$$\begin{aligned} &\text{minimize} \ \ \alpha_1 x_1 + \alpha_2 x_2 + \cdots + \alpha_n x_n \\ &\text{subject to} \ \ x_1^{\alpha_1} x_2^{\alpha_2} \cdots x_n^{\alpha_n} = 1, \qquad x_i > 0, \ \ i = 1, \ldots, n. \end{aligned}$$

Establish the arithmetic-geometric mean inequality

$$x_1^{\alpha_1} x_2^{\alpha_2} \cdots x_n^{\alpha_n} \leq \sum_{i=1}^{n} \alpha_i x_i,$$

for a set of positive numbers x_i, $i = 1, \ldots, n$. *Hint*: Use the change of variables $y_i = \ln(x_i)$.

3.1.7

Let $a_1, \ldots, a_m$ be given vectors in $\Re^n$, and consider the problem of minimizing $\sum_{j=1}^{m} \|x - a_j\|^2$ subject to $\|x\|^2 = 1$. Consider the center of gravity $\hat{a} = \frac{1}{m}\sum_{j=1}^{m} a_j$ of the vectors $a_1, \ldots, a_m$. Show that if $\hat{a} \neq 0$, the problem has a unique maximum and a unique minimum. What happens if $\hat{a} = 0$?

3.1.8

Show the angles x, y, and z of a triangle maximize $\sin x \sin y \sin z$ if and only if the triangle is equilateral.

3.1.9

(a) Show that of all triangles circumscribed about a given circle, the one possessing minimal area is the equilateral. *Hint*: Let x, y, and z be the lengths of the tangent lines from the vertices of the triangle to the circle, and let ρ be the radius of the circle. The area of the triangle is $A = \rho(x + y + z)$, and it can also be expressed as $A = \sqrt{xyz(x + y + z)}$ (a theorem by Heron of Alexandria). Consider the problem of minimizing $x + y + z$ subject to $xyz = \rho^2(x + y + z)$.

(b) Consider the unit circle and two points a and b in the plane. Find a point x on the circle such that the triangle with vertices a, b, and x has maximal area, by maximizing $\|x - \hat{x}\|^2$ over $\|x\|^2 = 1$, where $\hat{x}$ is the projection of x on the line that passes through a and b. Show that the line connecting x and $\hat{x}$ passes through the center of the circle.

(c) Show that of all triangles inscribed in a given circle, the one possessing maximal area is the equilateral. *Hint*: Use part (b).

(d) Show that among all polygons in the plane with a given number of sides that are inscribed in a given circle the one that has maximal area is regular (all its sides are equal). *Hint*: Use part (b).

3.1.10

Consider the unit circle and two points a and b in the plane.

(a) Find points x on the circle such that the triangle with vertices a, b, and x has minimal and maximal perimeter. Characterize the optimal solutions by relating the problem to Fermat's principle (Exercise 3.1.3). Use your analysis to show that of all triangles inscribed in a given circle, the one possessing maximal perimeter is the equilateral.

(c) Find points x on the circle such that the inner product $(x-a)'(x-b)$ is minimum or maximum.

3.1.11

Consider the problem $\min_{h(x)=0} f(x)$, suppose that x^* is a local minimum, that is a regular point, and let $I = \{i \mid \lambda_i^* \neq 0\}$, where λ^* is the corresponding Lagrange multiplier. Extend the proof based on the penalty approach of Section 3.1.1 to show that for each neighborhood N of x^*, there exists an $x \in N$ such that

$$\lambda_i^* h_i(x) > 0, \qquad \forall\, i \in I.$$

Hint: In the proof of Section 3.1.1, use the fact that $\lambda_i^k = k h_i(x^k)$ converges to λ_i^* for each i.

3.1.12

Consider the problem $\min_{h(x)=0} f(x)$, and suppose that x^* is a local minimum such that $\nabla f(x^*) \neq 0$. Show that x^* is a local minimum of the equality constrained problem

$$\min_{\|h(x)\|^2=0} f(x),$$

and that for this problem there is no Lagrange multiplier.

3.2 SUFFICIENT CONDITIONS AND SENSITIVITY ANALYSIS

As shown by Example 3.1.3 of the preceding section, the first order necessary condition may be satisfied by both local minima and local maxima (and possibly other vectors). The second order necessary condition is useful in narrowing down the field of candidates for local minima. To guarantee that a given vector is a local minimum, we need sufficient conditions for optimality, which are given by the following proposition.

Proposition 3.2.1: (Second Order Sufficiency Conditions) Assume that f and h are twice continuously differentiable, and let $x^* \in \Re^n$ and $\lambda^* \in \Re^m$ satisfy

$$\nabla_x L(x^*, \lambda^*) = 0, \qquad \nabla_\lambda L(x^*, \lambda^*) = 0, \tag{3.31}$$

$$y' \nabla^2_{xx} L(x^*, \lambda^*) y > 0, \qquad \text{for all } y \neq 0 \text{ with } \nabla h(x^*)' y = 0. \tag{3.32}$$

Then x^* is a strict local minimum of f subject to $h(x) = 0$. In fact, there exist scalars $\gamma > 0$ and $\epsilon > 0$ such that

$$f(x) \geq f(x^*) + \frac{\gamma}{2} \|x - x^*\|^2, \qquad \forall\ x \text{ with } h(x) = 0 \text{ and } \|x - x^*\| < \epsilon.$$

Note that the above sufficient conditions do not include regularity of the vector x^*. We will prove Prop. 3.2.1 in two different ways, in Sections 3.2.1 and 3.2.2, respectively. We first give an example.

Example 3.2.1

Consider the problem

$$\begin{aligned} &\text{minimize} \quad -(x_1 x_2 + x_2 x_3 + x_1 x_3) \\ &\text{subject to} \quad x_1 + x_2 + x_3 = 3. \end{aligned}$$

(If x_1, x_2, and x_3 represent the length, width, and height of a rectangular parallelepiped P, respectively, the problem can be interpreted as maximizing the surface area of P, subject to the sum of the edge lengths of P being fixed.) The first order necessary conditions are

$$-x_2^* - x_3^* + \lambda^* = 0,$$

$$-x_1^* - x_3^* + \lambda^* = 0,$$

$$-x_1^* - x_2^* + \lambda^* = 0,$$

$$x_1^* + x_2^* + x_3^* = 3,$$

which have the unique solution $x_1^* = x_2^* = x_3^* = 1$, $\lambda^* = 2$. The Hessian of the Lagrangian is

$$\nabla^2_{xx} L(x^*, \lambda^*) = \begin{pmatrix} 0 & -1 & -1 \\ -1 & 0 & -1 \\ -1 & -1 & 0 \end{pmatrix}.$$

We have for all $y \in V = \{y \mid \nabla h(x^*)'y = 0\} = \{y \mid y_1 + y_2 + y_3 = 0\}$ with $y \neq 0$,

$$\begin{aligned} y'\nabla^2_{xx}L(x^*, \lambda^*)y &= -y_1(y_2 + y_3) - y_2(y_1 + y_3) - y_3(y_1 + y_2) \\ &= y_1^2 + y_2^2 + y_3^2 > 0. \end{aligned}$$

Hence, the sufficient conditions of Prop. 3.2.1 are satisfied and x^* is a strict local minimum.

3.2.1 The Augmented Lagrangian Approach

We will first prove the sufficiency conditions, using penalty function concepts that will be used later in Section 4.2 as the basis for some important algorithms. An alternative proof, based on a feasible direction approach will be given in Section 3.2.2.

We begin with a useful lemma. Consider a symmetric matrix P and a positive semidefinite matrix Q. For each x that is not in the nullspace of Q, we have $x'Qx > 0$, so $x'(P + cQ)x > 0$ for sufficiently large scalars c. Thus, if we assume that $x'Px > 0$ for all $x \neq 0$ in the nullspace of Q, then for every $x \neq 0$, we have $x'(P + cQ)x > 0$, provided that c is sufficiently large. The essence of the lemma is that we can choose a threshold value of c that works for all x, so that $P + cQ$ is positive definite for all c greater than that threshold (cf. Fig. 3.2.1).

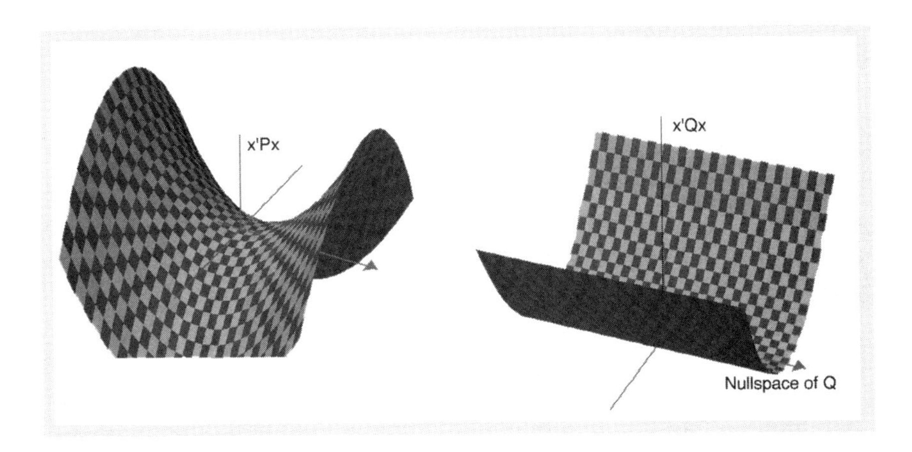

Figure 3.2.1. Illustration of Lemma 3.2.1. The possibly negative curvature of P outside the nullspace of Q is corrected by the positive curvature of Q appropriately magnified by the parameter c.

Lemma 3.2.1: Let P and Q be two symmetric matrices. Assume that Q is positive semidefinite and P is positive definite on the nullspace of Q, that is, $x'Px > 0$ for all $x \neq 0$ with $x'Qx = 0$. Then there exists a scalar $\bar{c}$ such that

$$P + cQ : \text{positive definite}, \qquad \forall\, c > \bar{c}.$$

Proof: Assume the contrary. Then for every integer k, there exists a vector x^k with $\|x^k\| = 1$ such that

$$x^{k\prime} P x^k + k x^{k\prime} Q x^k \leq 0.$$

Since $\{x^k\}$ is bounded, there is a subsequence $\{x^k\}_{k\in K}$ converging to some $\bar{x}$ [Prop. A.5(c) in Appendix A], and since $\|x^k\| = 1$ for all k, we have $\|\bar{x}\| = 1$. Taking the limit superior in the above inequality, we obtain

$$\bar{x}' P \bar{x} + \limsup_{k\to\infty,\, k\in K} (k x^{k\prime} Q x^k) \leq 0. \tag{3.33}$$

Since, by the positive semidefiniteness of Q, $x^{k\prime} Q x^k \geq 0$, we see that $\{x^{k\prime} Q x^k\}_{k\in K}$ must converge to zero, for otherwise the left-hand side of the above inequality would be $+\infty$. Therefore, $\bar{x}' Q \bar{x} = 0$ and from our hypothesis we obtain $\bar{x}' P \bar{x} > 0$. This contradicts Eq. (3.33). **Q.E.D.**

Let us introduce now the *augmented Lagrangian* function

$$L_c(x, \lambda) = f(x) + \lambda' h(x) + \frac{c}{2}\|h(x)\|^2,$$

where c is a scalar. This is the Lagrangian function for the problem

$$\begin{aligned} &\text{minimize} \;\; f(x) + \frac{c}{2}\|h(x)\|^2 \\ &\text{subject to} \;\; h(x) = 0, \end{aligned}$$

which has the same local minima as our original problem of minimizing $f(x)$ subject to $h(x) = 0$. The gradient and Hessian of L_c with respect to x are

$$\nabla_x L_c(x, \lambda) = \nabla f(x) + \nabla h(x)\big(\lambda + c h(x)\big),$$

$$\nabla^2_{xx} L_c(x, \lambda) = \nabla^2 f(x) + \sum_{i=1}^m \big(\lambda_i + c h_i(x)\big)\nabla^2 h_i(x) + c\nabla h(x) \nabla h(x)'.$$

In particular, if x^* and λ^* satisfy the sufficiency conditions of Prop. 3.2.1, we have

$$\nabla_x L_c(x^*, \lambda^*) = \nabla f(x^*) + \nabla h(x^*)\big(\lambda^* + c h(x^*)\big) = \nabla_x L(x^*, \lambda^*) = 0, \tag{3.34}$$

$$\begin{aligned}\nabla^2_{xx} L_c(x^*, \lambda^*) &= \nabla^2 f(x^*) + \sum_{i=1}^{m} \lambda_i^* \nabla^2 h_i(x^*) + c\nabla h(x^*)\nabla h(x^*)' \\ &= \nabla^2_{xx} L(x^*, \lambda^*) + c\nabla h(x^*)\nabla h(x^*)'.\end{aligned}$$

By the sufficiency condition (3.32), we have that $y'\nabla^2_{xx} L(x^*, \lambda^*)y > 0$ for all $y \neq 0$ such that $y'\nabla h(x^*)\nabla h(x^*)'y = 0$, so by applying Lemma 3.2.1 with $P = \nabla^2_{xx} L(x^*, \lambda^*)$ and $Q = \nabla h(x^*)\nabla h(x^*)'$, it follows that there exists a $\bar{c}$ such that

$$\nabla^2_{xx} L_c(x^*, \lambda^*) : \text{positive definite}, \qquad \forall\ c > \bar{c}. \tag{3.35}$$

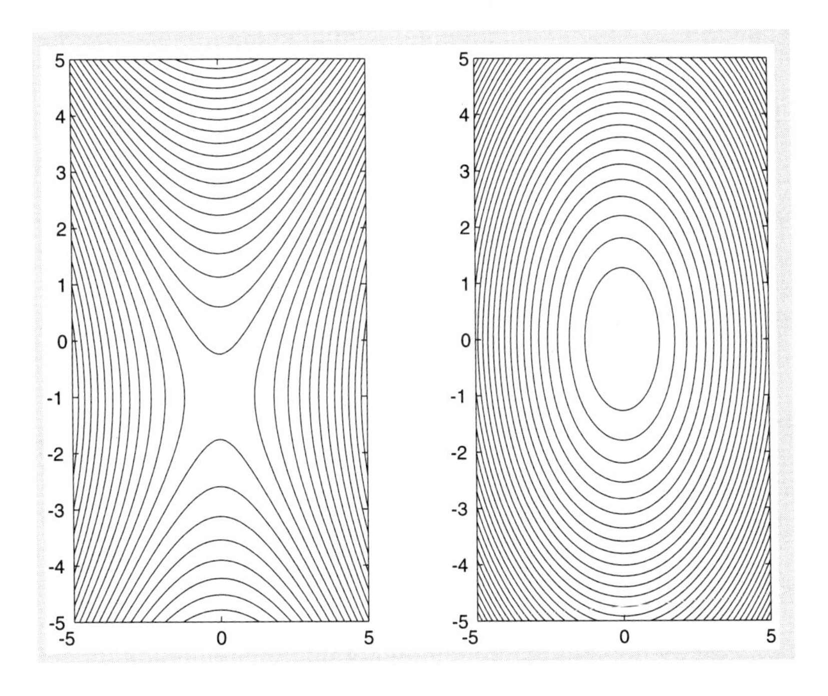

Figure 3.2.2. Illustration of how x^* is an unconstrained minimum of the augmented Lagrangian $L_c(\cdot, \lambda^*)$ for sufficiently large c. Consider the two-dimensional problem

$$\begin{aligned}&\text{minimize } f(x) = \tfrac{1}{2}\left(x_1^2 - x_2^2\right) - x_2 \\ &\text{subject to } x_2 = 0.\end{aligned}$$

Here, $x^* = (0, 0)$ is the unique global minimum, and $\lambda^* = 1$ is a corresponding Lagrange multiplier. The augmented Lagrangian function is

$$L_c(x, \lambda^*) = \tfrac{1}{2}\left(x_1^2 - x_2^2\right) - x_2 + \lambda^* x_2 + \tfrac{c}{2} x_2^2 = \tfrac{1}{2}\left(x_1^2 - x_2^2\right) + \tfrac{c}{2} x_2^2$$

and has $x^* = (0, 0)$ as its unique unconstrained minimum for $c > 1$. The figure shows the equal cost surfaces of f (left side) and of $L_c(\cdot, \lambda^*)$ for $c = 2$ (right side).

Using the sufficient optimality condition for unconstrained optimization (cf. Prop. 1.1.3), we conclude from Eqs. (3.34) and (3.35), that for $c > \bar{c}$, x^* is an unconstrained local minimum of $L_c(\cdot, \lambda^*)$. In particular, there exist $\gamma > 0$ and $\epsilon > 0$ such that

$$L_c(x, \lambda^*) \geq L_c(x^*, \lambda^*) + \frac{\gamma}{2}\|x - x^*\|^2, \qquad \forall\, x \text{ with } \|x - x^*\| < \epsilon.$$

Since for all x with $h(x) = 0$ we have $L_c(x, \lambda^*) = f(x)$, and by Eq. (3.31), $\nabla_\lambda L(x^*, \lambda^*) = h(x^*) = 0$, it follows that

$$f(x) \geq f(x^*) + \frac{\gamma}{2}\|x - x^*\|^2, \qquad \forall\, x \text{ with } h(x) = 0, \text{ and } \|x - x^*\| < \epsilon.$$

Thus x^* is a strict local minimum of f over $h(x) = 0$. The proof of Prop. 3.2.1 is complete.

The preceding analysis also shows that we can try to minimize f over $h(x) = 0$ by computing an unconstrained minimum of the augmented Lagrangian $L_c(\cdot, \lambda^*)$, as illustrated in Fig. 3.2.2. The difficulty here is that the Lagrange multiplier λ^* is unknown, but it turns out that by using in place of λ^*, readily computable approximations to λ^*, we can obtain useful algorithms; see Section 4.2.

3.2.2 The Feasible Direction Approach

Let us now prove the sufficiency conditions of Prop. 3.2.1 using a descent/feasible direction approach. In particular, we assume that there are points of lower cost arbitrarily close to x^*, focus on the directions leading to these points, and come to a contradiction. This method of proof can also be used to obtain fairly refined sufficiency conditions for problems with inequality constraints (see Exercise 3.3.7).

We first note that x^* is feasible, since, by Eq. (3.31), $\nabla_\lambda L(x^*, \lambda^*) = h(x^*) = 0$. Assume, to obtain a contradiction, that the conclusion of the proposition does not hold, so that there is a sequence $\{x^k\}$ such that $x^k \to x^*$, and for all k, $x^k \neq x^*$, $h(x^k) = 0$, and $f(x^k) < f(x^*) + (1/k)\|x^k - x^*\|^2$. Let us write $x^k = x^* + \delta^k y^k$, where

$$\delta^k = \|x^k - x^*\|, \qquad y^k = \frac{x^k - x^*}{\|x^k - x^*\|}.$$

The sequence $\{y^k\}$ is bounded, so it must have a subsequence converging to some y with $\|y\| = 1$. Without loss of generality, we assume that the whole sequence $\{y^k\}$ converges to y. By taking the limit as $\delta^k \to 0$ in the relation

$$0 = \frac{h_i(x^k) - h_i(x^*)}{\delta^k} = \frac{h_i(x^* + \delta^k y^k) - h_i(x^*)}{\delta^k} = \nabla h_i(x^*)' y^k + \frac{o(\delta^k)}{\delta^k},$$

we see that $\nabla h(x^*)'y = 0$.

We will now show that $y'\nabla^2_{xx}L(x^*, \lambda^*)y \leq 0$, thus coming to a contradiction [cf. Eq. (3.32)]. Since $x_k = x^* + \delta^k y^k$, by the mean value theorem [Prop. A.23(b) in Appendix A], we have

$$0 = h_i(x^k) - h_i(x^*) = \delta^k \nabla h_i(x^*)'y^k + \frac{(\delta^k)^2}{2} y^{k\prime} \nabla^2 h_i(\bar{\xi}_i^k) y^k, \tag{3.36}$$

$$\frac{1}{k}||x^k - x^*||^2 > f(x^k) - f(x^*) = \delta^k \nabla f(x^*)'y^k + \frac{(\delta^k)^2}{2} y^{k\prime} \nabla^2 f(\tilde{\xi}^k) y^k, \tag{3.37}$$

where all the vectors $\bar{\xi}_i^k$ and $\tilde{\xi}^k$ lie on the line segment joining x^* and x^k. Multiplying Eq. (3.36) by λ_i^* and adding Eq. (3.37) to it, we obtain

$$\begin{aligned} \frac{1}{k}||x^k - x^*||^2 > \delta^k &\left(\nabla f(x^*) + \sum_{i=1}^m \lambda_i^* \nabla h_i(x^*)\right)' y^k \\ &+ \frac{(\delta^k)^2}{2} y^{k\prime} \left(\nabla^2 f(\tilde{\xi}^k) + \sum_{i=1}^m \lambda_i^* \nabla^2 h_i(\bar{\xi}_i^k)\right) y^k. \end{aligned}$$

Since $\delta^k = ||x^k - x^*||$ and $\nabla f(x^*) + \sum_{i=1}^m \lambda_i^* \nabla h_i(x^*) = 0$, we obtain

$$\frac{2}{k} > y^{k\prime} \left(\nabla^2 f(\tilde{\xi}^k) + \sum_{i=1}^m \lambda_i^* \nabla^2 h_i(\bar{\xi}_i^k)\right) y^k.$$

By taking the limit as $k \to \infty$,

$$0 \geq y' \left(\nabla^2 f(x^*) + \sum_{i=1}^m \lambda_i^* \nabla^2 h_i(x^*)\right) y,$$

which contradicts the hypothesis (3.32). The proof of Prop. 3.2.1 is complete.

3.2.3 Sensitivity*

Lagrange multipliers frequently have an interesting interpretation in specific practical contexts. In economic applications they can often be interpreted as prices, while in other problems they represent quantities with concrete physical meaning. It turns out that within our mathematical framework, they can be viewed as rates of change of the optimal cost as the level of constraint changes. This is fairly easy to show for the case of linear constraints, as indicated in Fig. 3.2.3. The following proposition provides a formal proof for the case of nonlinear constraints.

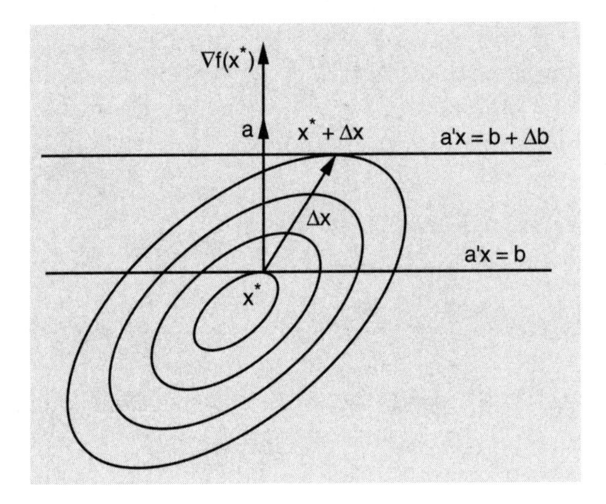

Figure 3.2.3. Illustration of the sensitivity theorem for a problem involving a single linear constraint,

$$\begin{aligned} &\text{minimize} \;\; f(x) \\ &\text{subject to} \;\; a'x = b. \end{aligned}$$

Here, x^* is a local minimum and λ^* is a corresponding Lagrange multiplier. If the level of constraint b is changed to $b + \Delta b$, the minimum x^* will change to $x^* + \Delta x$. Since $b + \Delta b = a'(x^* + \Delta x) = a'x^* + a'\Delta x = b + a'\Delta x$, we see that the variations Δx and Δb are related by

$$a'\Delta x = \Delta b.$$

Using the Lagrange multiplier condition $\nabla f(x^*) = -\lambda^* a$, the corresponding cost change can be written as

$$\Delta\text{cost} = f(x^* + \Delta x) - f(x^*) = \nabla f(x^*)'\Delta x + o(\|\Delta x\|) = -\lambda^* a'\Delta x + o(\|\Delta x\|).$$

By combining the above two relations, we obtain $\Delta\text{cost} = -\lambda^*\Delta b + o(\|\Delta x\|)$, so up to first order we have

$$\lambda^* = -\frac{\Delta\text{cost}}{\Delta b}.$$

Thus, the Lagrange multiplier λ^* gives the rate of optimal cost decrease as the level of constraint increases.

In the case where there are multiple constraints $a_i'x = b_i$, $i = 1, \ldots, m$, the preceding argument can be appropriately modified. In particular, we have

$$\begin{aligned} \Delta\text{cost} &= f(x^* + \Delta x) - f(x^*) \\ &= \nabla f(x^*)'\Delta x + o(\|\Delta x\|) \\ &= -\sum_{i=1}^m \lambda_i^* a_i'\Delta x + o(\|\Delta x\|), \end{aligned}$$

and $a_i'\Delta x = \Delta b_i$ for all i, so we obtain $\Delta\text{cost} = -\sum_{i=1}^m \lambda_i^*\Delta b_i + o(\|\Delta x\|)$.

Proposition 3.2.2: (Sensitivity Theorem) Let x^* and λ^* be a local minimum and Lagrange multiplier, respectively, satisfying the second order sufficiency conditions of Prop. 3.2.1. Consider the family of problems

$$\begin{aligned} &\text{minimize} \quad f(x) \\ &\text{subject to} \quad h(x) = u, \end{aligned} \tag{3.38}$$

parameterized by the vector $u \in \Re^m$. Then there exists an open sphere S centered at $u = 0$ such that for every $u \in S$, there is an $x(u) \in \Re^n$ and a $\lambda(u) \in \Re^m$, which are a local minimum-Lagrange multiplier pair of problem (3.38). Furthermore, $x(\cdot)$ and $\lambda(\cdot)$ are continuously differentiable functions within S and we have $x(0) = x^*$, $\lambda(0) = \lambda^*$. In addition, for all $u \in S$ we have

$$\nabla p(u) = -\lambda(u),$$

where $p(u)$ is the optimal cost parameterized by u, that is,

$$p(u) = f\big(x(u)\big).$$

Proof: Consider the system of equations

$$\nabla f(x) + \nabla h(x)\lambda = 0, \qquad h(x) = u. \tag{3.39}$$

For each fixed u, this system represents $n + m$ equations with $n + m$ unknowns – the vectors x and λ. For $u = 0$ the system has the solution (x^*, λ^*). The corresponding $(n + m) \times (n + m)$ Jacobian matrix with respect to (x, λ) is given by

$$J = \begin{pmatrix} \nabla^2_{xx} L(x^*, \lambda^*) & \nabla h(x^*) \\ \nabla h(x^*)' & 0 \end{pmatrix}.$$

Let us show that J is nonsingular. If it were not, some nonzero vector $(y', z')'$ would belong to the nullspace of J, that is,

$$\nabla^2_{xx} L(x^*, \lambda^*) y + \nabla h(x^*) z = 0, \tag{3.40}$$

$$\nabla h(x^*)' y = 0. \tag{3.41}$$

Premultiplying Eq. (3.40) by y' and using Eq. (3.41), we obtain

$$y' \nabla^2_{xx} L(x^*, \lambda^*) y = 0.$$

In view of Eq. (3.41), it follows that $y = 0$, for otherwise our second order sufficiency assumption would be violated [cf. Eq. (3.32)]. Since $y = 0$, Eq. (3.40) yields $\nabla h(x^*)z = 0$, which in view of the linear independence of the columns $\nabla h_i(x^*)$, $i = 1, \ldots, m$, of $\nabla h(x^*)$, yields $z = 0$. Thus, we obtain $y = 0$, $z = 0$, which is a contradiction. Hence, J is nonsingular.

Returning now to the system (3.39), it follows from the nonsingularity of J and the implicit function theorem (Prop. A.25 in Appendix A) that for all u in some open sphere S centered at $u = 0$, there exist $x(u)$ and $\lambda(u)$ such that $x(0) = x^*$, $\lambda(0) = \lambda^*$, the functions $x(\cdot)$ and $\lambda(\cdot)$ are continuously differentiable, and

$$\nabla f\big(x(u)\big) + \nabla h\big(x(u)\big)\lambda(u) = 0, \tag{3.42}$$

$$h\big(x(u)\big) = u.$$

For u sufficiently close to 0, the vectors $x(u)$ and $\lambda(u)$ satisfy the second order sufficiency conditions for problem (3.38), since they satisfy them by assumption for $u = 0$. This is straightforward to verify by using our continuity assumptions. [If it were not true, there would exist a sequence $\{u^k\}$ with $u^k \to 0$, and a sequence $\{y^k\}$ with $\|y^k\| = 1$ and $\nabla h\big(x(u^k)\big)' y^k = 0$ for all k, such that

$$y^{k\prime} \nabla^2_{xx} L\big(x(u^k), \lambda(u^k)\big) y^k \leq 0, \qquad \forall\ k.$$

By taking the limit along a convergent subsequence of $\{y^k\}$, we would obtain a contradiction of the second order sufficiency condition at (x^*, λ^*); cf. Eq. (3.32).] Hence, $x(u)$ and $\lambda(u)$ are a local minimum-Lagrange multiplier pair for problem (3.38).

There remains to show that $\nabla p(u) = \nabla_u \big\{f\big(x(u)\big)\big\} = -\lambda(u)$. By multiplying Eq. (3.42) by $\nabla x(u)$, we obtain

$$\nabla x(u) \nabla f\big(x(u)\big) + \nabla x(u) \nabla h\big(x(u)\big)\lambda(u) = 0.$$

By differentiating the relation $h\big(x(u)\big) = u$, it follows that

$$I = \nabla_u \big\{h\big(x(u)\big)\big\} = \nabla x(u) \nabla h\big(x(u)\big), \tag{3.43}$$

where I is the $m \times m$ identity matrix. Finally, by using the chain rule, we have

$$\nabla p(u) = \nabla_u \big\{f\big(x(u)\big)\big\} = \nabla x(u) \nabla f\big(x(u)\big).$$

Combining the above three relations, we obtain

$$\nabla p(u) + \lambda(u) = 0, \tag{3.44}$$

and the proof is complete. **Q.E.D.**

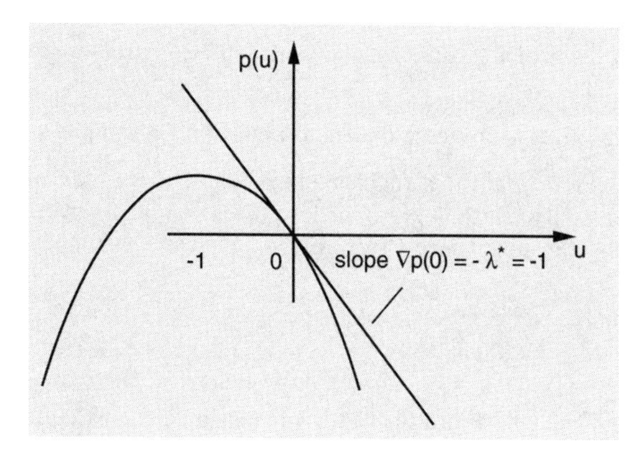

Figure 3.2.4. Illustration of the primal function $p(u) = f\big(x(u)\big)$ for the two-dimensional problem of Fig. 3.2.2

$$\text{minimize } f(x) = \tfrac{1}{2}\big(x_1^2 - x_2^2\big) - x_2$$
$$\text{subject to } h(x) = x_2 = 0.$$

Here,

$$p(u) = \min_{h(x)=u} f(x) = -\tfrac{1}{2}u^2 - u$$

and we have $\lambda^* = -\nabla p(0) = 1$ consistently with the sensitivity theorem.

The function $p(u) = f\big(x(u)\big)$ in the sensitivity theorem is called the *primal function* and plays an important role in contexts other than sensitivity. When we discuss duality theory later, in Section 3.4 and in Chapter 5, we will see that it is closely related to the, so-called, *dual function*. Within that context, the relation $\nabla p(0) = -\lambda^*$, illustrated in Fig. 3.2.4, will also turn out to be significant.

Note that the regularity of x^* is essential for the primal function to be defined within a sphere centered at 0. For example, for the problem of minimizing $\frac{1}{2}\big(x_1^2 - x_2^2\big) - x_2$ subject to $(x_2)^2 = 0$, which is equivalent to the problem of Fig. 3.2.4, the primal function is undefined for $u < 0$.

EXERCISES

3.2.1

Consider the problem

$$\text{maximize } x_1 + x_2$$
$$\text{subject to } x_1^2 + x_2^2 = 2.$$

(This is the problem of maximizing the perimeter of a rectangle inscribed in a given circle.)

(a) Show that it has a unique global maximum and a unique global minimum.

(b) Calculate the primal function corresponding to the maximum and the minimum and verify that its gradient is related to the Lagrange multiplier as specified by the sensitivity theorem.

3.2.2

Consider the problem of Example 3.1.3 in Section 3.1. Calculate the primal function corresponding to the minimum and verify that its gradient is related to the Lagrange multiplier as specified by the sensitivity theorem.

3.2.3

Use sufficiency conditions to verify optimality in the following problems:

(a) Among all rectangular parallelepipeds with given sum of lengths of edges, find one that has maximal volume.

(b) Among all rectangular parallelepipeds with given volume, find one that has minimal sum of lengths of edges.

(c) Among all rectangular parallelepipeds that are inscribed in an ellipsoid

$$\left\{ (x, y, z) \,\middle|\, \frac{x^2}{a^2} + \frac{y^2}{b^2} + \frac{z^2}{c^2} = 1 \right\},$$

and have sides that are parallel to the axes of the ellipsoid, find one that has maximal volume.

3.2.4

The purpose of this exercise is to show that under certain circumstances one may exchange the cost function with one of the constraints and obtain a new optimization problem whose solution is the same as the original. Consider the problem of minimizing $f(x)$ subject to $h_i(x) = 0$, $i = 1, \ldots, m$. Assume that x^* together with Lagrange multipliers λ_i^* satisfy the second order sufficiency conditions of Prop. 3.2.1. Let j be a constraint index such that $\lambda_j^* \neq 0$.

(a) Show that if $\lambda_j^* > 0$, then x^* is a local minimum of $h_j(x)$ subject to the constraints $f(x) = f(x^*)$ and $h_i(x) = 0$, $i = 1, \ldots, j-1, j+1, \ldots, m$.

(b) Show that if $\lambda_j^* < 0$, then x^* is a local maximum of $h_j(x)$ subject to the constraints $f(x) = f(x^*)$ and $h_i(x) = 0$, $i = 1, \ldots, j-1, j+1, \ldots, m$.

3.2.5 (Archimedes' Problem)

A spherical segment is the intersection of a sphere with one of the halfspaces corresponding to a 2-dimensional plane that meets the sphere. The area of a spherical segment is the area of the portion of its surface that is also part of the surface of the sphere. Note that if r is the radius of the sphere and h is the height of a spherical segment, the volume enclosed by the segment is $\pi h^2(r - h/3)$ and its spherical area is $2\pi rh$. Show that among all spherical segments with the same spherical area, the one that encloses the largest volume is a hemisphere.

3.2.6 (www)

Let x^* be a feasible point which is regular and together with some λ^* satisfies the first and second order necessary conditions of Prop. 3.1.1. Show that x^* and λ^* satisfy the second order sufficiency conditions of Prop. 3.2.1 if and only if the matrix

$$\begin{pmatrix} \nabla^2_{xx} L(x^*, \lambda^*) & \nabla h(x^*) \\ \nabla h(x^*)' & 0 \end{pmatrix}$$

is nonsingular. *Hint*: For the reverse assertion, see the proof of the sensitivity theorem. For the forward assertion, suppose that there exists a $\bar{y}$ such that $\nabla h(x^*)'\bar{y} = 0$ and $\bar{y}'\nabla^2_{xx} L(x^*, \lambda^*)\bar{y} = 0$. Use the fact that $\bar{y}$ minimizes $y'\nabla^2_{xx} L(x^*, \lambda^*)y$ over all y with $\nabla h(x^*)'y = 0$.

3.2.7 (Hessian of the Primal Function) (www)

Under the assumptions of the sensitivity theorem, show that for every scalar c for which the matrix

$$A_c(u) = \nabla^2_{xx} L\big(x(u), \lambda(u)\big) + c\nabla h\big(x(u)\big)\nabla h\big(x(u)\big)'$$

is invertible, we have

$$\nabla^2 p(u) = \Big(\nabla h\big(x(u)\big)' A_c(u)^{-1} \nabla h\big(x(u)\big)\Big)^{-1} - cI.$$

Hint: Differentiate Eqs. (3.42) and (3.44), and use Eq. (3.43).

3.3 INEQUALITY CONSTRAINTS

We now consider a problem involving both equality and inequality constraints

$$\begin{aligned} &\text{minimize} \quad f(x) \\ &\text{subject to} \quad h_1(x) = 0, \ldots, h_m(x) = 0, \\ &\qquad\qquad\quad g_1(x) \le 0, \ldots, g_r(x) \le 0, \end{aligned} \qquad \text{(ICP)}$$

where f, h_i, g_j are continuously differentiable functions from $\Re^n$ to $\Re$. More succinctly, we can write this problem as

$$\begin{aligned} &\text{minimize } f(x) \\ &\text{subject to } h(x) = 0, \qquad g(x) \leq 0, \end{aligned}$$

where $h : \Re^n \mapsto \Re^m$ and $g : \Re^n \mapsto \Re^r$ are the functions

$$h = (h_1, ..., h_m), \qquad g = (g_1, ..., g_r).$$

We will first use a simple approach to this problem that relies on the theory for equality constraints of the preceding sections. This leads to a straightforward development, which is however limited by regularity assumptions. Such assumptions are quite general for problems with nonlinear equality constraints, but are somewhat restrictive for problems with inequalities. The reason is that in many types of important problems (linear programs is an example) there may be many (possibly more than n) inequality constraints that are satisfied as equalities at a local minimum, and the corresponding constraint gradients are often dependent. With this motivation, we will develop later, starting in Section 3.5.5, a Lagrange multiplier theory that is not based on regularity-type assumptions. Still this theory is not too complicated and relies on the penalty approach that we have used for equality constraints in Section 3.1.1.

For any feasible point x, the *set of active inequality constraints* is denoted by

$$A(x) = \{j \mid g_j(x) = 0\}. \tag{3.45}$$

If $j \notin A(x)$, we say that the jth constraint is *inactive* at x. We note that if x^* is a local minimum of the inequality constrained problem (ICP), then x^* is also a local minimum for a problem identical to (ICP) except that the inactive constraints at x^* have been discarded. Thus, in effect, *inactive constraints at x^* don't matter*; they can be ignored in the statement of optimality conditions.

On the other hand, at a local minimum, *active inequality constraints can be treated to a large extent as equalities*. In particular, if x^* is a local minimum of the inequality constrained problem (ICP), then x^* is also a local minimum for the equality constrained problem

$$\begin{aligned} &\text{minimize } f(x) \\ &\text{subject to } h_1(x) = 0, \ldots, h_m(x) = 0, \qquad g_j(x) = 0, \quad \forall\, j \in A(x^*). \end{aligned} \tag{3.46}$$

Thus, if x^* is regular for the latter problem, there exist Lagrange multipliers $\lambda_1^*, \ldots, \lambda_m^*$, and μ_j^*, $j \in A(x^*)$ such that

$$\nabla f(x^*) + \sum_{i=1}^m \lambda_i^* \nabla h_i(x^*) + \sum_{j \in A(x^*)} \mu_j^* \nabla g_j(x^*) = 0.$$

Assigning zero Lagrange multipliers to the inactive constraints, we obtain

$$\nabla f(x^*) + \sum_{i=1}^{m} \lambda_i^* \nabla h_i(x^*) + \sum_{j=1}^{r} \mu_j^* \nabla g_j(x^*) = 0,$$

$$\mu_j^* = 0, \qquad \forall\, j \notin A(x^*),$$

which may be viewed as an analog of the first order optimality condition for the equality constrained problem.

There is one more important fact about the Lagrange multipliers μ_j^*: *they are nonnegative*. For a geometric illustration, see Fig. 3.3.1. For an algebraic argument, note that if the jth constraint is relaxed to $g_j(x) \leq u_j$, where $u_j > 0$, the optimal cost will tend to decrease because the constraint set will become larger. Therefore, the sensitivity interpretation of the Lagrange multiplier [$\mu_j^* = -(\Delta\text{cost due to } u_j)/u_j$, cf. Prop. 3.2.2] suggests that $\mu_j^* \geq 0$.

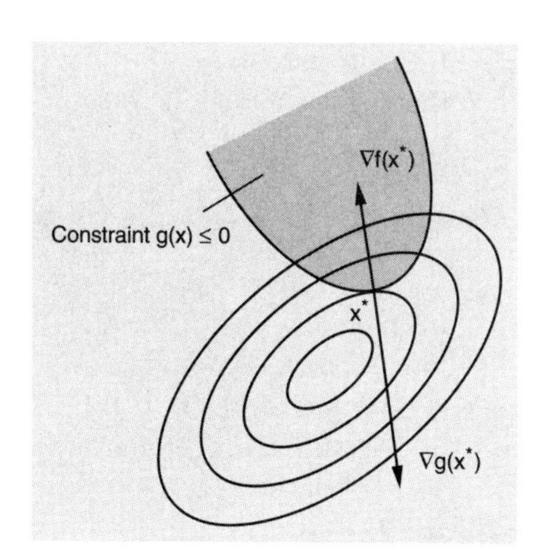

Figure 3.3.1. Illustration of the nonnegativity of the Lagrange multiplier for a problem with a single inequality constraint. If the constraint is inactive, then $\mu^* = 0$. Otherwise, $\nabla f(x^*)$ is normal to the constraint surface and points to the inside of the constraint set, while $\nabla g(x^*)$ is normal to the constraint surface and points to the outside of the constraint set. Thus $\nabla f(x^*)$ and $\nabla g(x^*)$ are collinear and have opposite signs, implying that the Lagrange multiplier is nonnegative.

3.3.1 Karush-Kuhn-Tucker Optimality Conditions

The preceding informal discussion will now be made rigorous. Our first line of development is based on an extended version of the notion of regularity that we used for equality constrained problems. This leads to a set of necessary conditions and sufficient conditions that are analogous to the ones for equality constraints. A more powerful line of development, to be given in Section 3.3.5, is based on the, so-called, Fritz John optimality conditions.

A feasible vector x is said to be *regular* if the equality constraint gradients $\nabla h_i(x)$, $i = 1, \ldots, m$, and the active inequality constraint gradients $\nabla g_j(x)$, $j \in A(x)$, are linearly independent. (We also say that x is regular in the exceptional case where there are no equality constraints and all the inequality constraints are inactive at x.) The following proposition generalizes the Lagrange multiplier theorem of Section 3.1. It will be stated in terms of the Lagrangian function defined by

$$L(x, \lambda, \mu) = f(x) + \sum_{i=1}^{m} \lambda_i h_i(x) + \sum_{j=1}^{r} \mu_j g_j(x). \tag{3.47}$$

Proposition 3.3.1: (Karush-Kuhn-Tucker Necessary Conditions) Let x^* be a local minimum of the problem

$$\begin{aligned} &\text{minimize} && f(x) \\ &\text{subject to} && h_1(x) = 0, \ldots, h_m(x) = 0, \\ & && g_1(x) \leq 0, \ldots, g_r(x) \leq 0, \end{aligned}$$

where f, h_i, g_j are continuously differentiable functions from $\Re^n$ to $\Re$, and assume that x^* is regular. Then there exist unique Lagrange multiplier vectors $\lambda^* = (\lambda_1^*, \ldots, \lambda_m^*)$, $\mu^* = (\mu_1^*, \ldots, \mu_r^*)$, such that

$$\nabla_x L(x^*, \lambda^*, \mu^*) = 0,$$

$$\mu_j^* \geq 0, \qquad j = 1, \ldots, r,$$

$$\mu_j^* = 0, \qquad \forall\ j \notin A(x^*),$$

where $A(x^*)$ is the set of active constraints at x^* [cf. Eq. (3.45)]. If in addition f, h, and g are twice continuously differentiable, there holds

$$y' \nabla_{xx}^2 L(x^*, \lambda^*, \mu^*) y \geq 0, \tag{3.48}$$

for all $y \in \Re^n$ such that

$$\nabla h_i(x^*)' y = 0,\ \forall\ i = 1, \ldots, m, \qquad \nabla g_j(x^*)' y = 0,\ \forall\ j \in A(x^*).$$

Proof: All the assertions of the proposition follow from the preceding discussion, and the first and second order optimality conditions for the equality constrained problem (3.46) (cf. Prop. 3.1.1), except for the assertion $\mu_j^* \geq 0$ for $j \in A(x^*)$. We give a proof of this assertion using the penalty approach of Section 3.1.1. We introduce the functions

$$g_j^+(x) = \max\{0, g_j(x)\}, \qquad j = 1, \ldots, r,$$

and for each $k = 1, 2, \ldots$, the "penalized" problem

$$\text{minimize} \quad F^k(x) \equiv f(x) + \frac{k}{2}||h(x)||^2 + \frac{k}{2}\sum_{j=1}^{r}\big(g_j^+(x)\big)^2 + \frac{\alpha}{2}||x - x^*||^2$$

$$\text{subject to} \quad x \in S,$$

where α is a fixed positive scalar, $S = \{x \mid ||x - x^*|| \leq \epsilon\}$, and $\epsilon > 0$ is such that $f(x^*) \leq f(x)$ for all feasible x with $x \in S$. Note that the function $\big(g_j^+(x)\big)^2$ is continuously differentiable with gradient $2g_j^+(x)\nabla g_j(x)$. If x^k minimizes $F^k(x)$ over S, a similar argument to the one used for the equality constraint case in Section 3.1.1 shows that $x^k \to x^*$, and that the Lagrange multipliers λ_i^* and μ_j^* are given by

$$\lambda_i^* = \lim_{k\to\infty} kh_i(x^k), \qquad i = 1, \ldots, m,$$

$$\mu_j^* = \lim_{k\to\infty} kg_j^+(x^k), \qquad j = 1, \ldots, r.$$

Since $g_j^+(x^k) \geq 0$, we obtain $\mu_j^* \geq 0$ for all j. **Q.E.D.**

One approach for using necessary conditions to solve inequality constrained problems is to consider separately all the possible combinations of constraints being active or inactive. The following example illustrates the process for the case of a single constraint.

Example 3.3.1

Consider the problem

$$\text{minimize} \quad \tfrac{1}{2}\big(x_1^2 + x_2^2 + x_3^2\big)$$
$$\text{subject to} \quad x_1 + x_2 + x_3 \leq -3.$$

Then for a local minimum x^*, the first order necessary condition [cf. Eq. (3.47)] yields

$$x_1^* + \mu^* = 0,$$

$$x_2^* + \mu^* = 0,$$

$$x_3^* + \mu^* = 0.$$

There are two possibilities:

(a) The constraint is inactive,

$$x_1^* + x_2^* + x_3^* < -3,$$

in which case $\mu^* = 0$. Then, we obtain $x_1^* = x_2^* = x_3^* = 0$, which contradicts the inequality $x_1^* + x_2^* + x_3^* < -3$. Hence, this possibility is excluded.

(b) The constraint is active,

$$x_1^* + x_2^* + x_3^* = -3.$$

Then we obtain $x_1^* = x_2^* = x_3^* = -1$ and $\mu^* = 1$, which satisfy all the necessary conditions for a local minimum. Since every point is regular it follows that $x^* = (-1, -1, -1)$ is the unique candidate for a local minimum. We may proceed further and verify that x^* and μ^* satisfy the second order sufficiency conditions for optimality (to be provided in Section 3.3.3) and conclude that x^* is indeed the unique local minimum for this problem. Alternatively, we can verify the sufficiency conditions based on convexity (to be provided in Section 3.3.4) and conclude that x^* is the unique global minimum.

The condition $\mu_j^* = 0$ for all $j \notin A(x^*)$ can also be succinctly written as

$$\mu_j^* g_j(x^*) = 0, \qquad j = 1, \ldots, r,$$

and is called *complementary slackness condition*. The name derives from the fact that for each j, whenever the constraint $g_j(x^*) \leq 0$ is slack [meaning that $g_j(x^*) < 0$], the constraint $\mu_j^* \geq 0$ must not be slack (meaning that $\mu_j^* = 0$), and reversely. A stronger form of the complementary slackness condition will be shown in Section 3.3.5 (see Prop. 3.3.5).

3.3.2 Conversion to the Equality Case*

We now provide an alternative proof of the Karush-Kuhn-Tucker conditions of Prop. 3.3.1, based on converting the inequality constrained problem (ICP) into an equality constrained problem and then using the corresponding necessary conditions of the previous section. This approach is straightforward and is also useful in other contexts. For example, it will be used in the next subsection to provide an easy proof of sufficiency conditions for inequality constraints. It is not as general, however, as our earlier approach in this section, because it requires twice differentiability of the problem functions.

Consider the equality constrained problem

$$\begin{aligned} &\text{minimize} \quad f(x) \\ &\text{subject to} \quad h_1(x) = 0, \ldots, h_m(x) = 0, \\ &\qquad\qquad\quad g_1(x) + z_1^2 = 0, \ldots, g_r(x) + z_r^2 = 0, \end{aligned} \tag{3.49}$$

where we have introduced additional variables $z_1, \ldots, z_r$.

Let x^* be a local minimum for our original problem (ICP). Then (x^*, z^*) is a local minimum for problem (3.49), where $z^* = (z_1^*, \ldots, z_r^*)$,

$$z_j^* = \big(-g_j(x^*)\big)^{1/2}, \qquad j = 1, \ldots, r.$$

It is straightforward to verify that x^* is regular for problem (ICP), if and only if (x^*, z^*) is regular for problem (3.49); we leave this as an exercise for the reader. By applying the first order necessary conditions for equality constraints (Prop. 3.1.1) to problem (3.49), we see that there exist Lagrange multipliers $\lambda_1^*, \ldots, \lambda_m^*, \mu_1^*, \ldots, \mu_r^*$ such that

$$\nabla f(x^*) + \sum_{i=1}^{m} \lambda_i^* \nabla h_i(x^*) + \sum_{j=1}^{r} \mu_j^* \nabla g_j(x^*) = 0,$$

$$2\mu_j^* z_j^* = 0, \qquad j = 1, \ldots, r.$$

Since $z_j^* = \big(-g_j(x^*)\big)^{1/2} > 0$ for $j \notin A(x^*)$, the last equation can also be written as

$$\mu_j^* = 0, \qquad \forall\, j \notin A(x^*). \tag{3.50}$$

Thus, to prove Prop. 3.3.1, there remains to show the nonnegativity condition $\mu_j^* \geq 0$ and the second order condition (3.48). For this purpose we use the second order necessary condition for the equivalent equality constrained problem (3.49). It yields [cf. Eq. (3.23) in Section 3.1]

$$\begin{pmatrix} y' & w' \end{pmatrix} \left(\begin{array}{c|cccc} \nabla_{xx}^2 L(x^*, \lambda^*, \mu^*) & & 0 & & \\ & 2\mu_1^* & 0 & \ldots & 0 \\ & 0 & 2\mu_2^* & \ldots & 0 \\ 0 & \vdots & \vdots & \vdots & \vdots \\ & 0 & 0 & \ldots & 2\mu_r^* \end{array} \right) \begin{pmatrix} y \\ w \end{pmatrix} \geq 0, \tag{3.51}$$

for all $y \in \Re^n$, $w \in \Re^r$ satisfying

$$\nabla h(x^*)' y = 0, \qquad \nabla g_j(x^*)' y + 2 z_j^* w_j = 0, \qquad j = 1, \ldots, r. \tag{3.52}$$

We will now try different pairs (y, w) satisfying Eq. (3.52) and obtain the desired results from Eq. (3.51). First let us select w by

$$w_j = \begin{cases} 0 & \text{if } j \in A(x^*), \\ -\dfrac{\nabla g_j(x^*)' y}{2 z_j^*} & \text{if } j \notin A(x^*). \end{cases}$$

Then $\nabla g_j(x^*)' y + 2 z_j^* w_j = 0$ for all $j \notin A(x^*)$, $z_j^* w_j = 0$ for all $j \in A(x^*)$, and by taking into account the fact $\mu_j^* = 0$ for $j \notin A(x^*)$ [cf. Eq. (3.50)], we also have $\mu_j^* w_j = 0$ for all j. Thus we obtain from Eqs. (3.51) and (3.52)

$$y' \nabla_{xx}^2 L(x^*, \lambda^*, \mu^*) y \geq 0,$$

for all y such that $\nabla h(x^*)' y = 0$ and $\nabla g_j(x^*)' y = 0$ for all $j \in A(x^*)$, thereby verifying the second order condition (3.48).

Next let us select, for every $j \in A(x^*)$, a vector (y, w) with $y = 0$, $w_j \neq 0$, $w_k = 0$ for all $k \neq j$. Such a vector satisfies the condition of Eq. (3.52). By using such a vector in Eq. (3.51), we obtain $2\mu_j^* w_j^2 \geq 0$, and

$$\mu_j^* \geq 0, \qquad \forall\, j \in A(x^*).$$

Thus, the conversion to the equivalent equality constrained problem (3.49) has yielded all the Karush-Kuhn-Tucker conditions for the inequality constrained problem.

3.3.3 Second Order Sufficiency Conditions and Sensitivity*

We now derive sufficiency conditions for the inequality constrained problem (ICP) using the transformation to an equality constrained problem described in Section 3.3.2.

Proposition 3.3.2: (Second Order Sufficiency Conditions) Assume that f, h, and g are twice continuously differentiable, and let $x^* \in \Re^n$, $\lambda^* \in \Re^m$, and $\mu^* \in \Re^r$ satisfy

$$\nabla_x L(x^*, \lambda^*, \mu^*) = 0, \qquad h(x^*) = 0, \qquad g(x^*) \leq 0,$$

$$\mu_j^* \geq 0, \qquad j = 1, \ldots, r,$$

$$\mu_j^* = 0, \qquad \forall\, j \notin A(x^*),$$

$$y' \nabla_{xx}^2 L(x^*, \lambda^*, \mu^*) y > 0,$$

for all $y \neq 0$ such that

$$\nabla h_i(x^*)'y = 0,\ \forall\, i = 1, \ldots, m, \qquad \nabla g_j(x^*)'y = 0,\ \forall\, j \in A(x^*). \tag{3.53}$$

Assume also that

$$\mu_j^* > 0, \qquad \forall\, j \in A(x^*). \tag{3.54}$$

Then x^* is a strict local minimum of f subject to $h(x) = 0$, $g(x) \leq 0$.

Proof: (*Abbreviated*) It is straightforward to verify using our assumptions, that for the equivalent equality constrained problem (3.49), the second order sufficiency conditions are satisfied by $(x^*, z_1^*, \ldots, z_r^*)$, λ^*, μ^*, where $z_j^* = \big(-g_j(x^*)\big)^{1/2}$. Hence $(x^*, z_1^*, \ldots, z_r^*)$ is a strict local minimum for problem (3.49) and the result follows. **Q.E.D.**

The condition (3.54) is known as the *strict complementary slackness condition*. Exercise 3.3.7 provides a sharper version of the preceding sufficiency theorem, where the strict complementary slackness condition is somewhat weakened. However, this assumption cannot be entirely discarded as the following example shows.

Example 3.3.2

Consider the two-dimensional problem

$$\begin{aligned} &\text{minimize} \ \ \tfrac{1}{2}\big(x_1^2 - x_2^2\big) \\ &\text{subject to} \ \ x_2 \leq 0. \end{aligned}$$

It can be verified that $x^* = (0,0)$ and $\mu^* = 0$ satisfy all the conditions of Prop. 3.3.2 except for the strict complementary slackness condition. It is seen that x^* is not a local minimum, because the cost can be decreased by moving from x^* along the feasible direction $(0,-1)$. This direction does not satisfy the condition (3.53). The problem here is that the constraint is active but *degenerate*, in the sense that its Lagrange multiplier is zero (see Exercise 3.3.7).

Using the transformation to an equality constrained problem of Section 3.3.2, together with the corresponding sensitivity result of Prop. 3.2.2, we can also obtain the following proposition. The proof is left for the reader.

Proposition 3.3.3: (Sensitivity) Let x^* and λ^* be a local minimum and Lagrange multiplier, respectively, satisfying the second order sufficiency conditions of Prop. 3.3.2. Consider the family of problems

$$\begin{aligned} &\text{minimize} \quad f(x) \\ &\text{subject to} \quad h(x) = u, \qquad g(x) \le v, \end{aligned} \tag{3.55}$$

parameterized by the vectors $u \in \Re^m$ and $v \in \Re^r$. Then there exists an open sphere S centered at $(u,v) = (0,0)$ such that for every $(u,v) \in S$ there is an $x(u,v) \in \Re^n$ and $\lambda(u,v) \in \Re^m$, $\mu(u,v) \in \Re^r$, which are a local minimum and associated Lagrange multiplier vectors of problem (3.55). Furthermore, $x(\cdot)$, $\lambda(\cdot)$, and $\mu(\cdot)$ are continuously differentiable in S and we have $x(0,0) = x^*$, $\lambda(0,0) = \lambda^*$, $\mu(0,0) = \mu^*$. In addition, for all $(u,v) \in S$, there holds

$$\nabla_u p(u,v) = -\lambda(u,v),$$

$$\nabla_v p(u,v) = -\mu(u,v),$$

where $p(u,v)$ is the optimal cost parameterized by (u,v),

$$p(u,v) = f\big(x(u,v)\big).$$

3.3.4 Sufficiency Conditions and Lagrangian Minimization*

The sufficiency conditions that we have discussed so far involve second derivatives and Hessian positive definiteness assumptions. Our experience with unconstrained problems suggests that the first order Lagrange multiplier conditions together with convexity assumptions should also be sufficient for optimality. Indeed this is so, as we will demonstrate shortly. In fact we will not need to impose convexity or even differentiability assumptions explicitly. A minimization condition on the Lagrangian function turns out to be sufficient.

Proposition 3.3.4: (General Sufficiency Condition) Consider the problem

$$\begin{aligned} &\text{minimize } f(x) \\ &\text{subject to } x \in X, \qquad g_j(x) \leq 0, \quad j = 1, \ldots, r, \end{aligned}$$

where f and g_j are real valued functions on $\Re^n$ and X is a given subset of $\Re^n$. Let x^* be a feasible vector which together with a vector $\mu^* = (\mu_1^*, \ldots, \mu_r^*)$, satisfies

$$\mu_j^* \geq 0, \qquad j = 1, \ldots, r,$$

$$\mu_j^* = 0, \qquad \forall\, j \notin A(x^*),$$

and minimizes the Lagrangian function $L(x, \mu^*)$ over $x \in X$:

$$x^* = \arg\min_{x \in X} L(x, \mu^*).$$

Then x^* is a global minimum of the problem.

Proof: We have

$$\begin{aligned} f(x^*) &= f(x^*) + \mu^{*\prime} g(x*) \\ &= \min_{x \in X} \big\{ f(x) + \mu^{*\prime} g(x) \big\} \\ &\leq \min_{x \in X,\, g(x) \leq 0} \big\{ f(x) + \mu^{*\prime} g(x) \big\} \\ &\leq \min_{x \in X,\, g(x) \leq 0} f(x), \end{aligned}$$

where the first equality follows from the hypothesis, which implies that $\mu^{*\prime} g(x*) = 0$, and the last inequality follows from the nonnegativity of μ^*. **Q.E.D.**

Note that if f and g_j are differentiable convex functions and $X = \Re^n$, the Lagrangian function $L(x, \mu^*)$ is convex as a function of x, so the Lagrangian minimization condition is equivalent to the first order condition

$$\nabla f(x^*) + \sum_{j=1}^{r} \mu_j^* \nabla g_j(x^*) = 0.$$

Thus, *in the presence of these convexity assumptions, the first order optimality conditions are also sufficient.*

However, the preceding proposition provides a more general sufficiency condition because it does not require differentiability and convexity of f and g_j, while X may be a strict subset of $\Re^n$.

3.3.5 Fritz John Optimality Conditions*

We will now develop a generalization of the Karush-Kuhn-Tucker necessary conditions that does not require a regularity assumption. With the help of this generalization, we will derive all the necessary conditions obtained so far in this chapter, plus quite a few new ones. To get a sense of the main idea, consider the equality constrained problem

$$\begin{aligned} &\text{minimize} \ \ f(x) \\ &\text{subject to} \ \ h_1(x) = 0, \dots, h_m(x) = 0, \end{aligned}$$

and note that an equivalent statement of the Lagrange multiplier theorem of Prop. 3.1.1 is that at a local minimum x^* (regular or not), there exist scalars $\mu_0^*, \lambda_1^*, \dots, \lambda_m^*$, not all equal to 0, such that $\mu_0^* \geq 0$ and

$$\mu_0^* \nabla f(x^*) + \sum_{i=1}^{m} \lambda_i^* \nabla h_i(x^*) = 0. \tag{3.56}$$

To see the equivalence, note that there are two possibilities:

(a) x^* is regular, in which case Eq. (3.56) is satisfied with $\mu_0^* = 1$.

(b) x^* is not regular, in which case the gradients $\nabla h_i(x^*)$ are linearly dependent, and there exist $\lambda_1^*, \dots, \lambda_m^*$, not all equal to 0, such that

$$\sum_{i=1}^{m} \lambda_i^* \nabla h_i(x^*) = 0,$$

so that Eq. (3.56) is satisfied with $\mu_0^* = 0$.

Necessary conditions that involve a nonnegative multiplier for the cost gradient, such as the one of Eq. (3.56), are known as *Fritz John necessary conditions*, and were first proposed in 1948 by John [Joh48]. One approach to deriving such conditions for the more general inequality-constrained case is to convert the inequality constraints to equalities, as in Section 3.3.1, and use the condition (3.56), but this approach is not fully satisfactory because it does not establish the nonnegativity of the Lagrange multipliers that correspond to the inequality constraints. On the other hand, a simple adaptation of the penalty-based proof of the Lagrange multiplier theorem (cf. Section 3.1.1, and Props. 3.1.1, 3.3.1) works well. This is shown in the following proposition, which also provides a more precise characterization of the constraints that correspond to nonzero multipliers.

Proposition 3.3.5: (Fritz John Necessary Conditions) Let x^* be a local minimum of the problem

$$\begin{aligned} &\text{minimize} \quad f(x) \\ &\text{subject to} \quad h_1(x) = 0, \ldots, h_m(x) = 0, \\ &\qquad\qquad\quad g_1(x) \leq 0, \ldots, g_r(x) \leq 0, \end{aligned}$$

where f, h_i, g_j are continuously differentiable functions from $\Re^n$ to $\Re$. Then there exists a scalar μ_0^* and multipliers $\lambda_1^*, \ldots, \lambda_m^*$ and $\mu_1^*, \ldots, \mu_r^*$, satisfying the following conditions:

(i)

$$\mu_0^* \nabla f(x^*) + \sum_{i=1}^{m} \lambda_i^* \nabla h_i(x^*) + \sum_{j=1}^{r} \mu_j^* \nabla g_j(x^*) = 0.$$

(ii)

$$\mu_j^* \geq 0, \qquad j = 0, 1, \ldots, r.$$

(iii) $\mu_0^*, \lambda_1^*, \ldots, \lambda_m^*, \mu_1^*, \ldots, \mu_r^*$ are not all equal to 0.

(iv) In every neighborhood N of x^* there is an $x \in N$ such that $\lambda_i^* h_i(x) > 0$ for all i with $\lambda_i^* \neq 0$ and $\mu_j^* g_j(x) > 0$ for all j with $\mu_j^* \neq 0$.

Note: Condition (iv) implies complementary slackness [$\mu_j^* g_j(x^*) = 0$ for all j], since if $\mu_j^* > 0$, the corresponding jth inequality constraint must be violated arbitrarily close to x^* implying that $g_j(x^*) = 0$. Another condition that is equivalent to condition (iv) is given in Exercise 3.3.12.

Proof: We follow the line of argument of the penalty approach used in Section 3.1.1 and in Prop. 3.3.1. Consider the functions

$$g_j^+(x) = \max\{0, g_j(x)\}, \qquad j = 1, \ldots, r,$$

and for each $k = 1, 2, \ldots$, the "penalized" problem

$$\text{minimize} \quad F^k(x) \equiv f(x) + \frac{k}{2} \sum_{i=1}^{m} \big(h_i(x)\big)^2 + \frac{k}{2} \sum_{j=1}^{r} \big(g_j^+(x)\big)^2 + \frac{1}{2} ||x - x^*||^2$$

$$\text{subject to} \quad x \in S,$$

where $S = \{x \mid ||x - x^*|| \leq \epsilon\}$, and $\epsilon > 0$ is such that $f(x^*) \leq f(x)$ for all feasible x with $x \in S$. The function $\big(g_j^+(x)\big)^2$ is continuously differentiable with gradient $2g_j^+(x)\nabla g_j(x)$, and if x^k minimizes $F^k(x)$ over S, an argument similar to the one used for the equality constraint casc in Scction 3.1.1 shows that $x^k \to x^*$ and that x^k is an interior point of S for all k greater than some $\bar{k}$. For such k, we have $\nabla F^k(x^k) = 0$, or

$$\nabla f(x^k) + \sum_{i=1}^{m} \xi_i^k \nabla h_i(x^k) + \sum_{j=1}^{r} \zeta_j^k \nabla g_j(x^k) + (x^k - x^*) = 0, \tag{3.57}$$

where

$$\xi_i^k = k h_i(x^k), \qquad \zeta_j^k = k g_j^+(x^k). \tag{3.58}$$

Denote,

$$\delta^k = \sqrt{1 + \sum_{i=1}^{m} (\xi_i^k)^2 + \sum_{j=1}^{r} (\zeta_j^k)^2}, \tag{3.59}$$

$$\mu_0^k = \frac{1}{\delta^k}, \qquad \lambda_i^k = \frac{\xi_i^k}{\delta^k}, \quad i = 1, \ldots, m, \qquad \mu_j^k = \frac{\zeta_j^k}{\delta^k}, \quad j = 1, \ldots, r. \tag{3.60}$$

Then by dividing Eq. (3.57) with δ^k, we obtain

$$\mu_0^k \nabla f(x^k) + \sum_{i=1}^{m} \lambda_i^k \nabla h_i(x^k) + \sum_{j=1}^{r} \mu_j^k \nabla g_j(x^k) + \frac{1}{\delta^k}(x^k - x^*) = 0. \tag{3.61}$$

Since by construction we have

$$(\mu_0^k)^2 + \sum_{i=1}^{m} (\lambda_i^k)^2 + \sum_{j=1}^{r} (\mu_j^k)^2 = 1, \tag{3.62}$$

the sequence $\{\mu_0^k, \lambda_1^k, \ldots, \lambda_m^k, \mu_1^k, \ldots, \mu_r^k\}$ is bounded and must contain a subsequence that converges to some limit $\{\mu_0^*, \lambda_1^*, \ldots, \lambda_m^*, \mu_1^*, \ldots, \mu_r^*\}$. From Eq. (3.61), we see that this limit must satisfy condition (i), from Eqs. (3.58) and (3.60), it must satisfy condition (ii), and from Eq. (3.62), it must satisfy condition (iii). Finally, to show that condition (iv) is satisfied, let

$$I = \{i \mid \lambda_i^* \neq 0\}, \qquad J = \{j \mid \mu_j^* > 0\}.$$

Then, for all sufficiently large k within the index set of the convergent subsequence, we must have $\lambda_i^* \lambda_i^k > 0$ for all $i \in I$ and $\mu_j^* \mu_j^k > 0$ for all $j \in J$. Therefore, for these k, from Eqs. (3.58) and (3.60), we must have $\lambda_i^* h_i(x^k) > 0$ for all $i \in I$ and $\mu_j^* g_j(x^k) > 0$ for all $j \in J$. Since every neighborhood of x^* must contain some x^k from the subsequence, this proves condition (iv). **Q.E.D.**

Example 3.3.3

Consider the problem

$$\begin{aligned} &\text{minimize} \quad f(x) = x_1 + x_2 \\ &\text{subject to} \quad g_1(x) = (x_1 - 1)^2 + x_2^2 - 1 \leq 0, \\ &\qquad\qquad\quad g_2(x) = -(x_1 - 2)^2 - x_2^2 + 4 \leq 0. \end{aligned}$$

This is an inequality constrained version of Example 3.1.1, whose geometry is illustrated in Fig. 3.1.2. It can be seen from this figure that the only feasible solution is $x^* = (0, 0)$, which is thus the global minimum. The condition

$$\mu_0^* \nabla f(x^*) + \mu_1^* \nabla g_1(x^*) + \mu_2^* \nabla g_2(x^*) = 0$$

can be written as

$$\mu_0^* - 2\mu_1^* + 4\mu_2^* = 0, \qquad \mu_0^* = 0.$$

Thus, consistently with Example 3.1.1, where we found that no Lagrange multiplier exists, we have $\mu_0^* = 0$. It can be seen that any positive μ_1^* and μ_2^* with $\mu_1^* = 2\mu_2^*$ satisfy the Fritz John necessary conditions (i)-(iii), and from the geometry of the problem as shown in Fig. 3.1.2, it can be verified that condition (iv) is satisfied as well.

Condition (iv) of Prop. 3.3.5 is somewhat unusual and has not been encountered so far. It asserts that *the constraints with nonzero multipliers can be simultaneously violated arbitrarily close to x^* in a way that the constraint violations have the same signs as the corresponding multipliers.* This condition replaces the weaker complementary slackness condition

$$\mu_j^* = 0, \qquad \forall\, j \notin A(x^*),$$

of Prop. 3.3.1, and narrows down the set of candidate multipliers that satisfy the Fritz John conditions.

From the Fritz John to the Karush-Kuhn-Tucker Conditions

To illustrate the use of the Fritz John conditions and condition (iv) in particular, suppose that we convert a problem with a single equality constraint, $\min_{h(x)=0} f(x)$, to the inequality constrained problem

$$\begin{aligned} &\text{minimize} \quad f(x) \\ &\text{subject to} \quad h(x) \leq 0, \quad -h(x) \leq 0. \end{aligned}$$

The Fritz John conditions assert the existence of nonnegative $\mu_0^*, \lambda^+, \lambda^-$, not all zero, such that

$$\mu_0^* \nabla f(x^*) + \lambda^+ \nabla h(x^*) - \lambda^- \nabla h(x^*) = 0. \tag{3.63}$$

The candidate multipliers that satisfy the above condition as well as the complementary slackness condition $\lambda^+ h(x^*) = \lambda^- h(x^*) = 0$, include those of the form $\mu_0^* = 0$ and $\lambda^+ = \lambda^- > 0$, which provide no relevant information about the problem. However, these multipliers fail the stronger condition (iv) of Prop. 3.3.5, showing that if $\mu_0^* = 0$, we must have either $\lambda^+ \neq 0$ and $\lambda^- = 0$ or $\lambda^+ = 0$ and $\lambda^- \neq 0$. For a regular x^*, this violates Eq. (3.63), so it follows that $\mu_0^* > 0$. Thus, by dividing Eq. (3.63) by μ_0^*, we recover the familiar first order condition $\nabla f(x^*) + \lambda^* \nabla h(x^*) = 0$ with $\lambda^* = (\lambda^+ - \lambda^-)/\mu_0^*$, under a regularity assumption on x^*. Note that this deduction would not have been possible without condition (iv).

Generalizing the above discussion, we note that if for the problem at hand we can verify that it is impossible to satisfy the Fritz John conditions (i)-(iv) with $\mu_0^* = 0$, then we obtain $\mu_0^* > 0$, and by dividing all λ_i^* and μ_j^* with μ_0^*, we recover the Karush-Kuhn-Tucker conditions with Lagrange multipliers equal to λ_i^*/μ_0^* and μ_j^*/μ_0^*. In fact, since the multipliers can be normalized by multiplication with a positive scalar, it is sufficient to consider just two possibilities for μ_0^*: either $\mu_0^* = 0$ or $\mu_0^* = 1$.

As an example, suppose that x^* is regular as per the assumption of Prop. 3.3.1. This implies that we cannot have

$$\sum_{i=1}^{m} \lambda_i^* \nabla h_i(x^*) + \sum_{j=1}^{r} \mu_j^* \nabla g_j(x^*) = 0$$

for some λ_i^*, $i = 1, \ldots, m$, and μ_j^*, $j = 1, \ldots, r$, that are not all zero, so the Fritz John conditions cannot be satisfied with $\mu_0^* = 0$. Hence, we can take $\mu_0^* = 1$ and there exist λ_i^*, $i = 1, \ldots, m$, and μ_j^*, $j = 1, \ldots, r$, such that

$$\nabla f(x^*) + \sum_{i=1}^{m} \lambda_i^* \nabla h_i(x^*) + \sum_{j=1}^{r} \mu_j^* \nabla g_j(x^*) = 0, \qquad \mu_j^* \geq 0, \quad j = 1, \ldots, r.$$

Furthermore, in every neighborhood N of x^* there is an $x \in N$ such that $\lambda_i^* h_i(x) > 0$ for all i with $\lambda_i^* \neq 0$ and $\mu_j^* g_j(x) > 0$ for all j with $\mu_j^* \neq 0$. We have thus shown the following proposition, which represents a slightly more powerful form of the Karush-Kuhn-Tucker conditions of Prop. 3.3.1 [condition (iv) of Prop. 3.3.5 has replaced the weaker complementary slackness condition, $\mu_j^* = 0$ if $j \notin A(x^*)$, of Prop. 3.3.1].

Proposition 3.3.6 Let x^* be a local minimum of the problem

$$\begin{aligned} &\text{minimize} && f(x) \\ &\text{subject to} && h_1(x) = 0, \ldots, h_m(x) = 0, \\ & && g_1(x) \leq 0, \ldots, g_r(x) \leq 0, \end{aligned}$$

where f, h_i, and g_j are continuously differentiable functions from $\Re^n$ to $\Re$, and assume that x^* is regular. Then there exist Lagrange multipliers $\lambda_1^*, \ldots, \lambda_m^*$ and $\mu_1^*, \ldots, \mu_r^*$, satisfying the following conditions:

(i)

$$\nabla f(x^*) + \sum_{i=1}^{m} \lambda_i^* \nabla h_i(x^*) + \sum_{j=1}^{r} \mu_j^* \nabla g_j(x^*) = 0.$$

(ii)

$$\mu_j^* \geq 0, \qquad j = 1, \ldots, r.$$

(iii) In every neighborhood N of x^* there is an $x \in N$ such that $\lambda_i^* h_i(x) > 0$ for all i with $\lambda_i^* \neq 0$ and $\mu_j^* g_j(x) > 0$ for all j with $\mu_j^* \neq 0$.

As another example, we consider in the following proposition the case where there are *linear* equality and *concave* (or, as a special case, *linear*) inequality constraints. The proof of the proposition relies strongly on the Fritz John condition (iv) of Prop. 3.3.5, the key fact being that if at least one λ_i^* or μ_j^* is nonzero, there exists an x arbitrarily close to x^* such that $\sum_{i=1}^{m} \lambda_i^* h_i(x) + \sum_{j=1}^{r} \mu_j^* g_j(x) > 0$.

Proposition 3.3.7: (Linear/Concave Constraints) Let x^* be a local minimum of the problem

$$\begin{aligned} &\text{minimize} \quad f(x) \\ &\text{subject to} \quad h_1(x) = 0, \ldots, h_m(x) = 0, \\ &\qquad\qquad\quad g_1(x) \leq 0, \ldots, g_r(x) \leq 0, \end{aligned}$$

where f, h_i, and g_j are continuously differentiable functions from $\Re^n$ to $\Re$. Assume that all the functions h_i are linear and all the functions g_j are concave. Then x^* satisfies the necessary conditions of Prop. 3.3.6.

Proof: We consider a set of λ_i^*, $i = 1, \ldots, m$, and μ_j^*, $j = 0, 1, \ldots, r$, satisfying the Fritz John conditions, and we assume that $\mu_0^* = 0$. We will come to a contradiction, implying that we can take $\mu_0^* = 1$, thereby proving the proposition. Indeed, for any $x \in \Re^n$, we have, by the linearity of h_i and the concavity of g_j,

$$h_i(x) = h_i(x^*) + \nabla h_i(x^*)'(x - x^*), \qquad i = 1, \ldots, m,$$

$$g_j(x) \leq g_j(x^*) + \nabla g_j(x^*)'(x - x^*), \qquad j = 1, \ldots, r.$$

By multiplying these two relations with λ_i^* and μ_j^*, and by adding over i and j, respectively, we obtain

$$\begin{aligned}\sum_{i=1}^m \lambda_i^* h_i(x) + \sum_{j=1}^r \mu_j^* g_j(x) &\le \sum_{i=1}^m \lambda_i^* h_i(x^*) + \sum_{j=1}^r \mu_j^* g_j(x^*) \\ &\quad + \left(\sum_{i=1}^m \lambda_i^* \nabla h_i(x^*) + \sum_{j=1}^r \mu_j^* \nabla g_j(x^*)\right)'(x - x^*) \\ &= 0,\end{aligned} \tag{3.64}$$

where the last equality holds because we have $\lambda_i^* h_i(x^*) = 0$ for all i and $\mu_j^* g_j(x^*) = 0$ for all j [by the Fritz John condition (iv)], and

$$\sum_{i=1}^m \lambda_i^* \nabla h_i(x^*) + \sum_{j=1}^r \mu_j^* \nabla g_j(x^*) = 0$$

[by the Fritz John condition (i)]. On the other hand, by the Fritz John condition (iii), we know that since $\mu_0^* = 0$, there exists some i for which $\lambda_i^* \neq 0$ or some j for which $\mu_j^* > 0$. By the Fritz John condition (iv), there is an x satisfying $\lambda_i^* h_i(x) > 0$ for all i with $\lambda_i^* \neq 0$ and $\mu_j^* g_j(x) > 0$ for all j with $\mu_j^* > 0$. For this x, we have $\sum_{i=1}^m \lambda_i^* h_i(x) + \sum_{j=1}^r \mu_j^* g_j(x) > 0$, contradicting Eq. (3.64). **Q.E.D.**

We next consider conditions that guarantee the existence of Lagrange multipliers in cases where the equality constraints are nonlinear and/or the inequality constraints are not concave. There are a number of such conditions, which are often referred to as *constraint qualifications*. The following proposition was shown by Arrow, Hurwitz, and Uzawa [AHU61] for the case of inequality constraints only, and was generalized to the case where there are equality constraints as well by Mangasarian and Fromovitz [MaF67]. (An alternative proof is given in Exercise 3.3.5 for the case where there are no equality constraints; see also Exercise 3.3.18.)

Proposition 3.3.8: (Mangasarian-Fromovitz Constraint Qualification) Let x^* be a local minimum of the problem

$$\begin{aligned}&\text{minimize } f(x)\\ &\text{subject to } h_1(x) = 0, \ldots, h_m(x) = 0,\\ &\qquad\qquad g_1(x) \le 0, \ldots, g_r(x) \le 0,\end{aligned}$$

where f, h_i, and g_j are continuously differentiable functions from $\Re^n$ to $\Re$. Assume that the gradients $\nabla h_i(x^*)$, $i = 1, \ldots, m$, are linearly independent, and that there exists a vector d such that

$$\nabla h_i(x^*)'d = 0, \quad \forall\, i = 1, \ldots, m, \qquad \nabla g_j(x^*)'d < 0, \quad \forall\, j \in A(x^*).$$

Then x^* satisfies the necessary conditions of Prop. 3.3.6.

Proof: We consider a set of multipliers λ_i^*, $i = 1, \ldots, m$, μ_j^*, $j = 0, 1, \ldots, r$, satisfying the Fritz John conditions, and we assume that $\mu_0^* = 0$, so that

$$\sum_{i=1}^{m} \lambda_i^* \nabla h_i(x^*) + \sum_{j=1}^{r} \mu_j^* \nabla g_j(x^*) = 0. \tag{3.65}$$

Since not all the λ_i^* and μ_j^* can be zero, we conclude that $\mu_j^* > 0$ for at least one $j \in A(x^*)$; otherwise we would have $\sum_{i=1}^{m} \lambda_i^* \nabla h_i(x^*) = 0$ with not all the λ_i^* equal to zero, and the linear independence of the gradients $\nabla h_i(x^*)$ would be violated. Since $\mu_j^* \geq 0$ for all j, with $\mu_j^* = 0$ for $j \notin A(x^*)$ and $\mu_j^* > 0$ for at least one j, we obtain

$$\sum_{i=1}^{m} \lambda_i^* \nabla h_i(x^*)'d + \sum_{j=1}^{r} \mu_j^* \nabla g_j(x^*)'d < 0,$$

where d is the vector of the hypothesis. This contradicts Eq. (3.65). Therefore, we must have $\mu_0^* > 0$, so that we can take $\mu_0^* = 1$. **Q.E.D.**

Note that the existence of a vector d satisfying the condition of Prop. 3.3.8 is equivalent to the more easily verifiable hypothesis that for each $j \in A(x^*)$, there exists a vector d^j such that

$$\nabla h_i(x^*)'d^j = 0, \quad i = 1, \ldots, m,$$

$$\nabla g_j(x^*)'d^j < 0, \qquad \nabla g_{\bar{j}}(x^*)'d^j \leq 0, \quad \forall\, \bar{j} \in A(x^*),\ \bar{j} \neq j.$$

Indeed, the vector $d = \sum_{j \in A(x^*)} d^j$ satisfies the condition of Prop. 3.3.8. Figure 3.3.2 illustrates the character of the Mangasarian-Fromovitz constraint qualification.

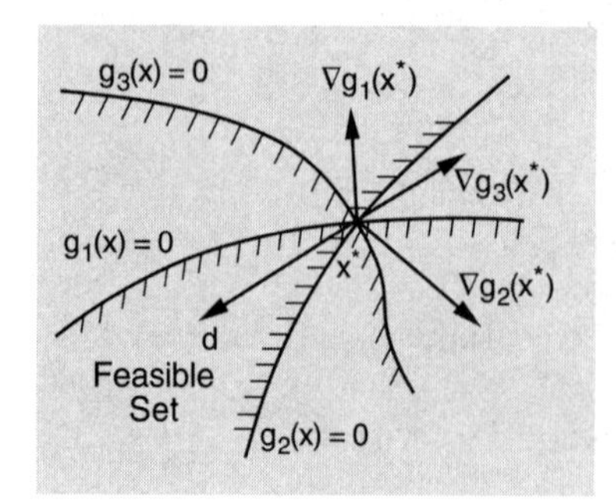

Figure 3.3.2. Illustration of the Mangasarian-Fromovitz constraint qualification. There exists a vector d such that $\nabla g_j(x^*)'d < 0$ for all $j \in A(x^*)$. As a result, if at least one of the μ_j^* is positive, we cannot have

$$\sum_{j=1}^{r} \mu_j^* \nabla g_j(x^*) = 0.$$

The following proposition, due to Slater [Sla50], is really a special case of the preceding one, and was one of the first shown for inequality constraints.

Proposition 3.3.9: (Slater Constraint Qualification for Convex Inequalities) Let x^* be a local minimum of the problem

$$\begin{aligned} &\text{minimize} \quad f(x) \\ &\text{subject to} \quad h_1(x) = 0, \ldots, h_m(x) = 0, \\ &\qquad\qquad\quad g_1(x) \leq 0, \ldots, g_r(x) \leq 0, \end{aligned}$$

where f and g_j are continuously differentiable functions from $\Re^n$ to $\Re$, and the functions h_i are linear. Assume that the functions g_j are convex and that there exists a feasible vector $\bar{x}$ satisfying

$$g_j(\bar{x}) < 0, \qquad \forall\, j \in A(x^*).$$

Then x^* satisfies the necessary conditions of Prop. 3.3.6.

Proof: In view of the linearity of h_i, we can assume with no loss of generality that the gradients $\nabla h_i(x^*)$ are linearly independent; otherwise we can eliminate a sufficient number of equalities, while assigning to them zero Lagrange multipliers, to obtain a set of linearly independent equality constraints. Using the convexity of g_j,

$$0 > g_j(\bar{x}) \geq g_j(x^*) + \nabla g_j(x^*)'(\bar{x} - x^*) = \nabla g_j(x^*)'(\bar{x} - x^*), \qquad \forall\, j \in A(x^*),$$

and by the feasibility of $\bar{x}$ and the linearity of h_i,

$$0 = \nabla h_i(x^*)'(\bar{x} - x^*), \qquad i = 1, \ldots, m.$$

Now let $d = \bar{x} - x^*$ and apply the result of Prop. 3.3.8. **Q.E.D.**

Minimization problems where the cost function is the maximum of several functions can be easily transformed to inequality constrained optimization problems for which the Lagrange multiplier theory developed so far applies. In this way we obtain the following proposition.

Proposition 3.3.10: (Minimax Problems) Let x^* be a local minimum of the problem

$$\begin{aligned} &\text{minimize} \quad \max\{f_1(x), \ldots, f_p(x)\} \\ &\text{subject to} \quad h_1(x) = 0, \ldots, h_m(x) = 0, \end{aligned}$$

where f_j and h_i are continuously differentiable functions from $\Re^n$ to $\Re$. Assume that the gradients $\nabla h_i(x^*)$, $i = 1, \ldots, m$, are linearly independent. Then there exists a vector $\lambda^* = (\lambda_1^*, \ldots, \lambda_m^*)$ and a vector $\mu^* = (\mu_1^*, \ldots, \mu_p^*)$ such that

(i)

$$\sum_{j=1}^{p} \mu_j^* \nabla f_j(x^*) + \sum_{i=1}^{m} \lambda_i^* \nabla h_i(x^*) = 0.$$

(ii)

$$\mu^* \geq 0, \qquad \sum_{j=1}^{p} \mu_j^* = 1.$$

(iii) For all $j = 1, \ldots, p$, if $\mu_j^* > 0$, then

$$f_j(x^*) = \max\{f_1(x^*), \ldots, f_p(x^*)\}.$$

(iv) In every neighborhood N of x^*, there is an $x \in N$ such that $\lambda_i^* h_i(x) > 0$ for all i with $\lambda_i^* \neq 0$.

Proof: We introduce an additional scalar variable z, and the problem

$$\begin{aligned} &\text{minimize } z \\ &\text{subject to } h_1(x) = 0, \ldots, h_m(x) = 0, \\ &\qquad f_1(x) \leq z, \ldots, f_p(x) \leq z. \end{aligned} \tag{3.66}$$

Let

$$z^* = \max\{f_1(x^*), \ldots, f_p(x^*)\}.$$

It can be seen that (x^*, z^*) is a local minimum of the above problem, and satisfies the Mangasarian-Fromovitz constraint qualification of Prop. 3.3.8 with $d = (0, 1)$. The conclusion of that proposition, as applied to this problem, implies the desired assertions. **Q.E.D.**

The preceding proposition can be easily extended to the case where there are additional inequality constraints $g_j(x) \leq 0$, satisfying $\nabla g_j(x^*)'d < 0$ for some d (cf. Prop. 3.3.8). By adding these constraints to the ones of problem (3.66), and by applying Prop. 3.3.8, we obtain a corresponding Lagrange multiplier theorem.

Additional Convex Set Constraints

Let us now consider a more general version of the equality/inequality constrained problem, whereby there is the additional constraint $x \in X$, where

X is a closed convex set. Then, the Fritz John conditions can be appropriately generalized. We have the following extension of Prop. 3.3.5.

Proposition 3.3.11: (Fritz John Conditions for an Additional Convex Set Constraint) Let x^* be a local minimum of the problem

$$\begin{aligned} \text{minimize} \quad & f(x) \\ \text{subject to} \quad & h_1(x) = 0, \ldots, h_m(x) = 0, \\ & g_1(x) \leq 0, \ldots, g_r(x) \leq 0, \quad x \in X, \end{aligned}$$

where f, h_i, g_j are continuously differentiable functions from $\Re^n$ to $\Re$, and X is a closed convex set. Then there exists a scalar μ_0^* and Lagrange multipliers $\lambda_1^*, \ldots, \lambda_m^*$ and $\mu_1^*, \ldots, \mu_r^*$, satisfying the following conditions:

(i) For all $x \in X$, we have

$$\left(\mu_0^* \nabla f(x^*) + \sum_{i=1}^{m} \lambda_i^* \nabla h_i(x^*) + \sum_{j=1}^{r} \mu_j^* \nabla g_j(x^*) \right)' (x - x^*) \geq 0.$$

(ii)

$$\mu_j^* \geq 0, \qquad j = 0, 1, \ldots, r.$$

(iii) $\mu_0^*, \lambda_1^*, \ldots, \lambda_m^*, \mu_1^*, \ldots, \mu_r^*$ are not all equal to 0.

(iv) In every neighborhood N of x^* there is an $x \in N \cap X$ such that $\lambda_i^* h_i(x) > 0$ for all i with $\lambda_i^* \neq 0$ and $\mu_j^* g_j(x) > 0$ for all j with $\mu_j^* \neq 0$.

Proof: The proof is very similar to the one of Prop. 3.3.5. We follow the same penalty approach. The penalized cost $F^k(x)$ is minimized over $x \in X \cap S$, where $S = \{x \mid ||x - x^*|| \leq \epsilon\}$, and $\epsilon > 0$ is such that $f(x^*) \leq f(x)$ for all feasible x with $x \in S$, yielding a vector x^k. Similar to Prop. 3.3.5, we have that $x^k \to x^*$. For sufficiently large k, x^k is an interior point of S, so that $\nabla F^k(x^k)'(x - x^k) \geq 0$ for all $x \in X$, or

$$\left(\nabla f(x^k) + \sum_{i=1}^{m} \xi_i^k \nabla h_i(x^k) + \sum_{j=1}^{r} \zeta_j^k \nabla g_j(x^k) + (x^k - x^*) \right)' (x - x^k) \geq 0,$$

for all $x \in X$, where

$$\xi_i^k = k h_i(x^k), \qquad \zeta_j^k = k g_j^+(x^k).$$

The proof from this point is essentially identical to the one of Prop. 3.3.5. **Q.E.D.**

The following proposition gives an extension of the Mangasarian-Fromovitz constraint qualification, which guarantees that $\mu_0^* > 0$ in the above Fritz John conditions.

Proposition 3.3.12: (Mangasarian-Fromovitz Constraint Qualification for an Additional Convex Set Constraint) Let x^* be a local minimum of the problem of Prop. 3.3.11, and assume that the following two conditions hold:

(1) There does not exist a nonzero vector $\lambda = (\lambda_1, \ldots, \lambda_m)$ such that

$$\left(\sum_{i=1}^{m} \lambda_i \nabla h_i(x^*)\right)'(x - x^*) \geq 0, \qquad \forall\, x \in X.$$

(2) There exists a feasible direction d of X at x^* such that

$$\nabla h_i(x^*)'d = 0, \quad i = 1, \ldots, m, \qquad \nabla g_j(x^*)'d < 0, \quad \forall\, j \in A(x^*).$$

Then the Fritz John conditions of Prop. 3.3.11 hold with $\mu_0^* = 1$.

Proof: We consider a set of multipliers λ_i^*, $i = 1, \ldots, m$, μ_j^*, $j = 0, 1, \ldots, r$, satisfying the Fritz John conditions of Prop. 3.3.11, and we assume that $\mu_0^* = 0$, so that

$$\left(\sum_{i=1}^{m} \lambda_i^* \nabla h_i(x^*) + \sum_{j=1}^{r} \mu_j^* \nabla g_j(x^*)\right)'(x - x^*) \geq 0, \qquad \forall\, x \in X. \tag{3.67}$$

Since not all the λ_i^* and μ_j^* can be zero, we conclude that $\mu_j^* > 0$ for at least one $j \in A(x^*)$; otherwise assumption (1) would be violated. Since $\mu_j^* \geq 0$ for all j, with $\mu_j^* = 0$ for $j \notin A(x^*)$ and $\mu_j^* > 0$ for at least one j, we obtain

$$\sum_{i=1}^{m} \lambda_i^* \nabla h_i(x^*)'d + \sum_{j=1}^{r} \mu_j^* \nabla g_j(x^*)'d < 0,$$

where d is the vector of the hypothesis. This contradicts Eq. (3.67). Therefore, we must have $\mu_0^* > 0$, so that we can take $\mu_0^* = 1$. **Q.E.D.**

In the case where the functions h_i are linear and the functions g_j are convex, it is possible to show that the Fritz John conditions of Prop.

3.3.11 are satisfied with $\mu_0^* = 1$ under assumptions that resemble the Slater constraint qualification of Prop. 3.3.9. Basically, condition (2) of Prop. 3.3.12 is replaced by the condition that there exists a feasible vector $\bar{x}$ satisfying $g_j(\bar{x}) < 0$ for all $j \in A(x^*)$.

Example 3.3.4

Consider the two-dimensional problem

$$\text{minimize} \quad f(x) = x_1$$
$$\text{subject to} \quad h(x) = x_2 = 0, \quad x \in X = \{(x_1, x_2) \mid x_1^2 \le x_2\};$$

[see Fig. 3.3.3(a)]. The only feasible solution is $x^* = (0, 0)$, but it can be graphically verified that there is no Lagrange multiplier λ^* such that

$$\big(\nabla f(x^*) + \lambda^* \nabla h(x^*)\big)'(x - x^*) = x_1 + \lambda^* x_2 \ge 0, \qquad \forall\, x \in X.$$

Here, both conditions (1) and (2) of Prop. 3.3.12 are violated. By contrast, any scalars μ_0^*, λ^* such that $\mu_0^* = 0$ and $\lambda^* > 0$ satisfy

$$\lambda^* \nabla h(x^*)'(x - x^*) = \lambda^* x_2 \ge 0, \qquad \forall\, x \in X,$$

which is consistent with Prop. 3.3.11.

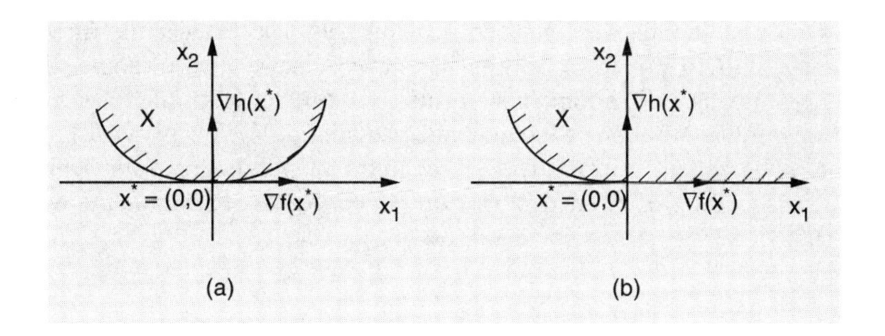

Figure 3.3.3. Constraint set X for the problems of Example 3.3.4.

In a variation of this example, suppose instead that the set X is given by

$$X = \{(x_1, x_2) \mid x_1^2 \le x_2,\, x_1 \le 0\} \cup \{(x_1, x_2) \mid x_1 \ge 0,\, x_2 \ge 0\};$$

[see Fig. 3.3.3(b)]. The optimal solution is $x^* = (0, 0)$, and again it can be seen that there is no Lagrange multiplier λ^* such that

$$\big(\nabla f(x^*) + \lambda^* \nabla h(x^*)\big)'(x - x^*) = x_1 + \lambda^* x_2 \ge 0, \qquad \forall\, x \in X.$$

Here, condition (2) of Prop. 3.3.12 is satisfied but condition (1) is violated.

3.3.6 Refinements*

In our derivation of optimality conditions for inequality-constrained optimization problems, we have used a penalty function approach: we have disregarded the constraints, while adding to the cost a penalty for their violation, and we have worked with the "penalized" unconstrained problem. By passing to the limit as the penalty increases to ∞, we have obtained all the necessary conditions thus far in this section. This approach has led to powerful results and simple proofs.

In this section, we consider an alternative line of analysis, which is based on a descent/feasible direction approach. The starting point here is the fact that at a local minimum there can be no cost improvement when traveling a small distance along a direction that leads to feasible points. This approach is straightforward when the constraint set is convex, as we have seen in Chapter 2 in the context of the feasible direction methodology. However, when nonlinear equality and/or nonconvex inequality constraints are involved, the notion of a feasible direction must be substantially modified before it can be used for analysis. With a fair amount of mathematical complication, this can be done and can be used to obtain optimality conditions as well as useful insight in many types of constrained problems, including some where the penalty approach does not work well.

Our aim in this section is to develop the descent/feasible direction approach, and to use it to obtain alternative derivations of some of our earlier results, as well as some new results, relating for example to problems with an infinite number of inequality constraints. We begin with Farkas' lemma, a basic fact relating to the geometry of polyhedral sets, which is useful in several constrained optimization contexts. Farkas' lemma says that the cone C generated by r vectors $a_1, \ldots, a_r$,

$$C = \left\{ c \;\middle|\; c = \sum_{j=1}^{r} \mu_j a_j,\ \mu_j \geq 0,\ j = 1, \ldots, r \right\},$$

consists of all vectors c satisfying

$$c'y \leq 0, \qquad \text{for all } y \text{ such that } a_j'y \leq 0,\ \forall\, j = 1, \ldots, r,$$

(see Fig. 3.3.4).

Farkas' lemma can be proved in several different ways. We give here a simple proof based on the Fritz John optimality conditions and the attendant Lagrange multiplier theorem for linear constraints (Prop. 3.3.7); an alternative proof, based on polyhedral convexity arguments, is given in Prop. B.16(d) in Appendix B.

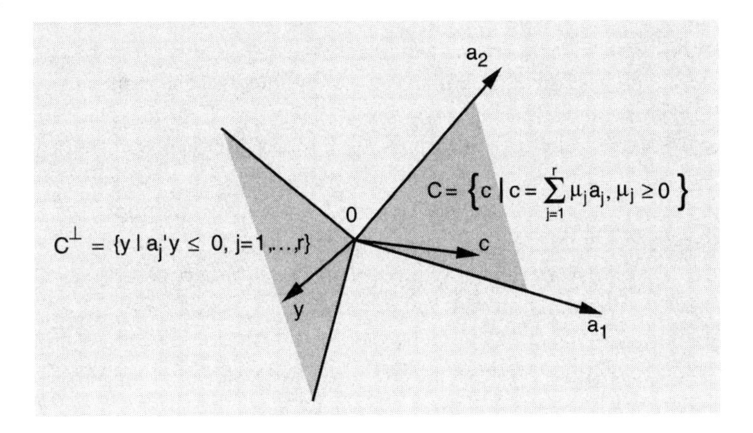

Figure 3.3.4. Geometric interpretation of Farkas' lemma. Consider the cone C generated by r vectors $a_1, \ldots, a_r$,

$$C = \left\{ c \;\middle|\; c = \sum_{j=1}^{r} \mu_j a_j,\ \mu_j \geq 0,\ j = 1, \ldots, r \right\},$$

and the cone

$$C^\perp = \{y \mid a_j' y \leq 0,\ j = 1, \ldots, r\}.$$

Farkas' lemma says that if c is such that

$$c'y \leq 0, \qquad \forall\ y \in C^\perp,$$

then c must belong to C (the reverse is also clearly true).

Proposition 3.3.13: (Farkas' Lemma) Consider r vectors $a_1, \ldots, a_r$ in $\Re^n$. Then a vector c satisfies

$$c'y \leq 0, \qquad \text{for all } y \text{ such that } a_j' y \leq 0,\ \forall\ j = 1, \ldots, r,$$

if and only if there exist nonnegative scalars $\mu_1, \ldots, \mu_r$ such that

$$c = \mu_1 a_1 + \cdots + \mu_r a_r.$$

Proof: If $c = \sum_{j=1}^{r} \mu_j a_j$ for some scalars $\mu_j \geq 0$, then for every y satisfying $a_j' y \leq 0$ for all j, we have $c'y = \sum_{j=1}^{r} \mu_j a_j' y \leq 0$. Reversely, if c satisfies $c'y \leq 0$ for all y such that $a_j' y \leq 0$ for all j, then $y^* = 0$ minimizes $-c'y$ subject to $a_j' y \leq 0$, $j = 1, \ldots, r$. The existence of nonnegative scalars $\mu_1, \ldots, \mu_r$ such that $c = \mu_1 a_1 + \cdots + \mu_r a_r$ follows from Prop. 3.3.7. **Q.E.D.**

Farkas' lemma, viewed as a mathematical result, is in fact equivalent to the existence of Lagrange multipliers in problems with differentiable cost and linear constraints. Indeed, the proof of Farkas' lemma given above was based on the existence of Lagrange multipliers for such problems, and it turns out that a reverse argument is possible, i.e., starting from Farkas' lemma, we can assert the existence of Lagrange multipliers for linearly constrained problems. This is illustrated in Fig. 3.3.5 for the case where there are inequality constraints only (linear equality constraints can be handled by conversion into two linear inequality constraints). In fact, in several textbooks, the construction given in this figure is the basis for proofs of Lagrange multiplier theorems, starting from Farkas' lemma.

Farkas' lemma can be used to interpret the existence of Lagrange multipliers for the problem

$$\begin{aligned} &\text{minimize} && f(x) \\ &\text{subject to} && h_1(x) = 0, \ldots, h_m(x) = 0, \\ &&& g_1(x) \leq 0, \ldots, g_r(x) \leq 0, \end{aligned} \tag{3.68}$$

where f, h_i, and g_j are continuously differentiable functions from $\Re^n$ to $\Re$. Let us consider the following two cones:

(a) The *cone of first order feasible variations* at x

$$V(x) = \big\{y \mid \nabla h_i(x)'y = 0,\ i = 1, \ldots, m,\ \nabla g_j(x)'y \leq 0,\ j \in A(x)\big\}.$$

(b) The *cone of descent directions* of f at x

$$D(x) = \{y \mid \nabla f(x)'y < 0\}.$$

We have the following:

Proposition 3.3.14: Let x^* be a local minimum of the equality and inequality constrained problem (3.68). Then there exist Lagrange multipliers $\lambda_1^*, \ldots, \lambda_m^*$ and $\mu_1^*, \ldots, \mu_r^*$, satisfying

$$\nabla f(x^*) + \sum_{i=1}^{m} \lambda_i^* \nabla h_i(x^*) + \sum_{j=1}^{r} \mu_j^* \nabla g_j(x^*) = 0,$$

$$\mu_j^* \geq 0, \quad j = 0, 1, \ldots, r, \qquad \mu_j^* = 0, \quad j \notin A(x^*),$$

if and only if there is no descent direction within the cone of first order feasible variations $V(x^*)$:

$$D(x^*) \cap V(x^*) = \emptyset.$$

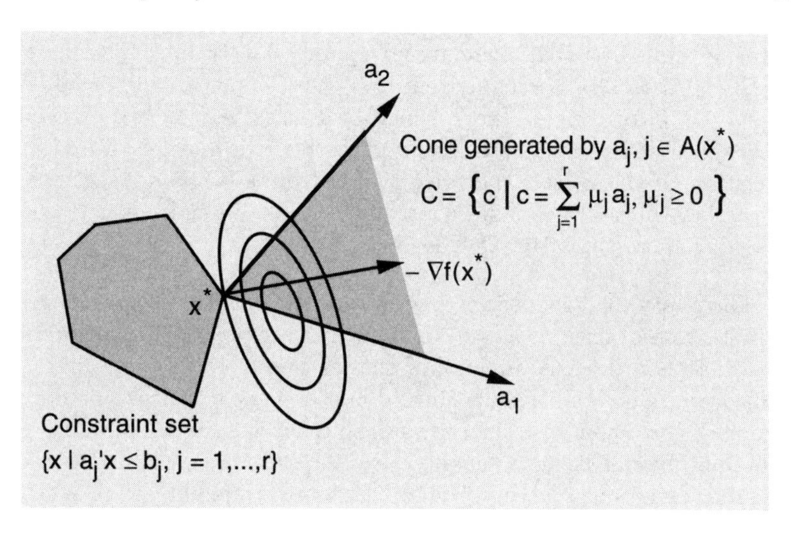

Figure 3.3.5. Proof of existence of Lagrange multipliers for linearly constrained problems, starting from Farkas' lemma. A local minimum x^* of f subject to $a_j'x \le b_j$, $j = 1, \ldots, r$, is also a local minimum for the problem

$$\begin{aligned} &\text{minimize} \quad f(x) \\ &\text{subject to} \quad a_j'x \le b_j, \qquad j \in A(x^*), \end{aligned}$$

where $A(x^*)$ is the set of active constraints at x^*. By Prop. 2.1.2 in Section 2.1, we have that

$$\nabla f(x^*)'(x - x^*) \ge 0, \qquad \forall\ x \text{ such that } a_j'x \le b_j,\ j \in A(x^*).$$

Since a constraint $a_j'x \le b_j$, $j \in A(x^*)$ can also be expressed as $a_j'(x - x^*) \le 0$, the preceding condition is equivalent to

$$\nabla f(x^*)'y \ge 0, \qquad \forall\ y \text{ such that } a_j'y \le 0,\ j \in A(x^*).$$

From Farkas' lemma, it follows that there exist nonnegative scalars μ_j^*, $j \in A(x^*)$, such that

$$\nabla f(x^*) + \sum_{j \in A(x^*)} \mu_j^* a_j = 0.$$

By letting $\mu_j^* = 0$ for all $j \notin A(x^*)$, the desired Lagrange multiplier conditions

$$\nabla f(x^*) + \sum_{j=1}^{r} \mu_j^* a_j = 0,$$

$$\mu_j^* = 0, \qquad \forall\ j \notin A(x^*),$$

follow.

Proof: Assume first that there are no equality constraints. The condition that $D(x^*) \cap V(x^*) = \emptyset$ is equivalent to having $\nabla f(x^*)'y \geq 0$ for all y with $\nabla g_j(x^*)'y \leq 0$ for all $j \in A(x^*)$, which by Farkas' lemma is equivalent to the existence of Lagrange multipliers μ_j^* with the properties stated in the proposition. If there are some equality constraints $h_i(x) = 0$, they can be converted to the two inequality constraints $h_i(x) \leq 0$ and $-h_i(x) \leq 0$, and the result follows similarly. **Q.E.D.**

The preceding proposition asserts that the problems where Lagrange multipliers do not exist at a local minimum x^* are precisely those problems where there is a descent direction $\bar{d}$ that is also a direction of first order feasible variation. As a result, since $\bar{d}$ is a descent direction, points near x^* along $\bar{d}$ have lower cost than x^*, and since x^* is a local minimum, these points must be infeasible. The only way this can happen is if the true set of feasible variations starting from x^* is "much different" than what the cone of first order feasible variations $V(x^*)$ suggests (see Example 3.1.1 and Fig. 3.1.2 for one such situation). In order to fully understand this phenomenon, we need a concept that allows us to describe more precisely the local character of the feasible set around x^*.

The Tangent Cone

Let us introduce another cone that is important in connection with optimality conditions.

Definition 3.3.1: Given a subset X of $\Re^n$ and a vector $x \in X$, a vector y is said to be a *tangent* of X at x if either $y = 0$ or there exists a sequence $\{x^k\} \subset X$ such that $x^k \neq x$ for all k and

$$x^k \to x, \qquad \frac{x^k - x}{\|x^k - x\|} \to \frac{y}{\|y\|}.$$

The set of all tangents of X at x is called the *tangent cone* of X at x, and is denoted by $T(x)$.

Thus a nonzero vector y is a tangent at x if it is possible to approach x with a feasible sequence $\{x^k\}$ such that the normalized direction sequence $(x^k - x)/\|x^k - x\|$ converges to $y/\|y\|$, the normalized direction of y. Figure 3.3.6 illustrates the tangent cone with some examples. It is easily verified that $T(x)$ is a cone. Furthermore, it can be shown that $T(x)$ is closed. In addition, if X is convex and does not consist of just the vector x, the cone $T(x)$ is equal to the closure of the set of feasible directions at x. We leave the verification of these facts for the reader (see Exercise 3.3.15).

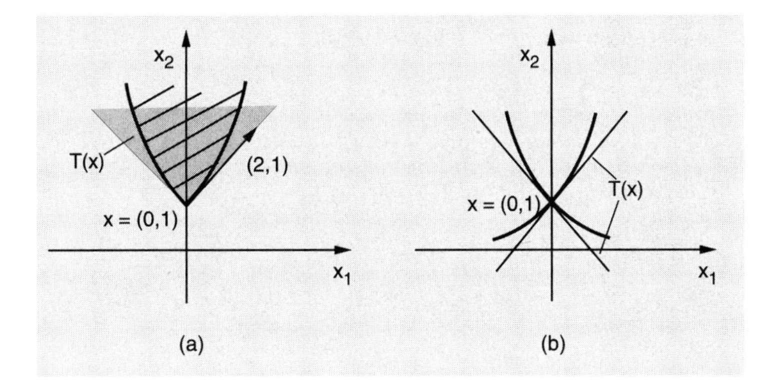

Figure 3.3.6. Examples of the tangent cone $T(x)$ of a set X at the vector $x = (0,1)$. In (a), we have

$$X = \big\{(x_1, x_2) \mid (x_1+1)^2 - x_2 \le 0,\ (x_1-1)^2 - x_2 \le 0\big\}.$$

Here X is convex and $T(x)$ is equal to the closure of the set of feasible directions at X. Note, however, that the vector $(2,1)$ belongs to $T(x)$ and also to the closure of the set of feasible directions, but is not a feasible direction. In (b), we have

$$X = \Big\{(x_1, x_2) \mid \big((x_1+1)^2 - x_2\big)\big((x_1-1)^2 - x_2\big) = 0\Big\}.$$

Here the set X is nonconvex, and $T(x)$ is closed but not convex.

The importance of the tangent cone is due to the following proposition:

Proposition 3.3.15: Let x^* be a local minimum of a continuously differentiable function f over a subset X of $\Re^n$. Then there is no descent direction within the tangent cone of X at x^*:

$$D(x^*) \cap T(x^*) = \emptyset.$$

Proof: Let y be a nonzero tangent of X at x^*. Then there exists a sequence $\{\xi^k\}$ and a sequence $\{x^k\} \subset X$ such that $x^k \neq x^*$ for all k,

$$\xi^k \to 0, \qquad x^k \to x^*,$$

and

$$\frac{x^k - x^*}{\|x^k - x^*\|} = \frac{y}{\|y\|} + \xi^k. \tag{3.69}$$

By the mean value theorem, we have for all k

$$f(x^k) = f(x^*) + \nabla f(\tilde{x}^k)'(x^k - x^*),$$

where $\tilde{x}^k$ is a vector that lies on the line segment joining x^k and x^*. Using Eq. (3.69), the last relation can be written as

$$f(x^k) = f(x^*) + \frac{\|x^k - x^*\|}{\|y\|} \nabla f(\tilde{x}^k)' y^k, \tag{3.70}$$

where

$$y^k = y + \|y\| \xi^k.$$

If the tangent y is a descent direction of f at x^*, then $\nabla f(x^*)'y < 0$, and since $\tilde{x}^k \to x^*$ and $y^k \to y$, we obtain for all sufficiently large k, $\nabla f(\tilde{x}^k)'y^k < 0$ and [from Eq. (3.70)] $f(x^k) < f(x^*)$. This contradicts the local optimality of x^*. **Q.E.D.**

In many problems the condition $D(x^*) \cap T(x^*) = \emptyset$ can be translated into more convenient necessary conditions. For example if X is a convex set, the feasible directions at x^* are all the vectors of the form $x - x^*$, where $x \in X$, and these vectors belong to $T(x^*)$. We thus obtain the familiar necessary condition

$$\nabla f(x^*)'(x - x^*) \geq 0, \qquad \forall\, x \in X,$$

which formed the basis for much of the analysis of Chapter 2.

We next explore the use of the tangent cone in the case where the constraint set X is specified by equality and inequality constraints.

Quasiregularity, Quasinormality, and Lagrange Multipliers

Consider the special case of the constraint set

$$X = \big\{x \mid h_1(x) = 0, \ldots, h_m(x) = 0, g_1(x) \leq 0, \ldots, g_r(x) \leq 0\big\},$$

where h_i, and g_j are continuously differentiable functions. A vector $x \in X$ is said to be a *quasiregular* point if the tangent cone at x is equal to the cone of first order feasible variations at x:

$$T(x) = V(x).$$

Figure 3.3.6(b) provides an example where there is a point that is not quasiregular [at the point $x = (0, 1)$, we have $\nabla h(x) = 0$, so $V(x)$ is equal to $\Re^2$, which strictly contains $T(x)$; every other vector in X is regular]. See also Fig. 3.1.2 for another similar example.

By combining Props. 3.3.14 and 3.3.15, we obtain the following proposition.

Proposition 3.3.16: Let x^* be a local minimum of the problem

$$\begin{aligned} &\text{minimize} \quad f(x) \\ &\text{subject to} \quad h_1(x) = 0, \ldots, h_m(x) = 0, \\ &\qquad\qquad\quad g_1(x) \le 0, \ldots, g_r(x) \le 0, \end{aligned}$$

where f, h_i, and g_j are continuously differentiable functions from $\Re^n$ to $\Re$. If x^* is a quasiregular point, there exist Lagrange multipliers $\lambda_1^*, \ldots, \lambda_m^*$ and $\mu_1^*, \ldots, \mu_r^*$, satisfying

$$\nabla f(x^*) + \sum_{i=1}^m \lambda_i^* \nabla h_i(x^*) + \sum_{j=1}^r \mu_j^* \nabla g_j(x^*) = 0,$$

$$\mu_j^* \ge 0, \quad j = 1, \ldots, r, \qquad \mu_j^* = 0, \quad j \notin A(x^*).$$

We will now develop a connection between the notion of quasiregularity and the Fritz John necessary conditions of Prop. 3.3.5. In particular, let us say that a feasible vector x^* is *quasinormal* if there is no set of scalars $\lambda_1, \ldots, \lambda_m, \mu_1, \ldots, \mu_r$, such that:

(i) $\sum_{i=1}^m \lambda_i \nabla h_i(x^*) + \sum_{j=1}^r \mu_j \nabla g_j(x^*) = 0$.

(ii) $\mu_j \ge 0, \qquad j = 1, \ldots, r$.

(iii) $\lambda_1, \ldots, \lambda_m, \mu_1, \ldots, \mu_r$ are not all equal to 0.

(vi) In every neighborhood N of x^* there is an $x \in N$ such that $\lambda_i h_i(x) > 0$ for all i with $\lambda_i \ne 0$ and $\mu_j g_j(x) > 0$ for all j with $\mu_j \ne 0$.

Thus if x^* is a quasinormal local minimum, the Fritz John conditions of Prop. 3.3.5 cannot be satisfied with $\mu_0^* = 0$, so that there exist corresponding Lagrange multipliers satisfying the standard necessary conditions implied by quasiregularity (cf. Prop. 3.3.16).

By reviewing the proofs of Props. 3.3.6-3.3.9, it can be seen that x^* is quasinormal under any one of the following conditions.

(a) x^* is a regular point.

(b) The constraint functions h_i and g_j are linear (or, more generally, the h_i are linear and the g_j are concave).

(c) x^* satisfies the Mangasarian-Fromovitz constraint qualification, as given in Prop. 3.3.8.

(d) x^* satisfies the Slater constraint qualification, as given in Prop. 3.3.9.

Traditional treatments of Lagrange multiplier theory revolve around establishing conditions that guarantee quasiregularity of the given local minimum x^* (see e.g., Luenberger [Lue84], p. 298, Hestenes [Hes75], p. 221, and Bazaraa, Sherali, and Shetty [BSS93], p. 191). In particular, it can be shown among others, that any one of the preceding conditions (a)-(d) guarantees quasiregularity of x^*. We will show next that quasinormality implies quasiregularity, thus providing an alternative proof that any one of the preceding conditions (a)-(d) implies quasiregularity of x^*.

Proposition 3.3.17: Assume that x^* satisfies the constraints

$$h_1(x) = 0, \ldots, h_m(x) = 0, \qquad g_1(x) \leq 0, \ldots, g_r(x) \leq 0,$$

where h_i and g_j are continuously differentiable functions from $\Re^n$ to $\Re$. If x^* is quasinormal, it is also quasiregular.

Proof: The proof is fairly intricate and relies on a simplification, developed in Exercise 3.3.21, which allows us to essentially treat inequality constraints as equalities (we work with the set $\overline{X}$ of that exercise). In particular, based on the result of Exercise 3.3.21(c), it is sufficient to consider the case where all the constraints are inequalities, which are active at x^* and are such that for each j, there is no vector $y \in V(x^*)$ with $\nabla g_j(x^*)'y < 0$.

We will first show that $T(x^*) \subset V(x^*)$ and then show the reverse inclusion. Indeed, let y be a nonzero tangent of X at x^*. Then there exists a sequence $\{\xi^k\}$ and a sequence $\{x^k\} \subset X$ such that $x^k \neq x^*$ for all k,

$$\xi^k \to 0, \qquad x^k \to x^*,$$

and

$$\frac{x^k - x^*}{\|x^k - x^*\|} = \frac{y}{\|y\|} + \xi^k.$$

By the mean value theorem, we have for all j and k

$$0 \geq g_j(x^k) = g_j(x^*) + \nabla g_j(\tilde{x}^k)'(x^k - x^*) = \nabla g_j(\tilde{x}^k)'(x^k - x^*),$$

where $\tilde{x}^k$ is a vector that lies on the line segment joining x^k and x^*. This relation can be written as

$$\frac{\|x^k - x^*\|}{\|y\|} \nabla g_j(\tilde{x}^k)'y^k \leq 0,$$

where $y^k = y + \xi^k\|y\|$, or equivalently

$$\nabla g_j(\tilde{x}^k)'y^k \leq 0, \qquad y^k = y + \xi^k\|y\|.$$

Taking the limit as $k \to \infty$, we obtain $\nabla g_j(x^*)'y \leq 0$ for all j, thus proving that $y \in V(x^*)$, and $T(x^*) \subset V(x^*)$.

To show that $V(x^*) \subset T(x^*)$, let y be a nonzero vector in $V(x^*)$. Then we have, using the assumption stated at the beginning of the proof,

$$\nabla g_j(x^*)'y = 0, \qquad j = 1, \ldots, r. \tag{3.71}$$

For each $t > 0$, choose an $x(t)$ that solves the problem

$$\begin{aligned} &\text{minimize} \ \ \frac{1}{2t}\|x - x^* - ty\|^2 \\ &\text{subject to} \ \ g_j(x) \leq 0, \ \ j = 1, \ldots, r. \end{aligned} \tag{3.72}$$

[Such a choice is possible since the cost function is coercive and the constraint set is closed; cf. Prop. A.8(b).] From the Fritz John conditions of Prop. 3.3.5 applied to problem (3.72), we see that there exist multipliers $\mu_0(t), \mu_1(t), \ldots, \mu_r(t)$ such that

$$\big(\mu_0(t)\big)^2 + \sum_{j=1}^r \big(\mu_j(t)\big)^2 = 1, \qquad \mu_j(t) \geq 0, \ \ j = 0, 1, \ldots, r, \tag{3.73}$$

$$\mu_0(t)\left(\frac{x(t) - x^*}{t} - y\right) + \sum_{j=1}^r \mu_j(t)\nabla g_j\big(x(t)\big) = 0, \tag{3.74}$$

and in every neighborhood of $x(t)$ there is an x such that if $\mu_j(t) > 0$ then $\mu_j(t)g_j(x) > 0$. We will now show that for some sequence t^k converging to 0, we have $(x(t^k) - x^*)/t^k \to y$, from which it will follow that $y \in T(x^*)$.

Indeed, we have

$$\|x(t) - x^* - ty\|^2 \leq \|x^* - x^* - ty\|^2 = \|ty\|^2,$$

from which we obtain $\|x(t) - x^*\| \leq 2\|ty\|$, and hence

$$\frac{\|x(t) - x^*\|}{t} \leq 2\|y\|, \qquad \forall\, t > 0.$$

Thus, we can select a sequence $\{t^k\}$ with $t^k \to 0$ such that the limit

$$\bar{y} = \lim_{k \to \infty} \frac{x(t^k) - x^*}{t^k}, \tag{3.75}$$

exists. Since $g_j\big(x(t^k)\big) \leq 0$, by using the mean value theorem, we have for all k, $\nabla g_j(\tilde{x}^k)'\big(x(t^k) - x^*\big) \leq 0$, where $\tilde{x}^k$ is a vector in the line segment connecting $x(t^k)$ and x^*. By taking the limit as $k \to \infty$, and using Eq. (3.75) and the assumption stated at the beginning of the proof, we obtain

$$\nabla g_j(x^*)'\bar{y} = 0, \qquad j = 1, \ldots, r. \tag{3.76}$$

In view of Eq. (3.73), we can assume without loss of generality that the sequences $\{\mu_j(t^k)\}$ converge to some μ_j, and we have

$$(\mu_0)^2 + \sum_{j=1}^{r}(\mu_j)^2 = 1, \qquad \mu_j \geq 0, \quad j = 0, 1, \ldots, r,$$

$$\mu_0(\bar{y} - y) + \sum_{j=1}^{r} \mu_j \nabla g_j(x^*) = 0, \tag{3.77}$$

and in every neighborhood of x^* there is an x such that if $\mu_j > 0$ then $\mu_j g_j(x) > 0$. By the quasinormality of x^*, we have $\mu_0 > 0$. Hence by multiplying Eq. (3.77) with y' and by using Eq. (3.71), we see that $y'(\bar{y} - y) = 0$. Similarly, by multiplying Eq. (3.77) with $\bar{y}'$ and by using Eq. (3.76), we see that $\bar{y}'(\bar{y} - y) = 0$. Thus,

$$y'(\bar{y} - y) = \bar{y}'(\bar{y} - y),$$

which implies that $\|\bar{y} - y\|^2 = 0$ or $\bar{y} = y$. From Eq. (3.75), we have $x(t^k) - x^* = t^k(y + \xi^k)$ for some sequence ξ^k with $\xi^k \to 0$. This implies that $\lim_{k\to\infty}\big(x(t^k) - x^*\big)/\|x(t^k) - x^*\| = \lim_{k\to\infty}(y + \xi^k)/\|y + \xi^k\| = y/\|y\|$. Hence $y \in T(x^*)$, and since y is an arbitrary nonzero vector in $V(x^*)$, we conclude that $V(x^*) \subset T(x^*)$. **Q.E.D.**

Figure 3.3.7 summarizes the relations of various conditions that guarantee existence of Lagrange multipliers corresponding to a local minimum of f subject to a finite number of equality and inequality constraints. Each of the implications in this figure is one-directional. In particular, while quasinormality implies quasiregularity, the reverse is not true (see Exercise 3.3.20 for a counterexample).

We will now extend the definition of quasiregularity and use the tangent cone to obtain a Lagrange multiplier theorem for problems involving an infinite number of constraints.

Semi-Infinite Programming

Consider the case of a problem with an infinite number of inequality constraints:

$$\begin{aligned} &\text{minimize} \;\; f(x) \\ &\text{subject to} \;\; x \in X, \end{aligned} \tag{3.78}$$

where

$$X = \{x \mid g_j(x) \leq 0,\ j \in J\},$$

and J is an index set that is a compact subset of some Euclidean space. Such problems are called *semi-infinite programming* problems (a finite number of variables but an infinite number of constraints). We assume that

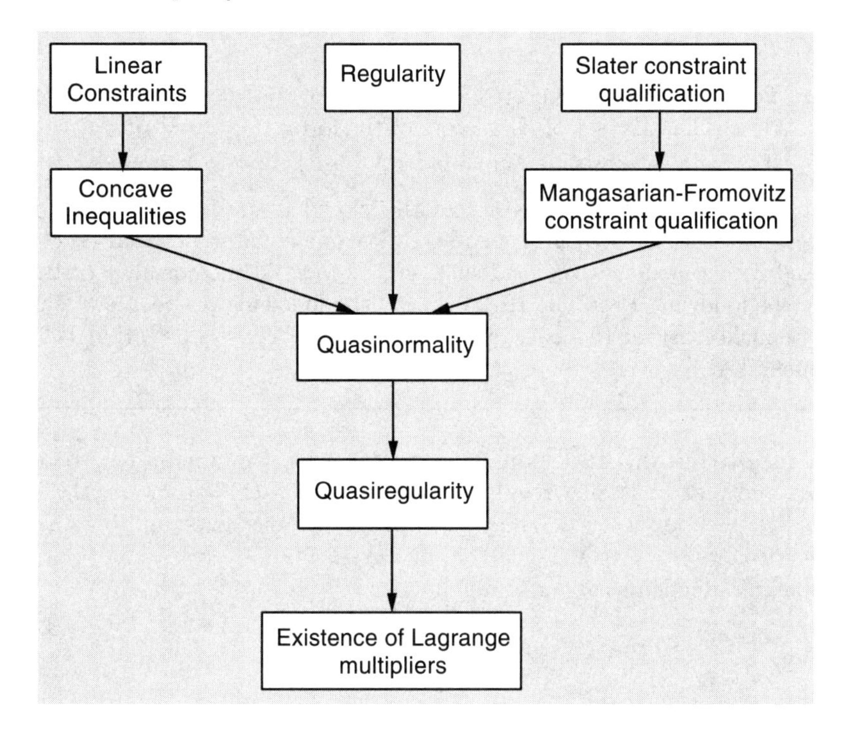

Figure 3.3.7. Relations between various conditions, which when satisfied at a local minimum x^*, guarantee existence of corresponding Lagrange multipliers.

f and g_j are continuously differentiable functions of x, and furthermore, $\nabla g_j(x)$ is continuous as a function of j for each feasible x.

For an important example of a semi-infinite programming problem, consider the case where the constraints $g_j(x) \leq 0$ are linear, i.e.,

$$X = \{x \mid a_j' x \leq b_j, j \in J\},$$

where for each $j \in J$, a_j is a given vector and b_j is a given scalar. In this case, X is a closed convex set, since it is the intersection of a collection of closed halfspaces. In fact any closed convex set X can be described as above for a suitable collection $\{(a_j, b_j) \mid j \in J\}$ (cf. Prop. B.15).

Let x^* be a local minimum and let $A(x^*)$ denote the set of active constraint indices at x^*:

$$A(x^*) = \{j \mid g_j(x^*) = 0,\ j \in J\}.$$

The cone of first order feasible variations is

$$V(x^*) = \{y \mid \nabla g_j(x^*)' y \leq 0,\ \forall\ j \in A(x^*)\}.$$

Let $T(x^*)$ denote the tangent cone of X at x^*. We say that x^* is a *quasiregular* point if the following two conditions are satisfied:

(a) $T(x^*) = V(x^*)$.

(b) The cone $G(x^*)$ generated by the gradients $\nabla g_j(x^*)$, $j \in A(x^*)$ is closed, i.e., the set of all vectors of the form $\sum_{j \in \bar{A}} \mu_j \nabla g_j(x^*)$, where $\bar{A}$ is a finite subset of $A(x^*)$ and $\mu_j \geq 0$ for all $j \in \bar{A}$, is closed.

Note that the cone $G(x^*)$ is necessarily closed if the index set $A(x^*)$ is finite; see Prop. B.16(b) in Appendix B. It can also be shown that $G(x^*)$ is closed if there exists a vector d such that $\nabla g_j(x^*)'d < 0$ for all $j \in A(x^*)$ (see the following Prop. 3.3.19). We have the following proposition, whose proof makes use of the background material on cones in Section B.3 of Appendix B.

Proposition 3.3.18: Assume that x^* is a local minimum of f over X, and that x^* is quasiregular. Then either $\nabla f(x^*) = 0$ or else there exists an index set $J^* \subset A(x^*)$, with at most n elements, and positive multipliers μ_j^*, $j \in J^*$, such that the gradients $\nabla g_j(x^*)$, $j \in J^*$, are linearly independent, and

$$\nabla f(x^*) + \sum_{j \in J^*} \mu_j^* \nabla g_j(x^*) = 0.$$

Proof: We first assert that we have

$$V(x^*)^\perp - G(x^*), \tag{3.79}$$

where $V(x^*)^\perp$ is the polar cone of $V(x^*)$ (see Section B.3 of Appendix B for the definition and various properties of polar cones). To see this, note that we have $V(x^*) = G(x^*)^\perp$, so that

$$V(x^*)^\perp = \big(G(x^*)^\perp\big)^\perp,$$

and since $G(x^*)$ is closed by quasiregularity, we have $G(x^*) = \big(G(x^*)^\perp\big)^\perp$ [Prop. B.16(a) on Appendix B].

Now, by Prop. 3.3.15, since x^* is a local minimum, we have $\nabla f(x^*)'y \geq 0$ for all $y \in T(x^*)$. By quasiregularity, we also have $T(x^*) = V(x^*)$, so that $\nabla f(x^*)'y \geq 0$ for all $y \in V(x^*)$. Hence, $-\nabla f(x^*)$ belongs to $V(x^*)^\perp$, and by Eq. (3.79), $-\nabla f(x^*)$ belongs to the cone $G(x^*)$. Thus, either $\nabla f(x^*) = 0$ or else, by Caratheodory's theorem for cones (Exercise B.1.7 in Appendix B), $-\nabla f(x^*)$ can be expressed as a positive combination of a collection of linearly independent vectors $\nabla g_j(x^*)$, $j \in J^*$. **Q.E.D.**

The following example illustrates the preceding proposition and highlights the significance of various assumptions, such as the compactness of the index set J.

Example 3.3.5

Consider the following problem in $\Re^2$:

$$\begin{array}{ll} \text{minimize} & -(x_1 + x_2) \\ \text{subject to} & x_1 \cos\phi + x_2 \sin\phi \le 1, \qquad \phi \in \Phi, \end{array}$$

where Φ is a given compact subset of the set $[0, 2\pi]$. To visualize the constraint set, note that for each ϕ, the equation $x_1 \cos\phi + x_2 \sin\phi = 1$ specifies a line that is tangent to the unit circle, as shown in Fig. 3.3.8(a). It can be seen that since there can be at most one active constraint at any feasible vector, quasiregularity is satisfied everywhere.

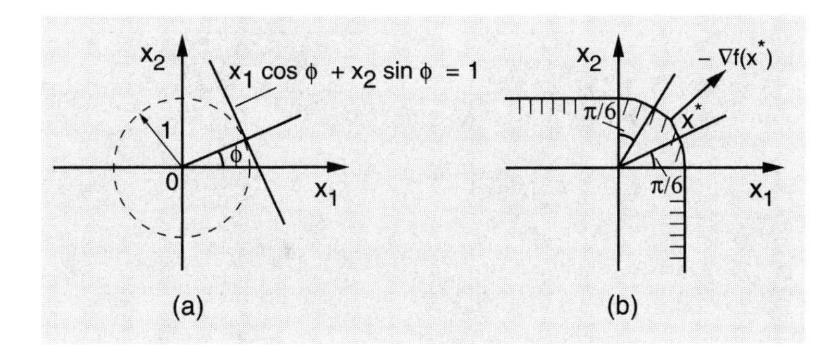

Figure 3.3.8. (a) The linear constraints in Example 3.3.5. (b) Constraint set and optimal solution for the case $\Phi = [0, \pi/6] \cup [\pi/3, \pi/2]$.

Consider first the case where $\Phi = [0, 2\pi]$. Then the constraint set of the problem is the unit circle. In this case, the optimal solution is $x^* = (\sqrt{2}/2, \sqrt{2}/2)$, and it can be seen that the only constraint that is active corresponds to $\phi = \pi/4$. From the condition $\nabla f(x^*) + \sum_{j \in J^*} \mu_j^* \nabla g_j(x^*) = 0$, we see that the corresponding Lagrange multiplier is

$$\mu_\phi^* = \begin{cases} 0 & \text{if } \phi \neq \pi/4, \\ \sqrt{2} & \text{if } \phi = \pi/4. \end{cases}$$

Consider next the case where $\Phi = [0, \pi/6] \cup [\pi/3, \pi/2]$. Then the constraint set is as shown in Fig. 3.3.8(b). It can be seen from this figure that the optimal solution is $x^* = \big(1/(1+\sqrt{3}), 1/(1+\sqrt{3})\big)$. There are two constraints that are active, corresponding to $\phi = \pi/6$ and $\phi = \pi/3$. The corresponding Lagrange multiplier is

$$\mu_\phi^* = \begin{cases} 0 & \text{if } \phi \neq \pi/6, \pi/3, \\ \dfrac{2}{1+\sqrt{3}} & \text{if } \phi = \pi/6 \text{ or } \phi = \pi/3. \end{cases}$$

It is interesting to observe that if we exclude the points $\pi/6$ and $\pi/3$ from Φ, so that Φ is the *noncompact* set $[0, \pi/6) \cup (\pi/3, \pi/2]$, the optimal solution x^*

will remain the same, but there will be no active constraints and corresponding Lagrange multipliers. This indicates the significance of the compactness assumption in the problem formulation.

Similarly, it can be seen that for any choice of the compact set Φ, there are at most two active constraints at the optimal solution, and the corresponding Lagrange multipliers are positive. We finally note that if we introduce some additional inequality constraints [such as for example the constraint $x_1 \geq x_2$ or the constraint $x_1 \leq 1/(1+\sqrt{3})$ in the case of Fig. 3.3.8(b)], it is still always possible to select a Lagrange multiplier vector, where at most two of the Lagrange multipliers are nonzero, even if the number of active constraints is larger than two. This is consistent with Prop. 3.3.18.

We now provide a condition that guarantees quasiregularity.

Proposition 3.3.19: Let x^* be a local minimum of f over X, and assume that there exists a vector d such that

$$\nabla g_j(x^*)'d < 0, \qquad \forall\, j \in A(x^*). \tag{3.80}$$

Then x^* is quasiregular and the conclusions of Prop. 3.3.18 hold.

Proof: For any vector $y \in V(x^*)$, the vector $y+\alpha(d-y)$, where $0 < \alpha \leq 1$, satisfies $\nabla g_j(x^*)'\big(y+\alpha(d-y)\big) < 0$ for all $j \in A(x^*)$, and based on this property, it can be shown that $y+\alpha(d-y)$ is a feasible direction at x^* so that it belongs to the tangent cone $T(x^*)$ (see Exercise 3.3.17). Hence $V(x^*)$ is a subset of the closure of $T(x^*)$, and since $T(x^*)$ is closed, we have $V(x^*) \subset T(x^*)$.

The reverse inclusion, $T(x^*) \subset V(x^*)$, is also true. To see this, let y be a nonzero tangent of X at x^*. Then there exists a sequence $\{\xi^k\}$ and a sequence $\{x^k\} \subset X$ such that $x^k \neq x^*$ for all k and

$$\xi^k \to 0, \qquad x^k \to x^*,$$

and

$$\frac{x^k - x^*}{\|x^k - x^*\|} = \frac{y}{\|y\|} + \xi^k.$$

By the mean value theorem, we have for all $j \in A(x^*)$ and k

$$0 \geq g_j(x^k) = g_j(x^*) + \nabla g_j(\tilde{x}^k)'(x^k - x^*) = \nabla g_j(\tilde{x}^k)'(x^k - x^*),$$

where $\tilde{x}^k$ is a vector that lies on the line segment joining x^k and x^*. This relation can be written as

$$0 \geq \frac{\|x^k - x^*\|}{\|y\|} \nabla g_j(\tilde{x}^k)'y^k,$$

or equivalently

$$\nabla g_j(\tilde{x}^k)'y^k \le 0,$$

where $y^k = y + \xi^k \|y\|$. Taking the limit as $k \to \infty$, we obtain

$$\nabla g_j(x^*)'y \le 0, \qquad \forall\ j \in A(x^*),$$

showing that $y \in V(x^*)$. Thus, we have $T(x^*) \subset V(x^*)$, completing the proof that $T(x^*) = V(x^*)$.

To show quasiregularity of x^*, there remains to show that the cone $G(x^*)$ generated by the set of gradients

$$R = \big\{\nabla g_j(x^*) \mid j \in A(x^*)\big\}$$

is closed. Let us first note that since $\nabla g_j(x^*)$ is continuous with respect to j, the set $A(x^*)$ is closed, and since J is compact, $A(x^*)$ is also compact. It follows that the set R, which is the image of the compact set $A(x^*)$ under a continuous mapping, is compact.

Consider any sequence $\{y^k\} \subset G(x^*)$ converging to some $y \neq 0$. We will show that $y \in G(x^*)$. Indeed, by Caratheodory's theorem (Prop. B.6 in Appendix B), for each k, there are indices $j_1^k, \ldots, j_{n+1}^k$ and nonnegative scalars $\beta_1^k, \ldots, \beta_{n+1}^k$ not all zero, such that

$$y^k = \beta_1^k \nabla g_{j_1^k}(x^*) + \cdots + \beta_{n+1}^k \nabla g_{j_{n+1}^k}(x^*).$$

Let

$$\beta^k = \beta_1^k + \cdots + \beta_{n+1}^k, \qquad \gamma_j^k = \frac{\beta_j^k}{\beta^k}, \quad j = j_1^k, \ldots, j_{n+1}^k.$$

Then we have

$$\frac{1}{\beta^k} y^k = \gamma_1^k \nabla g_{j_1^k}(x^*) + \cdots + \gamma_{n+1}^k \nabla g_{j_{n+1}^k}(x^*). \tag{3.81}$$

The sequences $\nabla g_{j_1^k}(x^*), \ldots, \nabla g_{j_{n+1}^k}(x^*)$ lie in the compact set R, while the sequences $\gamma_1^k, \ldots, \gamma_{n+1}^k$ are nonnegative and satisfy $\sum_{m=1}^{n+1} \gamma_m^k = 1$. Therefore, from Eq. (3.81), we see that there is a subsequence $\{y^k/\beta^k\}_{k\in\mathcal{K}}$ that converges to a vector of the form

$$g = \gamma_1 \nabla g_{j_1}(x^*) + \cdots + \gamma_{n+1} \nabla g_{j_{n+1}}(x^*),$$

where the scalars $\gamma_1, \ldots, \gamma_{n+1}$ are nonnegative with $\sum_{m=1}^{n+1} \gamma_m = 1$, and the indices $j_1, \ldots, j_{n+1}$ belong to $A(x^*)$. In view of the assumption (3.80), we must have $g \neq 0$. This together with the fact $y^k \to y$ imply that the subsequence $\{\beta^k\}_{k\subset\mathcal{K}}$ must converge to some $\beta > 0$, and that we must have $y = \beta g$. Since $g \in G(x^*)$ it follows that $y \in G(x^*)$. Therefore, $G(x^*)$ is closed. **Q.E.D.**

The assumption of the above proposition resembles the Mangasarian-Fromovitz constraint qualification of Prop. 3.3.8, and in fact reduces to it for the case where there is only a finite number of inequality constraints and no equality constraints. A related assumption that resembles the Slater constraint qualification of Prop. 3.3.9 is that the functions g_j are convex and there exists a feasible vector $\bar{x}$ such that

$$g_j(\bar{x}) < 0, \qquad \forall\, j \in A(x^*).$$

It can then be seen that the vector $d = \bar{x} - x^*$ satisfies the assumptions of Prop. 3.3.18, so x^* is quasiregular under this assumption as well.

Finally, suppose that the function f and the functions g_j, $j \in J$, are all convex over $\Re^n$, and suppose that we have a feasible vector x^*, a nonempty and finite index set $J^* \subset A(x^*)$, and positive multipliers μ_j^*, $j \in J^*$, such that

$$\nabla f(x^*) + \sum_{j \in J^*} \mu_j^* \nabla g_j(x^*) = 0.$$

Then by the convexity of f and g_j, x^* minimizes $f(x) + \sum_{j \in J^*} \mu_j^* g_j(x)$ over $x \in \Re^n$, and the proof of Prop. 3.3.4 can be used to assert that x^* is a global minimum of f subject to $g_j(x) \leq 0$, $j \in J$. Thus, under convexity assumptions on f and g_j, the necessary condition of Prop. 3.3.18 is also sufficient for optimality.

Example 3.3.6

Given a compact set C in $\Re^n$, consider the problem of finding the smallest sphere that contains C. We formulate this as the minimax problem

$$\begin{aligned} &\text{minimize} \ \max_{c \in C} \|x - c\|^2 \\ &\text{subject to} \ \ x \in \Re^n, \end{aligned}$$

where x is the unknown center of the sphere. Since the cost function is strictly convex and coercive, we conclude that there exists a unique optimal solution x^*. By introducing an additional variable z, the problem is equivalently written as

$$\begin{aligned} &\text{minimize} \ \ z \\ &\text{subject to} \ \ x \in \Re^n, \qquad \|x - c\|^2 \leq z, \quad \forall\, c \in C, \end{aligned}$$

which is a semi-infinite programming problem with convex cost and constraints. It can be seen, that the assumption (3.80) of Prop. 3.3.18 is satisfied with $d = (0, 1)$, and taking also into account the preceding discussion, we conclude that x^* is the optimal solution if and only if the optimality conditions of Prop. 3.3.18 hold. After a little calculation, it then follows that there exist elements $c_1, \ldots, c_s$ of C, with $s \leq n+1$, and corresponding positive multipliers $\mu_1, \ldots, \mu_s$, such that

$$\sum_{j=1}^{s} \mu_j^* = 1, \qquad x^* = \sum_{j=1}^{s} \mu_j^* c_j, \tag{3.82}$$

and the radius R of the optimal sphere satisfies

$$R = \|x^* - c_j\|, \qquad \forall\ j = 1, \ldots, s.$$

Let us now show that the radius satisfies the inequality

$$R \leq d\sqrt{\frac{s-1}{2s}}, \tag{3.83}$$

due to John [Joh48], where the scalar d is the diameter of the set $\{c_1, \ldots, c_s\}$:

$$d = \max_{i,j=1,\ldots,s} \|c_i - c_j\|.$$

Indeed, let

$$D_{ij} = \|c_i - c_j\|^2,$$

and note that

$$D_{ij} = \|c_i - c_j\|^2 = \|(c_i - x^*) - (c_j - x^*)\|^2 = 2R^2 - 2(c_i - x^*)'(c_j - x^*).$$

Multiplying this equation with $\mu_i^* \mu_j^*$, adding over i and j, and using the conditions (3.82), we have

$$\begin{aligned}\sum_{i,j} \mu_i^* \mu_j^* D_{ij} &= 2R^2 \left(\sum_{j=1}^{s} \mu_j^*\right)^2 - \left(\sum_{i=1}^{s} \mu_i^* (c_i - x^*)\right)' \left(\sum_{j=1}^{s} \mu_j^* (c_j - x^*)\right) \\ &= 2R^2.\end{aligned}$$

Since $D_{ij} \leq d^2$ and $D_{ii} = 0$, we thus obtain

$$2R^2 \leq d^2 \sum_{i \neq j} \mu_i^* \mu_j^*.$$

Since

$$\sum_{j=1}^{s} \mu_j^* = 1,$$

it can be seen that

$$\sum_{j=1}^{s} (\mu_j^*)^2 \geq 1/s,$$

so that

$$\sum_{i \neq j} \mu_i^* \mu_j^* = 1 - \sum_{j=1}^{s} (\mu_j^*)^2 \leq 1 - \frac{1}{s} = \frac{s-1}{s}.$$

By combining the last two relations, we obtain the desired bound (3.83).

EXERCISES

3.3.1

Solve the two-dimensional problem

$$\begin{aligned} &\text{minimize} \quad (x-a)^2 + (y-b)^2 + xy \\ &\text{subject to} \quad 0 \leq x \leq 1, \quad 0 \leq y \leq 1, \end{aligned}$$

for all possible values of the scalars a and b.

3.3.2

Given a vector y, consider the problem

$$\begin{aligned} &\text{maximize} \quad y'x \\ &\text{subject to} \quad x'Qx \leq 1, \end{aligned}$$

where Q is a positive definite symmetric matrix. Show that the optimal value is $\sqrt{y'Q^{-1}y}$ and use this fact to establish the inequality

$$(x'y)^2 \leq (x'Qx)(y'Q^{-1}y).$$

3.3.3

Solve Exercise 2.1.15 of Section 2.1 using Lagrange multipliers.

3.3.4

Solve Exercise 2.1.16 of Section 2.1 using Lagrange multipliers.

3.3.5 (Constraint Qualifications for Inequality Constraints) 

The purpose of this exercise is to explore a condition that implies all of the existence results for Lagrange multipliers that we have proved in this section for the case of inequality constraints only. Consider the problem

$$\begin{aligned} &\text{minimize} \;\; f(x) \\ &\text{subject to} \;\; g_j(x) \le 0, \qquad j = 1, \ldots, r. \end{aligned}$$

For a feasible x, let $F(x)$ be the set of all feasible directions at x defined by

$$F(x) = \big\{ d \mid d \neq 0, \text{ and for some } \overline{\alpha} > 0, \, g(x + \alpha d) \le 0 \text{ for all } \alpha \in [0, \overline{\alpha}] \big\}$$

and denote by $\overline{F}(x)$ the closure of $F(x)$. Let x^* be a local minimum. Show that:

(a)

$$\nabla f(x^*)' d \ge 0, \qquad \forall \, d \in \overline{F}(x^*).$$

(b) If we have $\overline{F}(x^*) = V(x^*)$, where

$$V(x^*) = \big\{ d \mid \nabla g_j(x^*)' d \le 0, \, \forall \, j \in A(x^*) \big\},$$

then there exists a Lagrange multiplier vector $\mu^* \ge 0$ such that

$$\nabla f(x^*) + \sum_{j=1}^{r} \mu_j^* \nabla g_j(x^*) = 0, \qquad \mu_j^* = 0, \qquad \forall \, j \notin A(x^*).$$

Hint: Use part (a) and Farkas' Lemma (Prop. 3.3.13).

(c) The condition of part (b) holds if any one of the following conditions holds:

1. The functions g_j are linear.

2. (Arrow, Hurwitz, Uzawa [AHU61]) There exists a vector d such that

$$\nabla g_j(x^*)' d < 0, \qquad \forall \, j \in A(x^*).$$

Hint: Let $\bar{d}$ satisfy $\nabla g_j(x^*)' \bar{d} \le 0$ for all $j \in A(x^*)$, and let $d_\gamma = \gamma d + (1 - \gamma)\bar{d}$. From the mean value theorem, we have for some $\epsilon \in [0, 1]$

$$\begin{aligned} g_j(x^* + \alpha d_\gamma) &= g_j(x^*) + \alpha \nabla g_j(x^* + \epsilon \alpha d_\gamma)' d_\gamma \\ &\le \alpha \big(\gamma \nabla g_j(x^* + \epsilon \alpha d_\gamma)' d + (1 - \gamma) \nabla g_j(x^* + \epsilon \alpha d_\gamma)' \bar{d} \big). \end{aligned}$$

For a fixed γ, there exists an $\overline{\alpha} > 0$ such that the right-hand side above is nonpositive for $\alpha \in [0, \overline{\alpha}]$. Thus $d_\gamma \in F(x^*)$, for all γ and $\lim_{\gamma \to 0} d_\gamma = \bar{d}$. Hence $\bar{d} \in \overline{F}(x^*)$.

3. (Slater [Sla50]) The functions g_j are convex and there exists a vector $\bar{x}$ satisfying

$$g_j(\bar{x}) < 0, \qquad \forall \, j \in A(x^*).$$

Hint: Let $d = \bar{x} - x^*$ and use condition 2.

4. The gradients $\nabla g_j(x^*)$, $j \in A(x^*)$, are linearly independent. *Hint*: Let d be such that $\nabla g_j(x^*)'d = -1$ for all $j \in A(x^*)$ and use condition 2.

(d) Consider the two-dimensional problem with two inequality constraints where

$$f(x_1, x_2) = x_1 + x_2,$$

$$g_1(x_1, x_2) = \begin{cases} -x_1^2 + x_2 & \text{if } x_1 \geq 0, \\ -x_1^4 + x_2 & \text{if } x_1 \leq 0, \end{cases}$$

$$g_2(x_1, x_2) = \begin{cases} x_1^4 - x_2 & \text{if } x_1 \geq 0, \\ x_1^2 - x_2 & \text{if } x_1 \leq 0. \end{cases}$$

Show that $x^* = (0, 0)$ is a local minimum, but each of the conditions 1-4 of part (c) is violated and there exists no Lagrange multiplier vector.

(e) Consider the equality constrained problem $\min_{h(x)=0} f(x)$, and suppose that x^* is a local minimum such that $\nabla f(x^*) \neq 0$. Repeat part (d) for the case of the equivalent inequality constrained problem

$$\min_{\|h(x)\|^2 \leq 0} f(x).$$

3.3.6 (Gordan's Theorem of the Alternative) (www)

Let A be an $m \times n$ matrix. Then exactly one of the following two conditions holds:

(i) There is an $x \in \Re^n$ such that $Ax < 0$ (all components of Ax are strictly negative).

(ii) There is a $\mu \in \Re^m$ such that $A'\mu = 0$, $\mu \neq 0$, and $\mu \geq 0$.

(A nonlinear version of this theorem is given in Exercise 5.3.8.) *Hint*: Assume that there exist x and μ satisfying simultaneously (i) and (ii), and reach a contradiction. Use Prop. 3.3.10 for a suitable minimax problem to show that if (i) fails to hold, then (ii) must hold.

3.3.7 (Sufficiency Without Strict Complementarity) (www)

Show that the sufficiency result of Prop. 3.3.2 holds if the strict complementary slackness condition (3.54) is removed, but the assumption on the Hessian of the Lagrangian is strengthened to read

$$y'\nabla^2_{xx}L(x^*, \lambda^*, \mu^*)y > 0$$

for all $y \neq 0$ satisfying

$$\nabla h_i(x^*)'y = 0, \qquad i = 1, \ldots, m,$$

$$\nabla g_j(x^*)'y = 0, \qquad \forall\, j \in A(x^*) \text{ with } \mu_j^* > 0,$$

$$\nabla g_j(x^*)'y \le 0, \qquad \forall\, j \in A(x^*) \text{ with } \mu_j^* = 0.$$

Hint: Extend the proof of Prop. 3.2.1, based on the descent approach, given in Section 3.2.2. Show that the vector y in that proof satisfies the above conditions.

3.3.8

Find $x_1^*, x_2^*, \mu_0^*, \mu_1^*, \mu_2^*$ satisfying the Fritz John conditions for the following two-dimensional problems with inequality constraints:

(a) $f(x_1, x_2) = -x_1, \quad g_1(x_1, x_2) = x_2 - (1 - x_1)^3, \quad g_2(x_1, x_2) = -x_2.$

(b) $f(x_1, x_2) = x_1 + x_2, \quad g_1(x_1, x_2) = x_1^2 + x_2^2 - 2, \quad g_2(x_1, x_2) = -x_1^2 - x_2^2 + 2.$

(c) $f(x_1, x_2) = x_1 + x_2, \quad g_1(x_1, x_2) = x_1^2 + x_2^2 - 2, \quad g_2(x_1, x_2) = x_1^2 + x_2^2 - 2.$

(d)

$$f(x_1, x_2) = x_1 + x_2,$$

$$g_1(x_1, x_2) = \begin{cases} -x_1^2 + x_2 & \text{if } x_1 \ge 0, \\ -x_1^4 + x_2 & \text{if } x_1 \le 0, \end{cases}$$

$$g_2(x_1, x_2) = \begin{cases} x_1^4 - x_2 & \text{if } x_1 \ge 0, \\ x_1^2 - x_2 & \text{if } x_1 \le 0. \end{cases}$$

3.3.9

Consider the equality constrained problem $\min_{h(x)=0} f(x)$, and suppose that x^* is a local minimum such that $h(x)$ is identically zero in a neighborhood of x^*. Use the Fritz John conditions to show that $\nabla f(x^*) = 0$.

3.3.10 (Relation of Fritz John and KKT multipliers) (www)

Consider the equality/inequality constrained problem

$$\begin{aligned} &\text{minimize } f(x) \\ &\text{subject to } h_1(x) = 0, \ldots, h_m(x) = 0, \\ &\qquad\qquad\quad g_1(x) \le 0, \ldots, g_r(x) \le 0, \end{aligned}$$

let x^* be a (not necessarily regular) local minimum and assume that there exist multipliers $\lambda_1^*, \ldots, \lambda_m^*$ and $\mu_1^*, \ldots, \mu_r^*$ that satisfy the Karush-Kuhn-Tucker conditions of Prop. 3.3.1.

(a) Show by example that $\mu_0^* = 1$ and $\lambda_1^*, \ldots, \lambda_m^*$ and $\mu_1^*, \ldots, \mu_r^*$ need not satisfy the Fritz John conditions of Prop. 3.3.5 (the last condition may be violated).

(b) Show that there exist some $\lambda_1, \ldots, \lambda_m$ and $\mu_1, \ldots, \mu_r$ (possibly other than λ_i^* and μ_j^*), which together with $\mu_0 = 1$ satisfy all the Fritz John conditions of Prop. 3.3.5, and are such that the vectors $\nabla h_i(x^*)$ with $\lambda_i \neq 0$, and $\nabla g_j(x^*)$ with $\mu_j > 0$ are linearly independent. *Hint*: Express $\nabla f(x^*)$ as a linear combination of a minimal number of constraint gradients and show that the coefficients of the linear combination have the desired property.

3.3.11 (www)

Consider the problem of minimizing $f(x)$ subject to $g_j(x) \leq 0$, $j = 1, \ldots, r$, and assume that $g_1, \ldots, g_r$ are convex over $\Re^n$. Let x^* be a local minimum which together with multipliers $\mu_0^*, \mu_1^*, \ldots, \mu_r^*$ satisfies the Fritz John conditions of Prop. 3.3.5 with $\mu_0^* = 0$. Show that $g_j(x) = 0$ for all feasible vectors x and all j with $\mu_j^* > 0$. In particular, there exists an index j such that $g_j(x) = 0$ for all feasible vectors x. *Hint*: Write $\mu_j^* g_j(x) \geq \mu_j^* g_j(x^*) + \mu_j^* \nabla g_j(x^*)'(x - x^*)$ and add over j.

3.3.12 (An Alternative Fritz John Condition) (www)

Consider a version of Prop. 3.3.5 where condition (iv) is replaced by the following: In every neighborhood N of x^* there is an $x \in N$ that simultaneously violates all the constraints corresponding to nonzero Lagrange multipliers. Show that this version of Prop. 3.3.5 is equivalent to the one given in the text, in the sense that each version of the proposition follows from the other. *Hint*: The condition above is clearly implied by condition (iv) in the Prop. 3.3.5 given in the text. To show the reverse, replace each equality constraint $h_i(x) = 0$ with the two constraints $h_i(x) \leq 0$ and $-h_i(x) \leq 0$.

3.3.13 (Fritz John Conditions for Nondifferentiable Cost) (www)

Let x^* be a local minimum of the problem

$$\begin{aligned} &\text{minimize} \quad f(x) \\ &\text{subject to} \quad h_1(x) = 0, \ldots, h_m(x) = 0, \\ &\qquad\qquad\quad g_1(x) \leq 0, \ldots, g_r(x) \leq 0, \end{aligned}$$

where f is a convex function, and h_i and g_j are continuously differentiable functions from $\Re^n$ to $\Re$.

(a) Adapt the proof of Prop. 3.3.5 to show that there exists a scalar μ_0^* and Lagrange multipliers $\lambda_1^*, \ldots, \lambda_m^*$ and $\mu_1^*, \ldots, \mu_r^*$, satisfying the following conditions:

(i) The vector

$$-\sum_{i=1}^m \lambda_i^* \nabla h_i(x^*) - \sum_{j=1}^r \mu_j^* \nabla g_j(x^*)$$

is a subgradient of the function $\mu_0^* f(\cdot)$ at x^*.

(ii)

$$\mu_j^* \geq 0, \qquad j = 0, 1, \ldots, r.$$

(iii) $\mu_0^*, \lambda_1^*, \ldots, \lambda_m^*, \mu_1^*, \ldots, \mu_r^*$ are not all equal to 0.

(iv) In every neighborhood N of x^* there is an $x \in N$ such that $\lambda_i^* h_i(x) > 0$ for all i with $\lambda_i^* \neq 0$ and $\mu_j^* g_j(x) > 0$ for all j with $\mu_j^* \neq 0$.

Hint: Verify that if x^* is an unconstrained local minimum of $f_1(x) + f_2(x)$, where f_1 is convex and f_2 is continuously differentiable, then $-\nabla f_2(x^*)$ is a subgradient of f_1 at x^*. Use Prop. B.24.

(b) Prove that one can take $\mu_0^* = 1$ if the constraint functions h_i and g_j are linear (cf. Prop. 3.3.7).

(c) Prove that one can take $\mu_0^* = 1$ under conditions that are analogous to the Mangasarian-Fromovitz and Slater constraint qualifications (cf. Props. 3.3.8 and 3.3.9).

3.3.14 (www)

Consider the problem of finding a sphere of minimum radius that contains given vectors $y_1, \ldots, y_p$ in $\Re^n$. Formulate this problem as the minimax problem

$$\begin{aligned} &\text{minimize} \quad \max\left\{\|x - y_1\|^2, \ldots, \|x - y_p\|^2\right\} \\ &\text{subject to} \quad x \in \Re^n, \end{aligned}$$

and write the corresponding optimality conditions of Prop. 3.3.10. Characterize the optimal solution and the Lagrange multipliers in the special case where $p = 3$.

3.3.15 (www)

Consider a subset X of $\Re^n$, and let x be a point in X. Show that the tangent cone $T(x)$ of X at x has the following properties:

(a) $T(x)$ is a closed cone.

(b) If X is convex and contains at least one feasible direction at x, $T(x)$ is equal to the closure of the set of feasible directions at x.

3.3.16 (www)

Consider the subset of $\Re^n$ defined by

$$X = \left\{x \mid g_j(x) \leq 0,\ j = 1, \ldots, r\right\},$$

where $g_j : \Re^n \mapsto \Re$ are concave and differentiable functions over $\Re^n$. Show that every point in X is quasiregular.

3.3.17 (www)

Consider the set

$$X = \{x \mid g_j(x) \leq 0,\ j \in J\},$$

where J is an index set that is a compact subset of some Euclidean space, and assume that the g_j are continuously differentiable functions of x, and furthermore, $\nabla g_j(x)$ is continuous as a function of j. Let x^* be a vector in X, and let $A(x^*) = \{j \mid g_j(x^*) = 0\}$. Let y be a vector such that $\nabla g_j(x^*)'y < 0$ for all $j \in A(x^*)$. Show that y is a feasible direction of X at x^* and belongs to the tangent cone $T(x^*)$. *Hint*: By the continuity of $\nabla g_j(x^*)$, there exists a neighborhood N of x^* and a neighborhood A of $A(x^*)$ (relative to J) such that $\nabla g_j(x)'y < 0$ for all $x \in N$ and $j \in A$. Furthermore, N can be chosen so that $g_j(x) < 0$ for all $x \in N$ and j in the compact set $J - A$. Choose $\delta > 0$ so that $x^* + \alpha y$ is in N when $0 \leq \alpha \leq \delta$. Show that $g_j(x^* + \alpha y) < 0$ for all α with $0 < \alpha \leq \delta$.

3.3.18 (www)

Use the elimination approach to prove the existence of Lagrange multipliers under the Mangasarian-Fromovitz and Slater constraint qualifications (cf. Props. 3.3.8 and 3.3.9) starting from the special case of this result where there are no inequality constraints. *Hint*: As in Section 3.1.2, use the implicit function theorem to express m of the variables in terms of the remaining variables, while eliminating the equality constraints.

3.3.19 (Boundedness of the Set of Lagrange Multipliers [Gau77], [KlH98]) (www)

Let x^* be a local minimum of f subject to $h_i(x) = 0$, $i = 1, \ldots, m$, $g_j(x) \leq 0$, $j = 1, \ldots, r$, where f, h_i, and g_j are continuously differentiable. Consider the set of pairs (λ^*, μ^*) that are Lagrange multipliers [i.e., satisfy $\mu^* \geq 0$, $\mu_j^* g_j(x^*) = 0$ for all j, and $\nabla L(x^*, \lambda^*, \mu^*) = 0$]. Show that this set is nonempty and bounded if and only if the Mangasarian-Fromovitz constraint qualification (cf. Prop. 3.3.8) holds. *Hint*: Suppose that $\{\lambda^k, \mu^k\}$ is a sequence of Lagrange multipliers with $\|\lambda^k\| + \|\mu^k\| \to \infty$. Define

$$\beta^k = \frac{\lambda^k}{\|\lambda^k\|}, \qquad \gamma^k = \frac{\mu^k}{\|\mu^k\|},$$

and show that $\nabla h(x^*)\beta + \nabla g(x^*)\gamma = 0$ for all limit points (β, γ) of $\{\beta^k, \gamma^k\}$. Show that this is a contradiction of the Mangasarian-Fromovitz constraint qualification. For the converse, use the theorem of the alternative of Exercise 3.3.6.
Note: A more general version of this result which applies to the case of an additional convex set constraint $x \in X$ (cf. Prop. 3.3.12) is given in Exercises 5.3.1 and 5.3.3.

3.3.20 (www)

Consider the subset $X = \{x \mid h_1(x) = 0,\ h_2(x) = 0\}$ of $\Re^2$, where

$$h_1(x) = x_2, \qquad h_2(x) = \begin{cases} x_1^4 \sin\left(\frac{1}{x_1}\right) - x_2 & \text{if } x_1 \neq 0, \\ 0 & \text{if } x_1 = 0. \end{cases}$$

Show that h_1 and h_2 are continuously differentiable, and that the vector $x^* = (0,0)$ is a quasiregular but not a quasinormal vector of X.

3.3.21 (www)

Consider the set

$$X = \{x \mid h_i(x) = 0,\ i = 1,\ldots,m,\ g_j(x) \leq 0,\ j = 1,\ldots,r\},$$

where h_i, and g_j are continuously differentiable functions, and let x^* be a vector in X. Let $\overline{J}$ be the set of indices $j \in A(x^*)$ with the property that $\nabla g_j(x^*)'y = 0$ for all y in the cone of first order feasible variations $V(x^*)$. Define

$$\overline{X} = \{x \mid h_i(x) = 0, i = 1,\ldots,m,\ g_j(x) \leq 0, j \in \overline{J}\}.$$

(In the degenerate case where there are no equalities and the set $\overline{J}$ is empty, we have $\overline{X} = \Re^n$ by convention. In this case, every vector is a quasiregular and a quasinormal vector of $\overline{X}$.) Show that:

(a) x^* is a quasinormal vector of $\overline{X}$ if and only if it is a quasinormal vector of X. *Hint*: For the hints of this exercise, we assume without loss of generality that all inequality constraints are active at x^*, and that there are no equality constraints [it can be seen that x^* is quasinormal with respect to the constraints $h_i(x) = 0$, $g_j(x) \leq 0$ if and only if it is quasinormal with respect to the constraints $h_i(x) \leq 0$, $-h_i(x) \leq 0$, $g_j(x) \leq 0$, so each equality constraint can be converted to two inequalities for the purposes of the proof]. Complete the details of the following argument. If x^* is a quasinormal vector of X, it is clearly a quasinormal vector of $\overline{X}$, based on the definition of quasinormality. Conversely, suppose, to arrive at a contradiction, that x^* is a quasinormal vector of $\overline{X}$, but not a quasinormal vector of X. Then there exist $\mu_j \geq 0$, not all zero, satisfying

$$\sum_{j=1}^{r} \mu_j \nabla g_j(x^*) = 0, \tag{3.84}$$

and such that there exist x arbitrarily close to x^* satisfying $\mu_j g_j(x) > 0$ for all j with $\mu_j > 0$. We must have $\mu_{\bar{j}} > 0$ for some $\bar{j} \notin \overline{J}$, since otherwise x^* would not be a quasinormal vector of $\overline{X}$. Take a $y \in V(x^*)$ such that $\nabla g_{\bar{j}}(x^*)'y < 0$, and multiply Eq. (3.84) with y to reach a contradiction.

(b) If x^* is a quasiregular vector of $\overline{X}$ it is also a quasiregular vector of X. *Hint*: Complete the details of the following argument. In general, $T(x^*)$, the tangent cone of X at x^*, is a closed cone and satisfies $T(x^*) \subset V(x^*)$. From these facts it follows that x^* is a quasiregular vector of X if and only if $\tilde{V}(x^*) \subset T(x^*)$, where

$$\tilde{V}(x^*) = \{y \mid \nabla g_j(x^*)'y = 0, j \in \overline{J},\, \nabla g_j(x^*)'y < 0, j \notin \overline{J}\}$$

[the closure of $\tilde{V}(x^*)$ is $V(x^*)$]. The hypothesis that x^* is a quasiregular vector of $\overline{X}$, implies that $\tilde{V}(x^*) \subset \overline{V}(x^*) = \overline{T}(x^*)$, where $\overline{V}(x^*)$ and $\overline{T}(x^*)$ are the cone of first order feasible variations and the tangent cone of $\overline{X}$ at x^*, respectively. We complete the proof by showing that $\tilde{V}(x^*) \subset T(x^*)$ as follows. Take any $y \in \tilde{V}(x^*)$ with $y \neq 0$. Since $y \in \overline{T}(x^*)$, there exists a sequence $\{x^k\} \subset \overline{X}$ such that $x^k \neq x^*$ for all k and

$$x^k \to x^*, \qquad \frac{x^k - x^*}{\|x^k - x^*\|} \to \frac{y}{\|y\|}.$$

We have for all j,

$$\lim_{k\to\infty} \frac{g_j(x^k) - g_j(x^*)}{\|x^k - x^*\|} = \lim_{k\to\infty} \frac{\nabla g_j(x^*)'(x^k - x^*)}{\|x^k - x^*\|} = \frac{\nabla g_j(x^*)'y}{\|y\|}$$

from which we see that for all $j \notin \overline{J}$, we have $g_j(x^k) < 0$ for sufficiently large k. Hence $x^k \in X$ for all sufficiently large k, and it follows that y is a tangent of X at x^*. Thus, we have $\tilde{V}(x^*) \subset T(x^*)$, which as shown earlier, implies that x^* is a quasiregular vector of X.

(c) Combine parts (a) and (b) to conclude that if every quasinormal vector of $\overline{X}$ is a quasiregular vector of $\overline{X}$, then every quasinormal vector of X is a quasiregular vector of X.

3.3.22 (An Alternative Version of the Mangasarian-Fromovitz Constraint Qualification) (www)

Let x^* be a local minimum of the problem

$$\begin{aligned} &\text{minimize } f(x) \\ &\text{subject to } h_1(x) = 0, \ldots, h_m(x) = 0, \\ &\qquad g_1(x) \le 0, \ldots, g_r(x) \le 0, \end{aligned}$$

where h_i are linear functions, and f and g_j are continuously differentiable functions. Let J be the subset of indices $j \in A(x^*)$ such that g_j is a concave function. Assume that there exists a vector d such that

$$\begin{aligned} \nabla h_i(x^*)'d = 0, &\quad i = 1, \ldots, m, \\ \nabla g_j(x^*)'d \le 0, &\quad \forall\, j \in J, \\ \nabla g_j(x^*)'d < 0, &\quad \forall\, j \in A(x^*),\, j \notin J. \end{aligned}$$

Then x^* satisfies the necessary conditions of Prop. 3.3.6. *Hint*: Combine the proofs of Props. 3.3.7 and 3.3.8.

3.4 LINEAR CONSTRAINTS AND DUALITY*

Problems with linear constraints have a remarkable property, which we derived in Prop. 3.3.7: they possess Lagrange multipliers always, even for local minima which are not regular. In this section we use this property to develop a duality theory for a problem with differentiable convex cost and linear constraints. In particular, we show that there is another problem, called *dual*, that has the same optimal value and has as optimal solutions the Lagrange multipliers of the original. This is an important relationship, which will be discussed extensively and generalized considerably in Chapter 5. Here, we develop some of the simpler aspects of duality theory, which follow easily from the theory developed so far in this chapter.

3.4.1 Convex Cost Functions and Linear Constraints

As one expects based on the corresponding results of Chapter 2 and Prop. 3.3.4, if f is convex, the necessary optimality conditions for linear constraints are also sufficient. In fact a sharper necessary and sufficient condition can be derived, involving minimization of a Lagrangian function. This minimization may involve *any* subset of the inequality constraints, while the remaining constraints are taken into account by the Lagrange multipliers, as shown in the following proposition. The flexibility of assigning Lagrange multipliers to only some of the constraints, while dealing with the other constraints explicitly is often very useful.

Proposition 3.4.1: (Optimality Conditions for Convex Cost and Linear Constraints) Consider the problem

$$
\begin{aligned}
&\text{minimize} && f(x) \\
&\text{subject to} && e_i'x = d_i, \quad i = 1, \ldots, m, \\
& && a_j'x \le b_j, \quad j = 1, \ldots, r,
\end{aligned}
$$

where e_i, a_j, and d_i, b_j are given vectors and scalars, respectively, and $f : \Re^n \mapsto \Re$ is convex and continuously differentiable. Let I be a subset of the index set $\{1, \ldots, m\}$, and J be a subset of the index set $\{1, \ldots, r\}$. Then x^* is a global minimum if and only if x^* is feasible and there exist scalars λ_i^*, $i \in I$, and μ_j^*, $j \in J$, such that

$$\mu_j^* \ge 0, \qquad j \in J, \tag{3.85}$$

$$\mu_j^* = 0, \qquad \forall\, j \in J \text{ with } j \notin A(x^*), \tag{3.86}$$

$$x^* = \arg \min_{\substack{e_i'x = d_i,\ i \notin I \\ a_j'x \leq b_j,\ j \notin J}} \left\{ f(x) + \sum_{i \in I} \lambda_i^*(e_i'x - d_i) + \sum_{j \in J} \mu_j^*(a_j'x - b_j) \right\}. \tag{3.87}$$

Proof: For simplicity, we assume that all the constraints are inequalities. Assume that x^* is a global minimum. Then by Prop. 3.3.7, there exist nonnegative scalars $\mu_1^*, \ldots, \mu_r^*$ such that

$$\mu_j^*(a_j'x^* - b_j) = 0, \qquad j = 1, \ldots, r, \tag{3.88}$$

and

$$\nabla f(x^*) + \sum_{j=1}^{r} \mu_j^* a_j = 0.$$

Using the convexity of f, the last relation implies that

$$x^* = \arg \min_{x \in \Re^n} \left\{ f(x) + \sum_{j=1}^{r} \mu_j^*(a_j'x - b_j) \right\}. \tag{3.89}$$

Combining Eqs. (3.88) and (3.89), we obtain

$$f(x^*) = \min_{x \in \Re^n} \left\{ f(x) + \sum_{j=1}^{r} \mu_j^*(a_j'x - b_j) \right\}. \tag{3.90}$$

Since $\mu_j^* \geq 0$, it follows that

$$\mu_j^*(a_j'x - b_j) \leq 0 \qquad \text{if} \qquad a_j'x - b_j \leq 0,$$

so Eq. (3.90) implies that

$$\begin{aligned} f(x^*) &\leq \min_{\substack{a_j'x \leq b_j \\ j \notin J}} \left\{ f(x) + \sum_{j=1}^{r} \mu_j^*(a_j'x - b_j) \right\} \\ &= \min_{\substack{a_j'x \leq b_j \\ j \notin J}} \left\{ f(x) + \sum_{j \in J} \mu_j^*(a_j'x - b_j) + \sum_{j \notin J} \mu_j^*(a_j'x - b_j) \right\} \\ &\leq \min_{\substack{a_j'x \leq b_j \\ j \notin J}} \left\{ f(x) + \sum_{j \in J} \mu_j^*(a_j'x - b_j) \right\} \\ &\leq f(x^*) + \sum_{j \in J} \mu_j^*(a_j'x^* - b_j) \\ &= f(x^*), \end{aligned} \tag{3.91}$$

where the last equality holds by Eq. (3.88). Thus, x^* attains the minimum in the above relations, proving the desired Eq. (3.87).

Conversely, assume that x^* is feasible and there exist scalars μ_j^*, $j \in J$, satisfying conditions (3.85)-(3.87). Then, using Prop. 3.3.4, it follows that x^* is a global minimum. **Q.E.D.**

Note that given x^*, the vectors λ^* and μ^* satisfying conditions (3.85)-(3.87) in the preceding proof are independent of the index subsets I and J; they are the vectors λ^* and μ^* corresponding to $I = \{1, \ldots, m\}$ and $J = \{1, \ldots, r\}$, respectively, but they satisfy condition (3.87) for any subsets I and J. This is not surprising in view of the sensitivity interpretation of Lagrange multipliers.

Example 3.4.1 (Optimization Over a Simplex)

Consider the case where f is convex and the constraint set is the simplex

$$\left\{ x \;\middle|\; x \geq 0,\ \sum_{i=1}^{n} x_i = r \right\},$$

where $r > 0$ is a given scalar. We will use the optimality condition of Prop. 3.4.1 to recover the conditions obtained in Example 2.1.2 of Section 2.1.

By Prop. 3.4.1, if x^* is a global minimum, there exists a scalar λ^* such that

$$x^* = \arg\min_{x \geq 0} \left\{ f(x) + \lambda^* \left(r - \sum_{i=1}^{n} x_i \right) \right\}. \tag{3.92}$$

By applying the optimality conditions for an orthant constraint (cf. Example 2.1.1 in Section 2.1), we obtain from Eq. (3.92)

$$\frac{\partial f(x^*)}{\partial x_i} \geq \lambda^*, \qquad i = 1, \ldots, n, \tag{3.93}$$

$$\frac{\partial f(x^*)}{\partial x_i} = \lambda^*, \qquad \text{if } x_i^* > 0. \tag{3.94}$$

We see therefore that all positive coordinates x_i^* must have partial cost derivatives which are minimal and equal to the Lagrange multiplier λ^*. By Prop. 3.4.1, this condition is also sufficient for a feasible vector x^* to be optimal.

3.4.2 Duality Theory: A Simple Form for Linear Constraints

Consider the problem

$$\begin{aligned} &\text{minimize } f(x) \\ &\text{subject to } e_i'x = d_i, \quad i = 1, \ldots, m, \\ &\qquad a_j'x \leq b_j, \quad j = 1, \ldots, r, \quad x \in X, \end{aligned} \tag{P}$$

where e_i, a_j, and d_i, b_j are given vectors and scalars, respectively, $f : \Re^n \to \Re$ is a convex continuously differentiable function, and X is a polyhedral set, i.e., a set specified by a finite collection of linear equality and inequality constraints. We refer to problem (P) as the *primal problem.*

Define the Lagrangian function

$$L(x, \lambda, \mu) = f(x) + \sum_{i=1}^{m} \lambda_i (e_i' x - d_i) + \sum_{j=1}^{r} \mu_j (a_j' x - b_j).$$

Consider also the *dual function* defined by

$$q(\lambda, \mu) = \inf_{x \in X} L(x, \lambda, \mu). \tag{3.95}$$

The *dual problem* is

$$\begin{aligned} &\text{maximize} \quad q(\lambda, \mu) \\ &\text{subject to} \quad \lambda \in \Re^m, \quad \mu \geq 0. \end{aligned} \tag{D}$$

Note that if the polyhedral set X is bounded, then it is also compact (every polyhedral set is closed since it is the intersection of closed subspaces), and the infimum in Eq. (3.95) is attained for all (λ, μ) by Weierstrass' theorem (Prop. A.8 in Appendix A). Thus if X is bounded, the dual function takes real values. In general, however, $q(\lambda, \mu)$ can take the value $-\infty$. Thus in effect, the constraint set of the dual problem (D) is the set

$$Q = \big\{(\lambda, \mu) \mid \mu \geq 0,\ q(\lambda, \mu) > -\infty\big\}.$$

Figure 3.4.1 illustrates the definition of the dual function for the case of a problem involving a single constraint. As the figure illustrates, the optimal primal value is equal to the optimal dual value, and the optimal dual solutions are Lagrange multipliers of the primal problem. This is the essence of the following basic result.

Proposition 3.4.2: (Duality Theorem)

(a) If the primal problem (P) has an optimal solution, the dual problem (D) also has an optimal solution and the corresponding optimal values are equal.

(b) In order for x^* to be an optimal primal solution and (λ^*, μ^*) to be an optimal dual solution, it is necessary and sufficient that x^* is primal feasible, $\mu^* \geq 0$, $\mu_j^* = 0$ for all $j \notin A(x^*)$, and

$$x^* = \arg\min_{x \in X} L(x, \lambda^*, \mu^*). \tag{3.96}$$

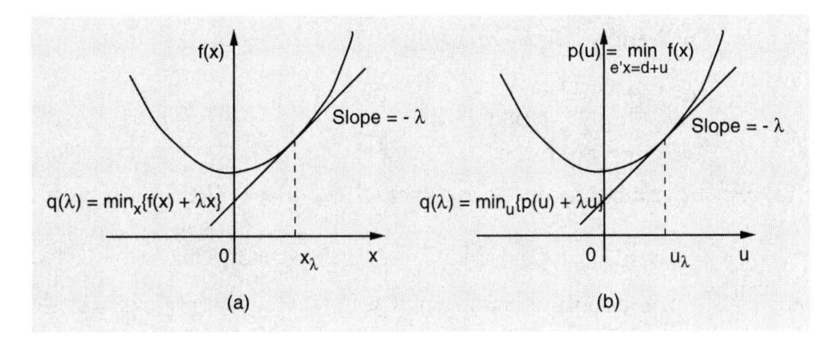

Figure 3.4.1. (a) Illustration of the dual function for the one-dimensional (trivial) problem

$$\begin{aligned} &\text{minimize} \quad f(x) \\ &\text{subject to} \quad x = 0, \end{aligned}$$

where x is a scalar. For the given λ, the dual value $q(\lambda) = \inf_x \{f(x) + \lambda x\}$ is obtained by determining the point x_λ attaining the infimum and calculating $q(\lambda)$ as the point of interception of the vertical axis and the corresponding line of slope $-\lambda$ supporting the graph of f. It can be seen that the maximum value of $q(\lambda)$ is obtained when λ is equal to $-\nabla f(0)$, which is the Lagrange multiplier of the problem.

(b) Illustration of the dual function for the case of a single-constraint problem

$$\begin{aligned} &\text{minimize} \quad f(x) \\ &\text{subject to} \quad e'x = d_i, \end{aligned}$$

where $x \in \Re^n$. This problem is equivalent to the one-dimensional problem

$$\begin{aligned} &\text{minimize} \quad p(u) \\ &\text{subject to} \quad u = 0, \end{aligned}$$

where

$$p(u) = \min_{\{x \in \Re^n \mid e'x = d_i + u\}} f(x).$$

(This is essentially the primal function; see the sensitivity theorem, Prop. 3.2.2.) The dual function can now be interpreted in terms of this problem as in (a). For the given λ, the dual value $q(\lambda)$ is obtained as the point of interception of the vertical axis and the corresponding line of slope $-\lambda$ supporting the graph of p. It can be seen that the maximum value of $q(\lambda)$ is obtained when λ is equal to $-\nabla p(0)$, which is the Lagrange multiplier of the problem (cf. the sensitivity theorem, Prop. 3.2.2).

Proof: (a) For all primal feasible x, and all $\lambda \in \Re^m$ and $\mu \geq 0$, we have

$\lambda_i(e_i'x - d_i) = 0$ and $\mu_j'(a_j'x - b_j) \leq 0$, so that

$$q(\lambda, \mu) \leq f(x) + \sum_{i=1}^{m} \lambda_i(e_i'x - d_i) + \sum_{j=1}^{r} \mu_j(a_j'x - b_j) \leq f(x). \tag{3.97}$$

By taking the minimum of the right-hand side over all primal feasible x, we obtain

$$q(\lambda, \mu) \leq f(x^*), \qquad \forall\, \lambda \in \Re^m,\ \mu \geq 0, \tag{3.98}$$

where x^* is a primal optimal solution. By Prop. 3.4.1, there exist $\lambda^* \in \Re^m$ and $\mu^* \geq 0$ such that $\mu_j^*(a_j'x^* - b_j) = 0$ for all j, and

$$x^* = \arg\min_{x \in X} L(x, \lambda^*, \mu^*),$$

so by using also the definition (3.95) of q, we have

$$\begin{aligned} q(\lambda^*, \mu^*) &= \inf_{x \in X} L(x, \lambda^*, \mu^*) \\ &= L(x^*, \lambda^*, \mu^*) \\ &= f(x^*) + \sum_{i=1}^{m} \lambda_i^*(e_i'x^* - d_i) + \sum_{j=1}^{r} \mu_j^*(a_j'x^* - b_j) \\ &= f(x^*). \end{aligned} \tag{3.99}$$

By combining Eqs. (3.98) and (3.99), we see that (λ^*, μ^*) is a dual optimal solution and that $q(\lambda^*, \mu^*) = f(x^*)$.

(b) If x^* is primal optimal and (λ^*, μ^*) is dual optimal, then $\mu^* \geq 0$ since (λ^*, μ^*) must be dual feasible. Furthermore, by part (a) we obtain

$$f(x^*) = q(\lambda^*, \mu^*),$$

which when combined with Eq. (3.97), yields

$$f(x^*) = L(x^*, \lambda^*, \mu^*) = q(\lambda^*, \mu^*) = \min_{x \in X} L(x, \lambda^*, \mu^*).$$

Conversely, the given conditions imply the above relation. Hence, since x^* is primal feasible and $\mu^* \geq 0$, Eq. (3.97) implies that x^* is primal optimal and (λ^*, μ^*) is dual optimal. **Q.E.D.**

The duality assertions of Prop. 3.4.2 were proved under hypotheses which are stronger than necessary. For example, the differentiability assumption on f is not needed – it turns out that convexity is sufficient. Furthermore, the existence of a primal optimal solution is not needed for the conclusion of Prop. 3.4.2(a) – it turns out that finiteness of the infimum of the primal cost is sufficient. However, to prove these and other extensions of the duality theorem we need an approach that is not calculus-based like the one of this chapter but is based instead on the geometry of convex sets; see Chapter 5.

Example 3.4.2 (The Dual of a Linear Program)

Consider the linear program

$$\begin{aligned}&\text{minimize } c'x\\ &\text{subject to } e_i'x = d_i, \quad i = 1,\ldots,m, \qquad x \geq 0,\end{aligned} \tag{LP}$$

where c and e_i are given vectors in $\Re^n$, and d_i are given scalars. We consider the dual function

$$q(\lambda) = \inf_{x \geq 0} \left\{ \sum_{j=1}^{n} \left(c_j - \sum_{i=1}^{m} \lambda_i e_{ij} \right) x_j + \sum_{i=1}^{m} \lambda_i d_i \right\},$$

where e_{ij} is the jth coordinate of the vector e_i. It is seen that if $c_j - \sum_{i=1}^{m} \lambda_i e_{ij} \geq 0$ for all j, the infimum above is attained for $x = 0$, and we have $q(\lambda) = \sum_{i=1}^{m} \lambda_i d_i$. On the other hand, if $c_j - \sum_{i=1}^{m} \lambda_i e_{ij} < 0$ for some j, we can make the expression in braces above arbitrarily small by taking x_j sufficiently large, so that we have $q(\lambda) = -\infty$ in this case. Thus, the dual problem is

$$\begin{aligned}&\text{maximize } \sum_{i=1}^{m} \lambda_i d_i\\ &\text{subject to } \sum_{i=1}^{m} \lambda_i e_{ij} \leq c_j, \qquad j = 1,\ldots,n.\end{aligned} \tag{DLP}$$

By the optimality conditions of Prop. 3.4.2, (x^*, λ^*) is a primal and dual optimal solution pair if and only if x^* is primal feasible, and the Lagrangian optimality condition

$$x^* = \arg\min_{x \geq 0} \left\{ \left(c - \sum_{i=1}^{m} \lambda_i^* e_i \right)' x + \sum_{i=1}^{m} \lambda_i^* d_i \right\}$$

holds [cf. Eq. (3.96)]. The above condition is equivalent to $c \geq \sum_{i=1}^{m} \lambda_i^* e_i$ (otherwise the minimum above would not be achieved at x^*), $x^* \geq 0$, and the following two relations

$$x_j^* > 0 \quad \Rightarrow \quad \sum_{i=1}^{m} \lambda_i^* e_{ij} = c_j, \qquad j = 1,\ldots,n, \tag{3.100}$$

$$\sum_{i=1}^{m} \lambda_i^* e_{ij} < c_j \quad \Rightarrow \quad x_j^* = 0, \qquad j = 1,\ldots,n, \tag{3.101}$$

known as the *complementary slackness conditions for linear programming.*

Let us consider now the dual of the dual problem (DLP). We first convert this program into the equivalent minimization problem

$$\begin{aligned} &\text{minimize} \quad \sum_{i=1}^{m} (-d_i)\lambda_i \\ &\text{subject to} \quad \sum_{i=1}^{m} \lambda_i e_{ij} \le c_j, \qquad j = 1, \ldots, n. \end{aligned}$$

Assigning a Lagrange multiplier x_j to the jth inequality constraint, the dual function of this problem is given by

$$\begin{aligned} p(x) &= \min_{\lambda \in \Re^m} \left\{ \sum_{i=1}^{m} \left(\sum_{j=1}^{n} e_{ij} x_j - d_i \right) \lambda_i - \sum_{j=1}^{n} c_j x_j \right\} \\ &= \begin{cases} -c'x & \text{if } e_i'x = d_i, \quad i = 1, \ldots, m, \\ -\infty & \text{otherwise.} \end{cases} \end{aligned}$$

The corresponding dual problem is

$$\begin{aligned} &\text{maximize} \quad p(x) \\ &\text{subject to} \quad x \ge 0, \end{aligned}$$

or equivalently

$$\begin{aligned} &\text{minimize} \quad c'x \\ &\text{subject to} \quad e_i'x = d_i, \quad i = 1, \ldots, m, \qquad x \ge 0, \end{aligned}$$

which is identical to the primal problem (LP). We have thus shown that the duality is symmetric, that is, *the dual of the dual linear program (DLP) is the primal problem (LP).*

The pair of primal and dual linear programs (LP) and (DLP) can also be written compactly in terms of the $m \times n$ matrix E having rows $e_1', \ldots, e_m'$, and the vector d having coordinates $d_1, \ldots, d_m$;

$$\min_{Ex = d,\ x \ge 0} c'x \qquad \Longleftrightarrow \qquad \max_{E'\lambda \le c} d'\lambda.$$

Other linear programming duality relations that can be verified by the reader include the following:

$$\min_{A'x \ge b} c'x \qquad \Longleftrightarrow \qquad \max_{A\mu = c,\ \mu \ge 0} b'\mu,$$

$$\min_{A'x \ge b,\ x \ge 0} c'x \qquad \Longleftrightarrow \qquad \max_{A\mu \le c,\ \mu \ge 0} b'\mu.$$

Example 3.4.3 (The Dual of a Quadratic Program)

Consider the quadratic programming problem

$$\begin{aligned} &\text{minimize} \quad \tfrac{1}{2}x'Qx + c'x \\ &\text{subject to} \quad Ax \le b, \end{aligned} \qquad \text{(QP)}$$

where Q is a given $n \times n$ positive definite symmetric matrix, A is a given $r \times n$ matrix, and $b \in \Re^r$ and $c \in \Re^n$ are given vectors. The dual function is

$$q(\mu) = \inf_{x \in \Re^n} \left\{ \tfrac{1}{2}x'Qx + c'x + \mu'(Ax - b) \right\}.$$

The infimum is attained for $x = -Q^{-1}(c + A'\mu)$, and a straightforward calculation after substituting this expression in the preceding relation, yields

$$q(\mu) = -\tfrac{1}{2}\mu' AQ^{-1}A'\mu - \mu'(b + AQ^{-1}c) - \tfrac{1}{2}c'Q^{-1}c.$$

The dual problem, after dropping the constant $\frac{1}{2}c'Q^{-1}c$ and changing the minus sign to convert the maximization to a minimization, can be written as

$$\begin{aligned} &\text{minimize} \quad \tfrac{1}{2}\mu'P\mu + t'\mu \\ &\text{subject to} \quad \mu \ge 0, \end{aligned} \qquad \text{(DQP)}$$

where

$$P = AQ^{-1}A', \qquad t = b + AQ^{-1}c.$$

If μ^* is any dual optimal solution, then the optimal solution of the primal problem is

$$x^* = -Q^{-1}(c + A'\mu^*).$$

Note that the dual problem is also a quadratic program, but it has simpler constraints than the primal. Furthermore, if the row dimension r of A is smaller than its column dimension n, the dual problem is defined on a space of smaller dimension than the primal, and this can be algorithmically significant.

EXERCISES

3.4.1

A company has available for sale a quantity of Q units of a certain product that will be sold in n market outlets. For each $i = 1, \ldots, n$, the quantity d_i, which is demanded at outlet i, and the price of sale p_i are known. The company wishes to determine the quantities s_i^* with $0 \le s_i^* \le d_i$ to be sold at each outlet that maximize the revenue $\sum_{i=1}^n p_i s_i$ from the sale. Assuming that $d_i > 0$, $p_i > 0$, and $\sum_{i=1}^n d_i \ge Q$, show that there exists a cutoff price level y such that for each i, if $p_i > y$, then $s_i^* = d_i$, and if $p_i < y$, then $s_i^* = 0$. What happens if $p_i = y$? Describe a procedure for obtaining s_i^*.

3.4.2

Consider a problem of finding the best way to place bets totalling A dollars ($A > 0$) in a race involving n horses. Assume that we know the probability p_i, that the ith horse wins, and the amount $s_i > 0$ that the rest of the public is betting on the ith horse. The track keeps a proportion $1 - c$ of the total amount ($0 < 1 - c < 1$) and distributes the rest among the public in proportion to the amounts bet on the winning horse. Thus, if we bet x_i on the ith horse, we receive

$$c\left(A + \sum_{i=1}^{n} s_i\right) \frac{x_i}{s_i + x_i}$$

if the ith horse wins. The problem is to find x_i^*, $i = 1, \ldots, n$, which maximize the expected net return

$$c\left(A + \sum_{i=1}^{n} s_i\right) \left\{\sum_{i=1}^{n} \frac{p_i x_i}{s_i + x_i}\right\} - A,$$

or, equivalently,

$$\sum_{i=1}^{n} \frac{p_i x_i}{s_i + x_i}$$

subject to $\sum_{i=1}^{n} x_i = A$, $x_i \geq 0$, $i = 1, \ldots, n$.

Assume that

$$\frac{p_1}{s_1} > \frac{p_2}{s_2} > \cdots > \frac{p_n}{s_n},$$

and show that there exists a scalar λ^* such that the optimal solution is

$$x_i^* = \begin{cases} \sqrt{\dfrac{s_i p_i}{\lambda^*}} - s_i & \text{for } i = 1, \ldots, m^*, \\ 0 & \text{for } i = m^* + 1, \ldots, n, \end{cases}$$

where m^* is the largest index m for which $\frac{p_m}{s_m} \geq \lambda^*$.

3.4.3 (www)

Verify the linear programming duality relations

$$\min_{A'x \geq b} c'x \quad \Longleftrightarrow \quad \max_{A\mu = c,\ \mu \geq 0} b'\mu,$$

$$\min_{A'x \geq b,\ x \geq 0} c'x \quad \Longleftrightarrow \quad \max_{A\mu \leq c,\ \mu \geq 0} b'\mu,$$

show that they are symmetric, and derive the corresponding complementary slackness conditions [cf. Eqs. (3.100) and (3.101)].

3.4.4 (Duality for Transportation Problems) (www)

Suppose that a quantity of a certain material must be shipped from m supply points to n demand points so as to minimize the total transportation cost. The supply at point i is denoted α_i and the demand at point j is denoted β_j. The unit transportation cost from i to j is a_{ij}. Letting x_{ij} denote the quantity shipped from supply point i to demand point j, the problem is

$$\text{minimize} \quad \sum_{i,j} a_{ij}x_{ij}$$

$$\text{subject to}$$

$$\sum_{j=1}^{n} x_{ij} = \alpha_i, \qquad i = 1, \ldots, m,$$

$$\sum_{i=1}^{m} x_{ij} = \beta_j, \qquad j = 1, \ldots, n,$$

$$0 \le x_{ij}, \qquad \forall\, i,\, j,$$

where α_i and β_j are positive scalars, which for feasibility must satisfy

$$\sum_{i=1}^{m} \alpha_i = \sum_{j=1}^{n} \beta_j.$$

(a) Assign Lagrange multipliers to the equality constraints and derive the corresponding dual problem.

(b) Introduce a price p_j that demand point j will pay per unit delivered at j. Show that if x^* is an optimal solution of the transportation problem, there is a set of prices $\{p_j^* \mid j = 1, \ldots, n\}$ such that if $x_{ij}^* > 0$ then j offers maximum net profit for i, i.e.,

$$p_j^* - a_{ij} = \max_{k=1,\ldots,n} \{p_k^* - a_{ik}\}.$$

(c) Relate prices to dual feasible solutions, and show that every dual optimal solution has a property of the type described in (b) above.

3.5 NOTES AND SOURCES

Section 3.1: The material on Lagrange multipliers is classical. For further analysis using penalty functions, see Hestenes [Hes75]. For an alternative proof of the Lagrange multiplier theorem, using an approach based on differentiable curves, see Hestenes [Hes75] and Luenberger [Lue84]. In this approach, differentiable curves $x(t)$, parameterized by a scalar t, are constructed using the implicit function theorem so that they lie on the constraint surface and pass through x^*, i.e., $x^* = x(0)$. The unconstrained optimality condition $df\big(x(t)\big)/dt|_{t=0} = 0$ is then applied to yield the first order Lagrange multiplier condition. This approach is insightful, but is considerably more complicated than the penalty approach we have followed.

Section 3.2: For a textbook treatment of sensitivity, see Fiacco [Fia78]. For sensitivity analysis and related topics under weaker assumptions than the second order sufficiency conditions that we have assumed, see Dontchev and Tongen [DoT86], Robinson [Rob87], Gauvin and Janin [GaJ88], Shapiro [Sha88], Auslender and Cominetti [AuC90], Kyparisis [Kyp90], Bonnans [Bon92], King and Rockafellar [KiR92], Ioffe [Iof94], Bonnans and Shapiro [BoS98], and Bonnans, Cominetti, and Shapiro [BCS99].

Section 3.3: Lagrange multiplier theorems for inequality constraints became available considerably later than their equality constraint counterparts. Important early works are those of Karush [Kar39] (an unpublished MS thesis), John [Joh48], and Kuhn and Tucker [KuT51]. The survey by Kuhn [Kuh76] gives a historical account of the development of the subject. There has been considerable effort to derive optimality conditions under weaker assumptions than regularity. Such conditions are generically called *constraint qualifications*. Important examples are those of Arrow, Hurwicz, and Uzawa [AHU61], Dubovitskii and Milyutin [DuM65], Abadie [Aba67], Mangasarian and Fromovitz [MaF67], and Guignard [Gui69]. For textbook treatments see Mangasarian [Man69] and Hestenes [Hes75]. There has been much subsequent work on the subject, some of which addresses nondifferentiable problems, e.g., Gould and Tolle [GoT71], [GoT72], Bazaraa, Goode, and Shetty [BGS72], Ben-Tal [Ben80], Ben-Tal and Zowe [BeZ82], Clarke [Cla83], Demjanov and Vasilév [DeV85], Mordukhovich [Mor88], and Rockafellar [Roc93]. Our treatment of Fritz John theory in Section 3.3.5, and its use in obtaining Lagrange multiplier theorems differs significantly from standard treatments. It is largely based on the work of Hestenes [Hes75], which has not received much attention thus far. Hestenes in turn relied on the penalty-based proof of the Karush-Kuhn-Tucker theorem given by McShane [McS73]. The material of Section 3.3.6 is also based on Hestenes [Hes75]. For further material on optimality conditions and computational methods for semi-infinite programming, see the articles in the edited volume by Reemtsen and Ruckman [ReR98], and the book by Polak [Pol97].

Section 3.4: Duality theory has its origins in the work of von Neuman on zero sum games. The proof of linear programming duality was given by Gale, Kuhn, and Tucker [GKT51]. Most of the research on duality has been done under the weaker assumptions and the geometrical framework of Chapter 5; see the references in Section 5.5.

4

Lagrange Multiplier Algorithms

Contents

In this chapter, we consider several computational methods for problems with equality and inequality constraints. All of these methods use some form of Lagrange multiplier estimates and typically provide in the limit not just stationary points of the original problem but also associated Lagrange multipliers. Additional methods using Lagrange multipliers will also be discussed in the next two chapters after the development of duality theory.

The methods of this chapter are based on one of the two following ideas:

(a) *Using a penalty or a barrier function.* Here a constrained problem is converted into a sequence of unconstrained problems, which involve an added high cost either for infeasibility or for approaching the boundary of the feasible region via its interior. These methods are discussed in Sections 4.1-4.3 and include interior point linear programming methods based on the logarithmic barrier function.

(b) *Solving the necessary optimality conditions*, viewing them as a system of equations and/or inequalities in the problem variables and the associated Lagrange multipliers. These methods, discussed in Section 4.4, guarantee only local convergence in their pure form; that is, they converge only when a good solution estimate is initially available. However, their convergence region can be enlarged by using various penalty and barrier functions.

Generally, the methods of this chapter are particularly well-suited for nonlinear constraints, because, contrary to the feasible direction methods of Chapter 2, they do not involve projections or direction finding subproblems, which tend to become more difficult when the constraints are nonlinear.

4.1 BARRIER AND INTERIOR POINT METHODS

Barrier methods apply to inequality constrained problems of the form

$$\begin{aligned}&\text{minimize } f(x)\\&\text{subject to } x \in X, \qquad g_j(x) \le 0, \quad j = 1, \ldots, r,\end{aligned} \tag{4.1}$$

where f and g_j are continuous real-valued functions, and X is a closed set. The interior (relative to X) of the set defined by the inequality constraints is

$$S = \{x \in X \mid g_j(x) < 0,\ j = 1, \ldots, r\}.$$

We assume that S is nonempty and that any feasible point that is not in S can be approached arbitrarily closely by a vector from S; that is, given any feasible x and any $\delta > 0$, there exists $\tilde{x} \in S$ such that $\|\tilde{x} - x\| \le \delta$. It can be seen, using the line segment principle [Prop. B.7(b) in Appendix

B], that the latter property holds automatically if the constraint functions g_j are convex.

In barrier methods, we add to the cost a function $B(x)$ that is defined in the interior set S. This function, called the *barrier function*, is continuous and goes to ∞ as any one of the constraints $g_j(x)$ approaches 0 from negative values. The two most common examples of barrier functions are:

$$B(x) = -\sum_{j=1}^{r} \ln\{-g_j(x)\}, \qquad \text{logarithmic,}$$

$$B(x) = -\sum_{j=1}^{r} \frac{1}{g_j(x)}, \qquad \text{inverse.}$$

Note that both of these barrier functions are convex if all the constraint functions g_j are convex. Figure 4.1.1 illustrates the form of $B(x)$.

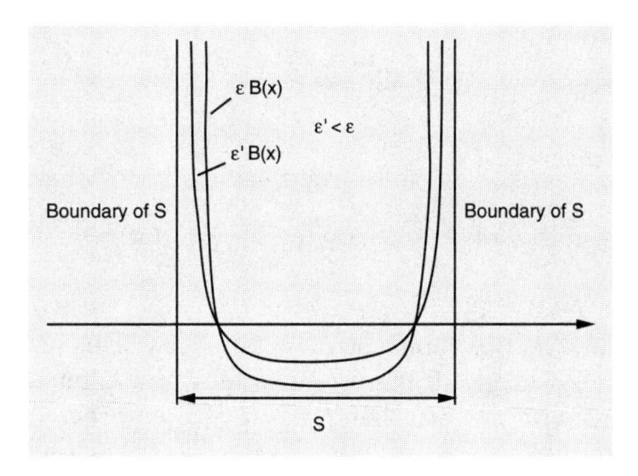

Figure 4.1.1. Form of a barrier function.

The barrier method is defined by introducing a parameter sequence $\{\epsilon^k\}$ with

$$0 < \epsilon^{k+1} < \epsilon^k, \quad k = 0, 1, \ldots, \qquad \epsilon^k \to 0.$$

It consists of finding

$$x^k = \arg\min_{x \in S} \{f(x) + \epsilon^k B(x)\}, \qquad k = 0, 1, \ldots$$

Since the barrier function is defined only on the interior set S, the successive iterates of any method used for this minimization must be interior points. If $X = \Re^n$, one may use unconstrained methods such as Newton's method with the stepsize properly selected to ensure that all iterates lie in S; an initial interior point can be obtained as discussed in Section 2.2. Note that the barrier term $\epsilon^k B(x)$ goes to zero for all interior points $x \in S$ as $\epsilon^k \to 0$. Thus the barrier term becomes increasingly inconsequential as far as interior points are concerned, while progressively allowing x^k to get closer to the boundary of S (as it should if the solutions of the original constrained problem lie on the boundary of S). Figure 4.1.2 illustrates the convergence process, and the following proposition gives the main convergence result.

Proposition 4.1.1: Every limit point of a sequence $\{x^k\}$ generated by a barrier method is a global minimum of the original constrained problem (4.1).

Proof: Let $\{\bar{x}\}$ be the limit of a subsequence $\{x^k\}_{k\in K}$. If $\bar{x} \in S$, we have $\lim_{k\to\infty,\, k\in K} \epsilon^k B(x^k) = 0$, while if $\bar{x}$ lies on the boundary of S, we have by assumption $\lim_{k\to\infty,\, k\in K} B(x^k) = \infty$. In either case we obtain

$$\liminf_{k\to\infty} \epsilon^k B(x^k) \geq 0,$$

which implies that

$$\liminf_{k\to\infty,\, k\in K} \{f(x^k) + \epsilon^k B(x^k)\} = f(\bar{x}) + \liminf_{k\to\infty,\, k\in K} \{\epsilon^k B(x^k)\} \geq f(\bar{x}). \quad (4.2)$$

The vector $\bar{x}$ is a feasible point of the original problem (4.1), since $x^k \in S$ and X is a closed set. If $\bar{x}$ were not a global minimum, there would exist a feasible vector x^* such that $f(x^*) < f(\bar{x})$ and therefore also (using our assumption that x^* can be approached arbitrarily closely through the interior set S) an interior point $\tilde{x} \in S$ such that $f(\tilde{x}) < f(\bar{x})$. We now have by the definition of x^k,

$$f(x^k) + \epsilon^k B(x^k) \leq f(\tilde{x}) + \epsilon^k B(\tilde{x}), \qquad k = 0, 1, \ldots,$$

which by taking the limit as $k \to \infty$ and $k \in K$, implies together with Eq. (4.2), that $f(\bar{x}) \leq f(\tilde{x})$. This is a contradiction, thereby proving that $\bar{x}$ is a global minimum of the original problem. **Q.E.D.**

The logarithmic barrier function has been central to much research on methods that generate successive iterates lying in the interior set S. These methods are generically referred to as *interior point methods*, and have been extensively applied to linear and quadratic programming problems following the work of Karmarkar [Kar84]. We proceed to discuss the linear programming case in detail.

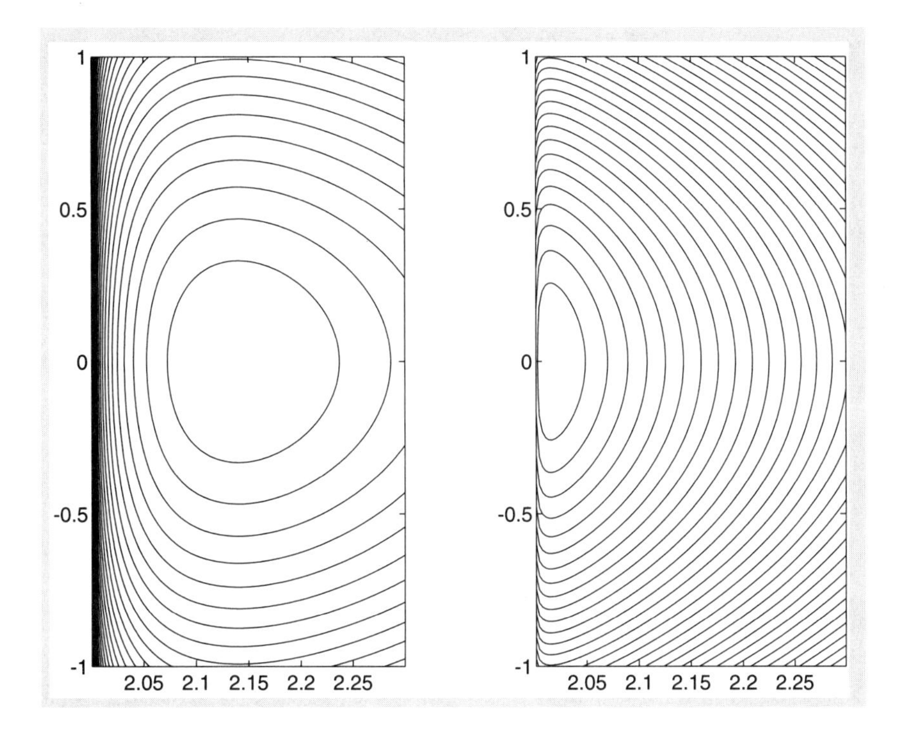

Figure 4.1.2. The convergence process of the barrier method for the two-dimensional problem

$$\begin{aligned}&\text{minimize} \ \ f(x) = \tfrac{1}{2}\big(x_1^2 + x_2^2\big)\\ &\text{subject to} \ \ 2 \leq x_1,\end{aligned}$$

with optimal solution $x^* = (2,0)$. For the case of the logarithmic barrier function $B(x) = -\ln(x_1 - 2)$, we have

$$x^k = \arg\min_{x_1 > 2}\left\{\tfrac{1}{2}\big(x_1^2 + x_2^2\big) - \epsilon^k \ln(x_1 - 2)\right\} = \left(1 + \sqrt{1 + \epsilon^k}\,, 0\right),$$

so as ϵ^k is decreased, the unconstrained minimum x^k approaches the constrained minimum $x^* = (2,0)$. The figure shows the equal cost surfaces of $f(x) + \epsilon B(x)$ for $\epsilon = 0.3$ (left side) and $\epsilon = 0.03$ (right side).

4.1.1 Linear Programming and the Logarithmic Barrier*

Let us apply the logarithmic barrier method to the linear programming problem

$$\begin{aligned}&\text{minimize} \ \ c'x\\ &\text{subject to} \ \ Ax = b, \qquad x \geq 0,\end{aligned} \tag{LP}$$

where $c \in \Re^n$ and $b \in \Re^m$ are given vectors, and A is an $m \times n$ matrix of rank m. We assume that the problem has at least one optimal solution.

From the theory of Section 3.4.2, we have that the dual problem, given by

$$\begin{aligned} &\text{maximize } \ b'\lambda \\ &\text{subject to } \ A'\lambda \le c, \end{aligned} \tag{DP}$$

also has an optimal solution. Furthermore, the optimal values of the primal and the dual problem are equal.

The method involves finding for various $\epsilon > 0$,

$$x(\epsilon) = \arg\min_{x \in S} F_\epsilon(x), \tag{4.3}$$

where

$$F_\epsilon(x) = c'x - \epsilon \sum_{i=1}^{n} \ln x_i,$$

and S is the interior set

$$S = \{x \mid Ax = b,\, x > 0\},$$

where $x > 0$ means that all the coordinates of x are strictly positive. We assume that S is nonempty and bounded. Since $-\ln x_i$ grows to ∞ as $x_i \to 0$, this assumption can be used together with Weierstrass' theorem (Prop. A.8 in Appendix A) to show that there exists at least one global minimum of $F_\epsilon(x)$ over S, which must be unique because the function F_ϵ can be seen to be strictly convex. Therefore, for each $\epsilon > 0$, $x(\epsilon)$ is uniquely defined by Eq. (4.3).

The Central Path

For given A, b, and c, as ϵ is reduced towards 0, $x(\epsilon)$ follows a trajectory that is known as the *central path*. Figure 4.1.3 illustrates the central path for various values of the cost vector c. Note the following:

(a) For fixed A and b, the central paths corresponding to different cost vectors c start at the same vector x_∞. This is the unique minimizing point over S of

$$-\sum_{i=1}^{n} \ln x_i,$$

corresponding to $\epsilon = \infty$, and is known as the *analytic center* of S.

(b) If c is such that (LP) has a unique optimal solution x^*, the central path ends at x^* [that is, $\lim_{\epsilon \to 0} x(\epsilon) = x^*$]. This follows from Prop. 4.1.1, which implies that for every sequence $\{\epsilon^k\}$ with $\epsilon^k \to 0$, the corresponding sequence $\{x(\epsilon^k)\}$ converges to x^*.

(c) If c is such that (LP) has multiple optimal solutions, it can be shown that the central path ends at one of the optimal solutions [that is,

$\lim_{\epsilon \to 0} x(\epsilon)$ exists and is equal to some optimal solution of (LP)]. We will not prove this fact. For a heuristic explanation note that from the definition of $x(\epsilon)$ we have

$$x(\epsilon) = \arg\min_{x \in S_\epsilon} \left\{ -\sum_{i=1}^{n} \ln x_i \right\},$$

where S_ϵ is the "slice" of the interior set S of points with the same cost value as $x(\epsilon)$, that is,

$$S_\epsilon = \big\{x \mid Ax = b,\ c'x = c'x(\epsilon),\ x > 0\big\}.$$

Thus $x(\epsilon)$ is the analytic center of S_ϵ. Since S_ϵ "converges" to the set of optimal solutions X^* of the linear program (LP) as $\epsilon \to 0$, it is natural to conclude that $x(\epsilon)$ must converge to the analytic center of the "interior" of X^*; see Fig. 4.1.4. (This argument can be made rigorous with a more precise definition of the analytic center of a polyhedral set; see the references.)

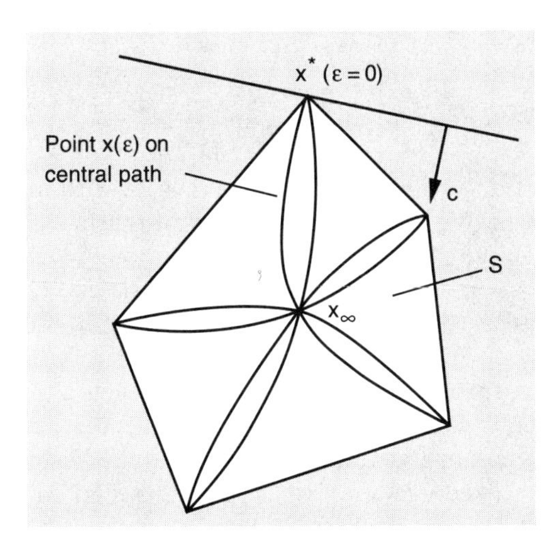

Figure 4.1.3. Central paths corresponding to ten different values of the cost vector c. All central paths start at the same vector, the *analytic center* x_∞, which corresponds to $\epsilon = \infty$,

$$x_\infty = \arg\min_{x \in S} \left\{ -\sum_{i=1}^{n} \ln x_i \right\},$$

and end at optimal solutions of (LP).

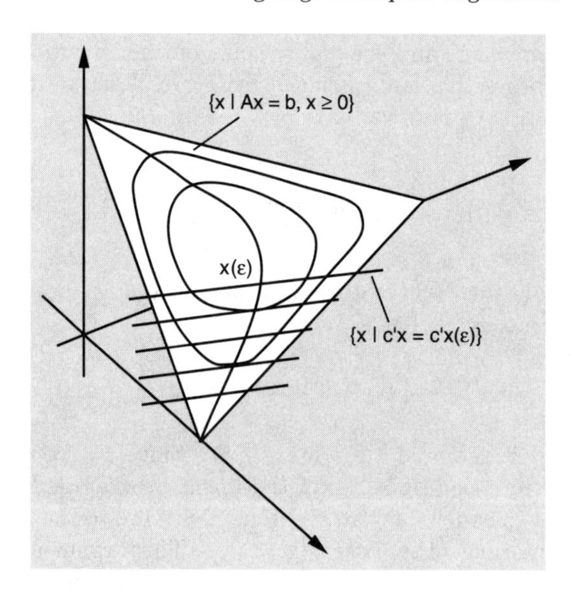

Figure 4.1.4. Illustration of how the central path $x(\epsilon)$ converges to an optimal solution as $\epsilon \to 0$. Each point $x(\epsilon)$ is the analytic center of the slice S_ϵ of the interior set S of points with the same cost value as $x(\epsilon)$.

Following Approximately the Central Path

The most straightforward way to implement the logarithmic barrier method is to use some iterative algorithm to minimize the function F_{ϵ^k} for some decreasing sequence $\{\epsilon^k\}$ with $\epsilon^k \to 0$. This is equivalent to finding a sequence $\{x(\epsilon^k)\}$ of points on the central path. However, this approach is inefficient because it requires an infinite number of iterations to compute each point $x(\epsilon^k)$.

It turns out that a far more efficient approach is possible, whereby each minimization is done approximately through a few iterations (possibly only one) of the constrained version of Newton's method that was given in Section 2.3. For a fixed ϵ and a given $x \in S$, this method replaces x by

$$\tilde{x} = x + \alpha(\bar{x} - x),$$

where $\bar{x}$ is the pure Newton iterate defined as the optimal solution of the quadratic program in the vector $z \in \Re^n$

$$\begin{aligned} &\text{minimize} \quad \nabla F_\epsilon(x)'(z - x) + \tfrac{1}{2}(z - x)'\nabla^2 F_\epsilon(x)(z - x) \\ &\text{subject to} \quad Az = b \end{aligned} \tag{4.4}$$

and α is a stepsize selected by some rule. We have

$$\nabla F_\epsilon(x) = c - \epsilon x^{-1}, \qquad \nabla^2 F_\epsilon(x) = \epsilon X^{-2},$$

where x^{-1} denotes the vector with coordinates $(x_i)^{-1}$ and X denotes the diagonal matrix with the coordinates x_i along the diagonal:

$$X = \begin{pmatrix} x_1 & 0 & \cdots & 0 \\ 0 & x_2 & \cdots & 0 \\ \cdots & \cdots & \cdots & \cdots \\ 0 & 0 & \cdots & x_n \end{pmatrix}.$$

We can obtain an expression for $\bar{x}$ by using the formula for the projection on a linear manifold given in Section 2.1, but we can also obtain the same expression by observing that the quadratic program (4.4) is the same as the one used in the affine scaling method of Section 2.6, except that c is replaced by $c - \epsilon x^{-1}$, and the stepsize s^k is taken to be $1/\epsilon$. By using the affine scaling formulas obtained in Section 2.6, we have that the pure Newton iterate is

$$\bar{x} = x - \epsilon^{-1} X^2 \left(c - \epsilon x^{-1} - A'\lambda\right), \tag{4.5}$$

where

$$\lambda = (AX^2A')^{-1}AX^2 \left(c - \epsilon x^{-1}\right). \tag{4.6}$$

These formulas can also be written as

$$\bar{x} = x - Xq(x, \epsilon), \tag{4.7}$$

where

$$q(x, \epsilon) = \frac{Xz}{\epsilon} - e, \tag{4.8}$$

with e and z being the vectors

$$e = \begin{pmatrix} 1 \\ \vdots \\ 1 \end{pmatrix}, \qquad z = c - A'\lambda, \tag{4.9}$$

and

$$\lambda = (AX^2A')^{-1}AX\big(Xc - \epsilon e\big). \tag{4.10}$$

Based on Eq. (4.7), we have $q(x, \epsilon) = X^{-1}(\bar{x} - x)$, so we may view the vector $q(x, \epsilon)$ as a transformed version of the Newton increment $(x - \bar{x})$ using the transformation matrix X^{-1} that maps the vector x into the vector e. Since $\bar{x}$ is the Newton step approximation to $x(\epsilon)$, we can consider $\|q(x, \epsilon)\|$ as a *a measure of proximity* of the current point x to the point $x(\epsilon)$ on the central path. In particular, it can be seen that we have $q(x, \epsilon) = 0$ if and only if $x = x(\epsilon)$.

The key result to be shown shortly is that for convergence of the logarithmic barrier method, it is sufficient to stop the minimization of F_{ϵ^k} and decrease ϵ^k to ϵ^{k+1} once the current iterate x^k satisfies

$$\|q(x^k, \epsilon^k)\| < 1.$$

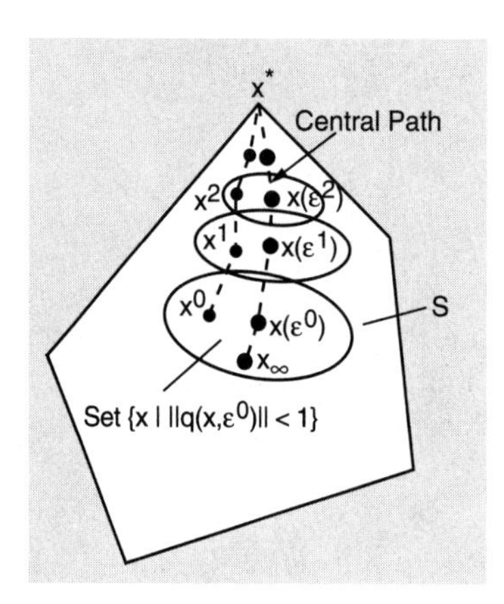

Figure 4.1.5. Following approximately the central path. For each ϵ^k, it is sufficient to carry out the minimization of F_{ϵ^k} up to the point where $\|q(x, \epsilon^k)\| < 1$.

Another way to phrase this result is that if a sequence of pairs $\{(x^k, \epsilon^k)\}$ satisfies

$$\|q(x^k, \epsilon^k)\| < 1, \qquad 0 < \epsilon^{k+1} < \epsilon^k, \quad k = 0, 1, \ldots, \qquad \epsilon^k \to 0,$$

then every limit point of $\{x^k\}$ is an optimal solution of (LP); see Fig. 4.1.5. The following proposition establishes this result.

Proposition 4.1.2: If $x > 0$, $Ax = b$, and $\|q(x, \epsilon)\| < 1$, then

$$c'x - f^* \leq c'x - b'\lambda \leq \epsilon\big(n + \|q(x, \epsilon)\|\sqrt{n}\big) \leq \epsilon\big(n + \sqrt{n}\big), \tag{4.11}$$

where λ is given by Eq. (4.10), and f^* is the optimal value of (LP), that is,

$$f^* = \min_{Ay=b,\, y \geq 0} c'y.$$

Proof: Using the definition (4.8)-(4.10) of q, we can write the hypothesis $\|q(x, \epsilon)\| < 1$ as

$$\left\| \frac{X(c - A'\lambda)}{\epsilon} - e \right\| < 1. \tag{4.12}$$

Thus the coordinates of $\big(X(c - A'\lambda)/\epsilon\big) - e$ must lie between -1 and 1, implying that the coordinates of $X(c - A'\lambda)$ are positive. Since the diagonal elements of X are positive, it follows that the coordinates of $c - A'\lambda$ are also positive. Hence $c \geq A'\lambda$, and for any optimal solution x^* of (LP), we obtain (using the fact $x^* \geq 0$)

$$f^* = c'x^* \geq \lambda' A x^* = \lambda' b. \tag{4.13}$$

On the other hand, since $||e|| = \sqrt{n}$, we have using Eq. (4.12),

$$e'\left(\frac{X(c-A'\lambda)}{\epsilon} - e\right) \leq ||e|| \left\| \frac{X(c-A'\lambda)}{\epsilon} - e \right\| = \sqrt{n}\,\|q(x,\epsilon)\| \leq \sqrt{n}, \tag{4.14}$$

and by using also Eq. (4.13),

$$e'\left(\frac{X(c-A'\lambda)}{\epsilon} - e\right) = \frac{x'(c-A'\lambda)}{\epsilon} - n = \frac{c'x - b'\lambda}{\epsilon} - n \geq \frac{c'x - f^*}{\epsilon} - n. \tag{4.15}$$

By combining Eqs. (4.14) and (4.15), the result follows. **Q.E.D.**

Note that from Eq. (4.11), $c'x - b'\lambda$ provides a readily computable upper bound to the (unknown) cost error $c'x - f^*$. What is happening here is that x and λ are feasible solutions to the primal and dual problems (LP) and (DP), respectively, and the common optimal value f^* lies between the corresponding primal and dual costs $c'x$ and $b'\lambda$.

Path-Following by Using Newton's Method

Since in order to implement the termination criterion $\|q(x,\epsilon)\| < 1$, we must calculate the pure Newton iterate $\bar{x} = x - Xq(x,\epsilon)$, it is natural to use a convergent version of Newton's method for approximate minimization of F_ϵ. This method replaces x by

$$\tilde{x} = x + \alpha(\bar{x} - x),$$

where α is a stepsize selected by the minimization rule or the Armijo rule (with unit initial stepsize) over the range of positive stepsizes such that $\tilde{x}$ is an interior point. We expect that for x sufficiently close to $x(\epsilon)$, the stepsize α can be taken equal to 1, so that the pure form of the method is used and a quadratic rate of convergence is obtained. The following proposition shows that the "termination set" $\big\{x \mid \|q(x,\epsilon)\| < 1\big\}$ is part of the region of quadratic convergence of the pure form of Newton's method.

Proposition 4.1.3: If $x > 0$, $Ax = b$, and $\|q(x,\epsilon)\| < 1$, then the pure Newton iterate $\bar{x} = x - Xq(x,\epsilon)$ is an interior point, i.e., $\bar{x} \in S$. Furthermore, we have $\|q(\bar{x},\epsilon)\| < 1$ and in fact

$$\|q(\bar{x},\epsilon)\| \leq \|q(x,\epsilon)\|^2. \tag{4.16}$$

Proof: Let us define $p = Xz/\epsilon = X(c - A'\lambda)/\epsilon$, so that $q(x,\epsilon) = p - e$ [cf. Eqs. (4.7) and (4.8)]. Since $\|p - e\| < 1$, we see that the coordinates of p satisfy $0 < p_i < 2$ for all i. We have $\bar{x} = x - X(p - e)$, so that $\bar{x}_i = (2 - p_i)x_i > 0$ for all i, and since also $A\bar{x} = b$, it follows that $\bar{x}$ is an interior point.

It can be shown (Exercise 4.1.3) that the vector $\bar{\lambda}$ corresponding to $\bar{x}$ in the manner of Eq. (4.10) satisfies

$$\bar{\lambda} = \arg\min_{\xi \in \Re^m} \left\| \frac{\bar{X}(c - A'\xi)}{\epsilon} - e \right\|,$$

where $\bar{X}$ is the diagonal matrix with $\bar{x}_i$ along the diagonal. Hence,

$$\|q(\bar{x}, \epsilon)\| = \left\| \frac{\bar{X}(c - A'\bar{\lambda})}{\epsilon} - e \right\| \leq \left\| \frac{\bar{X}(c - A'\lambda)}{\epsilon} - e \right\| = \|\bar{X}X^{-1}p - e\|.$$

Since $\bar{x} = 2x - Xp$, we have

$$\bar{X}X^{-1}p = (2X - XP)X^{-1}p = 2p - Pp,$$

where P is the diagonal matrix with p_i along the diagonal. The last two relations yield

$$\begin{aligned} \|q(\bar{x}, \epsilon)\|^2 &\leq \|2p - Pp - e\|^2 \leq \sum_{i-1}^{n} (2p_i - p_i^2 - 1)^2 = \sum_{i=1}^{n} (p_i - 1)^4 \\ &\leq \left(\sum_{i=1}^{n} (p_i - 1)^2 \right)^2 = \|p - e\|^4 = \|q(x, \epsilon)\|^4. \end{aligned}$$

This proves the result. **Q.E.D.**

The preceding proposition shows that if $\|q(x, \epsilon)\|$ is substantially less than 1, then a single pure Newton step, changing x to $\bar{x}$, reduces $\|q(x, \epsilon)\|$ by a substantial factor [cf. Eq. (4.16)]. Thus, we expect that if $\bar{\epsilon}$ is not much smaller than ϵ and $\|q(x, \epsilon)\|$ is substantially less than 1, then $\|q(\bar{x}, \bar{\epsilon})\|$ will also be substantially less than 1. This means that by carefully selecting the ϵ-reduction factor $\epsilon^{k+1}/\epsilon^k$ in combination with an appropriately small termination tolerance for the first minimization of F_ϵ $(k = 0)$, we can execute all the subsequent approximate minimizations of F_{ϵ^k} $(k \geq 1)$ in a single pure Newton step; see Fig. 4.1.6. One possibility is, given ϵ^k and x^k such that $\|q(x^k, \epsilon^k)\| \leq 1$, to obtain x^{k+1} by a single Newton step and then to select ϵ^{k+1} so that $\|q(x^{k+1}, \epsilon^{k+1})\|$ is minimized. This minimization can be done in closed form because $\|q(x, \epsilon)\|$ is quadratic in $1/\epsilon$ [cf. Eq. (4.8)]. Another possibility is shown in the next proposition.

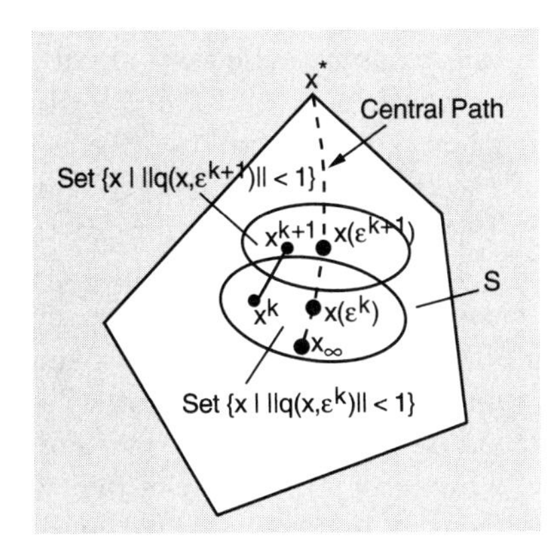

Figure 4.1.6. Following approximately the central path by using a single Newton step for each ϵ^k. If ϵ^k is close to ϵ^{k+1} and x^k is close to the central path, one expects that x^{k+1} obtained from x^k by a single pure Newton step will also be close to the central path.

Proposition 4.1.4: Suppose that $x > 0$, $Ax = b$, and that $\|q(x,\epsilon)\| \leq \gamma$ for some $\gamma < 1$. For any $\delta \in (0, n^{1/2})$, let $\bar{\epsilon} = (1 - \delta n^{-1/2})\epsilon$. Then

$$\|q(\bar{x}, \bar{\epsilon})\| \leq \frac{\gamma^2 + \delta}{1 - \delta n^{-1/2}}.$$

In particular, if

$$\delta \leq \frac{\gamma(1-\gamma)}{1+\gamma}, \tag{4.17}$$

we have $\|q(\bar{x}, \bar{\epsilon})\| \leq \gamma$.

Proof: Let $\theta = \delta n^{-1/2}$. We have using Eq. (4.8)

$$q(\bar{x}, \bar{\epsilon}) = \frac{\bar{X}z}{\bar{\epsilon}} - e = \frac{\bar{X}z}{(1-\theta)\epsilon} - e = \frac{q(\bar{x}, \epsilon) + e}{1-\theta} - e = \frac{1}{1-\theta}\big(q(\bar{x}, \epsilon) + \theta e\big).$$

Thus, using also Eq. (4.16),

$$\begin{aligned} \|q(\bar{x}, \bar{\epsilon})\| &\leq \frac{1}{1-\theta}\big(\|q(\bar{x}, \epsilon)\| + \theta\|e\|\big) \\ &= \frac{1}{1-\theta}\big(\|q(\bar{x}, \epsilon)\| + \theta n^{1/2}\big) \\ &\leq \frac{1}{1-\theta}\big(\|q(x, \epsilon)\|^2 + \delta\big) \\ &\leq \frac{\gamma^2 + \delta}{1-\theta}. \end{aligned}$$

Finally, Eq. (4.17) can be written as $(\gamma^2 + \delta)/(1 - \delta) \leq \gamma$, which, in combination with the relation just proved, implies that $\|q(\bar{x}, \bar{\epsilon})\| \leq \gamma$. **Q.E.D.**

Note that in the preceding proposition one can maintain x very close to the central path ($\gamma << 1$) provided one takes δ to be very small [cf. Eq. (4.17)], or equivalently, one uses an ϵ-reduction factor $1 - \delta n^{-1/2}$ that is very close to 1. Unfortunately, even when γ is close to 1, in order to guarantee the single-step attainment of the tolerance $\|q(x, \epsilon)\| < \gamma$, it is still necessary to decrease ϵ very slowly. In particular, since we must take $\delta < 1$ in order for $\|q(\bar{x}, \bar{\epsilon})\| < \gamma$ [cf. Eq. (4.17)], the reduction factor $\bar{\epsilon}/\epsilon$ must exceed $1 - n^{-1/2}$, which is very close to 1. This means that, even though each approximate minimization after the first will require a single Newton step, a very large number of approximate minimizations will be needed to attain an acceptable accuracy. Thus, it may be more efficient in practice to decrease ϵ^k at a faster rate, while accepting the possibility of multiple Newton steps before switching from ϵ^k to ϵ^{k+1}, as illustrated in Fig. 4.1.7.

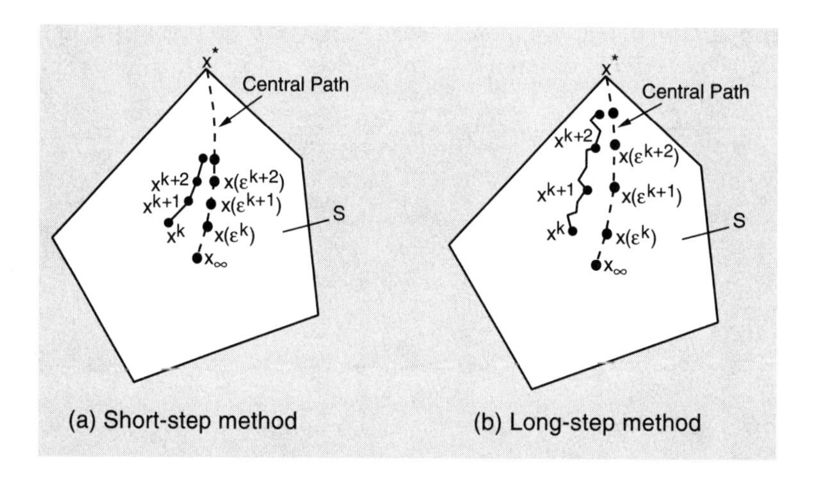

Figure 4.1.7. Following approximately the central path by decreasing ϵ^k slowly as in (a) or quickly as in (b). In (a) a single Newton step is required in each approximate minimization at the expense of a large number of approximate minimizations.

The preceding results form the basis for worst-case estimates of the number of Newton iterations required to reduce the error $c'x^k - f^*$ below some tolerance, where x^k is obtained by approximate minimization of F_{ϵ^k} using different termination criteria and reduction factors $\epsilon^{k+1}/\epsilon^k$. Exercise 4.1.5 provides a sample of this type of analysis.

We would like to caution the reader, however, that while many researchers consider a low estimate of number of iterations a good indicator of algorithmic performance, the worst-case estimates that have been obtained for interior point methods are so unrealistically high that they are entirely meaningless if taken literally. One may hope that these estimates are meaningful in a comparative sense, i.e., the practical performance of two algorithms would compare consistently with the corresponding worst-case estimates of required numbers of iterations. Unfortunately, this does not turn out to be true in practice. Thus one must view these worst-case analyses of the required number of iterations with skepticism; they should be primarily considered as an analytical vehicle for understanding better the corresponding methods. As an indication of the unreliability of the complexity-based conclusions, we note that the lowest estimates of the required number of iterations have been obtained for the so-called *short-step methods*, where ϵ^k is reduced very slowly so that the corresponding approximate minimization can be done in a single Newton step (cf. Prop. 4.1.4). The best practical performance, however, has been obtained with the so-called *long-step methods*, where ϵ^k is reduced at a much faster rate.

Decomposition of the Newton Direction

There is an interesting interpretation of the Newton direction for minimizing the barrier function F_ϵ. Using Eqs. (4.5) and (4.6), we can write this direction as

$$d_N = -\epsilon^{-1}X^2\big(c - \epsilon x^{-1} - A'(\lambda_a - \epsilon\lambda_c)\big),$$

where

$$\lambda_a = (AX^2A')^{-1}AX^2c,$$

$$\lambda_c = (AX^2A')^{-1}AXe.$$

Equivalently, we have

$$d_N = \epsilon^{-1}d_a + d_c, \tag{4.18}$$

where

$$d_a = -X^2(c - A\lambda_a),$$

$$d_c = -Xe + X^2A'\lambda_c. \tag{4.19}$$

It can be seen that d_a is the *affine scaling direction* defined in Section 2.6. We claim that d_c is the Newton direction for the problem

$$\begin{aligned} &\text{minimize} \quad -\sum_{i=1}^{n} \ln x_i \\ &\text{subject to} \quad x \in S, \end{aligned}$$

whose unique solution is the analytic center x_∞ of S; this is called the *centering direction*. Indeed this Newton direction is the one that solves the quadratic program

$$\begin{aligned} &\text{minimize} \quad \nabla C(x)'d + \tfrac{1}{2}d'\nabla^2 C(x)d \\ &\text{subject to} \quad Ad = 0, \end{aligned}$$

where C is the function

$$C(x) = -\sum_{i=1}^{n} \ln x_i.$$

Using the results of Example 2.1.5 in Section 2.1, the optimal solution can be written as

$$-\nabla^2 C(x)^{-1}\left(\nabla C(x) - A'\big(A\nabla^2 C(x)^{-1}A'\big)^{-1}A\nabla^2 C(x)^{-1}\nabla C(x)\right).$$

By making the substitutions

$$\nabla C(x) = x^{-1}, \qquad \nabla^2 C(x) = X^{-2},$$

it is straightforward to verify that this solution is equal to the direction d_c of Eq. (4.19).

We thus conclude from Eq. (4.18) that the barrier/Newton direction d_N can be decomposed into an affine scaling component d_a that aims at reducing the linear cost $c'x$, and a centering component d_c that aims at bringing the next iterate near the analytic center x_∞, thereby keeping it away from the constraint boundary; see Fig. 4.1.8. Note that *the centering component does not depend at all on the cost vector* c and that none of the two components depends on the barrier parameter ϵ. The role of ϵ is to control the relative weight of the centering component: as $\epsilon \to 0$, the affine scaling component dominates, and when $\epsilon = 0$ the Newton/barrier iteration coincides with the affine scaling iteration.

Many of the interior point methods that have been proposed use directions that combine the affine scaling direction and the centering direction with some weights. Any such method can be viewed as just the Newton/barrier method with a specific rule for choosing the relative weight parameter ϵ at each iteration. For further discussion of alternative methods and the corresponding practical implementation issues, we refer to the sources at the end of the chapter.

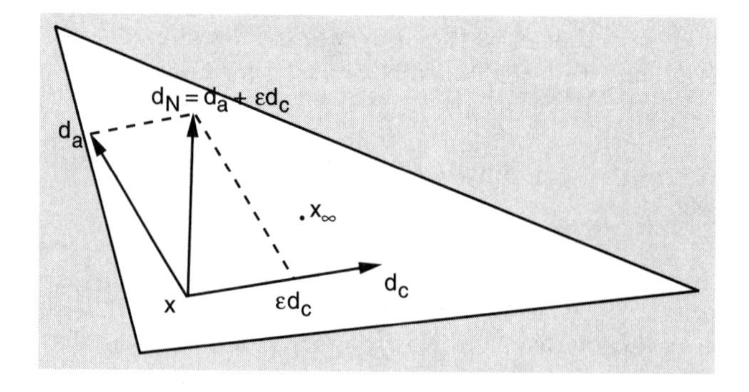

Figure 4.1.8. Decomposition of the Newton direction d_N into an affine scaling component d_a and a centering component d_c. The centering component is the Newton direction aimed at finding the analytic center x_∞.

Quadratic and Convex Programming

The logarithmic barrier method in conjunction with Newton's method can also be fruitfully applied to the convex programming problem

$$\begin{aligned} &\text{minimize } f(x) \\ &\text{subject to } Ax = b, \qquad x \geq 0, \end{aligned}$$

where $f : \Re^n \mapsto \Re$ is a convex function. The implementation of the method benefits from the extensive experience that has been accumulated from the linear programming case. For the special case of the quadratic programming problem

$$\begin{aligned} &\text{minimize } c'x + \tfrac{1}{2}x'Qx \\ &\text{subject to } Ax = b, \qquad x \geq 0, \end{aligned}$$

the performance and the analysis of the method are similar to that for linear programs. We refer to the references at the end of this chapter for a detailed treatment.

EXERCISES

4.1.1

Consider the linear program

$$\begin{aligned} &\text{minimize } x_1 + 2x_2 + 3x_3 \\ &\text{subject to } x_1 + x_2 + x_3 = 1, \qquad x \geq 0. \end{aligned}$$

Sketch on paper the central path. Write a computer program to implement a short-step and a long-step path-following method based on Newton's method for this problem. Compare the number of Newton steps for a given solution accuracy for the starting points $x^0 = (.8, .15, .05)$ and $x^0 = (.1, .2, .7)$.

4.1.2

Given x, show how to find an $\epsilon > 0$ that minimizes $\|q(x, \epsilon)\|$ [cf. Eq. (4.8)]. How would you use this idea to accelerate convergence in a short-step path-following method?

4.1.3

Show that the vector λ of Eq. (4.10) satisfies

$$\lambda = \arg\min_{\xi \in \Re^m} \left\| \frac{X(c - A'\xi)}{\epsilon} - e \right\|.$$

4.1.4

Let $\delta = \|q(x,\epsilon)\|$, where $q(x,\epsilon)$ is the scaled Newton step defined by Eq. (4.8), and assume that $\delta < 1$. Show that we have

$$\|X^{-1}\big(x - x(\epsilon)\big)\| \leq \frac{\delta}{1-\delta},$$

$$F_\epsilon(x) - F_\epsilon\big(x(\epsilon)\big) \leq \frac{\delta^2}{1-\delta^2},$$

$$|c'x - c'x(\epsilon)| \leq \frac{\delta(1+\delta)\epsilon}{1-\delta}\sqrt{n}.$$

4.1.5 (Complexity of the Short-Step Method)

The purpose of this exercise is to show that the number of iterations required by the short-step logarithmic barrier method to achieve a given accuracy is proportional to $\sqrt{n}$. Consider the linear programming problem (LP), a vector $x^0 \in S$, and a sequence $\{\epsilon^k\}$ such that $\|q(x^0,\epsilon^0)\| \leq 1/2$ and $\epsilon^{k+1} = (1-\theta)\epsilon^k$, where $\theta = 1/(6n^{1/2})$ (cf. Prop. 4.1.4). Let x^{k+1} be generated from x^k by a pure Newton step aimed at minimizing F_{ϵ^k}. For a given integer r, let $\bar{k}$ be the smallest integer k such that $-\ln(n\epsilon^k) \geq r$ and let $r^0 = -\ln(n\epsilon^0)$. Show that $\bar{k} \leq 6(r - r^0)\sqrt{n}$ and

$$c'x^{\bar{k}} - f^* \leq \frac{3}{2}e^{-r}.$$

Note: We have assumed here that a vector x^0 with $\|q(x^0,\epsilon^0)\| \leq 1/2$ is available. It is possible to show that such a point can be found in a number of Newton steps that is proportional to $\sqrt{n}$.

4.1.6

(a) Show that if the interior set $\{x \mid Ax = b,\ x > 0\}$ of (LP) is nonempty, then it is bounded if and only if the interior set $\{\lambda \mid A'\lambda < c\}$ of the dual problem (DP) is nonempty.

(b) Show that if the interior set $\{\lambda \mid A'\lambda < c\}$ of (DP) is nonempty, then it is bounded if and only if the interior set $\{x \mid Ax = b,\ x > 0\}$ of (LP) is nonempty.

4.1.7 (The Dual Problem as an Equality Constrained Problem)

Consider the dual problem

$$\begin{aligned} &\text{maximize} \;\; b'\lambda \\ &\text{subject to} \;\; A'\lambda \leq c \end{aligned} \tag{DP}$$

and its equivalent version

$$\begin{aligned} &\text{maximize} \;\; b'\lambda \\ &\text{subject to} \;\; A'\lambda + z = c, \quad z \geq 0, \end{aligned}$$

that involves the vector of additional variables z. Let P_A be the matrix that projects a vector x onto the nullspace of the matrix A, and note that using the analysis of Example 2.1.5 in Section 2.1, we have

$$P_A = I - A'(AA')^{-1}A.$$

Show that the dual linear program (DP) is equivalent to the linear program

$$\begin{aligned} &\text{minimize} \;\; \bar{x}'z \\ &\text{subject to} \;\; P_A z = P_A c, \quad z \geq 0, \end{aligned} \tag{4.20}$$

where $\bar{x}$ is any primal feasible vector, that is, $A\bar{x} = b$, $\bar{x} \geq 0$.

4.1.8 (Dual Central Path)

Consider the dual problem (DP). Using its equivalent reformulation (4.20) of Exercise 4.1.7, it is seen that the appropriate definition of the central path of the dual problem is

$$z(\epsilon) = \arg \min_{P_A z = P_A c,\, z > 0} \left\{ \bar{x}'z - \sum_{i=1}^{n} \ln z_i \right\},$$

where $\bar{x}$ is any primal feasible vector. Show that the primal and dual central paths are related by

$$z(\epsilon) = \epsilon x(\epsilon)^{-1},$$

and that the corresponding duality gap satisfies

$$c'x(\epsilon) - b'\lambda(\epsilon) = n\epsilon,$$

where $\lambda(\epsilon)$ is any vector such that $A'\lambda(\epsilon) = c - z(\epsilon)$.

4.2 PENALTY AND AUGMENTED LAGRANGIAN METHODS

The basic idea in penalty methods is to eliminate some or all of the constraints and add to the cost function a penalty term that prescribes a high cost to infeasible points. Associated with these methods is a penalty parameter c that determines the severity of the penalty and as a consequence, the extent to which the resulting unconstrained problem approximates the original constrained problem. As c takes higher values, the approximation becomes increasingly accurate. We focus attention primarily on the popular quadratic penalty function. Some other penalty functions, including the exponential, are discussed in Section 4.2.5.

Consider first the equality constrained problem

$$\begin{aligned} &\text{minimize} \quad f(x) \\ &\text{subject to} \quad h(x) = 0, \qquad x \in X, \end{aligned} \tag{4.21}$$

where $f : \Re^n \mapsto \Re$, $h : \Re^n \mapsto \Re^m$ are given functions, and X is a given subset of $\Re^n$.

Much of our analysis in this section will focus on the case where $X = \Re^n$, and x^* together with a Lagrange multiplier vector λ^* satisfies the sufficient optimality conditions of Prop. 3.2.1. At the center of our development is the *augmented Lagrangian function* $L_c : \Re^n \times \Re^m \mapsto \Re$, introduced in Section 3.2 and given by

$$L_c(x, \lambda) = f(x) + \lambda' h(x) + \frac{c}{2}\|h(x)\|^2,$$

where c is a positive penalty parameter.

There are two mechanisms by which unconstrained minimization of $L_c(\cdot, \lambda)$ can yield points close to x^*:

(a) *By taking λ close to λ^*.* Indeed, as shown in Section 3.2.1, if c is higher than a certain threshold, then for some $\gamma > 0$ and $\epsilon > 0$ we have

$$L_c(x, \lambda^*) \geq L_c(x^*, \lambda^*) + \frac{\gamma}{2}\|x - x^*\|^2, \qquad \forall\, x \text{ with } \|x - x^*\| < \epsilon,$$

so that x^* is a strict local minimum of the augmented Lagrangian $L_c(\cdot, \lambda^*)$ corresponding to λ^*. This suggests that if λ is close to λ^*, a good approximation to x^* can be found by unconstrained minimization of $L_c(\cdot, \lambda)$.

(b) *By taking c very large.* Indeed for high c, there is high cost for infeasibility, so the unconstrained minima of $L_c(\cdot, \lambda)$ will be nearly feasible. Since $L_c(x, \lambda) = f(x)$ for feasible x, we expect that $L_c(x, \lambda) \approx f(x)$ for nearly feasible x. Therefore, we can also expect to obtain a good approximation to x^* by unconstrained minimization of $L_c(\cdot, \lambda)$ when c is high.

Example 4.2.1

Consider the two-dimensional problem

$$\begin{aligned} &\text{minimize} \;\; f(x) = \tfrac{1}{2}(x_1^2 + x_2^2) \\ &\text{subject to} \;\; x_1 = 1, \end{aligned}$$

with optimal solution $x^* = (1, 0)$ and corresponding Lagrange multiplier $\lambda^* = -1$. The augmented Lagrangian is

$$L_c(x, \lambda) = \tfrac{1}{2}(x_1^2 + x_2^2) + \lambda(x_1 - 1) + \frac{c}{2}(x_1 - 1)^2,$$

and by setting its gradient to zero we can verify that its unique unconstrained minimum $x(\lambda, c)$ has coordinates given by

$$x_1(\lambda, c) = \frac{c - \lambda}{c + 1}, \qquad x_2(\lambda, c) = 0. \tag{4.22}$$

Thus, we have for all $c > 0$,

$$\lim_{\lambda \to \lambda^*} x_1(\lambda, c) = x_1(-1, c) = 1 = x_1^*, \qquad \lim_{\lambda \to \lambda^*} x_2(\lambda^*, c) = 0 = x_2^*,$$

showing that as λ is chosen close to λ^*, the unconstrained minimum of $L_c(x, \lambda)$ approaches the constrained minimum (see Fig. 4.2.1).

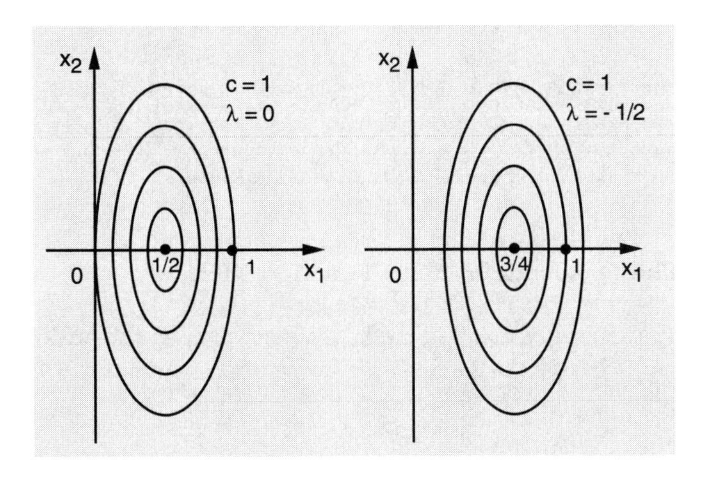

Figure 4.2.1. Equal cost surfaces of the augmented Lagrangian

$$L_c(x, \lambda) = \tfrac{1}{2}(x_1^2 + x_2^2) + \lambda(x_1 - 1) + \frac{c}{2}(x_1 - 1)^2,$$

of Example 4.2.1, for $c = 1$ and two different values of λ. The unconstrained minimum of $L_c(x, \lambda)$ approaches the constrained minimum $x^* = (1, 0)$ as $\lambda \to \lambda^* = -1$.

Using Eq. (4.22), we also have for all λ,

$$\lim_{c\to\infty} x_1(\lambda, c) = 1 = x_1^*, \qquad \lim_{c\to\infty} x_2(\lambda, c) = 0 = x_2^*,$$

showing that as c increases, the unconstrained minimum of $L_c(x, \lambda)$ approaches the constrained minimum (see Fig. 4.2.2).

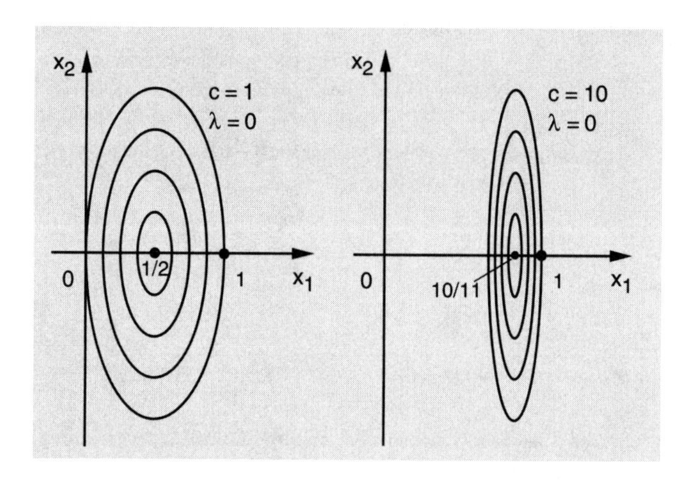

Figure 4.2.2. Equal cost surfaces of the augmented Lagrangian

$$L_c(x, \lambda) = \tfrac{1}{2}(x_1^2 + x_2^2) + \lambda(x_1 - 1) + \frac{c}{2}(x_1 - 1)^2,$$

of Example 4.2.1, for $\lambda = 0$ and two different values of c. The unconstrained minimum of $L_c(x, \lambda)$ approaches the constrained minimum $x^* = (1, 0)$ as $c \to \infty$.

4.2.1 The Quadratic Penalty Function Method

The quadratic penalty function method is motivated by the preceding considerations. It consists of solving a sequence of problems of the form

$$\begin{aligned} &\text{minimize} \;\; L_{c^k}(x, \lambda^k) \\ &\text{subject to} \;\; x \in X, \end{aligned}$$

where $\{\lambda^k\}$ is a sequence in $\Re^m$ and $\{c^k\}$ is a positive penalty parameter sequence.

In the original version of the penalty method introduced in the early sixties, the multipliers λ^k were taken to be equal to zero,

$$\lambda^k = 0, \qquad \forall\ k = 0, 1, \ldots$$

The idea of using λ^k that are "good" approximations to a Lagrange multiplier vector was not known at that time. In our development here we allow λ^k to change in the course of the algorithm but for the moment, we give no rule for updating λ^k. Thus the method depends for its validity on increasing c^k to ∞. The following proposition is the basic convergence result.

Proposition 4.2.1: Assume that f and h are continuous functions, that X is a closed set, and that the constraint set $\{x \in X \mid h(x) = 0\}$ is nonempty. For $k = 0, 1, \ldots$, let x^k be a global minimum of the problem

$$\begin{aligned} &\text{minimize } L_{c^k}(x, \lambda^k) \\ &\text{subject to } x \in X, \end{aligned}$$

where $\{\lambda^k\}$ is bounded, $0 < c^k < c^{k+1}$ for all k, and $c^k \to \infty$. Then every limit point of the sequence $\{x^k\}$ is a global minimum of the original problem (4.21).

Proof: Let $\bar{x}$ be a limit point of $\{x^k\}$. We have by definition of x^k

$$L_{c^k}(x^k, \lambda^k) \leq L_{c^k}(x, \lambda^k), \qquad \forall\, x \in X. \tag{4.23}$$

Let f^* denote the optimal value of the original problem (4.21). We have

$$\begin{aligned} f^* &= \inf_{h(x)=0,\, x\in X} f(x) \\ &= \inf_{h(x)=0,\, x\in X} \left\{ f(x) + \lambda^{k\prime} h(x) + \frac{c^k}{2}\|h(x)\|^2 \right\} \\ &= \inf_{h(x)=0,\, x\in X} L_{c^k}(x, \lambda^k). \end{aligned}$$

Hence, by taking the infimum of the right-hand side of Eq. (4.23) over $x \in X$, $h(x) = 0$, we obtain

$$L_{c^k}(x^k, \lambda^k) = f(x^k) + \lambda^{k\prime} h(x^k) + \frac{c^k}{2}\|h(x^k)\|^2 \leq f^*.$$

The sequence $\{\lambda^k\}$ is bounded and hence it has a limit point $\bar{\lambda}$. Without loss of generality, we may assume that $\lambda^k \to \bar{\lambda}$. By taking the limit superior in the relation above and by using the continuity of f and h, we obtain

$$f(\bar{x}) + \bar{\lambda}' h(\bar{x}) + \limsup_{k\to\infty} \frac{c^k}{2}\|h(x^k)\|^2 \leq f^*. \tag{4.24}$$

Since $\|h(x^k)\|^2 \geq 0$ and $c^k \to \infty$, it follows that $h(x^k) \to 0$ and

$$h(\bar{x}) = 0, \tag{4.25}$$

for otherwise the left-hand side of Eq. (4.24) would equal ∞, while $f^* < \infty$ (since the constraint set is assumed nonempty). Since X is a closed set, we also obtain that $\bar{x} \in X$. Hence, $\bar{x}$ is feasible, and since from Eqs. (4.24) and (4.25) we have $f(\bar{x}) \le f^*$, it follows that $\bar{x}$ is optimal. **Q.E.D.**

Lagrange Multiplier Estimates – Inexact Minimization

Proposition 4.2.1 assumes implicitly that the minimum of the augmented Lagrangian is found exactly. On the other hand, unconstrained minimization methods are usually terminated when the cost gradient is sufficiently small, but not necessarily zero. In particular, when $X = \Re^n$, and f and h are differentiable, the algorithm for solving the unconstrained problem

$$\begin{aligned} &\text{minimize } \ L_{c^k}(x, \lambda^k) \\ &\text{subject to } \ x \in \Re^n \end{aligned}$$

will typically be terminated at a point x^k satisfying

$$\|\nabla_x L_{c^k}(x^k, \lambda^k)\| \le \epsilon^k,$$

where ϵ^k is some small scalar. We address this situation in the next proposition, where we show in addition that we can usually obtain a Lagrange multiplier vector as a by-product of the computation.

Proposition 4.2.2: Assume that $X = \Re^n$, and f and h are continuously differentiable. For $k = 0, 1, \ldots$, let x^k satisfy

$$\|\nabla_x L_{c^k}(x^k, \lambda^k)\| \le \epsilon^k,$$

where $\{\lambda^k\}$ is bounded, and $\{\epsilon^k\}$ and $\{c^k\}$ satisfy

$$0 < c^k < c^{k+1}, \quad \forall\, k, \qquad c^k \to \infty,$$

$$0 \le \epsilon^k, \quad \forall\, k, \qquad \epsilon^k \to 0.$$

Assume that a subsequence $\{x^k\}_K$ converges to a vector x^* such that $\nabla h(x^*)$ has rank m. Then

$$\{\lambda^k + c^k h(x^k)\}_K \to \lambda^*,$$

where λ^* is a vector satisfying, together with x^*, the first order necessary conditions

$$\nabla f(x^*) + \nabla h(x^*)\lambda^* = 0, \qquad h(x^*) = 0.$$

Proof: Without loss of generality we assume that the entire sequence $\{x^k\}$ converges to x^*. Define for all k

$$\tilde{\lambda}^k = \lambda^k + c^k h(x^k).$$

We have

$$\nabla_x L_{c^k}(x^k, \lambda^k) = \nabla f(x^k) + \nabla h(x^k)\big(\lambda^k + c^k h(x^k)\big) = \nabla f(x^k) + \nabla h(x^k)\tilde{\lambda}^k. \tag{4.26}$$

Since $\nabla h(x^*)$ has rank m, $\nabla h(x^k)$ has rank m for all k that are sufficiently large. Without loss of generality, we assume that $\nabla h(x^k)$ has rank m for all k. Then, by multiplying Eq. (4.26) with

$$\big(\nabla h(x^k)'\nabla h(x^k)\big)^{-1}\nabla h(x^k)',$$

we obtain

$$\tilde{\lambda}^k = \big(\nabla h(x^k)'\nabla h(x^k)\big)^{-1}\nabla h(x^k)'\big(\nabla_x L_{c^k}(x^k, \lambda^k) - \nabla f(x^k)\big). \tag{4.27}$$

The hypothesis implies that $\nabla_x L_{c^k}(x^k, \lambda^k) \to 0$, so Eq. (4.27) yields

$$\tilde{\lambda}^k \to \lambda^*,$$

where

$$\lambda^* = -\big(\nabla h(x^*)'\nabla h(x^*)\big)^{-1}\nabla h(x^*)'\nabla f(x^*).$$

Using again the fact $\nabla_x L_{c^k}(x^k, \lambda^k) \to 0$ and Eq. (4.26), we see that

$$\nabla f(x^*) + \nabla h(x^*)\lambda^* = 0.$$

Since $\{\lambda^k\}$ is bounded and $\lambda^k + c^k h(x^k) \to \lambda^*$, it follows that $\{c^k h(x^k)\}$ is bounded. Since $c^k \to \infty$, we must have $h(x^k) \to 0$ and we conclude that $h(x^*) = 0$. **Q.E.D.**

Practical Behavior – Ill-Conditioning

Let us now consider the practical behavior of the quadratic penalty method. Assume that the kth unconstrained minimization of $L_{c^k}(x, \lambda^k)$ is terminated when

$$\|\nabla_x L_{c^k}(x^k, \lambda^k)\| \le \epsilon^k,$$

where $\epsilon^k \to 0$. There are three possibilities:

(a) The method breaks down because an x^k satisfying $\|\nabla_x L_{c^k}(x^k, \lambda^k)\| \le \epsilon^k$ cannot be found.

(b) A sequence $\{x^k\}$ with $\|\nabla_x L_{c^k}(x^k, \lambda^k)\| \leq \epsilon^k$ for all k is obtained, but it either has no limit points, or for each of its limit points x^* the matrix $\nabla h(x^*)$ has linearly dependent columns.

(c) A sequence $\{x^k\}$ with $\|\nabla_x L_{c^k}(x^k, \lambda^k)\| \leq \epsilon^k$ for all k is found and it has a limit point x^* such that $\nabla h(x^*)$ has rank m. Then, by Prop. 4.2.2, x^* together with λ^* [the corresponding limit point of $\{\lambda^k + c^k h(x^k)\}$] satisfies the first order necessary conditions for optimality.

Possibility (a) usually occurs when $L_{c^k}(\cdot, \lambda^k)$ is unbounded below as discussed following Prop. 4.2.1.

Possibility (b) usually occurs when $L_{c^k}(\cdot, \lambda^k)$ is bounded below, but the original problem has no feasible solution. Typically then the penalty term dominates as $k \to \infty$, and the method usually converges to an infeasible vector x^*, which is a stationary point of the function $\|h(x)\|^2$. This means that

$$\nabla h(x^*) h(x^*) = \tfrac{1}{2} \nabla \{\|h(x^*)\|^2\} = 0,$$

implying that $\nabla h(x^*)$ has linearly dependent columns.

Possibility (c) is the normal case, where the unconstrained minimization algorithm terminates successfully for each k and $\{x^k\}$ converges to a feasible vector, which is also regular. It is of course possible (although rare in practice) that $\{x^k\}$ converges to a local minimum x^*, which is not regular. Then, if there is no Lagrange multiplier vector corresponding to x^*, the sequence $\{\lambda^k + c^k h(x^k)\}$ diverges and has no limit point.

Extensive practical experience shows that the penalty function method is on the whole quite reliable and usually converges to at least a local minimum of the original problem. Whenever it fails, this is usually due to the increasing difficulty of the minimization

$$\begin{aligned} &\text{minimize} \;\; L_{c^k}(x, \lambda^k) \\ &\text{subject to} \;\; x \in X \end{aligned}$$

as $c^k \to \infty$. In particular, let us assume that $X = \Re^n$, and f and h are twice differentiable. Then, according to the convergence rate analysis of Section 1.3, the degree of difficulty depends on the ratio of largest to smallest eigenvalue (the condition number) of the Hessian matrix $\nabla^2_{xx} L_{c^k}(x^k, \lambda^k)$, and this ratio tends to increase with c^k. We illustrate this by means of an example. A proof is outlined in Exercise 4.2.8.

Example 4.2.2

Consider the problem of Example 4.2.1

$$\begin{aligned} &\text{minimize} \;\; f(x) = \tfrac{1}{2}(x_1^2 + x_2^2) \\ &\text{subject to} \;\; x_1 = 1. \end{aligned}$$

The augmented Lagrangian is

$$L_c(x, \lambda) = \tfrac{1}{2}(x_1^2 + x_2^2) + \lambda(x_1 - 1) + \frac{c}{2}(x_1 - 1)^2,$$

and its Hessian is

$$\nabla_{xx}^2 L_c(x, \lambda) = I + c \begin{pmatrix} 1 \\ 0 \end{pmatrix} (\, 1 \quad 0 \,) = \begin{pmatrix} 1+c & 0 \\ 0 & 1 \end{pmatrix}.$$

The ratio of largest to smallest eigenvalue of the Hessian is $1+c$ and tends to ∞ as $c \to \infty$. The associated ill-conditioning can also be observed from the narrow level sets of the augmented Lagrangian for large c in Fig. 4.2.2.

To overcome ill-conditioning, it is recommended to use a Newton-like method for minimizing $L_{c^k}(\cdot, \lambda^k)$, as well as double precision arithmetic to deal with roundoff errors. It is common to adopt as a starting point for minimizing $L_{c^k}(\cdot, \lambda^k)$ the last point x^{k-1} of the previous minimization. In order, however, for x^{k-1} to be near a minimizing point of $L_{c^k}(\cdot, \lambda^k)$, it is necessary that c^k is close to c^{k-1}. This in turn requires that the rate of increase of the penalty parameter c^k should be relatively small. There is a basic tradeoff here. If c^k is increased at a fast rate, then $\{x^k\}$ converges faster, but the likelihood of ill-conditioning is greater. Usually, a sequence $\{c^k\}$ generated by $c^{k+1} = \beta c^k$ with β in the range $[4, 10]$ works well if a Newton-like method is used for minimizing $L_{c^k}(\cdot, \lambda^k)$; otherwise, a smaller value of β may be needed. Some trial and error may be needed to choose the initial penalty parameter c^0, since there is no safe guideline on how to determine this value. For an indication of this, note that if the problem data f and h are scaled by multiplication with a scalar $s > 0$, then c^0 should be divided by s to maintain the same condition number for the Hessian of the augmented Lagrangian.

Inequality Constraints*

The simplest way to treat inequality constraints in the context of the quadratic penalty method, is to convert them to equality constraints by using squared additional variables. We have already used this device in our discussion of optimality conditions for inequality constraints in Section 3.3.2.

Consider the problem

$$\begin{aligned} &\text{minimize} \quad f(x) \\ &\text{subject to} \quad h_1(x) = 0, \ldots, h_m(x) = 0, \\ &\qquad\qquad\quad g_1(x) \le 0, \ldots, g_r(x) \le 0. \end{aligned} \tag{4.28}$$

As discussed in Section 3.3.2, we can convert this problem to the equality constrained problem

$$\begin{aligned} &\text{minimize} \quad f(x) \\ &\text{subject to} \quad h_1(x) = 0, \ldots, h_m(x) = 0, \\ &\qquad\qquad\quad g_1(x) + z_1^2 = 0, \ldots, g_r(x) + z_r^2 = 0, \end{aligned} \tag{4.29}$$

where $z_1, \ldots, z_r$ are additional variables. The quadratic penalty method for this problem involves unconstrained minimizations of the form

$$\min_{x,z} \bar{L}_c(x, z, \lambda, \mu) = f(x) + \lambda' h(x) + \frac{c}{2}\|h(x)\|^2 + \sum_{j=1}^{r} \left\{ \mu_j \left(g_j(x) + z_j^2\right) + \frac{c}{2}|g_j(x) + z_j^2|^2 \right\},$$

for various values of λ, μ, and c. This type of minimization can be done by first minimizing $\bar{L}_c(x, z, \lambda, \mu)$ with respect to z, obtaining

$$L_c(x, \lambda, \mu) = \min_z \bar{L}_c(x, z, \lambda, \mu),$$

and then by minimizing $L_c(x, \lambda, \mu)$ with respect to x. A key observation is that *the first minimization with respect to z can be carried out in closed form for each fixed x*, thereby yielding a closed form expression for $L_c(x, \lambda, \mu)$.

Indeed, we have

$$\min_z \bar{L}_c(x, z, \lambda, \mu) = f(x) + \lambda' h(x) + \frac{c}{2}\|h(x)\|^2 + \sum_{j=1}^{r} \min_{z_j} \left\{ \mu_j \left(g_j(x) + z_j^2\right) + \frac{c}{2}|g_j(x) + z_j^2|^2 \right\}, \tag{4.30}$$

and the minimization with respect to z_j in the last term is equivalent to

$$\min_{u_j \geq 0} \left\{ \mu_j \left(g_j(x) + u_j\right) + \frac{c}{2}|g_j(x) + u_j|^2 \right\}.$$

The function in braces above is quadratic in u_j. Its constrained minimum is $u_j^* = \max\{0, \hat{u}_j\}$, where $\hat{u}_j$ is the unconstrained minimum at which the derivative $\mu_j + c\big(g_j(x) + \hat{u}_j\big)$ is zero. Thus,

$$u_j^* = \max\left\{0, -\left(\frac{\mu_j}{c} + g_j(x)\right)\right\}.$$

Denoting

$$g_j^+(x, \mu, c) = \max\left\{g_j(x), -\frac{\mu_j}{c}\right\}, \tag{4.31}$$

we have $g_j(x) + u_j^* = g_j^+(x, \mu, c)$. Substituting this expression in Eq. (4.30), we obtain a closed form expression for $L_c(x, \lambda, \mu) = \min_z \bar{L}_c(x, z, \lambda, \mu)$ given by

$$L_c(x, \lambda, \mu) = f(x) + \lambda' h(x) + \frac{c}{2}\|h(x)\|^2 + \sum_{j=1}^{r} \left\{ \mu_j g_j^+(x, \mu, c) + \frac{c}{2}\left(g_j^+(x, \mu, c)\right)^2 \right\}. \tag{4.32}$$

After some calculation, left for the reader, we can also write this expression as

$$
\begin{aligned}
L_c(x,\lambda,\mu) = {} & f(x) + \lambda' h(x) + \frac{c}{2}\|h(x)\|^2 \\
& + \frac{1}{2c}\sum_{j=1}^{r}\Big\{\big(\max\{0,\mu_j + c g_j(x)\}\big)^2 - \mu_j^2\Big\},
\end{aligned}
\tag{4.33}
$$

and we can view it as the augmented Lagrangian function for the inequality constrained problem (4.28).

Note that the penalty term

$$
\frac{1}{2c}\big(\max\{0,\mu_j + c g_j(x)\}\big)^2 - \mu_j^2
$$

corresponding to the jth inequality constraint in Eq. (4.33) is continuously differentiable in x if g_j is continuously differentiable (see Fig. 4.2.3). However, its Hessian matrix is discontinuous for all x such that $g_j(x) = -\mu_j/c$; this may cause some difficulties in the minimization of $L_c(x,\lambda,\mu)$ and motivates alternative augmented Lagrangian methods for inequality constraints (see Section 4.2.5).

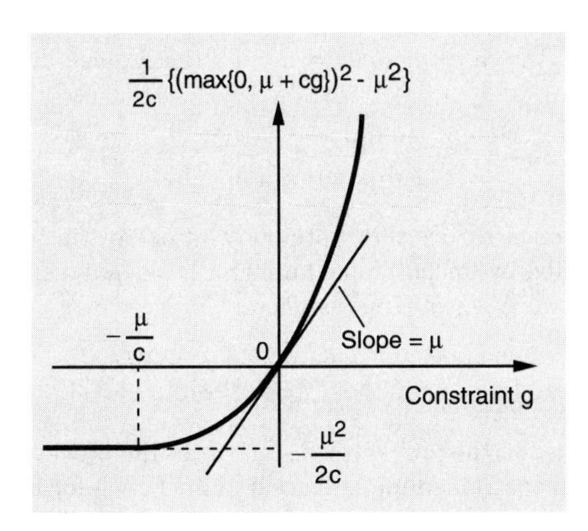

Figure 4.2.3. Form of the quadratic penalty function for inequality constraints.

To summarize, the quadratic penalty method for the inequality constrained problem (4.28) consists of a sequence of minimizations of the form

$$
\begin{aligned}
&\text{minimize} \quad L_{c^k}(x,\lambda^k,\mu^k) \\
&\text{subject to} \quad x \in X,
\end{aligned}
$$

where $L_c(x, \lambda^k, \mu^k)$ is given by Eq. (4.32) or Eq. (4.33), $\{\lambda^k\}$ and $\{\mu^k\}$ are sequences in $\Re^m$ and $\Re^r$, with coordinates denoted by λ_i^k and μ_j^k, respectively, and $\{c^k\}$ is a positive penalty parameter sequence. Since this method is equivalent to the equality-constrained method applied to the corresponding equality-constrained problem (4.29), our convergence result of Prop. 4.2.1 applies with the obvious modifications.

Furthermore, if $X = \Re^n$, and the generated sequence $\{x^k\}$ converges to a local minimum x^* which is also regular, application of Prop. 4.2.2 to the equivalent equality constrained problem (4.29) shows that the sequences

$$\{\lambda_i^k + c^k h_i(x^k)\}, \qquad \max\{0, \mu_j^k + c^k g_j(x^k)\} \tag{4.34}$$

converge to the corresponding Lagrange multipliers λ_i^* and μ_j^* [for the jth inequality constraint, the Lagrange multiplier estimate is

$$\mu_j^k + c^k g_j^+(x^k, \mu^k, c^k),$$

which is equal to $\max\{0, \mu_j^k + c^k g_j(x^k)\}$ in view of the form (4.31) for g_j^+].

4.2.2 Multiplier Methods – Main Ideas

Let us return to the case where $X = \Re^n$ and the problem has only equality constraints,

$$\begin{aligned} &\text{minimize } f(x) \\ &\text{subject to } h(x) = 0. \end{aligned}$$

We mentioned earlier that optimal solutions of this problem can be well approximated by unconstrained minima of the augmented Lagrangian $L_c(\cdot, \lambda)$ under two types of circumstances:

(a) The vector λ is close to a Lagrange multiplier.

(b) The penalty parameter c is large.

We analyzed in the previous subsection the quadratic penalty method consisting of unconstrained minimization of $L_{c^k}(\cdot, \lambda^k)$ for a sequence $c^k \to \infty$. No assumptions on the sequence $\{\lambda^k\}$ were made other than boundedness. Still, we found that, under minimal assumptions on f and h (continuity), the method has satisfactory convergence properties (Prop. 4.2.1).

We now consider intelligent ways to update λ^k so that it tends to a Lagrange multiplier. We will see that under some reasonable assumptions, this approach is workable even if c^k is not increased to ∞, thereby alleviating much of the difficulty with ill-conditioning. Furthermore, even when c^k is increased to ∞, the rate of convergence is significantly enhanced by using good updating schemes for λ^k.

The Method of Multipliers

A first update formula for λ^k in the quadratic penalty method is

$$\lambda^{k+1} = \lambda^k + c^k h(x^k). \tag{4.35}$$

The rationale is provided by Prop. 4.2.2, which shows that, if the generated sequence $\{x^k\}$ converges to a local minimum x^* which is regular, then $\{\lambda^k + c^k h(x^k)\}$ converges to the corresponding Lagrange multiplier λ^*.

The quadratic penalty method with the preceding update formula for λ^k is known as the *method of multipliers*. There are a number of interesting convergence and rate of convergence results regarding this method, which will be given shortly. We first illustrate the method with some examples.

Example 4.2.3

Consider again the problem of Examples 4.2.1 and 4.2.2

$$\begin{aligned} &\text{minimize} \ \ f(x) = \tfrac{1}{2}(x_1^2 + x_2^2) \\ &\text{subject to} \ \ x_1 = 1, \end{aligned}$$

with optimal solution $x^* = (1,0)$ and Lagrange multiplier $\lambda^* = -1$. The augmented Lagrangian is

$$L_c(x,\lambda) = \tfrac{1}{2}(x_1^2 + x_2^2) + \lambda(x_1 - 1) + \frac{c}{2}(x_1 - 1)^2.$$

The vectors x^k generated by the method of multipliers minimize $L_{c^k}(\cdot, \lambda^k)$ and are given by

$$x^k = \left(\frac{c^k - \lambda^k}{c^k + 1}, 0 \right).$$

Using this expression, the multiplier updating formula (4.35) can be written as

$$\lambda^{k+1} = \lambda^k + c^k \left(\frac{c^k - \lambda^k}{c^k + 1} - 1 \right) = \frac{\lambda^k}{c^k + 1} - \frac{c^k}{c^k + 1},$$

or by introducing the Lagrange multiplier $\lambda^* = -1$,

$$\lambda^{k+1} - \lambda^* = \frac{\lambda^k - \lambda^*}{c^k + 1}.$$

From this formula, it can be seen that

(a) $\lambda^k \to \lambda^* = -1$ and $x^k \to x^* = (1,0)$ for every nondecreasing sequence $\{c^k\}$ [since the scalar $1/(c^k+1)$ multiplying $\lambda^k - \lambda^*$ in the above formula is always less than one].

(b) The convergence rate becomes faster as c^k becomes larger; in fact $\{|\lambda^k - \lambda^*|\}$ converges superlinearly if $c^k \to \infty$.

Note that it is not necessary to increase c^k to ∞, although doing so results in a better convergence rate.

Example 4.2.4

Consider the problem

$$\begin{aligned}&\text{minimize } \tfrac{1}{2}(-x_1^2 + x_2^2)\\&\text{subject to } x_1 = 1\end{aligned}$$

with optimal solution $x^* = (1, 0)$ and Lagrange multiplier $\lambda^* = 1$. The augmented Lagrangian is given by

$$L_c(x, \lambda) = \tfrac{1}{2}(-x_1^2 + x_2^2) + \lambda(x_1 - 1) + \frac{c}{2}(x_1 - 1)^2.$$

The vector x^k minimizing $L_{c^k}(x, \lambda^k)$ is given by

$$x^k = \left(\frac{c^k - \lambda^k}{c^k - 1}, 0\right). \tag{4.36}$$

For this formula to be correct, however, it is necessary that $c^k > 1$; for $c^k < 1$ the augmented Lagrangian has no minimum, and the same is true for $c^k = 1$ unless $\lambda^k = 1$. The multiplier updating formula (4.35) can be written using Eq. (4.36) as

$$\lambda^{k+1} = \lambda^k + c^k \left(\frac{c^k - \lambda^k}{c^k - 1} - 1\right) = -\frac{\lambda^k}{c^k - 1} + \frac{c^k}{c^k - 1},$$

or by introducing the Lagrange multiplier $\lambda^* = 1$,

$$\lambda^{k+1} - \lambda^* = -\frac{\lambda^k - \lambda^*}{c^k - 1}. \tag{4.37}$$

From this iteration, it can be seen that similar conclusions to those of the preceding example can be drawn. In particular, it is not necessary to increase c^k to ∞ to obtain convergence, although doing so results in a better convergence rate. However, there is a difference: whereas in the preceding example, convergence was guaranteed for all positive sequences $\{c^k\}$, in the present example, the minimizing points exist only if $c^k > 1$. Furthermore, it is seen from Eq. (4.37) that to obtain convergence, the penalty parameter c^k must eventually exceed 2 [so that the scalar $-1/(c^k - 1)$ multiplying λ^k has absolute value less than one]. Thus in the present example, there is a threshold value that the penalty parameter must exceed in order for the method to work. We will see later that this threshold is due to the nonconvexity of the cost function.

Geometric Interpretation of the Method of Multipliers*

The conclusions from the preceding two examples hold in considerable generality. We first provide a geometric interpretation. Assume that f and h are twice differentiable and let x^* be a local minimum of f over $h(x) = 0$. Assume also that x^* is regular and together with its associated Lagrange multiplier vector λ^* satisfies the second order sufficiency conditions of Prop. 3.2.1. Then, the assumptions of the sensitivity theorem (Prop. 3.2.2) are satisfied and we can consider the *primal function*

$$p(u) = \min_{h(x)=u} f(x),$$

defined for u in an open sphere centered at $u = 0$. [The minimization above is understood to be local in an open sphere within which x^* is the unique local minimum of f over $h(x) = 0$ (cf. Prop. 3.2.2).] Note that we have

$$p(0) = f(x^*), \qquad \nabla p(0) = -\lambda^*,$$

(cf. Prop. 3.2.2). The primal function is illustrated in Fig. 4.2.4.

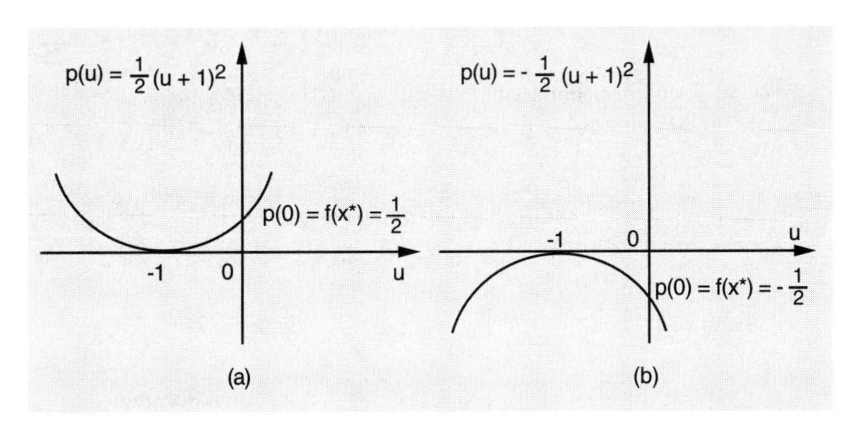

Figure 4.2.4. Illustration of the primal function. In (a) we show the primal function

$$p(u) = \min_{x_1 - 1 = u} \tfrac{1}{2}(x_1^2 + x_2^2)$$

for the problem of Example 4.2.3. In (b) we show the primal function

$$p(u) = \min_{x_1 - 1 = u} \tfrac{1}{2}(-x_1^2 + x_2^2)$$

for the problem of Example 4.2.4. The latter primal function is not convex because the cost function is not convex on the subspace that is orthogonal to the constraint set (this observation can be generalized).

We can break down the minimization of $L_c(\cdot, \lambda)$ into two stages, first minimizing over all x such that $h(x) = u$ with u fixed, and then minimizing over all u. Thus,

$$\begin{aligned} \min_x L_c(x, \lambda) &= \min_u \min_{h(x)=u} \left\{ f(x) + \lambda' h(x) + \frac{c}{2}\|h(x)\|^2 \right\} \\ &= \min_u \left\{ p(u) + \lambda' u + \frac{c}{2}\|u\|^2 \right\}, \end{aligned}$$

where the minimization above is understood to be local in a neighborhood of $u = 0$. This minimization can be interpreted as shown in Fig. 4.2.5. The minimum is attained at the point $u(\lambda, c)$ for which the gradient of $p(u) + \lambda' u + \frac{c}{2}\|u\|^2$ is zero, or, equivalently,

$$\nabla \left\{ p(u) + \frac{c}{2}\|u\|^2 \right\} \Big|_{u=u(\lambda,c)} = -\lambda.$$

Thus, the minimizing point $u(\lambda, c)$ is obtained as shown in Fig. 4.2.5. We also have

$$\min_x L_c(x, \lambda) - \lambda' u(\lambda, c) = p\big(u(\lambda, c)\big) + \frac{c}{2}\|u(\lambda, c)\|^2,$$

so the tangent hyperplane to the graph of $p(u) + \frac{c}{2}\|u\|^2$ at $u(\lambda, c)$ (which has "slope" $-\lambda$) intersects the vertical axis at the value $\min_x L_c(x, \lambda)$ as shown in Fig. 4.2.5. It can be seen that if c is sufficiently large, then the function

$$p(u) + \lambda' u + \frac{c}{2}\|u\|^2$$

is convex in a neighborhood of the origin. Furthermore, for λ close to λ^* and large c, the value $\min_x L_c(x, \lambda)$ is close to $p(0) = f(x^*)$.

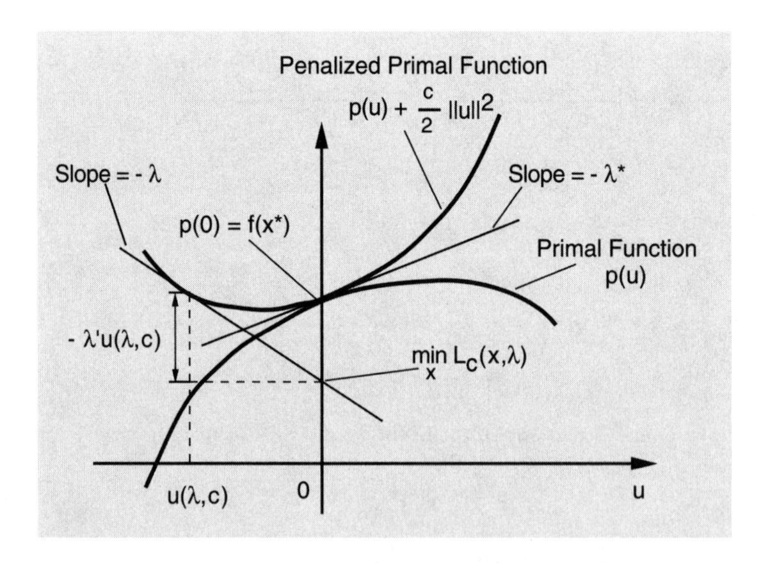

Figure 4.2.5. Geometric interpretation of minimization of the augmented Lagrangian.

Figure 4.2.6 provides a geometric interpretation of the multiplier iteration

$$\lambda^{k+1} = \lambda^k + c^k h(x^k).$$

To understand this figure, note that if x^k minimizes $L_{c^k}(\cdot, \lambda^k)$, then by the preceding analysis the vector u^k given by $u^k = h(x^k)$ minimizes $p(u) + \lambda^{k\prime} u + \frac{c^k}{2}\|u\|^2$. Hence,

$$\nabla \left\{ p(u) + \frac{c^k}{2}\|u\|^2 \right\} \Big|_{u=u^k} = -\lambda^k,$$

and

$$\nabla p(u^k) = -(\lambda^k + c^k u^k) = -\big(\lambda^k + c^k h(x^k)\big).$$

It follows that the next multiplier λ^{k+1} is

$$\lambda^{k+1} = \lambda^k + c^k h(x^k) = -\nabla p(u^k),$$

as shown in Fig. 4.2.6. The figure shows that if λ^k is sufficiently close to λ^* and/or c^k is sufficiently large, the next multiplier λ^{k+1} will be closer to λ^* than λ^k is. Furthermore, c^k need not be increased to ∞ in order to obtain convergence; it is sufficient that c^k eventually exceeds some threshold level. The figure also shows that if $p(u)$ is linear, convergence to λ^* will be achieved in one iteration.

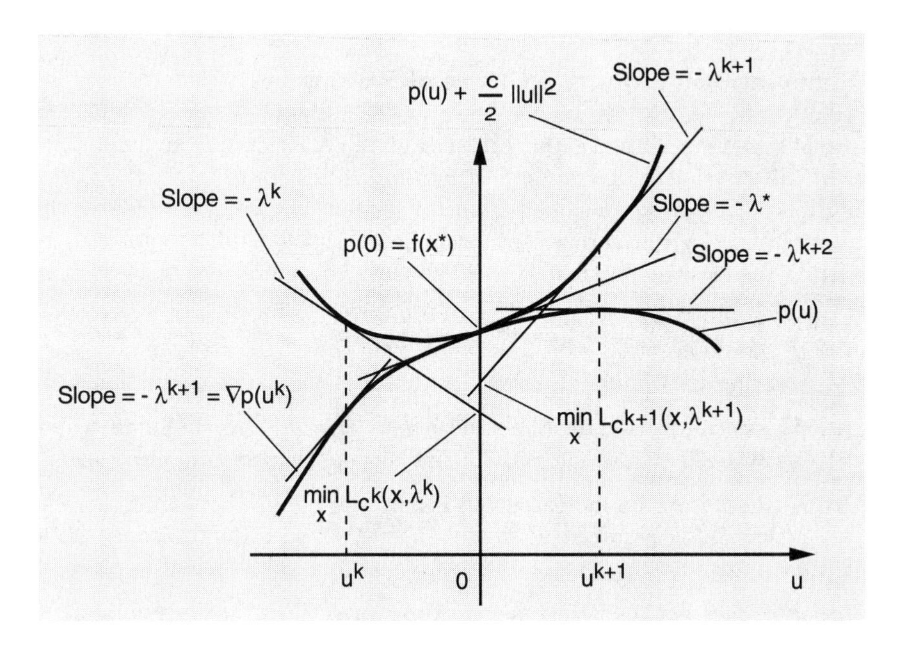

Figure 4.2.6. Geometric interpretation of the first order multiplier iteration.

In summary, the geometric interpretation of the method of multipliers just presented suggests the following:

(a) If c is large enough so that $cI + \nabla^2 p(0)$ is positive definite, then the "penalized primal function"

$$p(u) + \frac{c}{2}\|u\|^2$$

is convex within a sphere centered at $u = 0$. Furthermore, a local minimum of the augmented Lagrangian $L_c(x, \lambda)$ that is near x^* exists if λ is close enough to λ^*. The reason is that $\nabla^2_{xx} L_c(x^*, \lambda^*)$ is positive definite if and only if $cI + \nabla^2 p(0)$ is positive definite, a fact the reader may wish to verify as an exercise.

(b) If c^k is sufficiently large [the threshold can be shown to be twice the value of c needed to make $cI + \nabla^2 p(0)$ positive definite; see Exercise 4.2.4], then

$$\|\lambda^{k+1} - \lambda^*\| \leq \|\lambda^k - \lambda^*\|$$

and $\lambda^k \to \lambda^*$.

(c) Convergence can be obtained even if c^k is not increased to ∞.

(d) As c^k is increased, the rate of convergence of λ^k improves.

(e) If $\nabla^2 p(0) = 0$, the convergence is very fast.

These conclusions will be formalized in Section 4.2.3.

Computational Aspects Choice of Parameters

In addition to addressing the problem of ill-conditioning, an important practical question in the method of multipliers is how to select the initial multiplier λ^0 and the penalty parameter sequence. Clearly, in view of the interpretations given earlier, any prior knowledge should be exploited to select λ^0 as close as possible to λ^*. The main considerations to be kept in mind for selecting the penalty parameter sequence are the following:

(a) c^k should eventually become larger than the threshold level necessary to bring to bear the positive features of the multiplier iteration.

(b) The initial parameter c^0 should not be too large to the point where it causes ill-conditioning at the first unconstrained minimization.

(c) c^k should not be increased too fast to the point where too much ill-conditioning is forced upon the unconstrained minimization routine too early.

(d) c^k should not be increased too slowly, at least in the early minimizations, to the extent that the multiplier iteration has poor convergence rate.

A good practical scheme is to choose a moderate value c^0 (if necessary by preliminary experimentation), and then increase c^k via the equation $c^{k+1} = \beta c^k$, where β is a scalar with $\beta > 1$. In this way, the threshold level for multiplier convergence will eventually be exceeded. If the method used for augmented Lagrangian minimization is powerful, such as a Newton-like method, fairly large values of β (say $\beta \in [5, 10]$) are recommended; otherwise, smaller values of β may be necessary, depending on the method's ability to deal with ill-conditioning.

Another reasonable parameter adjustment scheme is to increase c^k by multiplication with a factor $\beta > 1$ only if the constraint violation as measured by $\|h(x(\lambda^k, c^k))\|$ is not decreased by a factor $\gamma < 1$ over the previous minimization; i.e.,

$$c^{k+1} = \begin{cases} \beta c^k & \text{if } \|h(x^k)\| > \gamma \|h(x^{k-1})\|, \\ c^k & \text{if } \|h(x^k)\| \leq \gamma \|h(x^{k-1})\|. \end{cases}$$

A choice such as $\gamma = 0.25$ is typically recommended.

Still another possibility is to use a different penalty parameter c_i^k for each constraint $h_i(x) = 0$, and to increase by a certain factor the penalty parameters of the constraints that are violated most. For example, increase c_i^k if the constraint violation as measured by $|h_i(x^k)|$ is not decreased by a certain factor over $|h_i(x^{k-1})|$.

Inexact Minimization of the Augmented Lagrangian*

In practice the minimization of $L_{c^k}(x, \lambda^k)$ is typically terminated early. For example, it may be terminated at a point x^k satisfying

$$\|\nabla_x L_{c^k}(x^k, \lambda^k)\| \leq \epsilon^k,$$

where $\{\epsilon^k\}$ is a positive sequence converging to zero. Then it is still appropriate to use the multiplier update

$$\lambda^{k+1} = \lambda^k + c^k h(x^k),$$

although in theory, some of the linear convergence rate results to be given shortly will not hold any more. This deficiency does not seem to be important in practice, but can also be corrected by using the alternative termination criterion

$$\|\nabla_x L_{c^k}(x^k, \lambda^k)\| \leq \min\{\epsilon^k, \gamma^k \|h(x^k)\|\},$$

where $\{\epsilon^k\}$ and $\{\gamma^k\}$ are positive sequences converging to zero; for an analysis see [Ber75b], [Ber76a], and [Ber82a].

In Section 4.4, we will see that with certain safeguards, it is possible to terminate the minimization of the augmented Lagrangian after a few Newton steps (possibly only one), and follow it by a second order multiplier update of the type that will be discussed later in this section. Such algorithmic strategies give rise to some of the most effective methods using Lagrange multipliers.

Inequality Constraints*

To treat inequality constraints $g_j(x) \leq 0$ in the context of the method of multipliers, we convert them into equality constraints $g_j(x) + z_j^2 = 0$, using the additional variables z_j [cf. problems (4.28) and (4.29)]. In particular, the multiplier update formulas are

$$\lambda^{k+1} = \lambda^k + c^k h(x^k),$$

$$\mu_j^{k+1} = \max\left\{0, \mu_j^k + c^k g_j(x^k)\right\},$$

[cf. Eq. (4.34)], where x^k minimizes the augmented Lagrangian

$$L_{c^k}(x, \lambda^k, \mu^k) = f(x) + \lambda^{k\prime} h(x) + \frac{c}{2}\|h(x)\|^2 + \frac{1}{2c^k}\sum_{j=1}^{r}\left\{\left(\max\{0, \mu_j^k + c^k g_j(x)\}\right)^2 - (\mu_j^k)^2\right\}.$$

Many problems encountered in practice involve two-sided constraints of the form

$$\alpha_j \leq g_j(x) \leq \beta_j,$$

where α_j and β_j are given scalars. Each two-sided constraint could of course be separated into two one-sided constraints. This would require, however, the assignment of two multipliers per two-sided constraint, and is somewhat wasteful, since we know that at a solution at least one of the two multipliers will be zero. It turns out that there is an alternative approach that requires only *one multiplier per two-sided constraint* (see Exercise 4.2.7).

Partial Elimination of Constraints*

In the preceding multiplier algorithms, all the equality and inequality constraints are eliminated by means of a penalty. In some cases it is convenient to eliminate only part of the constraints, while retaining the remaining constraints explicitly. A typical example is a problem of the form

$$\begin{aligned} &\text{minimize } f(x) \\ &\text{subject to } h(x) = 0, \qquad x \geq 0. \end{aligned}$$

While, in addition to $h(x) = 0$, it is possible to eliminate by means of a penalty the bound constraints $x \geq 0$, it is often desirable to handle these constraints explicitly by a gradient projection method of the type discussed in Sections 2.3 and 2.4.

More generally, the method of multipliers with partial elimination of constraints for the problem

$$\begin{aligned} &\text{minimize } f(x) \\ &\text{subject to } h(x) = 0, \qquad g(x) \leq 0, \end{aligned}$$

consists of finding x^k that solves the problem

$$\begin{aligned} &\text{minimize } f(x) + \lambda^{k\prime} h(x) + \frac{c}{2}\|h(x)\|^2 \\ &\text{subject to } g(x) \leq 0, \end{aligned}$$

followed by the multiplier iteration

$$\lambda^{k+1} = \lambda^k + c^k h(x^k).$$

It is not essential that just the equality constraints are eliminated by means of a penalty above. Any mixture of equality and inequality constraints can be eliminated by means of a penalty and a multiplier, while the remaining constraints can be explicitly retained. For a detailed treatment of partial elimination of constraints, we refer to [Ber77], [Ber82a], [Dun91a], and [Dun93b].

4.2.3 Convergence Analysis of Multiplier Methods*

We now discuss the convergence properties of multiplier methods and substantiate the conclusions derived informally earlier. We focus attention throughout on the equality constrained problem

$$\begin{aligned} &\text{minimize } f(x) \\ &\text{subject to } h(x) = 0, \end{aligned}$$

and on a particular local minimum x^*. We assume that *x^* is regular and together with a Lagrange multiplier vector λ^* satisfies the second order sufficiency conditions of Prop. 3.2.1.* In view of our earlier treatment of inequality constraints by conversion to equalities, our analysis readily carries over to the case of mixed equality and inequality constraints, under the second order sufficiency conditions of Prop. 3.3.2.

The convergence results described in this section can be strengthened considerably under additional convexity assumptions on the problem. This is discussed further in Section 5.4.6; see also [Ber82a].

Existence of Local Minima of the Augmented Lagrangian

A first basic issue is whether local minima x^k of the augmented Lagrangian $L_{c^k}(\cdot, \lambda^k)$ exist, so that the method itself is well-defined. We have shown that for the local minimum-Lagrange multiplier pair (x^*, λ^*) there exist scalars $\bar{c} > 0$, $\gamma > 0$, and $\epsilon > 0$, such that

$$L_c(x, \lambda^*) \geq L_c(x^*, \lambda^*) + \frac{\gamma}{2}\|x - x^*\|^2, \qquad \forall\, x \text{ with } \|x - x^*\| < \epsilon, \text{ and } c \geq \bar{c},$$

(cf. the discussion following Lemma 3.2.1). It is thus reasonable to infer that if λ is close to λ^*, there should exist a local minimum of $L_c(\cdot, \lambda)$ close to x^* for every $c \geq \bar{c}$. More precisely, for a fixed $c \geq \bar{c}$, we consider the system of equations

$$\nabla_x L_c(x, \lambda) = \nabla f(x) + \nabla h(x)\big(\lambda + ch(x)\big) = 0,$$

and we use the implicit function theorem in a neighborhood of (x^*, λ^*). [This can be done because the Jacobian of the system with respect to x is $\nabla^2_{xx} L_c(x^*, \lambda^*)$, and is positive definite since $c \geq \bar{c}$.] Thus, for λ sufficiently close to λ^*, there is an unconstrained local minimum $x(\lambda, c)$ of $L_c(\cdot, \lambda)$, which is defined via the equation

$$\nabla f\big(x(\lambda, c)\big) + \nabla h\big(x(\lambda, c)\big)\Big(\lambda + ch\big(x(\lambda, c)\big)\Big) = 0.$$

A closer examination of the preceding argument shows that for application of the implicit function theorem it is not essential that λ be close to λ^* but rather that the vector $\lambda + ch\big(x(\lambda, c)\big)$ be close to λ^*. Proposition 4.2.2 indicates that for any λ, if c is sufficiently large and $x(\lambda, c)$ minimizes $L_c(x, \lambda)$, the vector $\lambda + ch\big(x(\lambda, c)\big)$ is close to λ^*. This suggests that *there should exist a local minimum of $L_c(\cdot, \lambda)$ close to x^* even for λ that are far from λ^*, provided c is sufficiently large.* This can indeed be shown, and in fact it turns out that for existence of the local minimum $x(\lambda, c)$, what is really important is that the ratio $\|\lambda - \lambda^*\|/c$ be sufficiently small. However, proving this simultaneously for the entire range of values $c \in [\bar{c}, \infty)$ is not easy. The following proposition, due to [Ber82a], provides a precise mathematical statement of the existence result, together with some error estimates that quantify the rate of convergence. The proof requires the introduction of the variables $\tilde{\lambda} = \lambda + ch(x)$, and $t = (\lambda - \lambda^*)/c$, together with the system of equations $\nabla_x L_0(x, \tilde{\lambda}) = 0$ and an analysis based on a more advanced form of the implicit function theorem that the one given in Appendix A.

Proposition 4.2.3: Let $\bar{c}$ be a positive scalar such that

$$\nabla^2_{xx} L_{\bar{c}}(x^*, \lambda^*) > 0.$$

There exist positive scalars δ, ϵ, and M such that:

(a) For all (λ, c) in the set $D \subset \Re^{m+1}$ defined by

$$D = \big\{(\lambda, c) \mid \|\lambda - \lambda^*\| < \delta c,\ \bar{c} \le c\big\}, \tag{4.38}$$

the problem

$$\begin{aligned} &\text{minimize} \quad L_c(x, \lambda) \\ &\text{subject to} \quad \|x - x^*\| < \epsilon \end{aligned}$$

has a unique solution denoted $x(\lambda, c)$. The function $x(\cdot, \cdot)$ is continuously differentiable in the interior of D, and for all $(\lambda, c) \in D$, we have

$$\|x(\lambda, c) - x^*\| \le M \frac{\|\lambda - \lambda^*\|}{c}.$$

(b) For all $(\lambda, c) \in D$, we have

$$\|\tilde{\lambda}(\lambda, c) - \lambda^*\| \le M \frac{\|\lambda - \lambda^*\|}{c},$$

where

$$\tilde{\lambda}(\lambda, c) = \lambda + ch\big(x(\lambda, c)\big).$$

(c) For all $(\lambda, c) \in D$, the matrix $\nabla^2_{xx} L_c\big(x(\lambda, c), \lambda\big)$ is positive definite and the matrix $\nabla h\big(x(\lambda, c)\big)$ has rank m.

Proof: See [Ber82a], p. 108.

Figure 4.2.7 shows the set D of pairs (λ, c) within which the conclusions of Prop. 4.2.3 are valid [cf. Eq. (4.38)]. It can be seen that, for any λ, there exists a c_λ such that (λ, c) belongs to D for every $c \ge c_\lambda$. The estimate δc on the allowable distance of λ from λ^* grows linearly with c [compare with Eq. (4.38)]. In particular problems, the actual allowable distance may grow at a higher than linear rate, and in fact it is possible that for every λ and $c > 0$ there exists a unique global minimum of $L_c(\cdot, \lambda)$. (Take for instance the scalar problem $\min\{x^2 \mid x = 0\}$.) However, it is shown by example in [Ber82a], p. 111, that the estimate of a linear order of growth cannot be improved.

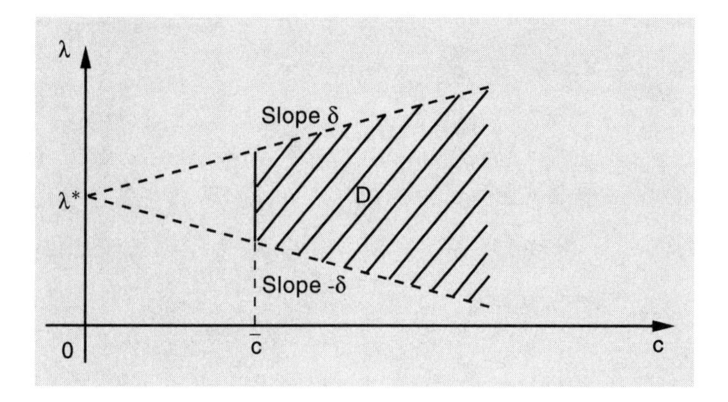

Figure 4.2.7. Illustration of the set

$$D = \big\{(\lambda, c) \mid \|\lambda - \lambda^*\| < \delta c,\, \bar{c} \le c\big\}$$

within which the conclusions of Prop. 4.2.3 are valid.

Convergence and Rate of Convergence

Proposition 4.2.3 yields both a convergence and a convergence rate result for the multiplier iteration

$$\lambda^{k+1} = \lambda^k + c^k h(x^k).$$

It shows that if the generated sequence $\{\lambda^k\}$ is bounded [this can be enforced if necessary by leaving λ^k unchanged if $\lambda^k + c^k h(x^k)$ does not belong to a prespecified bounded open set known to contain λ^*], the penalty parameter c^k is sufficiently large after a certain index [so that $(\lambda^k, c^k) \in D$], and after that index, minimization of $L_{c^k}(\cdot, \lambda^k)$ yields the local minimum $x^k = x(\lambda^k, c^k)$ closest to x^*, then we obtain $x^k \to x^*$, $\lambda^k \to \lambda^*$. Furthermore, the rate of convergence of the error sequences $\{\|x^k - x^*\|\}$ and $\{\|\lambda^k - \lambda^*\|\}$ is linear, and it is superlinear if $c^k \to \infty$.

It is possible to conduct a more refined convergence and rate of convergence analysis that supplements Prop. 4.2.3. This analysis quantifies the threshold level of the penalty parameter for convergence to occur and gives a precise estimate of the convergence rate. We refer to [Ber82a] for an extensive discussion; see also Exercise 4.2.4.

4.2.4 Duality and Second Order Multiplier Methods*

Let $\bar{c}$, δ, and ϵ be as in Prop. 4.2.3, and define for (λ, c) in the set

$$D = \big\{(\lambda, c) \mid \|\lambda - \lambda^*\| < \delta c,\, \bar{c} \le c\big\}$$

the *dual function* q_c by

$$q_c(\lambda) = \min_{\|x-x^*\|<\epsilon} L_c(x,\lambda) = L_c\big(x(\lambda,c),\lambda\big). \tag{4.39}$$

Since $x(\cdot,c)$ is continuously differentiable (Prop. 4.2.3), the same is true for q_c.

Calling q_c a dual function is not inconsistent with the duality theory already formulated in Section 3.4 and further developed in Chapter 5. Indeed q_c is the dual function for the problem

$$\begin{aligned} &\text{minimize} \;\; f(x) + \frac{c}{2}\|h(x)\|^2 \\ &\text{subject to} \;\; \|x - x^*\| < \epsilon, \quad h(x) = 0, \end{aligned}$$

which for $c \geq \bar{c}$, has x^* as its unique optimal solution and λ^* as the corresponding Lagrange multiplier.

We compute the gradient of q_c with respect to λ. From Eq. (4.39), we have

$$\nabla q_c(\lambda) = \nabla_\lambda x(\lambda,c)\nabla_x L_c\big(x(\lambda,c),\lambda\big) + h\big(x(\lambda,c)\big).$$

Since $\nabla_x L_c\big(x(\lambda,c),\lambda\big) = 0$, we obtain

$$\nabla q_c(\lambda) = h\big(x(\lambda,c)\big), \tag{4.40}$$

and since $x(\cdot,c)$ is continuously differentiable, the same is true for ∇q_c.

Next we compute the Hessian $\nabla^2 q_c$. Differentiating ∇q_c as given by Eq. (4.40) with respect to λ, we obtain

$$\nabla^2 q_c(\lambda) = \nabla_\lambda x(\lambda,c)\nabla h\big(x(\lambda,c)\big). \tag{4.41}$$

We also have, for all (λ,c) in the set D,

$$\nabla_x L_c\big(x(\lambda,c),\lambda\big) = 0.$$

Differentiating with respect to λ, we obtain

$$\nabla_\lambda x(\lambda,c)\nabla^2_{xx} L_c\big(x(\lambda,c),\lambda\big) + \nabla^2_{\lambda x} L_c\big(x(\lambda,c),\lambda\big) = 0,$$

and since

$$\nabla^2_{\lambda x} L_c\big(x(\lambda,c),\lambda\big) = \nabla h\big(x(\lambda,c)\big)',$$

we obtain

$$\nabla_\lambda x(\lambda,c) = -\nabla h\big(x(\lambda,c)\big)'\big\{\nabla^2_{xx} L_c\big(x(\lambda,c),\lambda\big)\big\}^{-1}.$$

Substitution in Eq. (4.41) yields the formula

$$\nabla^2 q_c(\lambda) = -\nabla h\big(x(\lambda,c)\big)'\big\{\nabla^2_{xx} L_c\big(x(\lambda,c),\lambda\big)\big\}^{-1}\nabla h\big(x(\lambda,c)\big). \tag{4.42}$$

Since $\nabla^2_{xx} L_c\big(x(\lambda,c),\lambda\big)$ is positive definite and $\nabla h\big(x(\lambda,c)\big)$ has rank m for $(\lambda,c)\in D$ (cf. Prop. 4.2.3), it follows from Eq. (4.42) that $\nabla^2 q_c(\lambda)$ is negative definite for all $(\lambda,c)\in D$, so that q_c is concave within the set $\{\lambda \mid \|\lambda-\lambda^*\|\le \delta c\}$. Furthermore, using Eq. (4.40), we have, for all $c\ge\bar{c}$,

$$\nabla q_c(\lambda^*) = h\big(x(\lambda^*,c)\big) = h(x^*) = 0.$$

Thus, *for every* $c\ge\bar{c}$, λ^* *maximizes* $q_c(\lambda)$ *over the set* $\{\lambda \mid \|\lambda-\lambda^*\| < \delta c\}$. Also in view of Eq. (4.40), the multiplier iteration can be written as

$$\lambda^{k+1} = \lambda^k + c^k \nabla q_{c^k}(\lambda^k), \tag{4.43}$$

so *it is a steepest ascent iteration for maximizing* q_{c^k}. When $c^k = c$ for all k, then Eq. (4.43) is the constant stepsize steepest ascent method

$$\lambda^{k+1} = \lambda^k + c\nabla q_c(\lambda^k)$$

for maximizing q_c.

The Second Order Method of Multipliers

In view of the interpretation of the multiplier iteration as a steepest ascent method, it is natural to consider the Newton-like iteration

$$\lambda^{k+1} = \lambda^k - \big(\nabla^2 q_{c^k}(\lambda^k)\big)^{-1}\nabla q_{c^k}(\lambda^k),$$

for maximizing the dual function. In view of Eqs. (4.40) and (4.42), this iteration can be written as

$$\lambda^{k+1} = \lambda^k + (B^k)^{-1} h(x^k),$$

where

$$B^k = \nabla h\big(x^k\big)'\big\{\nabla^2_{xx} L_{c^k}(x^k,\lambda^k)\big\}^{-1}\nabla h\big(x^k\big)$$

and x^k minimizes $L_c(\cdot,\lambda^k)$.

An alternative form, which turns out to be more appropriate when the minimization of the augmented Lagrangian is inexact is given by

$$\lambda^{k+1} = \lambda^k + (B^k)^{-1}\Big(h(x^k) - \nabla h(x^k)'\big(\nabla^2_{xx} L_{c^k}(x^k,\lambda^k)\big)^{-1}\nabla_x L_{c^k}(x^k,\lambda^k)\Big). \tag{4.44}$$

When $\nabla_x L_{c^k}(x^k,\lambda^k) = 0$, the two forms are equivalent.

To provide motivation for iteration (4.44), let us consider Newton's method for solving the system of necessary conditions

$$\nabla_x L_c(x,\lambda) = \nabla f(x) + \nabla h(x)\big(\lambda + ch(x)\big) = 0, \qquad h(x) = 0.$$

In this method, we linearize the above system around the current iterate (x^k, λ^k), and we obtain the next iterate (x^{k+1}, λ^{k+1}) from the solution of the linearized system

$$\begin{pmatrix} \nabla^2_{xx} L_{c^k}(x^k,\lambda^k) & \nabla h(x^k) \\ \nabla h(x^k)' & 0 \end{pmatrix} \begin{pmatrix} x^{k+1} - x^k \\ \lambda^{k+1} - \lambda^k \end{pmatrix} = - \begin{pmatrix} \nabla_x L_{c^k}(x^k,\lambda^k) \\ h(x^k) \end{pmatrix}.$$

It is straightforward to verify (a derivation will be given in Section 4.4) that λ^{k+1} is given by Eq. (4.44), while

$$x^{k+1} = x^k - \big(\nabla^2_{xx} L_{c^k}(x^k,\lambda^k)\big)^{-1} \nabla_x L_{c^k}\big(x^k,\lambda^{k+1}\big).$$

This justifies the use of the extra term

$$\nabla h(x^k)'(\nabla^2_{xx} L_{c^k}(x^k,\lambda^k))^{-1} \nabla_x L_{c^k}(x^k,\lambda^k)$$

in Eq. (4.44) when the minimization of the augmented Lagrangian is inexact.

4.2.5 The Exponential Method of Multipliers*

One of the drawbacks of the method of multipliers when applied to inequality constrained problems is that the corresponding augmented Lagrangian function is not twice differentiable even if the cost and constraint functions are. As a result, serious difficulties can arise when Newton-like methods are used to minimize the augmented Lagrangian, particularly when the problem is a linear program. This motivates alternative twice differentiable augmented Lagrangians to handle inequality constraints, which we now describe.

Consider the problem

$$\begin{aligned} &\text{minimize} \;\; f(x) \\ &\text{subject to} \;\; g_1(x) \leq 0, \ldots, g_r(x) \leq 0. \end{aligned}$$

We introduce a method of multipliers characterized by a twice differentiable penalty function $\psi : \Re \mapsto \Re$ with the following properties:

(i) $\nabla^2 \psi(t) > 0$ for all $t \in \Re$,

(ii) $\psi(0) = 0$, $\nabla \psi(0) = 1$,

(iii) $\lim_{t \to -\infty} \psi(t) > -\infty$,

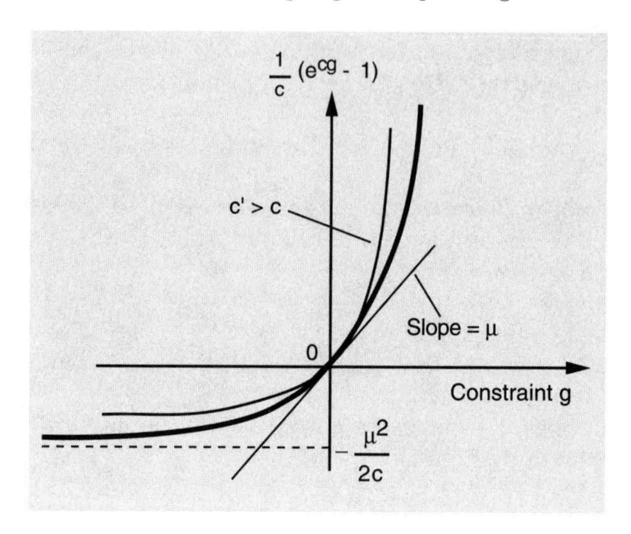

Figure 4.2.8. The penalty term of the exponential method of multipliers.

(iv) $\lim_{t\to-\infty} \nabla\psi(t) = 0$ and $\lim_{t\to\infty} \nabla\psi(t) = \infty$.

A simple and interesting special case is the *exponential penalty function*

$$\psi(t) = e^t - 1,$$

(see Fig. 4.2.8).

The method consists of the sequence of unconstrained minimizations

$$x^k \in \arg\min_{x\in\Re^n} \left\{ f(x) + \sum_{j=1}^m \frac{\mu_j^k}{c_j^k} \psi\big(c_j^k g_j(x)\big) \right\},$$

followed by the multiplier iterations

$$\mu_j^{k+1} = \mu_j^k \nabla\psi\big(c_j^k g_j(x^k)\big), \qquad j = 1,\ldots,r. \tag{4.45}$$

Here $\{c_j^k\}$ is a positive penalty parameter sequence for each j, and the initial multipliers μ_j^0 are arbitrary positive numbers.

Note that for fixed $\mu_j^k > 0$, the "penalty" term

$$\frac{\mu_j^k}{c_j^k} \psi\big(c_j^k g_j(x)\big)$$

tends (as $c_j^k \to \infty$) to ∞ for all infeasible x $[g_j(x) > 0]$ and to zero for all feasible x $[g_j(x) \le 0]$. To see this, note that by convexity of ψ, we have $\psi(ct) \ge \psi(ct/2) + (ct/2)\nabla\psi(ct/2)$, from which we obtain

$$\frac{1}{c}\psi(ct) \ge \frac{1}{c}\psi(ct/2) + \frac{t}{2}\nabla\psi(ct/2).$$

Thus if $t > 0$, the assumptions $\psi(ct/2) > 0$ and $\lim_{\tau\to\infty} \nabla\psi(\tau) = \infty$ imply that $\lim_{c\to\infty}(1/c)\psi(ct) = \infty$. Also, if $t < 0$, we have $\inf_{c>0} \psi(ct) = \inf_{\tau<0} \psi(\tau) > -\infty$, so that $\lim_{c\to\infty}(1/c)\psi(ct) = 0$.

On the other hand, for fixed c_j^k, as $\mu_j^k \to 0$ (which is expected to occur if the jth constraint is inactive at the optimum), the penalty term goes to zero for all x, feasible or infeasible. This is contrary to what happens in the quadratic penalty and augmented Lagrangian methods, and turns out to be a complicating factor in the analysis.

For the exponential penalty function $\psi(t) = e^t - 1$, the multiplier iteration (4.45) takes the form

$$\mu_j^{k+1} = \mu_j^k e^{c_j^k g_j(x^k)}, \qquad j = 1, \ldots, r.$$

Another interesting method, known as the *modified barrier method*, is based on the following modified version of the logarithmic barrier function

$$\psi(t) = -\ln(1 - t),$$

for which the multiplier iteration (4.45) takes the form

$$\mu_j^{k+1} = \frac{\mu_j^k}{1 - c_j^k g_j(x^k)}, \qquad j = 1, \ldots, r.$$

This method is not really a special case of the generic method because the penalty function ψ is defined only on the set $(-\infty, 1)$, but it shares the same qualitative characteristics as the generic method.

Two practical points regarding the exponential method are worth mentioning. The first is that the exponential terms in the augmented Lagrangian function can easily become very large with an attendant computer overflow. [The modified barrier method has a similar and even more serious difficulty: it tends to ∞ as $c_j^k g_j(x^k) \to 1$.] One way to deal with this disadvantage is to define $\psi(t)$ as the exponential $e^t - 1$ only for t in an interval $(-\infty, A]$, where A is such that e^A is within the floating point range of the computer; outside the interval $(-\infty, A]$, $\psi(t)$ can be defined as any function such that the properties required of ψ, including twice differentiability, are maintained over the entire real line. For example ψ can be a quadratic function with parameters chosen so that $\nabla^2\psi$ is continuous at the splice point A.

The second point is that it makes sense to introduce a different penalty parameter c_j^k for the jth constraint and to let c_j^k depend on the current values of the corresponding multiplier μ_j^k via

$$c_j^k = \frac{w^k}{\mu_j^k}, \qquad \forall\, j, \tag{4.46}$$

where $\{w^k\}$ is a positive scalar sequence with $w^k \leq w^{k+1}$ for all k. The reason can be seen by using the series expansion of the exponential term to write

$$\frac{\mu_j}{c}\left(e^{cg_j(x)} - 1\right) = \frac{\mu_j}{c}\left(cg_j(x) + \frac{c^2}{2}\big(g_j(x)\big)^2 + \frac{c^3}{3!}\big(g_j(x)\big)^3 + \cdots\right).$$

If the jth constraint is active at the eventual limit, the terms of order higher than quadratic can be neglected and we can write approximately

$$\frac{\mu_j}{c}\left(e^{cg_j(x)} - 1\right) \approx \mu_j g_j(x) + \frac{c\mu_j}{2} g_j(x)^2.$$

Thus the exponential augmented Lagrangian term becomes similar to the quadratic term, except that the role of the penalty parameter is played by the product $c\mu_j$. This motivates the use of selection rules such as Eq. (4.46). A similar penalty selection rationale applies also to other penalty functions in the class.

Generally the convergence analysis of the exponential method of multipliers and other methods in the class, under second order sufficiency conditions, turns out to be not much more difficult than for the quadratic method (see the references). The convergence results available are as powerful as for the quadratic method. The practical performances of the exponential and the quadratic method of multipliers are roughly comparable for problems where second order differentiability of the augmented Lagrangian function turns out to be of no concern. The exponential method has an edge for problems where the lack of second order differentiability in the quadratic method causes difficulties.

EXERCISES

4.2.1

Consider the problem

$$\begin{aligned} &\text{minimize} \;\; f(x) = \tfrac{1}{2}\left(x_1^2 - x_2^2\right) - 3x_2 \\ &\text{subject to} \;\; x_2 = 0. \end{aligned}$$

(a) Calculate the optimal solution and the Lagrange multiplier.

(b) For $k = 0, 1, 2$ and $c^k = 10^{k+1}$ calculate and compare the iterates of the quadratic penalty method with $\lambda^k = 0$ for all k and the method of multipliers with $\lambda^0 = 0$.

(c) For this problem, draw the figure that interprets geometrically the methods (cf. Figs. 4.2.5 and 4.2.6), and plot the iterates of the two methods on this figure for $k = 0, 1, 2$.

(d) Suppose that c is taken to be constant in the method of multipliers. For what values of c would the augmented Lagrangian have a minimum and for what values of c would the method converge?

4.2.2

Consider the problem

$$\begin{aligned} &\text{minimize} \;\; f(x) = \tfrac{1}{2}\big(x_1^2 + |x_2|^\rho\big) + 2x_2 \\ &\text{subject to} \;\; x_2 = 0, \end{aligned}$$

where $\rho > 1$.

(a) Calculate the optimal solution and the Lagrange multiplier.

(b) Write a computer program to calculate the iterates of the multiplier method with $\lambda^0 = 0$, and $c^k = 1$ for all k. Show computationally that the rate of convergence is sublinear if $\rho = 1.5$, linear if $\rho = 2$, and superlinear if $\rho = 3$.

(c) Give a heuristic argument why the rate of convergence is sublinear if $\rho < 2$, linear if $\rho = 2$, and superlinear if $\rho > 2$. What happens in the limit where $\rho = 1$?

4.2.3

Consider the problem of Exercise 4.2.1. Verify that the second order method of multipliers converges in one iteration provided c is sufficiently large, and estimate the threshold value for c.

4.2.4 (Convergence Threshold and Convergence Rate of the Method of Multipliers) (www)

Consider the quadratic problem

$$\begin{aligned} &\text{minimize} \;\; \tfrac{1}{2}x'Qx \\ &\text{subject to} \;\; Ax = b, \end{aligned}$$

where Q is symmetric and A is an $m \times n$ matrix of rank m. Let f^* be the optimal value of the problem and assume that the problem has a unique minimum x^* with associated Lagrange multiplier λ^*. Verify that for sufficiently large c, the penalized dual function is

$$q_c(\lambda) = -\tfrac{1}{2}(\lambda - \lambda^*)'A(Q + cA'A)^{-1}A'(\lambda - \lambda^*) + f^*.$$

(Use the quadratic programming duality theory of Section 3.4 to show that q_c is a quadratic function and to derive the Hessian matrix of q_c. Then use the fact that q_c is maximized at λ^* and that its maximum value is f^*.) Consider the first order method of multipliers.

(a) Use the theory of Section 1.3 to show that for all k

$$\|\lambda^{k+1} - \lambda^*\| \leq r^k \|\lambda^k - \lambda^*\|,$$

where

$$r^k = \max\{|1 - c^k E_{c^k}|, |1 - c^k e_{c^k}|\}$$

and E_c and e_c denote the maximum and the minimum eigenvalues of the matrix $A(Q + cA'A)^{-1}A'$.

(b) Using the matrix identity $\left(I + c^k AQ^{-1}A'\right)^{-1} = I - c^k A(Q + c^k A'A)^{-1}A'$ (cf. Section A.3 in Appendix A), relate the eigenvalues of the matrix $A(Q + c^k A'A)^{-1}A'$ with those of the matrix $AQ^{-1}A'$. Show that if $\gamma_1, \ldots, \gamma_m$ are the eigenvalues of $(AQ^{-1}A')^{-1}$, we have

$$r^k = \max_{i=1,\ldots,m} \left| \frac{\gamma_i}{\gamma_i + c^k} \right|.$$

(c) Show that the method converges to λ^* if $c > \bar{c}$, where $\bar{c} = 0$ if $\gamma_i \geq 0$ for all i, and $\bar{c} = -2\min\{\gamma_1, \ldots, \gamma_m\}$ otherwise.

4.2.5 (Stepsize Analysis of the Method of Multipliers) 

Consider the problem of Exercise 4.2.4. Use the results of that exercise to analyze the convergence and rate of convergence of the generalized method of multipliers

$$\lambda^{k+1} = \lambda^k + \alpha^k(Ax^k - b),$$

where α^k is a positive stepsize. Show in particular that if Q is positive definite and $c^k = c$ for all k, convergence is guaranteed if $\delta \leq \alpha^k \leq 2c$ for all k, where δ is some positive scalar. (For a solution and further analysis, see [Ber82a], p. 126.)

4.2.6

A weakness of the quadratic penalty method is that the augmented Lagrangian may not have a global minimum. As an example, show that the scalar problem

$$\begin{aligned} &\text{minimize} \quad -x^4 \\ &\text{subject to} \quad x = 0 \end{aligned}$$

has the unique global minimum $x^* = 0$ but its augmented Lagrangian

$$L_{c^k}(x, \lambda^k) = -x^4 + \lambda^k x + \frac{c^k}{2}x^2$$

has no global minimum for every c^k and λ^k. To overcome this difficulty, consider a penalty function of the form

$$\frac{c}{2}\|h(x)\|^2 + \|h(x)\|^\rho,$$

where $\rho > 4$, instead of $(c/2)\|h(x)\|^2$. Show that $L_{c^k}(x, \lambda^k)$ has a global minimum for every λ^k and $c^k > 0$.

4.2.7 (Two-Sided Inequality Constraints [Ber77], [Ber82a])

The purpose of this exercise is to show how to treat two-sided inequality constraints by using a single multiplier per constraint. Consider the problem

$$\begin{aligned} &\text{minimize} \ \ f(x) \\ &\text{subject to} \ \ \alpha_j \le g_j(x) \le \beta_j, \qquad j = 1, \ldots, r, \end{aligned}$$

where $f : \Re^n \mapsto \Re$ and $g_j : \Re^n \mapsto \Re$ are given functions, and α_j and β_j, $j = 1, \ldots, r$, are given scalars with $\alpha_j < \beta_j$. The method consists of sequential minimizations of the form

$$\begin{aligned} &\text{minimize} \ \ f(x) + \sum_{j=1}^{r} P_j\big(g_j(x), \mu_j^k, c^k\big) \\ &\text{subject to} \ \ x \in \Re^n, \end{aligned}$$

where

$$P_j\big(g_j(x), \mu_j^k, c^k\big) = \min_{u_j \in [g_j(x) - \beta_j,\, g_j(x) - \alpha_j]} \left\{ \mu_j^k u_j + \frac{c^k}{2}|u_j|^2 \right\}.$$

Each of these minimizations is followed by the multiplier iteration

$$\mu_j^{k+1} = \begin{cases} \mu_j^k + c^k\big(g_j(x^k) - \beta_j\big) & \text{if } \mu_j^k + c^k\big(g_j(x^k) - \beta_j\big) > 0, \\ \mu_j^k + c^k\big(g_j(x^k) - \alpha_j\big) & \text{if } \mu_j^k + c^k\big(g_j(x^k) - \alpha_j\big) < 0, \\ 0 & \text{otherwise,} \end{cases}$$

where x^k is a minimizing vector. Justify the method by introducing artificial variables u_j, by converting the problem to the equivalent form

$$\begin{aligned} &\text{minimize} \ \ f(x) \\ &\text{subject to} \ \ \alpha_j \le g_j(x) - u_j \le \beta_j, \qquad u_j = 0, \quad j = 1, \ldots, r, \end{aligned}$$

and by applying a multiplier method for this problem, where only the constraints $u_j = 0$ are eliminated by means of a quadratic penalty function (partial elimination of constraints).

4.2.8 (Proof of Ill-Conditioning as $c^k \to \infty$)

Consider the quadratic penalty method ($c^k \to \infty$) for the equality constrained problem of minimizing $f(x)$ subject to $h(x) = 0$, and assume that the generated sequence converges to a local minimum x^* that is also a regular point. Show that the condition number of the Hessian $\nabla^2_{xx} L_{c^k}(x^k, \lambda^k)$ tends to ∞. *Hint*: We have

$$\nabla^2_{xx} L_{c^k}(x^k, \lambda^k) = \nabla^2_{xx} L_0(x^k, \tilde{\lambda}^k) + c^k \nabla h(x^k) \nabla h(x^k)'$$

where $\tilde{\lambda}^k = \lambda^k + c^k h(x^k)$. The minimum eigenvalue $m(x^k, \lambda^k, c^k)$ of this matrix satisfies

$$\begin{aligned} m(x^k, \lambda^k, c^k) &= \min_{\|z\|=1} z' \nabla^2_{xx} L_{c^k}(x^k, \lambda^k) z \\ &\le \min_{\|z\|=1, \nabla h(x^k)' z = 0} z' \nabla^2_{xx} L_{c^k}(x^k, \lambda^k) z \\ &= \min_{\|z\|=1, \nabla h(x^k)' z = 0} z' \nabla^2_{xx} L_0(x, \tilde{\lambda}^k) z. \end{aligned}$$

The maximum eigenvalue $M(x^k, \lambda^k, c^k)$ satisfies

$$\begin{aligned} M(x^k, \lambda^k, c^k) &= \max_{\|z\|=1} z' \nabla^2_{xx} L_{c^k}(x^k, \lambda^k) z \\ &\ge \min_{\|z\|=1} z' \nabla^2_{xx} L_0(x^k, \tilde{\lambda}^k) z + c^k \max_{\|z\|=1} z' \nabla h(x^k) \nabla h(x^k)' z. \end{aligned}$$

Use Prop. 4.2.2 to show that

$$\lim_{c^k \to \infty} \frac{M(x^k, \lambda^k, c^k)}{m(x^k, \lambda^k, c^k)} = \infty.$$

4.2.9 (www)

Let $\{x^k\}$ be a sequence generated by the logarithmic barrier method. Formulate conditions under which the sequences $\{-\epsilon^k / g_j(x^k)\}$ converge to corresponding Lagrange multipliers. *Hint*: Compare with the corresponding result of Prop. 4.2.2 for the quadratic penalty function.

4.2.10

State and prove analogs of Props. 4.2.1 and 4.2.2 for the case where the penalty function $(c/2)\|h(x)\|^2 + \|h(x)\|^\rho$ with $\rho > 1$ is used in place of the quadratic $(c/2)\|h(x)\|^2$.

4.2.11 (Primal-Dual Methods not Using a Penalty) 

Consider the equality constrained problem

$$\begin{aligned} &\text{minimize} \ \ f(x) \\ &\text{subject to} \ \ h(x) = 0, \end{aligned}$$

where f and h are twice continuously differentiable, and let x^* be a local minimum that is regular. Let λ^* be the associated Lagrange multiplier vector, and assume that the Hessian $\nabla^2_{xx} L(x^*, \lambda^*)$ of the (ordinary) Lagrangian

$$L(x, \lambda) = f(x) + \lambda' h(x)$$

is positive definite. (Note that this is stronger than what is required by the second order sufficiency conditions.) Consider the iteration

$$\lambda^{k+1} = \lambda^k + \alpha h(x^k),$$

where α is a positive scalar stepsize and x^k minimizes $L(x, \lambda^k)$ (the minimization is local in a suitable neighborhood of x^*). Show that there exists a threshold $\bar{\alpha} > 0$ and a sphere centered at λ^* such that if λ^0 belongs to this sphere and $\alpha < \bar{\alpha}$, then λ^k converges to λ^*. Consider first the case where f is quadratic and h is linear, and sketch an analysis for the more general case. *Hint*: Even though there is no penalty parameter here, the method can be viewed as a first order method of multipliers for the artificial problem

$$\begin{aligned} &\text{minimize} \ \ f(x) - \frac{\alpha}{2} \|h(x)\|^2 \\ &\text{subject to} \ \ h(x) = 0. \end{aligned}$$

Use the threshold of Exercise 4.2.4(c) to verify that $\bar{\alpha}$ can be twice the minimum eigenvalue of $\nabla h(x^*)' \big(\nabla^2_{xx} L(x^*, \lambda^*)\big)^{-1} \nabla h(x^*)$.

4.3 EXACT PENALTIES – SEQUENTIAL QUADRATIC PROGRAMMING*

In this section we consider penalty methods that are *exact* in the sense that they require only one unconstrained minimization to obtain an optimal solution of the original constrained problem.

To get a sense of how this is possible, consider the equality constrained problem

$$\begin{aligned} &\text{minimize} \ \ f(x) \\ &\text{subject to} \ \ h_i(x) = 0, \qquad i = 1, \ldots, m, \end{aligned} \tag{4.47}$$

where f and h_i are continuously differentiable, and let

$$L(x,\lambda) = f(x) + \lambda' h(x)$$

be the corresponding Lagrangian function. Then by minimizing the function

$$P(x,\lambda) = \|\nabla_x L(x,\lambda)\|^2 + \|h(x)\|^2 \tag{4.48}$$

over $(x,\lambda) \in \Re^{n+m}$ we can obtain local minima-Lagrange multiplier pairs (x^*,λ^*) satisfying the first order necessary conditions

$$\nabla_x L(x^*,\lambda^*) = 0, \qquad h(x^*) = 0.$$

We may view $P(x,\lambda)$ as an *exact penalty function*, that is, a function whose unconstrained minima are (or strongly relate to) optimal solutions and/or Lagrange multipliers of a constrained problem.

The exact penalty function $P(x,\lambda)$ of Eq. (4.48) has been used extensively in the special case where $m = n$ and the problem is to solve the system of constraint equations $h(x) = 0$ (in this case, any cost function f may be used). However, both in this special case and more generally, $P(x,\lambda)$ has significant drawbacks because it does not discriminate between local minima and local maxima, and it may also have local minima $(\bar{x},\bar{\lambda})$ that are not global and do not satisfy the necessary optimality conditions, i.e., $P(\bar{x},\bar{\lambda}) > 0$. There are, however, more sophisticated exact penalty functions that do not have these drawbacks.

We may distinguish between *differentiable* and *nondifferentiable* exact penalty functions. The former have the advantage that they can be minimized by the unconstrained methods we have already studied in Chapter 1. The latter involve nondifferentiabilities, so the methods of Chapter 1 are not directly applicable. We will develop special algorithms, called *sequential quadratic programming methods*, for their minimization.

Nondifferentiable exact penalty methods have been more popular in practice than their differentiable counterparts, and they will receive most of our attention. On the other hand, differentiable exact penalty methods have some interesting advantages; see [Ber82a], which treats extensively both types of methods.

4.3.1 Nondifferentiable Exact Penalty Functions

We will show that solutions of the equality constrained problem (4.47) are related to solutions of the (nondifferentiable) unconstrained problem

$$\begin{aligned} &\text{minimize} \;\; f(x) + cP(x) \\ &\text{subject to} \;\; x \in \Re^n, \end{aligned}$$

where $c > 0$ and P is the nondifferentiable penalty function defined by

$$P(x) = \max_{i=1,\ldots,m} |h_i(x)|.$$

Indeed let x^* be a local minimum, which is a regular point and satisfies together with a corresponding Lagrange multiplier vector λ^*, the second order sufficiency conditions of Section 3.2 (Prop. 3.2.1). Consider also the primal function p defined in a neighborhood of the origin by

$$p(u) = \min\{f(x) \mid h(x) = u, \|x - x^*\| < \epsilon\},$$

where $\epsilon > 0$ is some scalar; see the sensitivity theorem (Prop. 3.2.2). Then if we locally minimize $f + cP$ around x^*, we can split the minimization in two: first minimize over all x satisfying $h(x) = u$ and then minimize over all possible u. We have

$$\begin{aligned} \inf_{\|x-x^*\|<\epsilon} & \left\{ f(x) + c \max_{i=1,\ldots,m} |h_i(x)| \right\} \\ &= \inf_{u \in U_\epsilon} \inf_{\{x \mid h(x)=u,\, \|x-x^*\|<\epsilon\}} \left\{ f(x) + c \max_{i=1,\ldots,m} |h_i(x)| \right\} \\ &= \inf_{u \in U_\epsilon} p_c(u), \end{aligned}$$

where

$$U_\epsilon = \{u \mid h(x) = u \text{ for some } x \text{ with } \|x - x^*\| < \epsilon\},$$

and

$$p_c(u) = p(u) + c \max_{i=1,\ldots,m} |u_i|.$$

Since, according to the sensitivity theorem, we have $\nabla p(0) = -\lambda^*$, we can use the mean value theorem to write for each u in a neighborhood of the origin

$$p(u) = p(0) - u'\lambda^* + \tfrac{1}{2} u' \nabla^2 p(\bar{\alpha} u) u,$$

where $\bar{\alpha}$ is some scalar in $[0, 1]$. Thus

$$p_c(u) = p(0) - \sum_{i=1}^m u_i \lambda_i^* + c \max_{i=1,\ldots,m} |u_i| + \tfrac{1}{2} u' \nabla^2 p(\bar{\alpha} u) u. \tag{4.49}$$

Assume that c is sufficiently large so that for some $\gamma > 0$,

$$c \geq \sum_{i=1}^m |\lambda_i^*| + \gamma.$$

Then it follows that

$$\begin{aligned} c \max_{i=1,\ldots,m} |u_i| &\geq \left(\sum_{i=1}^m |\lambda_i^*| + \gamma \right) \max_{i=1,\ldots,m} |u_i| \\ &\geq \sum_{i=1}^m u_i \lambda_i^* + \gamma \max_{i=1,\ldots,m} |u_i|. \end{aligned}$$

Using this relation in Eq. (4.49), we obtain

$$p_c(u) \geq p(0) + \gamma \max_{i=1,\ldots,m} |u_i| + \tfrac{1}{2} u' \nabla^2 p(\bar{\alpha} u) u.$$

For u sufficiently close to zero, the last term is dominated by the next to last term, so we have

$$p_c(u) > p(0) = p_c(0)$$

for all $u \neq 0$ in a neighborhood N of the origin. Hence $u = 0$ is a strict local minimum of p_c as shown in Fig. 4.3.1. Since we have

$$f(x) + cP(x) \geq p(u) + c \max_{i=1,\ldots,m} |u_i| = p_c(u)$$

for all $u \neq 0$ in the neighborhood N and x such that $h(x) = u$ with $\|x - x^*\| < \epsilon$, and we also have $p_c(0) = f(x^*)$, we obtain that for all x in a neighborhood of x^* with $P(x) > 0$,

$$f(x) + cP(x) > f(x^*).$$

Since x^* is a strict local minimum of f over all x such that $P(x) = 0$, it follows that x^* is a strict local minimum of $f + cP$, provided that $c > \sum_{i=1}^m |\lambda_i^*|$.

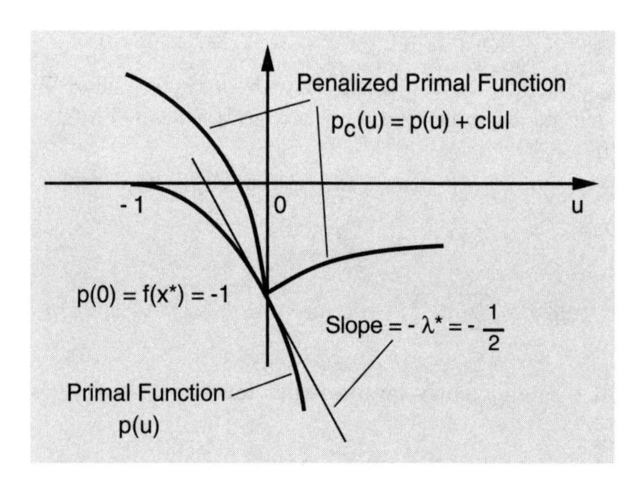

Figure 4.3.1. Illustration of how for c large enough, $u = 0$ is a strict local minimum of $p_c(u) = p(u) + c \max_{i=1,\ldots,m} |u_i|$. The figure corresponds to the two-dimensional problem where $f(x) = x_1$ and $h(x) = x_1^2 + x_2^2 - 1$. The optimal solution is $x^* = (-1, 0)$ and $\lambda^* = 1/2$. The primal functional is defined for $u \geq -1$ and is given by

$$p(u) = \min_{x_1^2 + x_2^2 - 1 = u} x_1 = -\sqrt{1+u}.$$

Note that $\nabla p(0) = \lambda^*$ and that we must have $c > \lambda^*$ in order for the nondifferentiable penalty function to be exact.

The above argument can be extended to the case where there are additional inequality constraints of the form $g_j(x) \le 0$, $j = 1, \ldots, r$. These constraints can be converted to the equality constraints

$$g_j(x) + z_j^2 = 0, \qquad j = 1, \ldots, r,$$

by introducing the squared slack variables z_j as in Section 3.3. The slack variables can be eliminated from the corresponding exact penalty function

$$f(x) + c\max\bigl\{|g_1(x) + z_1^2|, \ldots, |g_r(x) + z_r^2|, |h_1(x)|, \ldots, |h_m(x)|\bigr\} \qquad (4.50)$$

by explicit minimization. In particular, we have

$$\min_{z_j} |g_j(x) + z_j^2| = \max\bigl\{0, g_j(x)\bigr\},$$

so minimization of the exact penalty function (4.50) is equivalent to minimization of the function

$$f(x) + c\max\bigl\{0, g_1(x), \ldots, g_r(x), |h_1(x)|, \ldots, |h_m(x)|\bigr\}.$$

Thus, by repeating the earlier argument given for equality constraints, we have the following proposition.

Proposition 4.3.1: Let x^* be a local minimum of the problem

$$\begin{array}{ll} \text{minimize} & f(x) \\ \text{subject to} & h_i(x) = 0, \quad i = 1, \ldots, m, \qquad g_j(x) \le 0, \quad j = 1, \ldots, r, \end{array}$$

which is regular and satisfies together with corresponding Lagrange multiplier vectors λ^* and μ^*, the second order sufficiency conditions of Prop. 3.3.2. Then, if

$$c > \sum_{i=1}^{m} |\lambda_i^*| + \sum_{j=1}^{r} \mu_j^*,$$

the vector x^* is a strict unconstrained local minimum of $f + cP$, where

$$P(x) = \max\bigl\{0, g_1(x), \ldots, g_r(x), |h_1(x)|, \ldots, |h_m(x)|\bigr\}.$$

An example illustrating the above proposition is given in Fig. 4.3.2. The proof of the proposition was relatively simple but made assumptions that are stronger than necessary. There are related results that do not require second order differentiability assumptions. In particular, it is shown in Exercise 4.3.4 that if c is sufficiently large, a regular local minimum x^* is a "stationary" point of the exact penalty function $f + cP$ (in a sense to be made precise shortly). The reverse is not necessarily true. In particular, there may exist local minima of $f + cP$ that do not correspond to constrained local minima of f regardless of the value of c; see Exercise 4.3.1. Under convexity assumptions the situation is more satisfactory; see the theory developed in Section 5.4.5.

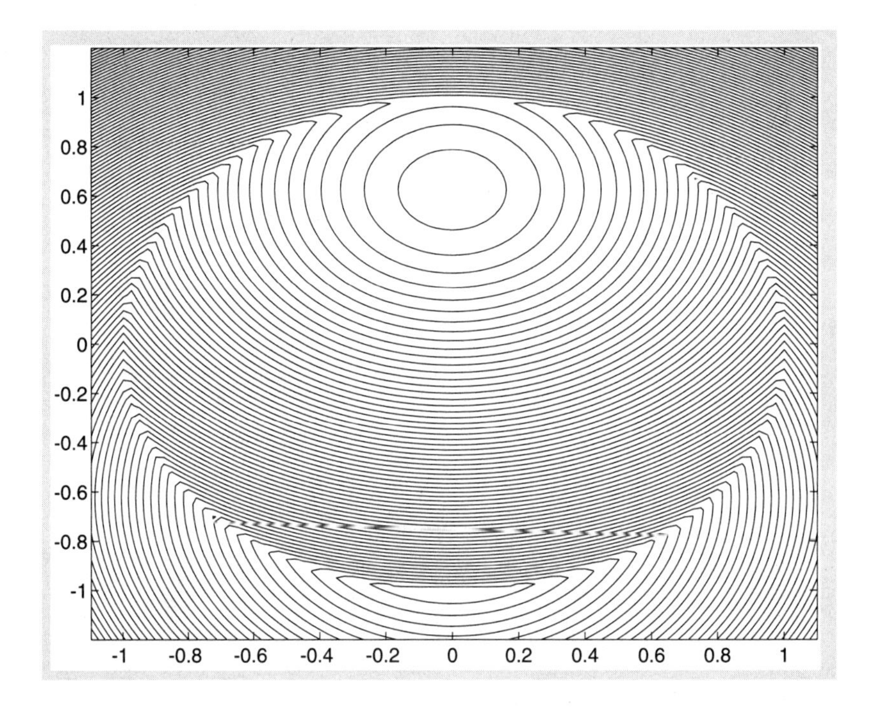

Figure 4.3.2. Contours of the function $f + cP$ for the two-dimensional problem where

$$f(x) = x_1, \qquad h(x) = x_1^2 + x_2^2 - 1$$

(cf. Fig. 4.3.1). For c greater than the Lagrange multiplier $\lambda^* = 1/2$, the optimal solution $x^* = (-1, 0)$ is a local minimum of $f + cP$. This is not so for $c < \lambda^*$. The figure corresponds to $c = 0.8$.

Local Minima and Stationary Points of Exact Penalties

In order to simplify notation in what follows, we will assume that all the constraints are inequalities. All the analysis and algorithms to be given admit simple extensions to the equality constrained case simply by converting each equality constraint into two inequalities. We assume throughout that the cost and constraint functions are at least once continuously differentiable.

We will develop an algorithm for minimizing $f + cP$, where $c > 0$ and

$$P(x) = \max\{g_0(x), g_1(x), \ldots, g_r(x)\}, \qquad \forall\, x \in \Re^n, \tag{4.51}$$

$$g_0(x) = 0, \qquad \forall\, x \in \Re^n. \tag{4.52}$$

We first introduce some definitions and develop some preliminary results.

For $x \in \Re^n$, $d \in \Re^n$, and $c > 0$, we introduce the index set

$$J(x) = \{j \mid g_j(x) = P(x),\ j = 0, 1, \ldots, r\},$$

and we denote

$$\theta_c(x; d) = \max\{\nabla f(x)'d + c\nabla g_j(x)'d \mid j \in J(x)\}.$$

The function θ_c plays the role that the gradient would play if the function $f + cP$ were differentiable. In particular,

$$f(x) + cP(x) + \theta_c(x; d)$$

may be viewed as a linear approximation of $f + cP$ for variations d around x; see Fig. 4.3.3.

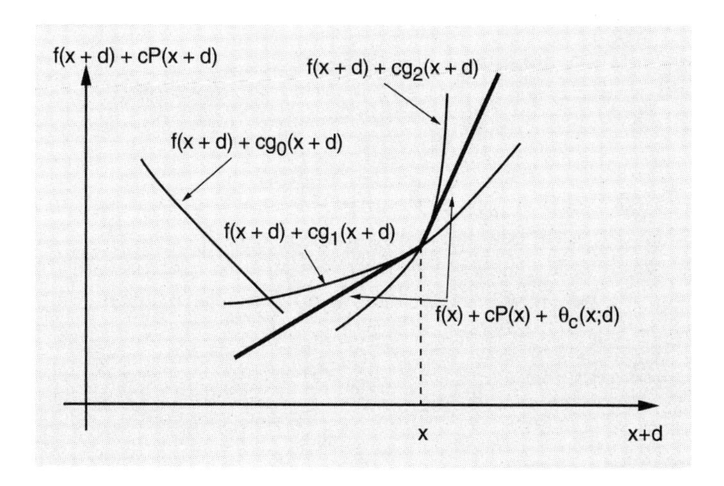

Figure 4.3.3. Illustration of $\theta_c(x; d)$ at a point x. It is the first order estimate of the variation

$$f(x + d) + cP(x + d) - f(x) - cP(x)$$

of $f + cP$ around x.

Since at an unconstrained local minimum x^*, $f+cP$ cannot decrease along any direction, the preceding interpretation of θ_c motivates us to call a vector x^* a *stationary point* of $f+cP$ if for all $d \in \Re^n$ there holds

$$\theta_c(x^*; d) \geq 0.$$

The following proposition shows that local minima of $f+cP$ must be stationary points (in fact global minima under convexity assumptions). Furthermore, descent directions of $f+cP$ at a given point x can be found only if x is nonstationary. Such directions can be obtained from the following (convex) quadratic program, in $(d,\xi) \in \Re^{n+1}$,

$$\begin{aligned} &\text{minimize} \quad \nabla f(x)'d + \tfrac{1}{2}d'Hd + c\xi \\ &\text{subject to} \quad (d,\xi) \in \Re^{n+1}, \qquad g_j(x) + \nabla g_j(x)'d \leq \xi, \quad j = 0,1,\ldots,r, \end{aligned} \tag{4.53}$$

where $c > 0$ and H is a positive definite symmetric matrix. Note that for a fixed d, the minimum with respect to ξ is obtained at

$$\xi = \max_{j=0,1,\ldots,r} \{g_j(x) + \nabla g_j(x)'d\},$$

so by eliminating the variable ξ, and by adding to the cost the constant term $f(x)$, we can write the quadratic program (4.53) in the alternative form

$$\begin{aligned} &\text{minimize} \quad \max_{j=0,1,\ldots,r} \{f(x) + cg_j(x) + \nabla f(x)'d + c\nabla g_j(x)'d\} + \tfrac{1}{2}d'Hd \\ &\text{subject to} \quad d \in \Re^n. \end{aligned} \tag{4.54}$$

For small $\|d\|$, the maximum over j is attained for $j \in J(x)$, so we can substitute $P(x)$ in place of $g_j(x)$. The cost function then takes the form

$$f(x) + cP(x) + \theta_c(x;d) + \tfrac{1}{2}d'Hd,$$

so locally, for d near zero, the problem (4.54) can be viewed as minimization of a quadratic approximation of $f+cP$ around x; see Fig. 4.3.4. Also since the cost function of problem (4.54) is strictly convex in d, its optimal solution is unique, implying also that the quadratic program (4.53) has a unique optimal solution (d,ξ).

Proposition 4.3.2:

(a) A local minimum of $f+cP$ is a stationary point. Furthermore, for all $x \in \Re^n$, $d \in \Re^n$, and $\alpha > 0$,

$$f(x+\alpha d) + cP(x+\alpha d) - f(x) - cP(x) = \alpha\theta_c(x;d) + o(\alpha), \tag{4.55}$$

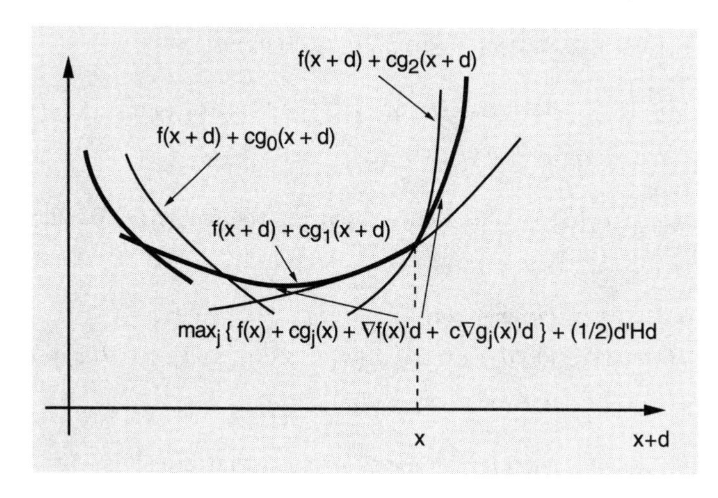

Figure 4.3.4. Illustration of the cost function

$$\max_{j=0,1,\ldots,r} \left\{ f(x) + cg_j(x) + \nabla f(x)'d + c\nabla g_j(x)'d \right\} + \tfrac{1}{2}d'Hd$$

of the quadratic program (4.54). For small $\|d\|$ this function takes the form

$$f(x) + cP(x) + \theta_c(x;d) + \tfrac{1}{2}d'Hd,$$

and is a quadratic approximation of $f + cP$ around x. It can be seen that by minimizing this function over d we obtain a direction of descent of $f + cP$ at x.

where $\lim_{\alpha \to 0^+} o(\alpha)/\alpha = 0$. As a result, if $\theta_c(x;d) < 0$, then d is a descent direction; that is, there exists $\bar{\alpha} > 0$ such that

$$f(x + \alpha d) + cP(x + \alpha d) < f(x) + cP(x), \qquad \forall\, \alpha \in (0, \bar{\alpha}].$$

(b) If f and g_j are convex functions, then a stationary point x^* of $f + cP$ is also a global minimum of $f + cP$.

(c) For any $x \in \Re^n$ and positive definite symmetric H, if (d, ξ) is the optimal solution of the quadratic program (4.53), then

$$\theta_c(x;d) \leq -d'Hd. \tag{4.56}$$

(d) A vector x is a stationary point of $f + cP$ if and only if the quadratic program (4.53) has $\{d = 0, \xi = P(x)\}$ as its optimal solution.

Proof: (a) We have for all $\alpha > 0$ and $j \in J(x)$,

$$f(x+\alpha d) + cg_j(x+\alpha d) = f(x) + \alpha \nabla f(x)'d + c\big(g_j(x) + \alpha \nabla g_i(x)'d\big) + o_j(\alpha),$$

where $\lim_{\alpha \to 0^+} o_j(\alpha)/\alpha = 0$. Hence, by using the fact $g_j(x) = P(x)$ for all $j \in J(x)$,

$$\begin{aligned} f(x+\alpha d) &+ c \max\{g_j(x+\alpha d) \mid j \in J(x)\} \\ &= f(x) + \alpha \nabla f(x)'d + c \max\{g_j(x) + \alpha \nabla g_j(x)'d \mid j \in J(x)\} + o(\alpha) \\ &= f(x) + cP(x) + \alpha \theta_c(x; d) + o(\alpha), \end{aligned}$$

where $\lim_{\alpha \to 0^+} o(\alpha)/\alpha = 0$. We have, for all α that are sufficiently small,

$$\begin{aligned} \max\{g_j(x+\alpha d) \mid j \in J(x)\} &= \max\{g_j(x+\alpha d) \mid j = 0, 1, \ldots, r\} \\ &= P(x+\alpha d). \end{aligned}$$

Combining the two above relations, we obtain

$$f(x+\alpha d) + cP(x+\alpha d) = f(x) + cP(x) + \alpha \theta_c(x; d) + o(\alpha),$$

which is Eq. (4.55).

If x^* is a local minimum of $f + cP$, then by Eq. (4.55), we have, for all d and $\alpha > 0$ such that $\|d\|$ and α are sufficiently small,

$$\alpha \theta_c(x^*; d) + o(\alpha) \geq 0.$$

Dividing by α and taking the limit as $\alpha \to 0$, we obtain $\theta_c(x^*; d) \geq 0$, so x^* is stationary.

(b) By convexity, we have (cf. Prop. B.3 in Appendix B) for all j and $x \in \Re^n$,

$$f(x) + cg_j(x) \geq f(x^*) + cg_j(x^*) + \big(\nabla f(x^*) + c\nabla g_j(x^*)\big)'(x - x^*).$$

Taking the maximum over j, we obtain

$$f(x) + cP(x) \geq \max_{j=0,1,\ldots,r} \big\{f(x^*) + cg_j(x^*) + \big(\nabla f(x^*) + c\nabla g_j(x^*)\big)'(x - x^*)\big\}.$$

For a sufficiently small scalar ϵ and for all x with $\|x - x^*\| < \epsilon$, the maximum above is attained for some $j \in J(x^*)$. Since $g_j(x^*) = P(x^*)$ for all $j \in J(x^*)$, we obtain for all x with $\|x - x^*\| < \epsilon$,

$$f(x) + cP(x) \geq f(x^*) + cP(x^*) + \theta_c(x^*; x - x^*) \geq f(x^*) + cP(x^*),$$

where the last inequality holds because x^* is a stationary point of $f + cP$. Hence x^* is a local minimum of the function $f + cP$, and in view of the convexity of $f + cP$, x^* is a global minimum.

(c) We have $g_j(x) + \nabla g_j(x)'d \leq \xi$ for all j. Since $g_j(x) = P(x)$ for all $j \in J(x)$, it follows that $\nabla g_j(x)'d \leq \xi - P(x)$ for all $j \in J(x)$ and therefore using the definition of θ_c we have

$$\theta_c(x; d) \leq \nabla f(x)'d + c\big(\xi - P(x)\big). \tag{4.57}$$

Let $\{\mu_j\}$ be a set of Lagrange multipliers for the quadratic program (4.53). The optimality conditions yield

$$\nabla f(x) + Hd + \sum_{j=0}^{r} \mu_j \nabla g_j(x) = 0, \tag{4.58}$$

$$c - \sum_{j=0}^{r} \mu_j = 0, \tag{4.59}$$

$$g_j(x) + \nabla g_j(x)'d \leq \xi, \qquad \mu_j \geq 0, \qquad j = 0, 1, \ldots, r,$$

$$\mu_j\big(g_j(x) + \nabla g_j(x)'d - \xi\big) = 0, \qquad j = 0, 1, \ldots, r.$$

By adding the last equation over all j and using Eq. (4.59), we have

$$\begin{aligned}
\sum_{j=0}^{r} \mu_j \nabla g_j(x)'d &= \sum_{j=0}^{r} \mu_j \xi - \sum_{j=0}^{r} \mu_j g_j(x) \\
&\geq \sum_{j=0}^{r} \mu_j \big(\xi - \max_{m=0,1,\ldots,r} g_m(x)\big) \\
&= \sum_{j=0}^{r} \mu_j \big(\xi - P(x)\big) \\
&= c\big(\xi - P(x)\big).
\end{aligned}$$

Combining this equation with Eq. (4.58) we obtain

$$\nabla f(x)'d + d'Hd + c\big(\xi - P(x)\big) \leq 0, \tag{4.60}$$

which in conjunction with Eq. (4.57) yields

$$\theta_c(x; d) + d'Hd \leq 0,$$

thus proving Eq. (4.56).

(d) x is stationary if and only if $\theta_c(x; d) \geq 0$ for all d, which by Eq. (4.56) is true if and only if $\big\{d = 0,\ \xi = P(x)\big\}$ is the optimal solution of the quadratic program (4.53). **Q.E.D.**

The Linearization Algorithm

We now introduce an iterative descent algorithm for minimizing the exact penalty function $f + cP$. It is called the *linearization algorithm* or *sequential quadratic programming*, and like the gradient projection method, it calculates the descent direction by solving a quadratic programming subproblem. It is given by

$$x^{k+1} = x^k + \alpha^k d^k,$$

where α^k is a nonnegative scalar stepsize, and d^k is a direction obtained by solving the quadratic program in (d, ξ)

$$\begin{aligned} &\text{minimize} \quad \nabla f(x^k)'d + \tfrac{1}{2} d'H^k d + c\xi \\ &\text{subject to} \quad g_j(x^k) + \nabla g_j(x^k)'d \leq \xi, \qquad j = 0, 1, \ldots, r. \end{aligned} \tag{4.61}$$

We require that H^k is a positive definite symmetric matrix. Proposition 4.3.2(c) implies that the solution d is a descent direction of $f + cP$ at x^k. The initial vector x^0 is arbitrary and the stepsize α^k is chosen by any one of the stepsize rules listed below:

(a) *Minimization rule*: Here α^k is chosen so that

$$f(x^k + \alpha^k d^k) + cP(x^k + \alpha^k d^k) = \min_{\alpha \geq 0} \big\{ f(x^k + \alpha d^k) + cP(x^k + \alpha d^k) \big\}.$$

(b) *Limited minimization rule*: Here a fixed scalar $s > 0$ is selected and α^k is chosen so that

$$f(x^k + \alpha^k d^k) = \min_{\alpha \in [0,s]} f(x^k + \alpha d^k).$$

(c) *Armijo rule*: Here fixed scalars s, β, and σ with $s > 0$, $\beta \in (0, 1)$, and $\sigma \in (0, \frac{1}{2})$, are selected, and we set $\alpha^k = \beta^{m_k} s$, where m_k is the first nonnegative integer m for which

$$f(x^k) + cP(x^k) - f(x^k + \beta^m s d^k) - cP(x^k + \beta^m s d^k) \geq \sigma \beta^m s d^{k\prime} H^k d^k. \tag{4.62}$$

It can be shown that if $d^k \neq 0$, the Armijo rule will yield a stepsize after a finite number of arithmetic operations. To see this, note that by Prop. 4.3.2(a) and Eq. (4.56), we have for all $\alpha > 0$,

$$\begin{aligned} f(x^k) + cP(x^k) - f(x^k + \alpha d^k) - cP(x^k + \alpha d^k) &= -\alpha \theta_c(x^k; d^k) + o(\alpha) \\ &\geq \alpha d^{k\prime} H^k d^k + o(\alpha). \end{aligned}$$

We then obtain

$$f(x^k) + cP(x^k) - f(x^k + \alpha d^k) - cP(x^k + \alpha d^k) \geq \sigma \alpha d^{k\prime} H^k d^k, \qquad \forall\, \alpha \in (0, \bar{\alpha}],$$

where $\bar{\alpha} > 0$ is such that for all $\alpha \in (0, \bar{\alpha}]$ we have $(1-\sigma)\alpha d^{k\prime} H^k d^k + o(\alpha) \geq 0$. Therefore, if $d^k \neq 0$, there is an integer m such that the Armijo test (4.62) is passed, while if $d^k = 0$, by Prop. 4.3.2(d), x^k is a stationary point of $f + cP$.

We have the following convergence result. Its proof is patterned after the corresponding proof for gradient methods for unconstrained minimization (cf. Prop. 1.2.1 in Section 1.2), but is considerably more complicated due to the constraints.

Proposition 4.3.3: Let $\{x^k\}$ be a sequence generated by the linearization algorithm, where the stepsize α^k is chosen by the minimization rule, the limited minimization rule, or the Armijo rule. Assume that there exist positive scalars γ and Γ such that

$$\gamma \|z\|^2 \leq z' H^k z \leq \Gamma \|z\|^2, \qquad \forall\, z \in \Re^n, \qquad k = 0, 1, \ldots,$$

(this condition corresponds to the assumption of a gradient-related direction sequence in unconstrained optimization). Then every limit point of $\{x^k\}$ is a stationary point of $f + cP$.

Proof: We argue by contradiction. Assume that a subsequence $\{x^k\}_K$ generated by the algorithm using the Armijo rule converges to a vector $\bar{x}$ that is not a stationary point of $f + cP$. Since $f(x^k) + cP(x^k)$ is monotonically decreasing, we have

$$f(x^k) + cP(x^k) \to f(\bar{x}) + cP(\bar{x})$$

and hence also

$$f(x^k) + cP(x^k) - f(x^{k+1}) - cP(x^{k+1}) \to 0.$$

By the definition of the Armijo rule, we have

$$f(x^k) + cP(x^k) - f(x^{k+1}) - cP(x^{k+1}) \geq \sigma \alpha^k d^{k\prime} H^k d^k.$$

Hence

$$\alpha^k d^{k\prime} H^k d^k \to 0. \tag{4.63}$$

Since for $k \in K$, d^k is the optimal solution of the quadratic program (4.61), we must have for some set of Lagrange multipliers $\{\mu_j^k\}$ and all $k \in K$,

$$\nabla f(x^k) + \sum_{j=0}^{r} \mu_j^k \nabla g_j(x^k) + H^k d^k = 0, \qquad c = \sum_{j=0}^{r} \mu_j^k, \tag{4.64}$$

$$\mu_j^k \geq 0, \qquad \mu_j^k\big(g_j(x^k) + \nabla g_j(x^k)'d^k - \xi^k\big) = 0, \qquad j = 0, 1, \ldots, r, \tag{4.65}$$

where

$$\xi^k = \max_{j=0,1,\ldots,r} \{g_j(x^k) + \nabla g_j(x^k)'d^k\}.$$

The relations $c = \sum_{j=0}^r \mu_j^k$ and $\mu_j^k \geq 0$ imply that the subsequences $\{\mu_j^k\}$ are bounded. Hence, without loss of generality, we may assume that for some μ_j, $j = 0, 1, \ldots, r$, we have

$$\{\mu_j^k\}_K \to \bar{\mu}_j, \qquad j = 0, 1, \ldots, r. \tag{4.66}$$

Using the assumption $\gamma\|z\|^2 \leq z'H^k z \leq \Gamma\|z\|^2$, we may also assume without loss of generality that

$$\{H^k\}_K \to \bar{H} \tag{4.67}$$

for some positive definite matrix $\bar{H}$.

Now from the fact $\alpha^k d^{k'} H^k d^k \to 0$ [cf. Eq. (4.63)], it follows that there are two possibilities. Either

$$\liminf_{k\to\infty,\, k\in K} \|d^k\| = 0, \tag{4.68}$$

or else

$$\liminf_{k\to\infty,\, k\in K} \alpha^k = 0, \qquad \liminf_{k\to\infty,\, k\in K} \|d^k\| > 0. \tag{4.69}$$

If Eq. (4.68) holds, then we may assume without loss of generality that $\{d^k\}_K \to 0$, and by taking the limit in Eqs. (4.64) and (4.65), and using Eq. (4.66), we have

$$\nabla f(\bar{x}) + \sum_{j=0}^r \bar{\mu}_j \nabla g_j(\bar{x}) = 0, \qquad c = \sum_{j=0}^r \bar{\mu}_j,$$

$$\bar{\mu}_j \geq 0, \qquad \bar{\mu}_j\big(g_j(\bar{x}) - \xi\big) = 0, \qquad j = 0, 1, \ldots, r,$$

where $\xi = \max_{j=0,1,\ldots,r} g_j(\bar{x})$. Hence the quadratic program (4.53) corresponding to $\bar{x}$ has $\{d = 0,\ \xi = P(\bar{x})\}$ as its optimal solution. From Prop. 4.3.2(d), it follows that $\bar{x}$ is a stationary point of $f + cP$, thus contradicting the hypothesis made earlier.

It will thus suffice to arrive at a contradiction assuming that Eq. (4.69) holds. We may assume without loss of generality that

$$\{\alpha^k\}_K \to 0.$$

Since Eqs. (4.64), (4.66), and (4.67) show that $\{d^k\}_K$ is a bounded sequence, we may also assume without loss of generality that

$$\{d^k\}_K \to \bar{d},$$

where $\bar{d}$ is some vector which cannot be zero in view of Eq. (4.69).

Since $\{\alpha^k\}_K \to 0$, it follows, in view of the definition of the Armijo rule, that the initial stepsize s will be reduced at least once for all $k \in K$ after some index $\bar{k}$. This means that for all $k \in K$, $k \geq \bar{k}$,

$$f(x^k) + cP(x^k) - f(x^k + \bar{\alpha}^k d^k) - cP(x^k + \bar{\alpha}^k d^k) < \sigma \bar{\alpha}^k d^{k\prime} H^k d^k, \quad (4.70)$$

where $\bar{\alpha}^k = \alpha^k/\beta$.

Define for all k and d

$$\zeta^k(d) = \nabla f(x^k)'d + c \max_{j \in J(x^k)} \{g_j(x^k) + \nabla g_j(x^k)'d\} - cP(x^k),$$

and restrict attention to $k \in K$, $k \geq \bar{k}$, that are sufficiently large so that $\bar{\alpha}^k \leq 1$, $J(x^k) \subset J(\bar{x})$, and $J(x^k + \bar{\alpha}^k d^k) \subset J(\bar{x})$. It will be shown that

$$f(x^k)+cP(x^k)-f(x^k+\bar{\alpha}^k d^k)-cP(x^k+\bar{\alpha}^k d^k) = -\zeta^k(\bar{\alpha}^k d^k)+o(\bar{\alpha}^k), \quad (4.71)$$

where

$$\lim_{k\to\infty,\, k\in K} \frac{o(\bar{\alpha}^k)}{\bar{\alpha}^k} = 0. \quad (4.72)$$

Indeed, we have

$$f(x^k + \bar{\alpha}^k d^k) = f(x^k) + \bar{\alpha}^k \nabla f(x^k)'d^k + o_0(\bar{\alpha}^k \|d^k\|)$$

$$g_j(x^k + \bar{\alpha}^k d^k) = g_j(x^k) + \bar{\alpha}^k \nabla g_j(x^k)'d^k + o_j(\bar{\alpha}^k \|d^k\|), \qquad j \in J(x^k),$$

where $o_j(\cdot)$ are functions satisfying $\lim_{k\to\infty} o_j(\bar{\alpha}^k\|d^k\|)/\bar{\alpha}^k = 0$. Adding and taking the maximum over $j \in J(x)$, and using the fact $J(x^k + \bar{\alpha}^k d^k) \subset J(\bar{x})$ [implying that $P(x^k + \bar{\alpha}^k d^k) = \max_j g_j(x^k + \bar{\alpha}^k d^k)$], we obtain for sufficiently large k,

$$\begin{aligned} f(x^k + \bar{\alpha}^k d^k) + cP(x^k + \bar{\alpha}^k d^k) &= f(x^k) + \bar{\alpha}^k \nabla f(x^k)'d^k \\ &\quad + c \max_{j \in J(x^k)} \{g_j(x^k) + \bar{\alpha}^k \nabla g_j(x^k)'d^k\} + o(\bar{\alpha}^k \|d^k\|) \\ &= f(x^k) + cP(x^k) + \zeta^k(\bar{\alpha}^k d^k) + o(\bar{\alpha}^k), \end{aligned}$$

thus proving Eq. (4.71).

We also claim that

$$-\frac{\zeta^k(\bar{\alpha}^k d^k)}{\bar{\alpha}^k} \geq -\zeta^k(d^k) \geq d^{k\prime} H^k d^k. \quad (4.73)$$

Indeed, let (d^k, ξ^k) be the optimal solution of the quadratic program

$$\begin{aligned} &\text{minimize} \quad \nabla f(x^k)'d + \tfrac{1}{2} d'H^k d + c\xi \\ &\text{subject to} \quad g_j(x^k) + \nabla g_j(x^k)'d \leq \xi, \quad j = 0, 1, \ldots, r. \end{aligned}$$

We have

$$\xi^k = \max_{j=0,1,\ldots,r} \{g_j(x^k) + \nabla g_j(x^k)'d^k\} \geq \max_{j\in J(x)} \{g_j(x^k) + \nabla g_j(x^k)'d^k\}$$
$$= \frac{\zeta^k(d^k) - \nabla f(x^k)'d^k}{c} + P(x^k).$$

On the other hand, in the proof of Prop. 4.3.2(c) we showed [cf. Eq. (4.60)] that

$$c\big(\xi^k - P(x^k)\big) - \nabla f(x^k)'d^k \geq d^{k\prime} H^k d^k.$$

The last two equations, together with the relation $\zeta^k(\bar{\alpha}^k d) \leq \bar{\alpha}^k \zeta^k(d)$, which follows from the convexity of $\zeta^k(\cdot)$, prove Eq. (4.73).

By dividing Eq. (4.70) with $\bar{\alpha}^k$ and by combining it with Eq. (4.71), we obtain

$$\sigma d^{k\prime} H^k d^k > -\frac{\zeta^k(\bar{\alpha}^k d)}{\bar{\alpha}^k} + \frac{o(\bar{\alpha}^k)}{\bar{\alpha}^k},$$

which in view of Eq. (4.73), yields

$$(1-\sigma) d^{k\prime} H^k d^k + \frac{o(\bar{\alpha}^k)}{\bar{\alpha}^k} < 0.$$

Since $\{H^k\}_K \to \bar{H}$, $\{d^k\}_K \to \bar{d}$, $\bar{H}$ is positive definite, $\bar{d} \neq 0$, and $o(\bar{\alpha}^k)/\bar{\alpha}^k \to 0$ [cf. Eq. (4.72)], we obtain a contradiction. This completes the proof of the proposition for the case of the Armijo rule.

Consider now the minimization rule and let $\{x^k\}_K$ converge to a vector $\bar{x}$, which is not a stationary point of $f + cP$. Let $\tilde{x}^{k+1}$ be the point that would be generated from x^k via the Armijo rule and let $\tilde{\alpha}^k$ be the corresponding stepsize. We have

$$f(x^k) - f(x^{k+1}) \geq f(x^k) - f(\tilde{x}^{k+1}) \geq \sigma \tilde{\alpha}^k d^{k\prime} H^k d^k.$$

By replacing α^k by $\tilde{\alpha}^k$ in the arguments of the earlier proof, we obtain a contradiction. This line of argument establishes that any stepsize rule that gives a larger reduction in the value of $f + cP$ at each iteration than the Armijo rule inherits its convergence properties, so it also proves the proposition for the limited minimization rule. **Q.E.D.**

Application to Constrained Optimization Problems

Given the inequality constrained problem

$$\begin{aligned} &\text{minimize} \ \ f(x) \\ &\text{subject to} \ \ g_j(x) \leq 0, \qquad j = 1, \ldots, r, \end{aligned} \tag{4.74}$$

one can attempt its solution by using the linearization algorithm to minimize the corresponding exact penalty function $f + cP$ for a value of c that exceeds the threshold $\sum_{j=1}^r \mu_j^*$ (cf. Prop. 4.3.1).

There are a number of complex implementation issues here. One difficulty is that we may not know a suitable threshold value for c. Under these circumstances, a possible approach is to choose an initial value c^0 for c and increase it as necessary at each iteration k if the algorithm indicates that the current value c^k is inadequate. An important question is to decide on the conditions that would prompt an increase of c^k. The most common approach here is based on trying to solve the quadratic program

$$\begin{aligned} &\text{minimize} \quad \nabla f(x^k)'d + \tfrac{1}{2}d'H^k d \\ &\text{subject to} \quad g_j(x^k) + \nabla g_j(x^k)'d \le 0, \qquad j = 1, \ldots, r, \end{aligned} \tag{4.75}$$

which differs from the direction finding quadratic program

$$\begin{aligned} &\text{minimize} \quad \nabla f(x^k)'d + \tfrac{1}{2}d'H^k d + c\xi \\ &\text{subject to} \quad g_j(x^k) + \nabla g_j(x^k)'d \le \xi, \qquad j = 0, 1, \ldots, r, \end{aligned} \tag{4.76}$$

of the linearization method in that ξ has been set to zero. If program (4.75) has a feasible solution, then it must have a unique optimal solution d^k and a set of Lagrange multipliers $\mu_1^k, \ldots, \mu_r^k$, since its cost function is a strictly convex quadratic and its constraints are linear. It can then be verified by checking the corresponding optimality conditions (Exercise 4.3.3) that for all $c > \sum_{j=1}^r \mu_j^k$, the pair $\{d^k, \xi = 0\}$ is the optimal solution of the quadratic program (4.76). Thus the direction d^k can be used as a direction of descent for minimizing $f + c^k P$, where $\sum_{j=1}^r \mu_j^k$ provides an underestimate for the appropriate value for c^k. The penalty parameter may be updated by

$$c^k = \max\left\{c^{k-1}, \sum_{j=1}^r \mu_j^k + \gamma\right\},$$

where γ is some positive scalar. Note that by the optimality conditions of the quadratic program (4.75) we have approximately, for small $\|d^k\|$,

$$\nabla f(x^k) + \sum_{j=1}^r \mu_j^k \nabla g_j(x^k) \approx 0, \qquad \mu_j^k g_j(x^k) \approx 0, \quad j = 1, \ldots, r,$$

so near convergence, the scalars μ_j^k are approximately equal to Lagrange multipliers of the constrained problem (4.74). This is consistent with the strategy of setting c^k at a somewhat higher value than $\sum_{j=1}^r \mu_j^k$. If on the other hand, the quadratic program (4.75) has no feasible solution, one can set $c^k = c^{k-1}$ and obtain a direction of descent d^k for the quadratic program (4.76).†

One of the drawbacks of this approach is that the value of the penalty parameter c^k may increase rapidly during the early stages of the algorithm, while during the final stage of the algorithm a much smaller value of c^k may be adequate. A large value of c^k results in very sharp corners of the surfaces of equal cost of the penalized cost $f+c^kP$ along the boundary of the constraint set, and can have a substantial adverse effect on the effectiveness of the stepsize procedure and thus on algorithmic progress (see Fig. 4.3.2). In this connection, it is interesting to note that if the system

$$g_j(x^k) + \nabla g_j(x^k)'d \leq 0, \qquad j = 1, \ldots, r,$$

is feasible, then the direction d^k obtained from the quadratic program (4.75) is independent of c^k, while the stepsize α^k depends strongly on c^k. For this reason, it may be important to provide schemes that allow for the reduction of c^k if circumstances appear to be favorable. The details of this can become quite complicated and we refer to [Ber82a] for a discussion of some possibilities.

An important question relates to the choice of the matrices H^k. In unconstrained minimization, one tries to employ a stepsize $\alpha^k = 1$ together with matrices H^k that approximate the Hessian of the cost function at a solution. A natural analog for the constrained case would be to choose H^k close to the Hessian of the Lagrangian function

$$L(x, \mu) = f(x) + \mu' g(x),$$

evaluated at (x^*, μ^*); a justification for this is provided in the next section, where it is shown that the direction d^k calculated by the linearization algorithm can be viewed as a Newton step.

There are two difficulties relating to such an approach. The first is that $\nabla^2_{xx} L(x^*, \mu^*)$ may not be positive definite. Actually this is not as serious as might appear. As we discuss more fully in the next section, what is important is that H^k approximate closely $\nabla^2_{xx} L(x^*, \mu^*)$ only on the

---

† It is possible that because of the constraint nonlinearities the quadratic program (4.75) has no feasible solution. This will not happen if the constraint functions g_j are convex and the original inequality constrained problem (4.74) has at least one feasible solution, say $\bar{x}$; it can be seen that the vector $\bar{d}^k = \bar{x} - x^k$ is a feasible solution of the quadratic program (4.75) since

$$g_j(x^k) + \nabla g_j(x^k)'(\bar{x} - x^k) \leq g_j(\bar{x}) \leq 0$$

by Prop. B.3 of Appendix B. Usually, even if the constraints are nonconvex, the quadratic problem (4.75) is feasible, provided the constrained problem (4.74) is feasible.

subspace tangent to the active constraints. Under second order sufficiency assumptions on (x^*, μ^*), this can be done with positive definite H^k, since then $\nabla^2_{xx} L(x^*, \mu^*)$ is positive definite on this subspace.

The second difficulty relates to the fact that even if we were to choose H^k equal to the (generally unknown) matrix $\nabla^2_{xx} L(x^*, \mu^*)$ and even if this matrix is positive definite, it may happen that arbitrarily close to x^* a stepsize $\alpha^k = 1$ is not acceptable by the algorithm because it does not decrease the value of $f + cP$; this can happen even for very simple problems (see Exercise 4.3.9). Reference [Ber82a] discusses this point in detail and introduces modifications to the basic linearization algorithm that allow a superlinear convergence rate.

Extension to Equality Constraints

The development given earlier for inequality constraints can be extended to the case of additional equality constraints simply by converting each equality constraint $h_i(x) = 0$ to the two inequalities

$$h_i(x) \le 0, \qquad -h_i(x) \le 0.$$

For example, the direction finding quadratic program of the linearization method is

$$\begin{aligned} \text{minimize} \quad & \nabla f(x^k)'d + \tfrac{1}{2} d' H^k d + c\xi \\ \text{subject to} \quad & g_j(x^k) + \nabla g_j(x^k)'d \le \xi, \quad j = 0, 1, \ldots, r, \\ & |h_i(x^k) + \nabla h_i(x^k)'d| \le \xi, \quad i = 1, \ldots, m. \end{aligned}$$

This program yields a descent direction for the exact penalty function

$$f(x) + c \max\{0, g_1(x), \ldots, g_r(x), |h_1(x)|, \ldots, |h_m(x)|\},$$

and can be used as a basis for an algorithm similar to the one developed for inequality constraints.

4.3.2 Differentiable Exact Penalty Functions

We now discuss briefly differentiable exact penalty functions for the equality constrained problem

$$\begin{aligned} \text{minimize} \quad & f(x) \\ \text{subject to} \quad & h_i(x) = 0, \qquad i = 1, \ldots, m. \end{aligned} \tag{4.77}$$

We assume that f and h_i are twice continuously differentiable. Furthermore, we assume that the matrix $\nabla h(x)$ has rank m for all x, although much of the following analysis can be conducted assuming $\nabla h(x)$ has rank m in a suitable open subset of $\Re^n$. Motivated by the exact penalty function

$$||\nabla_x L(x, \lambda)||^2 + \|h(x)\|^2$$

discussed earlier, we consider the function

$$P_c(x, \lambda) = L(x, \lambda) + \tfrac{1}{2}||W(x)\nabla_x L(x, \lambda)||^2 + \frac{c}{2}\|h(x)\|^2, \tag{4.78}$$

where

$$L(x, \lambda) = f(x) + \lambda' h(x)$$

is the Lagrangian function, c is a positive parameter, and $W(x)$ is any continuously differentiable $m \times n$ matrix such that the $m \times m$ matrix $W(x)\nabla h(x)$ is nonsingular for all x.

The idea here is that by introducing the Lagrangian $L(x, \lambda)$ in the cost, we build a preference towards local minima rather than local maxima. The use of the matrix function $W(x)$ cannot be motivated easily, but will be justified by subsequent developments. Two examples of choices of $W(x)$ that turn out to be useful are

$$W(x) = \nabla h(x)', \tag{4.79}$$

$$W(x) = \big(\nabla h(x)'\nabla h(x)\big)^{-1}\nabla h(x)'. \tag{4.80}$$

Reference [Ber82a] discusses in greater detail the role of the matrix $W(x)$ and also considers a different type of method whereby $W(x)$ is taken equal to the identity matrix.

Let us write $W(x)$ in the form

$$W(x) = \begin{pmatrix} w_1(x)' \\ \vdots \\ w_m(x)' \end{pmatrix},$$

where $w_i : \Re^n \mapsto \Re^n$ are some functions, and let $e_1, \ldots, e_m$ be the columns of the $m \times m$ identity matrix. It is then straightforward to verify that

$$\nabla_x P_c = \nabla_x L + \left(\nabla^2_{xx} L W' + \sum_{i=1}^m \nabla w_i \nabla_x L e_i'\right) W \nabla_x L + c\nabla h h, \tag{4.81}$$

$$\nabla_\lambda P_c = h + \nabla h' W' W \nabla_x L, \tag{4.82}$$

where all functions and gradients in the above expressions are evaluated at the typical pair (x, λ).

It can be seen that if (x^*, λ^*) is a local minimum-Lagrange multiplier pair of the original problem (4.77), then (x^*, λ^*) is also a stationary point of $P_c(x, \lambda)$, that is,

$$\nabla_x P_c(x^*, \lambda^*) = 0, \qquad \nabla_\lambda P_c(x^*, \lambda^*) = 0.$$

Under appropriate conditions, the reverse assertions are possible, namely that stationary points (x^*, λ^*) of $P_c(x, \lambda)$ satisfy the first order necessary conditions for the original constrained optimization problem. There are several results of this type, of which the following is typical. We outline the proof in Exercise 4.3.7 and we also refer to [Ber82a] for an extensive analysis.

Proposition 4.3.4: For every compact subset $X \times \Lambda$ of $\Re^{n+m}$ there exists a $\bar{c} > 0$ such that for all $c \geq \bar{c}$, every stationary point (x^*, λ^*) of P_c that belongs to $X \times \Lambda$ satisfies the first order necessary conditions

$$\nabla_x L(x^*, \lambda^*) = 0, \qquad \nabla_\lambda L(x^*, \lambda^*) = 0.$$

Differentiable Exact Penalty Functions Depending Only on x

One approach to minimizing $P_c(x, \lambda)$ is to first minimize it with respect to λ and then minimize it with respect to x. To simplify the subsequent formulas, let us focus on the function

$$W(x) = \big(\nabla h(x)'\nabla h(x)\big)^{-1}\nabla h(x)'$$

of Eq. (4.80). For this function, $W(x)\nabla h(x)$ is equal to the identity matrix and we have

$$P_c(x, \lambda) = f(x) + \lambda' h(x) + \tfrac{1}{2}\|W(x)\nabla f(x) + \lambda\|^2 + \frac{c}{2}\|h(x)\|^2. \tag{4.83}$$

We can minimize explicitly this function with respect to λ by setting

$$\nabla_\lambda P_c = h(x) + W(x)\nabla f(x) + \lambda = 0.$$

Substituting λ from this equation into Eq. (4.83), we obtain

$$\min_\lambda P_c(x, \lambda) = f(x) + \hat{\lambda}(x)' h(x) + \frac{c-1}{2}\|h(x)\|^2,$$

where

$$\hat{\lambda}(x) = -W(x)\nabla f(x).$$

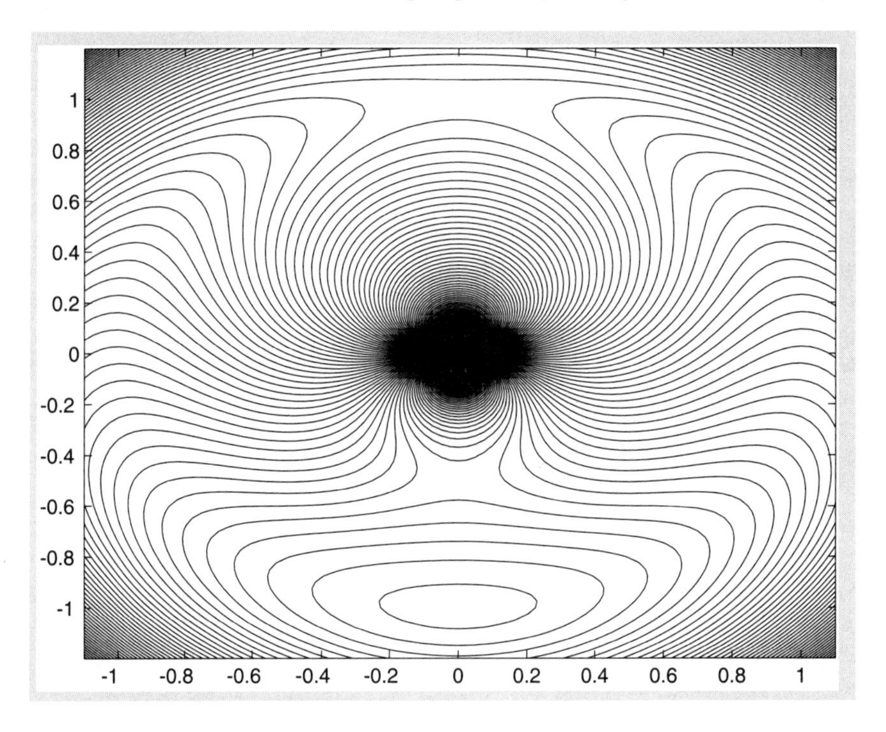

Figure 4.3.5. Contours of the differentiable exact penalty function $\hat{P}_c(x)$ for the two-dimensional problem where

$$f(x) = x_1, \qquad h(x) = x_1^2 + x_2^2 - 1$$

(cf. Figs. 4.3.1 and 4.3.2). The figure corresponds to $c = 2$. Note that there is a singularity at $(0, 0)$, which is a nonregular point at which $\hat{\lambda}(x)$ is undefined. The function $\hat{P}_c(x)$ takes arbitrarily large and arbitrarily small values sufficiently close to $(0, 0)$. This type of singularity can be avoided by using a modification of the exact penalty function (see Exercise 4.3.8).

Replacing $c - 1$ by c, it is seen that the function

$$\hat{P}_c(x) = f(x) + \hat{\lambda}(x)'h(x) + \frac{c}{2}\|h(x)\|^2 \tag{4.84}$$

is an exact penalty function, inheriting its properties from the exact penalty function $P_c(x, \lambda)$. Figure 4.3.5 illustrates the function $\hat{P}_c(x)$ for the same example problem that we used to illustrate the nondifferentiable exact penalty function $f + cP$ in Fig. 4.3.2. It can be seen that the two exact penalty functions $\hat{P}_c$ and $f + cP$ have quite different structures. For a detailed analysis of the convergence issues, extensions to inequality constraints, and methods to choose the penalty parameter c, we refer to [Ber82a].

EXERCISES

4.3.1

Consider a one-dimensional problem with two inequality constraints where $f(x) = 0$, $g_1(x) = -x$, $g_2(x) = 1 - x^2$. Show that for all c, $x = (1/2)(1 - \sqrt{5})$ and $x = 0$ are stationary points of $f + cP$, where P is the nondifferentiable exact penalty function (4.51)-(4.52), but are infeasible for the constrained problem. Plot $P(x)$ and discuss the behavior of the linearization method for this problem.

4.3.2

Let H be a positive definite symmetric matrix. Show that the pair (x^*, μ^*) satisfies the first order necessary conditions of Prop. 3.3.1 for the problem

$$\begin{aligned} &\text{minimize} \ \ f(x) \\ &\text{subject to} \ \ g_j(x) \leq 0, \qquad j = 1, \ldots, r, \end{aligned}$$

if and only if $(0, \mu^*)$ is a global minimum-Lagrange multiplier pair of the quadratic program

$$\begin{aligned} &\text{minimize} \ \ \nabla f(x^*)'d + \tfrac{1}{2} d'Hd \\ &\text{subject to} \ \ g_j(x^*) + \nabla g_j(x^*)'d \leq 0, \qquad j = 1, \ldots, r. \end{aligned}$$

(See [Ber82a], Section 4.1 for a solution.)

4.3.3

Show that if (d, μ) is a global minimum-Lagrange multiplier pair of the quadratic program

$$\begin{aligned} &\text{minimize} \ \ \nabla f(x)'d + \tfrac{1}{2} d'Hd \\ &\text{subject to} \ \ g_j(x) + \nabla g_j(x)'d \leq 0, \qquad j = 1, \ldots, r, \end{aligned}$$

where H is positive definite symmetric, and

$$c \geq \sum_{j=1}^{r} \mu_j,$$

then $(d, \xi = 0, \bar{\mu})$ is a global minimum-Lagrange multiplier pair of the quadratic program

$$\begin{aligned} &\text{minimize} \ \ \nabla f(x)'d + \tfrac{1}{2} d'Hd + c\xi \\ &\text{subject to} \ \ (x, \xi) \in \Re^{n+1}, \qquad g_j(x) + \nabla g_j(x)'d \leq \xi, \quad j = 0, 1, \ldots, r, \end{aligned}$$

where $\bar{\mu}_j = \mu_j$ for $j = 1, \ldots, r$, $\bar{\mu}_0 = c - \sum_{j=1}^{r} \mu_j$, and $g_0(x) \equiv 0$. (See [Ber82a], Section 4.1 for a solution.)

4.3.4

Show that if the pair (x^*, μ^*) satisfies the first order necessary conditions of Prop. 3.3.1 for the problem

$$\begin{aligned} &\text{minimize} \;\; f(x) \\ &\text{subject to} \;\; g_j(x) \le 0, \quad j = 1, \dots, r, \end{aligned}$$

then x^* is a stationary point of $f + cP$ for all $c \ge \sum_{j=1}^r \mu_j^*$. *Hint*: Combine the results of Exercises 4.3.2 and 4.3.3.

4.3.5

Show that when the constraints are linear, the linearization method based on the quadratic program (4.75) is equivalent to one of the gradient projection methods of Section 2.3.

4.3.6

For the one-dimensional problem of minimizing $(1/6)x^3$ subject to $x = 0$, consider the differentiable exact penalty function $P_c(x, \lambda)$ of Eq. (4.78) with $W(x)$ given by Eq. (4.79) or Eq. (4.80). Show that it has two stationary points: the pairs $(0, 0)$ and $\big(c-1, (1-c^2)/2\big)$. Are both of these local minima of $P_c(x, \lambda)$? Discuss how your analysis is consistent with Prop. 4.3.4.

4.3.7

Prove Prop. 4.3.4. *Hint*: By Eq. (4.82), the condition $\nabla_\lambda P_c = 0$ at some point of $X \times \Lambda$ implies $W \nabla_x L = -(\nabla h' W')^{-1} h$. If at this point $\nabla_x P_c = 0$ also holds, we obtain after some calculation

$$\begin{aligned} 0 &= W \nabla_x P_c \\ &= \left\{ cW\nabla h - \left(I + W \left(\nabla_{xx}^2 L W' + \sum_{i=1}^m \nabla w_i \nabla_x L e_i' \right) \right) (\nabla h' W')^{-1} \right\} h. \end{aligned}$$

Show that there exists $\bar{c} > 0$ such that for all $c \ge \bar{c}$ and stationary points within $X \times \Lambda$, the matrix within braces is nonsingular, implying that at such points $h = 0$. Conclude that we also have $W \nabla_x L = 0$ so that from Eq. (4.81), $\nabla_x L = 0$.

4.3.8 (Dealing with Singularities [Ber82a], p. 215)

A difficulty with the penalty function $P_c(x, \lambda)$ of Eq. (4.78) is the assumption that the matrix $\nabla h(x)$ has rank m for all x. When this assumption is violated, the λ-dependent terms of P_c may be unbounded below. Furthermore, the function $\hat{P}_c$ of Eq. (4.84) is undefined at some points and singularities of the type shown in Fig. 4.3.5 at $x = 0$ may arise. To deal with this difficulty, introduce the following modified version of P_c:

$$P_{c,\tau}(x, \lambda) = L(x, \lambda) + \tfrac{1}{2}\|\nabla h(x) \nabla_x L(x, \lambda)\|^2 + \frac{c + \tau\|\lambda\|^2}{2}\|h(x)\|^2,$$

where τ is an additional positive parameter.

(a) Show that $P_{c,\tau}(x, \lambda)$ is bounded from below if the function $f(x)+(c/2)\|h(x)\|^2$ is bounded from below.

(b) Obtain a corresponding differentiable penalty function depending only on x, by minimizing $P_{c,\tau}(x, \lambda)$ with respect to λ.

(c) Plot the contours of this function for the problem of Fig. 4.3.5 and verify that the singularity exhibited in that figure does not occur.

4.3.9 (Maratos' Effect [Mar78])

This example illustrates a fundamental difficulty in attaining superlinear convergence using the nondifferentiable exact penalty function for monitoring descent. (This difficulty does not arise for differentiable exact penalty functions; see [Ber82a], pp. 271-277.) Consider the problem

$$\begin{aligned} &\text{minimize} \quad f(x) = x_1 \\ &\text{subject to} \quad h(x) = x_1^2 + x_2^2 - 1 = 0 \end{aligned}$$

with optimal solution $x^* = (-1, 0)$ and Lagrange multiplier $\lambda^* = 1/2$ (see Figs. 4.3.1, 4.3.2, and 4.3.5). For any x, let (d, λ) be an optimal solution-Lagrange multiplier pair of the problem

$$\begin{aligned} &\text{minimize} \quad \nabla f(x) + \tfrac{1}{2} d' \nabla^2 L(x^*, \lambda^*) d \\ &\text{subject to} \quad h(x) + \nabla h(x)' d = 0. \end{aligned}$$

(Note that d is the Newton direction; see also the next section.) Show that for all c,

$$f(x + d) + c|h(x + d)| - f(x) - c|h(x)| = \lambda h(x) - c|h(x)| + (c - \lambda^*)\|d\|^2.$$

Conclude that for $c > 2\lambda^*$, there are points x arbitrarily close to x^* for which the exact penalty function $f(x) + c|h(x)|$ is not reduced by a pure Newton step. (For a solution of the exercise and for a broader discussion of this phenomenon, see [Ber82a], p. 290.)

4.4 LAGRANGIAN AND PRIMAL-DUAL INTERIOR POINT METHODS*

The algorithms of this section may be viewed as methods for solving the system of necessary optimality conditions of the equality constrained minimization problem

$$\begin{aligned} &\text{minimize} \ \ f(x) \\ &\text{subject to} \ \ h(x) = 0. \end{aligned}$$

Thus the necessary optimality conditions

$$\nabla f(x) + \nabla h(x)\lambda = 0, \qquad h(x) = 0, \tag{4.85}$$

are treated as a system of $(n+m)$ nonlinear equations with $(n+m)$ unknowns, the vectors x and λ.

This system, referred to as the *Lagrangian system*, can be solved by Newton's method as discussed in Section 1.4. The difficulty with Newton's method, as well as with most other methods for solving nonlinear systems of equations, is that they guarantee only local convergence, that is, convergence from a starting point that is sufficiently close to a solution. To enlarge the region of convergence, it is necessary to use some type of line search based on the improvement of some merit function. While the existence of such a function for the Lagrangian system (4.85) is not obvious, it turns out that a number of functions, such as the augmented Lagrangian function, the exact penalty functions of Section 4.3, and other functions can serve as the basis for globally convergent versions of Newton's method.

In what follows in this section, we consider various methods for solving the Lagrangian system (4.85), starting with a first order method, which does not require second derivatives. We then examine Newton-like methods. We focus initially on local convergence properties. Our presentation, however, is geared towards preparing the ground for modified versions with improved global convergence properties, which are discussed in Section 4.4.3 for the case of general nonlinear programs, and in Section 4.4.4 for the case of linear programs.

4.4.1 First Order Methods

We will consider algorithms for solving the Lagrangian system

$$\nabla f(x) + \nabla h(x)\lambda = 0, \qquad h(x) = 0.$$

These algorithms, called *Lagrangian methods*, have the generic form

$$x^{k+1} = G(x^k, \lambda^k), \qquad \lambda^{k+1} = H(x^k, \lambda^k), \tag{4.86}$$

where $G : \Re^{n+m} \mapsto \Re^n$ and $H : \Re^{n+m} \mapsto \Re^n$ are continuously differentiable functions. Since the above iteration can only converge to a pair (x^*, λ^*) such that

$$x^* = G(x^*, \lambda^*), \qquad \lambda^* = H(x^*, \lambda^*),$$

the functions G and H must be chosen so that local minima-Lagrange multiplier pairs satisfy the above equations.

The simplest Lagrangian method (also known as the *first order Lagrangian method*) is given by

$$x^{k+1} = x^k - \alpha \nabla_x L(x^k, \lambda^k), \tag{4.87}$$

$$\lambda^{k+1} = \lambda^k + \alpha h(x^k), \tag{4.88}$$

where L is the Lagrangian function

$$L(x, \lambda) = f(x) + \lambda' h(x)$$

and $\alpha > 0$ is a scalar stepsize. To motivate this method, consider the function

$$P(x, \lambda) = \tfrac{1}{2}\|\nabla_x L(x, \lambda)\|^2 + \tfrac{1}{2}\|h(x)\|^2.$$

This function is minimized at a local minimum-Lagrange multiplier pair, so it can be viewed as an exact penalty function (cf. the discussion of Section 4.3.2).

Let us consider the direction

$$d(x^k, \lambda^k) = \big(-\nabla_x L(x^k, \lambda^k), h(x^k)\big)$$

used in the first order Lagrangian iteration (4.87)-(4.88) and derive conditions under which it is a descent direction of the exact penalty function $P(x, \lambda)$. We have

$$\nabla P(x, \lambda) = \begin{pmatrix} \nabla^2_{xx} L(x, \lambda) \nabla_x L(x, \lambda) + \nabla h(x) h(x) \\ \nabla h(x)' \nabla_x L(x, \lambda) \end{pmatrix},$$

so that

$$\begin{aligned} d(x^k, \lambda^k)' \nabla P(x^k, \lambda^k) &= -\nabla_x L(x^k, \lambda^k)' \big(\nabla^2_{xx} L(x^k, \lambda^k) \nabla_x L(x^k, \lambda^k) \\ &\qquad + \nabla h(x^k) h(x^k)\big) + h(x^k)' \nabla h(x^k)' \nabla_x L(x^k, \lambda^k) \\ &= -\nabla_x L(x^k, \lambda^k)' \nabla^2_{xx} L(x^k, \lambda^k) \nabla_x L(x^k, \lambda^k). \end{aligned}$$

If the Hessian of the Lagrangian $\nabla^2_{xx} L(x^k, \lambda^k)$ is positive definite, we see that $d(x^k, \lambda^k)$ is not an ascent direction of the exact penalty function P [and is a descent direction when $\nabla_x L(x^k, \lambda^k) \neq 0$]. Note, however, that positive definiteness of $\nabla^2_{xx} L(x^k, \lambda^k)$ (which is essential for the preceding argument) is a stronger condition than what is required by the second order sufficiency conditions of Section 3.2, even near (x^*, λ^*).

To analyze the convergence of the first order Lagrangian iteration (4.87)-(4.88), we cannot quite use the convergence methodology for descent methods of Chapter 1 because $d(x^k, \lambda^k)$ is not always a descent direction of the exact penalty function P. Thus we need some new tools, and to this end, we develop a general result on the local convergence of methods for solving systems of nonlinear equations. A pair (x^*, λ^*) is said to be a *point of attraction* of the iteration (4.86) if there exists an open set $S \subset \Re^{n+m}$ such that if $(x^0, \lambda^0) \in S$, then the sequence $\big\{(x^k, \lambda^k)\big\}$ generated by the iteration belongs to S and converges to (x^*, λ^*). The following proposition is very useful for our purposes.

Proposition 4.4.1: Let $G : \Re^{n+m} \mapsto \Re^n$ and $H : \Re^{n+m} \mapsto \Re^n$ be continuously differentiable functions. Assume that (x^*, λ^*) satisfies

$$x^* = G(x^*, \lambda^*), \qquad \lambda^* = H(x^*, \lambda^*),$$

and that all eigenvalues of the $(n+m) \times (n+m)$ matrix

$$R^* = \begin{pmatrix} \nabla_x G(x^*, \lambda^*) & \nabla_x H(x^*, \lambda^*) \\ \nabla_\lambda G(x^*, \lambda^*) & \nabla_\lambda H(x^*, \lambda^*) \end{pmatrix} \tag{4.89}$$

lie strictly within the unit circle of the complex plane. Then (x^*, λ^*) is a point of attraction of the iteration

$$x^{k+1} = G(x^k, \lambda^k), \qquad \lambda^{k+1} = H(x^k, \lambda^k), \tag{4.90}$$

and when the generated sequence $\big\{(x^k, \lambda^k)\big\}$ converges to (x^*, λ^*), the rate of convergence of $\|x^k - x^*\|$ and $\|\lambda^k - \lambda^*\|$ is linear.

Proof: Denote $y = (x, \lambda)$, $y^k = (x^k, \lambda^k)$, $y^* = (x^*, \lambda^*)$, and consider the function $M : \Re^{n+m} \mapsto \Re^{n+m}$ given by $M(y) = \big(G(x, \lambda), H(x, \lambda)\big)$. By the mean value theorem, we have for any two vectors y and $\tilde{y}$,

$$M(\tilde{y}) - M(y) = R'(\tilde{y} - y),$$

where R is the matrix having as ith column the gradient $\nabla M_i(\hat{y}^i)$ of the ith component of M evaluated at some vector $\hat{y}^i$ on the line segment connecting y and $\tilde{y}$. By taking $\tilde{y}$ and y sufficiently close to y^*, we can make R as close to the matrix R^* of Eq. (4.89) as desired, and therefore we can make the eigenvalues of the transpose R' lie within the unit circle [the eigenvalues of R and R' coincide by Prop. A.13(f) of Appendix A]. It follows from Prop. A.15 of Appendix A that there exists a norm $\|\cdot\|$ and an open sphere S with respect to that norm centered at (x^*, λ^*) such that, within S, the induced matrix norm of R' is less than $1 - \epsilon$ where ϵ is some positive scalar. Since

$$\|M(\tilde{y}) - M(y)\| \le \|R'\|\,\|\tilde{y} - y\|,$$

it follows that within the sphere S, M is a contraction mapping as defined in Appendix A. The result then follows from the contraction mapping theorem (Prop. A.26 in Appendix A). **Q.E.D.**

We now prove the local convergence of the first order Lagrangian iteration (4.87)-(4.88).

Proposition 4.4.2: Assume that f and h are twice continuously differentiable, and let (x^*, λ^*) be a local minimum-Lagrange multiplier pair. Assume also that x^* is regular and that the matrix $\nabla^2_{xx} L(x^*, \lambda^*)$ is positive definite. Then there exists $\bar{\alpha} > 0$, such that for all $\alpha \in (0, \bar{\alpha}]$, (x^*, λ^*) is a point of attraction of iteration (4.87) and (4.88), and if the generated sequence $\{(x^k, \lambda^k)\}$ converges to (x^*, λ^*), then the rate of convergence of $\|x^k - x^*\|$ and $\|\lambda^k - \lambda^*\|$ is linear.

Proof: The proof consists of showing that, for α sufficiently small, the hypothesis of Prop. 4.4.1 is satisfied. Indeed for $\alpha > 0$, consider the mapping $M_\alpha : \Re^{n+m} \mapsto \Re^{n+m}$ defined by

$$M_\alpha(x, \lambda) = \begin{pmatrix} x - \alpha \nabla_x L(x, \lambda) \\ \lambda + \alpha \nabla_\lambda L(x, \lambda) \end{pmatrix}.$$

Clearly $(x^*, \lambda^*) = M_\alpha(x^*, \lambda^*)$, and we have

$$\nabla M_\alpha(x^*, \lambda^*)' = I - \alpha B, \tag{4.91}$$

where

$$B = \begin{pmatrix} \nabla^2_{xx} L(x^*, \lambda^*) & \nabla h(x^*) \\ -\nabla h(x^*)' & 0 \end{pmatrix}. \tag{4.92}$$

We will show that the real part of each eigenvalue of B is strictly positive, and then the result will follow from Eq. (4.91) by using Prop. 4.4.1. For any complex vector y, denote by $\hat{y}$ its complex conjugate, and for any complex number γ, denote by $Re(\gamma)$ its real part. Let β be an eigenvalue of B, and let $(z, w) \neq 0$ be a corresponding eigenvector where z and w are complex vectors of dimension n and m, respectively. We have

$$Re\left\{ (\hat{z}' \quad \hat{w}') B \begin{pmatrix} z \\ w \end{pmatrix} \right\} = Re\left\{ \beta (\hat{z}' \quad \hat{w}') \begin{pmatrix} z \\ w \end{pmatrix} \right\} = Re(\beta) \big(\|z\|^2 + \|w\|^2 \big), \tag{4.93}$$

while at the same time, by using Eq. (4.92),

$$Re\left\{ (\hat{z}' \quad \hat{w}') B \begin{pmatrix} z \\ w \end{pmatrix} \right\} = Re\big\{ \hat{z}' \nabla^2_{xx} L(x^*, \lambda^*) z + \hat{z}' \nabla h(x^*) w - \hat{w}' \nabla h(x^*)' z \big\}. \tag{4.94}$$

Since for any real $n \times m$ matrix M, we have

$$Re\{\hat{z}'M'w\} = Re\{\hat{w}'Mz\},$$

it follows from Eqs. (4.93) and (4.94) that

$$Re\{\hat{z}'\nabla^2_{xx}L(x^*,\lambda^*)z\} = Re\left\{(\hat{z}' \quad \hat{w}')B\begin{pmatrix} z \\ w \end{pmatrix}\right\} = Re(\beta)\big(\|z\|^2 + \|w\|^2\big). \tag{4.95}$$

Since for any positive definite matrix A, we have

$$Re\{\hat{z}'Az\} > 0, \qquad \forall\ z \neq 0,$$

it follows from Eq. (4.95) and the positive definiteness assumption on $\nabla^2_{xx}L(x^*,\lambda^*)$ that either $Re(\beta) > 0$ or else $z = 0$. But if $z = 0$, the equation

$$B\begin{pmatrix} z \\ w \end{pmatrix} = \beta\begin{pmatrix} z \\ w \end{pmatrix}$$

yields

$$\nabla h(x^*)w = 0.$$

Since $\nabla h(x^*)$ has rank m, it follows that $w = 0$. This contradicts our earlier assumption that $(z, w) \neq 0$. Consequently, we must have $Re(\beta) > 0$. **Q.E.D.**

By appropriately scaling the vectors x and λ, we can show that the result of Prop. 4.4.2 holds also for the more general iteration

$$x^{k+1} = x^k - \alpha D\nabla_x L(x^k,\lambda^k), \qquad \lambda^{k+1} = \lambda^k + \alpha E\nabla_\lambda L(x^k,\lambda^k),$$

where D and E are any positive definite symmetric matrices of appropriate dimension. However, the restrictive positive definiteness assumption on $\nabla^2_{xx}L(x^*,\lambda^*)$ is essential for the conclusion to hold.

4.4.2 Newton-Like Methods for Equality Constraints

Let us write the Lagrangian system

$$\nabla f(x) + \nabla h(x)\lambda = 0, \qquad h(x) = 0,$$

as

$$\nabla L(x,\lambda) = 0.$$

Newton's method for solving this system is given by

$$x^{k+1} = x^k + \Delta x^k, \qquad \lambda^{k+1} = \lambda^k + \Delta\lambda^k, \tag{4.96}$$

where $(\Delta x^k, \Delta\lambda^k) \in \Re^{n+m}$ is obtained by solving the system of equations

$$\nabla^2 L(x^k, \lambda^k) \begin{pmatrix} \Delta x^k \\ \Delta\lambda^k \end{pmatrix} = -\nabla L(x^k, \lambda^k). \tag{4.97}$$

We say that (x^{k+1}, λ^{k+1}) is *well-defined* by the Newton iteration (4.96), (4.97) if the matrix $\nabla^2 L(x^k, \lambda^k)$ is invertible. Note that if x^* is a local minimum that is regular and together with a Lagrange multiplier λ^* satisfies the second order sufficiency condition of Prop. 3.2.1, then $\nabla^2 L(x^*, \lambda^*)$ is invertible; this was shown as part of the proof of the sensitivity theorem (Prop. 3.2.2). As a result, $\nabla^2 L(x, \lambda)$ is invertible in a neighborhood of (x^*, λ^*), and within this neighborhood, points generated by the Newton iteration are well-defined. In the subsequent discussion, when stating various local convergence properties of the Newton iteration in connection with such a pair, we implicitly restrict the iteration within a neighborhood where it is well-defined.

The local convergence properties of the method can be inferred from the results of Section 1.4. For purposes of convenient reference, we provide the corresponding result in the following proposition.

Proposition 4.4.3: Let x^* be a strict local minimum that is regular and satisfies together with a corresponding Lagrange multiplier vector λ^* the second order sufficiency conditions of Prop. 3.2.1. Then (x^*, λ^*) is a point of attraction of the Newton iteration (4.96), (4.97). Furthermore, if the generated sequence converges to (x^*, λ^*), the rate of convergence of $\{||(x^k, \lambda^k) - (x^*, \lambda^*)||\}$ is superlinear (at least order two if $\nabla^2 f$ and $\nabla^2 h_i$, $i = 1, \ldots, m$, are Lipschitz continuous in a neighborhood of x^*).

Proof: See Prop. 1.4.1 in Section 1.4. **Q.E.D.**

The Newton iteration (4.96), (4.97) has a rich structure that can be used to provide interesting implementations, three of which we discuss below.

A First Implementation of Newton's Method

Let us write the Hessian of the Lagrangian function as

$$\nabla^2 L(x^k, \lambda^k) = \begin{pmatrix} H^k & N^k \\ N^{k\prime} & 0 \end{pmatrix}, \qquad \nabla L(x^k, \lambda^k) = \begin{pmatrix} \nabla_x L(x^k, \lambda^k) \\ h(x^k) \end{pmatrix},$$

where

$$H^k = \nabla^2_{xx} L(x^k, \lambda^k), \qquad N^k = \nabla h(x^k).$$

Thus, the system (4.97) takes the form

$$\begin{pmatrix} H^k & N^k \\ N^{k'} & 0 \end{pmatrix} \begin{pmatrix} \Delta x^k \\ \Delta \lambda^k \end{pmatrix} = - \begin{pmatrix} \nabla_x L(x^k, \lambda^k) \\ h(x^k) \end{pmatrix}. \tag{4.98}$$

Let us assume that H^k *is invertible and* N^k *has rank* m. Then we can provide a more explicit expression for the Newton iteration. Indeed the Newton system (4.98) can be written as

$$H^k \Delta x^k + N^k \Delta \lambda^k = -\nabla_x L(x^k, \lambda^k), \tag{4.99}$$

$$N^{k'} \Delta x^k = -h(x^k). \tag{4.100}$$

By multiplying the first equation with $N^{k'}(H^k)^{-1}$ and by using the second equation, it follows that

$$-h(x^k) + N^{k'}(H^k)^{-1} N^k \Delta \lambda^k = -N^{k'}(H^k)^{-1} \nabla_x L(x^k, \lambda^k).$$

Since N^k has rank m, the matrix $N^{k'}(H^k)^{-1}N^k$ is nonsingular, and we obtain

$$\lambda^{k+1} - \lambda^k = \Delta \lambda^k = \big(N^{k'}(H^k)^{-1}N^k\big)^{-1}\big(h(x^k) - N^{k'}(H^k)^{-1}\nabla_x L(x^k, \lambda^k)\big). \tag{4.101}$$

We have

$$\begin{aligned} \nabla_x L(x^k, \lambda^k) &= \nabla f(x^k) + N^k \lambda^k \\ &= \nabla f(x^k) + N^k \lambda^{k+1} - N^k \Delta \lambda^k \\ &= \nabla_x L(x^k, \lambda^{k+1}) - N^k \Delta \lambda^k, \end{aligned}$$

and by using this equation to substitute $\nabla_x L(x^k, \lambda^k)$ in Eqs. (4.99) and (4.101), we finally obtain

$$\lambda^{k+1} = \big(N^{k'}(H^k)^{-1}N^k\big)^{-1}\big(h(x^k) - N^{k'}(H^k)^{-1}\nabla f(x^k)\big), \tag{4.102}$$

$$x^{k+1} = x^k - (H^k)^{-1}\nabla_x L(x^k, \lambda^{k+1}). \tag{4.103}$$

These equations provide a first basic implementation of the Newton iteration (under the assumption that H^k is invertible and N^k has rank m). Note that the solution of systems of dimension at most n is required [as opposed to $n+m$, which is the dimension of $\nabla^2 L(x^k, \lambda^k)$].

A Second Implementation of Newton's Method

Another way to write the system of equations (4.99) and (4.100) is based on the observation that, for every scalar c, we have from Eq. (4.100),

$$cN^k N^{k'} \Delta x = -cN^k h(x^k),$$

and adding to Eq. (4.99) yields

$$(H^k + cN^kN^{k'})\Delta x^k + N^k\Delta\lambda^k = -\nabla_x L(x^k, \lambda^k) - cN^k h(x^k) \\ - \nabla_x L\big(x^k, \lambda^k + ch(x^k)\big).$$

Thus, if $(H^k + cN^kN^k)^{-1}$ exists, we obtain by the same type of calculation used to obtain Eqs. (4.102) and (4.103)

$$x^{k+1} = x^k - (H^k + cN^kN^{k'})^{-1}\nabla_x L(x^k, \hat{\lambda}^{k+1}),$$

$$\lambda^{k+1} = \hat{\lambda}^{k+1} - ch(x^k), \tag{4.104}$$

where

$$\hat{\lambda}^{k+1} = \big(N^{k'}(H^k + cN^kN^{k'})^{-1}N^k\big)^{-1}\big(h(x^k) - N^{k'}(H^k + cN^kN^{k'})^{-1}\nabla f(x^k)\big) \tag{4.105}$$

By observing that

$$\nabla_x L\big(x^k, \lambda^{k+1} + ch(x^k)\big) = \nabla_x L_c\big(x^k, \lambda^{k+1}\big),$$

we see that an alternative way to write the update equation for x^k is

$$x^{k+1} = x^k - (H^k + cN^kN^{k'})^{-1}\nabla_x L_c\big(x^k, \lambda^{k+1}\big). \tag{4.106}$$

An advantage that the implementation of Eqs. (4.104)-(4.106) may offer over the one of Eqs. (4.102) and (4.103) (which correspond to $c = 0$) is that the matrix H^k may not be invertible while $H^k + cN^kN^{k'}$ may be invertible for some values of c. For example, for c sufficiently large, we have that $H^k + cN^kN^{k'}$ is not only invertible but also positive definite near (x^*, λ^*) (see Lemma 3.2.1 in Section 3.2). An additional advantage offered by this property is that it allows us to differentiate between local minima and local maxima (near a local maximum-Lagrange multiplier pair, $H^k + cN^kN^{k'}$ is not likely to be positive definite for any positive value of c). Note that positive definiteness of $H^k + cN^kN^{k'}$ can be easily detected if the Cholesky factorization method is used for solving the various linear systems of equations in Eqs. (4.105) and (4.106) (compare with the discussion of Section 1.4).

Another property, which is particularly important for enlarging the region of convergence of Newton's method (see Section 4.4.3), can be inferred from Eq. (4.106): if c is large enough so that $H^k + cN^kN^{k'}$ is positive definite, $x^{k+1} - x^k$ is a descent direction of the augmented Lagrangian function $L_c\big(\cdot, \lambda^{k+1}\big)$ at x^k.

An Implementation of Newton's Method Based on Quadratic Programming

A third implementation of the Newton iteration is based on the observation that the Newton system (4.99)-(4.100) can be written as

$$\nabla f(x^k) + H^k \Delta x^k + N^k \lambda^{k+1} = 0, \qquad h(x^k) + N^{k\prime} \Delta x^k = 0,$$

which are the necessary optimality conditions for $(\Delta x^k, \lambda^{k+1})$ to be a global minimum-Lagrange multiplier pair of the quadratic program

$$\begin{aligned} &\text{minimize} \;\; \nabla f(x^k)'\Delta x + \tfrac{1}{2}\Delta x' H^k \Delta x \\ &\text{subject to} \;\; h(x^k) + N^{k\prime}\Delta x = 0. \end{aligned} \tag{4.107}$$

Thus we can obtain $(\Delta x^k, \lambda^{k+1})$ by solving this problem.

This implementation is not particularly useful for practical purposes but provides an interesting connection with the linearization method. This relation can be made more explicit by noting that the solution Δx^k of the quadratic program (4.107) is unaffected if H^k is replaced by any matrix of the form $H^k + cN^kN^{k\prime}$, where c is a scalar, thereby obtaining the program

$$\begin{aligned} &\text{minimize} \;\; \nabla f(x^k)'\Delta x + \tfrac{1}{2}\Delta x'(H^k + cN^kN^{k\prime})\Delta x \\ &\text{subject to} \;\; h(x^k) + N^{k\prime}\Delta x = 0. \end{aligned} \tag{4.108}$$

To see that problems (4.107) and (4.108) have the same solution Δx^k, simply note that they have the same constraints while their cost functions differ by the term $c\Delta x' N^k N^{k\prime}\Delta x$, which is equal to $c\|h(x^k)\|^2$ and is therefore constant. Near a local minimum-Lagrange multiplier pair (x^*, λ^*) satisfying the sufficiency conditions, we have that $H^k + cN^kN^{k\prime}$ is positive definite if c is sufficiently large (Lemma 3.2.1 in Section 3.2), and the quadratic program (4.108) is positive definite.

We see therefore that, under these circumstances, the Newton iteration can be viewed in effect as a special case of the linearization method of Section 4.3 with a constant unity stepsize, and scaling matrix

$$\bar{H}^k = H^k + cN^kN^{k\prime},$$

where c is any scalar for which $\bar{H}^k$ is positive definite.

Merit Functions and Descent Properties of Newton's Method

Since we would like to improve the global convergence properties of Newton's method, it is interesting to look for appropriate merit functions, that is, functions for which $(x^{k+1} - x^k)$ is a descent direction at x^k or

$(x^{k+1} - x^k, \lambda^{k+1} - \lambda^k)$ is a descent direction at (x^k, λ^k). By this, we mean functions F such that for a sufficiently small positive scalar $\bar{\alpha}$, we have

$$F\big(x^k + \alpha(x^{k+1} - x^k)\big) < F(x^k), \qquad \forall\ \alpha \in (0, \bar{\alpha}],$$

or

$$F\big(x^k + \alpha(x^{k+1} - x^k), \lambda^k + \alpha(\lambda^{k+1} - \lambda^k)\big) < F(x^k, \lambda^k), \qquad \forall\ \alpha \in (0, \bar{\alpha}].$$

The following proposition shows that there are several possible merit functions.

Proposition 4.4.4: Let x^* be a local minimum that is a regular point and satisfies together with a corresponding Lagrange multiplier vector λ^* the second order sufficiency conditions of Prop. 3.2.1. There exists a neighborhood S of (x^*, λ^*) such that if $(x^k, \lambda^k) \in S$ and $x^k \neq x^*$, then (x^{k+1}, λ^{k+1}) is well-defined by the Newton iteration and the following hold:

(a) There exists a scalar $\bar{c}$ such that for all $c \geq \bar{c}$, the vector $(x^{k+1} - x^k)$ is a descent direction at x^k for the exact penalty function

$$f(x) + c \max_{i=1,\ldots,m} |h_i(x)|. \tag{4.109}$$

(b) The vector $(x^{k+1} - x^k, \lambda^{k+1} - \lambda^k)$ is a descent direction at (x^k, λ^k) for the exact penalty function

$$P(x, \lambda) = \tfrac{1}{2}\|\nabla L(x, \lambda)\|^2.$$

Furthermore, given any scalar $r > 0$, there exists a $\delta > 0$ such that if

$$\|(x^k - x^*, \lambda^k - \lambda^*)\| < \delta,$$

we have

$$P(x^{k+1}, \lambda^{k+1}) \leq rP(x^k, \lambda^k). \tag{4.110}$$

(c) For every scalar c such that $H^k + cN^kN^{k\prime}$ is positive definite, the vector $(x^{k+1} - x^k)$ is a descent direction at x^k of the augmented Lagrangian function $L_c(\cdot, \lambda^{k+1})$.

Proof: (a) Take $\bar{c} > 0$ sufficiently large and a neighborhood S of (x^*, λ^*), which is sufficiently small, so that for $(x^k, \lambda^k) \in S$, the matrix $H^k + \bar{c}N^kN^{k\prime}$ is positive definite. Since Δx^k is the solution of the quadratic program

(4.108), it follows from Prop. 4.3.2 of Section 4.3 that if $x^k \neq x^*$, then Δx^k is a descent direction of the exact penalty function (4.109) for all $c \geq \bar{c}$.

(b) We have

$$\begin{pmatrix} x^{k+1} - x^k \\ \lambda^{k+1} - \lambda^k \end{pmatrix} = -\nabla^2 L(x^k, \lambda^k)^{-1} \nabla L(x^k, \lambda^k)$$

and

$$\nabla P(x^k, \lambda^k) = \nabla^2 L(x^k, \lambda^k) \nabla L(x^k, \lambda^k).$$

So

$$\big((x^{k+1} - x^k)', (\lambda^{k+1} - \lambda^k)'\big) \nabla P(x^k, \lambda^k) = -\|\nabla L(x^k, \lambda^k)\|^2 < 0,$$

and the descent property follows.

From Prop. 1.4.1 in Section 1.4, we have that, given any $\bar{r} > 0$, there exists a $\bar{\delta} > 0$ such that for $\|(x^k - x^*, \lambda^k - \lambda^*)\| < \bar{\delta}$, we have

$$\|(x^{k+1} - x^*, \lambda^{k+1} - \lambda^*)\| \leq \bar{r} \|(x^k - x^*, \lambda^k - \lambda^*)\|. \tag{4.111}$$

For every (x, λ), we have, by the mean value theorem,

$$\nabla L(x, \lambda) = B \begin{pmatrix} x - x^* \\ \lambda - \lambda^* \end{pmatrix},$$

where each row of B is the corresponding row of $\nabla^2 L$ evaluated at a point between (x, λ) and (x^*, λ^*). Since $\nabla^2 L(x^*, \lambda^*)$ is invertible, it follows that there is an $\epsilon > 0$ and scalars $\mu > 0$ and $M > 0$ such that for $\|(x - x^*, \lambda - \lambda^*)\| < \epsilon$, we have

$$\mu \|(x - x^*, \lambda - \lambda^*)\| \leq \|\nabla L(x, \lambda)\| \leq M \|(x - x^*, \lambda - \lambda^*)\|. \tag{4.112}$$

From Eqs. (4.111) and (4.112), it follows that for each $\bar{r} > 0$ there exists $\delta > 0$ such that, for $\|(x^k - x^*, \lambda^k - \lambda^*)\| < \delta$,

$$\|\nabla L(x^{k+1}, \lambda^{k+1})\| \leq (M\bar{r}/\mu) \|\nabla L(x^k, \lambda^k)\|.$$

or, equivalently,

$$P(x^{k+1}, \lambda^{k+1}) \leq (M^2 \bar{r}^2 / \mu^2) P(x^k, \lambda^k).$$

Given $r > 0$, we take $\bar{r} = (\mu/M)\sqrt{r}$ in the relation above, and Eq. (4.110) follows.

(c) We have shown that $x^{k+1} - x^k = -(H^k + cN^k N^{k\prime})^{-1} \nabla_x L_c\big(x^k, \lambda^{k+1}\big)$ [cf. Eq. (4.106)], which implies the conclusion. **Q.E.D.**

It is also possible to use the differentiable exact penalty functions of Section 4.3.2 as merit functions for Newton's method. The verification of this is somewhat tedious, so we refer to [Ber82a], p. 219, and [Ber82c] for an analysis. These references also show that differentiable exact penalty functions, while more complicated, have an interesting advantage over the nondifferentiable penalty function (4.109): they are not susceptible to the Maratos' effect discussed in Exercise 4.3.9 of Section 4.3, and they allow superlinear convergence of Newton-like methods without any complex modifications, as first shown in [Ber80] (see also [Ber82a], pp. 271-277).

Variations of Newton's Method

There are a number of variations of Newton's method, which are obtained by introducing some extra terms in the left-hand side of the Newton system. These variations have the general form

$$x^{k+1} = x^k + \Delta x^k, \qquad \lambda^{k+1} = \lambda^k + \Delta\lambda^k,$$

where

$$\big(\nabla^2 L(x^k, \lambda^k) + V^k(x^k, \lambda^k)\big)\begin{pmatrix}\Delta x^k \\ \Delta\lambda^k\end{pmatrix} = -\nabla L(x^k, \lambda^k),$$

where the extra term $V^k(x^k, \lambda^k)$ is "small" enough relative to $\nabla^2 L(x^k, \lambda^k)$, so that the eigenvalues of the matrix

$$I - \big(\nabla^2 L(x^k, \lambda^k) + V^k(x^k, \lambda^k)\big)^{-1}\nabla^2 L(x^k, \lambda^k)$$

are within the unit circle and the convergence result of Prop. 4.4.1 applies. In the case where the extra term $V^k(x^k, \lambda^k)$ converges to zero, superlinear convergence is attained; otherwise, the rate of convergence is linear (cf. Prop. 4.4.1).

An interesting approximation of Newton's method is obtained by adding a term $-(1/c^k)\Delta\lambda^k$ in the left-hand side of the equation $N^{k\prime}\Delta x^k = -h(x^k)$, where c^k is a positive parameter, so that Δx^k and $\Delta\lambda^k$ are obtained by solving the system

$$H^k\Delta x^k + N^k\Delta\lambda^k = -\nabla_x L(x^k, \lambda^k), \tag{4.113}$$

$$N^{k\prime}\Delta x^k - (1/c^k)\Delta\lambda^k = -h(x^k). \tag{4.114}$$

As $c^k \to \infty$, the system asymptotically becomes identical to the one corresponding to Newton's method.

We can show that the system (4.113), (4.114) has a unique solution if either $(H^k)^{-1}$ or $(H^k + c^k N^k N^{k\prime})^{-1}$ exists. Indeed when $(H^k)^{-1}$ exists, we can write explicitly the solution. By multiplying Eq. (4.113) by $N^{k\prime}(H^k)^{-1}$ and by using Eq. (4.114), we obtain

$$(1/c^k)\Delta\lambda^k - h(x^k) + N^{k\prime}(H^k)^{-1}N^k\Delta\lambda^k = -N^{k\prime}(H^k)^{-1}\nabla_x L(x^k, \lambda^k),$$

from which

$$\Delta\lambda^k = \big((1/c^k)I + N^{k\prime}(H^k)^{-1}N^k\big)^{-1}\big(h(x^k) - N^{k\prime}(H^k)^{-1}\nabla_x L(x^k, \lambda^k)\big)$$

and

$$\lambda^{k+1} = \lambda^k + \big((1/c^k)I + N^{k\prime}(H^k)^{-1}N^k\big)^{-1}\big(h(x^k) - N^{k\prime}(H^k)^{-1}\nabla_x L(x^k, \lambda^k)\big).$$

From Eq. (4.113), we then obtain

$$x^{k+1} = x^k - (H^k)^{-1} \nabla_x L(x^k, \lambda^{k+1}).$$

Also if $(H^k + c^k N^k N^{k'})^{-1}$ exists, by multiplying Eq. (4.114) with $c^k N^k$ and adding the resulting equation to Eq. (4.113), we obtain

$$(H^k + c^k N^k N^{k'}) \Delta x^k = -\nabla_x L(x^k, \lambda^k) - c^k N^k h(x^k),$$

and finally,

$$x^{k+1} = x^k - (H^k + c^k N^k N^{k'})^{-1} \nabla L_{c^k}(x^k, \lambda^k), \tag{4.115}$$

where L_{c^k} is the augmented Lagrangian function. Furthermore, from Eq. (4.114), we obtain

$$\lambda^{k+1} = \lambda^k + c^k \big(h(x^k) + N^{k'} (x^{k+1} - x^k) \big). \tag{4.116}$$

Note that the preceding development shows that N^k *need not have rank m in order for the system (4.113), (4.114) to have a unique solution, while this is not true for the Newton iteration.* Thus by introducing the term $(1/c^k)\Delta\lambda^k$ in the second equation, we avoid potential difficulties due to linear dependence of the constraint gradients.

The preceding analysis suggests that if c^k is taken sufficiently large, then the approximate Newton iteration (4.115), (4.116) should converge locally to a local minimum-Lagrange multiplier pair (x^*, λ^*) under the same conditions as the exact Newton iteration (cf. Prop. 4.4.3). Furthermore, the rate of convergence should be superlinear if $c^k \to \infty$. The proof of this, using Prop. 4.4.1, is straightforward, but is tedious and will not be given; see [Ber82a], pp. 240-243, where some variations of the method of Eqs. (4.115) and (4.116) are also discussed.

There is another type of approximate Newton's method, which is obtained by introducing extra terms in the right-hand side (rather than the left-hand side) of the Newton system. For local convergence to (x^*, λ^*), it is essential that the extra terms tend to zero. We will discuss such variations in Section 4.4.4 in connection with primal-dual methods for linear programming.

Connection with the First Order Method of Multipliers

From Eq. (4.115) it is seen that if $H^k + c^k N^k N^{k'}$ is positive definite, then $(x^{k+1} - x^k)$ is a descent direction for the augmented Lagrangian function $L_{c^k}(\cdot, \lambda^k)$. Furthermore, if the constraint functions h_i are linear, then Eq. (4.116) can be written as

$$\lambda^{k+1} = \lambda^k + c^k h(x^{k+1}), \tag{4.117}$$

while if in addition f is quadratic and $H^k + c^k N^k N^{k\prime}$ is positive definite, then from Eq. (4.115), x^{k+1} is the unique minimizing point of the augmented Lagrangian $L_{c^k}(\cdot, \lambda^k)$. Hence, it follows that if the constraints are linear and the cost function is quadratic, then the iteration (4.115), (4.116) is equivalent to the first order method of multipliers of Section 4.1.

Since near a local minimum-Lagrange multiplier pair the more general problem admits a quadratic approximation, it is natural to consider the iteration

$$x^{k+1} = x^k - \nabla^2_{xx} L_{c^k}(x^k, \lambda^k)^{-1} \nabla_x L_{c^k}(x^k, \lambda^k), \tag{4.118}$$

followed by the first order multiplier iteration

$$\lambda^{k+1} = \lambda^k + c^k h(x^{k+1}). \tag{4.119}$$

This is simply the first order multiplier iteration where the minimization of the augmented Lagrangian is replaced by a *single* pure Newton step, a method known as the *diagonalized method of multipliers*; see Tapia [Tap76]. For c^k large and x^k close to x^*, the Hessian $\nabla^2_{xx} L_{c^k}(x^k, \lambda^k)$ is nearly equal to $H^k + c^k N^k N^{k\prime}$, and $h(x^{k+1})$ is nearly equal to $h(x^k) + N^{k\prime}(x^{k+1} - x^k)$. Thus the diagonalized method of multipliers (4.118), (4.119) can be viewed as an approximation to the variation of Newton's method (4.115), (4.116) discussed earlier. This suggests that if c^k is taken larger than some threshold for all k, then the method should converge locally to a local minimum-Lagrange multiplier pair (x^*, λ^*) under the conditions of Prop. 4.4.3. The proof of this may be found in [Tap76] and [Ber82a], pp. 241-243, where it is also shown that the rate of convergence is superlinear if $c^k \to \infty$.

Extension to Inequality Constraints

Let us consider the inequality constrained problem

$$\begin{aligned} &\text{minimize} \quad f(x) \\ &\text{subject to} \quad g_j(x) \le 0, \qquad j = 1, \ldots, r, \end{aligned}$$

and focus on a local minimum x^* that is regular and together with a Lagrange multiplier μ^*, satisfies the second order sufficiency conditions of Prop. 3.3.2.

We can develop a Newton method for this problem, which is an extension of the quadratic programming implementation given earlier for equality constraints [cf. Eq. (4.107)]. This method is also similar to the constrained version of Newton's method for convex constraint sets given in Section 2.3 (in fact the two methods coincide when all the constraints are linear). In particular, given (x^k, μ^k), we obtain (x^{k+1}, μ^{k+1}) as an optimal solution-Lagrange multiplier pair of the quadratic program

$$\begin{aligned} &\text{minimize} \quad \nabla f(x^k)'(x - x^k) + \tfrac{1}{2}(x - x^k)' \nabla^2_{xx} L(x^k, \mu^k)(x - x^k) \\ &\text{subject to} \quad g_j(x^k) + \nabla g_j(x^k)'(x - x^k) \le 0, \qquad j = 1, \ldots, r. \end{aligned}$$

It is possible to show that there exists a neighborhood S of (x^*, μ^*) such that if (x^k, μ^k) is within S, then (x^{k+1}, μ^{k+1}) is uniquely defined as an optimal solution-Lagrange multiplier pair within S (an application of the implicit function theorem is needed to formalize this statement). Furthermore, (x^k, μ^k) converges to (x^*, μ^*) superlinearly. The details of this development are quite complex, and we refer to [Ber82a], Section 4.4.3, and the literature cited at the end of the chapter for further material.

4.4.3 Global Convergence

In order to enlarge the region of convergence of Lagrangian methods, it is necessary to combine them with some other method that has satisfactory global convergence properties. We refer to such a method as a *global method*. The main ideas here are similar to those underlying modifications of Newton's method for unconstrained minimization (compare with Section 1.4), although the resulting implementations tend to be somewhat more complex. Basically, we would like to have a combined method that when sufficiently close to a local minimum satisfying the second order sufficiency conditions switches automatically to a superlinearly convergent Lagrangian method, while when far away from such a point switches automatically to the global method, which is designed to make steady progress towards approaching the set of optimal solutions. Prime candidates for use as global methods are multiplier methods and exact penalty methods.

There are many possibilities for combining global and Lagrangian methods, and the suitability of any one of these depends strongly on the problem at hand. For this reason, our main purpose in this section is not to develop and recommend specific algorithms, but rather to focus on the main guidelines for harmoniously interfacing global and Lagrangian methods while retaining the advantages of both.

Once a global and a Lagrangian method have been selected, the main issue to be settled is the choice of what we shall call the *switching rule* and the *acceptance rule*. The switching rule determines at each iteration, on the basis of certain tests, whether a switch should be made to the Lagrangian method. The tests depend on the information currently available, and their purpose is to decide whether an iteration of the Lagrangian method has a reasonable chance of success. As an example, such tests might include verification that ∇h has rank m and that $\nabla^2_{xx} L$ is positive definite on the subspace $\{y \mid \nabla h' y = 0\}$. We hasten to add here that these tests should not require excessive computational overhead. In some cases a switch might be made without any test at all, subject only to the condition that the Lagrangian iteration is well-defined.

The acceptance rule determines whether the results of the Lagrangian iteration will be accepted as they are, whether they will be modified, or whether they will be rejected completely and a switch will be made back to the global method. Typically, acceptance of the results of the Lagrangian iteration is based on reducing the value of some exact penalty function.

Combinations with Multiplier Methods

One possibility for enlarging the region of convergence of Lagrangian methods is to combine them with the first or second order methods of multipliers discussed in Section 4.2. The resulting methods tend to be very reliable since they inherit the robustness of the method of multipliers. At the same time they typically require fewer iterations to converge within the same accuracy than pure methods of multipliers (see Glad [Gla79] for analysis and computational results).

The simplest possibility is to switch to a Lagrangian method at the beginning (or at the end) of each (perhaps approximate) unconstrained minimization of a method of multipliers and continue using the Lagrangian method as long as the value of the exact penalty function $\|\nabla L\|^2$ is being decreased by a certain factor at each iteration. If satisfactory progress in decreasing $\|\nabla L\|^2$ is not observed, a switch back to the method of multipliers is made. Another possibility is to attempt a switch to a Lagrangian method at each iteration. As an example, consider the following method for solving the equality constrained problem, which combines Newton's method for unconstrained minimization of the augmented Lagrangian together with the approximate Newton/Lagrangian iterations (4.115)-(4.119), which correspond to the first order method of multipliers.

At iteration k, we have x^k, λ^k, and a penalty parameter c^k. We also have a positive scalar w^k, which represents a target value of the exact penalty function $\|\nabla L\|^2$ that must be attained in order to accept the Lagrangian iteration, and a positive scalar ϵ^k that controls the accuracy of the unconstrained minimization of the method of multipliers. At the kth iteration, we determine x^{k+1}, λ^{k+1}, w^{k+1}, and ϵ^{k+1} as follows:

We form the Cholesky factorization $L^k L^{k'}$ of $\nabla^2_{xx} L_{c^k}(x^k, \lambda^k)$ as in Section 1.4. In the process, we modify $\nabla^2_{xx} L_{c^k}(x^k, \lambda^k)$ if it is not "sufficiently positive definite" (compare with Section 1.4). We then find the Newton direction

$$d^k = -(L^k L^{k'})^{-1} \nabla_x L_{c^k}(x^k, \lambda^k), \tag{4.120}$$

and if $\nabla^2_{xx} L_{c^k}(x^k, \lambda^k)$ was found "sufficiently positive definite" during the factorization process, we also carry out the Lagrangian iteration [compare with Eqs. (4.118) and (4.119)]:

$$\bar{x}^k = x^k + d^k, \tag{4.121}$$

$$\bar{\lambda}^k = \lambda^k + c^k h(\bar{x}^k). \tag{4.122}$$

[The analog of Eq. (4.116) could also be used in place of Eq. (4.122).]

If

$$\|\nabla L(\bar{x}^k, \bar{\lambda}^k)\|^2 \leq w^k,$$

then we accept the Lagrangian iteration and we set

$$x^{k+1} = \bar{x}^k, \qquad \lambda^{k+1} = \bar{\lambda}^k, \qquad c^{k+1} = c^k, \qquad \epsilon^{k+1} = \epsilon^k,$$

$$w^{k+1} = \gamma\|\nabla L(\bar{x}^k, \bar{\lambda}^k)\|^2,$$

where γ is a fixed scalar with $0 < \gamma < 1$.

Otherwise, we do not accept the results of the Lagrangian iteration, that is we do not update λ^k. Instead we revert to minimization of the augmented Lagrangian $L_{c^k}(\cdot, \lambda^k)$ by performing an Armijo-type line search. In particular, we set

$$x^{k+1} = x^k + \alpha^k d^k,$$

where the stepsize is obtained as

$$\alpha^k = \beta^{m_k},$$

where m_k is the first nonnegative integer m such that

$$L_{c^k}(x^k, \lambda^k) - L_{c^k}(x^k + \beta^m d^k, \lambda^k) \geq -\sigma\beta^m d^{k\prime}\nabla_x L_{c^k}(x^k, \lambda^k)$$

and β and σ are fixed scalars with $\beta \in (0, 1)$ and $\sigma \in (0, \frac{1}{2})$. If

$$\|\nabla_x L_{c^k}(x^{k+1}, \lambda^k)\| \leq \epsilon^k,$$

implying termination of the current unconstrained minimization, we do the ordinary first order multiplier iteration, setting

$$\lambda^{k+1} = \lambda^k + c^k h(x^k), \tag{4.123}$$

$$\epsilon^{k+1} = \gamma\epsilon^k, \qquad c^{k+1} = rc^k, \qquad w^{k+1} = \gamma\|\nabla L(x^{k+1}, \lambda^{k+1})\|^2,$$

where r is a fixed scalar with $r > 1$. If

$$\|\nabla_x L_{c^k}(x^{k+1}, \lambda^k)\| > \epsilon^k,$$

we set

$$\lambda^{k+1} = \lambda^k, \qquad \epsilon^{k+1} = \epsilon^k, \qquad c^{k+1} = c^k, \qquad w_{k=1} = w^k,$$

and proceed with the next iteration.

An alternative combined algorithm is obtained by using, in place of the first order iteration (4.122), the second order iteration

$$\bar{\lambda}^k = \left(N^{k\prime}(H^k + c^k N^k N^{k\prime})^{-1} N^k\right)^{-1}$$
$$\left(h(\tilde{x}^k) - N^{k\prime}(H^k + c^k N^k N^{k\prime})^{-1}\nabla f(\tilde{x}^k)\right) - c^k h(\tilde{x}^k),$$

where $\tilde{x}^k$ is obtained by a pure Newton step

$$\tilde{x}^k = x^k + d^k = x^k - (L^k L^{k'})^{-1} \nabla_x L(x^k, \lambda^k),$$

[cf. Eq. (4.120)]. This corresponds to the second implementation of Newton's method of Eqs. (4.104)-(4.106)]. One could then obtain the vector $\bar{x}^k$ by a line search on the augmented Lagrangian $L_{c^k}(\cdot, \bar{\lambda}^k)$ along the direction d^k. The first order multiplier update (4.123) could also be replaced by a second order update. This combination of Newton's method and the second order multiplier method has outstanding rate of convergence properties, particularly if relatively good starting points are known. The combination given earlier based on the first order multiplier updates (4.122) and (4.123) is simpler, particularly if second derivatives are hard to compute and/or a Quasi-Newton approximation is used in Eq. (4.120) in place of the inverse Hessian of the augmented Lagrangian $(L^k L^{k'})^{-1}$.

4.4.4 Primal-Dual Interior Point Methods

We now focus on the linear program

$$\begin{aligned} &\text{minimize } c'x \\ &\text{subject to } Ax = b, \qquad x \geq 0, \end{aligned} \tag{LP}$$

where $c \in \Re^n$ and $b \in \Re^m$ are given vectors, and A is an $m \times n$ matrix of rank m. The dual problem, derived in Section 3.4, is given by

$$\begin{aligned} &\text{maximize } b'\lambda \\ &\text{subject to } A'\lambda \leq c. \end{aligned} \tag{DP}$$

As shown in Section 3.4, (LP) has an optimal solution if and only if (DP) has an optimal solution. Furthermore, when optimal solutions to (LP) and (DP) exist, the corresponding optimal values are equal.

Recall that the logarithmic barrier method, discussed in Section 4.1.1, involves finding for various $\epsilon > 0$,

$$x(\epsilon) = \arg\min_{x \in S} F_\epsilon(x),$$

where

$$F_\epsilon(x) = c'x - \epsilon \sum_{i=1}^{n} \ln x_i,$$

and S is the interior set

$$S = \{x \mid Ax = b,\ x > 0\}.$$

We assume that S is nonempty and bounded.

In Section 4.1.1 we discussed the approach of following the central path $\{x(\epsilon) \mid \epsilon > 0\}$ by minimizing approximately the barrier function $F_{\epsilon^k}(\cdot)$ over S for a sequence $\{\epsilon^k\}$ that converges to zero. Each minimization requires one or more Newton iterations, aimed at iteratively decreasing $F_{\epsilon^k}(\cdot)$.

We now consider a Lagrangian approach, where we apply Newton's method for solving the system of optimality conditions for the problem of minimizing $F_{\epsilon^k}(\cdot)$ over S. The salient features of this approach are:

(a) Only one Newton/Lagrangian iteration is carried out for each value of ϵ^k.

(b) For every k, the pair (x^k, λ^k) is such that x^k is an interior point of the positive orthant, that is, $x^k > 0$, while λ^k is an interior point of the dual feasible region, that is,

$$c - A'\lambda^k > 0.$$

(However, x^k need not be primal-feasible, that is, it need not satisfy the equation $Ax = b$ as it does in the path-following approach of Section 4.1.1.)

(c) Global convergence is enforced by using as the merit function the expression

$$P^k = x^{k\prime} z^k + \|Ax^k - b\|, \tag{4.124}$$

where z^k is the vector of slack variables

$$z^k = c - A'\lambda^k.$$

The expression (4.124) consists of two nonnegative terms: the first term is $x^{k\prime} z^k$, which is positive (since $x^k > 0$ and $z^k > 0$) and can be written as

$$x^{k\prime} z^k = x^{k\prime}(c - A'\lambda^k) = c'x^k - b'\lambda^k + (b - Ax^k)'\lambda^k.$$

Thus when x^k is primal-feasible ($Ax^k = b$), $x^{k\prime} z^k$ is equal to the duality gap, that is, the difference between the primal and the dual costs, $c'x^k - b'\lambda^k$. The second term is the norm of the primal constraint violation $\|Ax^k - b\|$. In the method to be described, neither of the terms $x^{k\prime} z^k$ and $\|Ax^k - b\|$ may increase at each iteration, so that $P^{k+1} \le P^k$ (and typically $P^{k+1} < P^k$) for all k. If we can show that $P^k \to 0$, then asymptotically both the duality gap and the primal constraint violation will be driven to zero. Thus every limit point of $\{(x^k, \lambda^k)\}$ will be a pair of primal and dual optimal solutions, in view of the duality relation

$$\min_{Ax=b,\, x\ge 0} c'x = \max_{A'\lambda \le c} b'\lambda,$$

shown in Section 3.4.2.

Let us write the necessary and sufficient conditions for (x, λ) to be a (global) minimum-Lagrange multiplier pair for the problem of minimizing the barrier function $F_\epsilon(x)$ subject to $Ax = b$. They are

$$c - \epsilon x^{-1} - A'\lambda = 0, \qquad Ax = b, \tag{4.125}$$

where x^{-1} denotes the vector with coordinates $(x_i)^{-1}$. Let z be the vector of slack variables

$$z = c - A'\lambda.$$

Note that λ is dual feasible if and only if $z \geq 0$.

Using the vector z, we can write the first condition of Eq. (4.125) as $z - \epsilon x^{-1} = 0$ or, equivalently, $XZ = \epsilon e$, where X and Z are the diagonal matrices with the coordinates of x and z, respectively, along the diagonal, and e is the vector with unit coordinates,

$$X = \begin{pmatrix} x_1 & 0 & \cdots & 0 \\ 0 & x_2 & \cdots & 0 \\ \cdots & \cdots & \cdots & \cdots \\ 0 & 0 & \cdots & x_n \end{pmatrix}, \quad Z = \begin{pmatrix} z_1 & 0 & \cdots & 0 \\ 0 & z_2 & \cdots & 0 \\ \cdots & \cdots & \cdots & \cdots \\ 0 & 0 & \cdots & z_n \end{pmatrix}, \quad e = \begin{pmatrix} 1 \\ 1 \\ \vdots \\ 1 \end{pmatrix}.$$

Thus the optimality conditions (4.125) can be written in the equivalent form

$$XZe = \epsilon e, \tag{4.126}$$

$$Ax = b, \tag{4.127}$$

$$z + A'\lambda = c. \tag{4.128}$$

Given (x, λ, z) satisfying $z + A'\lambda = c$, and such that $x > 0$ and $z > 0$, a Newton iteration for solving this system is

$$x(\alpha, \epsilon) = x + \alpha \Delta x, \tag{4.129}$$

$$\lambda(\alpha, \epsilon) = \lambda + \alpha \Delta \lambda,$$

$$z(\alpha, \epsilon) = z + \alpha \Delta z,$$

where α is a stepsize such that $0 < \alpha \leq 1$ and

$$x(\alpha, \epsilon) > 0, \qquad z(\alpha, \epsilon) > 0,$$

and the pure Newton step $(\Delta x, \Delta \lambda, \Delta z)$ solves the linearized version of the system (4.126)-(4.128)

$$X\Delta z + Z\Delta x = -v, \tag{4.130}$$

$$A\Delta x = b - Ax, \tag{4.131}$$

$$\Delta z + A'\Delta\lambda = 0, \tag{4.132}$$

with v defined by

$$v = XZe - \epsilon e. \tag{4.133}$$

After a straightforward calculation, similar to the ones given earlier, the solution of the linearized system (4.130)-(4.132) can be written as

$$\Delta\lambda = \big(AZ^{-1}XA'\big)^{-1}\big(AZ^{-1}v + b - Ax\big), \tag{4.134}$$

$$\Delta z = -A'\Delta\lambda, \tag{4.135}$$

$$\Delta x = -Z^{-1}v - Z^{-1}X\Delta z.$$

Note that $\lambda(\alpha,\epsilon)$ is dual feasible, since from Eq. (4.132) and the condition $z + A'\lambda = c$, we see that $z(\alpha,\epsilon) + A'\lambda(\alpha,\epsilon) = c$. Note also that if $\alpha = 1$, that is, a pure Newton step is used, $x(\alpha,\epsilon)$ is primal feasible, since from Eq. (4.131) we have $A(x + \Delta x) = b$.

Merit Function Improvement

We will now evaluate the changes in the constraint violation and the merit function induced by the Newton iteration.

By using Eqs. (4.129)and (4.131), the new constraint violation is given by

$$Ax(\alpha,\epsilon) - b = Ax + \alpha A\Delta x - b = Ax + \alpha(b - Ax) - b = (1-\alpha)(Ax - b). \tag{4.136}$$

Thus, since $0 < \alpha \leq 1$, the new norm of constraint violation $\|Ax(\alpha,\epsilon) - b\|$ is always no larger than the old one. Furthermore, if x is primal-feasible ($Ax = b$), the new iterate $x(\alpha,\epsilon)$ is also primal-feasible.

The inner product

$$g = x'z \tag{4.137}$$

after the iteration becomes

$$\begin{aligned} g(\alpha,\epsilon) &= x(\alpha,\epsilon)'z(\alpha,\epsilon) \\ &= (x + \alpha\Delta x)'(z + \alpha\Delta z) \\ &= x'z + \alpha(x'\Delta z + z'\Delta x) + \alpha^2\Delta x'\Delta z. \end{aligned} \tag{4.138}$$

From Eqs. (4.131) and (4.135) we have

$$\Delta x'\Delta z = (Ax - b)'\Delta\lambda,$$

while by premultiplying Eq. (4.130) with e' and using the definition (4.133) for v, we obtain

$$x'\Delta z + z'\Delta x = -e'v = n\epsilon - x'z.$$

By substituting the last two relations in Eq. (4.138) and by using also the expression (4.137) for g, we see that

$$g(\alpha, \epsilon) = g - \alpha(g - n\epsilon) + \alpha^2 (Ax - b)'\Delta\lambda. \tag{4.139}$$

Let us now denote by P and $P(\alpha, \epsilon)$ the value of the merit function (4.124) before and after the iteration, respectively. We have by using the expressions (4.136) and (4.139),

$$\begin{aligned} P(\alpha, \epsilon) &= g(\alpha, \epsilon) + \|Ax(\alpha, \epsilon) - b\| \\ &= g - \alpha(g - n\epsilon) + \alpha^2 (Ax - b)'\Delta\lambda + (1 - \alpha)\|Ax - b\|, \end{aligned}$$

or

$$P(\alpha, \epsilon) = P - \alpha\big(g - n\epsilon + \|Ax - b\|\big) + \alpha^2 (Ax - b)'\Delta\lambda.$$

Thus if ϵ is chosen to satisfy

$$\epsilon < \frac{g}{n}$$

and α is chosen to be small enough so that the second order term $\alpha^2 (Ax - b)'\Delta\lambda$ is dominated by the first order term $\alpha(g - n\epsilon)$, the merit function will be improved as a result of the iteration.

A General Class of Primal-Dual Algorithms

Let us consider now the general class of algorithms of the form

$$x^{k+1} = x(\alpha^k, \epsilon^k), \qquad \lambda^{k+1} = \lambda(\alpha^k, \epsilon^k), \qquad z^{k+1} = z(\alpha^k, \epsilon^k),$$

where α^k and ϵ^k are positive scalars such that

$$x^{k+1} > 0, \qquad z^{k+1} > 0, \qquad \epsilon^k < \frac{g^k}{n},$$

where g^k is the inner product

$$g^k = x^{k\prime} z^k + (Ax^k - b)'\lambda^k,$$

and α^k is such that the merit function P^k is reduced. Initially we must have $x^0 > 0$, and $z^0 = c - A'\lambda^0 > 0$ (such a point can often be easily found; otherwise an appropriate reformulation of the problem is necessary for which we refer to the specialized literature). These methods have been called *primal-dual*, although we have seen that they are really Newton/Lagrangian methods supplemented by a stepsize procedure, which guarantees that the merit function P^k is improved at each iteration.

It can be shown that it is possible to choose α^k and ϵ^k so that the merit function is not only reduced at each iteration, but also converges to zero. Furthermore, with suitable choices of α^k and ϵ^k, algorithms with good theoretical properties, such as polynomial complexity and superlinear convergence, can be derived. The main convergence analysis ideas rely on the primal-dual version of the central path discussed in Section 4.1, and some of the associated path following concepts. We refer to the research monograph [Wri97] for a detailed discussion.

With properly chosen sequences α^k and ϵ^k, and appropriate implementation, the practical performance of the primal-dual methods has been shown to be excellent. The choice

$$\epsilon^k = \frac{g^k}{n^2},$$

leading to the relation

$$g^{k+1} = (1 - \alpha^k + \alpha^k/n)g^k$$

for feasible x^k, has been suggested as a good practical rule. Usually, when x^k has already become feasible, α^k is chosen as $\theta\tilde{\alpha}^k$, where θ is a factor very close to 1 (say 0.999), and $\tilde{\alpha}^k$ is the maximum stepsize α that guarantees that $x(\alpha, \epsilon^k) \geq 0$ and $z(\alpha, \epsilon^k) \geq 0$

$$\tilde{\alpha}^k = \min\left\{\min_{i=1,\ldots,n}\left\{\frac{x_i^k}{-\Delta x_i} \,\Big|\, \Delta x_i < 0\right\}, \min_{i=1,\ldots,n}\left\{\frac{z_i^k}{-\Delta z_i} \,\Big|\, \Delta z_i < 0\right\}\right\}.$$

When x^k is not feasible, the choice of α^k must also be such that the merit function is improved. In some works, a different stepsize for the x update than for the (λ, z) update has been suggested. The stepsize for the x update is near the maximum stepsize α that guarantees $x(\alpha, \epsilon^k) \geq 0$, and the stepsize for the (λ, z) update is near the maximum stepsize α that guarantees $z(\alpha, \epsilon^k) \geq 0$. There are a number of additional practical issues related to implementation, for which we refer to the specialized literature.

Predictor-Corrector Variants

We mentioned briefly in Section 1.4 the variation of Newton's method where the Hessian is evaluated periodically every $p > 1$ iterations in order to economize in iteration overhead. When $p = 2$ and the problem is to solve the system $g(x) = 0$, where $g : \Re^n \mapsto \Re^n$, this variation of Newton's method takes the form

$$\hat{x}^k = x^k - \left(\nabla g(x^k)'\right)^{-1} g(x^k), \tag{4.140}$$

$$x^{k+1} = \hat{x}^k - \left(\nabla g(x^k)'\right)^{-1} g(\hat{x}^k). \tag{4.141}$$

Thus, given x^k, this iteration performs a regular Newton step to obtain $\hat{x}^k$, and then an approximate Newton step from $\hat{x}^k$, using, however, the already available Jacobian inverse $\left(\nabla g(x^k)'\right)^{-1}$. It can be shown that if $x^k \to x^*$, the order of convergence of the error $\|x^k - x^*\|$ is cubic, that is,

$$\limsup_{k\to\infty} \frac{\|x^{k+1} - x^*\|}{\|x^k - x^*\|^3} < \infty,$$

under the same assumptions that the ordinary Newton's method ($p = 1$) attains a quadratic order of convergence; see [OrR70], p. 315. Thus, the price for the 50% saving in Jacobian evaluations and inversions is a small degradation of the convergence rate over the ordinary Newton's method (which attains a quartic order of convergence when two successive ordinary Newton steps are counted as one).

Two-step Newton methods such as the iteration (4.140), (4.141), when applied to the system of optimality conditions (4.126)-(4.128) for the linear program (LP) are known as *predictor-corrector* methods (the name comes from their similarity with predictor-corrector methods for solving differential equations). They operate as follows:

Given (x, z, λ) with $x > 0$, and $z = c - A'\lambda > 0$, the *predictor iteration* [cf. Eq. (4.140)], solves for $(\Delta\hat{x}, \Delta\hat{z}, \Delta\hat{\lambda})$ the system

$$X\Delta\hat{z} + Z\Delta\hat{x} = -\hat{v}, \tag{4.142}$$

$$A\Delta\hat{x} = b - Ax, \tag{4.143}$$

$$\Delta\hat{z} + A'\Delta\hat{\lambda} = 0, \tag{4.144}$$

with $\hat{v}$ defined by

$$\hat{v} = XZe - \hat{\epsilon}e, \tag{4.145}$$

[cf. Eqs. (4.130)-(4.133)].

The *corrector iteration* [cf. Eq. (4.141)], solves for $(\Delta\bar{x}, \Delta\bar{z}, \Delta\bar{\lambda})$ the system

$$X\Delta\bar{z} + Z\Delta\bar{x} = -\bar{v}, \tag{4.146}$$

$$A\Delta\bar{x} = b - A(x + \Delta\hat{x}), \tag{4.147}$$

$$\Delta\bar{z} + A'\Delta\bar{\lambda} = 0, \tag{4.148}$$

with $\bar{v}$ defined by

$$\bar{v} = (X + \Delta\hat{X})(Z + \Delta\hat{Z})e - \bar{\epsilon}e, \tag{4.149}$$

where $\Delta\hat{X}$ and $\Delta\hat{Z}$ are the diagonal matrices corresponding to $\Delta\hat{x}$ and $\Delta\hat{z}$, respectively. Here $\hat{\epsilon}$ and $\bar{\epsilon}$ are the barrier parameters corresponding to the two iterations.

The *composite Newton direction* is

$$\Delta x = \Delta\hat{x} + \Delta\bar{x},$$

$$\Delta z = \Delta\hat{z} + \Delta\bar{z},$$

$$\Delta\lambda = \Delta\hat{\lambda} + \Delta\bar{\lambda},$$

and the corresponding iteration is

$$x(\alpha, \epsilon) = x + \alpha\Delta x,$$

$$\lambda(\alpha, \epsilon) = \lambda + \alpha\Delta\lambda,$$

$$z(\alpha, \epsilon) = z + \alpha\Delta z,$$

where α is a stepsize such that $0 < \alpha \leq 1$ and

$$x(\alpha, \epsilon) > 0, \qquad z(\alpha, \epsilon) > 0.$$

Adding Eqs. (4.142)-(4.144)and Eqs. (4.146)-(4.148), we obtain

$$X(\Delta\hat{z} + \Delta\bar{z})z + Z(\Delta\hat{x} + \Delta\bar{x}) = -\hat{v} - \bar{v}, \tag{4.150}$$

$$A(\Delta\hat{x} + \Delta\bar{x})x = b - Ax + b - A(x + \Delta\hat{x}), \tag{4.151}$$

$$\Delta\hat{z} + \Delta\bar{z} + A'(\Delta\hat{\lambda} + \Delta\bar{\lambda}) = 0, \tag{4.152}$$

We now use the fact

$$b - A(x + \Delta\hat{x}) = 0$$

[cf. Eq. (4.143)], and we also use Eqs. (4.149) and (4.142) to write

$$\begin{aligned}
\bar{v} &= (X + \Delta\hat{X})(Z + \Delta\hat{Z})e - \bar{\epsilon}e \\
&= XZe + \Delta\hat{X}Ze + X\Delta\hat{Z}e + \Delta\hat{X}\Delta\hat{Z}e - \bar{\epsilon}e \\
&= XZe + Z\Delta\hat{x} + X\Delta\hat{z} + \Delta\hat{X}\Delta\hat{Z}e - \bar{\epsilon}e \\
&= XZe - \hat{v} + \Delta\hat{X}\Delta\hat{Z}e - \bar{\epsilon}e.
\end{aligned}$$

Substituting in Eqs. (4.150)-(4.152) we obtain the following system of equations for the composite Newton direction $(\Delta x, \Delta z, \Delta\lambda) = (\Delta\hat{x} + \Delta\bar{x}, \Delta\hat{z} + \Delta\bar{z}, \Delta\hat{\lambda} + \Delta\bar{\lambda})$:

$$X\Delta z + Z\Delta x = -XZe - \Delta\hat{X}\Delta\hat{Z}e + \bar{\epsilon}e, \tag{4.153}$$

$$A\Delta x = b - Ax, \tag{4.154}$$

$$\Delta z + A'\Delta\lambda = 0. \tag{4.155}$$

To implement the predictor-corrector method, we need to solve the system (4.142)-(4.145) for some value of $\hat{\epsilon}$ to obtain $(\Delta\hat{X}, \Delta\hat{Z})$, and then to solve the system (4.153)-(4.155) for some value of $\bar{\epsilon}$ to obtain $(\Delta x, \Delta z, \Delta\lambda)$. It is important to note here that most of the work needed for the first system, namely the factorization of the matrix

$$AZ^{-1}XA'$$

in Eq. (4.134), need not be repeated when solving the second system, so that solving both systems requires relatively little extra work over solving the first one.

In an implementation proposed in [Meh92], which has proved very successful in practice, $\hat{\epsilon}$ is taken to be zero, so that the solution of the first system, $(\Delta\hat{x}, \Delta\hat{z}, \Delta\hat{\lambda})$, is the affine scaling primal-dual direction. Furthermore, $\bar{\epsilon}$ is chosen on the basis of the solution of the first system according to the formula

$$\bar{\epsilon} = \left(\frac{\hat{g}}{x'z}\right)^2 \frac{\hat{g}}{n},$$

where $\hat{g}$ is the duality gap that would result from a feasibility-restricted primal-dual affine scaling step, as given by

$$\hat{g} = (x + \alpha_P \Delta\hat{x})'(z + \alpha_D \Delta\hat{z}),$$

where

$$\alpha_P = \theta \min_{i=1,\ldots,n} \left\{ \frac{x_i^k}{-\Delta x_i} \;\Big|\; \Delta x_i < 0 \right\},$$

$$\alpha_D = \theta \min_{i=1,\ldots,n} \left\{ \frac{z_i^k}{-\Delta z_i} \;\Big|\; \Delta z_i < 0 \right\}.$$

and θ is a factor very close to 1 (say 0.999). Note that this formula chooses a small $\bar{\epsilon}$ when good progress can be made in the affine scaling direction and a large value of $\bar{\epsilon}$ otherwise. We refer to the specialized literature for further details [Meh92], [LMS92], [Wri97], [Ye97].

4.4.5 Comparison of Various Methods

Quite a few barrier, penalty, and Lagrange multiplier methods were given in this chapter, so it is worth reflecting on their suitability for different types of problems. Even though it is hard to provide reliable guidelines, one may at least delineate the relative strengths and the weaknesses of the various methods in specific practical contexts.

The barrier methods of Section 4.1, generally must solve a sequence of minimization problems that are increasingly ill-conditioned. This is a disadvantage relative to the multiplier methods of Section 4.2, whose sequence of minimization problems need not be ill-conditioned, and also relative to the exact penalty methods of Section 4.1, which have to solve only one minimization problem. However, for linear and for quadratic programs there is special structure that makes the logarithmic barrier and also the primal-dual interior point methods of Section 4.4.4 preferable to multiplier and exact penalty methods, from the theoretical and apparently the practical point of view. Whether, there are other important classes of problems for which this is also true, is currently an open question.

Multiplier methods are excellent general purpose constrained optimization methods. In fact some of the most popular software packages for solving nonlinear programming problems are based on multiplier methods. The main advantages of multiplier methods are simplicity and robustness. They take advantage of the well-developed unconstrained optimization technology, and they require fewer assumptions for their validity relative to their competitors. In particular, they can deal with nonlinear equality constraints, and they do not rely on the existence of second derivatives and on regularity of the generated iterates (although they can be made more efficient when second derivatives can be used and when the iterates are regular).

The main disadvantage of multiplier methods relative to exact penalty methods is that they require a sequence of unconstrained minimizations as opposed to a single minimization. This disadvantage can be ameliorated by making the minimizations inexact or by combining the multiplier method with a Lagrangian method as described in Section 4.4.3. Still, practice has shown that minimization of an exact penalty function by a Newton-like method can require substantially fewer iterations relative to a multiplier method. Note, however, that each of these iterations may require a potentially costly subproblem (as in the linearization method) or may require complex calculations (as in differentiable exact penalty methods).

Generally, both multiplier methods and exact penalty methods require some trial and error to obtain appropriate values of the penalty parameter and also to ensure that there are no difficulties with ill-conditioning. However, multiplier methods typically are easier to "tune" than exact penalty methods, and deal more comfortably with the absence of a good starting point. Thus, if only a limited number of optimization runs are required in a given practical problem after development of the optimization code, one is typically better off with a method of multipliers than with an exact penalty method. If on the other hand, repetitive solution of the same problem with minor variations is envisioned, solution time is an issue, and the associated overhead per iteration is reasonable, one may prefer to use an exact penalty method.

E X E R C I S E S

4.4.1

Consider the problem

$$\begin{aligned} &\text{minimize} \quad -x_1x_2 \\ &\text{subject to} \quad x_1 + x_2 = 2. \end{aligned}$$

Write the implementation (4.102)-(4.103) of Newton's method, and show that it finds the optimal solution in a single iteration, regardless of the starting point. Write also the approximate implementation (4.115)-(4.116) for the starting point $x^0 = (0, 0)$, $\lambda^0 = 0$, and for the two values $c = 10^{-2}$ and $c = 10^2$. How does the error after a single iteration depend on c?

4.4.2

Consider the linear program

$$\begin{aligned} &\text{minimize} \;\; x_1 + 2x_2 + 3x_3 \\ &\text{subject to} \;\; x_1 + x_2 + x_3 = 1, \qquad x \geq 0. \end{aligned}$$

Write a computer program to implement a primal-dual interior point method and its predictor-corrector variant, and solve the problem for the starting points $x^0 = (.8, .15, .05)$ and $x^0 = (.1, .2, .7)$, and $\lambda^0 = 0$.

4.4.3

Use Prop. 4.4.1 to derive a local convergence result for the approximate implementation (4.115)-(4.116) of Newton's method. Do the same for iteration (4.118)-(4.119).

4.5 NOTES AND SOURCES

Section 4.1: The logarithmic barrier method dates to the work of M. R. Frisch in the middle 50's [Fri56]. Other barrier methods have been proposed by Carroll [Car61]. An important early reference on penalty and barrier methods is Fiacco and McCormick [FiM68]. Properties of the central path are investigated by McLinden [McL80], Sonnevend [Son86], Bayer and Lagarias [BaL89], and Guler [Gul94]. For surveys of logarithmic barrier interior point methods for linear programming, which give many additional references, see Gonzaga [Gon92] and den Hertog [Her94]. The line of analysis that we use is due to Tseng [Tse89], which also gives a computational complexity result along the lines of Exercise 4.1.5 (see also Gonzaga [Gon91]).

There has been a lot of effort to efficiently apply interior point methods to broader classes of problems, such as quadratic programming (Anstreicher, den Hertog, Roos, and Terlaky [AHR93], Wright [Wri96]), linear complementarity problems (Kojima, Meggido, Noma, and Yoshise [KMN91], Tseng [Tse92], Wright [Wri93c]), matrix inequalities (Alizadeh [Ali92], [Ali95], Nesterov and Nemirovskii [NeN94], Vandenberghe and Boyd [VaB95]), general convex programming problems (Wright [Wri92], Kortanek and Zhu [KoZ93], Anstreicher and Vial [AnV94], Jarre and Saunders [JaS95], Kortanek and Zhu [KoZ95]), and optimal control (Wright [Wri93b]). The research monographs by Nesterov and Nemirovskii [NeN94], Wright [Wri97], and Ye [Ye97] are devoted to interior point methods for linear, quadratic, and convex programming.

Section 4.2: The method of multipliers for equality constraints was independently proposed by Hestenes [Hes69], Powell [Pow69], and Haarhoff and Buys [HaB70]. These references contained very little analysis and empirical evidence, but much subsequent work established the convergence properties of the method and proposed various extensions. Surveys by Bertsekas [Ber76b] and Rockafellar [Roc76] summarize the work up to 1976, and the author's research monograph [Ber82a] provides a detailed analysis and many references. The global convergence of the method under a variety of conditions is discussed by Poljak and Tretjakov [PoT73b], Bertsekas [Ber76a], Polak and Tits [PoT80], Bertsekas [Ber82a], and Conn, Gould, and Toint [CGT91].

The class of nonquadratic penalty methods given in Section 4.2.5 was introduced by Kort and Bertsekas [KoB72], with a special focus on the exponential method of multipliers. The convergence properties of the sequence $\{x^k\}$ generated by this method are quite intricate and are discussed by Bertsekas [Ber82a], Tseng and Bertsekas [TsB93], and Iusem [Ius98] for convex problems. For nonconvex problems under second order sufficiency conditions, the convergence analysis follows the pattern of the corresponding analysis for the quadratic method of multipliers (see Nguyen and Strodiot [NgS79]). The exponential method was applied to the solution of systems of nonlinear inequalities by Bertsekas [Ber82a], Section 5.1.3, and Schnabel [Sch82]. The modified barrier method was proposed and developed by Polyak; see [Pol92] and the references given therein.

There has been considerable recent research interest on the exponential penalty, the modified barrier, and other related methods that use nonquadratic penalty functions; see Freund [Fre91], Guler [Gul92], Teboulle [Teb92], Chen and Teboulle [ChT93], Tseng and Bertsekas [TsB93], Eckstein [Eck94a], Iusem, Svaiter, and Teboulle [IST94], Iusem and Teboulle [IuT95], Ben-Tal and Zibulevsky [BeZ97], Kiwiel [Kiw97a], Polyak and Teboulle [PoT97], Doljansky and Teboulle [DoT98], Iusem [Ius98], and Wei, Qi, and Birge [WQB98].

Section 4.3: Nondifferentiable exact penalty methods were first proposed by Zangwill [Zan67b]; see also Han and Mangasarian [HaM79], who survey the subject and give many references. The linearization method is due to Pschenichny [Psc70] (see also Pschenichny and Danilin [PsD76]), and was also derived independently by Han [Han77]. Convergence rate issues and modifications to improve the convergence rate of sequential quadratic programming algorithms and to avoid the Maratos' effect (Exercise 4.3.9) are discussed in Boggs, Tolle, and Wang [BTW82], Coleman and Conn [CoC82a], [CoC82b], Gabay [Gab82], Mayne and Polak [MaP82], Panier and Tits [PaT91], Bonnans, Panier, Tits, and Zhou [BPT92], and Bonnans [Bon89b], [Bon94]. Combinations of the linearization method and the two-metric projection method discussed in Section 2.4 have been proposed by Heinkenschloss [Hei96]. Note that since the linearization method can minimize the nondifferentiable exact penalty function $f(x) + cP(x)$, it can also be used to minimize

$$P(x) = \max\{g_0(x), g_1(x), \ldots, g_r(x)\},$$

which is a typical version of a minimax problem.

Exact differentiable penalty methods involving only x were introduced by Fletcher [Fle70b]. Exact differentiable penalty methods involving both x and λ were introduced by DiPillo and Grippo [DiG79]. The relation between these two types of methods, their utility for sequential quadratic programming, and a number of variations were derived by Bertsekas [Ber82c] (see also [Ber82a], which contains a detailed convergence analysis). Extensions to inequality constraints are given by Glad and Polak [GlP79], and Bertsekas [Ber82a]; see also Mukai and Polak [MuP75], Boggs and Tolle [BoT80], Bertsekas [Ber82c], and DiPillo and Grippo [DiG89]. Differentiable exact penalty functions are used by Nazareth [Naz96], and Nazareth and Qi [NaQ96] to extend the region of convergence of Newton-like methods for solving systems of nonlinear equations.

Section 4.4: First order Lagrangian methods were introduced by Arrow, Hurwicz, and Uzawa [AHU59]. They were also analyzed by Poljak [Pol70], whom we follow in our presentation.

A Newton-like Lagrangian method for inequality constraints was proposed by Wilson [Wil63]. For this method, a superlinear convergence rate was established by Robinson [Rob74] under second order sufficiency conditions, including strict complementarity. Superlinear convergence results of a variant of the method were shown under weaker conditions by Wright [Wri98] and Hager [Hag99]. An extensive discussion of Newton-like Lagrangian methods is given in the author's research monograph [Ber82a].

Among interior point algorithms, primal-dual methods are widely considered as generally the most effective; see e.g., McShane, Monma, and Shanno [MMS91]. They were introduced for linear programming through the study of the primal-dual central path by Megiddo [Meg88]. They were turned into path-following algorithms in the subsequent papers by Kojima, Mizuno, and Yoshise [KMY89], and Monteiro and Adler [MoA89a]. Their convergence, rate of convergence, and computational complexity are discussed in Zhang and Tapia [ZhT92], [ZTP93], [ZhT93], Potra [Pot94], and Tapia, Zhang, and Ye [TZY95]. The predictor-corrector variant was proposed by Mehrotra [Meh92]; see also Lustig, Marsten, and Shanno [LMS92]. There have been several extensions to broader classes of problems, including quadratic programming (Monteiro and Adler [MoA89b]), and linear complementarity (Wright [Wri94] and Tseng [Tse95b]). The research monograph by Wright [Wri97] provides an extensive development of primal-dual interior point methods.

5

Duality and Convex Programming

Contents

$\alpha\gamma\epsilon\omega\mu\epsilon\tau\rho\eta\tau o s\ \mu\eta\delta\epsilon\iota s\ \epsilon\iota\sigma\iota\tau\omega$

(No one ignorant of geometry shall enter)

Door inscription at Plato's Academy

Our analysis of earlier chapters was primarily calculus-based, comparing local minima to their close neighbors using first and second derivatives of the cost and the constraints. In this chapter we use a fundamentally different line of analysis. In particular, we will be concerned exclusively with global minima, as opposed to local minima, and at least in the beginning, we will make no differentiability or even continuity assumptions on the cost and constraint functions.

An important consequence of the generality of our framework is that much of the theory that we will develop is very broadly applicable. For example, it is useful for discrete optimization problems, such as integer programming, as we will show in Section 5.5. On the other hand, the power of the theory is enhanced as the problem becomes more structured, and the existence of Lagrange multipliers can be guaranteed. In particular, the strongest results of this chapter apply only when the cost and constraints are convex.

The central notion of this chapter is *duality*. We have already encountered a simple version of duality in Section 3.4, which was developed under differentiability assumptions on the cost function and linearity assumptions on the constraints. Our analysis in this chapter will be deeper and more powerful, making full use of intuitive geometrical notions such as hyperplanes, and their convex set support and separation properties. Because our development is geometry-based, the duality results that we develop and their proofs can be easily visualized. As an example, we provide two abstract problems, dual to each other, that will prove particularly relevant in the context of the subsequent analysis. Let S be a subset of $\Re^n$ and consider the following two problems.

(a) *Min Common Point Problem*: Here, among all points that are common to both S and the nth axis, we want to find the one whose nth coordinate is minimum.

(b) *Max Intercept Point Problem*: Here we consider all hyperplanes that intersect the nth axis and support the set S from "below" (see Fig. 5.1.1). We want to find the maximum point of interception of the nth axis with such a hyperplane.

Figure 5.1.1 shows that the optimal value of the max intercept point problem is no larger than the optimal value of the min common point problem, and that under favorable circumstances the two optimal values are equal.

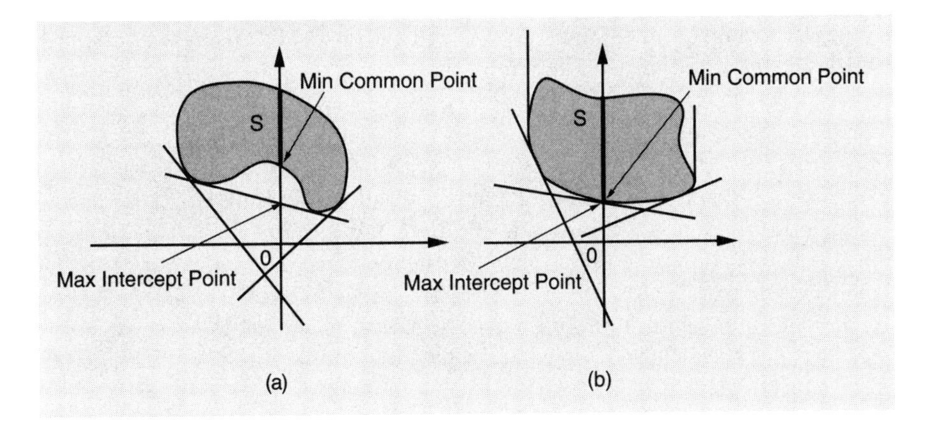

Figure 5.1.1. Illustration of the optimal values of the min common point and max intercept point problems. In (a), the two optimal values are not equal. In (b), the set S, when "extended upwards" along the nth axis, yields the set

$$\bar{S} = \{\bar{x} \mid \text{for some } x \in S,\ \bar{x}_n \geq x_n,\ \bar{x}_i = x_i, \forall\ i = 1, \ldots, n-1\},$$

which is convex. As a result, the two optimal values are equal. This fact, when suitably formalized, will form the basis for a number of duality results in Sections 5.2 and 5.3.

5.1 THE DUAL PROBLEM

We consider the problem

$$\begin{aligned} &\text{minimize} \quad f(x) \\ &\text{subject to} \quad x \in X, \qquad g_j(x) \leq 0, \quad j = 1, \ldots, r, \end{aligned}$$

where $f : \Re^n \mapsto \Re$, $g_j : \Re^n \mapsto \Re$ are given functions, and X is a subset of $\Re^n$. We also denote

$$g(x) = \big(g_1(x), \ldots, g_r(x)\big),$$

and we write the constraints $g_j(x) \leq 0$ compactly as $g(x) \leq 0$. We refer to this problem as the *primal problem* and we denote by f^* its optimal value:

$$f^* = \inf_{\substack{x \in X \\ g_j(x) \leq 0,\, j=1,\ldots,r}} f(x).$$

Unless the opposite is clearly stated, we will assume throughout this chapter the following:

Assumption 5.1.1: (Feasibility and Boundedness) There exists at least one feasible solution for the primal problem and the cost is bounded from below, i.e., $-\infty < f^* < \infty$.

At several points, however, we will pause to discuss what happens when this assumption is violated.

5.1.1 Lagrange Multipliers

We want to define a notion of a Lagrange multiplier that is not tied to a specific local or global minimum, and does not assume differentiability of the cost and constraint functions. To this end, we draw motivation from the case where $X = \Re^n$, and f and g_j are convex and differentiable. Then, if x^* is a global minimum and a regular point, there exists a vector $\mu^* = (\mu_1^*, \ldots, \mu_r^*)$ such that $\mu_j^* \geq 0$ and $\mu_j^* g_j(x^*) = 0$ for all j, and $\nabla_x L(x^*, \mu^*) = 0$ or equivalently,

$$f^* = f(x^*) = \min_{x \in \Re^n} L(x, \mu^*),$$

where $L : \Re^{n+r} \mapsto \Re$ is the Lagrangian function

$$L(x, \mu) = f(x) + \sum_{j=1}^{r} \mu_j g_j(x) = f(x) + \mu' g(x).$$

We are thus led to the following definition:†

Definition 5.1.1: A vector $\mu^* = (\mu_1^*, \ldots, \mu_r^*)$ is said to be a *Lagrange multiplier vector* (or simply *Lagrange multiplier*) for the primal problem if

$$\mu_j^* \geq 0, \qquad j = 1, \ldots, r,$$

and

$$f^* = \inf_{x \in X} L(x, \mu^*).$$

To visualize the definition of a Lagrange multiplier, as well as other concepts related to duality, it is useful to consider hyperplanes in the space of constraint-cost pairs $\big(g(x), f(x)\big)$ (viewed as vectors in $\Re^{r+1}$). We recall

† From here until the end of the book, we will use the term "Lagrange multiplier" in accordance with Definition 5.1.1. This differs somewhat from our earlier usage of the term. In particular, a vector that qualifies as Lagrange multiplier in the local, calculus-oriented context of Chapter 3 (cf. Prop. 3.1.1) may not qualify as Lagrange multiplier under the global, geometrically-motivated Definition 5.1.1 of this chapter, and reversely. It would have been more proper to use a different term here than in Chapter 3, but we have chosen to use a single term for multipliers to emphasize their common essential use, which is to facilitate the solution process by eliminating some or all of the constraints.

(see Section B.2 in Appendix B) that a hyperplane in $\Re^{r+1}$ is specified by a linear equation involving a nonzero vector (μ, μ_0) (called the *normal vector of H*), where $\mu \in \Re^r$ and $\mu_0 \in \Re$, and by a constant c as follows:

$$H = \{(z, w) \mid z \in \Re^r,\ w \in \Re,\ \mu_0 w + \mu' z = c\}.$$

Any vector $(\bar{z}, \bar{w})$ that belongs to the hyperplane H specifies the constant c as

$$c = \mu_0 \bar{w} + \mu' \bar{z}.$$

Thus the hyperplane with given normal (μ, μ_0) that passes through a given vector $(\bar{z}, \bar{w})$ is the set of (z, w) that satisfy the equation

$$\mu_0 w + \mu' z = \mu_0 \bar{w} + \mu' \bar{z}.$$

This hyperplane defines two halfspaces: the *positive halfspace*

$$H^+ = \{(z, w) \mid \mu_0 w + \mu' z \geq \mu_0 \bar{w} + \mu' \bar{z}\}$$

and the *negative halfspace*

$$H^- = \{(z, w) \mid \mu_0 w + \mu' z \leq \mu_0 \bar{w} + \mu' \bar{z}\}.$$

Hyperplanes with normals (μ, μ_0) where $\mu_0 \neq 0$ are referred to as *nonvertical* (their normal has a nonzero last component). A nonvertical hyperplane can be *normalized* by dividing its normal vector by μ_0, and assuming this is done, we have $\mu_0 = 1$.

The above definitions will now be used to interpret Lagrange multipliers as normalized nonvertical hyperplanes with a certain orientation to the set of all constraint-cost pairs as x ranges over X, i.e., the subset of $\Re^{r+1}$

$$S = \{(g(x), f(x)) \mid x \in X\}. \tag{5.1}$$

We have the following lemma, which is graphically illustrated in Fig. 5.1.2.

Visualization Lemma:

(a) The hyperplane with normal $(\mu, 1)$ that passes through a vector $(g(x), f(x))$ intercepts the vertical axis $\{(0, w) \mid w \in \Re\}$ at the level $L(x, \mu)$.

(b) Among all hyperplanes with normal $(\mu, 1)$ that contain in their positive halfspace the set S of Eq. (5.1), the highest attained level of interception of the vertical axis is $\inf_{x \in X} L(x, \mu)$.

(c) μ^* is a Lagrange multiplier (as per Definition 5.1.1) if and only if $\mu^* \geq 0$ and among all hyperplanes with normal $(\mu^*, 1)$ that contain in their positive halfspace the set S, the highest attained level of interception of the vertical axis is f^*.

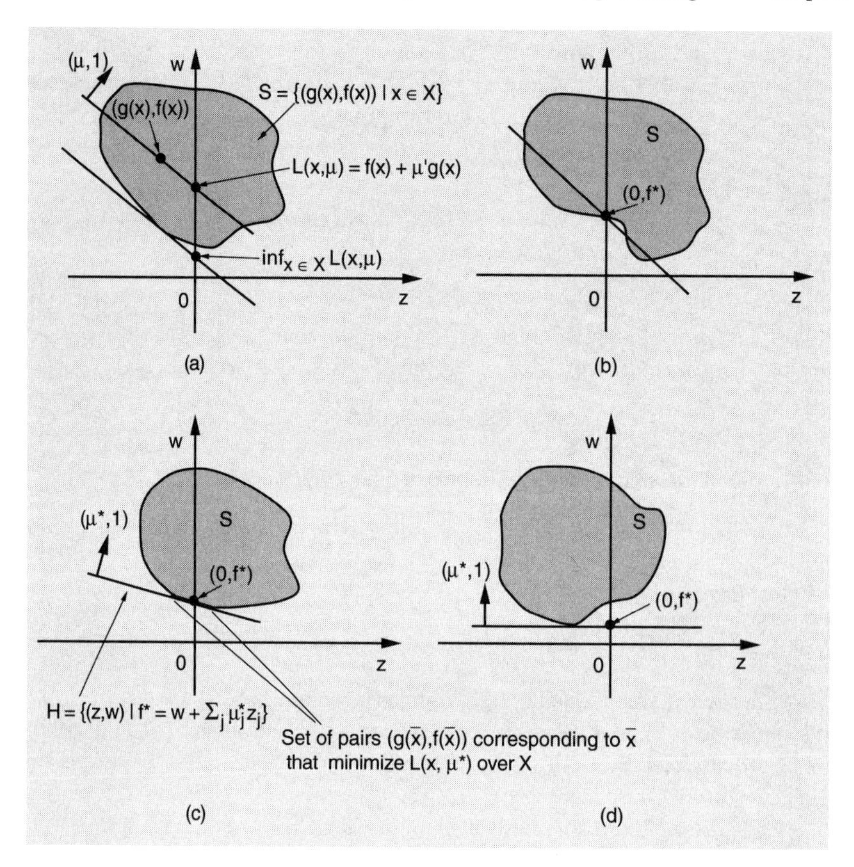

Figure 5.1.2. (a) Geometrical interpretation of the Lagrangian value $L(x, \mu)$ as the level of interception of the vertical axis with the hyperplane with normal $(\mu, 1)$ that passes through $(g(x), f(x))$. (b) A case where there is no Lagrange multiplier. Here, there is no hyperplane that passes through the point $(0, f^*)$ and contains S in its positive halfspace. (c) and (d) Illustration of a Lagrange multiplier vector μ^* in the cases $\mu^* \neq 0$ and $\mu^* = 0$, respectively. It defines a hyperplane H that has normal $(\mu^*, 1)$, passes through $(0, f^*)$, and contains S in its positive halfspace. Note that the common points of H and S (if any) are the pairs $(g(\bar{x}), f(\bar{x}))$ corresponding to points $\bar{x}$ that minimize the Lagrangian $L(x, \mu^*)$ over $x \in X$. In (c), the point $(0, f^*)$ belongs to S and corresponds to optimal primal solutions for which the constraint is active (as well as to some infeasible primal solutions). In (d), the point $(0, f^*)$ does not belong to S and corresponds to an optimal primal solution for which the constraint is inactive.

Proof: (a) By the preceding discussion, the hyperplane with normal $(\mu, 1)$ that passes through $\big(g(x), f(x)\big)$ is the set of (z, w) satisfying

$$w + \mu' z = f(x) + \mu' g(x) = L(x, \mu).$$

The only vector in the vertical axis satisfying this equation is $\big(0, L(x, \mu)\big)$.

(b) The hyperplane with normal $(\mu, 1)$ that intercepts the vertical axis at level c is the set of vectors (z, w) that satisfy the equation

$$w + \mu' z = c,$$

and this hyperplane contains S in its positive halfspace if and only if

$$L(x, \mu) = f(x) + \mu' g(x) \geq c, \qquad \forall\, x \in X.$$

Therefore the maximum point of interception is $c^* = \inf_{x \in X} L(x, \mu)$.

(c) Follows from the definition of a Lagrange multiplier and part (b). **Q.E.D.**

The hyperplane with normal $(\mu, 1)$ that attains the highest level of interception of the vertical axis, as in part (b) of the visualization lemma, is seen to "support" the set S (the notion of a supporting hyperplane is defined more precisely in Section B.2 of Appendix B). Figure 5.1.3 gives some examples where there exist one or more Lagrange multipliers. Figure 5.1.4 shows cases where there is no Lagrange multiplier.

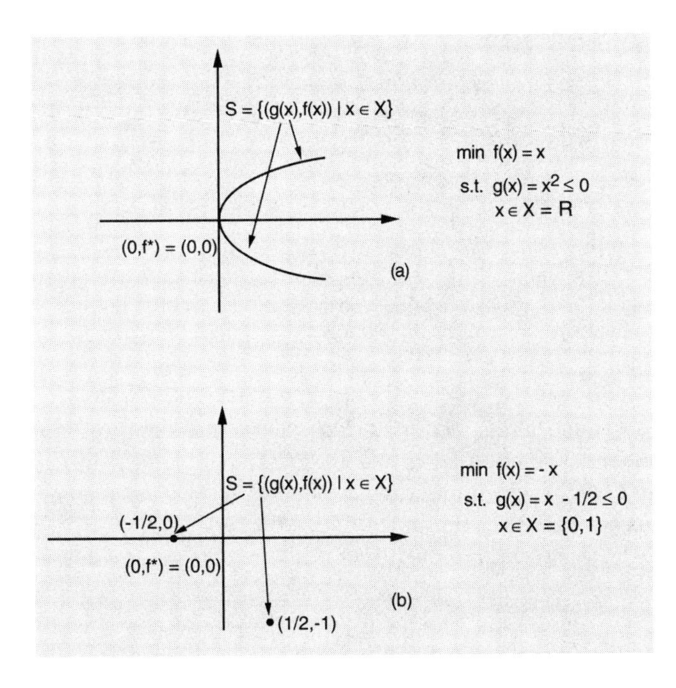

Figure 5.1.4. Examples where there exists no Lagrange multiplier. In (a), a Lagrange multiplier does not exist because the only hyperplane that supports S at $(0, f^*)$ is vertical. In (b), we have an integer programming problem, where the set X is discrete. Here there is no Lagrange multiplier because there is no hyperplane that supports S and passes through $(0, f^*)$.

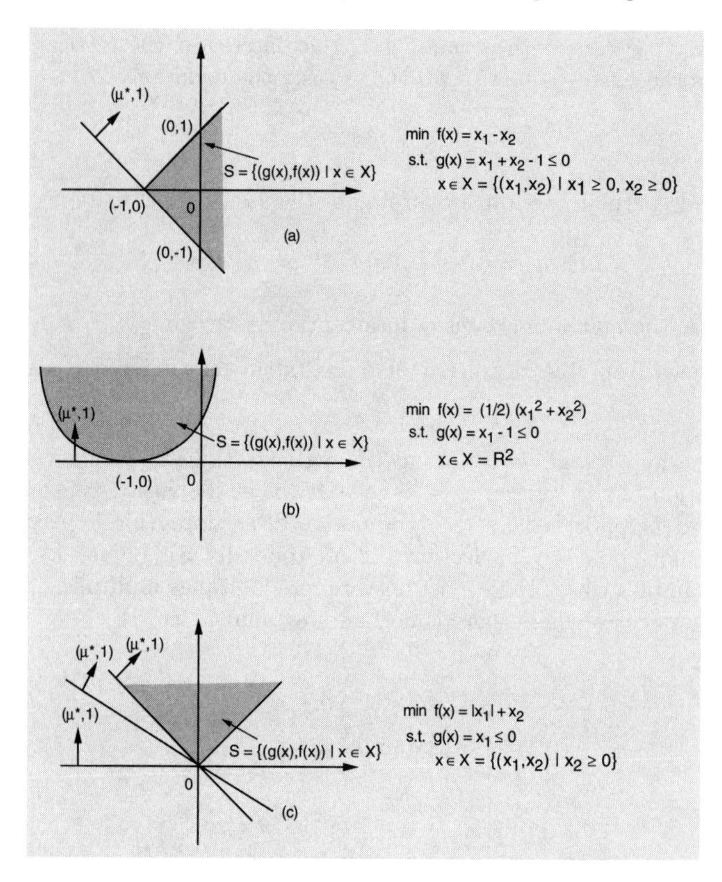

Figure 5.1.3. Examples where there exists at least one Lagrange multiplier. In (a), there is a unique Lagrange multiplier, $\mu^* = 1$. In (b), there is a unique Lagrange multiplier, $\mu^* = 0$. In (c), the set of Lagrange multipliers is the interval $[0, 1]$.

If a Lagrange multiplier μ^* is known, then all optimal solutions x^* can be obtained by minimizing the Lagrangian $L(x, \mu^*)$ over $x \in X$, as indicated in Fig. 5.1.2(c) and shown in the following proposition. However, there may be vectors that minimize $L(x, \mu^*)$ over $x \in X$ but do not satisfy the inequality constraints $g(x) \leq 0$ [cf. Figs. 5.1.2(c) and 5.1.3(a)].

Proposition 5.1.1: Let μ^* be a Lagrange multiplier. Then x^* is a global minimum of the primal problem if and only if x^* is feasible and

$$x^* = \arg\min_{x \in X} L(x, \mu^*), \qquad \mu_j^* g_j(x^*) = 0, \qquad j = 1, \ldots, r. \tag{5.2}$$

Proof: If x^* is a global minimum, then x^* is feasible and furthermore,

$$f^* = f(x^*) \geq f(x^*) + \sum_{j=1}^{r} \mu_j^* g_j(x^*) = L(x^*, \mu^*) \geq \inf_{x \in X} L(x, \mu^*), \tag{5.3}$$

where the first inequality follows from the definition of a Lagrange multiplier ($\mu^* \geq 0$) and the feasibility of x^* $[g(x^*) \leq 0]$. Using again the definition of Lagrange multiplier, we have $f^* = \inf_{x \in X} L(x, \mu^*)$, so that equality holds throughout in Eq. (5.3), implying Eq. (5.2).

Conversely, if x^* is feasible and Eq. (5.2) holds, we have, using also the definition of a Lagrange multiplier,

$$f(x^*) = f(x^*) + \sum_{j=1}^{r} \mu_j^* g_j(x^*) = L(x^*, \mu^*) = \min_{x \in X} L(x, \mu^*) = f^*,$$

so x^* is a global minimum. **Q.E.D.**

5.1.2 The Weak Duality Theorem

We consider the *dual function* q defined for $\mu \in \Re^r$ by

$$q(\mu) = \inf_{x \in X} L(x, \mu).$$

This definition is illustrated in Fig. 5.1.5, where $q(\mu)$ is interpreted as the highest point of interception with the vertical axis over all hyperplanes with normal $(\mu, 1)$, which contain the set S in their positive halfspace [compare also with the visualization lemma and Fig. 5.1.2(a)]. The *dual problem* is

$$\text{maximize } q(\mu)$$
$$\text{subject to } \mu \geq 0,$$

and corresponds to finding the maximum point of interception, over all hyperplanes with normal $(\mu, 1)$ where $\mu \geq 0$.

Note that $q(\mu)$ may be equal to $-\infty$ for some μ. In this case, effectively we have the additional constraint

$$\mu \in D,$$

in the dual problem, where D, called the *domain* of q, is the set of μ for which $q(\mu)$ is finite:

$$D = \{\mu \mid q(\mu) > -\infty\}.$$

In fact, we may have $q(\mu) = -\infty$ for all $\mu \geq 0$, in which case the dual optimal value

$$q^* = \sup_{\mu \geq 0} q(\mu),$$

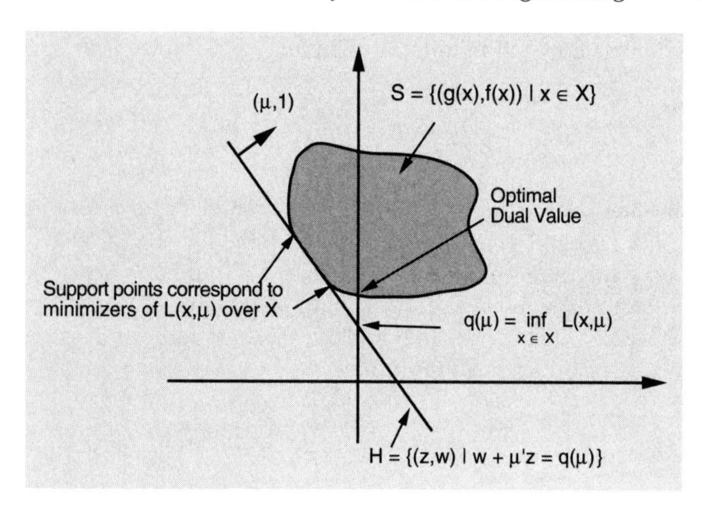

Figure 5.1.5. Geometrical interpretation of the dual function.

is equal to $-\infty$ (situations where this can happen will be discussed later).

Regardless of the structure of the cost and constraints of the primal problem, the dual problem has nice convexity properties, as shown by the following proposition.

Proposition 5.1.2: The domain D of the dual function q is convex and q is concave over D.

Proof: For any x, μ, $\bar{\mu}$, and $\alpha \in [0,1]$, we have

$$L\big(x, \alpha\mu + (1-\alpha)\bar{\mu}\big) = \alpha L(x,\mu) + (1-\alpha)L(x,\bar{\mu}).$$

Taking the infimum over all $x \in X$, we obtain

$$\inf_{x \in X} L\big(x, \alpha\mu + (1-\alpha)\bar{\mu}\big) \geq \alpha \inf_{x \in X} L(x,\mu) + (1-\alpha) \inf_{x \in X} L(x,\bar{\mu})$$

or

$$q\big(\alpha\mu + (1-\alpha)\bar{\mu}\big) \geq \alpha q(\mu) + (1-\alpha)q(\bar{\mu}).$$

Therefore if μ and $\bar{\mu}$ belong to D, the same is true for $\alpha\mu + (1-\alpha)\bar{\mu}$, so D is convex. Furthermore, q is concave over D. **Q.E.D.**

The concavity of q can also be verified by observing that q is defined as the infimum of a collection of the concave (in fact linear) functions $L(x,\cdot)$ parameterized by $x \in X$; such a function is concave by Prop. B.2(d) in Appendix B.

Another important property is that the optimal dual value is always an underestimate of the optimal primal value. This is evident from the geometric interpretation of Fig. 5.1.5 and is formally shown in the next proposition.

Proposition 5.1.3: (Weak Duality Theorem) We have

$$q^* \leq f^*.$$

Proof: For all $\mu \geq 0$, and $x \in X$ with $g(x) \leq 0$, we have

$$q(\mu) = \inf_{z \in X} L(z, \mu) \leq f(x) + \sum_{j=1}^{r} \mu_j g_j(x) \leq f(x),$$

so

$$q^* = \sup_{\mu \geq 0} q(\mu) \leq \inf_{x \in X,\, g(x) \leq 0} f(x) = f^*.$$

Q.E.D.

If $q^* = f^*$ we say that *there is no duality gap* and if $q^* < f^*$ we say that *there is a duality gap*. Note that if there exists a Lagrange multiplier μ^*, the weak duality theorem ($q^* \leq f^*$) and the definition of a Lagrange multiplier [$f^* = q(\mu^*) \leq q^*$] imply that there is no duality gap. However, the converse is not true. In particular, it is possible that no Lagrange multiplier exists even though there is no duality gap [cf. Fig. 5.1.4(a)]; in this case the dual problem does not have an optimal solution, as implied by the following proposition.

Proposition 5.1.4:

(a) If there is no duality gap, the set of Lagrange multipliers is equal to the set of optimal dual solutions.

(b) If there is a duality gap, the set of Lagrange multipliers is empty.

Proof: By definition, a vector $\mu^* \geq 0$ is a Lagrange multiplier if and only if $f^* = q(\mu^*) \leq q^*$, which by the weak duality theorem, holds if and only if there is no duality gap and μ^* is a dual optimal solution. **Q.E.D.**

The preceding results are illustrated in Figs. 5.1.6 and 5.1.7 for the problems of Figs. 5.1.3 and 5.1.4, respectively.

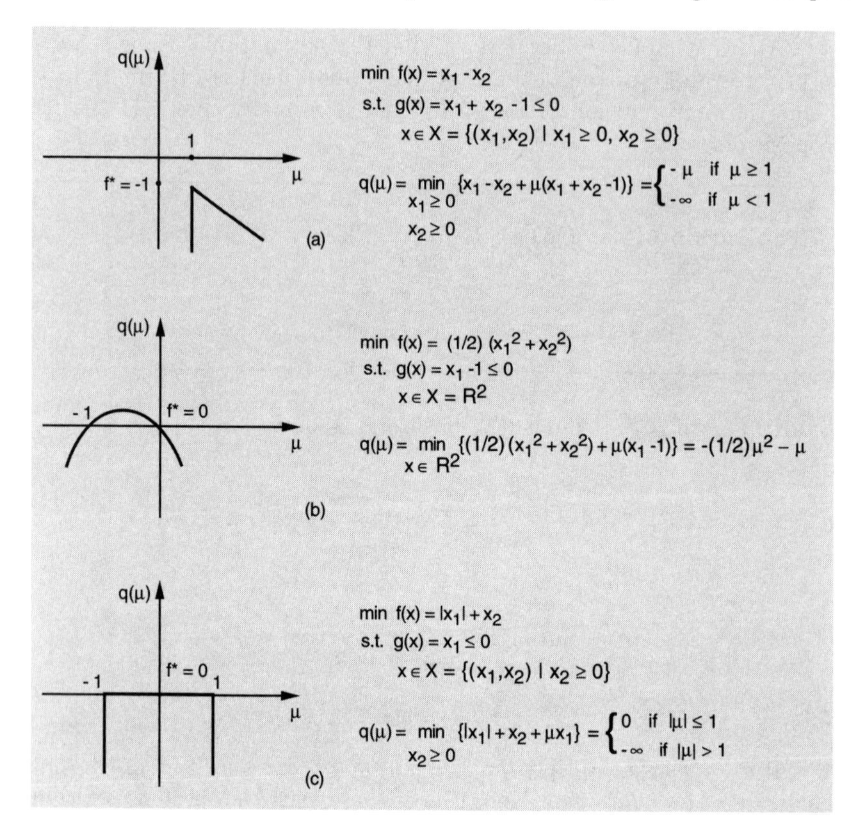

Figure 5.1.6. The duals of the problems of Fig. 5.1.3. In all these problems, there is no duality gap and the set of Lagrange multipliers is equal to the set of dual optimal solutions (cf. Prop. 5.1.3). In (a), there is a unique dual optimal solution $\mu^* = 1$. In (b), there is a unique dual optimal solution $\mu^* = 0$; at the corresponding primal optimal solution ($x^* = 0$) the constraint is inactive. In (c), the set of dual optimal solutions is $\{\mu^* \mid 0 \leq \mu^* \leq 1\}$.

Duality theory is most useful when there is no duality gap. To guarantee that there is no duality gap and that a Lagrange multiplier exists, it is typically necessary to impose various types of convexity conditions on the cost and the constraints of the primal problem; we will develop some of these conditions in Sections 5.2-5.4. However, even with a duality gap, the dual problem can be useful, as indicated by the following example.

Example 5.1.1 (Integer Programming, and Branch-and-Bound)

Many important practical optimization problems of the form

$$\begin{aligned} &\text{minimize } f(x) \\ &\text{subject to } x \in X, \qquad g_j(x) \leq 0, \quad j = 1, \ldots, r, \end{aligned}$$

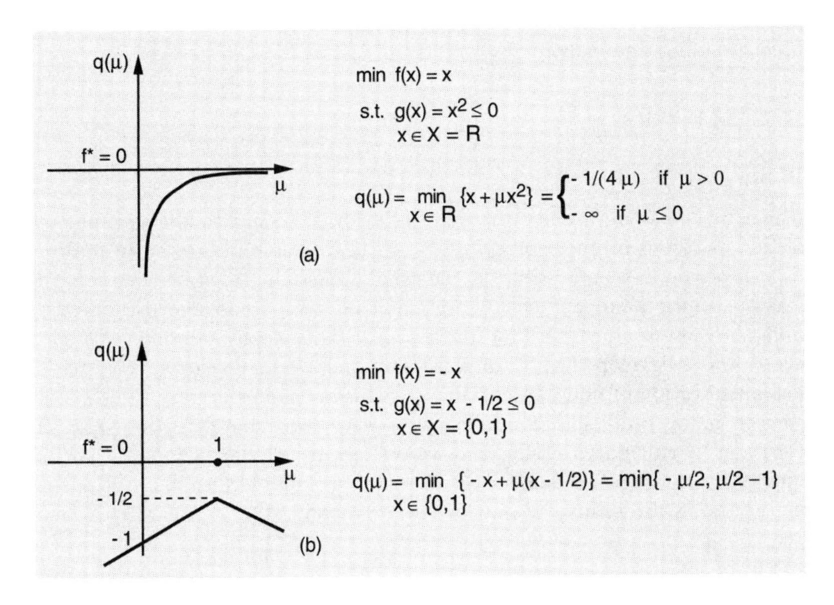

Figure 5.1.7. The duals of the problems of Fig. 5.1.4. In these problems, there is no Lagrange multiplier. In (a), there is no duality gap and the dual problem has no optimal solution (cf. Prop. 5.1.3). In (b), there is a duality gap ($f^* - q^* = 1/2$). The dual problem has a unique optimal solution $\mu^* = 1$, which is the Lagrange multiplier of a "convexified" version of the problem, where the constraint set $X = \{0, 1\}$ is replaced by its convex hull, the interval $[0, 1]$ (this observation can be generalized).

have a finite constraint set X. An example is *integer programming*, where the coordinates of x must be integers from a bounded range (usually 0 or 1). An important special case is the linear 0-1 integer programming problem

$$\begin{aligned} &\text{minimize} \;\; c'x \\ &\text{subject to} \;\; Ax \leq b, \qquad x_i = 0 \text{ or } 1, \quad i = 1, \ldots, n. \end{aligned}$$

A principal approach for solving such problems is the *branch-and-bound method*, which will be described in Section 5.5. This method relies on obtaining lower bounds to the optimal cost of restricted problems of the form

$$\begin{aligned} &\text{minimize} \;\; f(x) \\ &\text{subject to} \;\; x \in \tilde{X}, \qquad g_j(x) \leq 0, \quad j = 1, \ldots, r, \end{aligned}$$

where $\tilde{X}$ is a subset of X; for example in the 0-1 integer case where X specifies that all x_i should be 0 or 1, $\tilde{X}$ may be the set of all 0-1 vectors x such that one or more coordinates x_i are restricted to satisfy $x_i = 0$ for all $x \in \tilde{X}$ or $x_i = 1$ for all $x \in \tilde{X}$. These lower bounds can often be obtained by finding a dual-feasible (possibly dual-optimal) solution μ of this problem and the corresponding dual value

$$q(\mu) = \inf_{x \in \tilde{X}} \left\{ f(x) + \sum_{j=1}^{r} \mu_j g_j(x) \right\},$$

which by the weak duality theorem, is a lower bound to the optimal value of the restricted problem $\min_{x \in \tilde{X},\, g(x) \leq 0} f(x)$. One is interested in finding as tight lower bounds as possible, so the usual approach is to start with some dual feasible solution and iteratively improve it by using some algorithm. A major difficulty here is that the dual function $q(\mu)$ is typically nondifferentiable, so the methods developed so far cannot be used. We will develop special methods for optimization of nondifferentiable cost functions in the next chapter. Note that in many problems of interest that have favorable structure, the value $q(\mu)$ can be calculated easily, and we will see in the next chapter that other quantities that are needed for application of nondifferentiable optimization methods are also easily obtained together with $q(\mu)$.

5.1.3 Characterization of Primal and Dual Optimal Solutions

There are powerful characterizations of primal and dual optimal solution pairs, given in the following two propositions. Note, however, that these characterizations are useful only if there is no duality gap, since otherwise there is no Lagrange multiplier [cf. Prop. 5.1.4(b)], even if the dual problem has an optimal solution.

Proposition 5.1.5: (Optimality Conditions) (x^*, μ^*) is an optimal solution-Lagrange multiplier pair if and only if

$$x^* \in X, \quad g(x^*) \leq 0, \qquad \text{(Primal Feasibility)}, \tag{5.4}$$

$$\mu^* \geq 0, \qquad \text{(Dual Feasibility)}, \tag{5.5}$$

$$x^* = \arg\min_{x \in X} L(x, \mu^*), \qquad \text{(Lagrangian Optimality)}, \tag{5.6}$$

$$\mu_j^* g_j(x^*) = 0, \quad j = 1, \ldots, r, \qquad \text{(Complementary Slackness)}. \tag{5.7}$$

Proof: If (x^*, μ^*) is an optimal solution-Lagrange multiplier pair, then x^* is primal feasible and μ^* is dual feasible. Equations (5.6) and (5.7) follow from Prop. 5.1.1.

Conversely, using Eqs. (5.4)-(5.7), we obtain

$$f^* \leq f(x^*) = L(x^*, \mu^*) = \min_{x \in X} L(x, \mu^*) = q(\mu^*) \leq q^*.$$

Using the weak duality theorem (Prop. 5.1.3), we see that equality holds throughout in the preceding relation. It follows that x^* is primal optimal and μ^* is dual optimal, while there is no duality gap. **Q.E.D.**

Proposition 5.1.6: (Saddle Point Theorem) (x^*, μ^*) is an optimal solution-Lagrange multiplier pair if and only if $x^* \in X$, $\mu^* \geq 0$, and (x^*, μ^*) is a saddle point of the Lagrangian, in the sense that

$$L(x^*, \mu) \leq L(x^*, \mu^*) \leq L(x, \mu^*), \qquad \forall\, x \in X,\ \mu \geq 0. \tag{5.8}$$

Proof: If (x^*, μ^*) is an optimal solution-Lagrange multiplier pair, then x^* is primal optimal and μ^* is dual optimal, so that $x^* \in X$, $\mu^* \geq 0$, and Eq. (5.6) holds, thereby proving the right-hand side of Eq. (5.8). Furthermore, for all $\mu \geq 0$, using the fact $g(x^*) \leq 0$, we have $L(x^*, \mu) \leq f(x^*)$ and in view of Eqs. (5.6) and (5.7), we obtain

$$L(x^*, \mu) \leq f(x^*) = L(x^*, \mu^*),$$

proving the left-hand side of Eq. (5.8).

Conversely, assume that $x^* \in X$, $\mu^* \geq 0$, and Eq. (5.8) holds. We will show that the four conditions (5.4)-(5.7) of Prop. 5.1.5 hold. We have

$$\sup_{\mu \geq 0} L(x^*, \mu) = \sup_{\mu \geq 0} \left\{ f(x^*) + \sum_{j=1}^{r} \mu_j g_j(x^*) \right\} = \begin{cases} f(x^*) & \text{if } g(x^*) \leq 0, \\ +\infty & \text{otherwise.} \end{cases}$$

Therefore, from the left-hand side of Eq. (5.8), we obtain that $g(x^*) \leq 0$, and $f(x^*) = L(x^*, \mu^*)$. Thus, Eqs. (5.4) and (5.5) hold, and we have $\sum_{j=1}^{r} \mu_j^* g_j(x^*) = 0$. Since $\mu^* \geq 0$ and $g(x^*) \leq 0$, we obtain $\mu_j^* g_j(x^*) = 0$ for all j, and taking into account the right-hand side of Eq. (5.8), we see that Eq. (5.6) holds. From Prop. 5.1.5 it follows that (x^*, μ^*) is an optimal solution-Lagrange multiplier pair. **Q.E.D.**

5.1.4 The Case of an Infeasible or Unbounded Primal Problem

We now consider what happens when our standing Assumption 5.1.1 (feasibility and boundedness) does not hold.

Suppose that the primal problem is unbounded, i.e., $f^* = -\infty$. Then, it is seen that the proof of the weak duality theorem still applies and that $q(\mu) = -\infty$ for all $\mu \geq 0$. As a result the dual problem is infeasible.

Suppose now that X is nonempty but the primal problem is infeasible, i.e., the set $\{x \in X \mid g(x) \leq 0\}$ is empty. Then, by convention, we write

$$f^* = \infty.$$

The dual function $q(\mu) = \inf_{x \in X} L(x, \mu)$ satisfies $q(\mu) < \infty$ for all μ, but it is possible that $q^* = \infty$, in which case the dual problem is unbounded. It is also possible, however, that $q^* < \infty$ or even that $q^* = -\infty$, as the examples of Fig. 5.1.8 show. For linear programs, it will be shown in the next section that $-\infty < q^* < \infty$ implies $-\infty < f^* < \infty$. However, even for linear programs, it is possible that both the primal and the dual are infeasible, i.e., $f^* = \infty$ and $q^* = -\infty$ (see Exercise 5.1.5).

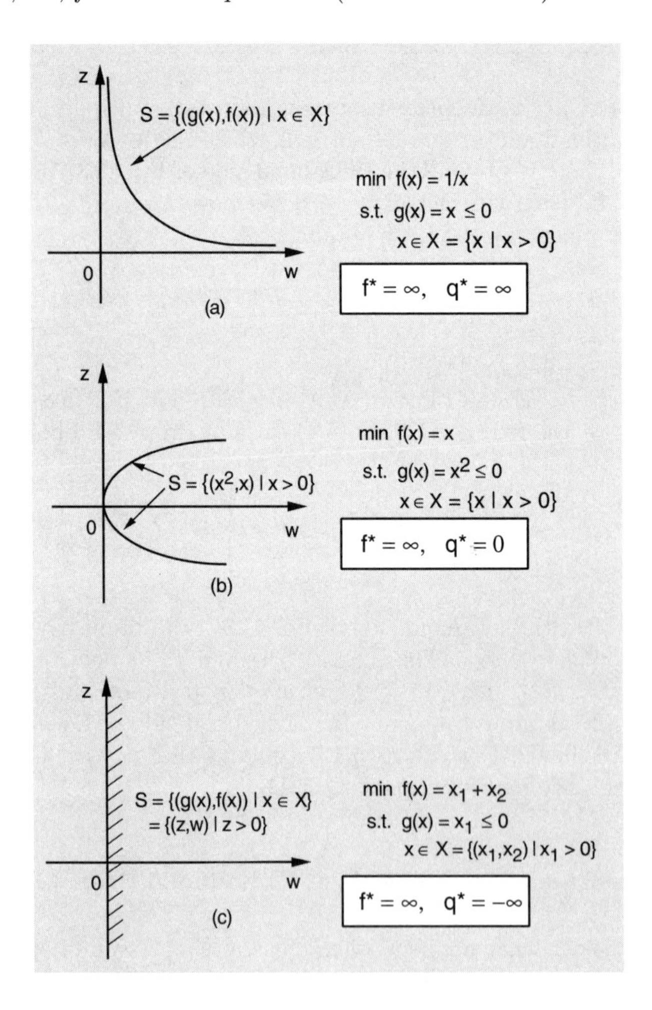

Figure 5.1.8. Examples where X is nonempty but the primal problem is infeasible. In (a), we have $f^* = q^* = \infty$. In (b), we have $f^* = \infty$ and $-\infty < q^* < \infty$. In (c), we have $f^* = \infty$ and $q^* = -\infty$.

5.1.5 Treatment of Equality Constraints

The theory developed so far in this section can be extended to handle additional equality constraints of the form $h_i(x) = 0$, $i = 1, \ldots, m$. A constraint of this type can be converted into the two inequality constraints

$$h_i(x) \le 0, \qquad -h_i(x) \le 0.$$

In particular, consider the problem

$$\begin{aligned} &\text{minimize } \ f(x) \\ &\text{subject to } \ x \in X, \qquad g_j(x) \le 0, \ j = 1, \ldots, r, \qquad h_i(x) = 0, \ i = 1, \ldots m. \end{aligned}$$

If we replace each equality constraint by two inequalities as above, we see that a vector $(\mu_1^*, \ldots, \mu_r^*, \lambda_1^+, \lambda_1^-, \ldots, \lambda_m^+, \lambda_m^-)$ is a Lagrange multiplier if $\mu_j^* \ge 0$ for all j, $\lambda_i^+ \ge 0$ and $\lambda_i^- \ge 0$, for all i, and

$$f^* = \inf_{x \in X} \left\{ f(x) + \sum_{j=1}^{r} \mu_j^* g_j(x) + \sum_{i=1}^{m} (\lambda_i^+ - \lambda_i^-) h_i(x) \right\},$$

where λ_i^+ corresponds to the constraint $h_i(x) \le 0$ and λ_i^- corresponds to the constraint $-h_i(x) \le 0$. By replacing $\lambda_i^+ - \lambda_i^-$ with a single scalar λ_i^* with unrestricted sign in the above expression, we are led to introduce a Lagrangian function of the form

$$L(x, \mu, \lambda) = f(x) + \sum_{j=1}^{r} \mu_j g_j(x) + \sum_{i=1}^{m} \lambda_i h_i(x).$$

We say that (μ^*, λ^*) is a Lagrange multiplier, if $\mu^* \ge 0$ and

$$f^* = \inf_{x \in X} L(x, \mu^*, \lambda^*).$$

We also define the dual problem

$$\begin{aligned} &\text{maximize } \ q(\mu, \lambda) \\ &\text{subject to } \ \mu \ge 0, \qquad \lambda \in \Re^m, \end{aligned}$$

where the dual function q is defined by

$$q(\mu, \lambda) = \inf_{x \in X} L(x, \mu, \lambda).$$

It is straightforward to verify that Props. 5.1.1 through 5.1.6 have analogs where the sign of λ_i is unrestricted.

5.1.6 Separable Problems and their Geometry

Suppose that x has m components $x_1, \ldots, x_m$ of dimensions $n_1, \ldots, n_m$, respectively, and the problem has the form

$$\begin{aligned} &\text{minimize} \ \sum_{i=1}^{m} f_i(x_i) \\ &\text{subject to} \ \sum_{i=1}^{m} g_{ij}(x_i) \leq 0, \qquad j = 1, \ldots, r, \\ &\qquad x_i \in X_i, \qquad i = 1, \ldots, m, \end{aligned}$$

where $f_i : \Re^{n_i} \mapsto \Re$ and $g_{ij} : \Re^{n_i} \mapsto \Re$ are given functions, and X_i are given subsets of $\Re^{n_i}$. We note that if the constraints $\sum_{i=1}^{m} g_{ij}(x_i) \leq 0$ were not present, it would be possible to decompose this problem into m independent subproblems. This motivates us to consider the following dual problem that involves Lagrange multipliers for these constraints:

$$\begin{aligned} &\text{maximize} \ q(\mu) \\ &\text{subject to} \ \mu \geq 0, \end{aligned}$$

where the dual function is given by

$$q(\mu) = \inf_{\substack{x_i \in X_i \\ i=1,\ldots,m}} \left\{ \sum_{i=1}^{m} \left(f_i(x_i) + \sum_{j=1}^{r} \mu_j g_{ij}(x_i) \right) \right\} = \sum_{i=1}^{m} q_i(\mu),$$

and

$$q_i(\mu) = \inf_{x_i \in X_i} \left\{ f_i(x_i) + \sum_{j=1}^{r} \mu_j g_{ij}(x_i) \right\}, \qquad i = 1, \ldots, m.$$

Note that the minimization involved in the calculation of the dual function has been decomposed into m simpler minimizations. These minimizations are often conveniently done either analytically or computationally, in which case the dual function can be easily evaluated. This is the key advantageous structure of separable problems.

The separable structure is additionally helpful when the cost and/or the constraints are not convex. In particular, in this case *the duality gap turns out to be relatively small and can often be shown to diminish to zero relative to the optimal primal value as the number m of separable terms increases.* As a result, one can often obtain a near-optimal primal solution, starting from a dual-optimal solution, without resorting to costly branch-and-bound procedures (cf. Example 5.1.1).

The small duality gap size is a consequence of the structure of the set of constraint-cost pairs

$$S = \big\{ \big(g(x), f(x)\big) \mid x \in X \big\}.$$

In the case of the separable problem, this set can be written as a vector sum of m sets, one for each separable term, i.e.,

$$S = S_1 + \cdots + S_m,$$

where

$$S_i = \{(g_i(x_i), f_i(x_i)) \mid x_i \in X_i\},$$

and $g_i : \Re^{n_i} \mapsto \Re^r$ is the function $g_i(x_i) = (g_{i1}(x_i), \ldots, g_{im}(x_i))$. Generally, a set $S \subset \Re^n$ that is the vector sum of a large number of possibly nonconvex but roughly similar sets "tends to be convex" in the sense that any vector in the convex hull of S can be closely approximated by a vector in S. As a result, the duality gap tends to be relatively small, as illustrated in Fig. 5.1.9. An example is given in Exercise 5.1.7. The analytical substantiation is based on a theorem by Shapley and Folkman (see [Ber82a], Section 5.6.1). In particular, it can be shown under some reasonable assumptions that the duality gap satisfies

$$f^* - q^* \leq (r+1) \max_{i=1,\ldots,m} \rho_i,$$

where for each i, ρ_i is a nonnegative scalar that depends on the structure of the functions f_i, g_{ij}, $j = 1, \ldots, r$, and the set X_i. This estimate suggests that as $m \to \infty$, the duality gap is bounded. Thus, if $|f^*| \to \infty$ as $m \to \infty$, the "relative" duality gap $(f^* - q^*)/|f^*|$ diminishes to 0 as $m \to \infty$.

Let us close this section with a simple but important example of a separable problem.

Example 5.1.2 (Separable Problem with a Single Constraint)

Suppose that at least A units of a certain commodity (for example, electric power) must be produced by n independent production units at minimum cost. Let x_i denote the amount produced by the ith unit and let $f_i(x_i)$ be the corresponding production cost. The problem is

$$\begin{aligned} &\text{minimize} \quad \sum_{i=1}^{n} f_i(x_i) \\ &\text{subject to} \quad \sum_{i=1}^{n} x_i \geq A, \qquad \alpha_i \leq x_i \leq \beta_i, \qquad i = 1, \ldots, n, \end{aligned}$$

where α_i and β_i are known lower and upper bounds on the production capacity of the ith unit.

We assume that each f_i is continuously differentiable and convex, and that $\sum_{i=1}^{n} \beta_i \geq A$, so that the problem has at least one feasible solution. By Prop. 5.1.5, (x^*, μ^*) is an optimal solution-Lagrange multiplier pair if and

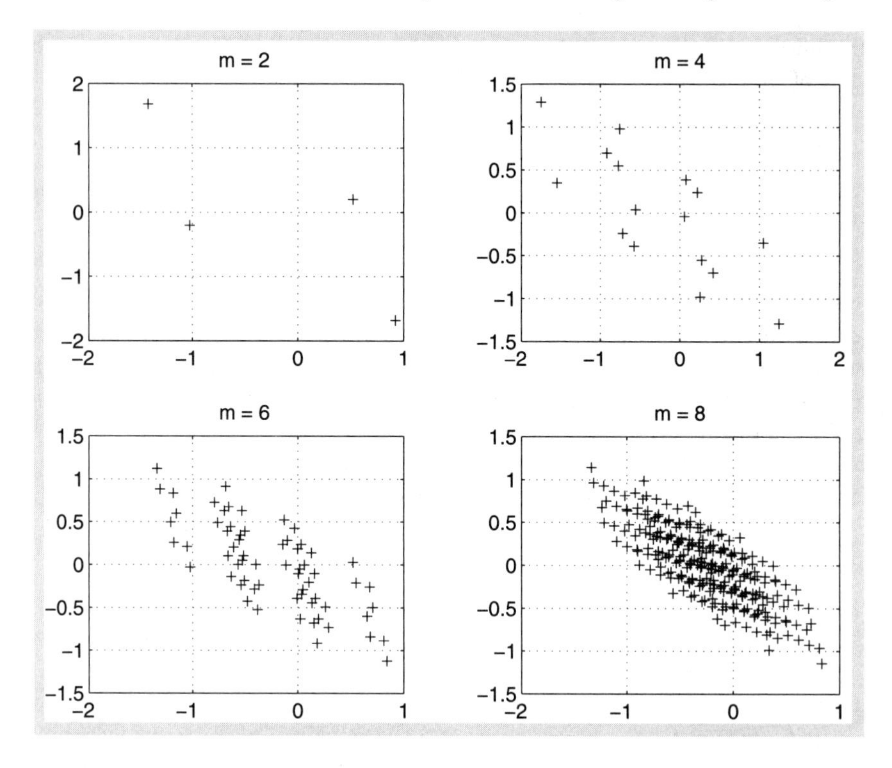

Figure 5.1.9. Illustration of the set of constraint-cost pairs

$$S = \{(g(x), f(x)) \mid x \in X\}$$

for the case of a separable problem. It can be expressed as the vector sum $S = S_1 + \cdots + S_m$, where

$$S_i = \{(g_i(x_i), f_i(x_i)) \mid x_i \in X_i\}.$$

The figure shows S for the integer programming problem

$$\begin{aligned} &\text{minimize} && \frac{1}{m}\sum_{i=1}^{m} r_i x_i \\ &\text{subject to} && \frac{1}{m}\sum_{i=1}^{m} w_i x_i \leq 0.25, \quad x_i = -1 \text{ or } 1, \end{aligned}$$

for various values of m, and for r_i and w_i chosen randomly from the range [-2,0] and [0,2], respectively, according to a uniform distribution. It can be seen that the duality gap diminishes relative to the optimal primal value. For an analytical substantiation, based on a theorem by Shapley and Folkman, see [Ber82a], Section 5.6.1.

only if x^* is primal feasible, $\mu^* \geq 0$, $\mu^*(\sum_{i=1}^n x_i^* - A) = 0$, and the Lagrangian optimality condition

$$x^* = \arg \min_{\substack{\alpha_i \leq x_i \leq \beta_i \\ i=1,\ldots,n}} \left\{ \sum_{i=1}^n \big(f_i(x_i) - \mu^* x_i\big) + \mu^* A \right\}$$

holds. Thanks to the separable structure of the problem, the last condition can be decomposed into the n conditions

$$x_i^* = \arg \min_{\alpha_i \leq x_i \leq \beta_i} \big\{ f_i(x_i) - \mu^* x_i \big\}, \qquad i = 1, \ldots, n.$$

Using the differentiability and convexity of f_i, it is seen that this condition is equivalent to the following three conditions

$$\nabla f_i(x_i^*) = \mu^*, \qquad \text{if } \alpha_i < x_i^* < \beta_i, \tag{5.9}$$

$$\nabla f_i(x_i^*) \geq \mu^*, \qquad \text{if } \alpha_i = x_i^*, \tag{5.10}$$

$$\nabla f_i(x_i^*) \leq \mu^*, \qquad \text{if } x_i^* = \beta_i. \tag{5.11}$$

The gradient $\nabla f_i(x_i)$ is also called the *marginal production cost* of the ith unit. Thus, at an optimum, all units not operating at the upper or lower bound must operate at points of equal marginal production cost.

The computational solution of the problem typically involves duality. For example, in many situations the dual problem can be written in a convenient closed form and it can be solved by a one-dimensional search (cf. Appendix C). In an alternative but related approach, one may try to find a solution (x^*, μ^*) of the Lagrangian optimality conditions of Eqs. (5.9)-(5.11) and the primal feasibility condition

$$\sum_{i=1}^n x_i^* \geq A, \qquad \alpha_i \leq x_i^* \leq \beta_i, \quad i = 1, \ldots, n.$$

Let us assume for convenience that each f_i is strictly convex on $[\alpha_i, \beta_i]$, so that for every μ^*, the equation $\nabla f_i(x_i^*) = \mu^*$ has a unique solution for x_i^*. Then, for each μ^*, Eqs. (5.9)-(5.11) yield a unique x^*. Furthermore, it is seen that $\sum_{i=1}^n x_i^*$ increases as μ^* increases, in view of the convexity of f_i. Thus, we may plot $\sum_{i=1}^n x_i^*$ versus μ^*. The point of interception of the resulting graph with the level A yields the optimal dual solution as shown in Fig. 5.1.10.

In the preceding example, we may view the dual solution method illustrated in Fig. 5.1.10 as a *coordination mechanism*. A coordinator sets the price of the commodity to be produced at the optimal dual value μ^*. The production units then respond by acting in their own self-interest, setting their production levels to the values

$$x_i^* = \arg \min_{\alpha_i \leq x_i \leq \beta_i} \big\{ f_i(x_i) - \mu^* x_i \big\}, \qquad i = 1, \ldots, n,$$

that minimize their net production cost (cost of production minus revenue from sale of the product). This interpretation of separable problem duality can be easily extended to the case where multiple commodities are being produced, and there is a demand inequality constraint for each commodity.

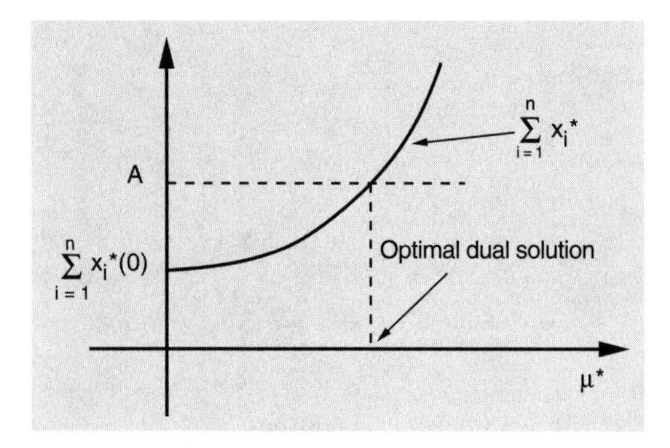

Figure 5.1.10. Graphical approach for solving the production allocation problem. We plot $\sum_{i=1}^n x_i^*$ as an increasing function of $\mu^* \geq 0$ based on the solution of Eqs. (5.9)-(5.11), and we find μ^* for which

$$\sum_{i=1}^{n} x_i^* = A.$$

The figure shows the case where the vector $x^*(0)$ corresponding to $\mu^* = 0$ satisfies $\sum_{i=1}^n x_i^*(0) < A$. If $\sum_{i=1}^n x_i^*(0) \geq A$, then $x^*(0)$ is optimal.

5.1.7 Additional Issues About Duality

Mathematical programming duality is a broadly applicable subject with a rich theory, and most of the remainder of the book revolves around its analytical and algorithmic aspects. Here is a preview and brief summary of some of the topics that we will address in the sequel:

(a) *Conditions under which there is no duality gap and there exists a Lagrange multiplier*. Convexity of the cost function and the constraints is essentially a prerequisite for this, but additional technical conditions, similar to the constraint qualifications encountered in Chapter 3, are needed. We focus on two cases, which parallel some of the corresponding conditions discussed in Section 3.3.5.

  (1) The constraints $g_j(x) \leq 0$ are linear and the constraint set X is polyhedral. This case is discussed in Section 5.2.

  (2) The constraints $g_j(x) \leq 0$ are convex and possess a common interior point $\bar{x} \in X$ satisfying $g_j(\bar{x}) < 0$ for all j (cf. the Slater constraint qualification discussed in Section 3.3.5). This case is discussed in Section 5.3.

(b) *Specially structured dual problems*. Here, specific assumptions are made on the form of the cost functions and the constraints, and strengthened versions of duality theorems and/or other specialized results are accordingly obtained. A principal example is Fenchel duality, developed in Section 5.4, which relies on the theory of conjugate functions to derive an alternative framework for obtaining duality results in specially structured problems. Examples of such problems include:

(1) *Monotropic programming.* This class of problems, discussed in Section 5.4.1, involve a separable cost and a subspace constraint, and include linear and quadratic programming problems. The duality results here are symmetric and very strong. In fact, monotropic programming problems form the largest class of nonlinear programming problems with a duality theory that is as sharp as the one for linear programs.

(2) *Network optimization problems* with a single commodity and no side constraints. This is a practically important class, including problems such as shortest path, assignment, max-flow, etc. The duality theory here is a special case of monotropic programming duality.

(c) *Exact penalty function methodology.* There is a strong and insightful relation between the nondifferentiable exact penalty functions discussed in Section 4.3 and Fenchel duality theory. This relation provides the connecting link between a seemingly disparate set of topics relating to penalty and augmented Lagrangian methods, sensitivity, and smoothing approaches for nondifferentiable optimization. These connections are explored in Sections 5.4.5 and 5.4.6.

(d) *Discrete optimization methodology.* Discrete optimization problems are often approached via duality, as indicated earlier in Example 5.1.1. In particular, the solution of dual problems provides lower bounds to the corresponding optimal primal values, which can be used in the context of the branch-and-bound technique to be discussed in Section 5.5.

(e) *Dual algorithmic methodology.* There are several important methods for large scale optimization that rely on duality. In effect these methods try to solve the dual problem directly, while exploiting the available special structure. Several generic dual approaches will be discussed in Chapter 6.

(f) *Infinite dimensional problems.* Since duality theory is based on geometrical constructions, it is not surprising that it has far-reaching application to problems with optimization variables taking values in infinite dimensional linear spaces, where convexity notions can be readily defined. Extensions of this type can be challenging and complicated, but the advanced reader can verify that the entire framework, analysis, and results presented so far in this section can be used in an infinite dimensional context with essentially no modification. In particular, the definitions and derivations given in this section for the problem

$$\begin{aligned} &\text{minimize } f(x) \\ &\text{subject to } x \in X, \qquad g_j(x) \leq 0, \quad j = 1, \ldots, r, \end{aligned}$$

apply nearly verbatim to the case where x belongs to an arbitrary linear vector space rather than the Euclidean space $\Re^n$. Furthermore, while we will not go into the development of the corresponding theory, the advanced reader can verify that the proof of the Lagrange multiplier theorem of Prop. 5.3.1 under the Slater constraint qualification (see Section 5.3) also carries through.

EXERCISES

5.1.1

Find the sets of all optimal solutions and all Lagrange multipliers, and sketch the dual function for the following two-dimensional convex programming problems:

$$\begin{aligned} &\text{minimize } x_1 \\ &\text{subject to } |x_1| + |x_2| \leq 1, \qquad x \in X = \Re^2, \end{aligned}$$

and

$$\begin{aligned} &\text{minimize } x_1 \\ &\text{subject to } |x_1| + |x_2| \leq 1, \qquad x \in X = \big\{x \mid |x_1| \leq 1,\ |x_2| \leq 1\big\}. \end{aligned}$$

5.1.2

Consider the problem

$$\begin{aligned} &\text{minimize } f(x) = 10x_1 + 3x_2 \\ &\text{subject to } 5x_1 + x_2 \geq 4, \quad x_1, x_2 = 0 \text{ or } 1. \end{aligned}$$

(a) Sketch the set of constraint-cost pairs

$$\big\{(4 - 5x_1 - x_2, 10x_1 + 3x_2) \mid x_1, x_2 = 0 \text{ or } 1\big\}.$$

(b) Sketch the dual function.

(c) Solve the problem and its dual, and relate the solutions to your sketch in part (a).

5.1.3

Derive the dual of the projection problem

$$\begin{aligned}&\text{minimize } \|z-x\|^2\\&\text{subject to } Ax=0,\end{aligned}$$

where the $m \times n$ matrix A and the vector $z \in \Re^n$ are given. Show that the dual problem is also a problem of projection on a subspace.

5.1.4 (www)

Use duality to show that in three-dimensional space, the (minimum) distance from the origin to a line is equal to the maximum over all (minimum) distances of the origin from planes that contain the line.

5.1.5

Show that the dual of the (infeasible) linear program

$$\begin{aligned}&\text{minimize } x_1-x_2\\&\text{subject to } x \in X=\{x \mid x_1 \geq 0, x_2 \geq 0\},\ \ x_1+1 \leq 0,\ 1-x_1-x_2 \leq 0\end{aligned}$$

is the (infeasible) linear program

$$\begin{aligned}&\text{maximize } \mu_1+\mu_2\\&\text{subject to } \mu_1 \geq 0,\ \mu_2 \geq 0,\ -\mu_1+\mu_2-1 \leq 0,\ \mu_2+1 \leq 0.\end{aligned}$$

5.1.6

Consider the optimal control problem of minimizing $\sum_{i=1}^{N-1} |u_i|^2$ subject to

$$x_{i+1}=A_i x_i+B_i u_i, \qquad i=0,\ldots,N-1,$$

$$x_0 : \text{given},$$

$$x_N \geq c,$$

where c is a given vector. Show that a dual problem is of the form

$$\begin{aligned}&\text{minimize } \mu' Q\mu+\mu' d\\&\text{subject to } \mu \geq 0,\end{aligned}$$

where Q is an appropriate $n \times n$ matrix (n is the dimension of x_N) and $d \in \Re^n$ is an appropriate vector.

5.1.7 (Duality Gap of the Knapsack Problem) (www)

Given objects $i = 1, \ldots, n$ with positive weights w_i and values v_i, we want to assemble a subset of the objects so that the sum of the weights of the subset does not exceed a given $A > 0$, and the sum of the values of the subset is maximized.

(a) Show that this problem can be written as

$$\begin{aligned} &\text{maximize} \quad \sum_{i=1}^{n} v_i x_i \\ &\text{subject to} \quad \sum_{i=1}^{n} w_i x_i \leq A, \qquad x_i \in \{0, 1\}, \quad i = 1, \ldots, n. \end{aligned}$$

(b) Use a graphical procedure to solve the dual problem where a Lagrange multiplier is assigned to the constraint $\sum_{i=1}^{n} w_i x_i \leq A$.

(c) Let f^* and q^* be the optimal values of the primal and dual problems, respectively. Show that

$$0 \leq q^* - f^* \leq \max_{i=1,\ldots,n} v_i.$$

(d) Consider the problem where A is multiplied by a positive integer k and each object is replaced by k replicas of itself, while the object weights and values stay the same. Let $f^*(k)$ and $q^*(k)$ be the corresponding optimal primal and dual values. Show that

$$\frac{q^*(k) - f^*(k)}{f^*(k)} \leq \frac{1}{k} \frac{\max_{i=1,\ldots,n} v_i}{f^*},$$

so that the relative value of the duality gap tends to 0 as $k \to \infty$.

5.1.8 (Sensitivity) (www)

Consider the class of problems

$$\begin{aligned} &\text{minimize} \quad f(x) \\ &\text{subject to} \quad x \in X, \qquad g_j(x) \leq u_j, \quad j = 1, \ldots, r, \end{aligned}$$

where $u = (u_1, \ldots, u_r)$ is a vector parameterizing the right-hand side of the constraints. Given two distinct values $\bar{u}$ and $\tilde{u}$ of u, let $\bar{f}$ and $\tilde{f}$ be the corresponding optimal values, and let $\bar{\mu}$ and $\tilde{\mu}$ be corresponding Lagrange multipliers. Assume that $-\infty < \bar{f} < \infty$ and $-\infty < \tilde{f} < \infty$. Show that

$$(\tilde{u} - \bar{u})'\tilde{\mu} \leq \bar{f} - \tilde{f} \leq (\tilde{u} - \bar{u})'\bar{\mu}.$$

5.1.9 (Upper Bounds to the Optimal Dual Value [Ber97]) (www)

Consider the problem

$$\begin{aligned} &\text{minimize} \;\; f(x) \\ &\text{subject to} \;\; x \in X, \qquad g_j(x) \leq 0, \qquad j = 1, \ldots, r, \end{aligned}$$

where X is a nonempty subset of $\Re^n$, and $f : \Re^n \mapsto \Re$, $g_j : \Re^n \mapsto \Re$ are given functions. Let f^* and q^* be the optimal primal and dual value, respectively:

$$f^* = \inf_{\substack{x \in X \\ g_j(x) \leq 0,\, j=1,\ldots,r}} f(x), \qquad q^* = \sup_{\mu \geq 0} q(\mu),$$

where $q : \Re^r \mapsto [-\infty, +\infty)$ is the dual function $q(\mu) = \inf_{x \in X} L(x, \mu)$. Assume that we have two vectors x_F and x_I from X such that:

(1) x_F is feasible, i.e., $g(x_F) \leq 0$.

(2) x_I is infeasible, i.e., $g_j(x_I) > 0$ for at least one j.

(3) $f(x_I) < f(x_F)$.

Let

$$\Gamma = \inf\big\{\gamma \geq 0 \mid g_j(x_I) \leq -\gamma g_j(x_F),\, j = 1, \ldots, r\big\}.$$

Show that

$$q^* \leq \frac{\Gamma}{\Gamma + 1} f(x_F) + \frac{1}{\Gamma + 1} f(x_I),$$

and that this inequality provides a tighter upper bound to q^* than $f(x_F)$.

5.2 CONVEX COST – LINEAR CONSTRAINTS*

We saw in Section 3.3.5 that problems with differentiable cost and linear constraints are special because they possess Lagrange multipliers, even for local minima that are not regular points. We traced this fact to properties of polyhedral convexity and we discussed a connection with Farkas' lemma. It turns out that these same properties are responsible for powerful duality results for the linearly constrained problem

$$\begin{aligned} &\text{minimize} \;\; f(x) \\ &\text{subject to} \;\; x \in X, \;\; e_i'x - d_i = 0, \;\; i = 1, \ldots, m, \\ &\qquad\qquad\quad a_j'x - b_j \leq 0, \;\; j = 1, \ldots, r, \end{aligned} \tag{5.12}$$

where $f : \Re^n \mapsto \Re$ is a convex (not necessarily differentiable) function and X is a polyhedral subset of $\Re^n$. We state below the main result of this section, and we prove it later in Section 5.2.1.

Proposition 5.2.1: (Strong Duality Theorem - Linear Constraints) Assume that problem (5.12) is feasible and its optimal value f^* is finite. Let also f be convex over $\Re^n$ and let X be polyhedral. Then there is no duality gap and there exists at least one Lagrange multiplier.

Actually, we have already shown Prop. 5.2.1 for the special case where f is continuously differentiable and there exists a primal optimal solution; see Prop. 3.4.2 in Section 3.4.2. On the other hand, proving Prop. 5.2.1 in the general case is considerably more challenging. The proof given in Section 5.2.1 first shows existence of an optimal primal solution for linear programs, and then uses this result to show a nonlinear version of Farkas' lemma, from which Prop. 5.2.1 follows.

We note that convexity of f over X is not enough for Prop. 5.2.1 to hold; it is essential that f be convex over the entire space $\Re^n$, as the following example shows.

Example 5.2.1 (A Convex Problem with a Duality Gap)

Consider the two-dimensional problem

$$\begin{aligned} &\text{minimize } f(x) \\ &\text{subject to } x_1 = 0, \qquad x \in X = \{x \mid x \geq 0\}, \end{aligned}$$

where

$$f(x) = e^{-\sqrt{x_1 x_2}}, \qquad \forall\, x \in X,$$

and $f(x)$ is arbitrarily defined for $x \notin X$. Here it can be verified that f is convex over X (its Hessian is positive definite in the interior of X). Since for feasibility, we must have $x_1 = 0$, we see that $f^* = 1$. On the other hand, for all $\mu \geq 0$ we have

$$q(\mu) = \inf_{x \geq 0} \left\{ e^{-\sqrt{x_1 x_2}} + \mu x_1 \right\} = 0,$$

since the expression in braces is nonnegative for $x \geq 0$ and can approach zero by taking $x_1 \to 0$ and $x_1 x_2 \to \infty$. It follows that $q^* = 0$. Thus, there is a duality gap, $f^* - q^* = 1$. The difficulty here is that $f(x)$ is not defined as a convex function over $\Re^2$.

Proposition 5.2.1 asserts that the dual problem has an optimal solution when the optimal value is finite and the constraint set is nonempty and polyhedral. If this constraint set is also bounded, then the primal problem has an optimal solution by Weierstrass' theorem (Prop. A.8 in Appendix A), since f is a convex function and therefore is also continuous (Prop. B.9 in Appendix B). In the important special case of a linear program, where the cost function f is linear, however, it turns out that this boundedness assumption is not needed to guarantee that the primal problem has an optimal solution. This is a fundamental result, which we state as a proposition and we prove in Section 5.2.1.

Proposition 5.2.2: (Linear Programming Duality) Assume that problem (5.12) is feasible and its optimal value f^* is finite. Let also f be a linear function and X be a polyhedral subset of $\Re^n$. Then, the primal and the dual problem have optimal solutions, and there is no duality gap.

5.2.1 Proofs of Duality Theorems

The proof of Prop. 5.2.1 is based on a nonlinear version of Farkas' lemma, which we will prove by using the following linear programming result.

Lemma 5.2.1: (Existence of Primal Optimal Solutions of Linear Programs) Assume that problem (5.12) is feasible and its optimal value f^* is finite. Let also f be linear and X be polyhedral. Then, problem (5.12) has at least one optimal solution.

Proof: We will first prove the result for the special case where the linear program is

$$\begin{aligned} &\text{minimize} \quad c'x \\ &\text{subject to} \quad Ax = b, \qquad x \geq 0, \end{aligned} \tag{LP}$$

where A is an $m \times n$ matrix, $c \in \Re^n$, and $b \in \Re^m$. We will subsequently generalize by showing that any linear program can be expressed in the form (LP).

We argue by contradiction. Assume that (LP) has no optimal solution. Then, we have

$$c'x > f^*, \qquad \forall\, x \geq 0 \text{ such that } Ax = b, \tag{5.13}$$

and thus there exists no $x \geq 0$ such that

$$\begin{pmatrix} f^* \\ -b \end{pmatrix} = \begin{pmatrix} c' \\ -A \end{pmatrix} x.$$

We now use Farkas' lemma [Prop. B.16(d) in Appendix B; in applying the lemma, identify $(f^*, -b')$, x, and the rows of $(c, -A')'$ with x, μ, and a_j of the lemma, respectively]. It follows that there exists some $\alpha \in \Re$ and some $\pi \in \Re^m$ such that

$$\alpha f^* - b'\pi < 0, \tag{5.14}$$

$$\alpha c - A'\pi \geq 0. \tag{5.15}$$

Since problem (LP) is assumed feasible, there exists $x \geq 0$ such that $Ax = b$. Taking inner product of $\alpha c - A'\pi$ with such an x preserves the sign of the inequality (5.15), since $x \geq 0$. Therefore, using also the equation $Ax = b$, we obtain

$$\alpha c'x - b'\pi \geq 0. \tag{5.16}$$

Combining this inequality with Eq. (5.14), we obtain

$$\alpha c'x > \alpha f^*,$$

which in view of Eq. (5.13), implies that $\alpha > 0$. Dividing Eqs. (5.14) and (5.15) with α, we see that

$$f^* < b'\bar{\pi}, \qquad A'\bar{\pi} \leq c, \tag{5.17}$$

where $\bar{\pi} = \pi/\alpha$. We now recall from Example 3.4.2 of Section 3.4.2 that the dual problem to (LP) is

$$\begin{aligned} &\text{maximize } \ b'\pi \\ &\text{subject to } \ A'\pi \leq c. \end{aligned} \tag{DLP}$$

Equation (5.17) implies that there is a dual feasible solution with dual value that is larger than the optimal primal value. In view of the weak duality theorem (Prop. 5.1.3), this is a contradiction, proving the result for the special case of the linear program (LP).

The general linear program

$$\begin{aligned} &\text{minimize } \ c'x \\ &\text{subject to } \ x \in X, \quad e_i'x - d_i = 0, \quad i = 1, \ldots, m, \\ &\qquad\qquad a_j'x - b_j \leq 0, \quad j = 1, \ldots, r, \end{aligned} \tag{LP$'$}$$

where X is a polyhedral set, can be converted to the form (LP). In particular, if X is expressed in terms of linear inequalities as

$$X = \{x \mid a_j'x - b_j \leq 0, \, j = r+1, \ldots, \bar{r}\},$$

we can write the program (LP$'$) in the equivalent form

$$\begin{aligned} &\text{minimize } \ c'(x_+ - x_-) \\ &\text{subject to } \ a_j'(x_+ - x_-) - b_j + z_j = 0, \quad j = 1, \ldots, \bar{r}, \\ &e_i'(x_+ - x_-) - d_i = 0, \quad i = 1, \ldots, m, \quad x_+ \geq 0, \; x_- \geq 0, \; z \geq 0, \end{aligned} \tag{LP$''$}$$

where x_+, x_-, and z are new variables. The program (LP$''$) is in the form (LP) and therefore by what has been proved so far, it has an optimal solution (x_+^*, x_-^*, z^*). It can be seen that $x^* = x_+^* - x_-^*$ is an optimal solution of problem (LP$'$). **Q.E.D.**

We now provide a generalization of Farkas' lemma. In this lemma, we use the notion of *relative interior* of a set, which is discussed in Section B.1. In particular, let C be a convex subset of $\Re^n$. We recall from Appendix B that the affine hull of C, denoted $aff(C)$ is the intersection of all linear manifolds containing C. The relative interior of C, denoted $ri(C)$, is the set of all $x \in C$ for which there exists an $\epsilon > 0$ such that

$$\big\{z \mid \|z - x\| < \epsilon,\ z \in aff(C)\big\} \subset C,$$

i.e., $ri(C)$ is the interior of C relative to $aff(C)$. Every nonempty convex set has a nonempty relative interior (Prop. B.7 in Appendix B).

Lemma 5.2.2: (Extended Farkas' Lemma) Let C be a convex subset of $\Re^n$ and let $F : C \mapsto \Re$ be a function that is convex over C. Let also $a_j \in \Re^n$ and $b_j \in \Re$, $j = 1, \ldots, r$, be given vectors and scalars, respectively. Assume that the set

$$S = \{x \mid a_j'x \le b_j,\ j = 1, \ldots, r\} \tag{5.18}$$

contains a point in the relative interior of C, and that

$$F(x) \ge 0, \qquad \forall\ x \in S \cap C.$$

Then, there exist scalars $\mu_j \ge 0$, $j = 1, \ldots, r$, such that

$$F(x) + \sum_{j=1}^{r} \mu_j (a_j'x - b_j) \ge 0, \qquad \forall\ x \in C. \tag{5.19}$$

Proof: Consider the convex subsets A_1 and A_2 of $\Re^{n+1}$ defined by

$$A_1 = \big\{(x, w) \mid x \in C,\ F(x) < w\big\}, \qquad A_2 = \big\{(x, 0) \mid x \in S\big\},$$

(cf. Fig. 5.2.1), and assume first that C has a nonempty interior.

In view of the hypothesis $F(x) \ge 0$ for all $x \in S \cap C$, we see that A_1 and A_2 are disjoint. Hence, there exists a hyperplane separating A_1 and A_2, i.e., a vector $(\xi, \beta) \ne (0, 0)$ such that

$$\xi'z \le \xi'y + \beta w, \qquad \forall\ (y, w) \in A_1,\ z \in S. \tag{5.20}$$

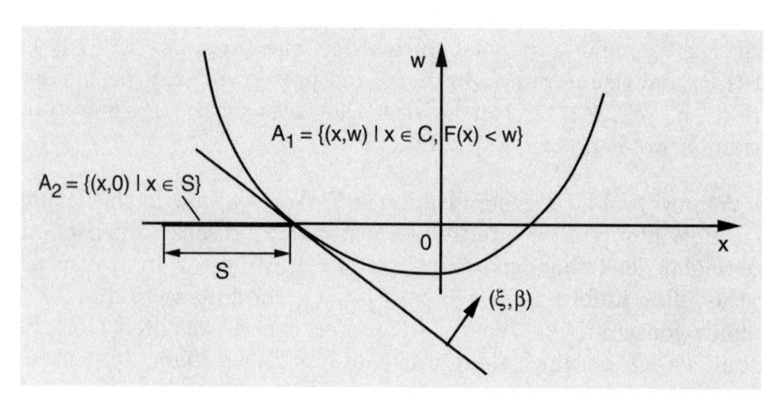

Figure 5.2.1. Separating hyperplane argument used for the proof of Lemma 5.2.2.

In view of the definition of A_1, it is seen that $\beta \geq 0$, since otherwise the right-hand side in Eq. (5.20) may be decreased without bound. If $\beta = 0$, then we must have $\xi \neq 0$, while Eq. (5.20) yields $0 \leq \xi'(y - z)$ for all $y \in C$ and $z \in S$. Let $\bar{x}$ be a point of S, which is in the interior of C. Then, since we have $0 \leq \xi'(y - \bar{x})$ for all $y \in C$, it follows that $\xi = 0$, which is a contradiction. Therefore, $\beta > 0$ and by normalizing (ξ, β) if necessary, we may assume that $\beta = 1$. Thus, using the definition (5.18) of the set S and Eq. (5.20), we obtain

$$\sup_{a_j'x - b_j \leq 0,\ j=1,\ldots,r} \xi'x \leq \inf_{x \in C} \{F(x) + \xi'x\}. \tag{5.21}$$

The linear program in the left-hand side of Eq. (5.21) is feasible and bounded, so by Lemma 5.2.1, it has an optimal solution x^*. By the Lagrange multiplier theorem for linear constraints (Prop. 3.4.1 in Section 3.4.1), it follows that there exist nonnegative scalars $\mu_1, \ldots, \mu_r$ such that

$$\xi = \sum_{j=1}^{r} \mu_j a_j, \tag{5.22}$$

$$\mu_j = 0, \qquad \forall\ j \text{ with } a_j'x^* < b_j. \tag{5.23}$$

Taking the inner product of both sides of Eq. (5.22) with x^* and using Eq. (5.23), we obtain

$$\xi'x^* = \sum_{j=1}^{r} \mu_j a_j'x^* = \sum_{j=1}^{r} \mu_j b_j. \tag{5.24}$$

Using Eqs. (5.22) and (5.24) in Eq. (5.21), we obtain

$$\sum_{j=1}^{r} \mu_j b_j \leq \inf_{x \in C} \left\{ F(x) + \left(\sum_{j=1}^{r} \mu_j a_j \right)' x \right\}$$

or equivalently,

$$0 \leq \inf_{x \in C} \left\{ F(x) + \sum_{j=1}^{r} \mu_j (a_j' x - b_j) \right\}.$$

This proves the desired relation (5.19).

If C has empty interior, a slight modification of the preceding argument works. In particular, we reformulate the problem so that it is defined over the subspace that is parallel to the affine hull $aff(C)$. The vector ξ in the preceding argument is taken to belong to this subspace, and the vector $\bar{x}$ is taken to be in the intersection of S and the relative interior of C. **Q.E.D.**

We are now ready to prove Prop. 5.2.1.

Proof of Prop. 5.2.1: Without loss of generality, we assume that there are no equality constraints, so we are dealing with the problem

$$\begin{aligned} &\text{minimize} \quad f(x) \\ &\text{subject to} \quad x \in X, \qquad a_j' x - b_j \leq 0, \quad j = 1, \ldots, r, \end{aligned}$$

(each equality constraint can be converted into two inequality constraints, as discussed in Section 5.1.5). Let X be expressed in terms of linear inequalities as

$$X = \{x \mid a_j' x - b_j \leq 0, \ j = r+1, \ldots, p\},$$

where p is an integer with $p > r$. By applying Lemma 5.2.2 with $C = \Re^n$, $S = \{x \mid a_j' x - b_j \leq 0, \ j = 1, \ldots, p\}$, and $F(x) = f(x) - f^*$, we see that there exist $\mu_1, \ldots, \mu_p$ with $\mu_j \geq 0$ for all j, such that

$$f^* \leq f(x) + \sum_{j=1}^{p} \mu_j (a_j' x - b_j), \qquad \forall \ x \in \Re^n.$$

Since for $x \in X$, we have $\mu_j (a_j' x - b_j) \leq 0$ for all $j = r+1, \ldots, p$, the above equation yields

$$f^* \leq f(x) + \sum_{j=1}^{r} \mu_j (a_j' x - b_j), \qquad \forall \ x \in X,$$

from which

$$f^* \leq \inf_{x \in X} L(x, \mu) = q(\mu) \leq q^*.$$

Using the weak duality theorem (Prop. 5.1.3), it follows that μ is a Lagrange multiplier and that there is no duality gap. **Q.E.D.**

The proof of Prop. 5.2.2 is obtained by combining Prop. 5.2.1 and Lemma 5.2.1. Note that Exercise 5.2.2 provides a version of the strong duality theorem of Prop. 5.2.1, which relaxes the requirement that X be a polyhedral set. Instead, X is assumed to be the intersection of a polyhedron and a general convex set C. However, there is a requirement that there exists a feasible solution, which is a relative interior point of C (this is needed in order to fulfill the relative interior point condition of the extended Farkas' lemma 5.2.2).

EXERCISES

5.2.1

Consider the problem

$$\begin{aligned} &\text{minimize} \quad \sum_{i=0}^{m} f_i(x) \\ &\text{subject to} \quad x \in X_i, \qquad i = 0, 1, \ldots, m, \end{aligned}$$

where $f_i : \Re^n \mapsto \Re$ are convex functions and X_i are bounded polyhedral subsets of $\Re^n$ with nonempty intersection. Show that a dual problem is given by

$$\begin{aligned} &\text{maximize} \quad q_0(\lambda_1 + \cdots + \lambda_m) + \sum_{i=1}^{m} q_i(\lambda_i) \\ &\text{subject to} \quad \lambda_i \in \Re^n, \qquad i = 1, \ldots, m, \end{aligned}$$

where the functions $q_i : \Re^n \mapsto \Re$ are given by

$$q_0(\lambda) = \min_{x \in X_0} \{ f_0(x) - \lambda' x \},$$

$$q_i(\lambda) = \min_{x \in X_i} \{ f_i(x) + \lambda' x \}, \qquad i = 1, \ldots, m.$$

Show that the primal and dual problems have optimal solutions, and that there is no duality gap.

5.2.2 (A Stronger Version of the Duality Theorem) (www)

For the problem (5.12) of this section with $f^* > -\infty$, show that the conclusions of Prop. 5.2.1 hold if the assumption that X is a polyhedron and $f : \Re^n \mapsto \Re$ is convex over $\Re^n$ is replaced by the following assumptions:

(1) X is the intersection of a polyhedron P and a convex set C.

(2) $f : \Re^n \mapsto \Re$ is convex over C.

(3) There exists a feasible solution of the problem that belongs to the relative interior of C.

Use Example 5.2.1 to demonstrate the need for assumption (3). *Hint*: Assume that there are no inequality constraints, and express P as

$$P = \{x \mid a_j'x - b_j \le 0,\ j = r+1, \ldots, p\},$$

where p is an integer with $p > r$. Apply Lemma 5.2.2 and proceed as in the proof of Prop. 5.2.1.

5.3 CONVEX COST – CONVEX CONSTRAINTS

We now consider the nonlinearly constrained problem

$$\begin{aligned} &\text{minimize } f(x) \\ &\text{subject to } x \in X, \qquad g_j(x) \le 0, \quad j = 1, \ldots, r, \end{aligned} \tag{5.25}$$

under convexity assumptions. In particular, we will assume the following.

Assumption 5.3.1: (Convexity and Interior Point) Problem (5.25) is feasible and its optimal value f^* is finite. Furthermore, the set X is a convex subset of $\Re^n$ and the functions $f : \Re^n \mapsto \Re$, $g_j : \Re^n \mapsto \Re$ are convex over X. In addition, there exists a vector $\bar{x} \in X$ such that

$$g_j(\bar{x}) < 0, \qquad \forall\ j = 1, \ldots, r. \tag{5.26}$$

The interior point condition (5.26) is known as the *Slater constraint qualification* [Sla50] (compare also with the analysis of Section 3.3.5). The following is the central result of this section.

Proposition 5.3.1: (Strong Duality Theorem - Inequality Constraints) Let the convexity and interior point Assumption 5.3.1 hold for problem (5.25). Then there is no duality gap and there exists at least one Lagrange multiplier.

Proof: Consider the subset of $\Re^{r+1}$ given by

$$A = \{(z_1, \ldots, z_r, w) \mid \text{there exists } x \in X \text{ such that } g_j(x) \le z_j,\ j = 1, \ldots, r,\ f(x) \le w\},$$

(cf. Fig. 5.3.1). We first show that A is convex. To this end, we consider vectors $(z, w) \in A$ and $(\tilde{z}, \tilde{w}) \in A$, and we show that their convex combinations lie in A. The definition of A implies that for some $x \in X$ and $\tilde{x} \in X$, we have

$$f(x) \le w, \qquad g_j(x) \le z_j, \qquad j = 1, \ldots, r,$$

$$f(\tilde{x}) \le \tilde{w}, \qquad g_j(\tilde{x}) \le \tilde{z}_j, \qquad j = 1, \ldots, r.$$

For any $\alpha \in [0, 1]$, we multiply these relations with α and $1-\alpha$, respectively, and add. By using the convexity of f and g_j, we obtain

$$f\big(\alpha x + (1-\alpha)\tilde{x}\big) \le \alpha f(x) + (1-\alpha) f(\tilde{x}) \le \alpha w + (1-\alpha)\tilde{w},$$

$$g_j\big(\alpha x + (1-\alpha)\tilde{x}\big) \le \alpha g_j(x) + (1-\alpha) g_j(\tilde{x}) \le \alpha z_j + (1-\alpha)\tilde{z}_j, \qquad j = 1, \ldots, r.$$

In view of the convexity of X, we have $\alpha x + (1-\alpha)\tilde{x} \in X$, so these equations imply that the convex combination of (z, w) and $(\tilde{z}, \tilde{w})$, i.e., $\big(\alpha z + (1-\alpha)\tilde{z}, \alpha w + (1-\alpha)\tilde{w}\big)$, belongs to A. This proves the convexity of A.

We next observe that $(0, f^*)$ is not an interior point of A; otherwise, for some $\epsilon > 0$, the point $(0, f^* - \epsilon)$ would belong to A, contradicting the definition of f^* as the primal optimal value. Therefore, by Prop. B.12 of Appendix B, there exists a hyperplane passing through $(0, f^*)$ and containing A in one of the two corresponding halfspaces. In particular, there exists a vector $(\mu, \beta) \ne (0, 0)$ such that

$$\beta f^* \le \beta w + \mu' z, \qquad \forall\ (z, w) \in A. \tag{5.27}$$

This equation implies that

$$\beta \ge 0, \qquad \mu_j \ge 0, \qquad \forall\ j = 1, \ldots, r, \tag{5.28}$$

since for each $(z, w) \in A$, we have that $(z, w + \gamma) \in A$ and $(z_1, \ldots, z_j + \gamma, \ldots, z_r, w) \in A$ for all $\gamma > 0$ and j.

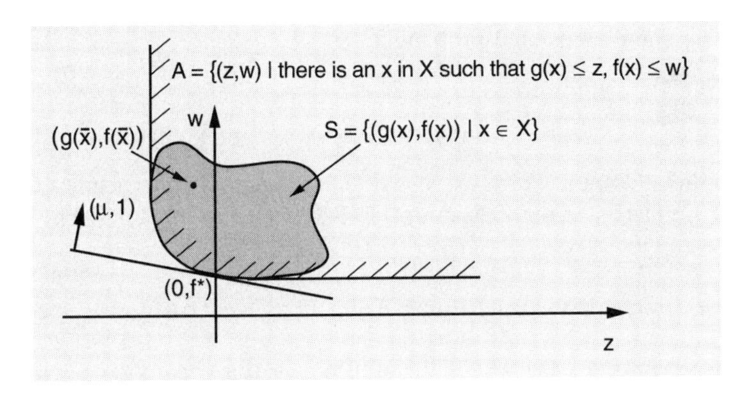

Figure 5.3.1. Illustration of the set

$$S = \{ \big(g(x), f(x)\big) \mid x \in X \}$$

and the set

$$A = \Big\{ (z_1, \ldots, z_r, w) \mid \text{there exists } x \in X \text{ such that} \\ g_j(x) \le z_j,\ j = 1, \ldots, r,\ f(x) \le w \Big\},$$

used in the proof of Prop. 5.3.1. The idea of the proof is to show that A is convex and that $(0, f^*)$ is not an interior point of A. A hyperplane passing through $(0, f^*)$ and supporting A is used to construct a Lagrange multiplier.

We now claim that $\beta > 0$. Indeed, if this were not so, we would have $\beta = 0$ and Eq. (5.27) would imply that $0 \le \mu' z$ for all $(z, w) \in A$. Since, $\big(g_1(\bar{x}), \ldots, g_r(\bar{x}), f(\bar{x})\big) \in A$, we would obtain

$$0 \le \sum_{j=1}^{r} \mu_j g_j(\bar{x}),$$

which in view of $\mu \ge 0$ [cf. Eq. (5.28)] and the assumption $g_j(\bar{x}) < 0$ for all j, implies that $\mu = 0$. This means, however, that $(\mu, \beta) = (0, 0)$, arriving at a contradiction. Thus, we must have $\beta > 0$ and by dividing if necessary the vector (μ, β) by β, we may assume that $\beta = 1$. Thus, since $\big(g(x), f(x)\big) \in A$ for all $x \in X$, Eq. (5.27) yields

$$f^* \le f(x) + \mu' g(x), \qquad \forall\, x \in X. \tag{5.29}$$

Taking the infimum over $x \in X$ and using the fact $\mu \ge 0$, we obtain

$$f^* \le \inf_{x \in X} \{ f(x) + \mu' g(x) \} = q(\mu) \le q^*,$$

where q^* is the optimal dual value. Using the weak duality theorem (Prop. 5.1.3), it follows that μ is a Lagrange multiplier for problem (5.25) and there is no duality gap. **Q.E.D.**

The one-dimensional problem

$$\begin{aligned} &\text{minimize} \quad f(x) = x \\ &\text{subject to} \quad g(x) = x^2 \leq 0, \qquad x \in X = \Re, \end{aligned}$$

[cf. Fig. 5.1.4(a)], for which there is no Lagrange multiplier, shows that the interior point condition of Assumption 5.3.1 is necessary for Prop. 5.3.1 to hold.

Convex Inequality and Linear Equality Constraints*

Consider now the case where in addition to the convex inequality constraints in problem (5.25), there are linear equality constraints

$$\begin{aligned} &\text{minimize} \quad f(x) \\ &\text{subject to} \quad x \in X, \quad g_j(x) \leq 0, \quad j = 1, \ldots, r, \quad e_i'x - d_i = 0, \quad i = 1, \ldots, m. \end{aligned} \tag{5.30}$$

Even after each equality constraint $e_i'x - d_i = 0$ is converted to the two inequality constraints $e_i'x - d_i \leq 0$ and $-e_i'x + d_i \leq 0$, Prop. 5.3.1 cannot be applied to this problem, because the interior point condition of Assumption 5.3.1 is violated. To cover this case, we develop a related result, which involves an assumption regarding the relative interior $ri(X)$ of the set X. The following assumption extends Assumption 5.3.1 to deal with additional equality constraints.

Assumption 5.3.2: (For the Equality Constrained Problem (5.30)) The set X is a convex subset of $\Re^n$ and the functions $f : \Re^n \mapsto \Re$, $g_j : \Re^n \mapsto \Re$ are convex over X. Furthermore, the optimal value f^* is finite and there exists a vector $\bar{x} \in ri(X)$ such that

$$g_j(\bar{x}) < 0, \qquad j = 1, \ldots, r,$$

$$e_i'\bar{x} - d_i = 0, \qquad i = 1, \ldots, m.$$

The proof of the following proposition parallels the one of Prop. 5.3.1. A slightly stronger version of the proposition, together with an alternative method of proof, based on the extended Farkas' lemma 5.2.2, is outlined in Exercise 5.3.7.

Proposition 5.3.2: (Strong Duality Theorem - Equality Constraints) Let Assumption 5.3.2 hold for problem (5.30). Then there is no duality gap and there exists at least one Lagrange multiplier.

Proof: (*Abbreviated*) We first assume in addition that X has a nonempty interior and that the vectors $e_1, \ldots, e_m$ are linearly independent. Under these assumptions, we prove the result with an argument that closely parallels the one of Prop. 5.3.1. In particular, we consider the set

$$A = \big\{(z_1, \ldots, z_r, y_1, \ldots, y_m, w) \mid \text{there exists } x \in X \text{ such that} \\ g_j(x) \leq z_j, \forall\, j,\ e_i'x - d_i = y_i, \forall\, i,\ f(x) \leq w\big\}.$$

Similar to the proof of Prop. 5.3.1, we show that A is convex and that the vector $(0, 0, f^*)$ is not an interior point of A. Therefore, there exists a vector (μ, λ, β) such that

$$\beta f^* \leq \beta w + \mu' z + \lambda' y, \qquad \forall\, (z, y, w) \in A. \tag{5.31}$$

As in the proof of Prop. 5.3.1, we argue that $\mu \geq 0$ and $\beta \geq 0$. If $\beta = 0$, then, using Eq. (5.31), the fact $e_i'\bar{x} = d_i$, and the definition of A it follows that

$$0 \leq \sum_{j=1}^{r} \mu_j g_j(x) + \sum_{i=1}^{m} \lambda_i e_i'(x - \bar{x}), \qquad \forall\, x \in X. \tag{5.32}$$

Applying this relation with $x = \bar{x}$, we obtain

$$0 \leq \sum_{j=1}^{r} \mu_j g_j(\bar{x}),$$

which in view of $\mu \geq 0$ and the assumption $g_j(\bar{x}) < 0$ for all j, implies that $\mu = 0$. Thus, Eq. (5.32) yields

$$0 \leq \left(\sum_{i=1}^{m} \lambda_i e_i\right)' (x - \bar{x}), \qquad \forall\, x \in X.$$

Since $\bar{x}$ is an interior point of X, it follows that $\sum_{i=1}^{m} \lambda_i e_i = 0$, which in view of the linear independence of $e_1, \ldots, e_m$, implies that $\lambda = 0$. This means that $(\beta, \mu, \lambda) = (0, 0, 0)$, arriving at a contradiction. Thus, we must have $\beta > 0$ and we may assume that $\beta = 1$. As in the proof of Prop. 5.3.1, it follows that

$$f^* = \inf_{x \in X} \left\{ f(x) + \sum_{j=1}^{r} \mu_j g_j(x) + \sum_{i=1}^{m} \lambda_i (e_i'x - d_i) \right\}, \tag{5.33}$$

so (μ, λ) is a Lagrange multiplier.

If the vectors $e_1, \ldots, e_m$ are linearly dependent, let $\{e_i \mid i \in I\}$ be a linearly independent set that spans the same space as $e_1, \ldots, e_m$. Then, since the equality constraints have at least one feasible solution, the equations $e_i'x - d_i = 0$, $i \notin I$, can be expressed in terms of the remaining equations. Thus, we can consider a modified problem where the equality constraints $e_i'x - d_i = 0$, $i \notin I$, have been eliminated. The proof for the case of linear independent vectors e_i applies to this problem and yields Lagrange multipliers $\mu \geq 0$, and λ_i, $i \in I$. Defining $\lambda_i = 0$ for $i \notin I$, we recover Eq. (5.33), implying that (μ, λ) is a Lagrange multiplier.

Finally, in the case where X has an empty interior, we reformulate the problem so that it is defined over the subspace which is parallel to the affine hull of X. Relative to this subspace, $\bar{x}$ is an interior point of X and the preceding proof shows that there exists a Lagrange multiplier. **Q.E.D.**

To see that Assumption 5.3.2 is necessary for Prop. 5.3.2 to hold, consider the problem of Example 5.2.1, where there is a duality gap. This problem satisfies all the conditions of Assumption 5.3.2, except that the feasible set and the relative interior of X have no common point. For another example, which involves linear cost and constraint functions (but nonpolyhedral set X), consider the following two-dimensional problem:

$$\begin{aligned} &\text{minimize} \ \ f(x) = x_1 \\ &\text{subject to} \ \ x_2 = 0, \qquad x \in X = \big\{(x_1, x_2) \mid x_1^2 \leq x_2\big\}. \end{aligned}$$

Here, the unique optimal solution is $x^* = (0,0)$ and the primal optimal value is $f^* = 0$. The dual function is given by

$$q(\lambda) = \inf_{x_1^2 \leq x_2} \{x_1 + \lambda x_2\} = \begin{cases} \inf_{x_1}\{x_1 + \lambda x_1^2\} & \text{if } \lambda > 0 \\ -\infty & \text{if } \lambda \leq 0 \end{cases} = \begin{cases} -\frac{1}{4\lambda} & \text{if } \lambda > 0, \\ -\infty & \text{if } \lambda \leq 0. \end{cases}$$

There is no duality gap here, but there is no dual optimal solution and therefore there is no Lagrange multiplier (cf. Prop. 5.1.4). The difficulty is that the feasible set and the relative interior of X have no common point.

E X E R C I S E S

5.3.1 (Boundedness of the Set of Lagrange Multipliers) (www)

Assume that X is convex, f and g_j are convex over X, and $-\infty < f^* < \infty$. Show that the set of Lagrange multipliers is nonempty and bounded if and only if there exists an $\bar{x} \in X$ such that $g_j(\bar{x}) < 0$ for all j. *Hint*: Show that if $\bar{x} \in X$ such that $g_j(\bar{x}) < 0$ for all j, then for every Lagrange multiplier μ, we have

$$\sum_{j=1}^{r} \mu_j \leq \frac{f(\bar{x}) - f^*}{\min_{j=1,\ldots,r}\big\{-g_j(\bar{x})\big\}}.$$

Conversely, assume that there is no vector $\bar{x} \in X$ such that $g_j(\bar{x}) < 0$ for all j. Use the line of proof of Prop. 5.3.1 to show that there exists a nonzero vector $\gamma \geq 0$ such that $\gamma' g(x) \geq 0$ for all $x \in X$, and conclude that the set of Lagrange multipliers must be unbounded.

5.3.2 (Optimality Conditions for Nonconvex Cost) (www)

Consider the problem

$$\begin{aligned} &\text{minimize } f(x) \\ &\text{subject to } x \in X, \qquad g_j(x) \leq 0, \quad j = 1, \ldots, r, \end{aligned}$$

where f is continuously differentiable, and $X \subset \Re^n$ is a closed convex set. Let x^* be a local minimum.

(a) Assume that the g_j are convex over X, and that there exists a vector $\bar{x} \in X$ such that $g_j(\bar{x}) < 0$ for all $j = 1, \ldots, r$. Show that there exists a vector $\mu^* \geq 0$ such that

$$x^* = \arg\min_{x \in X} \left\{ \nabla f(x^*)' x + \sum_{j=1}^{r} \mu_j^* g_j(x) \right\}, \qquad \mu_j^* g_j(x^*) = 0, \quad j = 1, \ldots, r.$$

(b) Assume that the g_j are continuously differentiable, and that there exists a feasible direction d of X at x^* such that

$$\nabla g_j(x^*)' d < 0, \quad \forall\, j \in A(x^*).$$

Show that there exists a vector $\mu^* \geq 0$ such that

$$x^* = \arg\min_{x \in X} \nabla_x L(x^*, \mu^*)' x, \qquad \mu_j^* g_j(x^*) = 0, \quad j = 1, \ldots, r.$$

5.3.3 (Boundedness of Lagrange Multipliers for Nonconvex Constraints) (www)

Consider the problem

$$\begin{aligned} &\text{minimize } f(x) \\ &\text{subject to } x \in X, \qquad g_j(x) \leq 0, \quad j = 1, \ldots, r, \end{aligned}$$

where f, and the g_j are continuously differentiable, and $X \subset \Re^n$ is a closed convex set. Let x^* be a local minimum, and let M^* be the set of vectors $\mu \geq 0$ such that

$$x^* = \arg\min_{x \in X} \nabla_x L(x^*, \mu)' x, \qquad \mu' g(x^*) = 0.$$

Show that M^* is nonempty and bounded if and only if there exists a feasible direction d of X at x^* such that

$$\nabla g_j(x^*)' d < 0, \qquad \forall\, j \in A(x^*).$$

Hint: Combine Prop. 3.3.12 and Exercise 5.3.1.

5.3.4 (Characterization of Pareto Optimality) (www)

A decisionmaker wishes to choose a vector $x \in X$, which keeps the values of *two* cost functions $f_1 : \Re^n \mapsto \Re$ and $f_2 : \Re^n \mapsto \Re$ reasonably small. Since a vector x^* minimizing simultaneously both f_1 and f_2 over X need not exist, he/she decides to settle for a *Pareto optimal solution*, i.e., a vector $x^* \in X$ with the property that there does not exist any vector $\bar{x} \in X$ that is strictly better than x^*, in the sense that either

$$f_1(\bar{x}) \le f_1(x^*), \qquad f_2(\bar{x}) < f_2(x^*)$$

or

$$f_1(\bar{x}) < f_1(x^*), \qquad f_2(\bar{x}) \le f_2(x^*).$$

(a) Show that if x^* is a vector in X, and λ_1^* and λ_2^* are two positive scalars such that

$$\lambda_1^* f_1(x^*) + \lambda_2^* f_2(x^*) = \min_{x \in X} \big\{ \lambda_1^* f_1(x) + \lambda_2^* f_2(x) \big\},$$

then x^* is a Pareto optimal solution.

(b) Assume that X is convex and f_1, f_2 are convex over X. Show that if x^* is a Pareto optimal solution, then there exist non-negative scalars λ_1^*, λ_2^*, not both zero, such that

$$\lambda_1^* f_1(x^*) + \lambda_2^* f_2(x^*) = \min_{x \in X} \big\{ \lambda_1^* f_1(x) + \lambda_2^* f_2(x) \big\}.$$

Hint: Consider the set

$$A = \big\{ (z_1, z_2) \mid \text{there exists } x \in X \text{ such that } f_1(x) \le z_1,\ f_2(x) \le z_2 \big\}$$

and show that it is a convex set. Use hyperplane separation arguments.

(c) Generalize the results of (a) and (b) to the case where there are m cost functions rather than two.

5.3.5 (Directional Convexity) (www)

Consider the following problem, which involves a pair of optimization variables (x, u):

$$\begin{aligned} &\text{minimize} \ \ f(x, u) \\ &\text{subject to} \ \ h(x, u) = 0, \qquad u \in U, \end{aligned}$$

where $f : \Re^{n+s} \mapsto \Re$ and $h : \Re^{n+s} \mapsto \Re^m$ are given functions, and U is a subset of $\Re^s$. Assume that:

(1) The following subset of $\Re^{m+1}$

$$\big\{ (z, w) \mid \text{there exists } (x, u) \in \Re^{n+s} \text{ such that } \\ h(x, u) = z,\ f(x, u) \le w,\ u \in U \big\}$$

is convex. (This property is known as *directional convexity*.)

(2) There exists an $\epsilon > 0$ such that for all $z \in \Re^m$ with $\|z\| < \epsilon$, we have $h(x, u) = z$ for some $u \in U$ and $x \in \Re^n$.

Show that there exists $\lambda \in \Re^m$ such that

$$\inf_{x \in \Re^n,\, u \in U} \big\{ f(x, u) + \lambda' h(x, u) \big\} = f^*,$$

where f^* is the optimal value of the problem. Furthermore, if (x^*, u^*) is an optimal solution, we have

$$u^* = \arg\min_{u \in U} \big\{ f(x^*, u) + \lambda' h(x^*, u) \big\}.$$

If in addition, f and h are continuously differentiable with respect to x for each $u \in U$, we have

$$\nabla_x f(x^*, u^*) + \nabla_x h(x^*, u^*) \lambda = 0.$$

Hint: Follow the proof of Prop. 5.3.1.

5.3.6 (Directional Convexity and Optimal Control) (www)

Consider finding sequences $u = (u_0, u_1, \ldots, u_{N-1})$ and $x = (x_1, x_2, \ldots, x_N)$, which minimize

$$g_N(x_N) + \sum_{i=0}^{N-1} g_i(x_i, u_i),$$

subject to the constraints

$$x_{i+1} = f_i(x_i, u_i), \quad i = 0, \ldots, N-1, \qquad x_0 : \text{ given},$$

$$u_i \in U_i \subset \Re^m, \qquad i = 0, \ldots, N-1.$$

We assume that the functions g_i and f_i are continuously differentiable with respect to x_i for each $u_i \in U_i$, and we also assume that the set

$$\left\{ (z, w) \;\middle|\; x_{i+1} = f_i(x_i, u_i) + z_i,\ g_N(x_N) + \sum_{i=0}^{N-1} g_i(x_i, u_i) \le w,\ u_i \in U_i \right\}$$

is convex; this is a directional convexity assumption (refer to the preceding exercise). Furthermore, there exists a $\epsilon > 0$ such that for all $z = (z_0, \ldots, z_{N-1})$ with $\|z\| < \epsilon$, we have $x_{i+1} = f_i(x_i, u_i) + z_i$ for some x_{i+1}, $u_i \in U_i$, and all i. Show that if u^* and x^* are optimal, then the minimum principle

$$u_i^* = \arg\min_{u_i \in U_i} H_i\big(x_i^*, u_i, p_{i+1}^*\big), \qquad \forall\, i = 0, \ldots, N-1,$$

holds, where

$$H_i(x_i, u_i, p_{i+1}) = g_i(x_i, u_i) + p_{i+1}' f_i(x_i, u_i)$$

is the Hamiltonian function, and the vectors $p_1^*, \ldots, p_N^*$ are obtained from the adjoint equation

$$p_i^* = \nabla_{x_i} H_i\big(x_i^*, u_i^*, p_{i+1}^*\big), \qquad i = 1, \ldots, N-1,$$

with the terminal condition

$$p_N^* = \nabla g_N(x_N^*).$$

5.3.7 (Strong Duality for Mixed Linear and Nonlinear Constraints) (www)

Consider the problem with mixed linear and nonlinear constraints

$$\begin{aligned}&\text{minimize} \;\; f(x)\\ &\text{subject to} \;\; x \in X, \;\; e_i'x - d_i = 0, \;\; i = 1, \ldots, m,\\ &\qquad\qquad g_j(x) \le 0, \;\; j = 1, \ldots, \bar{r}, \quad a_j'x - b_j \le 0, \;\; j = \bar{r}+1, \ldots, r.\end{aligned}$$

The set X is a convex subset of $\Re^n$ and the functions $f : \Re^n \mapsto \Re$, $g_j : \Re^n \mapsto \Re$ are convex over X. Furthermore, the optimal value f^* is finite, and there exists a feasible vector $\bar{x}$ that belongs to $ri(X)$ and is such that

$$g_j(\bar{x}) < 0, \qquad j = 1, \ldots, \bar{r}.$$

Show that there exists at least one Lagrange multiplier and that there is no duality gap. *Hint*: Use Prop. 5.3.1 to show that there exist $\mu_j^* \ge 0$, $j = 1, \ldots, r$, such that

$$f^* = \inf_{\substack{x \in X,\, a_j'x - b_j \le 0,\, j = \bar{r}+1, \ldots, r \\ e_i'x - d_i = 0,\, i = 1, \ldots, m}} \left\{ f(x) + \sum_{j=1}^{\bar{r}} \mu_j^* g_j(x) \right\}.$$

Apply the result of Exercise 5.2.2 to the optimization problem of the above equation.

5.3.8 (Inconsistent Convex Systems of Inequalities) 

Let $g_j : \Re^n \mapsto \Re$, $j = 1, \ldots, r$, be convex functions over the nonempty convex set $X \subset \Re^n$. Show that the system

$$g_j(x) < 0, \qquad j = 1, \ldots, r,$$

has no solution within X if and only if there exists a vector $\mu \in \Re^r$ such that

$$\sum_{j=1}^{r} \mu_j = 1, \qquad \mu \ge 0,$$

$$\mu' g(x) \ge 0, \qquad \forall\, x \in X.$$

Hint: Consider the convex program

$$\begin{aligned}&\text{minimize} \;\; y\\ &\text{subject to} \;\; x \in X, \quad y \in \Re, \qquad g_j(x) \le y, \quad j = 1, \ldots, r.\end{aligned}$$

5.4 CONJUGATE FUNCTIONS AND FENCHEL DUALITY*

In this section we consider the problem

$$\begin{aligned} &\text{minimize} \;\; f_1(x) - f_2(x) \\ &\text{subject to} \;\; x \in X_1 \cap X_2, \end{aligned} \tag{5.34}$$

where f_1 and f_2 are real-valued functions on $\Re^n$, and X_1 and X_2 are subsets of $\Re^n$. We assume throughout that this problem has a feasible solution and a finite optimal value, denoted f^*.†

We will explore a classical form of duality, which is useful in many contexts. This duality was developed by W. Fenchel, a Danish mathematician who pioneered in the 40s the use of geometric convexity methods in optimization. One way to derive Fenchel duality is to convert problem (5.34) to the following problem in the variables $y \in \Re^n$ and $z \in \Re^n$

$$\begin{aligned} &\text{minimize} \;\; f_1(y) - f_2(z) \\ &\text{subject to} \;\; z = y, \qquad y \in X_1, \qquad z \in X_2, \end{aligned} \tag{5.35}$$

and to dualize the constraint $z = y$. The dual function is

$$\begin{aligned} q(\lambda) &= \inf_{y \in X_1,\, z \in X_2} \big\{ f_1(y) - f_2(z) + (z-y)'\lambda \big\} \\ &= \inf_{z \in X_2} \big\{ z'\lambda - f_2(z) \big\} + \inf_{y \in X_1} \big\{ f_1(y) - y'\lambda \big\} \end{aligned}$$

or, equivalently,

$$q(\lambda) = g_2(\lambda) - g_1(\lambda), \tag{5.36}$$

where the functions $g_1 : \Re^n \mapsto (-\infty, \infty]$ and $g_2 : \Re^n \mapsto [-\infty, \infty)$ are defined by

$$g_1(\lambda) = \sup_{x \in X_1} \big\{ x'\lambda - f_1(x) \big\}, \tag{5.37}$$

$$g_2(\lambda) = \inf_{x \in X_2} \big\{ x'\lambda - f_2(x) \big\}. \tag{5.38}$$

The function g_1 is known as the *conjugate convex function* corresponding to the pair (f_1, X_1), while the function g_2 is known as the *conjugate concave function* corresponding to (f_2, X_2) (see Fig. 5.4.1). It is

---

† Much of the literature dealing with the material of the present section treats the functions f_1 and f_2 as extended real-valued functions, which are defined over $\Re^n$ but take the value ∞ outside the sets X_1 and X_2, respectively. There are notational advantages to this format, but it is simpler for our purposes, to maintain the framework of real-valued functions, while keeping explicit the (effective) domains of f_1, f_2, and other related functions (e.g., the conjugate functions g_1 and g_2 that will be introduced shortly).

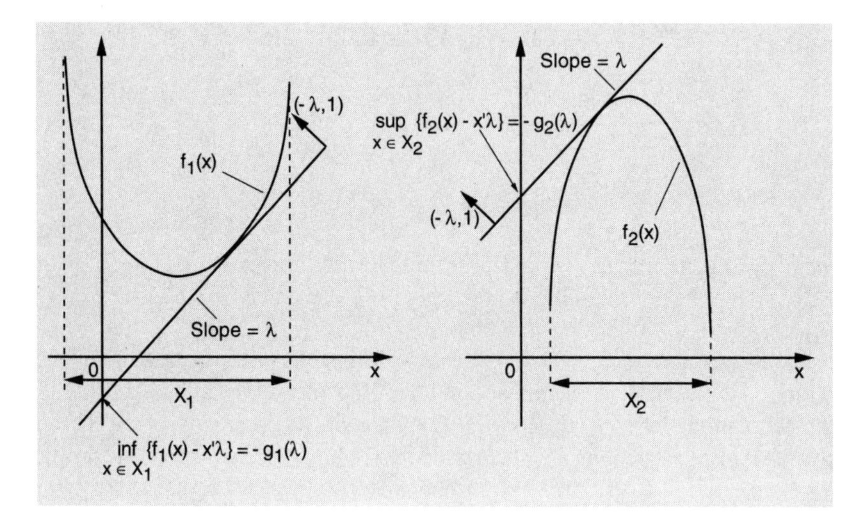

Figure 5.4.1. Visualization of the conjugate convex function

$$g_1(\lambda) = \sup_{x \in X_1} \big\{ x'\lambda - f_1(x) \big\}$$

and the conjugate concave function

$$g_2(\lambda) = \inf_{x \in X_2} \big\{ x'\lambda - f_2(x) \big\}.$$

The point of interception of the vertical axis with the hyperplane that has normal $(-\lambda, 1)$ and supports the graph of (f_1, X_1) is $\inf_{x \in X_1} \{ f_1(x) - x'\lambda \}$, which by definition is equal to $-g_1(\lambda)$. This is illustrated in the figure on the left. The figure on the right gives a similar interpretation of $g_2(\lambda)$.

straightforward to show the convexity of the sets Λ_1 and Λ_2 over which g_1 and g_2 are finite,

$$\Lambda_1 = \big\{ \lambda \mid g_1(\lambda) < \infty \big\}, \qquad \Lambda_2 = \big\{ \lambda \mid g_2(\lambda) > -\infty \big\}, \tag{5.39}$$

(compare with the proof of convexity of the domain of the dual function in Prop. 5.1.2). Furthermore, g_1 is convex over Λ_1 and g_2 is concave over Λ_2. Figure 5.4.2 shows some examples of conjugate convex functions.

The dual problem is given by [cf. Eqs. (5.36)-(5.39)]

$$\begin{aligned} &\text{maximize} \quad g_2(\lambda) - g_1(\lambda) \\ &\text{subject to} \quad \lambda \in \Lambda_1 \cap \Lambda_2. \end{aligned} \tag{5.40}$$

From Prop. 5.1.5, we see that there is no duality gap, while simultaneously, (x^*, λ^*) is an optimal primal and dual solution pair, if and only if

$$x^* \in X_1 \cap X_2, \qquad \text{(primal feasibility)}, \tag{5.41}$$

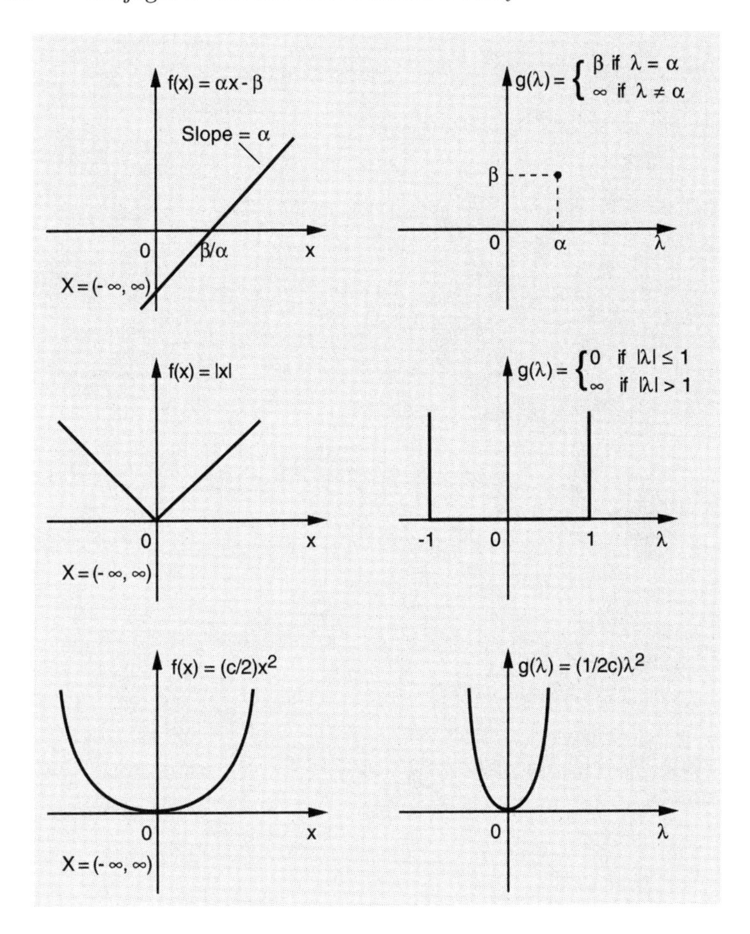

Figure 5.4.2. Some examples of conjugate convex functions.

$$\lambda^* \in \Lambda_1 \cap \Lambda_2, \qquad \text{(dual feasibility)}, \tag{5.42}$$

$$\begin{aligned} x^* &= \arg \max_{x \in X_1} \{x'\lambda^* - f_1(x)\} \\ &= \arg \min_{x \in X_2} \{x'\lambda^* - f_2(x)\}, \qquad \text{(Lagrangian optimality)}. \end{aligned} \tag{5.43}$$

The duality between the problems (5.34) and (5.40), and the Lagrangian optimality condition (5.43) are illustrated in Fig. 5.4.3.

As in Sections 5.2 and 5.3, to assert that there is no duality gap, we need convexity assumptions given below:

Assumption 5.4.1: The sets X_1, X_2 are convex. Furthermore, the function f_1 is convex over X_1, and the function f_2 is concave over X_2.

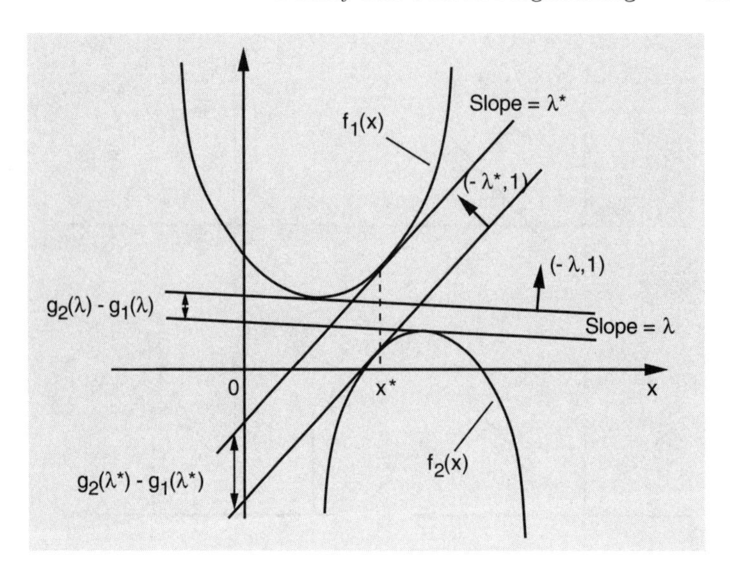

Figure 5.4.3. Illustration of Fenchel' s duality theory. The dual value $g_2(\lambda) - g_1(\lambda)$ becomes maximal when hyperplanes that are normal to $(-\lambda, 1)$ support the graphs of f_1 and f_2 at a common point x^*, which solves the primal problem [cf. the Lagrangian optimality condition (5.43)].

Under Assumption 5.4.1, by applying Prop. 5.2.1 and Prop. 5.3.2, together with the fact

$$ri(X_1 \times X_2) = ri(X_1) \times ri(X_2)$$

to problem (5.35), we obtain the following:

Proposition 5.4.1: Let Assumption 5.4.1 hold. Then the dual problem has an optimal solution, there is no duality gap, and we have

$$\inf_{x \in X_1 \cap X_2} \{f_1(x) - f_2(x)\} = \max_{\lambda \in \Lambda_1 \cap \Lambda_2} \{g_2(\lambda) - g_1(\lambda)\}, \tag{5.44}$$

if one of the following two conditions holds:

(1) The relative interiors of the sets X_1 and X_2 intersect.

(2) X_1 and X_2 are polyhedral and f_1 and f_2 are convex and concave over $\Re^n$, respectively, (rather than just over X_1 and X_2, respectively).

If Assumption 5.4.1 holds, then the epigraphs

$$\{(x, \gamma) \mid f_1(x) \leq \gamma\}, \qquad \{(x, \gamma) \mid f_2(x) \geq \gamma\}$$

are convex subsets of $\Re^{n+1}$. If these epigraphs are also closed, it is possible to show that the conjugate convex function of (g_1, Λ_1) is

$$\sup_{\lambda \in \Lambda_1} \{\lambda' x - g_1(\lambda)\} = \begin{cases} f_1(x) & \text{if } x \in X_1, \\ \infty & \text{if } x \notin X_1, \end{cases}$$

and the conjugate concave function of (g_2, X_2) is

$$\inf_{\lambda \in \Lambda_2} \{\lambda' x - g_2(\lambda)\} = \begin{cases} f_2(x) & \text{if } x \in X_2, \\ -\infty & \text{if } x \notin X_2. \end{cases}$$

A proof of this fact, which is fundamental in the theory of convex functions, is sketched in Exercise 5.4.7. It follows that under Assumption 5.4.1, the duality is symmetric; i.e., the dual problem is of the same type as the primal problem, and by dualizing the dual problem, we obtain the primal. Thus, by applying Prop. 5.4.1, we obtain the following:

Proposition 5.4.2: Let Assumption 5.4.1 hold. Then there exists an optimal primal solution and there is no duality gap if one of the following two conditions holds:

(1) The relative interiors of Λ_1 and Λ_2 intersect.

(2) Λ_1 and Λ_2 are polyhedral, and g_1 and g_2 can be extended to a real-valued convex and concave function, respectively, which are defined over the entire space $\Re^n$.

Furthermore, (x^*, λ^*) are a primal and dual optimal solution pair if and only if the primal and dual feasibility conditions (5.41) and (5.42) hold, together with the alternative Lagrangian optimality condition

$$\lambda^* = \arg\max_{\lambda \in \Lambda_1} \{\lambda' x^* - g_1(\lambda)\} = \arg\min_{\lambda \in \Lambda_2} \{\lambda' x^* - g_2(\lambda)\}. \tag{5.45}$$

We now discuss a few applications of Fenchel duality.

5.4.1 Monotropic Programming Duality

Monotropic programming problems have the generic form

$$\begin{aligned} &\text{minimize} \quad \sum_{i=1}^{n} f_i(x_i) \\ &\text{subject to} \quad x \in S, \qquad x_i \in X_i, \quad i = 1, \ldots, n, \end{aligned} \tag{MP}$$

where

(a) X_i are convex intervals of real numbers.

(b) S is a subspace of $\Re^n$.

The following analysis extends easily to the case where S is a linear manifold, i.e. the corresponding constraint is $Ax = b$ for some $m \times n$ matrix A and some vector $b \in \Re^n$ (Exercises 5.4.1 and 5.4.2). We assume throughout that this problem has at least one feasible solution and a finite optimal value, denoted f^*.

Let us apply Fenchel duality to problem (MP) with the following identifications:

(a) The functions f_1 and f_2 in the Fenchel duality framework are

$$\sum_{i=1}^{n} f_i(x_i)$$

and the identically zero function, respectively.

(b) The sets X_1 and X_2 in the Fenchel duality framework are the Cartesian product

$$X_1 \times \cdots \times X_n$$

and the subspace S, respectively.

The corresponding conjugate concave and convex functions g_2 and g_1 in the Fenchel duality framework are

$$\inf_{x \in S} \lambda' x = \begin{cases} 0 & \text{if } \lambda \in S^\perp, \\ -\infty & \text{if } \lambda \notin S^\perp, \end{cases}$$

where $S^\perp$ is the orthogonal subspace of S, and

$$\sup_{x_i \in X_i} \left\{ \sum_{i=1}^{n} \big(x_i \lambda_i - f_i(x_i) \big) \right\} = \sum_{i=1}^{n} g_i(\lambda_i),$$

where, for each i,

$$g_i(\lambda_i) = \sup_{x_i \in X_i} \{ x_i \lambda_i - f_i(x_i) \}$$

is the conjugate convex function of (f_i, X_i). After a sign change which converts it to a minimization problem, the Fenchel dual problem (5.40) becomes

$$\begin{aligned} &\text{minimize } \sum_{i=1}^{n} g_i(\lambda_i) \\ &\text{subject to } \lambda \in S^\perp, \qquad \lambda_i \in \Lambda_i, \;\; i = 1, \ldots, n, \end{aligned} \tag{DMP}$$

where Λ_i is the domain of g_i,

$$\Lambda_i = \{\lambda_i \mid g_i(\lambda_i) < \infty\}.$$

Here, the dual problem has an optimal solution and there is no duality gap if the functions f_i are convex and one of the following two conditions holds:

(1) The relative interiors of S and $X_1 \times \cdots \times X_n$ intersect.

(2) The intervals X_i are closed and the functions f_i are convex over the entire real line.

These conditions correspond to the two conditions for no duality gap in Prop. 5.4.1. At least one of these conditions is satisfied by the great majority of practical monotropic programs.

Finally, let us consider the Lagrangian optimality condition, which together with primal and dual feasibility, guarantees that x^* is primal optimal, λ^* is dual optimal, and there is no duality gap. This condition takes the form [cf. Eq. (5.43)]

$$x_i^* = \arg \max_{x_i \in X_i} \{x_i \lambda_i^* - f_i(x_i)\}, \qquad \forall\, i = 1, \ldots, n. \tag{5.46}$$

We can express this condition in terms of the right and left derivatives of the function f_i over the set X_i. Denote

$$f_i^+(x_i) = \begin{cases} \infty & \text{if } x_i \text{ is a right endpoint of } X_i, \\ \lim_{\gamma \downarrow 0} \dfrac{f_i(x_i+\gamma) - f_i(x_i)}{\gamma} & \text{otherwise,} \end{cases}$$

$$f_i^-(x_i) = \begin{cases} -\infty & \text{if } x_i \text{ is a left endpoint of } X_i, \\ \lim_{\gamma \uparrow 0} \dfrac{f_i(x_i+\gamma) - f_i(x_i)}{\gamma} & \text{otherwise.} \end{cases}$$

Then the Lagrangian optimality condition (5.46) can be written as

$$f_i^-(x_i^*) \le \lambda_i^* \le f_i^+(x_i^*), \qquad \forall\, i = 1, \ldots, n,$$

as illustrated in Fig. 5.4.4.

The theory of monotropic programming can be developed under more refined assumptions than the ones used in this section, and a sharper duality result can be obtained. In fact, monotropic programming problems form the largest class of nonlinear programming problems, where equality of the primal and dual optimal values can be shown under conditions as weak as for linear programs. We state the corresponding result below, and for the proof we refer to the books [Roc84] and [Ber98], which develop monotropic programming in detail.

Proposition 5.4.3: Assume that the subsets of $\Re^2$

$$\{(x_i, \gamma) \mid x \in X_i,\, f_i(x_i) \le \gamma\}$$

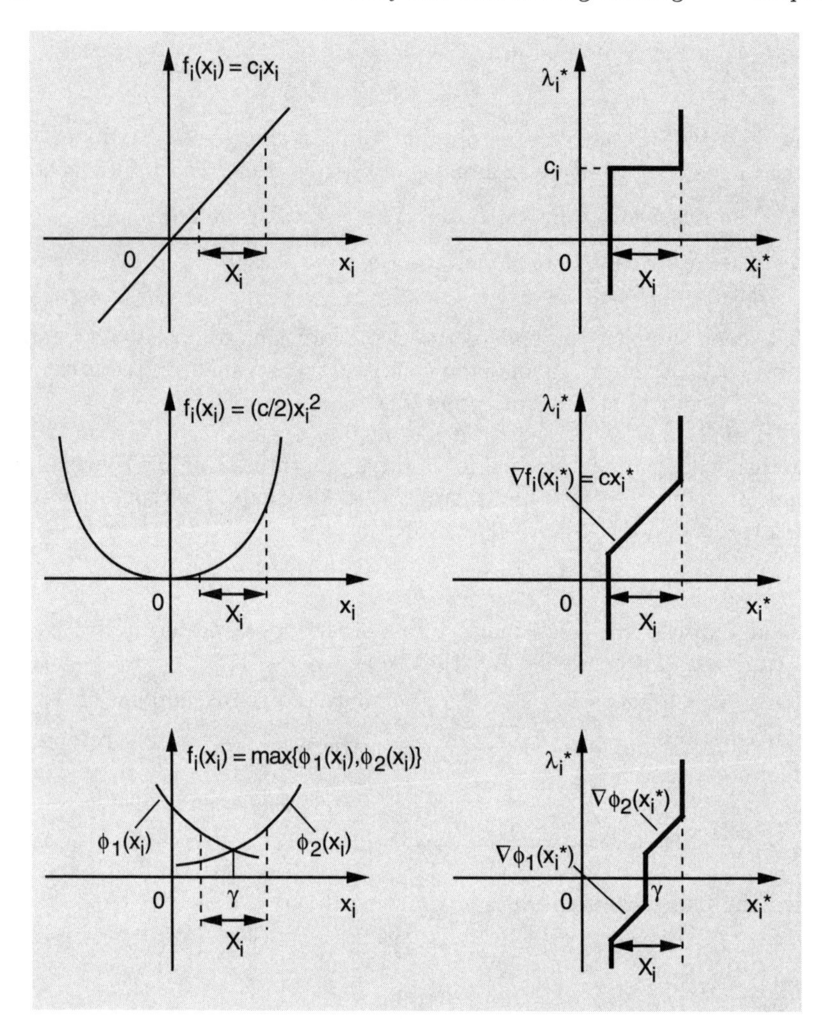

Figure 5.4.4. Illustration of the Lagrangian optimality condition

$$f_i^-(x_i^*) \leq \lambda_i^* \leq f_i^+(x_i^*), \qquad \forall\, i = 1, \ldots, n,$$

for several special cases of the function f_i. To satisfy this condition, the pair (x_i^*, λ_i^*) should lie on the thick-lined graph.

are closed and convex for all $i = 1, \ldots, n$. Then if there exists at least one feasible solution to the primal problem (MP), or at least one feasible solution to the dual problem (DMP), the optimal primal and dual costs are equal.

Note that part of the assertion of the above proposition is that if the primal problem is feasible but unbounded, then the dual problem is infeasible (the optimal costs of both problems are equal to $-\infty$), and that if the dual problem is feasible but unbounded, the primal problem is infeasible (the optimal costs of both problems are equal to ∞).

5.4.2 Network Optimization

We now consider an important special case of a monotropic programming problem. Consider a directed graph consisting of a set $\mathcal{N}$ of nodes and a set $\mathcal{A}$ of directed arcs. Letting x_{ij} be the flow of the arc (i,j), the problem is to minimize

$$\sum_{(i,j)\in\mathcal{A}} f_{ij}(x_{ij})$$

subject to the constraints

$$\sum_{\{j\mid(i,j)\in\mathcal{A}\}} x_{ij} - \sum_{\{j\mid(j,i)\in\mathcal{A}\}} x_{ji} = 0, \qquad \forall\, i\in\mathcal{N}, \tag{5.47}$$

and

$$b_{ij} \le x_{ij} \le c_{ij}, \qquad \forall\, (i,j)\in\mathcal{A}, \tag{5.48}$$

where b_{ij}, c_{ij} are given scalars, and $f_{ij} : \Re \mapsto \Re$ are convex functions.

The constraints (5.47), known as the *conservation of flow constraints*, specify that the total inflow into each node must be equal to the total outflow from that node. This formulation of the network optimization problem, where the right-hand side in the constraint (5.47) is zero for all nodes i, is called the *circulation* format. In a different formulation, the right-hand side of the constraint (5.47) may involve nonzero "supplies" or "demands" for some nodes i, but it is possible to convert the problem to the circulation format by introducing an artificial node to which all supplies and demands accumulate (see e.g., [Ber98]). In some formulations, the arc flow bounds (5.48) may not be present. The following development can be easily modified in this case, provided the directional derivatives of each arc cost function f_{ij} tend to $-\infty$ and ∞ as x_{ij} tends to $-\infty$ and ∞, respectively.

The problem is in the form of the monotropic program (MP), where the subspace S corresponds to the conservation of flow constraints (5.47). The orthogonal subspace $S^\perp$ is given by

$$S^\perp = \{\lambda \mid \lambda_{ij} = p_i - p_j \text{ for some scalars } p_i,\ i\in\mathcal{N}\},$$

so the dual problem is

$$\text{minimize} \sum_{(i,j)\in\mathcal{A}} g_{ij}(\lambda_{ij})$$

$$\text{subject to } \lambda_{ij} = p_i - p_j \text{ for some scalars } p_i,\ i\in\mathcal{N},$$

where

$$g_{ij}(\lambda_{ij}) = \max_{b_{ij} \le x_{ij} \le c_{ij}} \{\lambda_{ij} x_{ij} - f_{ij}(x_{ij})\}.$$

Based on the results for monotropic programming that we outlined in Section 5.4.1, it can be seen that if the network flow problem is feasible, there exists an optimal dual solution and there is no duality gap.

For network problems, it is customary to write the dual problem in terms of the variables p_i (also known as the node *prices* or *potentials*) in the following equivalent form:

$$\text{maximize} \sum_{(i,j)\in\mathcal{A}} q_{ij}(p_i - p_j)$$

$$\text{subject to no constraint on } p_i, \ i \in \mathcal{N},$$

where

$$q_{ij}(p_i - p_j) = \min_{b_{ij} \le x_{ij} \le c_{ij}} \{f_{ij}(x_{ij}) - (p_i - p_j) x_{ij}\}.$$

The form of the dual arc functions q_{ij} is illustrated in Fig. 5.4.5 for the case of a linear primal cost. Figure 5.4.6 illustrates the Lagrangian optimality condition

$$f_{ij}^-(x_{ij}^*) \le p_i^* - p_j^* \le f_{ij}^+(x_{ij}^*), \qquad \forall\ (i,j), \tag{5.49}$$

in the network context.

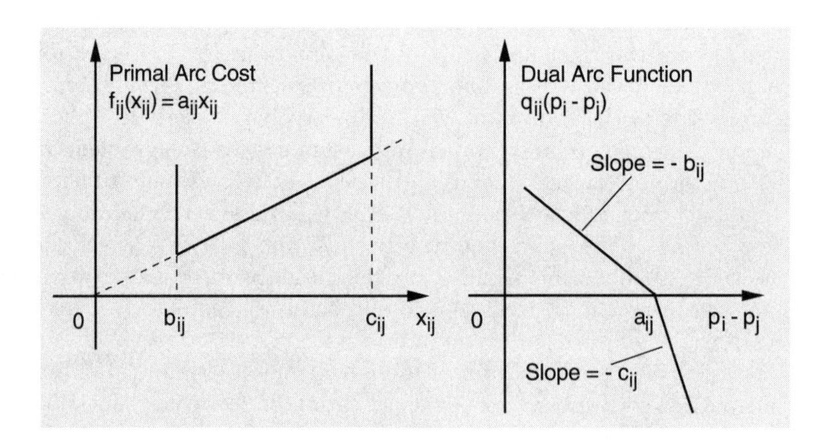

Figure 5.4.5. Illustration of primal and dual arc cost functions of the network optimization problem when the primal cost is linear.

An interesting interpretation is to view the graph as an electric circuit, where x_{ij} and p_i represents the current of arc (i, j) and the voltage of node i, respectively. In this context, the conservation of flow constraint (5.47) is equivalent to Kirchhoff's current law, while the Lagrangian optimality

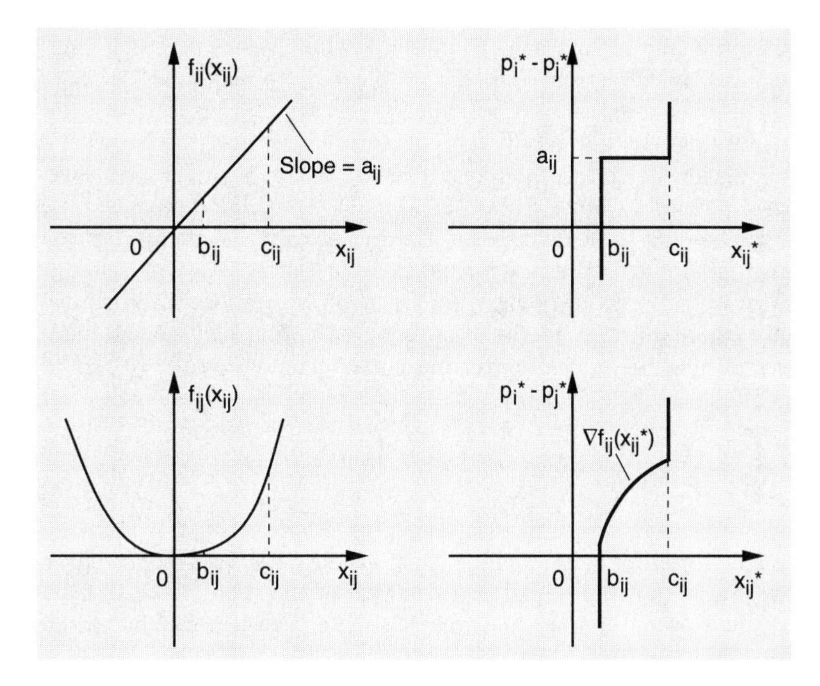

Figure 5.4.6. Illustration of the Lagrangian optimality condition

$$f_{ij}^-(x_{ij}^*) \le p_i^* - p_j^* \le f_{ij}^+(x_{ij}^*),$$

for some special cases of the network optimization problem. To satisfy this condition, the pair $(x_{ij}^*, p_i^* - p_j^*)$ should lie on the thick-lined graph.

condition (5.49) corresponds to Ohm's law, which specifies a monotonically increasing resistive relation between the current and the voltage differential of each arc.

The dual problem is particularly useful in the context of specialized numerical algorithms. We will discuss several general dual algorithmic approaches in Chapter 6. There are also several dual algorithms that are specialized to the network structure of the problem, which are both theoretically interesting and practically efficient. For a detailed account we refer to the author's network optimization textbook [Ber98].

5.4.3 Games and the Minimax Theorem

As another application of Fenchel duality, we develop a classical result of game theory. In the simplest type of zero sum game, there are two players: the first may choose one out of n moves and the second may choose one out of m moves. If moves i and j are selected by the first and the second player, respectively, the first player gives a specified amount a_{ij} to the second. The objective of the first player is to minimize the amount given to the other player, and the objective of the second player is to maximize this amount.

The players use mixed strategies, whereby the first player selects a probability distribution $x = (x_1, \ldots, x_n)$ over his n possible moves and the second player selects a probability distribution $z = (z_1, \ldots, z_m)$ over his m possible moves. Since the probability of selecting i and j is $x_i z_j$, the expected amount to be paid by the first player to the second is $\sum_{i,j} a_{ij} x_i z_j$ or $x'Az$, where A is the $n \times m$ matrix with elements a_{ij}. If each player adopts a worst case viewpoint, whereby he optimizes his choice against the worst possible selection by the other player, the first player must minimize $\max_z x'Az$ and the second player must maximize $\min_x x'Az$.

Let us denote

$$X = \left\{ x \;\middle|\; \sum_{i=1}^{n} x_i = 1,\ x \geq 0 \right\}, \quad Y = \left\{ y \;\middle|\; y = Az,\ \sum_{j=1}^{m} z_j = 1,\ z \geq 0 \right\}.$$

Since a selection of a probability distribution z by the second player has the same effect on the payoff as a selection of the vector $y = Az$ from the set Y, the first player's optimal value (or guaranteed upper bound on the expected amount he must pay) is

$$\min_{x \in X} \max_{y \in Y} x'y.$$

This is also the minimum amount that the first player would pay if the second player chooses his move optimally with knowledge of the first player's probability distribution x. The second player's optimal value (or guaranteed lower bound on the expected amount he will receive) is

$$\max_{y \in Y} \min_{x \in X} x'y.$$

This is also the maximum amount that the second player would obtain if the first player chooses his move optimally with knowledge of the second player's probability distribution z.

The main result here is that these two optimal values are equal, implying that there is an amount that can be meaningfully viewed as the value of the game for its participants. The following proposition proves and extends this result.

Proposition 5.4.4: (Minimax Theorem) Let X and Y be nonempty, convex, and compact subsets of $\Re^n$. Then

$$\min_{x \in X} \max_{y \in Y} x'y = \max_{y \in Y} \min_{x \in X} x'y.$$

Proof: We apply the Fenchel duality theory, with the identifications

$$f_1(x) = \max_{y \in Y} x'y, \qquad f_2(x) \equiv 0, \qquad X_1 = \Re^n, \qquad X_2 = X. \tag{5.50}$$

For each x, the maximum above is attained by Weierstrass' theorem (Prop. A.8 in Appendix A), since Y is compact. Hence $f_1(x)$ is finite for all x and convex over $\Re^n$ [Prop. B.2(d) in Appendix B], so that Assumption 5.4.1 is satisfied. Furthermore, the relative interiors of X_1 and X_2 intersect, because $ri(X_1) = \Re^n$ and $ri(X_2)$ is nonempty [the relative interior of every nonempty convex set is nonempty; see Prop. B.7(a)]. Therefore, there is no duality gap and we have [cf. Eq. (5.44)]

$$\min_{x \in X} \max_{y \in Y} x'y = \max_{\lambda \in \Lambda_1 \cap \Lambda_2} \big\{ g_2(\lambda) - g_1(\lambda) \big\}, \tag{5.51}$$

where g_1, g_2 and Λ_1, Λ_2 are given by Eqs. (5.37)-(5.39). Using the identifications (5.50), and the definitions of the conjugates [cf. Eqs. (5.37) and (5.38)], we have

$$g_2(\lambda) = \inf_{x \in X} \lambda' x, \tag{5.52}$$

$$\begin{aligned} g_1(\lambda) &= \sup_{x \in \Re^n} \big\{ \lambda' x - f_1(x) \big\} \\ &= \sup_{x \in \Re^n} \left\{ \lambda' x - \max_{y \in Y} x'y \right\} \\ &= - \inf_{x \in \Re^n} \left\{ \max_{y \in Y} x'y - x'\lambda \right\} \\ &= - \inf_{x \in \Re^n} \max_{y \in Y} x'(y - \lambda). \end{aligned} \tag{5.53}$$

If $\lambda \in Y$, then for every $x \in \Re^n$, we have

$$\max_{y \in Y} x'(y - \lambda) \geq 0, \qquad x \in \Re^n;$$

i.e. $\max_{y \in Y} x'(y - \lambda)$ is at least as large as zero, which is the value of $x'(y - \lambda)$ obtained for $y = \lambda$. Thus, the minimum of $\max_{y \in Y} x'(y - \lambda)$ is attained for $x = 0$ and we have

$$\inf_{x \in \Re^n} \max_{y \in Y} x'(y - \lambda) = 0, \qquad \text{if } \lambda \in Y. \tag{5.54}$$

If $\lambda \notin Y$, then because Y is convex and compact, there exists a hyperplane passing through λ, not intersecting Y, and such that Y is contained in one of the corresponding halfspaces; that is, there exists $\bar{x} \in \Re^n$ such that

$$\bar{x}'y < \bar{x}'\lambda, \qquad \forall \ y \in Y.$$

Therefore, since Y is compact,

$$\max_{y \in Y} \bar{x}'(y - \lambda) < 0,$$

so by taking $x = \alpha\bar{x}$ for α arbitrarily large, we see that

$$\inf_{x \in \Re^n} \max_{y \in Y} x'(y - \lambda) = -\infty, \qquad \text{if } \lambda \notin Y. \tag{5.55}$$

Combining Eqs. (5.53)-(5.55), we see that

$$g_1(\lambda) = \begin{cases} 0 & \text{if } \lambda \in Y, \\ \infty & \text{if } \lambda \notin Y. \end{cases}$$

Using this equation together with Eqs. (5.51) and (5.52), we obtain

$$\min_{x \in X} \max_{y \in Y} x'y = \max_{\lambda \in Y} \min_{x \in X} x'\lambda,$$

which by replacing λ with y, gives the desired result. **Q.E.D.**

5.4.4 The Primal Function

Let $f : \Re^n \mapsto \Re$ and $g_j :\mapsto \Re$ be given functions, and X be a given subset of $\Re^n$. Consider the family of problems

$$\begin{aligned} &\text{minimize } f(x) \\ &\text{subject to } x \in X, \qquad g_j(x) \le u_j, \quad j = 1, \ldots, r, \end{aligned}$$

parameterized by $u = (u_1, \ldots, u_r)$, and let $p(u)$ denote the corresponding optimal value, i.e.,

$$p(u) = \inf_{\substack{x \in X,\, g_j(x) \le u_j, \\ j = 1, \ldots, r}} f(x).$$

The function $p : \Re^r \mapsto [-\infty, \infty]$ is called the *primal function* of the problem

$$\begin{aligned} &\text{minimize } f(x) \\ &\text{subject to } x \in X, \qquad g_j(x) \le 0, \quad j = 1, \ldots, r, \end{aligned} \tag{5.56}$$

We encountered versions of this function in Chapter 3.

Let P be the *domain* of p, i.e., the set of all u for which the constraint set $\{x \in X \mid g_j(x) \le u_j,\, j = 1, \ldots, r\}$ is nonempty and, consequently, $p(u) < \infty$,

$$P = \{u \mid p(u) < \infty\}.$$

There are some interesting connections between the primal and dual functions. In particular, let us write for every $\mu \geq 0$,

$$
\begin{aligned}
q(\mu) &= \inf_{x \in X} \left\{ f(x) + \sum_{j=1}^{r} \mu_j g_j(x) \right\} \\
&= \inf_{\{(u,x) \mid u \in P,\, x \in X,\, g_j(x) \leq u_j,\, j=1,\ldots,r\}} \left\{ f(x) + \sum_{j=1}^{r} \mu_j g_j(x) \right\} \\
&= \inf_{\{(u,x) \mid u \in P,\, x \in X,\, g_j(x) \leq u_j,\, j=1,\ldots,r\}} \left\{ f(x) + \sum_{j=1}^{r} \mu_j u_j \right\} \\
&= \inf_{u \in P} \inf_{\substack{x \in X,\, g_j(x) \leq u_j, \\ j=1,\ldots,r}} \left\{ f(x) + \sum_{j=1}^{r} \mu_j u_j \right\},
\end{aligned}
$$

and finally

$$
q(\mu) = \inf_{u \in P} \{p(u) + \mu' u\}, \qquad \forall\, \mu \geq 0. \tag{5.57}
$$

Thus, we have $q(\mu) = -\sup_{u \in P}\big\{(-\mu)'u - p(u)\big\}$, implying that

$$
q(\mu) = -h(-\mu), \qquad \forall\, \mu \geq 0,
$$

where h is the conjugate convex function of (p, P).

Generally, neither the primal function nor its domain have any convexity properties. However, if we assume that X is convex, f and g_j are convex over X, and $p(u) > -\infty$ for all $u \in P$, it can be shown that P is a convex set and p is convex over P. The proof is the same as the one given in Prop. 3.4.3.

Under the preceding convexity assumptions, we can show that the structure of the primal function around $u = 0$ determines the existence of Lagrange multipliers. In particular, if μ^* is a Lagrange multiplier for problem (5.56), then we have $q(\mu^*) = f^* = p(0)$, and from Eq. (5.57) we see that

$$
p(0) \leq p(u) + u'\mu^*, \qquad \forall\, u \in \Re^r. \tag{5.58}
$$

This implies that $-\mu^*$ is a subgradient of $p(u)$ at 0. Conversely, if the above equation holds for some μ^*, then since $p(u)$ is monotonically nonincreasing with respect to the coordinates of u, we have $\mu^* \geq 0$ [otherwise the right side of Eq. (5.58) would be unbounded below]. Furthermore, from Eqs. (5.58) and (5.57), it follows that

$$
f^* = p(0) \leq \inf_{u \in P} \big\{p(u) + u'\mu^*\big\} = q(\mu^*),
$$

which, in view of the weak duality theorem (Prop. 5.1.3), implies that μ^* is a Lagrange multiplier. In conclusion, if the problem is feasible and p is convex and real-valued over P, we have

$$-\mu^* \text{ is a subgradient of } p \text{ at } u = 0$$
$$\iff \quad \mu^* \text{ is a Lagrange multiplier of the problem } \min_{x \in X,\, g(x) \le 0} f(x).$$

This result is consistent with the sensitivity analyses of Sections 3.2, 3.3, and 3.4, and characterizes the subdifferential of p at $u = 0$. More generally, by replacing the constraint $g(x) \le 0$ with a constraint $g(x) \le u$, where $u \in P$, we see that if p is convex and real-valued over P, we have

$$-\mu \text{ is a subgradient of } p \text{ at } u \in P$$
$$\iff \quad \mu \text{ is a Lagrange multiplier of the problem } \min_{x \in X,\, g(x) \le u} f(x).$$

The relation of the primal and the dual functions, and their subgradients is further discussed in Exercises 5.4.7 and 5.4.8.

5.4.5 A Dual View of Penalty Methods

In Section 4.3 we developed a local version of the theory of nondifferentiable exact penalty functions for nonconvex problems. In this section, we develop a generalized version of this theory for convex problems by using the Fenchel duality theorem.

We consider the problem

$$\begin{aligned} &\text{minimize } \ f(x) \\ &\text{subject to } \ x \in X, \qquad g_j(x) \le 0, \quad j = 1, \ldots, r, \end{aligned} \tag{5.59}$$

where $f : \Re^n \mapsto \Re$, $g_j : \Re^n \mapsto \Re$ are convex functions, and X is a convex subset of $\Re^n$. We denote

$$g(x) = \big(g_1(x), \ldots, g_r(x)\big).$$

We assume that the optimal value f^* is finite and that there is no duality gap. We also assume that the constraint set X is compact; at the end of this section, we will discuss what happens when X is not compact.

We consider penalty functions $P : \Re^r \mapsto \Re$ that are convex and satisfy

$$P(u) = 0, \qquad \forall\ u \le 0, \tag{5.60}$$

$$P(u) > 0, \qquad \text{if } u_j > 0 \text{ for some } j = 1, \ldots, r. \tag{5.61}$$

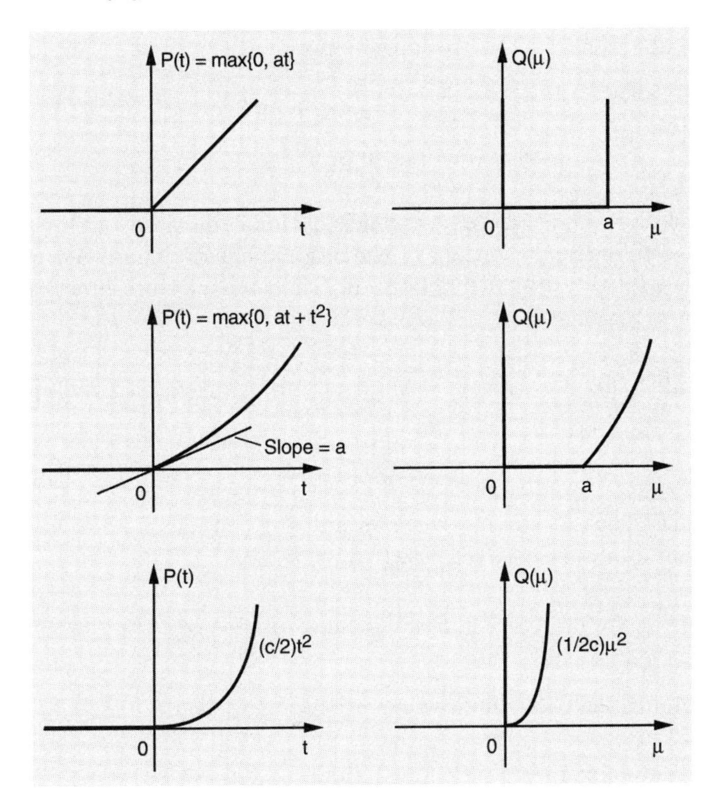

Figure 5.4.7. Illustration of conjugates of various penalty functions.

Such functions impose a penalty only for infeasible points. An example is the quadratic penalty function for the case where the multiplier vector μ is zero:

$$P(u) = \frac{c}{2} \sum_{j=1}^{r} \bigl(\max\{0, u_j\}\bigr)^2,$$

where c is the penalty parameter (see Section 4.2). The conjugate convex function of P is given by

$$Q(\mu) = \sup_{u \in \Re^r} \bigl\{u'\mu - P(u)\bigr\},$$

and it can be seen that

$$Q(\mu) \geq 0, \qquad \forall\ \mu \in \Re^r,$$

$$Q(\mu) = \infty, \qquad \text{if } \mu_j < 0 \text{ for some } j = 1, \ldots, r.$$

Some interesting penalty functions P are shown in Fig. 5.4.7, together with their conjugates.

We now consider the "penalized" problem

$$\begin{aligned} &\text{minimize} \;\; f(x) + P\big(g(x)\big) \\ &\text{subject to} \;\; x \in X. \end{aligned} \tag{5.62}$$

We will derive conditions under which optimal solutions of this problem are related to optimal solutions of the original constrained problem (5.59).

Let p be the primal function and let q be the dual function of the constrained problem (5.59). We have

$$\begin{aligned} \min_{x\in X}\big\{f(x) + P\big(g(x)\big)\big\} &= \min_{u\in\Re^r} \min_{x\in X,\, g(x)\le u} \big\{f(x) + P\big(g(x)\big)\big\} \\ &= \min_{u\in\Re^r} \big\{p(u) + P(u)\big\}, \end{aligned}$$

and we use Fenchel's duality theorem and the conjugacy relation between primal and dual functions [cf. Eq. (5.57)] to write

$$\min_{u\in\Re^r} \big\{p(u) + P(u)\big\} = \max_{\mu\ge 0}\big\{q(\mu) - Q(\mu)\big\}.$$

By combining the preceding two equations, we obtain

$$\min_{x\in X}\big\{f(x) + P\big(g(x)\big)\big\} = \max_{\mu\ge 0}\big\{q(\mu) - Q(\mu)\big\}; \tag{5.63}$$

see Fig. 5.4.8. Note that the conditions for application of the Fenchel duality theorem are satisfied, including the requirement that the optimal value of the problem on the left-hand side of Eq. (5.63) be finite, which is implied by the compactness of X. Note also that the maximum is attained for some $\mu \ge 0$ above because the domain of the penalty function P is the entire space $\Re^r$, so the relative interiors of the domains of p and P have nonempty intersection.

It can be seen from Fig. 5.4.8 that in order for the penalized problem (5.62) to have the same optimal value as the original constrained problem (5.59), the conjugate Q must be "flat" along a sufficiently large "area," the size of which is related to the "size" of the dual optimal solutions. In particular, we have the following proposition.

Proposition 5.4.5:

(a) In order for the penalized problem (5.62) and the original constrained problem (5.59) to have equal optimal values, it is necessary and sufficient that there exists a Lagrange multiplier μ^* such that

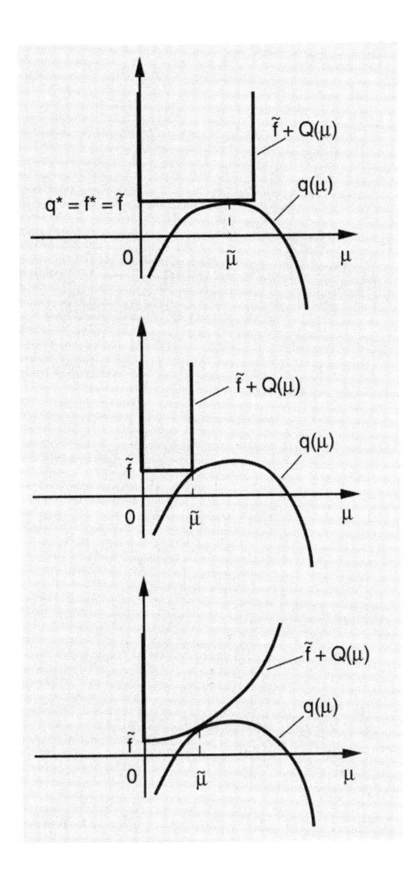

Figure 5.4.8. Illustration of the duality relation (5.63), and the optimal values of the penalized and the dual problem. Here f^* is the optimal value of the original problem, which is assumed to be equal to the optimal dual value q^*, while $\tilde{f}$ is the optimal value of the penalized problem,

$$\tilde{f} = \min_{x \in X} \big\{ f(x) + P\big(g(x)\big) \big\}.$$

The point of contact of the graphs of the functions $\tilde{f} + Q(\mu)$ and $q(\mu)$ corresponds to the vector $\tilde{\mu}$ that attains the maximum in the relation

$$\tilde{f} = \max_{\mu \geq 0} \big\{ q(\mu) - Q(\mu) \big\}.$$

$$u'\mu^* \leq P(u), \qquad \forall\, u \in \Re^r. \tag{5.64}$$

(b) In order for some optimal solution of the penalized problem (5.62) to be an optimal solution of the constrained problem (5.59), it is necessary that there exists a Lagrange multiplier μ^* such that

$$u'\mu^* \leq P(u), \qquad \forall\, u \in \Re^r.$$

(c) In order for the penalized problem (5.62) and the constrained problem (5.59) to have the same set of optimal solutions, it is sufficient that there exists a Lagrange multiplier μ^* such that

$$u'\mu^* < P(u), \qquad \forall\, u \in \Re^r \text{ with } u_j > 0 \text{ for some } j. \tag{5.65}$$

Proof: (a) Since $P(g(x)) = 0$ for all feasible solutions of the constrained problem (5.59), we have using the Fenchel duality relation (5.63)

$$f^* \geq \min_{x \in X} \big\{ f(x) + P\big(g(x)\big) \big\} = \max_{\mu \geq 0} \big\{ q(\mu) - Q(\mu) \big\}.$$

Thus $f^* = \min_{x \in X}\{f(x) + P(g(x))\}$ if and only if $f^* = q(\mu^*) - Q(\mu^*)$ for all vectors μ^* attaining the maximum above. Since $Q(\mu^*) \geq 0$ for all $\mu^* \geq 0$ and we have $f^* \geq q(\mu^*)$ with equality if and only if μ^* is a Lagrange multiplier, it follows that $f^* = \min_{x \in X}\{f(x) + P(g(x))\}$ if and only if there is a Lagrange multiplier μ^* satisfying $Q(\mu^*) = 0$. Using the definition of Q, it is seen that the relation $Q(\mu^*) = 0$ is equivalent to the desired relation (5.64).

(b) If x^* is a common optimal solution of problems (5.59) and (5.62), then by feasibility of x^*, we have $P(g(x^*)) = 0$, so these two problems have equal optimal values. The result then follows from part (a).

(c) If x^* is an optimal solution of the constrained problem (5.59), then $P(g(x^*)) = 0$, so we have

$$f^* = f(x^*) = f(x^*) + P(g(x^*)) \geq \min_{x \in X} \{f(x) + P(g(x))\}.$$

By part (a), the given condition (5.65) implies that equality holds throughout in the above relation, showing that x^* is also an optimal solution of the penalized problem (5.62).

Conversely, if x^* is an optimal solution of the penalized problem (5.62), then x^* is either feasible [satisfies $g(x^*) \leq 0$], in which case it is an optimal solution of the constrained problem (5.59) [in view of $P(g(x)) = 0$ for all feasible vectors x], or it is infeasible in which case $g_j(x^*) > 0$ for some j. In the latter case, by using the given condition (5.65), it follows that there exists an $\epsilon > 0$ such that

$$g(x^*)'\mu^* + \epsilon < P\big(g(x^*)\big).$$

Let $\tilde{x}$ be a feasible vector such that $f(\tilde{x}) \leq f^* + \epsilon$. Since $P(g(\tilde{x})) = 0$ and $f^* = \min_{x \in X}\{f(x) + g(x)'\mu^*\}$, we obtain

$$f(\tilde{x}) + P\big(g(\tilde{x})\big) = f(\tilde{x}) \leq f^* + \epsilon \leq f(\tilde{x}) + g(\tilde{x})'\mu^* + \epsilon.$$

By combining the last two equations, we obtain

$$f(\tilde{x}) + P\big(g(\tilde{x})\big) < f(x^*) + P\big(g(x^*)\big),$$

which contradicts the hypothesis that x^* is an optimal solution of the penalized problem (5.62). This completes the proof. **Q.E.D.**

Note that in the case where the necessary condition (5.64) holds but the sufficient condition (5.65) does not, it is possible that the constrained problem (5.59) has optimal solutions that are not optimal solutions of the penalized problem (5.62), even though the two problems have the same optimal value. This can be seen from Fig. 5.4.8 and corresponds to the case where the conjugate Q is "flat" over an area that includes some but not all Lagrange multipliers.

As an application of Prop. 5.4.5, consider the penalty function

$$P(u) = c \sum_{j=1}^{r} \max\{0, u_j\},$$

where $c > 0$. The condition $u'\mu^* \leq P(u)$ for all $u \in \Re^r$ [cf. Eq. (5.64)], is equivalent to

$$\mu_j^* \leq c, \qquad \forall\ j = 1, \ldots, r.$$

Similarly, the condition $u'\mu^* < P(u)$ for all $u \in \Re^r$ with $u_j > 0$ for some j [cf. Eq. (5.65)], is equivalent to

$$\mu_j^* < c, \qquad \forall\ j = 1, \ldots, r.$$

Finally, consider the penalty function

$$P(u) = c \max\{0, u_1, \ldots, u_r\}$$

that we used in Section 4.3 in connection with sequential quadratic programming. The condition

$$u'\mu^* \leq P(u), \qquad \forall\ u \in \Re^r$$

[cf. Eq. (5.64)], is equivalent to

$$\sum_{j=1}^{r} \mu_j^* u_j \leq c \max\{0, u_1, \ldots, u_r\}, \qquad \forall\ u \in \Re^r,$$

which with a little thought is seen to be equivalent to

$$\sum_{j=1}^{r} \mu_j^* \leq c.$$

Similarly, the condition (5.65) that guarantees that the optimal solution sets of the constrained problem (5.59) and the penalized problem (5.62) are the same, is equivalent to

$$\sum_{j=1}^{r} \mu_j^* < c.$$

This result is consistent with the results obtained in Section 4.3, but is considerably stronger, which is not unexpected since convexity of the cost and the constraints is assumed.

In the analysis of the present section we have assumed that X is a compact set because for application of the Fenchel duality theorem we need the optimal value

$$\tilde{f} = \inf_{x \in X} \{f(x) + P\big(g(x)\big)\} \tag{5.66}$$

of the penalized problem (5.62) to be finite. The compactness assumption is unnecessary, however, for parts (a) and (c) of Prop. 5.4.5. In particular, if X is nonempty but not compact and the other assumptions of this subsection hold, then we can consider separately the two cases $\tilde{f} > -\infty$ and $\tilde{f} = -\infty$. In the case $\tilde{f} > -\infty$, the proofs of parts (a) and (c) of Prop. 5.4.5 go through with essentially no change. In the case $\tilde{f} = -\infty$, we have

$$\sup_{\mu \geq 0} \big\{q(\mu) - Q(\mu)\big\} = -\infty, \tag{5.67}$$

because the two problems in Eqs. (5.66) and (5.67) are dual to each other and weak duality holds. From Eq. (5.67), it can be seen that for each $\mu \geq 0$ we must have either $q(\mu) = -\infty$ or $Q(\mu) = \infty$. Using this fact, it can be seen that parts (a) and (c) of Prop. 5.4.5 are correct as stated.

5.4.6 The Proximal and Entropy Minimization Algorithms

The duality correspondence developed in the preceding subsection can be generalized considerably by introducing more general penalty functions that do not satisfy the conditions (5.60) and (5.61). In particular, let us consider the problem

$$\begin{aligned} &\text{minimize} \quad f(x) \\ &\text{subject to} \quad x \in X, \qquad g_j(x) \leq 0, \quad j = 1, \ldots, r, \end{aligned} \tag{5.68}$$

under the convexity assumptions of the preceding subsection. Consider also the penalized problem that arises in the method of multipliers of Section 4.2

$$\begin{aligned} &\text{minimize} \quad L_{c^k}(x, \mu^k) = f(x) + \frac{1}{2c^k} \sum_{j=1}^{r} \Big\{ \big(\max\{0, \mu_j^k + c^k g_j(x)\}\big)^2 - (\mu_j^k)^2 \Big\} \\ &\text{subject to} \quad x \in X. \end{aligned} \tag{5.69}$$

This corresponds to the penalty function

$$P_{c^k}(u; \mu^k) = \frac{1}{2c^k} \sum_{j=1}^{r} \Big\{ \big(\max\{0, \mu_j^k + c^k u_j\}\big)^2 - (\mu_j^k)^2 \Big\}.$$

The conjugate of this function can be verified to be

$$Q_{c^k}(\mu;\mu^k) = \begin{cases} \frac{1}{2c^k}\|\mu-\mu^k\|^2 & \text{if } \mu \geq 0, \\ \infty & \text{otherwise.} \end{cases}$$

By using the Fenchel duality theorem similar to Eq. (5.63), the minimization of the augmented Lagrangian can be written as

$$\begin{aligned} \min_{x\in X} L_{c^k}(x,\mu^k) &= \min_{u\in\Re^r}\{p(u)+P_{c^k}(u;\mu^k)\} \\ &= \max_{\mu\geq 0}\left\{q(\mu)-\frac{1}{2c^k}\|\mu-\mu^k\|^2\right\}, \end{aligned} \tag{5.70}$$

where q is the dual function of problem (5.68). The maximization above can be visualized as in Fig. 5.4.9, and it can be seen that the optimal value is near the maximal value of the dual function if c^k is large and/or μ^k is near a Lagrange multiplier of problem (5.68). This is consistent with our conclusions regarding the quadratic augmented Lagrangian in Section 4.2.

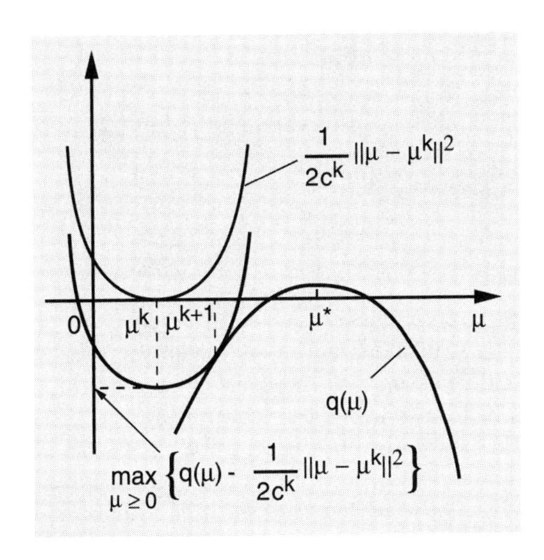

Figure 5.4.9. Dual view of the minimization of the augmented Lagrangian and the first order multiplier method.

Let x^k attain the minimum of the augmented Lagrangian in (5.69). Then $u^k = g(x^k)$ attains the minimum in Eq. (5.70), and from the Lagrangian optimality condition (5.45), the vector μ^{k+1} that attains the maximum in Eq. (5.70) satisfies

$$\mu^{k+1} = \arg\max_{\mu\geq 0}\left\{\mu' u^k - \frac{1}{2c^k}\|\mu-\mu^k\|^2\right\}.$$

Equivalently,

$$\mu_j^{k+1} = \max\{0, \mu_j^k + c^k u^k\} = \max\{0, \mu_j^k + c^k g_j(x^k)\}.$$

This is precisely the first order multiplier iteration for inequality constraints (see Section 4.2).

We have thus reached the conclusion that the first order multiplier method of Section 4.2 can equivalently be written as

$$\mu^{k+1} = \arg\max_{\mu \geq 0} \left\{ q(\mu) - \frac{1}{2c^k} \|\mu - \mu^k\|^2 \right\}. \tag{5.71}$$

Conversely, the iteration above when applied to any function q that is the dual function of a convex programming problem is equivalent to the first order method of multipliers applied to that problem.

The preceding observation motivates us also to consider the problem

$$\begin{aligned} &\text{minimize} \quad F(x) \\ &\text{subject to} \quad x \in X, \end{aligned}$$

where $F : \Re^n \mapsto \Re$ is a convex function, and X is a nonempty closed convex set, and the iteration

$$x^{k+1} = \arg\min_{x \in X} \left\{ F(x) + \frac{1}{2c^k} \|x - x^k\|^2 \right\}. \tag{5.72}$$

This is known as the *proximal minimization algorithm*, and we have seen that it is essentially equivalent to the first order method of multipliers [the only difference is the presence of the constraint $\mu \geq 0$ in Eq. (5.71), but if we consider equality constraints, the maximization in Eq. (5.71) becomes unconstrained]. We have also interpreted the algorithm as a gradient projection method (Exercise 2.3.8 of Section 2.3) and as a block coordinate descent method (Exercise 2.7.1 of Section 2.7), under various assumptions on f. The chief utility of the algorithm is that it regularizes the original problem in the sense that the quadratic term $\|x - x^k\|^2$ makes the cost function of problem (5.72) strictly convex.

The proximal minimization algorithm has a number of interesting convergence and rate of convergence properties (a detailed textbook analysis is given in Section 3.4 of [BeT89]). Chief among its properties is that it converges to some global minimum of $F(x)$ over X provided at least one such minimum exists. Furthermore, when F is linear and X is polyhedral (a linear program), the algorithm converges to an optimal solution in a finite number of iterations (see Exercise 5.4.16). Figure 5.4.10 interprets graphically the convergence process of the algorithm.

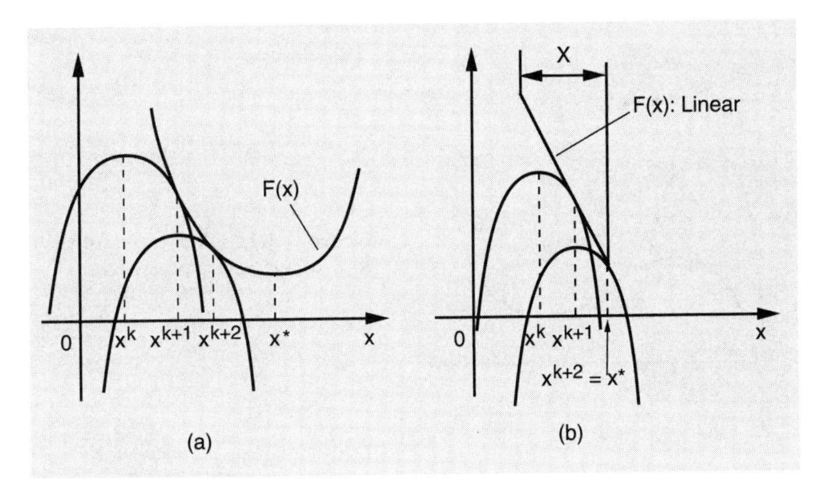

Figure 5.4.10. Illustration of the convergence of the proximal minimization algorithm. When F is linear and X is polyhedral as in (b), convergence occurs in a finite number of iterations.

The Entropy Minimization Algorithm

When the exponential penalty function is used in place of the quadratic, the preceding analysis can be used to show that the dual counterpart of the exponential method of multipliers is the so called *entropy minimization algorithm*, which is given by

$$x^{k+1} = \arg \min_{x \in X} \left\{ F(x) + \frac{1}{c^k} \sum_{i=1}^{n} x_i \left(\ln \left(\frac{x_i}{x_i^k} \right) - 1 \right) \right\}.$$

This algorithm is illustrated in Fig. 5.4.11, and can be used to solve the problem

$$\begin{aligned} &\text{minimize} \quad F(x) \\ &\text{subject to} \quad x \in X, \end{aligned}$$

where $F : \Re^n \mapsto \Re$ is a convex function, and X is a nonempty closed subset of the nonnegative orthant.

The convergence properties of the entropy minimization algorithm are quite similar to those of the proximal minimization algorithm. However, the analysis is considerably more difficult because when one of the coordinates x_i^k tends to zero, the fraction of the corresponding entropy term tends to infinity. We refer to the literature cited at the end of the chapter.

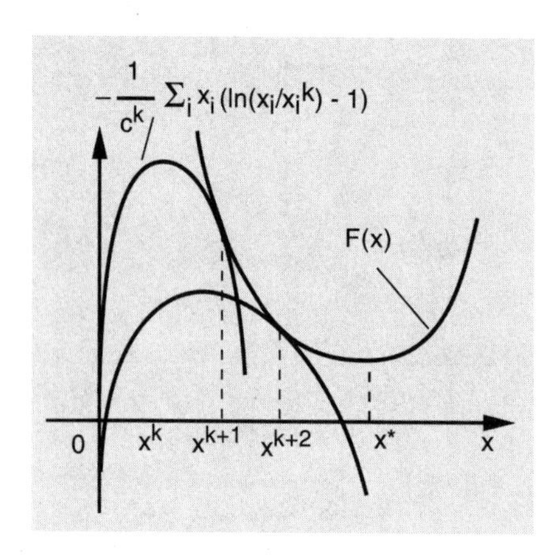

Figure 5.4.11. Illustration of the entropy minimization algorithm.

EXERCISES

5.4.1

Consider the problem

$$\begin{aligned} &\text{minimize} \quad \sum_{i=1}^{n} f_i(x_i) \\ &\text{subject to} \quad Ax = b, \qquad x_i \in X_i, \qquad i = 1, \ldots, n. \end{aligned}$$

Convert this problem into a monotropic programming problem by introducing an additional vector z that is constrained by the equation $z = b$. Show that a dual problem is given by

$$\begin{aligned} &\text{minimize} \quad \sum_{i=1}^{n} g_i(\lambda_i) - b'\xi \\ &\text{subject to} \quad \lambda = A'\xi, \qquad \lambda_i \in \Lambda_i, \qquad i = 1, \ldots, n, \end{aligned}$$

where $\Lambda_i = \{\lambda_i \mid g_i(\lambda_i) < \infty\}$.

5.4.2

Consider the variation of the monotropic programming problem

$$\begin{aligned} &\text{minimize} \quad \sum_{i=1}^{n} f_i(x_i) \\ &\text{subject to} \quad x \in M, \qquad x_i \in X_i, \quad i = 1, \ldots, n, \end{aligned}$$

where M is a linear manifold given by $M = \bar{x} + S$, where $\bar{x}$ is a given vector and S is a subspace. Derive a dual problem by translating the origin to the vector $\bar{x}$.

5.4.3 (www)

Consider the problem

$$\text{minimize } f(x)$$
$$\text{subject to } x \in X \cap C,$$

where X is a convex subset of $\Re^n$, C is a cone in $\Re^n$, and $f : \Re^n \mapsto \Re$ is convex over X. Derive conditions under which we have

$$\inf_{x \in X \cap C} f(x) = \max_{-\lambda \in C^\perp} -g(\lambda),$$

where $C^\perp$ is the polar cone of C and g is defined by

$$g(\lambda) = \sup_{x \in X} \{\lambda' x - f(x)\}.$$

5.4.4 (Extension of the Fenchel Duality Theorem [Roc70])

Consider the problem

$$\text{minimize } f_1(x) - f_2(Ax)$$
$$\text{subject to } x \in X_1, \qquad Ax \in X_2,$$

where A is an $m \times n$ matrix, X_1 and X_2 are convex subsets of $\Re^n$ and $\Re^m$, respectively, and f_1 and f_2 are real-valued functions that are convex over X_1 and X_2, respectively. Assume that this problem has a feasible solution and a finite optimal value, denoted f^*. Derive conditions under which we have

$$\min_{x \in X_1,\, Ax \in X_2} \{f_1(x) - f_2(Ax)\} = \max_{A'\lambda \in \Lambda_1,\, \lambda \in \Lambda_2} \{g_2(\lambda) - g_1(A'\lambda)\}, \tag{5.73}$$

where the functions $g_1 : \Re^m \mapsto (-\infty, \infty]$ and $g_2 : \Re^n \mapsto [-\infty, \infty)$ are defined by

$$g_1(\xi) = \sup_{x \in X_1} \{x'\xi - f_1(x)\},$$

$$g_2(\lambda) = \inf_{z \in X_2} \{z'\lambda - f_2(z)\},$$

and the sets Λ_1 and Λ_2 are given by

$$\Lambda_1 = \{\xi \mid g_1(\xi) < \infty\}, \qquad \Lambda_2 = \{\lambda \mid g_2(\lambda) > -\infty\}.$$

Derive also primal feasibility, dual feasibility, and Lagrangian optimality conditions, which are necessary and sufficient for x^* and λ^* to be optimal for the two problems in Eq. (5.73).

5.4.5 (Smoothing of Nondifferentiabilities [Ber75c], [Ber77], [Ber82a], [Pap81], [Pol79]) (www)

A simple and effective technique to handle nondifferentiabilities in the cost or the constraints of optimization problems is to replace them by smooth approximations and then use gradient-based algorithms. This exercise develops a general technique for deriving such approximations. Let $\gamma : \Re^n \mapsto \Re$ be a convex real-valued function with conjugate convex function given by

$$g(\lambda) = \sup_{u \in \Re^n} \{u'\lambda - \gamma(u)\}.$$

For each $t \in \Re^n$, define

$$P_c(t) = \inf_{u \in \Re^n} \{\gamma(t-u) + \tfrac{c}{2}\|u\|^2\},$$

where c is a positive constant.

(a) Use the Fenchel duality theorem to show that

$$P_c(t) = \sup_{\lambda \in \Re^n} \{t'\lambda - g(\lambda) - \tfrac{1}{2c}\|\lambda\|^2\}.$$

Show also that P_c is convex and differentiable, and that the gradient $\nabla P_c(t)$ is the unique vector λ attaining the supremum in the above equation. *Hint*: Interpret $P_c(t)$ as the primal function of a suitable problem, and use the theory of Section 5.4.4.

(b) Derive $g(\lambda)$ and $P_c(t)$ for the following functions, the first three of which are defined on $\Re$,

$$\gamma(t) = \max\{0, t\},$$

$$\gamma(t) = |t|,$$

$$\gamma(t) = \frac{s}{2}t^2, \qquad (s \text{ is a positive scalar}),$$

$$\gamma(t) = \max\{t_1, \ldots, t_n\}.$$

(c) Show that $P_c(t)$ is real-valued and that for all t, $\lim_{c \to \infty} P_c(t) = \gamma(t)$, so the accuracy of the approximation can be controlled using c. Verify graphically this property for the first three functions in (b).

(d) Repeat (a) and prove results analogous to (b) and (c), when P_c is instead defined by

$$P_c(t) = \inf_{u \in \Re^n} \{\gamma(t-u) + y'u + \tfrac{c}{2}\|u\|^2\},$$

where y is some fixed vector in $\Re^n$.

5.4.6 (Symmetry of the Conjugacy Operation) (www)

Let $f : \Re^n \mapsto \Re$ be a function and let X be a nonempty subset of $\Re^n$. Define the conjugate convex function

$$g(\lambda) = \sup_{x \in X} \big\{x'\lambda - f(x)\big\}$$

and the set D over which g is finite

$$D = \big\{\lambda \mid g(\lambda) < \infty\big\}.$$

Define also the conjugate of the conjugate

$$\tilde{f}(x) = \sup_{\lambda \in D} \big\{\lambda' x - g(\lambda)\big\}$$

and the set $\tilde{X}$ over which $\tilde{f}$ is finite

$$\tilde{X} = \big\{x \mid \tilde{f}(x) < \infty\big\}.$$

(a) Show that

$$\tilde{X} \supset X, \qquad \tilde{f}(x) \leq f(x), \quad \forall\, x \in X.$$

(b) Show that if the epigraph of (f, X), i.e., the subset of $\Re^{n+1}$

$$\big\{(x, \gamma) \mid x \in X,\, f(x) \leq \gamma\big\},$$

is closed and convex, then

$$\tilde{X} = X, \qquad \tilde{f}(x) = f(x), \quad \forall\, x \in X.$$

Hints: For a closed convex set C in $\Re^{n+1}$:

(1) Use the fact that C is the intersection of all the halfspaces that contain C (Prop. B.15), to show that if C contains no vertical lines [sets of the form $\{(z, w) \mid w \in \Re\}$, where z is a fixed vector in $\Re^n$], then C is contained in a halfspace corresponding to a nonvertical hyperplane; that is, there exists (ξ, ζ) such that $\xi \in \Re^n$, $\zeta \in \Re$, $\zeta \neq 0$, and $\beta \in \Re$ such that

$$\xi' z + \zeta w \geq \beta, \qquad \forall\, (z, w) \in C.$$

(2) Show that if C is closed and contains no vertical lines and $(\bar{z}, \bar{w})$ does not belong to C, there exists a nonvertical hyperplane strictly separating $(\bar{z}, \bar{w})$ from C. [Take a suitable convex combination with the nonvertical hyperplane of hint (1).]

5.4.7 (Symmetry of Duality) (www)

Consider the primal function

$$p(u) = \inf_{x \in X,\, g(x) \leq u} f(x)$$

of the problem

$$\begin{aligned} &\text{minimize} \;\; f(x) \\ &\text{subject to} \;\; x \in X, \qquad g_j(x) \leq 0, \quad j = 1, \ldots, r, \end{aligned} \tag{5.74}$$

Consider also the problem

$$\begin{aligned} &\text{minimize} \;\; p(u) \\ &\text{subject to} \;\; u \in P, \qquad u \leq 0, \end{aligned} \tag{5.75}$$

where P is the domain of p,

$$P = \big\{ u \mid \text{ there exists } x \in X \text{ with } g(x) \leq u \big\}.$$

Assume that $-\infty < p(0) < \infty$.

(a) Show that problems (5.74) and (5.75) have equal optimal values and equal sets of Lagrange multipliers.

(b) Consider the dual functions of problems (5.74) and (5.75) and show that they are equal on the positive orthant, i.e., for all $\mu \geq 0$,

$$q(\mu) = \inf_{x \in X} \left\{ f(x) + \sum_{j=1}^{r} \mu_j g_j(x) \right\} = \inf_{u \in P} \big\{ p(u) + \mu' u \big\}.$$

(c) Assume that the epigraph

$$\big\{ (u, \gamma) \mid p(u) \leq \gamma \big\}$$

is a closed and convex subset of $\Re^{r+1}$. Show that u^* is an optimal solution of problem (5.75) if and only if $-u^*$ is a Lagrange multiplier for the dual problem

$$\begin{aligned} &\text{maximize} \;\; q(\mu) \\ &\text{subject to} \;\; \mu \geq 0, \end{aligned}$$

in the sense that

$$q^* = \sup_{\mu \geq 0} \big\{ q(\mu) - \mu' u^* \big\}.$$

Hint: Use the result of the preceding exercise.

5.4.8 (www)

Consider the problem

$$\begin{aligned} &\text{minimize} \quad f(x) \\ &\text{subject to} \quad \|x\| \le 1, \end{aligned}$$

where the function $f : \Re^n \mapsto \Re$ is given by

$$f(x) = \max_{y \in Y} x'y,$$

and Y is a convex and compact set. Show that this problem can be solved as a problem of projection on the set Y.

5.4.9

Show that

$$\min_{x \in X} \max_{y \in Y} \big\{x'y + F(x) - G(y)\big\} = \max_{y \in Y} \min_{x \in X} \big\{x'y + F(x) - G(y)\big\},$$

where X and Y are convex and compact subsets of $\Re^n$, F is a continuous convex function over X, and G is a continuous concave function over Y.

5.4.10 (Network Optimization and Kirchhoff's Laws)

Consider a linear resistive electric network with node set $\mathcal{N}$ and arc set $\mathcal{A}$. Let v_i be the voltage of node i and let x_{ij} be the current of arc (i, j). Kirchhoff's current law says that for each node i, the total outgoing current is equal to the total incoming current

$$\sum_{\{j \mid (i,j) \in \mathcal{A}\}} x_{ij} = \sum_{\{j \mid (j,i) \in \mathcal{A}\}} x_{ji}.$$

Ohm's law says that the current x_{ij} and the voltage drop $v_i - v_j$ along each arc (i, j) are related by

$$v_i - v_j = R_{ij} x_{ij} - t_{ij},$$

where $R_{ij} \ge 0$ is a resistance parameter and t_{ij} is another parameter that is nonzero when there is a voltage source along the arc (i, j) (t_{ij} is positive if the voltage source pushes current in the direction from i to j).

(a) Consider the problem

$$\begin{aligned} &\text{minimize} \quad \sum_{(i,j) \in \mathcal{A}} \big(\tfrac{1}{2} R_{ij} x_{ij}^2 - t_{ij} x_{ij}\big) \\ &\text{subject to} \quad \sum_{\{j \mid (i,j) \in \mathcal{A}\}} x_{ij} = \sum_{\{j \mid (j,i) \in \mathcal{A}\}} x_{ji}, \quad \forall\, i \in \mathcal{N}. \end{aligned} \tag{5.76}$$

(The quadratic cost above has an energy interpretation.) Give a set of necessary and sufficient conditions for a set of variables $\{x_{ij} \mid (i,j) \in \mathcal{A}\}$ and $\{v_i \mid i \in \mathcal{N}\}$ to be an optimal solution-Lagrange multiplier pair for this problem.

(b) Show that if a set of currents $\{x_{ij} \mid (i,j) \in \mathcal{A}\}$ and voltages $\{v_i \mid i \in \mathcal{N}\}$, satisfy Kirchhoff's current law and Ohm's law, then $\{x_{ij} \mid (i,j) \in \mathcal{A}\}$ solve problem (5.76).

5.4.11

Consider the case where in Prop. 5.4.5 the function P has the form

$$P(z) = \sum_{j=1}^{r} P_j(z_j),$$

where $P_j : \Re \mapsto \Re$ are convex real-valued functions satisfying

$$P_j(z_j) = 0, \quad \forall\ z_j \le 0, \qquad P_j(z_j) > 0, \quad \forall\ z_j > 0.$$

Show that the conditions (5.64) and (5.65) of Prop. 5.4.5 are equivalent to

$$\mu_j^* \le \lim_{z_j \to 0+} \frac{P_j(z_j)}{z_j}, \qquad j = 1, \ldots, r,$$

and

$$\mu_j^* < \lim_{z_j \to 0+} \frac{P_j(z_j)}{z_j}, \qquad j = 1, \ldots, r,$$

respectively.

5.4.12 (Geometric Interpretation of the Exponential Method of Multipliers [KoB72]) (www)

Consider the primal function

$$p(u) = \inf_{\substack{g_j(x) \le u_j,\ j=1,\ldots,r, \\ x \in X}} f(x)$$

and the following "penalized" primal function

$$p_c(u) = \inf_{\substack{\frac{1}{c}\psi(cg_j(x)) \le u_j,\ j=1,\ldots,r, \\ x \in X}} f(x)$$

where $\psi(\cdot)$ belongs to the class of penalty functions of Section 4.2.5 (cf. the exponential method of multipliers), and $c > 0$ is the penalty parameter. Consider Fig. 5.4.12, which provides an interpretation of the multiplier iteration

$$\bar{\mu}_j = \mu_j \nabla\psi\big(cg_j(\bar{x})\big), \qquad j = 1, \ldots, r,$$

where $\bar{x}$ minimizes

$$L_c(x,\mu) = f(x) + \frac{1}{c}\sum_{j=1}^{r} \psi\big(cg_j(x)\big).$$

State conditions under which the primal functions p and p_c are convex as in the figure, and for which we have

$$q_c(\mu) < q(\bar{\mu}) < q_c(\bar{\mu}),$$

as indicated in the figure, where

$$q(\mu) = \inf_{x\in X} L(x,\mu), \qquad q_c(\mu) = \inf_{x\in X} L_c(x,\mu).$$

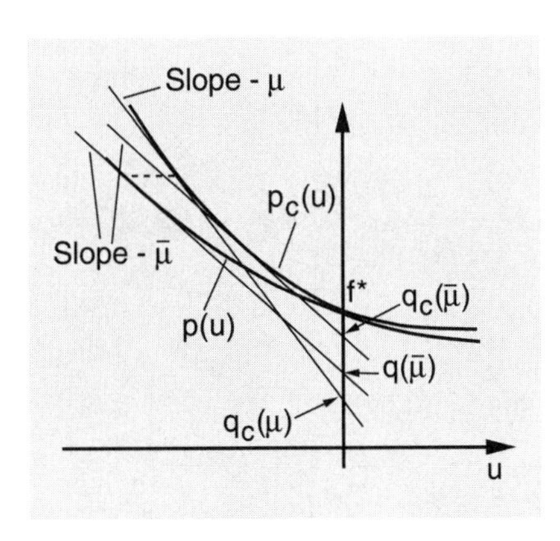

Figure 5.4.12. Geometric interpretation of a nonquadratic method of multipliers in terms of the primal function p and a "penalized" primal function p_c; compare with the geometric interpretations of the method of multipliers in Section 4.2.2.

5.4.13 (Convexification of Nonconvex Problems [Ber79a]) 

Consider the problem

$$\begin{aligned} &\text{minimize } f(x) \\ &\text{subject to } Ax = b \end{aligned}$$

where $f(x)$ is a quadratic but not necessarily positive semidefinite function, and suppose that it has a unique optimal solution x^*. Consider the iteration

$$x^{k+1} = \arg\min_{Ax=b} \left\{ f(x) + \frac{1}{2c}\|x - x^k\|^2 \right\}, \tag{5.77}$$

and show that it converges to x^*. What is the connection with the proximal minimization algorithm? *Note*: The idea of this exercise extends to problems where f is nonquadratic and the constraint is nonlinear. The algorithm (5.77) is useful if there is a particularly advantageous method for finding the minimum in Eq. (5.77), which, however, requires convexity of the cost function $f(x) + (1/2c)\|x - x^k\|^2$. This requirement can be met by choosing c sufficiently small. Prominent examples are methods that rely on separability of the cost and the constraints (see the next chapter). For an analysis in the case of nonquadratic cost and a nonlinear equality constraint, see [Ber79a]. Related methods are discussed in [StW75] and [TaM85].

5.4.14 (Partial Proximal Minimization Algorithm [BeT94]) 

For each $c > 0$, let F_c be the real-valued convex function on $\Re^n$ defined by

$$F_c(x) = \min_{y \in X} \left\{ f(y) + \frac{1}{2c} \|y - x\|^2 \right\},$$

where f is a convex function over the closed convex set X. Let I be a subset of the index set $\{1, \ldots, n\}$. For any $x \in \Re^n$, consider a vector $\bar{x}$ satisfying

$$\bar{x} \in \arg\min_{y \in X} \left\{ f(y) + \frac{1}{2c} \sum_{i \in I} |y_i - x_i|^2 \right\},$$

and let $\tilde{x}$ be the vector with components

$$\tilde{x}_i = \begin{cases} x_i & \forall\, i \in I, \\ \bar{x}_i & \forall\, i \notin I. \end{cases}$$

(a) Show that

$$\tilde{x} \in \arg \min_{\{y \mid y_i = x_i,\, i \in I\}} F_c(y),$$

$$\bar{x} = \arg\min_{y \in X} \left\{ f(y) + \frac{1}{2c} \|y - \tilde{x}\|^2 \right\}.$$

(b) Interpret $\bar{x}$ as the result of a coordinate descent step followed by a proximal minimization step, and show that $F_c(\bar{x}) \le f(\bar{x}) \le F_c(x) \le f(x)$.

5.4.15 (Convergence Rate of the Proximal Minimization Algorithm [KoB76])

Consider the problem of minimizing $F(x)$ over $x \in X$, where $F : \Re^n \mapsto \Re$ is a convex function, and X is a nonempty closed convex set. Assume that the set X^* of minimizing points is nonempty and that there exist $\beta > 0$, $\delta > 0$, and $\alpha \in (1, 2]$ such that

$$F^* + \beta\big(\rho(x)\big)^\alpha \le F(x), \qquad \forall\, x \in X \text{ with } \rho(x) \le \delta,$$

where $F^* = \min_{x \in X} F(x)$ and

$$\rho(x) = \min_{x^* \in X^*} \|x - x^*\|.$$

Let $\{x^k\}$ be a sequence generated by the proximal minimization algorithm

$$x^{k+1} = \arg\min_{x \in X} \left\{ F(x) + \frac{1}{2c^k} \|x - x^k\|^2 \right\},$$

and assume that $\lim_{k\to\infty} \rho(x^k) = 0$ and that $x^k \notin X^*$ for all k. Show that:

(a) If $\alpha < 2$ and for some $\underline{c} > 0$ we have $c^k \geq \underline{c}$ for all k, then

$$\limsup_{k\to\infty} \frac{\rho(x^{k+1})}{\big(\rho(x^k)\big)^{1/(\alpha-1)}} < \infty,$$

so the convergence of $\rho(x^k)$ is superlinear of order $1/(\alpha - 1)$. *Hint*: Part (a) and the following parts (b) and (c) are based on the relation

$$\rho(x^{k+1}) + \beta c^k \big(\rho(x^{k+1})\big)^{\alpha-1} \leq \rho(x^k). \tag{5.78}$$

To show this relation, let $\hat{x}$ denote the projection of any x on the closed and convex set X^* and let $d = \hat{x}^{k+1} - x^{k+1}$. Consider the function

$$H(\gamma) = F(x^{k+1} + \gamma d) + \frac{1}{2c^k} \|x^{k+1} + \gamma d - x^k\|^2.$$

Since H is minimized at $\gamma = 0$, its right derivative $H^+(0)$ is nonnegative, from which we obtain

$$\begin{aligned} 0 \leq H^+(0) &= F'(x^{k+1}; d) + \frac{1}{c^k}(x^{k+1} - x^k)'(\hat{x}^{k+1} - x^{k+1}) \\ &\leq F^* - F(x^{k+1}) + \frac{1}{c^k}(x^{k+1} - x^k)'(\hat{x}^{k+1} - x^{k+1}). \end{aligned}$$

Using the hypothesis, it follows that

$$\beta c^k \big(\rho(x^{k+1})\big)^{\alpha} \leq (x^{k+1} - x^k)'(\hat{x}^{k+1} - x^{k+1}),$$

for k sufficiently large. We now add to both sides $(x^{k+1} - \hat{x}^k)'(x^{k+1} - \hat{x}^{k+1})$ and we use the fact

$$\|x^{k+1} - \hat{x}^{k+1}\|^2 \leq (x^{k+1} - \hat{x}^k)'(x^{k+1} - \hat{x}^{k+1}),$$

(which follows from the projection theorem) to obtain

$$\|x^{k+1} - \hat{x}^{k+1}\|^2 + \beta c^k \big(\rho(x^{k+1})\big)^{\alpha} \leq \|x^k - \hat{x}^k\| \|x^{k+1} - \hat{x}^{k+1}\|,$$

from which the desired relation follows using the assumption $\rho(x^{k+1}) > 0$.

(b) If $\alpha = 2$ and $\lim_{k\to\infty} c^k = \bar{c} < \infty$ then

$$\limsup_{k\to\infty} \frac{\rho(x^{k+1})}{\rho(x^k)} \leq \frac{1}{1 + \beta\bar{c}}.$$

(c) If $\alpha = 2$ and $\lim_{k\to\infty} c^k = \infty$, $\rho(x^k)$ converges superlinearly.

(d) If F is a positive definite quadratic function, $X = \Re^n$, $\alpha = 2$ and $\lim_{k\to\infty} c^k = \bar{c} < \infty$, then

$$\limsup_{k\to\infty} \frac{\rho(x^{k+1})}{\rho(x^k)} \leq \frac{1}{1 + 2\beta\bar{c}}.$$

Show by example that this estimate is tight. *Hint*: Let x^* be the minimizing point of F over $x \in \Re^n$, and let $\tilde{y}$ denote the unique vector that minimizes $\beta\|x - x^*\|^2 + (1/2c)\|x - y\|^2$ over $x \in \Re^n$. Show that $y - x^* = (1 + 2\beta c)(\tilde{y} - x^*)$, and that $\|x(y,c) - x^*\| \leq \|\tilde{y} - x^*\|$, where $x(y,c)$ denotes the unique vector that minimizes $F(x) + (1/2c)\|x - y\|^2$ over $x \in \Re^n$.

5.4.16 (Finite Convergence of the Proximal Minimization Algorithm [Ber75a])

Consider the proximal minimization algorithm as in Exercise 5.4.15, and assume that X^* is nonempty. Assume further that there exists a scalar $\beta > 0$ such that

$$F^* + \beta\rho(x) \leq F(x), \qquad \forall\, x \in X,$$

where F^* and $\rho(x)$ are as in Exercise 5.4.15. (This assumption is satisfied if F is linear and X is polyhedral.) Show that if $\sum_{k=0}^{\infty} c^k = \infty$, the algorithm converges to X^* finitely [that is, there exists $\bar{k} > 0$ such that $x^k \in X^*$ for all $k \geq \bar{k}$]. Furthermore, for a given x^0, the algorithm converges in a single iteration if c^0 is sufficiently large. *Hint*: Prove the following analog of Eq. (5.78) for $\alpha = 1$ and $x^{k+1} \notin X^*$:

$$\rho(x^{k+1}) + \beta c^k \leq \rho(x^k).$$

5.4.17 (Second-Order Cone Programming) (www)

Consider the problem

$$\begin{aligned} &\text{minimize} \quad c'x \\ &\text{subject to} \quad \|A_j x + b_j\| \leq e_j' x + d_j, \quad j = 1, \ldots, r, \end{aligned}$$

where $x \in \Re^n$, and c, A_j, b_j, e_j, and d_j are given and have appropriate dimension. Assume that the problem is feasible. Consider the equivalent problem

$$\begin{aligned} &\text{minimize} \quad c'x \\ &\text{subject to} \quad \|u_j\| \leq t_j, \quad u_j = A_j x + b_j, \quad t_j = e_j' x + d_j, \quad j = 1, \ldots, r, \end{aligned} \tag{5.79}$$

where u_j and t_j are auxiliary optimization variables.

(a) Show that problem (5.79) has cone constraints of the type described in Exercise 5.4.3.

(b) Use the duality theory of Exercise 5.4.3 to show that a dual problem is given by

$$\begin{aligned} &\text{maximize} \quad -\sum_{j=1}^{r}(b_j' z_j + d_j w_j) \\ &\text{subject to} \quad \sum_{j=1}^{r}(A_j' z_j + e_j w_j) = c, \qquad \|z_j\| \leq w_j, \quad j = 1, \ldots, r. \end{aligned} \tag{5.80}$$

Furthermore, show that there is no duality gap if either there exists a feasible solution of problem (5.79) or a feasible solution of problem (5.80) satisfying strictly all the corresponding inequality constraints.

5.4.18 (Quadratically Constrained Quadratic Problems [LVB98]) (www)

Consider the quadratically constrained quadratic problem

$$\begin{aligned} &\text{minimize} \quad x' P_0 x + 2q_0' x + r_0 \\ &\text{subject to} \quad x' P_i x + 2q_i' x + r_i \leq 0, \quad i = 1, \ldots, p, \end{aligned}$$

where $P_0, P_1, \ldots, P_p$ are symmetric positive definite matrices. Show that the problem can be converted to one of the type described in Exercise 5.4.17 and derive the corresponding dual problem. *Hint*: Consider the equivalent problem

$$\begin{aligned} &\text{minimize} \quad \|P_0^{1/2} x + P_0^{-1/2} q_0\| \\ &\text{subject to} \quad \|P_i^{1/2} x + P_i^{-1/2} q_i\| \leq (r_i - q_i' P_i^{-1} q_i)^{1/2}, \quad i = 1, \ldots, p. \end{aligned}$$

5.4.19 (Minimizing the Sum or the Maximum of Norms [LVB98]) (www)

Consider the problems

$$\begin{aligned} &\text{minimize} \quad \sum_{i=1}^{p} \|F_i x + g_i\| \\ &\text{subject to} \quad x \in \Re^n, \end{aligned} \tag{5.81}$$

and

$$\begin{aligned} &\text{minimize} \quad \max_{i=1,\ldots,p} \|F_i x + g_i\| \\ &\text{subject to} \quad x \in \Re^n, \end{aligned} \tag{5.82}$$

where F_i and g_i are given matrices and vectors, respectively. Convert these problems to second-order cone programming problems (cf. Exercise 5.4.17) and derive the corresponding dual problems.

5.4.20 (Complex l_1 and l_∞ Approximation [LVB98]) 

Consider the complex l_1 approximation problem

$$\begin{aligned} &\text{minimize} \quad \|Ax - b\|_1 \\ &\text{subject to} \quad x \in \mathcal{C}^n, \end{aligned}$$

where $\mathcal{C}^n$ is the set of n-dimensional vectors with complex components. Show that it is a special case of problem (5.81) and derive the corresponding dual problem. Repeat for the complex l_∞ approximation problem

$$\begin{aligned} &\text{minimize} \quad \|Ax - b\|_\infty \\ &\text{subject to} \quad x \in \mathcal{C}^n. \end{aligned}$$

5.5 DISCRETE OPTIMIZATION AND DUALITY

Many optimization problems, in addition to the usual equality and inequality constraints, involve integer constraints. For example, the variables x_i may be constrained to be 0 or 1. Problems where some or all of the optimization variables take values in a finite set are generally referred to as *discrete optimization problems*. There is a large variety of practical problems of this type, arising for example in scheduling, resource allocation, and engineering design. The methodology for their solution is quite diverse, but an important subset of this methodology relies on the solution of continuous optimization subproblems, as well as on duality. In this section, we explore this interface between discrete and continuous optimization, and we illustrate it through analysis, algorithms, and paradigms.

We first discuss, in Section 5.5.1, some examples of discrete optimization problems that are suitable for the use of duality, and provide vehicles for illustrating the methodology to be discussed later. Then, in Section 5.5.2, we describe the branch-and-bound method, which is in principle capable of producing an optimal solution to an integer-constrained problem. This method relies on upper and lower bound estimates of the optimal cost of various problems that are derived from the given problem. Usually, the upper bounds are obtained with various heuristics, while the lower bounds are obtained through *integer constraint relaxation* (neglecting the integer constraints) or through *Lagrangian relaxation* (using the weak duality theorem).

The methods for obtaining lower bounds are elaborated on in Section 5.5.3, and the Lagrangian relaxation method is discussed in detail. This method requires the optimization of nondifferentiable functions, and some of the major relevant algorithms, subgradient and cutting plane methods, will be discussed in Chapter 6. The effectiveness of duality-based methods for discrete optimization is greatly enhanced when there is a small duality gap. As discussed in Section 5.1.6, separability, the structural property that most enhances duality, also tends to diminish the size of the duality gap in relative terms, particularly when the number of variables is much larger than the number of constraints that couple the variables.

The material in this section is fairly elementary. It is based on the main duality ideas of Section 5.1, but does not rely on the material of the subsequent Sections 5.2-5.4. However, it does require some background on directed graphs and associated optimization concepts, since many of the interesting discrete optimization problems involve a graph structure. The author's text on the subject [Ber98] provides an introductory treatment, as do several other sources given at the end of the chapter.

5.5.1 Examples of Discrete Optimization Problems

There is a very large variety of integer-constrained optimization problems. Furthermore, small changes in the problem formulation can often make a significant difference in the character of the solution. As a result, it is not easy to provide a taxonomy of the major problems of interest. It is helpful, however, to study in some detail a few representative examples that can serve as paradigms when dealing with other problems that have similar structure.

Example 5.5.1. Integer-Constrained Network Optimization

Let us consider network optimization problems, various forms of which we have already discussed several times so far. In particular, in Section 2.1, we discussed convex multicommodity flow problems arising in communication and transportation networks, and in Section 5.4.2 we considered a convex network problem that is a special case of a monotropic programming problem. Here we will focus on the combinatorial aspects of linear cost network flow problems where integer constraints are involved.

Consider a directed graph consisting of a set $\mathcal{N}$ of nodes and a set $\mathcal{A}$ of directed arcs. The flow x_{ij} of each arc (i,j) *is constrained to be an integer*. The problem is to minimize

$$\sum_{(i,j)\in\mathcal{A}} a_{ij}x_{ij} \tag{5.83}$$

subject to the constraints

$$\sum_{\{j\mid(i,j)\in\mathcal{A}\}} x_{ij} - \sum_{\{j\mid(j,i)\in\mathcal{A}\}} x_{ji} = s_i, \qquad \forall\, i\in\mathcal{N}, \tag{5.84}$$

and

$$b_{ij} \le x_{ij} \le c_{ij}, \qquad \forall\,(i,j)\in\mathcal{A}, \tag{5.85}$$

where a_{ij} are given scalars, and s_i, b_{ij}, and c_{ij} are given integers.

We refer to the constraints (5.84) and (5.85) as the *conservation of flow constraints*, and the *flow bound constraints*, respectively. The conservation of flow constraint at a node i expresses the requirement that the difference between the total flows coming into and out of i must be equal to the given amount s_i, which may be viewed as a *supply* provided by node i to the outside world.

For a typical application of the minimum cost flow problem, think of the nodes as locations (cities, warehouses, or factories) where a certain product is produced or consumed. Think of the arcs as transportation links between the locations, each with transportation cost a_{ij} per unit transported. The problem then is to move the product from the production points to the consumption points at minimum cost while observing the capacity constraints of the transportation links. However, the minimum cost flow problem has many applications that are well beyond the transportation context just described, as will be seen from the following examples. These examples illustrate how some important discrete/combinatorial problems can be modeled as minimum cost flow problems, and highlight the important connection between continuous and discrete network optimization.

Let us consider the classical *shortest path problem*. Here, to each arc (i, j) of a graph, we assign a scalar cost a_{ij} and we define the cost of a path consisting of a sequence of arcs $(i_1, i_2), (i_2, i_3), \ldots, (i_{n-1}, i_n)$ to be the sum of the costs of the arcs $\sum_{k=1}^{n_1} a_{i_k i_{k+1}}$. Given an origin node s and a destination node t, the shortest path problem is to find a path that starts at s, ends at t, and has minimum cost.

It is possible to cast this problem as the following network optimization problem with integer constraints:

$$\begin{aligned} &\text{minimize} \quad \sum_{(i,j)\in\mathcal{A}} a_{ij}x_{ij} \\ &\text{subject to} \quad \sum_{\{j|(i,j)\in\mathcal{A}\}} x_{ij} - \sum_{\{j|(j,i)\in\mathcal{A}\}} x_{ji} = \begin{cases} 1 & \text{if } i = s, \\ -1 & \text{if } i = t, \\ 0 & \text{otherwise,} \end{cases} \\ &\qquad\qquad x_{ij} = 0 \text{ or } 1, \qquad \forall\ (i,j) \in \mathcal{A}. \end{aligned} \tag{5.86}$$

Indeed, because of the form of the conservation of flow constraint, it can be seen that the feasible solutions x are associated with the paths P from s to t via the one-to-one correspondence

$$x_{ij} = \begin{cases} 1 & \text{if } (i,j) \text{ belongs to } P, \\ 0 & \text{otherwise.} \end{cases}$$

Furthermore, the cost of x is the length of the corresponding path P. Thus, the path corresponding to an optimal solution of the network optimization problem (5.86) is shortest.

Another interesting combinatorial optimization problem is the *assignment problem*. Here we have n persons and n objects, which we have to match on a one-to-one basis. There is a benefit or value a_{ij} for matching person i

with object j, and we want to assign persons to objects so as to maximize the total benefit. There is also a restriction that person i can be assigned to object j only if (i,j) belongs to a given set of pairs $\mathcal{A}$. Mathematically, we want to find a set of person-object pairs $(1,j_1),\ldots,(n,j_n)$ from $\mathcal{A}$ such that the objects $j_1,\ldots,j_n$ are all distinct, and the total benefit $\sum_{i=1}^n a_{ij_i}$ is maximized.

To formulate the assignment problem as a network optimization problem with integer constraints, we introduce the graph shown in Fig. 5.5.1. Here, there are $2n$ nodes divided into two groups: n corresponding to persons and n corresponding to objects. Also, for every possible pair $(i,j)\in\mathcal{A}$, there is an arc connecting person i with object j. The variable x_{ij} is the flow of arc (i,j), and is constrained to be either 1 or 0, indicating that person i is or is not assigned to person j, respectively. The constraint that each person/node i must be assigned to some object can be expressed as

$$\sum_{\{j\mid(i,j)\in\mathcal{A}\}} x_{ij}=1,$$

while the constraint that each object/node j must be assigned to some person can be expressed as

$$\sum_{\{i\mid(i,j)\in\mathcal{A}\}} x_{ij}=1.$$

Finally, we may view $(-a_{ij})$ as the cost coefficient of the arc (i,j) (by reversing the sign of a_{ij}, we convert the problem from a maximization to a minimization problem). Thus the problem is

$$\begin{aligned}
\text{minimize}\quad & \sum_{(i,j)\in\mathcal{A}} (-a_{ij})x_{ij}\\
\text{subject to}\quad & \sum_{\{j\mid(i,j)\in\mathcal{A}\}} x_{ij}=1, \qquad \forall\, i=1,\ldots,n,\\
& \sum_{\{i\mid(i,j)\in\mathcal{A}\}} x_{ij}=1, \qquad \forall\, j=1,\ldots,n,\\
& x_{ij}=0 \text{ or } 1, \qquad \forall\,(i,j)\in\mathcal{A},
\end{aligned} \tag{5.87}$$

which is a special case of the network optimization problem (5.83)-(5.85).

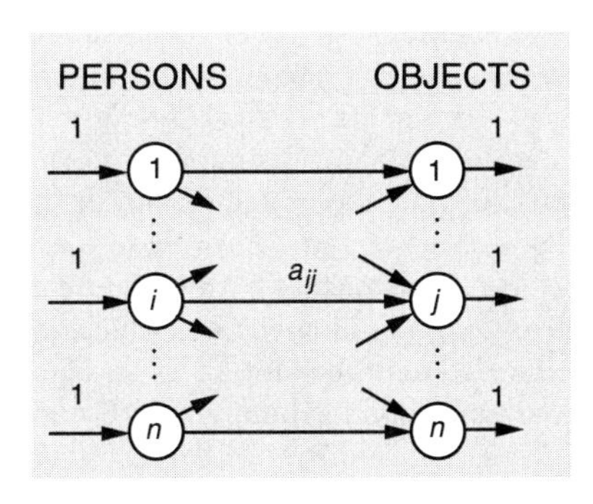

Figure 5.5.1. The graph representation of an assignment problem.

The most important property of the network optimization problem (5.83)-(5.85) is that *the integer constraints can be neglected.* By this we mean that the *relaxed* problem, i.e., the linear program of minimizing the cost $\sum_{(i,j)} a_{ij}x_{ij}$ subject to the conservation of flow and bound constraints (5.84) and (5.85), but *without* the integer constraints, has the same optimal value as the integer-constrained original. This remarkable fact will be shown as a consequence of a structural property of the network problem, called *unimodularity*, which is discussed in the following example.

Example 5.5.2. Unimodular Problems

An important fact about linear programs of the form

$$\begin{aligned} &\text{minimize } a'x \\ &\text{subject to } Ex = d, \quad b \leq x \leq c, \end{aligned}$$

is that if they attain an optimal solution, they attain one among the set of extreme points of the constraint polyhedron

$$P = \{x \mid Ex = d,\, b \leq x \leq c\}.$$

This is the fundamental theorem of linear programming, given as Prop. B.21(c) in Appendix B. Thus if we knew that all the extreme points of P are integer, we would be assured that the above linear program and its integer-constrained version, where the components of x are further constrained to be integer, share some optimal solutions. Furthermore, given the integer-constrained version of the problem, we could neglect the integer constraints, solve the resulting linear program with a method that finds an optimal extreme point (such as the simplex method), and thus solve the original integer-constrained version.

It turns out that the extreme points of a polyhedron $P = \{x \mid Ex = d,\, b \leq x \leq c\}$ are integer provided that the components of E, b, c, and d are integer, and the matrix E has a property called *total unimodularity*. To introduce this property, let us say that a square matrix A with integer components is *unimodular* if its determinant is 0, 1, or -1. Unimodularity can be used to assert the integrality of solutions of linear systems of equations. To see this, note that by Kramer's rule, if A is invertible and unimodular, then the inverse matrix A^{-1} has integer components. Therefore, the unique solution x of the system $Ax = b$ is integer for every integer vector b.

A rectangular matrix with integer components is called *totally unimodular* if each of its square submatrices is unimodular. Using the property of unimodular matrices just described, we can show the integrality of all the extreme points (vertices) of a polyhedron of the form $\{x \mid Ex = d,\, b \leq x \leq c\}$, where E is totally unimodular and b, c, and d are vectors with integer components. This follows from standard linear programming theory, which asserts that every extreme point $\hat{x}$ of this polyhedron is the solution of an equation of the form $\hat{E}\hat{x} = \hat{d}$, where $\hat{E}$ is a square invertible submatrix of E and $\hat{d}$ is a subvector of d (see Prop. B.21).

It turns out that the conservation of flow constraints (5.84) of the network optimization problem (5.83)-(5.85) involve a matrix that is totally unimodular, and as a result, the extreme points of the associated constraint polyhedron are integer. This matrix, denoted E is the, so-called, *arc incidence matrix* of the underlying graph. It has a row for each node and a column for each arc. The component corresponding to the ith row and a given arc is a 1 if the arc is outgoing from i, is a -1 if the arc is incoming to i, and is a 0 otherwise. Indeed, we can show that the determinant of each square submatrix of E is 0, 1, or -1 by induction on the dimension of the submatrix. In particular, the submatrices of dimension 1 of E are the scalar components of E, which are 0, 1, or -1. Suppose that the determinant of each square submatrix of dimension $n \geq 1$ is 0, 1, or -1. Consider a square submatrix of dimension $n+1$. If this matrix has a column with all components 0, the matrix is singular, and its determinant is 0. If the matrix has a column with a single nonzero component (a 1 or a -1), by expanding its determinant along that component and using the induction hypothesis, we see that the determinant is 0, 1, or -1. Finally, if each column of the matrix has two components (a 1 and a -1), the sum of its rows is 0, so the matrix is singular, and its determinant is 0.

The network optimization problem (5.83)-(5.85) is the most important example of an integer programming problem with polyhedral constraints that has the unimodular property. However, there are other interesting examples for which we refer to specialized sources, such as [NeW88] and [Sch86].

Example 5.5.3. Generalized Assignment and Facility Location Problems

Consider a problem of assigning m jobs to n machines. If job i is performed at machine j, it costs a_{ij} and requires t_{ij} time units. We want to find a minimum cost assignment of the jobs to the machines, given the total available time T_j at machine j. We assume that each job must be performed in its entirety at a single machine, and this introduces a combinatorial character to the problem.

Let us introduce for each job i and machine j, a variable x_{ij} which takes the value 1 or 0 depending on whether the job is or is not assigned to machine j, respectively. Then the problem takes the form

$$
\begin{aligned}
\text{minimize} \quad & \sum_{i=1}^{m} \sum_{j=1}^{n} a_{ij} x_{ij} \\
\text{subject to} \quad & \sum_{j=1}^{n} x_{ij} = 1, \qquad i = 1, \ldots, m, \\
& \sum_{i=1}^{m} t_{ij} x_{ij} \leq T_j, \qquad j = 1, \ldots, n, \\
& x_{ij} = 0 \text{ or } 1, \qquad i = 1, \ldots, m, \; j = 1, \ldots, n.
\end{aligned}
$$

The constraint $\sum_{j=1}^{n} x_{ij} = 1$ specifies that each job i must be assigned to some machine, and the constraint $\sum_{i=1}^{m} t_{ij} x_{ij} \leq T_j$ specifies that the total working time of machine j must not exceed the available T_j.

When there is only one machine, the index j is superfluous, and the problem becomes equivalent to the classical *knapsack problem* (cf. Exercise 5.1.7). Here we want to place in a knapsack the most valuable subcollection out of a given collection of objects, subject to a total weight constraint

$$\sum_{i=1}^{m} w_i x_i \leq T,$$

where T is a given total weight threshold, w_i is the weight of object i, and x_i is a variable which is 1 or 0 depending on whether the ith object is placed in the knapsack or not. The value to be maximized is $\sum_{i=1}^{m} v_i x_i$, where v_i is the value of the ith object.

For another variant of the generalized assignment problem, let us introduce a setup cost b_j for using machine j, that is, for having $\sum_{i=1}^{m} x_{ij} > 0$. We then obtain an integer-constrained problem of the form

$$\begin{aligned}
\text{minimize} \quad & \sum_{i=1}^{m} \sum_{j=1}^{n} a_{ij} x_{ij} + \sum_{j=1}^{n} b_j y_j \\
\text{subject to} \quad & \sum_{j=1}^{n} x_{ij} = 1, \qquad i = 1, \ldots, m, \\
& \sum_{i=1}^{m} t_{ij} x_{ij} \leq T_j y_j, \qquad j = 1, \ldots, n, \\
& x_{ij} = 0 \text{ or } 1, \qquad i = 1, \ldots, m,\ j = 1, \ldots, n, \\
& y_j = 0 \text{ or } 1, \qquad j = 1, \ldots, n,
\end{aligned}$$

where $y_j = 1$ indicates that the jth machine is used. When $t_{ij} = 1$ for all pairs (i, j), this problem is also known as the *facility location problem*. Within this context, we must select a subset of locations from a given candidate set, and place in each of these locations a "facility" that will serve the needs of certain "clients" up to a given capacity bound. The 0-1 decision variable y_j corresponds to selecting location j for facility placement. To make the connection with the generalized assignment problem, associate clients with jobs, and locations with machines.

Example 5.5.4. Traveling Salesman Problem

An important model for scheduling a sequence of operations is the classical traveling salesman problem. This is perhaps the most studied of all combinatorial optimization problems. In addition to its use as a practical model, it has served as a testbed for a large variety of formal and heuristic approaches in discrete optimization.

In a colloquial description of the problem, a salesman wants to find a minimum cost tour that visits each of N given cities exactly once and returns to the starting city. Let a_{ij} be the cost of going from city i to city j, and let x_{ij} be a variable that takes the value 1 if the salesman visits city j immediately following city i, and the value 0 otherwise. Then the problem is

$$
\begin{aligned}
\text{minimize} \quad & \sum_{i=1}^{N} \sum_{\substack{j=1,\ldots,N \\ j \neq i}} a_{ij} x_{ij} \\
\text{subject to} \quad & \sum_{\substack{j=1,\ldots,N \\ j \neq i}} x_{ij} = 1, \qquad i = 1, \ldots, N, \\
& \sum_{\substack{i=1,\ldots,N \\ i \neq j}} x_{ij} = 1, \qquad j = 1, \ldots, N, \\
& x_{ij} = 0 \text{ or } 1, \qquad \forall\, (i,j) \in \mathcal{A},
\end{aligned}
$$

plus the additional constraint that the set of arcs $\{(i,j) \mid x_{ij} = 1\}$ forms a connected tour. The last constraint can be expressed as

$$
\sum_{i \in S,\, j \notin S} (x_{ij} + x_{ji}) \geq 2, \qquad \forall \text{ nonempty proper subsets } S \text{ of cities.} \tag{5.88}
$$

It turns out that if this constraint were not present, the problem would be much easier. In particular, it would be an assignment problem (assign the N cities to the N successor cities in a salesman tour), which is a relatively easy problem as discussed in Example 5.5.1. Unfortunately, however, these constraints are essential, since without them, there would be feasible solutions involving multiple disconnected cycles, as illustrated in Fig. 5.5.2. Nonetheless, with the aid of duality, the corresponding assignment problem can form the basis for solution of the traveling salesman problem.

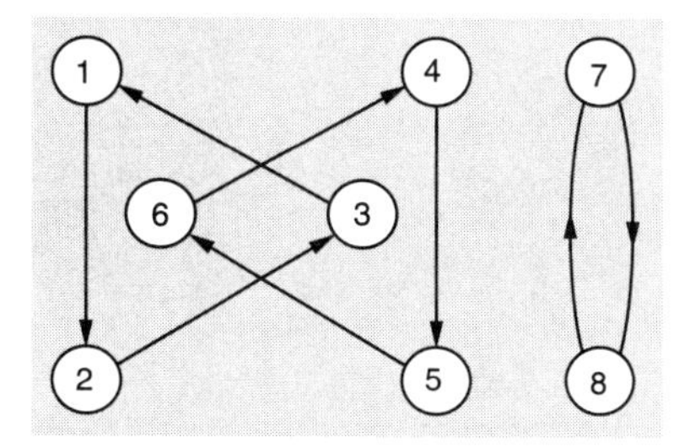

Figure 5.5.2. Example of an infeasible solution of a traveling salesman problem where all the constraints are satisfied except for the connectivity constraint (5.88). This solution may have been obtained by solving an $N \times N$ assignment problem and consists of multiple cycles [(1,2,3), (4,5,6), and (7,8) in the figure]. The arcs of the cycles correspond to the assigned pairs (i,j) in the assignment problem.

Example 5.5.5. Separable Resource Allocation Problems

Consider a problem of optimally producing a given amount of product using n production units. We assume that the production and cost characteristics of the units may change abruptly as the production level changes; for example there may be a setup or startup cost that is incurred at an arbitrarily small (but positive) level of production. This situation leads to a "mixed" discrete/continuous problem formulation.

Let x_i be the amount produced by the ith unit, where $i = 1, \ldots, n$. Let there be a finite set M_i of production modes associated with the ith production unit. The set of possible productions levels in production mode $m_i \in M_i$ of the ith unit is denoted $X_i(m_i)$, so that the amount produced by the ith unit must satisfy $x_i \in X_i(m_i)$ when the production mode n_i is selected (see Fig. 5.5.3).

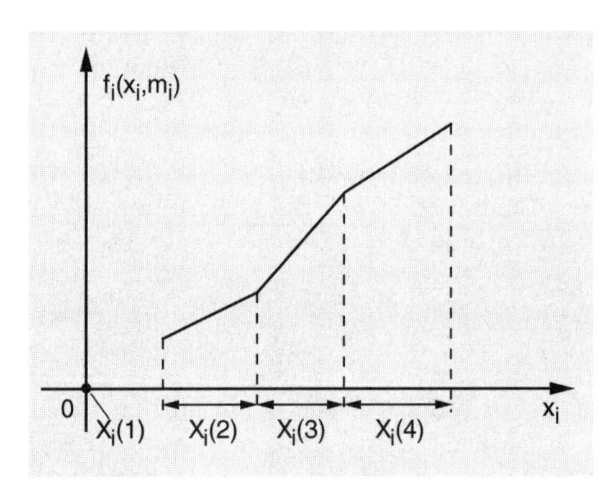

Figure 5.5.3. Production modes and cost function of a discrete resource allocation problem. Here there are four modes of production, $m_i = 1, 2, 3, 4$, and corresponding constraints, $x_i \in X_i(m_i)$. The choice $m_i = 1$ corresponds to no production ($x_i = 0$).

Let $f_i(x_i, m_i)$ be the cost of producing amount x_i via the production mode m_i. The problem then is to

$$\begin{aligned} &\text{minimize} \quad \sum_{i=1}^{n} f_i(x_i, m_i) \\ &\text{subject to} \quad m_i \in M_i, \qquad x_i \in X(m_i). \end{aligned}$$

There are also some additional constraints, which we generically represent in the form

$$\sum_{i=1}^{n} g_{ij}(x_i, m_i) \leq 0, \qquad j = 1, \ldots, r,$$

where the function g_{ij} represents the contribution of the ith unit in the jth constraint. These constraints specify that the units must collectively meet certain requirements (such as satisfying a given demand d, i.e., $\sum_{i=1}^{m} x_i \geq d$), security constraints (such as a lower bound on the number of units involved in production), etc.

An important characteristic of this problem is that it has a separable structure. As discussed in Section 5.1.6, this structure is particularly well-suited for the application of duality for two reasons: first, the computation of the dual function values is facilitated through decomposition, and second, the duality gap tends to be small in relative terms, particularly as the dimension n increases. In what follows, we will illustrate how these properties enhance the application of various solution algorithms.

5.5.2 Branch-and-Bound

The branch-and-bound method implicitly enumerates all the feasible solutions, using optimization problems that involve no integer constraints. The method can be very time-consuming, but is in principle capable of yielding an exactly optimal solution. The idea of the method is to partition the feasible set into smaller subsets, and then calculate certain bounds on the attainable cost within some of the subsets to eliminate from further consideration other subsets. This idea is encapsulated in the following simple observation.

Bounding Principle

Given the problem of minimizing $f(x)$ over $x \in X$, and two subsets $Y_1 \subset X$ and $Y_2 \subset X$, suppose that we have bounds

$$\underline{f}_1 \leq \min_{x \in Y_1} f(x), \qquad \bar{f}_2 \geq \min_{x \in Y_2} f(x).$$

Then, if $\bar{f}_2 \leq \underline{f}_1$, the solutions in Y_1 may be disregarded since their cost cannot be smaller than the cost of the best solution in Y_2.

The branch-and-bound method uses suitable upper and lower bounds, and the bounding principle to eliminate from consideration substantial portions of the feasible set. To describe the method, consider a general discrete optimization problem

$$\begin{aligned} &\text{minimize} \quad f(x) \\ &\text{subject to} \quad x \in X, \end{aligned}$$

where the feasible set X is a *finite* set. The branch-and-bound algorithm uses an acyclic graph known as the *branch-and-bound tree*, which corresponds to a progressively finer partition of X. In particular, the nodes of this graph correspond to a collection $\mathcal{X}$ of subsets of X, which is such that:

1. $X \in \mathcal{X}$ (i.e., the set of all solutions is a node).

2. If x is a feasible solution, then $\{x\} \in \mathcal{X}$ (i.e., each solution viewed as a singleton set is a node).

3. If a set $Y \in \mathcal{X}$ contains more than one solution $x \in X$, then there exist disjoint sets $Y_1, \ldots, Y_n \in \mathcal{X}$ such that

$$\bigcup_{i=1}^{n} Y_i = Y.$$

The set Y is called the *parent* of $Y_1, \ldots, Y_n$, and the sets $Y_1, \ldots, Y_n$ are called the *children* or *descendants* of Y.

4. Each set in $\mathcal{X}$ other than X has a parent.

The collection of sets $\mathcal{X}$ defines the branch-and-bound tree as in Fig. 5.5.4. In particular, this tree has the set of all feasible solutions X as its root node and the singleton solutions $\{x\}$, $x \in X$, as terminal nodes. The arcs of the graph are those that connect parents Y and their children Y_i.

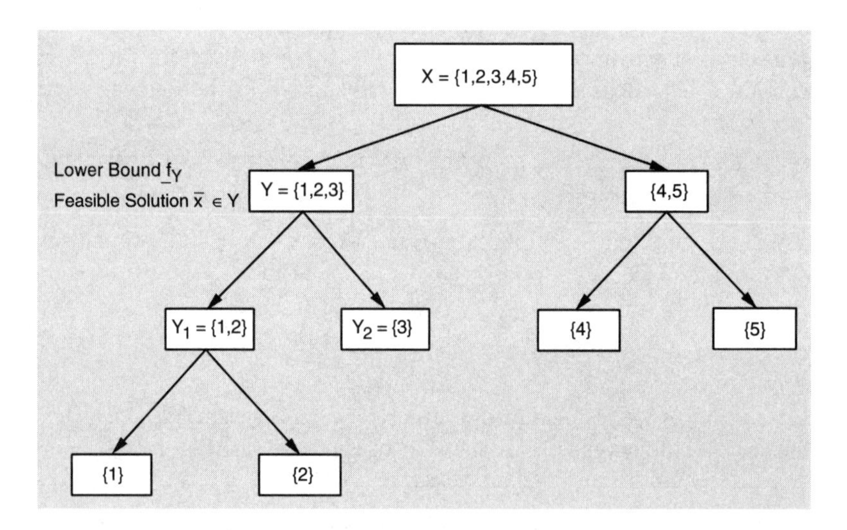

Figure 5.5.4. Illustration of a branch-and-bound tree. Each node Y (a subset of the feasible set X), except those consisting of a single solution, is partitioned into several other nodes (subsets) $Y_1, \ldots, Y_n$. The original feasible set is divided repeatedly into subsets until no more division is possible. For each node/subset Y of the tree, one may compute a lower bound $\underline{f}_Y$ to the optimal cost of the corresponding restricted subproblem $\min_{x \in Y} f(x)$, and a feasible solution $\bar{x} \in Y$, whose cost can serve as an upper bound to the optimal cost $\min_{x \in X} f(x)$ of the original problem. The idea is to use these bounds to economize computation by eliminating nodes of the tree that cannot contain an optimal solution.

The key assumption in the branch-and-bound method is that for every nonterminal node Y, there is an algorithm that calculates:

(a) A lower bound $\underline{f}_Y$ to the minimum cost over Y

$$\underline{f}_Y \leq \min_{x \in Y} f(x).$$

(b) A feasible solution $\bar{x} \in Y$, whose cost $f(\bar{x})$ can serve as an upper bound to the minimum cost over Y (as well as over X).

These bounds are used to save computation by discarding the nodes/subsets of the tree that have no chance of containing a solution that is better than the best currently available. In particular, the algorithm selects nodes Y from the branch-and-bound tree, and checks whether the lower bound $\underline{f}_Y$ exceeds the best available upper bound [the minimal cost $f(\bar{x})$ over all feasible solutions $\bar{x}$ found so far]. If this is so, we know that Y cannot contain an optimal solution, so all its descendant nodes in the tree need not be considered further.

To organize the search through the tree, the algorithm maintains a node list called OPEN, and also maintains a scalar called UPPER, which is equal to the minimal cost over feasible solutions found so far. Initially, OPEN contains just X, and UPPER is equal to ∞ or to the cost $f(\bar{x})$ of some feasible solution $\bar{x} \in X$.

Branch-and-Bound Algorithm

Step 1: Remove a node Y from OPEN. For each child Y_j of Y, do the following: Find the lower bound $\underline{f}_{Y_j}$ and a feasible solution $\bar{x} \in Y_j$. If

$$\underline{f}_{Y_j} < \text{UPPER},$$

place Y_j in OPEN. If in addition

$$f(\bar{x}) < \text{UPPER},$$

set

$$\text{UPPER} = f(\bar{x})$$

and mark $\bar{x}$ as the best solution found so far.

Step 2: (Termination Test) If OPEN is nonempty, go to step 1. Otherwise, terminate; the best solution found so far is optimal.

A node Y_j that is not placed in OPEN in Step 1 is said to be *fathomed*. Such a node cannot contain a better solution than the best solution found so far, since the corresponding lower bound $\underline{f}_{Y_j}$ is not smaller than UPPER. Therefore nothing is lost when we drop this node from further consideration and forego the examination of its descendants. Regardless of how many nodes are fathomed, the branch-and-bound algorithm is guaranteed to examine either explicitly or implicitly (through fathoming) all the terminal nodes, which are the singleton solutions. As a result, it will terminate with an optimal solution. Note that a small (near-optimal) value of UPPER and tight lower bounds $\underline{f}_{Y_j}$ contribute to the quick fathoming of large portions of the branch-and-bound tree, so it is important to obtain good bounds as early as possible.

In a popular variant of the algorithm, termination is accelerated at the expense of obtaining a solution that is suboptimal within some tolerance $\epsilon > 0$. In particular, if we replace the test

$$\underline{f}_{Y_j} < \text{UPPER}$$

with

$$\underline{f}_{Y_j} < \text{UPPER} - \epsilon$$

in Step 1, the best solution obtained upon termination is guaranteed to be within ϵ of optimality.

Other variations of branch-and-bound relate to the method for selecting a node from OPEN in Step 1. For example, a possible strategy is to choose the node with minimal lower bound; alternatively, one may choose the node containing the best solution found so far. In fact *it is neither practical nor necessary to generate a priori the branch-and-bound tree.* Instead, one may adaptively decide on the order and the manner in which the nodes are partitioned into descendants based on the progress of the algorithm.

Branch-and-bound typically uses continuous optimization problems (without integer constraints) to obtain lower bounds to the optimal costs of the restricted problems $\min_{x \in Y} f(x)$ and to construct corresponding feasible solutions. For example, suppose that our original problem has a convex cost function, and a feasible set X that consists of convex inequality constraints, *plus the additional constraint that all optimization variables must be 0 or 1.* Then a restricted subset Y may specify that the values of some given subset of variables are fixed at 0 or at 1, while the remaining variables may take either the value 0 or the value 1. A lower bound to the restricted optimal cost $\min_{x \in Y} f(x)$ is then obtained by relaxing the 0-1 constraint on the latter variables, thereby allowing them to take any value in the interval $[0, 1]$ and resulting in a convex problem with inequality constraints. Thus the solution by branch-and-bound of a problem with convex cost and inequality constraints *plus* additional integer constraints requires the solution of many convex problems with inequality constraints but *without* integer constraints.

Example 5.5.6. Facility Location Problems

Let us consider the facility location problem introduced in Example 5.5.3, which involves m clients and n locations. By $x_{ij} = 1$ (or $x_{ij} = 0$) we indicate that client i is assigned to location j at a cost a_{ij} (or is not assigned, respectively). We also introduce a 0-1 integer variable y_j to indicate (with $y_j = 1$) that a facility is placed at location j at a cost b_j. The problem is

$$
\begin{aligned}
\text{minimize} \quad & \sum_{i=1}^{m} \sum_{j=1}^{n} a_{ij} x_{ij} + \sum_{j=1}^{n} b_j y_j \\
\text{subject to} \quad & \sum_{j=1}^{n} x_{ij} = 1, \qquad i = 1, \ldots, m, \\
& \sum_{i=1}^{m} x_{ij} \le T_j y_j, \qquad j = 1, \ldots, n, \\
& x_{ij} = 0 \text{ or } 1, \qquad \forall\, (i,j), \\
& y_j = 0 \text{ or } 1, \qquad j = 1, \ldots, n,
\end{aligned}
$$

where T_j is the maximum number of customers that can be served by a facility at location j.

The solution of the problem by branch-and-bound involves the partition of the feasible set X into subsets. The choice of subsets is somewhat arbitrary, but it is convenient to select subsets of the form

$$X(J_0, J_1) = \big\{(x, y) : \text{ feasible} \mid y_j = 0, \, \forall\, j \in J_0, \, y_j = 1, \, \forall\, j \in J_1\big\},$$

where J_0 and J_1 are disjoint subsets of the index set $\{1, \ldots, n\}$ of facility locations. Thus, $X(J_0, J_1)$ is the subset of feasible solutions such that:

a facility is placed at the locations in J_1,

no facility is placed at the locations in J_0,

a facility may or may not be placed at the remaining locations.

For each node/subset $X(J_0, J_1)$, we may obtain a lower bound and a feasible solution by solving the linear program where all integer constraints are relaxed except that the variables y_j, $j \in J_0 \cup J_1$ are fixed at either 0 or 1.

As an illustration, let us work out the example shown in Fig. 5.5.5, which involves 3 clients and 2 locations. The facility capacities at the two locations are $T_1 = T_2 = 3$. The cost coefficients a_{ij} and b_j are shown next to the corresponding arcs. The optimal solution corresponds to $y_1 = 0$ and $y_2 = 1$, i.e., placing a facility only in location 2 and serving all the clients at that facility. The corresponding optimal cost is

$$f^* = 5.$$

Let us apply the branch-and-bound algorithm using the tree shown in Fig. 5.5.5. We first consider the top node $\big(J_0 = \varnothing, J_1 = \varnothing\big)$, where neither y_1 nor y_2 is fixed at 0 or 1. The lower bound $\underline{f}_Y$ is obtained by solving the (relaxed) linear program

$$
\begin{aligned}
\text{minimize} \quad & (2x_{11} + x_{12}) + (2x_{21} + x_{22}) + (x_{31} + 2x_{32}) + 3y_1 + y_2 \\
\text{subject to} \quad & x_{11} + x_{12} = 1, \qquad x_{21} + x_{22} = 1, \qquad x_{31} + x_{32} = 1, \\
& x_{11} + x_{21} + x_{31} \le 3y_1, \qquad x_{12} + x_{22} + x_{32} \le 3y_2, \\
& 0 \le x_{ij} \le 1, \qquad \forall\, (i,j), \\
& 0 \le y_1 \le 1, \qquad 0 \le y_2 \le 1.
\end{aligned}
$$

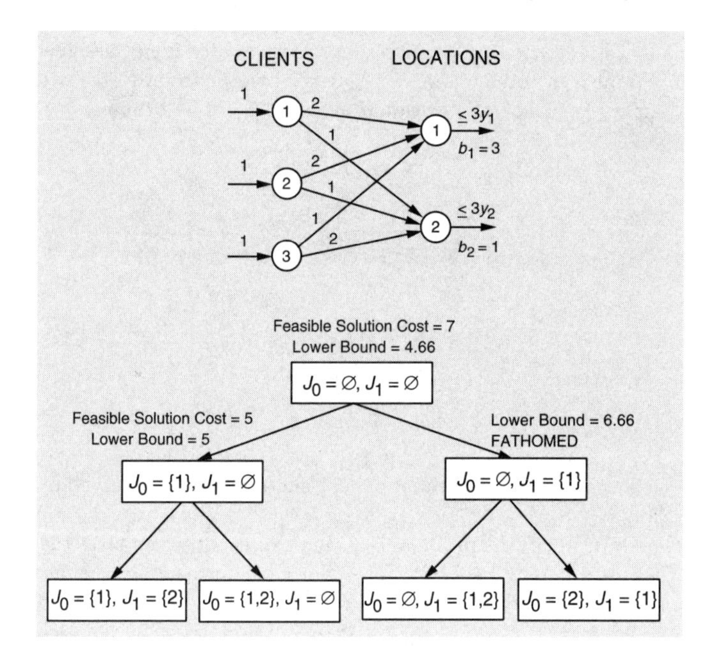

Figure 5.5.5. Branch-and-bound solution of a facility location problem with 3 clients and 2 locations. The facility capacities at the two locations are $T_1 = T_2 = 3$. The cost coefficients a_{ij} and b_j are shown next to the corresponding arcs. The relaxed problem for the top node ($J_0 = \emptyset, J_1 = \emptyset$), corresponding to relaxing all the integer constraints, is solved first, obtaining the lower and upper bounds shown. Then the relaxed problem corresponding to the left node ($J_0 = \{1\}, J_1 = \emptyset$) is solved, obtaining the lower and upper bounds shown. Finally, the relaxed problem corresponding to the right node ($J_0 = \emptyset, J_1 = \{1\}$) is solved, obtaining a lower bound that is higher than the current value of UPPER. As a result this node can be fathomed, and its descendants need not be considered further.

The optimal solution of this program can be obtained by a linear programming algorithm and can be verified to be

$$x_{ij} = \begin{cases} 1 & \text{if } (i,j) = (1,2), (2,2), (3,1), \\ 0 & \text{otherwise,} \end{cases}$$

$$y_1 = 1/3, \qquad y_2 = 2/3.$$

The corresponding optimal cost (lower bound) is

$$\underline{f}_Y = 4.66.$$

A feasible solution of the original problem is obtained by rounding the fractional values of y_1 and y_2 to

$$\bar{y}_1 = 1, \qquad \bar{y}_2 = 1,$$

and the associated cost is 7. Thus, we set

$$\text{UPPER} = 7,$$

and we place in OPEN the two descendants $\big(J_0 = \{1\}, J_1 = \emptyset\big)$ and $\big(J_0 = \emptyset, J_1 = \{1\}\big)$, corresponding to fixing y_1 at 0 and at 1, respectively.

We proceed with the left branch of the branch-and-bound tree, and consider the node $\big(J_0 = \{1\}, J_1 = \emptyset\big)$, corresponding to fixing y_1 as well as the corresponding variables x_{11}, x_{21}, and x_{31} to 0. The associated (relaxed) linear program is

$$\begin{aligned} \text{minimize} \quad & x_{12} + x_{22} + 2x_{32} + y_2 \\ \text{subject to} \quad & x_{12} = 1, \qquad x_{22} = 1, \qquad x_{32} = 1, \\ & x_{12} + x_{22} + x_{32} \leq 3y_2, \\ & 0 \leq x_{12} \leq 1, \quad 0 \leq x_{22} \leq 1, \quad 0 \leq x_{32} \leq 1, \\ & 0 \leq y_2 \leq 1. \end{aligned}$$

The optimal solution (in fact the only feasible solution) of this program is

$$x_{ij} = \begin{cases} 1 & \text{if } (i,j) = (1,2), (2,2), (3,2), \\ 0 & \text{otherwise,} \end{cases}$$

$$y_2 = 1,$$

and the corresponding optimal cost (lower bound) is

$$\underline{f}_Y = 5.$$

The optimal solution of the relaxed problem is integer, and its cost, 5, is lower than the current value of UPPER, so we set

$$\text{UPPER} = 5.$$

The two descendants, $\big(J_0 = \{1\}, J_1 = \{2\}\big)$ and $\big(J_0 = \{1,2\}, J_1 = \emptyset\big)$, corresponding to fixing y_2 at 1 and at 0, respectively, are placed in OPEN.

We proceed with the right branch of the branch-and-bound tree, and consider the node $\big(J_0 = \emptyset, J_1 = \{1\}\big)$, corresponding to fixing y_1 to 1. The associated (relaxed) linear program is

$$\begin{aligned} \text{minimize} \quad & (2x_{11} + x_{12}) + (2x_{21} + x_{22}) + (x_{31} + 2x_{32}) + 3 + y_2 \\ \text{subject to} \quad & x_{11} + x_{12} = 1, \qquad x_{21} + x_{22} = 1, \qquad x_{31} + x_{32} = 1, \\ & x_{11} + x_{21} + x_{31} \leq 3, \qquad x_{12} + x_{22} + x_{32} \leq 3y_2, \\ & 0 \leq x_{ij} \leq 1, \qquad \forall\ (i,j), \\ & 0 \leq y_2 \leq 1. \end{aligned}$$

The optimal solution of this program can be verified to be

$$x_{ij} = \begin{cases} 1 & \text{if } (i,j) = (1,2), (2,2), (3,1), \\ 0 & \text{otherwise,} \end{cases}$$

$$y_2 = 2/3,$$

and the corresponding optimal cost (lower bound) is

$$\underline{f}_Y = 6.66.$$

This is larger than the current value of UPPER, so the node can be fathomed, and its two descendants are not placed in OPEN.

We conclude that one of the two descendants of the left node, $(J_0 = \{1\}, J_1 = \{2\})$ and $(J_0 = \{1, 2\}, J_1 = \varnothing)$ (the only nodes in OPEN), contains the optimal solution. We can proceed to solve the relaxed linear programs corresponding to these two nodes, and obtain the optimal solution. However, there is also a shortcut here: since these are the only two remaining nodes and the upper bound corresponding to these nodes coincides with the lower bound, we can conclude that the lower bound is equal to the optimal cost and the corresponding integer solution $(y_1 = 0, y_2 = 1)$ is optimal.

Generally, for the success of the branch-and-bound approach it is important that the lower bounds are as tight as possible, because this facilitates the fathoming of nodes, and leads to fewer restricted problem solutions. On the other hand, the tightness of the bounds strongly depends on how the problem is formulated as an integer programming problem. There may be several possible formulations, some of which are "stronger" than others in the sense that they provide better bounds within the branch-and-bound context. The following example provides an illustration.

Example 5.5.7. Facility Location – Alternative Formulation

Consider the following alternative formulation of the preceding facility location problem

$$\begin{aligned}
\text{minimize} \quad & \sum_{i=1}^{m} \sum_{j=1}^{n} a_{ij} x_{ij} + \sum_{j=1}^{n} b_j y_j \\
\text{subject to} \quad & \sum_{j=1}^{n} x_{ij} = 1, \qquad i = 1, \ldots, m, \\
& \sum_{i=1}^{m} x_{ij} \leq T_j y_j, \qquad j = 1, \ldots, n, \\
& x_{ij} \leq y_j, \qquad \forall\ (i, j), \\
& x_{ij} = 0 \text{ or } 1, \qquad \forall\ (i, j), \\
& y_j = 0 \text{ or } 1, \qquad j = 1, \ldots, n.
\end{aligned}$$

This formulation involves a lot more constraints, but is in fact superior to the one given earlier (cf. Example 5.5.6). The reason is that, once we relax the 0-1 constraints on x_{ij} and y_j, the constraints $\sum_{i=1}^{m} x_{ij} \leq T_j y_j$ of Example 5.5.6 define a larger region than the constraints $\sum_{i=1}^{m} x_{ij} \leq T_j y_j$ and $x_{ij} \leq y_j$

of the present example. As a result, the lower bounds obtained by relaxing some of the 0-1 constraints are tighter in the alternative formulation just given, thereby enhancing the effectiveness of the branch-and-bound method. In fact, it can be verified that for the example of Fig. 5.5.5, by relaxing the 0-1 constraints in the stronger formulation of the present example, we obtain the correct optimal integer solution at the very first node of the branch-and-bound tree.

The preceding example illustrates an important fact: *it is possible to accelerate the branch-and-bound solution of a problem by adding constraints that do not affect the set of feasible integer solutions*. Such constraints, referred to as *valid inequalities*, may improve the lower bounds obtained by relaxing the 0-1 constraints. Basically, when the integer constraints are relaxed, one obtains a superset of the feasible set of integer solutions, so with more constraints, the corresponding superset becomes smaller and approximates better the true feasible set (see Fig. 5.5.6). Thus one should strive to select a problem formulation and/or try to generate additional valid inequalities so that when the integer constraints are relaxed, the feasible set is as small as possible. In particular, this leads to a variant of branch-and-bound, called *branch-and-cut*, which generates suitable valid inequalities at several nodes of the branch-and-bound tree, together with the associated bounds. We refer to the surveys [JRT95], [LuB96], and [CaF97] for accounts of recent progress on this subject.

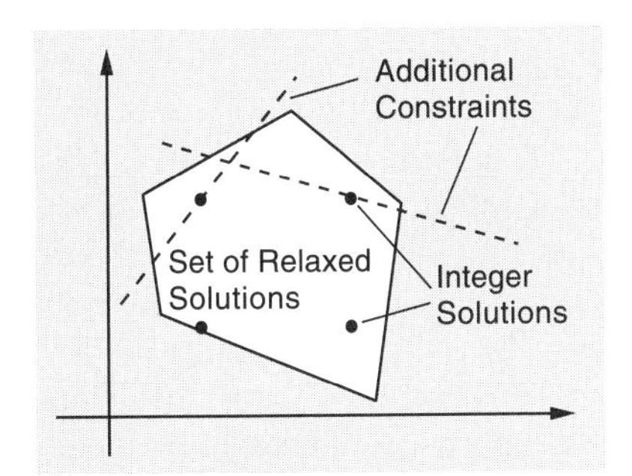

Figure 5.5.6. Illustration of the effect of additional constraints. They do not affect the set of feasible integer solutions, but they reduce the set of "relaxed solutions," that is, those x that satisfy all the constraints except for the integer constraints. This results in improved lower bounds and a faster branch-and-bound solution.

We note that the subject of characterizing the feasible set of an integer programming problem, and approximating it tightly with a polyhedral set has received extensive attention. In particular, there is a lot of theory and accumulated practical knowledge on characterizing the feasible set in specific problem contexts; see the references cited at the end of the chapter. A further discussion of branch-and-bound is beyond our scope. We refer to sources on linear and combinatorial optimization, such as Zoutendijk [Zou76], Papadimitriou and Steiglitz [PaS82], Schrijver [Sch86], Nemhauser and Wolsey [NeW88], Bertsimas and Tsitsiklis [BeT97], Cook, Cunningham, Pulleyblank, and Schrijver [CCP98], Wolsey [Wol98], which also describe many applications.

5.5.3 Lagrangian Relaxation

In this section, we consider an important approach for obtaining lower bounds to use in the branch-and-bound method. Let us consider the problem

$$\begin{aligned} &\text{minimize} \quad f(x) \\ &\text{subject to} \quad g_j(x) \le 0, \qquad j = 1, \ldots, r, \\ &\qquad\qquad\quad x \in X, \end{aligned}$$

where X is a *finite* set of vectors in $\Re^n$.

In the Lagrangian relaxation approach, we eliminate the constraints $g_j(x) \le 0$ by using nonnegative multipliers μ_j, and by forming the Lagrangian function

$$L(x, \mu) = f(x) + \sum_{j=1}^{r} \mu_j g_j(x).$$

As in Section 5.1, the dual function is

$$q(\mu) = \min_{x \in X} L(x, \mu)$$

and the dual problem is

$$\begin{aligned} &\text{maximize} \quad q(\mu) \\ &\text{subject to} \quad \mu \ge 0. \end{aligned}$$

A key fact here is that for any $\mu \ge 0$, *the dual value $q(\mu)$ provides a lower bound to the optimal primal value f^*, and so does the optimal dual value q^*.* This is the weak duality theorem (Prop. 5.1.3). Thus solving the dual problem (even approximately) provides a lower bound that can be used in the context of the branch-and-bound procedure.

The minimization of $L(x, \mu)$ is facilitated when the cost function f, the constraint functions g_j, and the set X have a special structure, such as a separable structure, since then one can take advantage of the associated decomposition, as discussed in Section 5.1.6. More generally, the minimization of $L(x, \mu)$ is facilitated when by eliminating the constraints $g_j(x) \le 0$ using multipliers, one obtains an easily solvable problem. A major case of this type is when the cost and constraint functions f and g_j are linear, and the finite set X is the constraint set of an integer-constrained network optimization problem discussed in Example 5.5.1. Then the minimization of $L(x, \mu)$ over X can be done with fast specialized network flow algorithms such as shortest path, max-flow, assignment, transportation, and minimum cost flow algorithms (see network optimization books, such as the author's [Ber98], for an extensive discussion).

Example 5.5.8. Facility Location – Lagrangian Relaxation

Consider the facility location problem as formulated in Example 5.5.7:

$$\begin{aligned}
&\text{minimize} && \sum_{i=1}^{m}\sum_{j=1}^{n} a_{ij}x_{ij} + \sum_{j=1}^{n} b_j y_j \\
&\text{subject to} && \sum_{j=1}^{n} x_{ij} = 1, \qquad i = 1,\ldots,m, \\
& && \sum_{i=1}^{m} x_{ij} \le T_j y_j, \qquad j = 1,\ldots,n, \\
& && x_{ij} \le y_j, \qquad \forall\,(i,j), \\
& && x_{ij} = 0 \text{ or } 1, \qquad \forall\,(i,j), \\
& && y_j = 0 \text{ or } 1, \qquad j = 1,\ldots,n.
\end{aligned}$$

The solution of the problem by branch-and-bound involves the partition of the feasible set into subsets whereby some of the integer variables are fixed at the 0 or 1 value, while the 0-1 constraints on the remaining variables are replaced by interval constraints. In particular, at the typical subproblem/node of the branch-and-bound tree we select two disjoint subsets J_0 and J_1 of the index set $\{1,\ldots,n\}$. These subsets correspond to:

placing a facility at the locations in J_1,

placing no facility at the locations in J_0,

leaving open the placement of a facility at the remaining locations.

We must then obtain a lower bound and a feasible solution of the restricted integer program where the variables y_j, $j \in J_0 \cup J_1$ have been fixed at either 0 or 1:

$$\begin{aligned}
&\text{minimize} && \sum_{i=1}^{m}\sum_{j=1}^{n} a_{ij}x_{ij} + \sum_{j=1}^{n} b_j y_j \\
&\text{subject to} && \sum_{j=1}^{n} x_{ij} = 1, \qquad i = 1,\ldots,m, \\
& && \sum_{i=1}^{m} x_{ij} \le T_j y_j, \qquad j \notin J_0, \\
& && x_{ij} \le y_j, \qquad j \notin J_0, \\
& && x_{ij} = 0 \text{ or } 1, \qquad \forall\,(i,j) \text{ with } j \notin J_0, \\
& && x_{ij} = 0, \qquad \forall\,(i,j) \text{ with } j \in J_0, \\
& && y_j = 0 \text{ or } 1, \qquad \forall\, j \notin J_0 \cup J_1, \\
& && y_j = 0, \quad \forall\, j \in J_0, \qquad y_j = 1, \quad \forall\, j \in J_1.
\end{aligned} \tag{5.89}$$

Here are three ways to obtain a lower bound. The first is based on constraint relaxation, as in Examples 5.5.6 and 5.5.7, and the last two are based on Lagrangian relaxation:

(a) We replace the 0-1 integer constraints on the variables x_{ij} and y_j by $[0, 1]$ interval constraints. The resulting linear problem is then solved by standard linear programming methods.

(b) We eliminate the constraints $\sum_{i=1}^m x_{ij} \le T_j y_j$, $j \notin J_0$, using multipliers $\mu_j \ge 0$, and the constraints $x_{ij} \le y_j$, $j \notin J_0$, using multipliers $\xi_{ij} \ge 0$. The corresponding Lagrangian function,

$$L(x, y, \mu, \xi) = \sum_{i=1}^m \sum_{j \notin J_0} (a_{ij} + \mu_j + \xi_{ij}) x_{ij} + \sum_{j \notin J_0} \left(b_j - \mu_j T_j - \sum_{i=1}^m \xi_{ij} \right) y_j,$$

is then minimized over the remaining constraints. This minimization is decoupled with respect to x and y. Minimization over y_j yields optimal values

$$\hat{y}_j = \begin{cases} 0 & \text{if } j \in J_0, \text{ or } b_j - \mu_j T_j - \sum_{i=1}^m \xi_{ij} \ge 0 \text{ and } j \notin J_1, \\ 1 & \text{otherwise.} \end{cases}$$

Minimization over x_{ij} can be done by finding for each i, an index $j_i \notin J_0$ such that

$$a_{ij_i} + \mu_{j_i} + \xi_{ij_i} = \arg \min_{j \notin J_0} \{a_{ij} + \mu_j + \xi_{ij}\},$$

and then obtaining the optimal values

$$\hat{x}_{ij} = \begin{cases} 0 & \text{if } j \ne j_i, \\ 1 & \text{if } j = j_i. \end{cases}$$

The value $L(\hat{x}, \hat{y}, \mu, \xi)$, which is the dual value corresponding to μ and ξ, is a lower bound to the optimal cost of the restricted integer program (5.89). This lower bound can be strengthened by maximizing $L(\hat{x}, \hat{y}, \mu, \xi)$ over $\mu \ge 0$ and $\xi \ge 0$, thereby obtaining the optimal dual value/lower bound (we discuss later the issues associated with this maximization).

(c) We eliminate only the constraints, $x_{ij} \le y_j$, $j \notin J_0$, using multipliers $\xi_{ij} \ge 0$. The corresponding Lagrangian function,

$$L(x, y, \xi) = \sum_{i=1}^m \sum_{j \notin J_0} (a_{ij} + \xi_{ij}) x_{ij} + \sum_{j \notin J_0} \left(b_j - \sum_{i=1}^m \xi_{ij} \right) y_j,$$

is then minimized over the remaining constraints for x. This minimization is decoupled with respect to x and y. Minimization over y_j yields optimal values

$$\hat{y}_j = \begin{cases} 0 & \text{if } j \in J_0, \text{ or } b_j - \sum_{i=1}^m \xi_{ij} \ge 0 \text{ and } j \notin J_1, \\ 1 & \text{otherwise.} \end{cases}$$

The minimization over x_{ij} does not decompose over the clients i as in case (b) above. Instead it has the form

$$\begin{aligned}
&\text{minimize} \quad \sum_{i=1}^{m} \sum_{j \notin J_0} (a_{ij} + \xi_{ij}) x_{ij} \\
&\text{subject to} \quad \sum_{j=1}^{n} x_{ij} = 1, \qquad i = 1, \ldots, m, \\
&\qquad \sum_{i=1}^{m} x_{ij} \le T_j, \qquad \forall\ j \text{ with } \hat{y}_j = 1, \\
&\qquad x_{ij} = 0 \text{ or } 1, \qquad \forall\ (i,j) \text{ with } \hat{y}_j = 1, \\
&\qquad x_{ij} = 0, \qquad \forall\ (i,j) \text{ with } \hat{y}_j = 0.
\end{aligned}$$

This problem is an integer-constrained network optimization problem (cf. Example 5.5.1), known as a *transportation problem*, which can be solved efficiently with specialized algorithms. The value $L(\hat{x}, \hat{y}, \xi)$, is the dual value corresponding to ξ, and is a lower bound to the optimal cost of the restricted integer program (5.89). The optimal lower bound is obtained by maximizing $L(\hat{x}, \hat{y}, \xi)$ over $\xi \ge 0$.

Note that the three lower bounds involve qualitatively different computations. In particular, in cases (b) and (c), which involve Lagrangian relaxation, an easy optimization problem is solved many times in the context of the algorithm that maximizes the dual function. By contrast, in constraint relaxation [case (a)], a single optimization problem is solved to obtain the lower bound, but the solution of this problem is complicated by the presence of extra constraints.

Regarding the quality of the three bounds, it turns out that the lower bound obtained by constraint relaxation, and the two lower bounds obtained by Lagrangian relaxation, are all equal thanks to the linearity of the cost function, as we will demonstrate shortly in some generality.

Example 5.5.9. Traveling Salesman Problem – Lagrangian Relaxation

Consider the traveling salesman problem of Example 5.5.4. Here, we want to find a minimum cost tour in a complete graph where the cost of arc (i, j) is denoted a_{ij}. We formulate this as the discrete optimization problem

$$\begin{aligned}
&\text{minimize} \quad \sum_{i=1}^{N} \sum_{\substack{j=1,\ldots,N \\ j \ne i}} a_{ij} x_{ij} \\
&\text{subject to} \quad \sum_{\substack{j=1,\ldots,N \\ j \ne i}} x_{ij} = 1, \qquad i = 1, \ldots, N,
\end{aligned} \tag{5.90}$$

$$\sum_{\substack{i=1,\ldots,N \\ i\neq j}} x_{ij} = 1, \qquad j = 1,\ldots,N, \tag{5.91}$$

$$x_{ij} = 0 \text{ or } 1, \qquad \forall\ (i,j) \in \mathcal{A}, \tag{5.92}$$

$$\text{the subgraph with node-arc set } \big(\mathcal{N}, \{(i,j) \mid x_{ij} = 1\}\big) \text{ is connected.} \tag{5.93}$$

We may express the connectivity constraint (5.93) in several different ways, leading to different Lagrangian relaxation and branch-and-bound algorithms. One of the most successful formulations is based on the notion of a *1-tree*, which consists of a tree that spans nodes $2,\ldots,N$, *plus* two arcs that are incident to node 1. Equivalently, a 1-tree is a connected subgraph that contains a single cycle passing through node 1 (see Fig. 5.5.7). Note that if the conservation of flow constraints (5.90) and (5.91), and the integer constraints (5.92) are satisfied, then the connectivity constraint (5.93) is equivalent to the constraint that the subgraph $\big(\mathcal{N}, \{(i,j) \mid x_{ij} = 1\}\big)$ is a 1-tree.

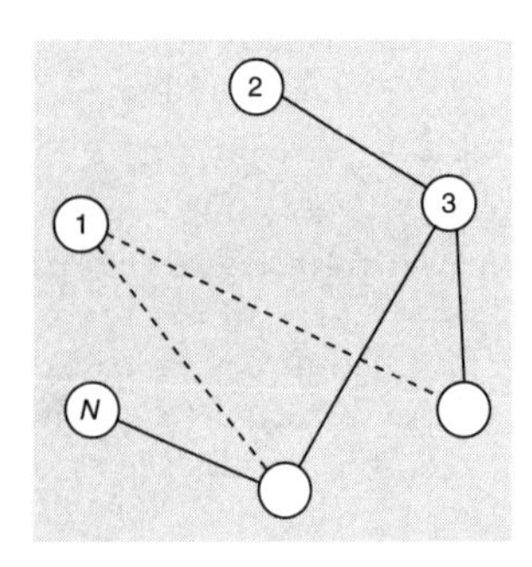

Figure 5.5.7. Illustration of a 1-tree. It consists of a tree that spans nodes $2,\ldots,N$, *plus* two arcs that are incident to node 1.

Let X_1 be the set of all x with $0-1$ components, and such that the subgraph $\big(\mathcal{N}, \{(i,j) \mid x_{ij} = 1\}\big)$ is a 1-tree. Let us consider a Lagrangian relaxation approach based on elimination of the conservation of flow equations. Assigning multipliers u_i and v_j to the constraints (5.90) and (5.91), respectively, the Lagrangian function is

$$L(x,u,v) = \sum_{i,j,i\neq j} (a_{ij} + u_i + v_j)x_{ij} - \sum_{i=1}^{N} u_i - \sum_{j=1}^{N} v_j.$$

The minimization of the Lagrangian is over all 1-trees, leading to the problem

$$\min_{x\in X_1} \left\{ \sum_{i,j,i\neq j} (a_{ij} + u_i + v_j)x_{ij} \right\}.$$

If we view $a_{ij} + u_i + v_j$ as a *modified cost* of arc (i,j), this minimization is quite easy. It is equivalent to obtaining a tree of minimum modified cost that spans the nodes $2,\ldots,N$, and then adding two arcs that are incident to node 1 and have minimum modified cost. The minimum cost spanning tree problem can be easily solved using a variety of efficient algorithms such as the Prim-Dijkstra algorithm (see e.g., [BeT97], [Ber98], [Mur92], [PaS82]).

The preceding examples illustrate an important characteristic of Lagrangian relaxation: *as we eliminate the troublesome constraints $g_j(x) \leq 0$, the problem is often simplified enough so that we also essentially eliminate the integer constraints.* This should be contrasted with the integer constraint relaxation approach, where we eliminate just the integer constraints, while leaving the other inequality constraints unaffected.

Comparing Lagrangian and Constraint Relaxation

Let us now compare the lower bounds obtained by relaxing the integer constraints and by dualizing the inequality constraints. We focus on the special case of a convex cost function $f(x)$ and the constraints

$$a_j'x \leq b_j, \qquad j = 1, \ldots, r,$$

$$x_i \in X_i, \qquad i = 1, \ldots, n,$$

where X_i are given finite subsets of the real line, and a_j and b_j are given vectors and scalars, respectively. Let f^* denote the optimal primal cost,

$$f^* = \inf_{\substack{a_j'x \leq b_j,\, j=1,\ldots,r \\ x_i \in X_i,\, i=1,\ldots,n}} f(x),$$

and let q^* denote the optimal dual cost,

$$q^* = \sup_{\mu \geq 0} q(\mu) = \sup_{\mu \geq 0} \inf_{x_i \in X_i,\, i=1,\ldots,n} L(x, \mu), \tag{5.94}$$

where

$$L(x, \mu) = f(x) + \sum_{j=1}^{r} \mu_j (a_j'x - b_j)$$

is the Lagrangian function. Let $\hat{X}_i$ denote the interval which is the convex hull of the set X_i, and denote by $\hat{f}$ the optimal cost of the problem, where each set X_i is replaced by $\hat{X}_i$,

$$\hat{f} = \inf_{\substack{a_j'x \leq b_j,\, j=1,\ldots,r \\ x_i \in \hat{X}_i,\, i=1,\ldots,n}} f(x). \tag{5.95}$$

Note that $\hat{f}$ is the lower bound obtained by constraint relaxation.

We will show that

$$\hat{f} \leq q^*,$$

so that *for a convex cost and linear inequality constraints, the lower bound* q^* *obtained by Lagrangian relaxation is no worse that the lower bound* $\hat{f}$ *obtained by constraint relaxation.* Indeed by using Prop. 5.2.1, we see that problem (5.95) has no duality gap, so $\hat{f}$ is equal to the corresponding dual cost, which is

$$\hat{q} = \sup_{\mu \geq 0} \inf_{x_i \in \hat{X}_i,\, i=1,\ldots,n} L(x,\mu). \tag{5.96}$$

By comparing Eqs. (5.94) and (5.96), we see that $\hat{q} \leq q^*$, so we conclude that $\hat{f} = \hat{q} \leq q^*$.

Let us now focus on the special case where f is linear. Then, the Lagrangian function $L(\cdot,\mu)$ is also linear and the minimization in the two dual problems (5.94) and (5.96) yields identical results, so that $\hat{f} = \hat{q} = q^*$. Thus, *when the cost and the inequality constraints are linear, the lower bounds obtained by Lagrangian and constraint relaxation are equal.* In fact, the preceding argument also shows that the same is true if the Lagrangian function $L(\cdot,\mu)$ is *concave* for each fixed $\mu \geq 0$.

Weaknesses of Lagrangian Relaxation

Finally, let us point out that the Lagrangian relaxation method has several weaknesses. First, the minimization of $L(x,\mu)$ over the set X may yield an x that violates some of the constraints $g_j(x) \leq 0$, so it may be necessary to adjust this x for feasibility using heuristics. There is not much one can say in generality about such heuristics because they are typically problem-dependent.

Second, the maximization of the dual function $q(\mu)$ over $\mu \geq 0$ may be quite nontrivial for a number of reasons, including the fact that q is typically nondifferentiable. In Chapter 6, we will discuss the algorithmic methodology for solving the dual problem, including the subgradient and cutting plane methods, which have enjoyed a great deal of popularity for nondifferentiable optimization.

EXERCISES

5.5.1

Apply branch-and-bound to solve the problem

$$\begin{aligned} &\text{minimize} \quad 5x_1 + 2x_2 + x_3 \\ &\text{subject to} \quad x_1 + 2x_2 + 2x_3 \geq 3, \quad x_1, x_2, x_3 \in \{0,1\}. \end{aligned}$$

5.5.2

Apply branch-and-bound to solve the problem of Exercise 5.1.2. Obtain lower bounds using Lagrangian relaxation and constraint relaxation, and verify that they are identical.

5.5.3 (Separable Problems with Integer/Simplex Constraints) (www)

Consider the problem

$$\begin{aligned} &\text{minimize} \quad \sum_{j=1}^{n} f_j(x_j) \\ &\text{subject to} \quad \sum_{j=1}^{n} x_j \leq A, \\ &\qquad x_j \in \{0,1,\ldots,m_j\}, \qquad j=1,\ldots,n, \end{aligned}$$

where A and $m_1,\ldots,m_n$ are given positive integers, and each function f_j is convex over the interval $[0,m_j]$. Consider an iterative algorithm (due to Ibaraki and Katoh [IbK88]) that starts at $(0,\ldots,0)$ and maintains a feasible vector $(x_1,\ldots,x_n)$. At the typical iteration, we consider the set of indices $J = \{j \mid x_j < m_j\}$. If J is empty or $\sum_{j=1}^{n} x_j = A$, the algorithm terminates. Otherwise, we find an index $\bar{j} \in J$ that maximizes $f_j(x_j) - f_j(x_j+1)$. If $f_{\bar{j}}(x_{\bar{j}}) - f_{\bar{j}}(x_{\bar{j}}+1) \leq 0$, the algorithm terminates. Otherwise, we increase $x_{\bar{j}}$ by one unit, and go to the next iteration. Show that upon termination, the algorithm yields an optimal solution. *Note*: The book by Ibaraki and Katoh [IbK88] contains a lot of material on this problem, and addresses the issues of efficient implementation.

5.5.4 (Monotone Discrete Problems [WuB99]) (www)

Consider the problem

$$\begin{aligned} &\text{minimize} \quad f(x) \\ &\text{subject to} \quad g_i(x) \leq 0, \qquad i=1,\ldots,n, \\ &\qquad x_i \in X_i, \qquad i=1,\ldots,n, \end{aligned}$$

where the functions $f : \Re^n \mapsto \Re$ and $g_i : \Re^n \mapsto \Re$ are given, and the sets X_i are *finite* subsets of real numbers. Let X denote the Cartesian product $X_1 \times \cdots \times X_n$ and let $e_i = (0,\ldots,0,1,0,\ldots,0)'$ denote the ith unit vector. Assume that for all $x = (x_1,\ldots,x_n) \in X$, all $i = 1,\ldots,n$, and all $\xi > x_i$, we have

$$f\big(x + (\xi - x_i)e_i\big) \geq f(x),$$

$$g_i\big(x + (\xi - x_i)e_i\big) < g_i(x),$$

$$g_j\big(x + (\xi - x_i)e_i\big) \geq g_j(x), \qquad \forall\, j \neq i.$$

[This assumption is satisfied for example if f and g_i are linear of the form

$$f(x) = \sum_{i=1}^{n} c_i x_i, \qquad g_i(x) = \sum_{j=1}^{n} a_{ij} x_j - b_i, \quad i = 1, \ldots, n,$$

where for all i, $c_i \geq 0$, $a_{ii} < 0$, and $a_{ij} \geq 0$, for $j \neq i$.]

(a) Assume that x^* is an optimal solution of the problem. Consider the following algorithm: Start with an $x^0 \in X$ satisfying $x^0 \leq x^*$ (for example, set $x_i^0 = \min X_i$). Given x^k, stop if $g_i(x^k) \leq 0$ for all $i = 1, \ldots, n$; otherwise, select any i such that $g_i(x^k) > 0$ and set

$$x^{k+1} = x^k + (\xi - x_i^k)e_i,$$

where ξ is the smallest element of X_i such that $g_i\big(x^k + (\xi - x_i^k)e_i\big) \leq 0$ (see Fig. 5.5.8). Show that the algorithm is well-defined and terminates after a finite number of iterations with an optimal solution. Under what conditions will this optimal solution be equal to x^*? *Hint*: Verify that $x^k \leq x^*$ for all k.

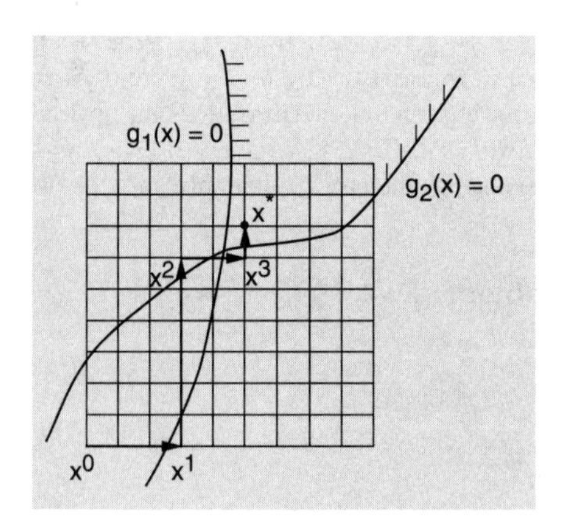

Figure 5.5.8. Illustration of the algorithm of Exercise 5.5.8 in two dimensions. The grid points correspond to the elements of $X_1 \times X_2$.

(b) Show that regardless of the starting point, the algorithm will terminate in a finite number of iterations even if the problem is infeasible. How should the starting point be chosen so that the algorithm can detect infeasibility?

(c) Consider the more general problem where there are inequality constraints $g_i(x) \leq 0$, $i \in I$, for only a subset I of the set of indices $\{1, \ldots, n\}$. Show how this case can be reduced to the case where $I = \{1, \ldots, n\}$.

5.5.5 (www)

Show that an $m \times n$ matrix E is totally unimodular if and only if every subset J of $\{1, \ldots, n\}$ can be partitioned into two subsets J_1 and J_2 such that

$$\left| \sum_{j \in J_1} e_{ij} - \sum_{j \in J_2} e_{ij} \right| \leq 1, \qquad \forall\, i = 1, \ldots, m.$$

5.5.6 (www)

Let E be a matrix with entries that are either 1 or -1 or 0. Suppose that E has at most two nonzero entries in each column. Show that E is totally unimodular.

5.5.7 (www)

Let E be a matrix with entries that are either 0 or 1. Suppose that in each column, all the entries that are equal to 1 appear consecutively. Show that E is totally unimodular.

5.5.8 (Birkhoff's Theorem for Doubly Stochastic Matrices)

A *doubly stochastic* $n \times n$ matrix $X = \{x_{ij}\}$ is a matrix such that the elements of each of its rows and columns are nonnegative, and add to one. Thus the set of doubly stochastic matrices can be identified with the polyhedron S of $\Re^{n^2}$ specified by the inequalities

$$\sum_{j=1}^{n} x_{ij} = 1, \qquad \forall\, i = 1, \ldots, n,$$

$$\sum_{i=1}^{n} x_{ij} = 1, \qquad \forall\, j = 1, \ldots, n,$$

$$x_{ij} \geq 0, \qquad \forall\, i,\, j = 1, \ldots, n.$$

A *permutation matrix* is a doubly stochastic matrix whose elements are either one or zero.

(a) Show that the extreme points of S are the permutation matrices.

(b) Verify that the matrix specifying the polyhedron S is totally unimodular.

(c) Show that every doubly stochastic matrix X can be written as $\sum_{i=1}^{k} \gamma_i X_i^*$, where X_i^* are permutation matrices and $\gamma_i \geq 0$, $\sum_{i=1}^{k} \gamma_i = 1$.

5.5.9 (Constructing Valid Inequalities [Gom58])

Consider the polyhedron

$$P = \left\{ x \geq 0 \;\middle|\; \sum_{i=1}^{n} a_i x_i \leq b \right\},$$

where $a_1, \ldots, a_n$ and b are given vectors in $\Re^n$, and $x_1, \ldots, x_n$ are the coordinates of x. Let u be any vector with nonnegative coordinates. Show that the following steps produce progressively tighter inequalities, which are satisfied by all $x \in P$ that are integer.

(1) $\sum_{j=1}^{n} u' a_j x_j \leq u' b$.

(2) $\sum_{j=1}^{n} \lfloor u' a_j \rfloor x_j \leq u' b$.

(3) $\sum_{j=1}^{n} \lfloor u' a_j \rfloor x_j \leq \lfloor u' b \rfloor$.

Note: This procedure is important both for theoretical representations of valid inequalities and for use in algorithms (see e.g., [Wol98], p. 120).

5.5.10 (www)

The purpose of this problem is to show that potentially better bounds can be obtained from Lagrangian relaxation by dualizing as few constraints as possible. Consider the problem

$$\begin{aligned} &\text{minimize } f(x) \\ &\text{subject to } x \in X, \qquad g_j(x) \leq 0, \quad j = 1, \ldots, r, \end{aligned} \tag{5.97}$$

where X is a subset of $\Re^n$, and $f : \Re^n \mapsto \Re$, $g_j : \Re^n \mapsto \Re$ are given functions. We assume that the problem is feasible and that its optimal value f^* satisfies $f^* > -\infty$.

Let $\bar{r}$ be an integer with $1 \leq \bar{r} < r$, consider the set

$$\bar{X} = \left\{ x \in X \mid g_{\bar{r}+1}(x) \leq 0, \ldots, g_r(x) \leq 0 \right\},$$

and the problem

$$\begin{aligned} &\text{minimize } f(x) \\ &\text{subject to } x \in \bar{X}, \qquad g_j(x) \leq 0, \quad j = 1, \ldots, \bar{r}. \end{aligned} \tag{5.98}$$

Let q^* and $\bar{q}^*$ be the optimal dual values of problems (5.97) and (5.98), respectively.

(a) Show that $q^* \leq \bar{q}^* \leq f^*$.

(b) Show that if problem (5.97) has no duality gap, the same is true for problem (5.98). Furthermore, if $(\mu_1^*, \ldots, \mu_r^*)$ is a Lagrange multiplier for problem (5.97), then $(\mu_1^*, \ldots, \mu_{\bar{r}}^*)$ is a Lagrange multiplier for problem (5.98).

(c) Construct an example to show that it is possible that $q^* < \bar{q}^* = f^*$.

5.5.11 (Integer Approximations of Feasible Solutions)

(a) Given a feasible solution x of the network optimization problem of Example 5.5.1, show that there exists an integer feasible solution $\bar{x}$ satisfying

$$|x_{ij} - \bar{x}_{ij}| < 1, \qquad \forall\ (i,j) \in \mathcal{A}.$$

Hint: For each arc (i,j), define the integer flow bounds

$$\bar{b}_{ij} = \lfloor x_{ij} \rfloor, \qquad \bar{c}_{ij} = \lceil x_{ij} \rceil.$$

(b) Generalize the result of part (a) for the case of a unimodular problem of Example 5.5.2.

5.6 NOTES AND SOURCES

Section 5.1: An extensive account of duality theory for convex programming is given by Rockafellar [Roc70]. Estimates of the duality gap for separable problems are given by Aubin and Ekeland [AuE76], and by Bertsekas [Ber82a], Section 5.6.1, which we follow in our discussion.

Section 5.3: For a recent weakening of the Slater constraint qualification (cf. Assumption 5.3.1), see Jeyakumar and Wolkowicz [JeW92]. The use of directional convexity in duality theory and optimality conditions for discrete-time optimal control (Exercise 5.3.6) is discussed by Cannon, Cullum, and Polak [CCP70].

Section 5.4: Fenchel duality theory and the theory of conjugate convex functions were developed by Fenchel [Fen49], [Fen51], where the relation with the classical Legendre transformation was noted. An extensive development, including extensions, is given by Rockafellar [Roc70]; see also Luenberger [Lue69].

Monotropic programming has been developed by Rockafellar [Roc67], [Roc69], following the work of Minty [Min60], who treated the special case where the constraint subspace is specified by conservation of flow constraints in a network (cf. Section 5.4.2). An extensive account of monotropic programming theory, associated algorithms, and application to network optimization are given in Rockafellar's book [Roc84]. A more condensed version of this theory, including a proof of the main duality result (Prop. 5.4.3), is given in the author's network optimization book [Ber98]. Primal-dual algorithms for monotropic programming are given by Rockafellar [Roc84], and Tseng and Bertsekas [TsB90]. The paper by Tseng [Tse98] gives a complexity analysis for monotropic programming based on a dual algorithm.

The exact penalty function analysis based on Fenchel's duality theorem is based on Bertsekas [Ber75a]; see also Yokohama [Yok92]. Finite convergence of the quadratic method of multipliers for linear programming problems (cf. Exercise 5.4.16) was shown by Poljak and Tretjakov [PoT74].

The proximal minimization algorithm has been proposed by Martinet [Mar70], [Mar72] and was extensively developed in a more general setting that applies to solution of equations and variational inequalities by Rockafellar [Roc76a]. The relation of the algorithm with the method of multipliers is discussed by Rockafellar [Roc76b], following his earlier work in [Roc73]. The convergence rate has been discussed under various conditions by Kort and Bertsekas [KoB76] (see Exercise 5.4.15), Rockafellar [Roc76a], and Luque [Luq84]. A survey is given by Lemaire [Lem89].

The entropy minimization algorithm together with its primal counterpart, the exponential method of multipliers, were introduced in the paper by Kort and Bertsekas [KoB72], and the thesis by Kort [Kor75], in the context of broad classes of methods involving nonquadratic augmented Lagrangians (see also Kort and Bertsekas [KoB76], and Bertsekas [Ber82a]). There has been much recent work in this area, directed at obtaining additional classes of methods, sharper convergence results, and an understanding of the properties that enhance computational performance; see Censor and Zenios [CeZ92], Guler [Gul92], Teboulle [Teb92], Chen and Teboulle [ChT93], [ChT94], Tseng and Bertsekas [TsB93], Bertsekas and Tseng [BeT94], Eckstein [Eck94a], Iusem, Svaiter, and Teboulle [IST94], Iusem and Teboulle [IuT95], Polyak and Teboulle [PoT95], Auslender, Cominetti, and Haddou [AHR97], Kiwiel [Kiw97a], Doljansky and Teboulle [DoT98], and Iusem [Ius98].

Section 5.5: There is a great variety of integer constrained optimization problems, and the associated methodological and applications literature is vast. For textbook treatments at various levels of sophistication, some of which are restricted to network optimization, see Lawler [Law76], Zoutendijk [Zou76], Papadimitriou and Steiglitz [PaS82], Minoux [Min86], Schrijver [Sch86], Nemhauser and Wolsey [NeW88], Bazaraa, Jarvis, and Sherali [BJS90], Bogart [Bog90], Murty [Mur92], Pulleyblank, Cook, Cunningham, and Schrijver [PCC93], Cameron [Cam94], Bertsimas and Tsitsiklis [BeT97], Bertsekas [Ber98], Cook, Cunningham, Pulleyblank, and Schrijver [CCP98], Wolsey [Wol98], and Sherali and Adams [ShA99].

Volumes 7 and 8 of the Handbooks in OR/MS, edited by Ball, Magnanti, Monma, and Nemhauser [BMM95a], [BMM95b], are devoted to network theory and applications, and include several excellent survey papers with large bibliographies. O'hEigeartaigh, Lenstra, and Rinnoy Kan [OLR85] provide an extensive bibliography on combinatorial optimization.

The paper by Dantzig, Fulkerson, and Johnson [DFJ54] introduced the branch-and-bound method in the context of the traveling salesman problem. This paper was followed by Croes [Cro58], Eastman [Eas58], and Land and Doig [LaD60], who considered versions of the branch-and-bound method in the context of various integer programming problems. Balas and Toth [BaT85], Nemhauser and Wolsey [NeW88], and Wolsey [Wol98] provide extensive surveys of branch-and-bound and its variations. Lagrangian relaxation was suggested in the context of discrete optimization by Held and Karp [HeK70], [HeK71], and was applied to the traveling salesman problem.

In many practical discrete optimization problems, the branch-and-bound method is too time-consuming for exact optimal solution, so it can only be used as an approximation scheme. There are other approaches, which do not offer the theoretical guarantees of branch-and-bound, but are much faster in practice. Local search methods, such as genetic algorithms, tabu search, and simulated annealing are among the most popular of such alternative possibilities. For some textbook references that also include bibliographies, see Aarts and Lenstra [AaL97], Bertsimas and Tsitsiklis [BeT97], Bertsekas [Ber98], Glover and Laguna [GlL97], Goldberg [Gol89], Korst, Aarts, and Korst [KAK89].

6

Dual Methods

Contents

In this chapter we consider dual methods, i.e., methods for solving dual problems. In particular, we focus on the primal problem

$$\begin{aligned} &\text{minimize } f(x) \\ &\text{subject to } x \in X, \qquad g_j(x) \le 0, \quad j = 1, \ldots, r, \end{aligned} \tag{P}$$

and its dual

$$\begin{aligned} &\text{maximize } q(\mu) \\ &\text{subject to } \mu \ge 0, \end{aligned} \tag{D}$$

where $f : \Re^n \mapsto \Re$, $g_j : \Re^n \mapsto \Re$ are given functions, X is a subset of $\Re^n$, and

$$q(\mu) = \inf_{x \in X} L(x, \mu) = \inf_{x \in X} \{f(x) + \mu' g(x)\}$$

is the dual function (cf. Section 5.1). We may also consider additional equality constraints in the primal problem, and on occasion we pause to discuss special issues regarding their treatment.

As we embark on the study of dual methods, it is worth reflecting on the potential incentives for solving the dual problem in place of the primal. These are:

(a) The dual is a concave problem (concave cost, convex constraint set). By contrast, the primal need not be convex.

(b) The dual may have smaller dimension and/or simpler constraints than the primal.

(c) If there is no duality gap and the dual is solved exactly to yield a Lagrange multiplier μ^*, all optimal primal solutions can be obtained by minimizing the Lagrangian $L(x, \mu^*)$ over $x \in X$ [however, there may be additional minimizers of $L(x, \mu^*)$ that are primal-infeasible]. Furthermore, if the dual is solved approximately to yield an approximate Lagrange multiplier μ, and x_μ minimizes $L(x, \mu)$ over $x \in X$, then it can be seen by applying Prop. 5.1.5 of Section 5.1, that x_μ also solves the problem

$$\begin{aligned} &\text{minimize } f(x) \\ &\text{subject to } x \in X, \qquad g_j(x) \le g_j(x_\mu), \quad j = 1, \ldots, r. \end{aligned}$$

Thus if the constraint violations $g_j(x_\mu)$ are not much larger than zero, x_μ may be an acceptable practical solution.

(d) Even if there is a duality gap, for every $\mu \ge 0$, the dual value $q(\mu)$ is a lower bound to the optimal primal value (the weak duality theorem; Prop. 5.1.3). This lower bound may be useful in the context of discrete optimization and branch-and-bound procedures (cf. Section 5.5).

We should also consider some of the difficulties in solving the dual problem. The most important ones are the following:

(a) To evaluate the dual function at any μ requires minimization of the Lagrangian $L(x, \mu)$ over $x \in X$. In effect, this restricts the utility of dual methods to problems where this minimization can either be done in closed form or else is relatively simple; for example, when there is special structure that allows decomposition, as in the separable problems of Section 5.1.6 and the monotropic programming problems of Section 5.4.1.

(b) In many types of problems, the dual function is nondifferentiable, in which case the algorithms of Chapters 1, 2, and 4 do not apply.

(c) Even if we find an optimal dual solution μ^*, it may be difficult to obtain a primal feasible vector x from the minimization of $L(x, \mu^*)$ over $x \in X$ as required by the primal-dual optimality conditions of Prop. 5.1.5, since this minimization can also yield primal-infeasible vectors.

Another important point regarding large-scale optimization problems is that there are several different ways to introduce duality in their solution. For example an alternative strategy to take advantage of separability, often called *partitioning*, is to divide the variables in two subsets, and minimize first with respect to one subset while taking advantage of whatever simplification may arise by fixing the variables in the other subset. In particular, the problem

$$\begin{aligned} &\text{minimize} \ \ F(x) + G(y) \\ &\text{subject to} \ \ Ax + By = c, \quad x \in X, \ \ y \in Y \end{aligned}$$

can be written as

$$\begin{aligned} &\text{minimize} \ \ F(x) + \inf_{By = c - Ax,\, y \in Y} G(y) \\ &\text{subject to} \ \ x \in X, \end{aligned}$$

or

$$\begin{aligned} &\text{minimize} \ \ F(x) + \tilde{p}(c - Ax) \\ &\text{subject to} \ \ x \in X, \end{aligned}$$

where $\tilde{p}(\cdot)$ is the primal function of the minimization problem involving y above:

$$\tilde{p}(u) = \inf_{By = u,\, y \in Y} G(y).$$

Assuming no duality gap, this primal function and its subgradients can be calculated using the corresponding dual function and associated Lagrange multipliers, as discussed in Section 5.4.4.

Naturally, the differentiability properties of dual functions are a very important determinant of the type of dual method that is appropriate for a given problem. We consequently develop these properties first in Section

6.1. In Section 6.2, we consider methods for the case where q is differentiable, and discuss the applicability of various gradient-based methods from Chapters 1 and 2. In Sections 6.3 and 6.4, we consider methods for coping with nondifferentiability in the dual cost, and we discuss specific methods for large problems with special structure.

6.1 DUAL DERIVATIVES AND SUBGRADIENTS*

In this section, we will use some properties of subgradients of a convex or a concave function, developed in Appendix B, to characterize various differentiability properties that are of interest in the context of duality. We have already characterized in Section 5.4.4 the subgradients of the primal function in terms of optimal solutions of the dual problem, assuming no duality gap. We now focus on differentiability properties of the dual function.

For a given $\mu \in \Re^r$, suppose that x_μ minimizes the Lagrangian $L(x, \mu)$ over $x \in X$,

$$x_\mu = \arg\min_{x \in X} L(x, \mu) = \arg\min_{x \in X} \big\{ f(x) + \mu' g(x) \big\}.$$

An important fact for our purposes is that *$g(x_\mu)$ is a subgradient of the dual function q at μ*, i.e.,

$$q(\bar{\mu}) \leq q(\mu) + (\bar{\mu} - \mu)' g(x_\mu), \qquad \forall\, \bar{\mu} \in \Re^r. \tag{6.1}$$

To see this, we use the definition of q and x_μ to write for all $\bar{\mu} \in \Re^r$,

$$\begin{aligned} q(\bar{\mu}) &= \inf_{x \in X} \big\{ f(x) + \bar{\mu}' g(x) \big\} \\ &\leq f(x_\mu) + \bar{\mu}' g(x_\mu) \\ &= f(x_\mu) + \mu' g(x_\mu) + (\bar{\mu} - \mu)' g(x_\mu) \\ &= q(\mu) + (\bar{\mu} - \mu)' g(x_\mu). \end{aligned}$$

Note that this calculation is valid for all $\mu \in \Re^r$ for which there is a minimizing vector x_μ, regardless of whether $\mu \geq 0$.

What is particularly important here is that we need to compute x_μ anyway in order to evaluate the dual function at μ, so *a subgradient $g(x_\mu)$ is obtained essentially at no cost*. All of the dual methods to be discussed solve the dual problem by computing the dual function value and a subgradient at a sequence of vectors $\{\mu^k\}$. It is not necessary to compute the set of *all* subgradients at μ^k in these methods; a single subgradient is sufficient.

Despite the fact that the full set of subgradients at a point is not needed for the application of the following methodology, it is still useful to have characterizations of this set. For example it is important to derive conditions under which q is differentiable. We know from our preceding discussion that if q is differentiable at μ, there can be at most one value of $g(x_\mu)$ corresponding to vectors $x_\mu \in X$ minimizing $L(x, \mu)$. This suggests that q is everywhere differentiable (as well real-valued and concave) if for all μ, $L(x, \mu)$ is minimized at a unique $x_\mu \in X$. Indeed this can be inferred under some assumptions from the convexity results developed in Appendix B. In particular, we have the following proposition.

Proposition 6.1.1: Let X be a compact set, and let f and g be continuous over X. Assume also that for every $\mu \in \Re^r$, $L(x, \mu)$ is minimized over $x \in X$ at a unique point x_μ. Then, q is everywhere continuously differentiable and

$$\nabla q(\mu) = g(x_\mu), \qquad \forall\, \mu \in \Re^r.$$

Proof: To assert the uniqueness of the subgradient of q at μ, apply Danskin's theorem [Prop. B.25(a) in Appendix B] with the identifications $x \sim z$, $X \sim Z$, $\mu \sim x$, and $-L(x, \mu) \sim \phi(x, z)$. The assumptions of this theorem are satisfied because X is compact, while $L(x, \mu)$ is continuous as a function of x and concave (in fact linear) as a function of μ. The continuity of the dual gradient ∇q follows from Prop. B.23. **Q.E.D.**

Note that if the constraint functions g_j are linear, X is convex and compact, and f is *strictly* convex, then the assumptions of Prop. 6.1.1 are satisfied and the dual function q is differentiable. We will focus on this and other related cases in the next section, where we discuss methods for differentiable dual functions.

In the case where X is a discrete set, as for example in integer programming, the continuity and compactness assumptions of Prop. 6.1.1 are satisfied, but there typically exist some μ for which $L(x, \mu)$ has multiple minima, leading to nondifferentiabilities. In fact, it can be shown that *if there exists a duality gap, the dual function is nondifferentiable at every dual optimal solution*; see Exercise 6.1.1. Thus, nondifferentiabilities tend to arise at the most interesting points and cannot be ignored in dual methods.

An important special case of a nondifferentiable dual function is when q is polyhedral, i.e., it has the form

$$q(\mu) = \min_{i \in I} \{a_i'\mu + b_i\}, \tag{6.2}$$

where I is a finite index set, and $a_i \in \Re^r$ and b_i are given vectors and scalars, respectively. This case arises, for example, when dealing with discrete problem where X is a finite set. The set of all subgradients of q is then the convex hull of the vectors a_i for which the minimum is attained in Eq. (6.2), as shown in the following proposition.

Proposition 6.1.2: Let the dual function q be polyhedral of the form (6.2), and for every $\mu \in \Re^r$, let I_μ be the set of indices attaining the minimum in Eq. (6.2),

$$I_\mu = \{ i \in I \mid a_i'\mu + b_i = q(\mu) \}.$$

The set of all subgradients of q at μ is given by

$$\partial q(\mu) = \left\{ g \;\middle|\; g = \sum_{i \in I_\mu} \xi_i a_i,\ \xi_i \geq 0,\ \sum_{i \in I_\mu} \xi_i = 1 \right\}.$$

Proof: Apply Danskin's theorem [Prop. B.25(b) in Appendix B]. **Q.E.D.**

Even though a subgradient may not be a direction of ascent at points μ where $q(\mu)$ is nondifferentiable, it still maintains an important property of the gradient: *it makes an angle less than 90 degrees with all ascent directions at* μ, i.e., all the vectors $\alpha(\bar{\mu} - \mu)$ such that $\alpha > 0$ and $q(\bar{\mu}) > q(\mu)$. In particular, *a small move from* μ *along any subgradient at* μ *decreases the distance to any maximizer* μ^* *of* q. This property follows from Eq. (6.1) and is illustrated in Fig. 6.1.1. It will form the basis for a number of dual methods that use subgradients (see Section 6.3).

Second Derivatives

Under particularly favorable circumstances, the dual function may be twice differentiable, which opens up the possibility of using Newton's method to solve the dual problem.

To illustrate this, let us assume the following:

(a) $X = \Re^n$.

(b) f and g_j are twice continuously differentiable convex functions.

(c) There exists a primal and dual optimal solution pair (x^*, μ^*) such that

$$\nabla^2_{xx} L(x^*, \mu^*) : \text{ positive definite.} \tag{6.3}$$

Consider the system of n equations

$$\nabla_x L(x, \mu) = 0, \tag{6.4}$$

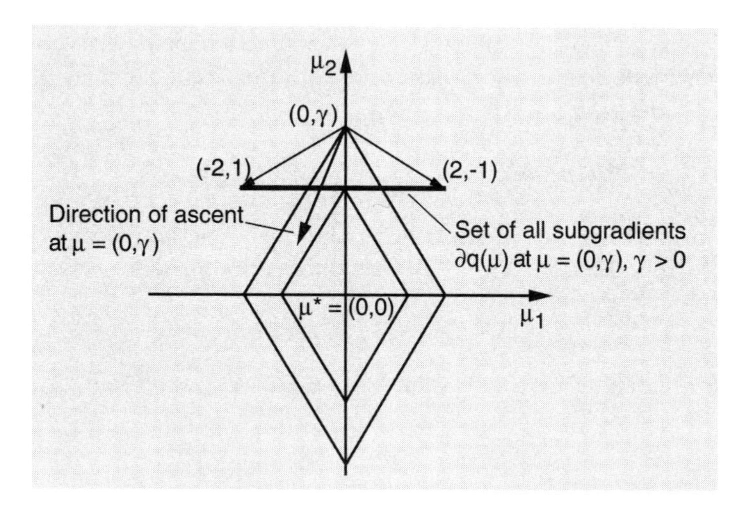

Figure 6.1.1. Illustration of the set of all subgradients $\partial q(\mu)$ at $\mu = (0,1)$ of the function

$$q(\mu) = -2|\mu_1| - |\mu_2|.$$

By expressing this function as

$$q(\mu) = \min\{-2\mu_1 - \mu_2, 2\mu_1 - \mu_2, -2\mu_1 + \mu_2, 2\mu_1 + \mu_2\},$$

we see that $\partial q(\mu)$ is the convex hull of the gradients of the functions attaining the above minimum (cf. Prop. 6.1.2). Thus, for $\mu = (0,\gamma)$, $\gamma > 0$, the minimum is attained by the first two functions having gradients $(-2,-1)$ and $(2,-1)$. As the result, $\partial q(\mu)$ is as shown in the figure. Note the characteristic property of a subgradient at μ; it makes an angle less than 90 degrees with every ascent direction at μ. As a result, a small move along a subgradient decreases the distance from the maximizing point $\mu^* = (0,0)$.

with the $(n+r)$ unknowns (x,μ). The pair (x^*,μ^*) is a solution of this system and the Jacobian with respect to x at this solution is $\nabla^2_{xx}L(x^*,\mu^*)$, which is positive definite by assumption (c). Therefore, the implicit function theorem (Prop. A.25 in Appendix A) applies and asserts that there exist neighborhoods S_{x^*} and S_{μ^*} of x^* and μ^*, respectively, such that Eq. (6.4) has a unique solution x_μ within S_{x^*} for every $\mu \in S_{\mu^*}$, i.e.,

$$\nabla_x L(x_\mu,\mu) = 0, \qquad \forall\ \mu \in S_{\mu^*}. \tag{6.5}$$

For $\mu \geq 0$, the Lagrangian $L(x,\mu)$ is convex as a function of x, so it follows that x_μ minimizes $L(x,\mu)$ over $x \in \Re^n$ for all $\mu \in S_{\mu^*}$ with $\mu \geq 0$, and the dual function satisfies

$$q(\mu) = \min_{x\in\Re^n} L(x,\mu) = L(x_\mu,\mu), \qquad \forall\ \mu \in S_{\mu^*} \cap \{\mu \mid \mu \geq 0\}. \tag{6.6}$$

We can derive the Hessian matrix of q by differentiating twice the above equation. Here, we will use the implicit function theorem to infer that x_μ is continuously differentiable as a function of μ with an $n \times r$ gradient matrix denoted ∇x_μ. By differentiating Eq. (6.6) and by using the fact $\nabla_x L(x_\mu, \mu) = 0$ [cf. Eq. (6.5)], we obtain

$$\nabla q(\mu) = \nabla x_\mu \nabla_x L(x_\mu, \mu) + g(x_\mu) = g(x_\mu),$$

which, incidentally, is consistent with the dual function differentiability result of Prop. 6.1.1 [cf. Eq. (6.2)]. By differentiating again this relation, we have

$$\nabla^2 q(\mu) = \nabla x_\mu \nabla g(x_\mu), \tag{6.7}$$

while by differentiating the relation $\nabla_x L(x_\mu, \mu) = 0$, we obtain

$$\nabla x_\mu \nabla^2_{xx} L(x_\mu, \mu) + \nabla g(x_\mu)' = 0. \tag{6.8}$$

By taking the neighborhood S_{μ^*} sufficiently small if needed, we can assert that

$$\nabla^2_{xx} L(x_\mu, \mu) : \text{ positive definite}, \qquad \forall\ \mu \in S_{\mu^*},$$

so that $\big(\nabla^2_{xx} L(x_\mu, \mu)\big)^{-1}$ exists. From Eq. (6.8) we see that

$$\nabla x_\mu = -\nabla g(x_\mu)' \big(\nabla^2_{xx} L(x_\mu, \mu)\big)^{-1},$$

which by substitution in Eq. (6.7), yields the desired Hessian of the dual function

$$\nabla^2 q(\mu) = -\nabla g(x_\mu)' \big(\nabla^2_{xx} L(x_\mu, \mu)\big)^{-1} \nabla g(x_\mu). \tag{6.9}$$

Example 6.1.1 (Separable Problems)

Consider the separable problem

$$\begin{aligned} &\text{minimize} \quad \sum_{i=1}^n f_i(x_i) \\ &\text{subject to} \quad \sum_{i=1}^n g_{ij}(x_i) \le 0, \qquad j = 1, \ldots, r, \end{aligned}$$

where $f_i : \Re \mapsto \Re$, $g_{ij} : \Re \mapsto \Re$, are twice continuously differentiable convex functions. An important consequence of separability is that the Hessian of the Lagrangian is a diagonal matrix. In particular, assuming that (x^*, μ^*) is a primal and dual optimal solution pair, and that the positive definiteness condition (6.3) is satisfied, we have using Eq. (6.9)

$$\nabla^2 q(\mu) = -\sum_{i=1}^n \frac{G_i' G_i}{\nabla^2_{x_i x_i} L} = -\sum_{i=1}^n \frac{G_i' G_i}{\nabla^2 f_i + \sum_{j=1}^r \mu_j \nabla^2 g_{ij}},$$

where G_i is the row vector $\left(\partial g_{i1}/\partial x_i, \ldots, \partial g_{ir}/\partial x_i\right)$, and all derivatives are evaluated at x_μ. Thus, since $\nabla^2_{x_i x_i} L$ is a scalar, the calculation of $\nabla^2 q$ does not involve a matrix inverse, thereby greatly simplifying the use of Newton's method for solving the dual problem.

EXERCISES

6.1.1

This exercise shows that nondifferentiabilities of the dual function often tend to arise at the most interesting points and thus cannot be ignored. Consider problem (P) and assume that for all $\mu \geq 0$, the infimum of the Lagrangian $L(x, \mu)$ over X is attained by at least one $x_\mu \in X$. Show that if there is a duality gap, then the dual function $q(\mu) = \inf_{x \in X} L(x, \mu)$ is nondifferentiable at every dual optimal solution. *Hint*: If q is differentiable at a dual optimal solution μ^*, by Prop. B.24(f), we must have $\partial q(\mu^*)/\partial \mu_j \leq 0$ and $\mu_j^* \partial q(\mu^*)/\partial \mu_j = 0$ for all j. Use this to show that any vector $x_{\mu^*} = \arg\min_{x \in X} L(x, \mu^*)$ together with μ^* satisfy the conditions for an optimal solution-Lagrange multiplier pair of Section 5.1.

6.1.2

Consider the separable problem of Example 6.1.1 for the case where each x_i is not a scalar coordinate but rather is an m-dimensional vector. Write a form of $\nabla^2 q$ that involves inversion of order m.

6.1.3

Consider the monotropic programming problem of Section 5.4.1

$$\begin{aligned} &\text{minimize} \quad \sum_{i=1}^{n} f_i(x_i) \\ &\text{subject to} \quad x \in S, \qquad x_i \in X_i, \quad i = 1, \ldots, n. \end{aligned}$$

Assume that the sets X_i are closed intervals, and that the functions $f_i : \Re \mapsto \Re$ are strictly convex and satisfy

$$\lim_{|x_i| \to \infty} f(x_i) = \infty.$$

Use Danskin's theorem (Prop. B.25 in Appendix B) to show that the dual function has the form

$$\sum_{i=1}^{n} g_i(\lambda_i)$$

and is everywhere real-valued and differentiable, where g_i is the conjugate of (f_i, X_i),

$$g_i(\lambda_i) = \sup_{x_i \in X_i} \big\{x_i\lambda_i - f_i(x_i)\big\}.$$

6.1.4

Let $g(\lambda)$ be the convex conjugate function of a pair (f, X) (cf. Section 5.4):

$$g(\lambda) = \sup_{x\in X}\big\{\lambda' x - f(x)\big\}, \qquad \lambda \in \Re^n.$$

Show that if x_λ attains the supremum above, then x_λ is a subgradient of g at λ.

6.2 DUAL ASCENT METHODS FOR DIFFERENTIABLE DUAL PROBLEMS*

If the dual problem is differentiable, we can apply the methodology for unconstrained and simply constrained problems of Chapters 1 and 2. For example, the use of scaled steepest ascent, gradient projection, and Newton-like methods is straightforward using the formulas for the first and the second derivatives developed in Section 6.1. (See Exercise 4.2.11 in Section 4.2 for a simple method of analysis of the steepest ascent method.) In some cases, even methods requiring line search, such as the conjugate gradient method, become possible thanks to the special structure of the dual function (see Exercise 6.2.1). In a dual context, coordinate ascent methods often become particularly interesting because of the structure of the dual problem. We first discuss such methods for the case of a strictly convex quadratic programming problem. We then consider various special structures and we delineate the type of assumptions that are needed in order to use methods that rely on differentiability.

6.2.1 Coordinate Ascent for Quadratic Programming

Consider the quadratic programming problem

$$\begin{aligned} &\text{minimize} \quad \tfrac{1}{2}x'Qx + c'x \\ &\text{subject to} \quad Ax \le b, \end{aligned} \tag{6.10}$$

where Q is a given $n \times n$ positive definite symmetric matrix, A is a given $m \times n$ matrix, and $b \in \Re^m$ and $c \in \Re^n$ are given vectors. Based on the duality theory developed in Section 3.4, the dual of the quadratic programming problem of Eq. (6.10) is given by

$$\begin{aligned} &\text{minimize} \quad \tfrac{1}{2}\mu' P \mu + t'\mu \\ &\text{subject to} \quad \mu \geq 0, \end{aligned}$$

where

$$P = AQ^{-1}A', \qquad t = b + AQ^{-1}c.$$

It is shown in Section 3.4 that if μ^* solves the dual problem, then $x^* = -Q^{-1}\big(c + A'\mu^*\big)$ solves the primal problem of Eq. (6.10). The dual problem has a simple constraint set, so it may be easier to solve than the primal. Furthermore, it turns out to be particularly well suited to the use of parallel algorithms; see e.g. Bertsekas and Tsitsiklis [BeT89], Censor and Zenios [CeZ97].

Let a_j denote the jth column of A'. We assume that a_j is nonzero for all j (if $a_j = 0$, then the corresponding constraint $a_j'x \leq b_j$ is meaningless and can be eliminated). Since Q is symmetric and positive definite, the jth diagonal element of P, given by $p_{jj} = a_j'Q^{-1}a_j$, is positive. This means that for every j, the dual cost function is strictly convex along the jth coordinate. Therefore, the strict convexity assumption of Prop. 2.7.1 in Section 2.7 is satisfied and it is possible to use the coordinate ascent method. Because the dual cost is quadratic, the minimization with respect to μ along each coordinate can be done analytically, and the iteration can be written explicitly.

The first partial derivative of the dual cost function with respect to μ_j is given by

$$t_j + \sum_{k=1}^{m} p_{jk}\mu_k,$$

where p_{jk} and t_j are the corresponding elements of the matrix P and the vector t, respectively. Setting the derivative to zero, we see that the unconstrained minimum of the dual cost along the jth coordinate starting from μ is attained at $\tilde{\mu}_j$ given by

$$\tilde{\mu}_j = -\frac{1}{p_{jj}}\left(t_j + \sum_{k \neq j} p_{jk}\mu_k\right) = \mu_j - \frac{1}{p_{jj}}\left(t_j + \sum_{k=1}^{m} p_{jk}\mu_k\right).$$

Taking into account the nonnegativity constraint $\mu_j \geq 0$, we see that the jth coordinate update has the form

$$\mu_j := \max\{0, \tilde{\mu}_j\} = \max\left\{0, \mu_j - \frac{1}{p_{jj}}\left(t_j + \sum_{k=1}^{m} p_{jk}\mu_k\right)\right\}. \tag{6.11}$$

The matrix A often has a sparse structure in practice, and one would like to take advantage of this structure. Unfortunately, the matrix $P = AQ^{-1}A'$ typically has a less advantageous sparsity structure than A. Furthermore, it may be undesirable to calculate and store the elements of P, particularly when m is large. It turns out that the coordinate ascent iteration (6.11) can be performed without explicit knowledge of the elements p_{jk} of the matrix P; only the elements of the matrix AQ^{-1} are needed instead. To see how this can be done, consider the vector

$$\nu = -A'\mu. \tag{6.12}$$

We have

$$P\mu = AQ^{-1}A'\mu = -AQ^{-1}\nu,$$

and the jth component of this vector equation yields

$$\sum_{k=1}^{m} p_{jk}\mu_k = -w_j'\nu, \tag{6.13}$$

where w_j' is the jth row of AQ^{-1}. We also have

$$p_{jj} = w_j'a_j, \tag{6.14}$$

where a_j is the jth column of A'. The coordinate ascent iteration (6.11) can now be written, using Eqs. (6.13) and (6.14), as

$$\mu_i := \begin{cases} \max\left\{0, \mu_j - \dfrac{1}{w_j'a_j}(t_j - w_j'\nu)\right\} & \text{if } i = j, \\ \mu_i & \text{if } i \neq j, \end{cases}$$

or, equivalently,

$$\mu := \mu - \min\left\{\mu_j, \frac{1}{w_j'a_j}(t_j - w_j'\nu)\right\} e_j, \tag{6.15}$$

where e_j is the jth unit vector (all its elements are 0 except for the jth, which is 1). The corresponding iteration for the vector ν of Eq. (6.12) is obtained by multiplication of Eq. (6.15) with $-A'$ yielding

$$\nu := \nu + \min\left\{\mu_j, \frac{1}{w_j'a_j}(t_j - w_j'\nu)\right\} a_j. \tag{6.16}$$

The coordinate ascent method can now be summarized as follows. Initially, μ is any vector in the nonnegative orthant and $\nu = -A'\mu$. At each iteration, a coordinate index j is chosen and μ and ν are iterated using Eqs. (6.15) and (6.16).

6.2.2 Decomposition and Primal Strict Convexity

Consider a strictly convex version of the separable problem discussed in Section 5.1.6:

$$
\begin{aligned}
&\text{minimize} \quad \sum_{i=1}^{m} f_i(x_i) \\
&\text{subject to} \quad \sum_{i=1}^{m} g_{ij}(x_i) \leq 0, \qquad j = 1, \ldots, r, \\
&\qquad\qquad x_i \in X_i, \qquad i = 1, \ldots, m,
\end{aligned}
\tag{6.17}
$$

where $f_i : \Re^{n_i} \mapsto \Re$ are strictly convex functions, x_i are the components of x, g_{ij} are given convex functions on $\Re^{n_i}$, and X_i are given convex and compact subsets of $\Re^{n_i}$. We derived the following dual problem that involves Lagrange multipliers for the inequality constraints:

$$
\begin{aligned}
&\text{maximize} \quad q(\mu) \\
&\text{subject to} \quad \mu \geq 0,
\end{aligned}
$$

where the dual function is given by

$$
q(\mu) = \min_{\substack{x_i \in X_i \\ i=1,\ldots,m}} \left\{ \sum_{i=1}^{m} \left(f_i(x_i) + \sum_{j=1}^{r} \mu_j g_{ij}(x_i) \right) \right\} = \sum_{i=1}^{m} q_i(\mu),
$$

and

$$
q_i(\mu) = \min_{x_i \in X_i} \left\{ f_i(x_i) + \sum_{j=1}^{r} \mu_j g_{ij}(x_i) \right\}, \qquad i = 1, \ldots, m. \tag{6.18}
$$

By applying Prop. 6.1.1, we see that strict convexity of f_i implies that the dual function is continuously differentiable, and that if the minimum in Eq. (6.18) is attained at the point $x_i(\mu)$, the partial derivative of q with respect to μ_j is given by

$$
\frac{\partial q(\mu)}{\partial \mu_j} = \sum_{i=1}^{m} g_{ij}\big(x_i(\mu)\big),
$$

where $x(\mu) = \big\{x_1(\mu), \ldots, x_m(\mu)\big\}$. Since the dual function is differentiable, we can apply methods considered in Chapters 1 and 2, such as coordinate ascent and gradient projection. The application of these methods is enhanced by the separable structure of the problem, which in turn greatly facilitates the evaluation of the dual function.

Example 6.2.1 (Price Coordination and Hierarchical Decomposition)

Ascent methods for solving the dual problem of the separable problem (6.17) admit an interesting interpretation in the context of optimization of a system consisting of m subsystems, where the cost function f_i is associated with the operations of the ith subsystem, and there are r resources to be shared by the subsystems. Here x_i is viewed as a vector of local decision variables that influences the cost of the ith subsystem only, and each of the constraints $\sum_{i=1}^{m} g_{ij}(x_i) \leq 0$ is viewed as a restriction of the availability of the jth resource. In particular, $g_{ij}(x_i)$ accounts for the portion of the jth resource consumed by the ith subsystem when its decision variables are x_i.

We may interpret the dual variable μ_j as a price of the jth resource, which is set by a coordinator. Given a price vector μ, the subsystems respond by setting their local decision variables to values $x_i(\mu)$ that minimize their total cost $f_i(x_i) + \sum_{j=1}^{r} \mu_j g_{ij}(x_i)$, including the cost of the resources at the given price vector μ. A dual ascent algorithm can be viewed as an iterative adjustment by the coordinator, which aims to set the prices at levels such that the subsystems, acting in their own best interest, minimize the overall system cost subject to the given resource constraints.

Unfortunately, in many convex separable problems the functions f_i are not strictly convex, so even though there is typically no duality gap, the dual function is not differentiable. In such cases it is possible to approach the problem via the proximal or entropy minimization algorithms of Section 5.4.6, and introduce strict convexity by adding to the primal cost function $\sum_{i=1}^{m} f_i(x_i)$ the corresponding strictly convex proximal term. Thus the problem is solved by solving a sequence of strictly convex separable problems, each of which has a differentiable dual problem. It turns out that this approach is additionally attractive when a parallel computer is available, because the differentiable dual problems often have structure that is suitable for parallel computation; see Bertsekas and Tsitsiklis [BeT89], Censor and Zenios [CeZ97], which describe a number of related algorithms.

6.2.3 Partitioning and Dual Strict Concavity

We now consider a different way to take advantage of separability, which is based on the partitioning approach outlined in the introduction to this chapter. Our starting point is the problem

$$\begin{aligned} &\text{minimize} \;\; F(x) + G(y) \\ &\text{subject to} \;\; Ax + By = c, \qquad x \in X, \; y \in Y. \end{aligned} \tag{6.19}$$

Here $F : \Re^n \mapsto \Re$ and $G : \Re^m \mapsto \Re$ are convex functions, X and Y are convex subsets of $\Re^n$ and $\Re^m$, respectively, and A is an $r \times n$ matrix, B is an $r \times m$ matrix, and $c \in \Re^r$ is a given vector. The optimization variables are x and y, and they are linked through the constraint $Ax + By = c$.

The idea here is to eliminate y by expressing its optimal value as a function of x. In particular, we first consider optimization with respect to y for a fixed value of x, i.e.,

$$\begin{aligned} &\text{minimize} \;\; G(y) \\ &\text{subject to} \;\; By = c - Ax, \qquad y \in Y, \end{aligned} \tag{6.20}$$

and then minimize with respect to x. Suppose that an optimal solution, denoted $y(x)$, of this problem exists for each $x \in X$. Then if x^* is an optimal solution of the problem

$$\begin{aligned} &\text{minimize} \;\; F(x) + G\big(y(x)\big) \\ &\text{subject to} \;\; x \in X, \end{aligned} \tag{6.21}$$

it is clear that $\big(x^*, y(x^*)\big)$ is an optimal solution of the original problem (6.19). We call problem (6.21) the *master problem*. The following proposition summarizes its differentiability properties.

Proposition 6.2.1: Assume that problem (6.20) has an optimal solution and at least one Lagrange multiplier for each $x \in X$.

(a) The master problem (6.21) can also be written as

$$\begin{aligned} &\text{minimize} \;\; F(x) + \tilde{p}(c - Ax) \\ &\text{subject to} \;\; x \in X, \end{aligned} \tag{6.22}$$

where $\tilde{p}(\cdot)$ is the primal function of problem (6.20)

$$\tilde{p}(u) = \inf_{By=u,\, y \in Y} G(y).$$

(b) The set of subgradients of $\tilde{p}$ at $c - Ax$ is the set of all $-\lambda$, where λ is a Lagrange multiplier of the problem

$$\begin{aligned} &\text{minimize} \;\; G(y) \\ &\text{subject to} \;\; By = c - Ax, \qquad y \in Y, \end{aligned}$$

corresponding to the constraint $By = c - Ax$.

(c) We have

$$\tilde{p}(c - Ax) = \max_{\lambda \in \Re^r} \big\{ \tilde{q}(\lambda) - \lambda'(c - Ax) \big\}, \tag{6.23}$$

where

$$\tilde{q}(\lambda) = \inf_{y \in Y} \big\{ G(y) + \lambda' By \big\}.$$

Furthermore, $\tilde{p}$ is differentiable at $c - Ax$ if $\tilde{q}$ is strictly concave over the set $\{\lambda \mid \tilde{q}(\lambda) > -\infty\}$.

Proof: Parts (a) and (b) follow from the analysis of Section 5.4.4. For part (c), we use the fact that $\tilde{q}(\lambda) - \lambda'(c - Ax)$ is the dual function corresponding to $\tilde{p}$, so $\tilde{p}$ is differentiable at $c - Ax$ if there is a unique λ attaining the maximum in Eq. (6.23). **Q.E.D.**

The preceding proposition suggests a solution approach whereby we first find an optimal solution x^* of the master problem (6.21), and then solve the corresponding problem (6.20) to obtain $y(x^*)$. An iterative method can be used for this; the required values and gradients or subgradients of Q at the iterates can be calculated by solving problems of the form (6.20), as shown by Prop. 6.2.1. In some special cases, the function $\tilde{p}$ can be explicitly calculated; see Exercise 6.2.2.

Example 6.2.2 (Coupling Variables – Stochastic Programming)

Suppose that $y = (y_1, \ldots, y_J)$, that G and Y have the separable form

$$G(y) = \sum_{j=1}^{J} G_j(y_j), \qquad Y = \{y \mid y_j \in Y_j,\ j = 1, \ldots, J\},$$

and that the constraint $Ax + By = c$ also has the separable form

$$A_j x + B_j y_j = c_j, \qquad j = 1, \ldots, J.$$

Thus the constraint matrix is block diagonal with the blocks being coupled by the entries A_j that correspond to x. We refer to x as the vector of *coupling variables*, since without this vector, the problem would be entirely decoupled into J subproblems, one for each y_j.

Separable problems involving coupling variables arise often in practice. A representative example is *stochastic programming*, which can also be viewed as a two-stage stochastic optimal control problem (see e.g. the textbooks by Kall and Wallace [KaW94], Prekopa [Pre95], and Birge and Louveaux [BiL97], and the survey by Ruszczynski [Rus97]). Here, once a vector $x \in X$ is selected, a random event occurs that has J possible outcomes $w_1, \ldots, w_J$. Another vector $y \in Y$ is then selected with knowledge of the outcome that occurred; thus for optimization purposes, we need to specify a different vector $y_j \in Y$ for each outcome w_j. The problem is then to minimize the expected cost

$$F(x) + \sum_{j=1}^{J} \pi_j G_j(y_j),$$

where $G_j(y_j)$ is the cost associated with the occurrence of w_j and π_j is the corresponding probability. If there are constraints of the form $A_jx + B_jy_j = c_j$, the stochastic programming problem becomes a special case of the problem with coupling variables formulated above.

The corresponding master problem may be solved by a gradient-based iterative method. The typical iteration at x involves solving the J subproblems

$$\begin{aligned} &\text{minimize} \quad G_j(y_j) \\ &\text{subject to} \quad B_jy_j = c_j - A_jx, \quad y_j \in Y_j, \end{aligned}$$

to obtain the value of $\tilde{p}(c-Ax)$. The gradient of $\tilde{p}(c-Ax)$ is given in terms of the vector of the Lagrange multipliers λ_j, $j = 1, \ldots, J$, of these subproblems, assuming that the functions

$$\tilde{q}_j(\lambda_j) = \inf_{y_j \in Y_j} \big\{G_j(y_j) + \lambda_j' B_j y_j\big\}$$

are strictly concave.

By Prop. 6.2.1(c), we can apply a differentiable descent method for solving the master problem (6.22) if F is differentiable and $\tilde{q}$ is strictly concave. If $\tilde{q}$ is not strictly concave we can consider adding to $\tilde{q}$ a strictly concave proximal quadratic term $-(1/2\gamma^k)\|\lambda - \lambda^k\|^2$, where $\gamma^k > 0$. By the theory of the proximal minimization algorithm of Section 5.4.6, this is equivalent to considering the augmented Lagrangian function

$$L_{\gamma^k}(x, y, \lambda^k) = G(y) + \lambda^{k\prime}(Ax + By - c) + \frac{\gamma^k}{2}\|Ax + By - c\|^2,$$

and replacing $\tilde{p}(c - Ax)$ in the master problem (6.22) by

$$\tilde{p}_{\gamma^k}(c - Ax, \lambda^k) = \min_{y \in Y} L_{\gamma^k}(x, y, \lambda^k).$$

However, once the solution x^k of the master problem

$$\begin{aligned} &\text{minimize} \quad F(x) + \tilde{p}_{\gamma^k}(c - Ax, \lambda^k) \\ &\text{subject to} \quad x \in X \end{aligned}$$

and the corresponding vector

$$y^k = \arg\min_{y \in Y} L_{\gamma^k}(x^k, y, \lambda^k)$$

are obtained, λ^k must be updated by

$$\lambda^{k+1} = \lambda^k + \gamma^k(Ax^k + By^k - c),$$

and a new master problem $\min_{x \in X}\big\{F(x) + \tilde{p}_{\gamma^{k+1}}(c - Ax, \lambda^{k+1})\big\}$ must be solved.

EXERCISES

6.2.1 (Line Search in Dual Ascent Methods [Pan84])

Consider the convex programming problem $\min_{Ax=b,\, x\in X} f(x)$ and its dual function $q(\lambda)$. Given a vector λ and a direction d, suppose that we want to find a stepsize α^* by the line search

$$\alpha^* = \arg\max_{\alpha\geq 0} q(\lambda + \alpha d).$$

Show that α^* is a Lagrange multiplier for the problem

$$\begin{aligned} &\text{minimize} \;\; f(x) + \lambda' Ax \\ &\text{subject to} \;\; d'(Ax - b) \leq 0, \quad x \in X. \end{aligned}$$

6.2.2 (Thevenin Decomposition [Ber93])

Consider problem (6.19), and assume that the matrix B has rank r, and that

$$G(y) = \tfrac{1}{2} y' R y + w' y, \qquad Y = \{y \mid Dy = d\},$$

where R is a positive definite symmetric $m \times m$ matrix, D is a given matrix, and d, w are given vectors. Assume further that the constraint set $\{y \mid By = c - Ax,\, Dy = d\}$ is nonempty for all x.

(a) Show that the function $\tilde{p}$ appearing in the master problem (6.22) is given by

$$\tilde{p}(c - Ax) = \min_{By = c - Ax,\, y \in Y} G(y) = \tfrac{1}{2}(Ax - b)' M (Ax - b) + \gamma,$$

where M is a $r \times r$ positive definite symmetric matrix, γ is a constant, and

$$b = c - B\bar{y},$$

with

$$\bar{y} = \arg\min_{y \in Y} G(y).$$

Furthermore, the vector

$$M(Ax - b)$$

is the unique Lagrange multiplier of the problem

$$\begin{aligned} &\text{minimize} \;\; G(y) \\ &\text{subject to} \;\; By = c - Ax, \qquad y \in Y, \end{aligned}$$

[cf. Eq. (6.20)] associated with the constraint $By = c - Ax$.

(b) Suppose that the matrix A has rank r. Show how to calculate the columns of M by solving r problems of the form (6.20) with x equal to each of r vectors such that the corresponding vectors $Ax - b$ are linearly independent.

6.3 NONDIFFERENTIABLE OPTIMIZATION METHODS*

We now consider algorithms for the case of a nondifferentiable dual function. We view the dual function value

$$q(\mu) = \inf_{x \in X} L(x, \mu) = \inf_{x \in X} \{f(x) + \mu' g(x)\}$$

as being defined for all $\mu \in \Re^r$, even though this value may be equal to $-\infty$ for some μ. We focus on the dual problem

$$\begin{aligned} &\text{maximize} \quad q(\mu) \\ &\text{subject to} \quad \mu \in M, \end{aligned}$$

where the constraint set M is given by

$$M = \{\mu \mid \mu \geq 0,\, q(\mu) > -\infty\}. \tag{6.24}$$

We assume throughout this section that for every $\mu \in M$, some vector x_μ that minimizes $L(x, \mu)$ over $x \in X$ can be calculated, yielding a subgradient $g(x_\mu)$ of q at μ [cf. Eq. (6.1)].

Note that if the dual function is real-valued [$q(\mu) > -\infty$ for all $\mu\Re^r$], then it is continuous, since it is concave [cf. Prop. B.9(a)]. In general, q need not be real-valued, but it is always upper-semicontinuous, i.e., for any sequence μ^k converging to a vector μ, we have

$$\limsup_{k \to \infty} q(\mu^k) \leq q(\mu).$$

To see this, note that we have

$$q(\mu^k) = \inf_{z \in X} L(z, \mu^k) \leq L(x, \mu^k), \qquad \forall\, x \in X,$$

so that

$$\limsup_{k \to \infty} q(\mu^k) \leq L(x, \mu), \qquad \forall\, x \in X.$$

Hence

$$\limsup_{k \to \infty} q(\mu^k) \leq \inf_{x \in X} L(x, \mu) = q(\mu).$$

The upper-semicontinuity of q implies that the domain $\{\mu \mid q(\mu) > -\infty\}$ of q is a closed set, and therefore the set M of Eq. (6.24) is also closed. While the discussion of this section focuses on the case where q is a dual function, *the analysis and the results apply to any function $q : \Re^r \mapsto [-\infty, \infty)$ that is upper-semicontinuous and concave, and for which we can calculate a subgradient at any μ in the set M of Eq. (6.24).*

We discuss two algorithms, both of which calculate a single subgradient at each iteration: the *subgradient method*, in Section 6.3.1, which at each iteration uses only the current subgradient, and the *cutting plane method*, in Section 6.3.3, which at each iteration uses all the subgradients previously calculated. Despite similarities, the philosophies underlying these two methods are quite different. The subgradient method may be viewed as an extension of the gradient and gradient projection methods, and uses subgradients as directions of improvement of the distance to the optimum. On the other hand, the cutting plane method uses subgradients in an approximation scheme, to construct increasingly accurate polyhedral approximations to the dual problem.

6.3.1 Subgradient Methods

The subgradient method generates a sequence of dual feasible points according to the iteration

$$\mu^{k+1} = \left[\mu^k + s^k g^k\right]^+, \tag{6.25}$$

where g^k is the subgradient $g(x_{\mu^k})$, $[\cdot]^+$ denotes projection on the closed convex set M, and s^k is a positive scalar stepsize. The iteration looks like the gradient projection method of Section 2.3, except that the subgradient g^k is used in place of the gradient (which may not exist). This difference is quite fundamental, however, because in contrast with the gradient projection method, the new iterate may not improve the dual cost for all values of the stepsize; that is, for some k we may have

$$q\big([\mu^k + sg^k]^+\big) < q(\mu^k), \qquad \forall\ s > 0;$$

see Fig. 6.3.1.

What makes the subgradient method work is that for sufficiently small stepsize s^k, the distance of the current iterate to the optimal solution set is reduced, as illustrated in Fig. 6.3.2. This is shown in the following proposition, which also provides an estimate for the range of appropriate stepsizes:

Proposition 6.3.1: If μ^k is not optimal, then for every dual optimal solution μ^*, we have

$$\|\mu^{k+1} - \mu^*\| < \|\mu^k - \mu^*\|,$$

for all stepsizes s^k such that

$$0 < s^k < \frac{2\big(q(\mu^*) - q(\mu^k)\big)}{\|g^k\|^2}. \tag{6.26}$$

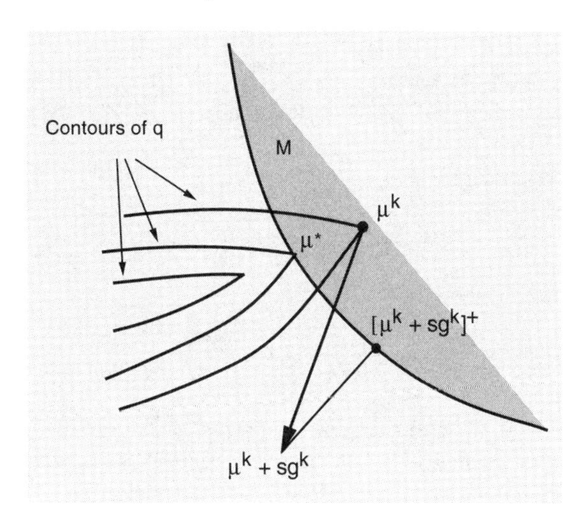

Figure 6.3.1. Illustration of how the iterate $[\mu^k + sg^k]^+$ may not improve the dual function with a particular choice of subgradient g^k, regardless of the value of the stepsize s.

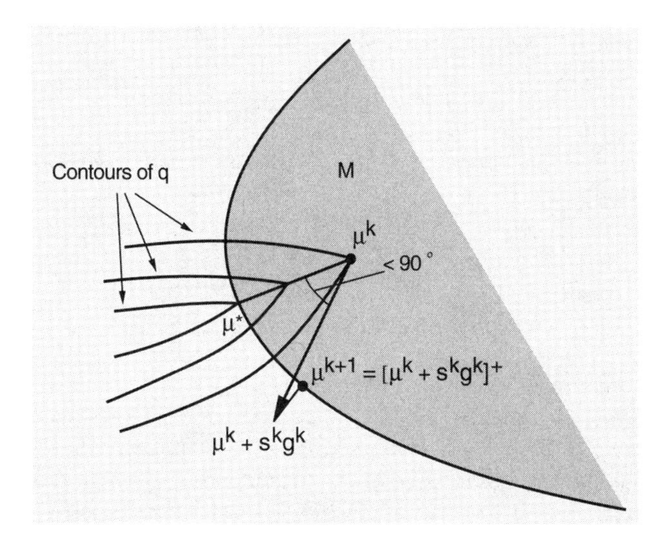

Figure 6.3.2. Illustration of how, given a nonoptimal μ^k, the distance to any optimal solution μ^* is reduced using a subgradient iteration with a sufficiently small stepsize. The crucial fact, which follows from the definition of a subgradient, is that the angle between the subgradient g^k and the vector $\mu^* - \mu^k$ is less than 90 degrees. As a result, the vector $\mu^k + s^k g^k$ is closer to μ^* than μ^k for s^k small enough. Through the projection on M, $[\mu^k + s^k g^k]^+$ gets even closer to μ^*.

Proof: We have

$$\|\mu^k + s^k g^k - \mu^*\|^2 = \|\mu^k - \mu^*\|^2 - 2s^k(\mu^* - \mu^k)'g^k + (s^k)^2\|g^k\|^2,$$

and by using the subgradient inequality,

$$(\mu^* - \mu^k)'g^k \geq q(\mu^*) - q(\mu^k),$$

we obtain

$$\|\mu^k + s^k g^k - \mu^*\|^2 \leq \|\mu^k - \mu^*\|^2 - 2s^k\big(q(\mu^*) - q(\mu^k)\big) + (s^k)^2\|g^k\|^2.$$

We can verify that for the range of stepsizes of Eq. (6.26) the sum of the last two terms in the above relation is negative. In particular, with a straightforward calculation, we can write this relation as

$$\|\mu^k + s^k g^k - \mu^*\|^2 \leq \|\mu^k - \mu^*\|^2 - \frac{\gamma^k(2-\gamma^k)\big(q(\mu^*) - q(\mu^k)\big)^2}{\|g^k\|^2}, \tag{6.27}$$

where

$$\gamma^k = \frac{s^k\|g^k\|^2}{q(\mu^*) - q(\mu^k)}.$$

If the stepsize s^k satisfies Eq. (6.26), then $0 < \gamma^k < 2$, so Eq. (6.27) yields

$$\|\mu^k + s^k g^k - \mu^*\| < \|\mu^k - \mu^*\|.$$

We now observe that since $\mu^* \in M$ and the projection operation is nonexpansive [Prop. 2.1.3(c) in Section 2.1], we have

$$\big\|\big[\mu^k + s^k g^k\big]^+ - \mu^*\big\| \leq \|\mu^k + s^k g^k - \mu^*\|.$$

By combining the last two inequalities, the result follows. **Q.E.D.**

The inequality (6.27) can also be used to establish convergence and rate of convergence results for the subgradient method with stepsize rules satisfying Eq. (6.26) (see Exercise 6.3.1). Unfortunately, however, unless we know the dual optimal value $q(\mu^*)$, which is rare, the range of stepsizes (6.26) is unknown. In practice, typically one uses the stepsize formula

$$s^k = \frac{\alpha^k\big(q^k - q(\mu^k)\big)}{\|g^k\|^2}, \tag{6.28}$$

where q^k is an approximation to the optimal dual value and

$$0 < \alpha^k < 2.$$

To estimate the optimal dual value from below, we can use the best current dual value

$$\hat{q}^k = \max_{0 \le i \le k} q(\mu^i).$$

As an upper bound, we can use any primal value $f(\bar{x})$ corresponding to a primal feasible solution $\bar{x}$; in many circumstances, primal feasible solutions are naturally obtained in the course of the algorithm (see also Exercise 5.1.9, which provides a method to obtain improved upper bounds from primal feasible solutions). Finally, the special structure of many problems can be exploited to yield improved bounds to the optimal dual value.

Here are two common ways to choose α^k and q^k in the stepsize formula (6.28):

(a) q^k is the best known upper bound to the optimal dual value at the kth iteration and α^k is a number, which is initially equal to one and is decreased by a certain factor (say, two) every few (say, five or ten) iterations. An alternative formula for α^k is

$$\alpha^k = \frac{1+m}{k+m},$$

where m is a fixed positive integer. This formula satisfies the usual conditions for a diminishing stepsize

$$\alpha^k \to 0, \qquad \sum_{k=0}^{\infty} \alpha^k = \infty,$$

and in fact, one can show various convergence results for subgradient methods with a diminishing stepsize (see Exercises 6.3.13 and 6.3.14).

(b) $\alpha^k = 1$ for all k and q^k is given by

$$q^k = \big(1 + \beta(k)\big)\hat{q}^k, \tag{6.29}$$

where $\hat{q}^k$ is the best current dual value $\hat{q}^k = \max_{0 \le i \le k} q(\mu^i)$. Furthermore, $\beta(k)$ is a number greater than zero, which is increased by a certain factor if the previous iteration was a "success", i.e., it improved the best current dual value, and is decreased by some other factor otherwise. This method requires that $\hat{q}^k > 0$. Also, if upper bounds $\tilde{q}^k$ to the optimal dual value are available as discussed earlier, then a natural improvement to Eq. (6.29) is

$$q^k = \min\big\{\tilde{q}^k, \big(1 + \beta(k)\big)\hat{q}^k\big\}.$$

A number of convergence properties of the subgradient method are developed in the exercises under various assumptions. For example a convergent variation is given in Exercise 6.3.2. However, the theoretical convergence properties of the schemes often used in practice, including the ones given above, are neither solid nor well understood. Clearly, there are problems where the subgradient method works very poorly; think of an ill-conditioned differentiable problem, in which case the method becomes a variant of gradient projection.

On the other hand, the subgradient method is simple and works well for many types of problems, yielding good approximate solutions within a few tens or hundreds of iterations. Also, frequently a good primal solution can be obtained thanks to effective heuristics, even with a fairly suboptimal dual solution. There is no clear explanation of these fortuitous phenomena; apparently they are due to the special structure of many of the important types of practical problems.

There are several variations of subgradient methods that aim to accelerate the convergence of the basic method (see e.g., [CFM75], [Sho85], [Min86], [Sho98], [ZhL98], [ZLW99]). We develop some of these variations in exercises, and we discuss in some detail one type of variation in the following section.

6.3.2 Approximate and Incremental Subgradient Methods

Similar to their gradient counterparts, discussed in Section 1.2, subgradient methods with errors arise naturally in a variety of contexts. For example, if for a given $\mu \in M$, the dual function value $q(\mu)$ is calculated by minimizing *approximately* $L(x, \mu)$ over $x \in X$, the subgradient obtained [as well as the value of $q(\mu)$] will involve an error.

To analyze such methods, it is useful to introduce a notion of approximate subgradient. In particular, given a scalar $\epsilon \geq 0$ and a vector $\bar{\mu}$ with $q(\bar{\mu}) > -\infty$, we say that g is an *ϵ-subgradient at $\bar{\mu}$* if

$$q(\mu) \leq q(\bar{\mu}) + \epsilon + g'(\mu - \bar{\mu}), \qquad \forall\ \mu \in \Re^r.$$

The set of all ϵ-subgradients at $\bar{\mu}$ is called the *ϵ-subdifferential at $\bar{\mu}$* and is denoted by $\partial_\epsilon q(\bar{\mu})$. Note that every subgradient at a given point is also an ϵ-subgradient for all $\epsilon \geq 0$. Generally, however, an ϵ-subgradient need not be a subgradient, unless $\epsilon = 0$:

$$\partial q(\bar{\mu}) = \partial_0 q(\bar{\mu}) \subset \partial_\epsilon q(\bar{\mu}), \qquad \forall\ \epsilon \geq 0.$$

An approximate subgradient method is defined by

$$\mu^{k+1} = [\mu^k + s^k g^k]^+, \tag{6.30}$$

where g^k is an ϵ^k-subgradient at μ^k. A convergence analysis for this method is outlined in Exercises 6.3.12 and 6.3.13, where it is shown that it converges to within 2ϵ of the optimal dual value, assuming that $\epsilon^k \to \epsilon$ and the stepsize is appropriately chosen.

For an interesting context where this method applies, suppose that we minimize *approximately* $L(x, \mu^k)$ over $x \in X$, thereby obtaining a vector $x^k \in X$ with

$$L(x^k, \mu^k) \leq \inf_{x \in X} L(x, \mu^k) + \epsilon^k. \tag{6.31}$$

We claim that the corresponding constraint vector, $g(x^k)$, is an ϵ^k-subgradient at μ^k. Indeed, we have for all $\mu \in \Re^r$

$$\begin{aligned} q(\mu) &= \inf_{x \in X} \{ f(x) + \mu' g(x) \} \\ &\leq f(x^k) + \mu' g(x^k) \\ &= f(x^k) + {\mu^k}' g(x^k) + g(x^k)'(\mu - \mu^k) \\ &= q(\mu^k) + \epsilon^k + g(x^k)'(\mu - \mu^k), \end{aligned}$$

where the last inequality follows from Eq. (6.31). Thus we can view a subgradient method with errors in the Lagrangian minimization, as in Eq. (6.31), as a special case of the approximate subgradient method (6.30).

Another interesting type of approximate subgradient method is an *incremental* variant that is patterned after the incremental gradient method, discussed in Section 1.5.2. Assume that the dual function has the form

$$q(\mu) = \sum_{i=1}^{m} q_i(\mu),$$

where the q_i are concave and continuous over M (i.e., the structure corresponding to separable problems). The idea of the incremental method is to sequentially take steps along the subgradients of the component functions q_i, with intermediate adjustment of μ after processing each component function. Similar to the incremental approach for least squares of Section 1.5.2, *an iteration is viewed as a cycle of m subiterations*. If μ^k is the vector obtained after k cycles, the vector μ^{k+1} obtained after one more cycle is

$$\mu^{k+1} = \psi_m,$$

where ψ_m is obtained after the m steps

$$\psi_i = \left[\psi_{i-1} + s^k g_i\right]^+, \qquad i = 1, \ldots, m, \tag{6.32}$$

with g_i being a subgradient of q_i at ψ_{i-1}. The cycle starts with

$$\psi_0 = \mu^k.$$

As in Section 1.5.2, the motivation for the incremental subgradient method is faster convergence. In particular, we hope that far from the solution, a single cycle of the incremental subgradient method will be as effective as several (as many as m) iterations of the ordinary subgradient method. This type of behavior is illustrated in the following example.

Example 6.3.1

Assume that μ is a scalar, and that the problem has the form

$$\begin{aligned} &\text{maximize} \quad q(\mu) = \sum_{i=1}^{2(M+N)} q_i(\mu) \\ &\text{subject to} \quad \mu \in \Re, \end{aligned}$$

where the component functions q_i are given by

$$q_i(\mu) = \begin{cases} -|\mu - 1| & \text{for } i = 1, \ldots, M, \\ -|\mu + 1| & \text{for } i = M+1, \ldots, 2M, \\ -|\mu| & \text{for } i = 2M+1, \ldots, 2(M+N), \end{cases}$$

and, to illustrate most prominently the effects of incrementalism, we assume that the integers M and N are large and comparable, in the sense that $M/N = O(1)$. For simplicity here, we assume that μ is unconstrained; a similar example can be constructed when there are constraints. The maximum of q is attained for $\mu^* = 0$.

Consider the incremental subgradient method with a constant stepsize $s \in (0,1)$. Then for μ outside the interval $[-1,1]$ that contains the maxima of the component functions q_i, the subgradient of q_i is

$$g_i = \begin{cases} 1 & \text{if } \mu < -1, \\ -1 & \text{if } \mu > 1. \end{cases}$$

Thus each of the steps

$$\psi_i = \psi_{i-1} + s g_i$$

of Eq. (6.32) makes progress towards the maximum $\mu^* = 0$ when $\mu \notin [-1,1]$, and in fact we have $|\psi_i| = |\psi_{i-1}| - s$.

However, once the method enters the interval $[-1,1]$, it exhibits the type of oscillatory behavior that is also characteristic of the incremental gradient method of Section 1.5.2. In particular, when a function q_i of the form $-|\mu - 1|$ (or $-|\mu+1|$) is processed, the method takes a step of size s towards 1 (or -1, respectively). Thus the method generically oscillates, and the asymptotic size of the oscillation witin a cycle is roughly proportional to s.

It is important to note that the size of the oscillation also depends substantially on the *order* in which the functions q_i are processed within a cycle. The maximum oscillation size occurs when N functions $-|\mu|$ are processed, followed by the M functions $-|\mu+1|$, followed by N functions $-|\mu|$, and followed by the M functions $-|\mu - 1|$. Then it can be seen that the size of the oscillation is of order Ms. The minimum oscillation size occurs when the processing of the M functions $-|\mu+1|$ is interleaved with the processing of the M functions $-|\mu-1|$. Then within the interval $[-1,1]$, the steps corresponding to the functions $-|\mu+1|$ are essentially canceled by the steps corresponding to the functions $-|\mu-1|$, and asymptotically the size of the oscillation is a small integer multiple of s (depending on the details of when the component functions $-|\mu|$ are processed).

The preceding example illustrates several characteristics of the incremental subgradient method, which tend to manifest themselves in some generality:

(a) When far from the solution, the method can make much faster progress than the nonincremental subgradient method, particularly if the number of component functions m is large. The rate of progress also depends on the stepsize.

(b) When close to the solution, the method oscillates and the size of the oscillation (from start to end of a cycle) is proportional to the stepsize. Thus there is a tradeoff between rapid initial convergence (large stepsize) and size of asymptotic oscillation (small stepsize). With a diminishing stepsize the method is capable of attaining convergence (no asymptotic oscillation).

(c) The size of the oscillation depends also on the order in which the component functions q_i are processed within a cycle.

To address the potentially detrimental effect of an unfavorable order of processing the component functions, one may consider randomization. Indeed, a popular technique for incremental gradient methods, discussed in Section 1.5.2, is to reshuffle randomly the order of the component functions after each cycle. A variation of this method is to pick randomly a function q_i at each step rather than to pick each q_i exactly once in each cycle according to a randomized order. This idea seems to be particularly effective in the nondifferentiable context. As an illustration, consider the dual function of Example 6.3.1, and a randomized variant of the method which uses a small constant stepsize s, and at each step, selects with equal probability $1/2(M+N)$ an index i from the range $[1, 2(M+N)]$ and executes the iteration

$$\bar{\mu} = \mu + s g_i.$$

If the starting point is an integer multiple of s, then all the points generated by the algorithm will be integer multiples of s. As a result, the algorithm can be modeled by a Markov chain of the random walk type. The stationary probability distribution of the chain can be easily calculated, and one can show that asymptotically, the expected value of μ will be the optimal $\mu^* = 0$, and that its variance will be within a small integer multiple of s, which *does not depend on the number of components* $m = 2(M+N)$. Thus, the effect on the oscillation size of a poor order of processing of the components q_i within a cycle is mitigated by randomization. It is interesting to compare the nature of this result with the corresponding result for differentiable cost functions (see Example 1.5.6), where we showed that randomization has a qualitatively different and less advantageous effect.

Let us now focus on the connection between the incremental and the approximate subgradient methods (6.30) and (6.32). An important fact here is that *if two vectors μ and $\bar{\mu}$ are "near" each other, then subgradients* *at* $\bar{\mu}$ *can be viewed as* ϵ*-subgradients at* μ, where ϵ is "small." In particular, if $g \in \partial q(\bar{\mu})$, we have for all $w \in \Re^r$,

$$\begin{aligned} q(w) &\leq q(\bar{\mu}) + g'(w - \bar{\mu}) \\ &\leq q(\mu) + g'(w - \mu) + q(\bar{\mu}) - q(\mu) + g'(\mu - \bar{\mu}) \\ &\leq q(\mu) + g'(w - \mu) + \epsilon, \end{aligned}$$

where

$$\epsilon = |q(\bar{\mu}) - q(\mu)| + \|g\| \cdot \|\bar{\mu} - \mu\|.$$

Thus, we have $g \in \partial_\epsilon q(\mu)$, and ϵ is small when $\bar{\mu}$ is near μ.

We now observe from Eq. (6.32) that the ith step within a cycle of the incremental subgradient method involves the direction g_i, which is a subgradient of q_i at the corresponding vector ψ_{i-1}. If the stepsize s^k is small, then ψ_{i-1} is close to the vector μ^k available at the start of the cycle, and hence g_i is an ϵ_i-subgradient at q_i at μ^k, where ϵ_i is small. Let us ignore the projection operation in Eq. (6.32), and let us also use the easily shown formula

$$\partial_{\epsilon_1} q(\mu) + \cdots + \partial_{\epsilon_m} q(\mu) \subset \partial_\epsilon q(\mu), \qquad \forall\ \mu \text{ with } q(\mu) > -\infty,$$

where $\epsilon = \epsilon_1 + \cdots + \epsilon_m$, to approximate the ϵ-subdifferential of the sum $q = \sum_{i=1}^m q_i$. Then, it can be seen that the incremental subgradient iteration can be viewed as an approximate subgradient iteration with $\epsilon^k = \epsilon_1 + \cdots + \epsilon_m$. The size of ϵ^k depends on the size of s^k, as well as the dual function q, and we generally have $\epsilon^k \to 0$ as $s^k \to 0$.

The view of the incremental subgradient method that we outlined above, parallels the view of the incremental gradient method for least squares problems as a gradient method with errors (see Section 1.5.2). The convergence analysis of the incremental subgradient method is similar in spirit to the one for the least squares case, but considerably more complicated in its details. Some representative analysis is outlined in Exercises 6.3.15 and 6.3.16, together with various methods for choosing the stepsize s^k. Computational experimentation with the method has been quite encouraging.

6.3.3 Cutting Plane Methods

Consider again the dual problem

$$\begin{aligned} &\text{maximize} \quad q(\mu) \\ &\text{subject to} \quad \mu \in M. \end{aligned}$$

The cutting plane method consists of solving at the kth iteration the problem

$$\begin{aligned} &\text{maximize} \quad Q^k(\mu) \\ &\text{subject to} \quad \mu \in M, \end{aligned}$$

where the dual function is replaced by a polyhedral approximation Q^k, constructed using the points μ^i generated so far and their subgradients $g(x_{\mu^i})$, which are denoted by g^i. In particular, for $k = 1, 2, \ldots$,

$$Q^k(\mu) = \min\{q(\mu^0) + (\mu - \mu^0)'g^0, \ldots, q(\mu^{k-1}) + (\mu - \mu^{k-1})'g^{k-1}\} \quad (6.33)$$

and

$$\mu^k = \arg\max_{\mu \in M} Q^k(\mu); \quad (6.34)$$

see Fig. 6.3.3. We assume that the maximum of $Q^k(\mu)$ above is attained for all k. For those k for which this is not guaranteed, artificial bounds may be placed on the coordinates of μ, so that the maximization will be carried out over a compact set and consequently the maximum will be attained.

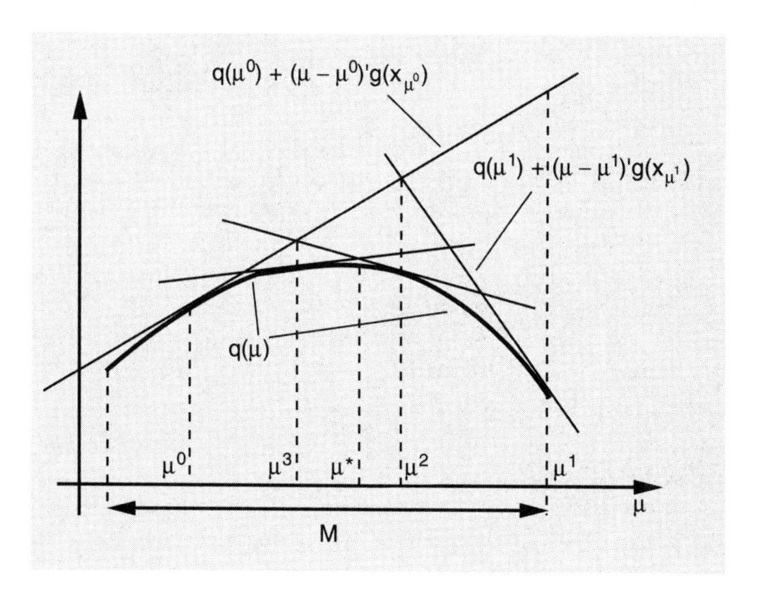

Figure 6.3.3. Illustration of the cutting plane method. With each new iterate μ^i, a new hyperplane $q(\mu^i) + (\mu - \mu^i)'g^i$ is added to the polyhedral approximation of the dual function.

The following proposition establishes the convergence properties of the cutting plane method. An important special case arises when the primal problem is a linear program, in which case the dual function is polyhedral and can be put in the form

$$q(\mu) = \min_{i \in I}\{a_i'\mu + b_i\}, \quad (6.35)$$

where I is a finite index set, and $a_i \in \Re^r$ and b_i are given vectors and scalars, respectively. Then, the subgradient g^k in the cutting plane method

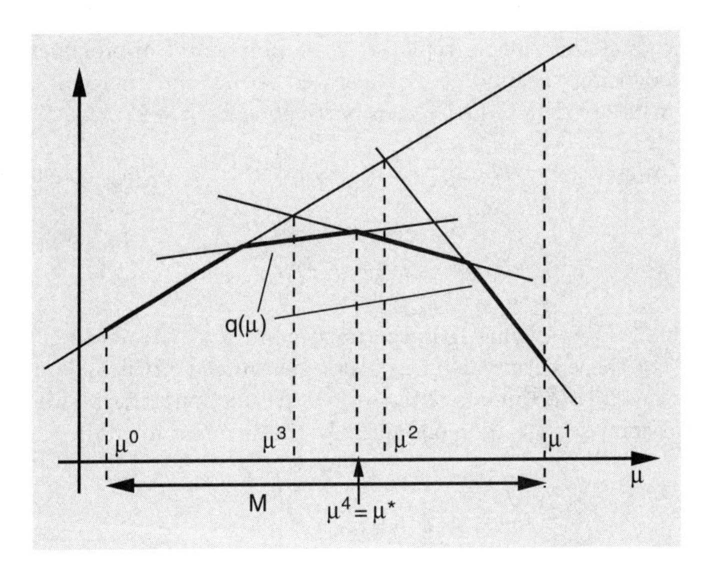

Figure 6.3.4. Illustration of the finite convergence property of the cutting plane method in the case where q is polyhedral. What happens here is that if μ^k is not optimal, a new cutting plane will be added at the corresponding iteration, and there can be only a finite number of cutting planes.

is a vector a_{i_k} for which the minimum in Eq. (6.35) is attained (cf. Prop. 6.1.2). In this case, the following proposition shows that the cutting plane converges finitely; see also Fig. 6.3.4.

Proposition 6.3.2:

(a) Assume that $\{g^k\}$ is a bounded sequence. Then every limit point of a sequence $\{\mu^k\}$ generated by the cutting plane method is a dual optimal solution.

(b) Assume that the dual function q is polyhedral of the form (6.35). Then the cutting plane method terminates finitely; that is, for some k, μ^k is a dual optimal solution.

Proof: (a) Since for all i, g^i is a subgradient of q at μ^i, we have

$$q(\mu^i) + (\mu - \mu^i)' g^i \geq q(\mu), \qquad \forall\, \mu \in M,$$

so from the definitions (6.33) and (6.34) of Q^k and μ^k, it follows that

$$Q^k(\mu^k) \geq Q^k(\mu) \geq q(\mu), \qquad \forall\, \mu \in M. \tag{6.36}$$

Suppose that a subsequence $\{\mu^k\}_K$ converges to $\bar{\mu}$. Then, since M is closed, we have $\bar{\mu} \in M$, and by using Eq. (6.36) and the definitions (6.33) and (6.34) of Q^k and μ^k, we obtain for all k and $i < k$,

$$q(\mu^i) + (\mu^k - \mu^i)'g^i \geq Q^k(\mu^k) \geq Q^k(\bar{\mu}) \geq q(\bar{\mu}). \tag{6.37}$$

We will now take the limit above as $i \to \infty$, $k \to \infty$, $i \in K$, $k \in K$. We have using the upper-semicontinuity of q,

$$\limsup_{i\to\infty,\, i\in K} q(\mu^i) \leq q(\bar{\mu}). \tag{6.38}$$

Also by using the assumption that the subgradient sequence $\{g^i\}$ is bounded, we have

$$\lim_{\substack{i\to\infty,\, k\to\infty,\\ i\in K,\, k\in K}} (\mu^k - \mu^i)'g^i = 0. \tag{6.39}$$

By combining Eqs. (6.37)-(6.39), we see that

$$q(\bar{\mu}) \geq \limsup_{k\to\infty,\, k\in K} Q^k(\mu^k) \geq \liminf_{k\to\infty,\, k\in K} Q^k(\mu^k) \geq q(\bar{\mu}),$$

and hence

$$\lim_{k\to\infty,\, k\in K} Q^k(\mu^k) = q(\bar{\mu}).$$

Combining this equation with Eq. (6.36), we obtain

$$q(\bar{\mu}) \geq q(\mu), \qquad \forall\ \mu \in M,$$

showing that $\bar{\mu}$ is a dual optimal solution.

(b) Let i^k be an index attaining the minimum in the equation

$$q(\mu^k) = \min_{i\in I}\{a_i'\mu^k + b_i\},$$

so that a_{i^k} is a subgradient at μ^k. From Eq. (6.36), we see that

$$Q^k(\mu^k) = q(\mu^k) \qquad \Longrightarrow \qquad \mu^k \text{ is dual optimal.}$$

Therefore, if μ^k is not dual optimal, we must have $Q^k(\mu^k) > q(\mu^k) = a_{i^k}'\mu^k + b_{i^k}$. Since

$$Q^k(\mu^k) = \min_{0\leq m\leq k-1}\{a_{i^m}'\mu^k + b_{i^m}\},$$

the pair (a_{i^k}, b_{i^k}) is not equal to any of the pairs $(a_{i^0}, b_{i^0}), \ldots, (a_{i^{k-1}}, b_{i^{k-1}})$. It follows that there can be only a finite number of iterations for which μ^k is not dual optimal. **Q.E.D.**

The reader may verify [using Prop. B.24(c)] that the boundedness assumption in Prop. 6.3.2(a) can be replaced by the assumption that $q(\mu)$

is real-valued for all $\mu \in \Re^r$, which can be ascertained if X is a finite set, or alternatively if f and g_j are continuous, and X is a compact set. Despite the finite convergence property shown in Prop. 6.3.2(b), the cutting plane method often tends to converge slowly, even for problems where the dual function is polyhedral. Indeed, typically one should base termination on the upper and lower bounds

$$Q^k(\mu^k) \geq \max_{\mu \in M} q(\mu) \geq \max_{0 \leq i \leq k-1} q(\mu^i),$$

[cf. Eq. (6.36)], rather that wait for finite termination to occur. Nonetheless, the method is often much better suited for solution of particular types of large problems than its competitors; see Section 6.4.

Linearly Constrained Versions

Consider the case where the constraint set M is polyhedral of the form

$$M = \{\mu \mid \gamma_i'\mu + \beta_i \leq 0,\ i \in I\},$$

where I is a finite set, and γ_i and β_i are given vectors and scalars, respectively. Let

$$w(\mu) = \max_{i \in I}\{\gamma_i'\mu + \beta_i\}.$$

It is then possible to consider a variation of the cutting plane method, where both functions q and w are replaced by polyhedral approximations Q^k and W^k, respectively. The method is

$$\mu^k = \arg \max_{W^k(\mu) \leq 0} Q^k(\mu).$$

As earlier,

$$Q^k(\mu) = \min\big\{q(\mu^0) + (\mu - \mu^0)'g^0, \ldots, q(\mu^{k-1}) + (\mu - \mu^{k-1})'g^{k-1}\big\},$$

with g^i being a subgradient of q at μ^i. The polyhedral approximation W^k is given by

$$W^k(\mu) = \max_{i \in I^k}\{\gamma_i{}'\mu + \beta^i\},$$

where I^k is a subset of I generated as follows: I^0 is an arbitrary subset of I, and I^k is obtained from I^{k-1} by setting $I^k = I^{k-1}$ if $w(\mu^k) \leq 0$, and by adding to I^{k-1} one or more of the indices $i \notin I^{k-1}$ such that $\gamma_i{}'\mu^k + \beta^i > 0$ otherwise.

The convergence properties of this method are very similar to the ones of the earlier method. In fact a proposition analogous to Prop. 6.3.2 can be formulated and proved.

Proximal Cutting Plane Methods

One of the drawbacks of the cutting plane method is that it can take large steps away from the optimum even when it is close to (or even at) the optimum (cf. Exercise 6.3.7). This phenomenon is referred to as *instability*, and has another undesirable effect, namely, that μ^{k-1} may not be a good starting point for the algorithm that minimizes $Q^k(\mu)$. A way to limit the effects of this phenomenon is to add to the polyhedral function approximation a quadratic term that penalizes large deviations from the current point. Thus in this method, μ^{k+1} is obtained as

$$\mu^{k+1} = \arg\max_{\mu \in M} \left\{ Q^{k+1}(\mu) - \frac{1}{2c^k} \|\mu - \mu^k\|^2 \right\}, \tag{6.40}$$

where $\{c^k\}$ is a positive nondecreasing scalar parameter sequence, and as in Eq. (6.33),

$$Q^{k+1}(\mu) = \min\{q(\mu^0) + (\mu - \mu^0)'g^0, \ldots, q(\mu^k) + (\mu - \mu^k)'g^k\}.$$

We recognize this as an approximate version of the proximal minimization algorithm of Section 5.4.6, where the polyhedral approximation Q^{k+1} is used in place of the dual function q. One advantage of this method is that the maximum in Eq. (6.40) is guaranteed to be attained regardless of the nature of the constraint set M. It can also be shown that the method maintains the finite termination property of Prop. 6.3.2(b) in the case where q is polyhedral (combine Prop. 6.3.2 with Exercise 5.4.16 in Section 5.4).

The proximal cutting plane method idea can be strengthened by introducing scaling in the proximal term, yielding the method

$$\mu^{k+1} = \arg\max_{\mu \in M}\{Q^{k+1}(\mu) - \tfrac{1}{2}(\mu - \mu^k)'H^k(\mu - \mu^k)\},$$

where H^k is a positive definite symmetric matrix. The subgradient information collected in the cutting plane iterations can be used to improve the quality of the scaling matrices H^k via quasi-Newton updating formulas. Some of the ideas underlying this process are described in Exercise 2.3.8 of Section 2.3; see the references cited at the end of the chapter.

Central Cutting Plane Methods

These methods maintain a polyhedral approximation

$$Q^k(\mu) = \min\{q(\mu^0) + (\mu - \mu^0)'g^0, \ldots, q(\mu^{k-1}) + (\mu - \mu^{k-1})'g^{k-1}\}$$

to the dual function q, but they generate the next vector μ^k by using a somewhat different mechanism. In particular, instead of maximizing Q^k as in Eq. (6.34), the methods obtain μ^k by finding a "central pair" (μ^k, z^k) within the subset

$$S^k = \{(\mu, z) \mid \mu \in M,\ \tilde{q}^k \le q(\mu),\ \tilde{q}^k \le z \le Q^k(\mu)\},$$

where $\tilde{q}^k$ is the best lower bound to the optimal dual value that has been found so far,

$$\tilde{q}^k = \max_{i=0,\ldots,k-1} q(\mu^i).$$

The set S^k is illustrated in Fig. 6.3.5.

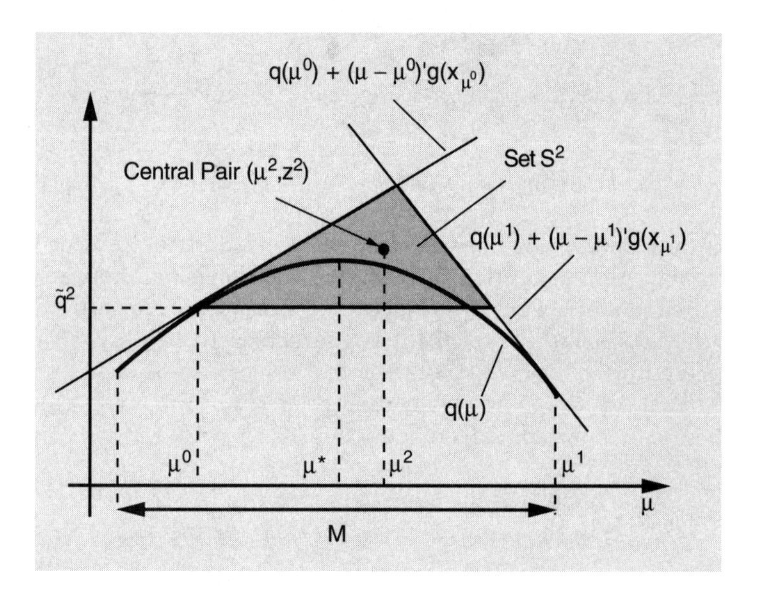

Figure 6.3.5. Illustration of the set

$$S^k = \{(\mu, z) \mid \mu \in M,\ \tilde{q}^k \le q(\mu),\ \tilde{q}^k \le z \le Q^k(\mu)\},$$

in the central cutting plane method.

There are several possible methods for finding the central pair (μ^k, z^k). Roughly, the idea is that the central pair should be "somewhere in the middle" of S. For example, consider the case where S is polyhedral with nonempty interior. Then (μ^k, z^k) could be the *analytic center* of S, where for any polyhedron $P = \{y \mid a_p' y \le c_p,\ p = 1, \ldots, m\}$ with nonempty interior, its analytic center is the unique maximizer of $\sum_{p=1}^m \ln(c_p - a_p' y)$ over $y \in P$. Another possibility is the *ball center* of S, i.e., the center of the largest inscribed sphere in S; for the generic polyhedron $P = \{y \mid a_p' y \le c_p,\ p = 1, \ldots, m\}$ with nonempty interior, the ball center can be obtained by solving the following problem with optimization variables (y, σ):

$$\begin{aligned} &\text{maximize } \ \sigma \\ &\text{subject to } \ a_p'(y + d) \leq c_p, \quad \forall\ \|d\| \leq \sigma,\ p = 1, \ldots, m. \end{aligned}$$

It can be seen that this problem is equivalent to the linear program

$$\begin{aligned} &\text{maximize } \ \sigma \\ &\text{subject to } \ a_p'y + \|a_p\|\sigma \leq c_p, \quad p = 1, \ldots, m. \end{aligned}$$

The convergence properties of central cutting plane methods are satisfactory, even though they do not terminate finitely in the case of a polyhedral q. Furthermore, these methods have benefited from advances in the implementation of interior point methods; see the references cited at the end of the chapter.

6.3.4 Ascent and Approximate Ascent Methods

The subgradient and cutting plane methods are not monotonic in the sense that they do not guarantee a dual function improvement at each iteration. We will now briefly discuss some of the difficulties in constructing ascent methods, i.e., methods where the dual function is increased at each iterate. For simplicity, we will restrict ourselves to the case where the dual function is real-valued, and the dual problem is unconstrained (i.e., $M = \Re^r$). Note, however, that constraints may be lumped into the dual function by using a nondifferentiable penalty function as discussed in Sections 4.3 and 5.4.

Coordinate Ascent

It is possible to consider the analog of the coordinate ascent method discussed in Section 1.8. Here we maximize the dual function successively along each coordinate direction. Unfortunately there is a fundamental difficulty. It is possible to get stuck at a corner from which it is impossible to make progress along any coordinate direction (see Fig. 6.3.6). This difficulty does not arise in some problem types (see Exercise 6.3.9), but may be hard to overcome in general.

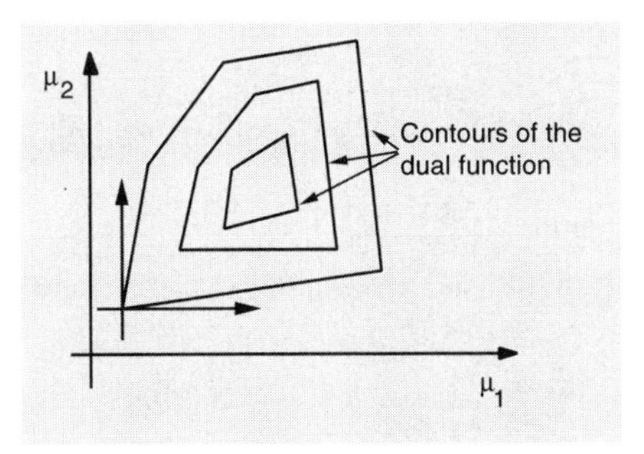

Figure 6.3.6. The basic difficulty with coordinate ascent for a nondifferentiable dual function. At some points it may be impossible to improve the dual function along any coordinate direction.

In some important special cases, however, such as when the primal problem is a network optimization problem of the type discussed in Sections 5.4.2 and 5.5, it is possible to modify the coordinate ascent method to make it workable. For such problems, two types of approaches have been introduced by the author:

(a) The *relaxation method*, first proposed for the assignment problem in [Ber81], and extended to linear cost network flow problems with and without arc gains [Ber85], [BHT87], [BeT88]. Here, ascent directions involving several coordinates are constructed, whenever no progress can be made along any coordinate direction.

(b) The *auction algorithm*, first proposed for the assignment problem in [Ber79b], and extended to linear cost network flow problems under the name *ϵ-relaxation*, [Ber86], [BeE88]. (See [BeT89], [Ber91], [Ber98] for textbook presentations of these methods, [Ber92] for a tutorial survey, and [BPT95], [TsB97] for extensions to the convex arc cost case with and without arc gains.) The main idea is to allow a single coordinate to change even if this worsens the dual function. When a coordinate is changed, however, it is set to ϵ plus the value that maximizes the dual function along that coordinate, where ϵ is a positive number. If ϵ is small enough, the algorithm can eventually approach the optimal solution as illustrated in Fig. 6.3.7. The proper choice of coordinate at each iteration is made using the notion of *ϵ-complementary slackness*, which is in turn also related to approximate subgradients (see Exercise 6.3.17 that can be used as the basis for approximate subgradient algorithms for separable problems).

Steepest Ascent and ϵ-Ascent Methods

The steepest ascent direction d_μ of a concave function $q : \Re^n \mapsto \Re$ at a vector μ is obtained by solving the problem

$$\begin{aligned} &\text{maximize} \quad q'(\mu; d) \\ &\text{subject to} \quad \|d\| \le 1, \end{aligned}$$

where

$$q'(\mu; d) = \lim_{\alpha \downarrow 0} \frac{q(\mu + \alpha d) - q(\mu)}{\alpha}$$

is the directional derivative of q at μ in the direction d. Using Prop. B.24 of Appendix B (and making an adjustment for the concavity of q), we have

$$q'(\mu; d) = \min_{g \in \partial q(\mu)} d'g,$$

and by the minimax theorem of Section 5.4.3 [using also the compactness of $\partial q(\mu)$; cf. Prop. B.24], we have

$$\max_{\|d\| \le 1} q'(\mu; d) = \max_{\|d\| \le 1} \min_{g \in \partial q(\mu)} d'g = \min_{g \in \partial q(\mu)} \max_{\|d\| \le 1} d'g = \min_{g \in \partial q(\mu)} \|g\|.$$

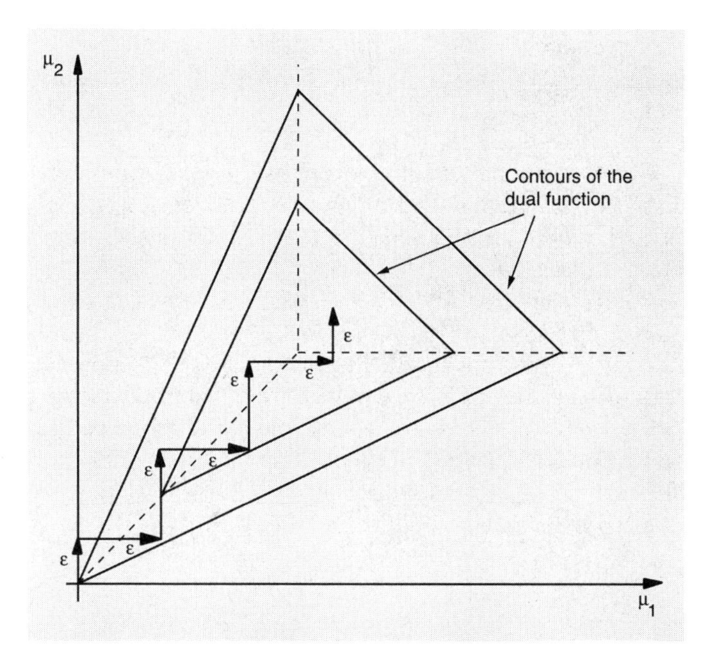

Figure 6.3.7. Path of the ϵ-relaxation method. When ϵ is small, it is possible to approach the optimal solution even if each step does not result in a dual function improvement. The method eventually reaches a small neighborhood of the optimal solution if the problem has a favorable (typically a network) structure.

It follows that the steepest ascent direction is

$$d_\mu = \frac{\bar{g}_\mu}{\|\bar{g}_\mu\|},$$

where $\bar{g}_\mu$ is the unique minimum-norm subgradient at μ,

$$\bar{g}_\mu = \arg \min_{g \in \partial q(\mu)} \|g\|. \tag{6.41}$$

The steepest ascent method has the form

$$\mu^{k+1} = \mu^k + \alpha^k \bar{g}^k,$$

where $\bar{g}^k$ is the vector of minimum norm in $\partial q(\mu^k)$, and α^k is a positive stepsize that guarantees that $q(\mu^{k+1}) > q(\mu^k)$ if μ^k is not optimal. One disadvantage of this method is that to calculate the steepest ascent direction, one needs to compute all subgradients at the current iterate. It is possible, however, to compute an approximate steepest ascent direction by employing approximations to the subdifferential that involve only a finite number of subgradients (see Exercise 6.3.4).

A more serious drawback of the steepest ascent method is that, depending on the stepsize rule, it may get stuck far from the optimum. This can happen even if the stepsize is chosen by line maximization; see the example of Exercise 6.3.8 and Fig. 6.3.8. The difficulty in this example is that at the limit, q is nondifferentiable and has subgradients that cannot be approximated by subgradients at any one of the iterates, arbitrarily close to the limit (this would not happen if q were continuously differentiable at the limit).

One way to construct a variant of steepest ascent that is convergent, is to obtain a direction by projecting in the manner of Eq. (6.41) on a larger set than the subdifferential. Such a set should include not only the subgradients at the current iterate, but also the subgradients of neighboring points. One possibility is to replace the subdifferential $\partial q(\mu)$ in the projection of Eq. (6.41) with the ϵ-subdifferential $\partial_\epsilon q(\mu)$ at μ, where ϵ is a positive scalar. The resulting method, called the *ϵ-ascent method*, and first introduced by Bertsekas and Mitter [BeM71], [BeM73], has the form

$$\mu^{k+1} = \mu^k + \alpha^k \tilde{g}^k,$$

where

$$\tilde{g}^k = \arg \min_{g \in \partial_\epsilon q(\mu^k)} \|g\|. \tag{6.42}$$

It can be implemented by either calculating explicitly the ϵ-subdifferential $\partial_\epsilon q(\mu)$ (in problems with favorable structure) or by approximating $\partial_\epsilon q(\mu)$ with a finite number of ϵ-subgradients, generated with a procedure described in Exercises 6.3.4 and 6.3.5. This procedure is due to Lemarechal [Lem74], who also developed in [Lem75] a class of methods, called *bundle methods*, which are closely related to the ϵ-ascent method. We refer to the specialized literature (particularly Hiriart-Urruty and Lemarechal [HiL93]) for further discussion.

An interesting variation of the ϵ-ascent method applies to the case where the dual function consists of the sum of several concave functions:

$$q(\mu) = q_1(\mu) + \cdots + q_m(\mu).$$

As we have seen, this is very common, particularly for problems with separable structure. Then, we can approximate the ϵ-subdifferential of q with the vector sum

$$\tilde{\partial}_\epsilon q(\mu) = \partial_\epsilon q_1(\mu) + \cdots + \partial_\epsilon q_m(\mu)$$

in the projection problem (6.42). The method (also proposed by Bertsekas and Mitter [BeM71]) consists of the iteration

$$\mu^{k+1} = \mu^k + \alpha^k \tilde{g}^k,$$

where

$$\tilde{g}^k = \arg \min_{g \in \tilde{\partial}_\epsilon q(\mu^k)} \|g\|. \tag{6.43}$$

In many cases, the approximation $\tilde{\partial}_\epsilon q(\mu)$ may be obtained much more easily than the exact ϵ-subdifferential $\partial_\epsilon q(\mu)$, thereby resulting in an algorithm that is much easier to implement. For an application of this algorithm to monotropic programming, we refer to Rockafellar [Roc81], [Roc84], and Bertsekas [Ber98]. The algorithm draws its validity from the fact

$$\partial_\epsilon q(\mu) \subset \tilde{\partial}_\epsilon q(\mu) \subset \partial_{m\epsilon} q(\mu),$$

in combination with the properties of ϵ-subdifferentials discussed in Exercise 6.3.5.

EXERCISES

6.3.1 (Convergence of the Subgradient Method [Pol69b]) (www)

Consider the subgradient method $\mu^{k+1} = [\mu^k + s^k g^k]^+$, where

$$s^k = \frac{q^* - q(\mu^k)}{\|g^k\|^2}$$

and q^* is the optimal dual value. Assume that there exists at least one optimal dual solution.

(a) Use Eq. (6.27) to show that $\{\mu^k\}$ is bounded.

(b) Assuming that $\{g^k\}$ is bounded, use Eq. (6.27) to show that $q(\mu^k) \to q^*$, and that $\{\mu^k\}$ converges to some optimal dual solution.

(c) Show that $\{g^k\}$ is bounded provided that $q(\mu)$ is real-valued for all $\mu \in \Re^r$. *Hint*: Use Prop. B.24 in Appendix B.

6.3.2 (A Convergent Variation of the Subgradient Method) (www)

Consider the subgradient method $\mu^{k+1} = [\mu^k + s^k g^k]^+$, where

$$s^k = \frac{\tilde{q} - q(\mu^k)}{\|g^k\|^2}.$$

(a) Suppose that $\tilde{q}$ is an *underestimate* of the optimal dual value q^* such that $q(\mu^k) < \tilde{q} < q^*$. [Here $\tilde{q}$ is fixed and the algorithm stops at μ^k if $q(\mu^k) \geq \tilde{q}$.] Assuming that $\{g^k\}$ is bounded, show that either for some $\bar{k}$ we have

$q(\mu^{\bar{k}}) \geq \tilde{q}$ or else $\{\mu^k\}$ converges to some $\tilde{\mu}$ with $q(\tilde{\mu}) \geq \tilde{q}$. *Hint*: Consider the function $\min\{q(\mu), \tilde{q}\}$ and use the results of Exercise 6.3.1.

(b) Suppose that $\tilde{q}$ is an *overestimate* of the optimal dual value, i.e., $\tilde{q} > q^*$. Assuming that $\{g^k\}$ is bounded, show that the length of the path traveled by the method is infinite, i.e.,

$$\sum_{k=0}^{\infty} s^k \|g^k\| = \sum_{k=0}^{\infty} \frac{\tilde{q} - q(\mu^k)}{\|g^k\|} = \infty.$$

Note: Parts (a) and (b) provide the basis for a method that uses an adjustable level $\tilde{q}$ in the stepsize formula. The method uses two positive scalars δ^0 and B, and operates in cycles during which $\tilde{q}$ is kept constant. Cycle 0 begins with the starting point μ^0. At the beginning of the typical cycle k, we have a vector μ^k and we set $\tilde{q} = q(\mu^k) + \delta^k$. Cycle k consists of successive subgradient iterations that start with μ^k and end when one of the following two occurs:

(1) The dual value exceeds $q(\mu^k) + \delta^k/2$.

(2) The length of the path traveled starting from μ^k exceeds B.

Then cycle $k+1$ begins with μ^{k+1} equal to the vector that has the highest dual value within cycle k, and either $\delta^{k+1} = \delta^k$ or $\delta^{k+1} = \delta^k/2$, depending on whether cycle k terminated with case (1) or case (2), respectively. This method was proposed by Brannlund [Bra93] who showed that it has satisfactory convergence properties; Goffin and Kiwiel [GoK99] provide a more refined convergence result.

6.3.3 (Convergence Rate of the Subgradient Method) 

Consider the subgradient method of Exercise 6.3.1 and assume that $\{g^k\}$ is a bounded sequence.

(a) Show that

$$\liminf_{k \to \infty} \sqrt{k}\big(q^* - q(\mu^k)\big) = 0.$$

Hint: Use Eq. (6.27) to show that $\sum_{k=0}^{\infty} \big(q^* - q(\mu^k)\big)^2 < \infty$. Assume that $\sqrt{k}\big(q^* - q(\mu^k)\big) \geq \epsilon$ for some $\epsilon > 0$ and arbitrarily large k, and reach a contradiction.

(b) Assume that for some $a > 0$ and all k, we have $q(\mu^*) - q(\mu^k) \geq a\|\mu^* - \mu^k\|$, where μ^* is some dual optimal solution. Use Eq. (6.27) to show that $\{\|\mu^k - \mu^*\|\}$ converges linearly. In particular, for all k we have

$$\|\mu^{k+1} - \mu^*\| \leq r\|\mu^k - \mu^*\|,$$

where $r = \sqrt{1 - a^2/b^2}$ and b is an upper bound for $\|g^k\|$.

6.3.4 (Generating Ascent Directions of Concave Functions [Lem74], [Lem75]) (www)

Let $q : \Re^n \mapsto \Re$ be a concave function and fix a vector $\mu \in \Re^n$. A vector $d \in \Re^n$ is said to be an ascent direction of q at μ if the corresponding directional derivative of q satisfies

$$q'(\mu; d) > 0.$$

This exercise provides a method for generating an ascent direction in circumstances where obtaining a single subgradient at μ is relatively easy.

Assume that μ does not maximize q and let g_1 be some subgradient of q at μ. For $k = 2, 3, \ldots$, let w^k be the vector of minimum norm in the convex hull of $g^1, \ldots, g^{k-1}$,

$$w^k = \arg \min_{g \in conv\{g^1, \ldots, g^{k-1}\}} \|g\|.$$

If w^k is an ascent direction stop; else let g^k be an element of $\partial q(\mu)$ such that

$$g^{k\prime} w^k = \min_{g \in \partial q(\mu)} g' w^k = q'(\mu; w^k),$$

[cf. Prop. B.24(b)]. Show that this process terminates in a finite number of steps with an ascent direction. *Hint*: If w^k is not an ascent direction, then $g^{i\prime} w^k \geq \|w^k\|^2 \geq \|g^*\|^2 > 0$ for all $i = 1, \ldots, k-1$, where g^* is the subgradient of minimum norm, while at the same time $g^{k\prime} w^k \leq 0$. Consider a limit point of $\{(w^k, g^k)\}$.

6.3.5 (ϵ-Ascent Method [BeM71], [BeM73]) (www)

Consider the unconstrained maximization of a concave function $q : \Re^r \mapsto \Re$, and for $\epsilon > 0$, consider the ϵ-subdifferential at a vector μ

$$\partial_\epsilon q(\mu) = \big\{ g \mid q(\bar{\mu}) \leq q(\mu) + g'(\bar{\mu} - \mu) + \epsilon, \ \forall \ \bar{\mu} \in \Re^r \big\}.$$

(a) Use the line of proof of Prop. B.24 [parts (a) and (b)] to show that $\partial_\epsilon q(\mu)$ is nonempty, convex, and compact, and that

$$\sup_{s>0} \frac{q(\mu + sd) - q(\mu) - \epsilon}{s} = \inf_{g \in \partial_\epsilon q(\mu)} d' g.$$

(b) Show that $0 \in \partial_\epsilon q(\mu)$ if and only if μ is an ϵ-optimal solution, i.e.,

$$\sup_{\bar{\mu} \in \Re^r} q(\bar{\mu}) - \epsilon \leq q(\mu).$$

(c) Show that if a direction d is such that $\inf_{g \in \partial_\epsilon q(\mu)} d' g > 0$, then

$$\sup_{s>0} q(\mu + sd) > q(\mu) + \epsilon.$$

(d) Show that if $0 \notin \partial_\epsilon q(\mu)$, then the direction

$$g_\mu = \arg \min_{g \in \partial_\epsilon q(\mu)} \|g\|$$

satisfies $g_\mu' g > 0$ for all $g \in \partial_\epsilon q(\mu)$.

(e) Consider the following procedure (first given in [Lem74]) that parallels the one of Exercise 6.3.4. Given μ, let g_1 be some ϵ-subgradient of q at μ. For $k = 2, 3, \ldots$, let w^k be the vector of minimum norm in the convex hull of $g^1, \ldots, g^{k-1}$,

$$w^k = \arg \min_{g \in conv\{g^1, \ldots, g^{k-1}\}} \|g\|.$$

If $w^k = 0$, stop; we have $0 \in \partial_\epsilon q(\mu)$, so $\sup_{\bar{\mu} \in \Re^r} q(\bar{\mu}) - \epsilon \leq q(\mu)$, by part (b). Otherwise, by a search along the line $\{\mu + s w^k \mid s \geq 0\}$, determine whether there exists a scalar $\bar{s}$ such that $q(\mu + \bar{s} w^k) > q(\mu) - \epsilon$. If such a $\bar{s}$ can be found, stop and replace μ with $\mu + \bar{s} w^k$ (the dual value has been improved by at least ϵ). Otherwise let g^k be an element of $\partial_\epsilon q(\mu)$ such that

$$g^{k\prime} w^k = \min_{g \in \partial_\epsilon q(\mu)} g' w^k,$$

[note that from part (a), we have $g^{k\prime} w^k \leq 0$]. Show that this process will terminate in a finite number of steps with either an improvement of the dual value by at least ϵ, or by confirmation that μ is an ϵ-optimal solution.

(f) Generalize parts (a)-(d) to the case of constrained maximization of the concave function q over a convex set M. *Hint*: Redefine the ϵ-subdifferential at a vector $\mu \in M$ to be the set

$$\big\{ g \mid q(\bar{\mu}) \leq q(\mu) + g'(\bar{\mu} - \mu) + \epsilon, \ \forall\, \bar{\mu} \in M \big\}.$$

6.3.6 (A Variation of the Subgradient Method [CFM75]) (www)

Consider the dual problem and the following variation of the subgradient method

$$\mu^{k+1} = [\mu^k + s^k d^k]^+,$$

where

$$d^k = \begin{cases} g^k & \text{if } k = 0, \\ g^k + \beta^k d^{k-1} & \text{if } k > 0, \end{cases}$$

s^k and β^k are scalars satisfying

$$0 < s^k \leq \frac{q(\mu^*) - q(\mu^k)}{\|d^k\|^2},$$

$$\beta^k = \begin{cases} -\gamma \dfrac{g^{k\prime} d^{k-1}}{\|d^{k-1}\|^2} & \text{if } g^{k\prime} d^{k-1} < 0, \\ 0 & \text{otherwise,} \end{cases}$$

with $\gamma \in [0, 2]$, and μ^* is an optimal dual solution. Assuming $\mu^k \neq \mu^*$, show that

$$\|\mu^* - \mu^{k+1}\| < \|\mu^* - \mu^k\|.$$

Furthermore,

$$\frac{(\mu^* - \mu^k)' d^k}{\|d^k\|} \geq \frac{(\mu^* - \mu^k)' g^k}{\|g^k\|},$$

i.e., the angle between d^k and $\mu^* - \mu^k$ is no larger than the angle between g^k and $\mu^* - \mu^k$.

6.3.7

Give an example of a one-dimensional problem where the cutting plane method is started at an optimal solution but does not terminate finitely.

6.3.8 (Counterexample for Steepest Ascent [Wol75])

Consider the maximization of the two-dimensional function

$$q(\mu_1, \mu_2) = \begin{cases} -5(9\mu_1^2 + 16\mu_2^2)^{1/2} & \text{if } \mu_1 > |\mu_2|, \\ -(9\mu_1 + 16|\mu_2|) & \text{if } \mu_1 \leq |\mu_2|, \end{cases}$$

using the steepest ascent method with the line maximization stepsize rule. Suppose that the algorithm starts anywhere in the region $\{(\mu_1, \mu_2) \mid \mu_1 > |\mu_2| > (9/16)^2 |\mu_1|\}$. Show that it converges to the nonoptimal point $(0, 0)$. See Fig. 6.3.8.

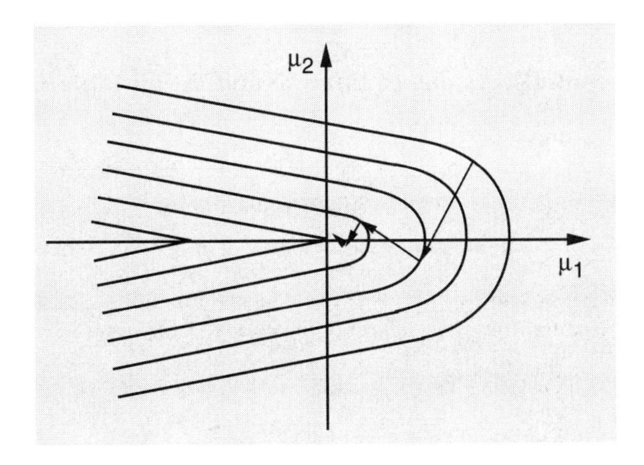

Figure 6.3.8. Contours and steepest ascent path for the function of Exercise 6.3.8.

6.3.9 (Coordinate Ascent for Problems with Special Structure)

Consider the maximization of a function $q : \Re^r \mapsto \Re$ subject to $\mu_j \geq 0$, $j = 1, \ldots, r$. Assume that q has the form

$$q(\mu) = Q(\mu) + \sum_{j=1}^{r} q_j(\mu_j),$$

where $Q : \Re^r \mapsto \Re$ is concave and differentiable, and the functions $q_j : \Re \mapsto \Re$ are concave (possibly nondifferentiable). Show that the coordinate ascent method improves the value of q at every μ that is not optimal.

6.3.10 (Solution of Linear Inequalities) (www)

Consider the problem of finding a point in the intersection of m halfspaces

$$C = \{\mu \mid a_i'\mu - b_i \geq 0,\ i = 1, \ldots, m\},$$

which is assumed to be nonempty. Consider the algorithm that stops if $\mu^k \in C$ and otherwise sets

$$\mu^{k+1} = \mu^k + s^k a_{i^k},$$

where i^k is the index of the inequality $a_i'\mu - b_i \geq 0$ that is most violated at μ^k, and

$$s^k = \frac{b_{i^k} - a_{i^k}'\mu^k}{\|a_{i^k}\|^2}.$$

Show that this algorithm is a subgradient method for a suitable function and prove its convergence to a point of C.

6.3.11 (Ascent Directions of Some Nondifferentiable Functions)

Consider the problem

$$\begin{aligned} &\text{maximize} \quad q(\mu) = q_1(\mu) + q_2(\mu) \\ &\text{subject to} \quad \mu \in M, \end{aligned}$$

where $q_1 : \Re^r \mapsto \Re$ and $q_2 : \Re^r \mapsto \Re$ are concave functions, and M is a closed convex set. Assume that q_1 is differentiable and that for some $L > 0$,

$$\|\nabla q_1(\bar{\mu}) - \nabla q_1(\mu)\| \leq L\|\bar{\mu} - \mu\|, \qquad \forall\ \bar{\mu},\ \mu \in M.$$

(a) Given $\mu^k \in M$, let μ^{k+1} be the optimal solution of the problem

$$\begin{aligned} &\text{maximize} \quad \nabla q_1(\mu^k)'(\mu - \mu^k) - \frac{1}{2s}\|\mu - \mu^k\|^2 + q_2(\mu) \\ &\text{subject to} \quad \mu \in M. \end{aligned}$$

Show that if $0 < s \le 1/L$, then either μ^k maximizes q over M, or else $q(\mu^{k+1}) > q(\mu^k)$. *Hint*: Use the descent lemma (Prop. A.24 in Appendix A).

(b) Suppose that q_2 and M have a separable structure. Construct a method based on part (a) that exploits separability.

6.3.12 (Approximate Subgradient Method) (www)

Let the dual function be real-valued [$q(\mu) > -\infty$ for all $\mu \in \Re^r$], and let μ^* be a dual optimal solution. Consider the subgradient iteration

$$\mu^{k+1} = \left[\mu^k + \frac{q(\mu^*) - q(\mu^k)}{\|g^k\|^2} g^k\right]^+,$$

with the difference that g^k is an ϵ-subgradient at μ^k [i.e., $g^k \in \partial_\epsilon q(\mu^k)$], where $\epsilon > 0$. Show that as long as μ^k is such that $q(\mu^k) < q(\mu^*) - 2\epsilon$, we have $\|\mu^{k+1} - \mu^*\| < \|\mu^k - \mu^*\|$. *Hint*: Similar to the proof of Prop. 6.3.1, show that if g^k is an ϵ-subgradient, then for all $\mu \in M$ we have

$$\|\mu^{k+1} - \mu\|^2 \le \|\mu^k - \mu\|^2 - 2s^k\big(q(\mu) - q(\mu^k) - \epsilon\big) + (s^k)^2\|g^k\|^2.$$

6.3.13 (Approximate Subgradient Method with Diminishing Stepsize [CoL94]) (www)

Let the dual function be real-valued [$q(\mu) > -\infty$ for all $\mu \in \Re^r$]. Consider the iteration

$$\mu^{k+1} = \left[\mu^k + s^k g^k\right]^+,$$

where g^k is an ϵ^k-subgradient at μ^k [i.e., $g^k \in \partial_{\epsilon^k} q(\mu^k)$], and we have

$$\sum_{k=0}^{\infty} s^k = \infty, \qquad s^k\|g^k\|^2 \to 0,$$

$$\epsilon^k \ge 0, \qquad \epsilon^k \to \epsilon.$$

(a) Use the hint of the preceding exercise to show that

$$\sup_{\mu \in M} q(\mu) - \epsilon \le \limsup_{k\to\infty} q(\mu^k) \le \sup_{\mu \in M} q(\mu).$$

(b) Assume that $\epsilon^k = 0$ for all k and that

$$\sum_{k=0}^{\infty} s^k = \infty, \qquad \sum_{k=0}^{\infty} (s^k)^2 < \infty.$$

Show that $\limsup_{k\to\infty} q(\mu^k) = \sup_{\mu\in M} q(\mu)$ and that μ^k converges to a maximizing point of q over M if there exists at least one such point.

6.3.14 (Subgradient Method with Fixed and with Diminishing Stepsize [Sho85]) (www)

Let the dual function be real-valued [$q(\mu) > -\infty$ for all $\mu \in \Re^r$]. Consider the subgradient iteration

$$\mu^{k+1} = \left[\mu^k + \alpha^k \frac{g^k}{\|g^k\|}\right]^+,$$

where α^k is a positive scalar.

(a) Assume that α^k is fixed ($\alpha^k = \alpha$ for all k). Show that for any $\epsilon > 0$ and any dual optimal solution μ^*, there exists an index $\bar{k}$ and a point $\bar{\mu} \in M$ such that $q(\bar{\mu}) = q(\mu^{\bar{k}})$ and $\|\bar{\mu} - \mu^*\| < \alpha(1+\epsilon)/2$.

(b) Assume that $\alpha^k \to 0$ and $\sum_{k=0}^{\infty} \alpha^k = \infty$, and that the set of dual optimal solutions M^* is nonempty and compact. Show that either $\mu^{\bar{k}} \in M^*$ for some index $\bar{k}$, or else

$$\lim_{k\to\infty} \min_{\mu^* \in M^*} \|\mu^k - \mu^*\| = 0, \qquad \lim_{k\to\infty} q(\mu^k) = \max_{\mu \in M} q(\mu).$$

6.3.15 (Incremental Subgradient Method with Diminishing Stepsize [GeB99]) (www)

Consider the problem of maximizing over the closed set M the function $q : \Re^r \to \Re$ given by

$$q(\mu) = \sum_{i=1}^{m} q_i(\mu),$$

where each $q_i : \Re^r \mapsto \Re$ is a concave function. The incremental subgradient method for this problem is given by

$$\mu^{k+1} = \psi^{m,k},$$

where $\psi^{m,k}$ is obtained after the m steps

$$\psi^{i,k} = \left[\psi^{i-1,k} + \alpha^k g^{i,k}\right]^+, \qquad g^{i,k} \in \partial q_i(\psi^{i-1,k}), \qquad i = 1, \ldots, m,$$

starting with

$$\psi^{0,k} = \mu^k.$$

Assume the following conditions:

(i) We have $\max_{\mu \in M} q_i(\mu) = q_i^* < \infty$, and the set $\{\mu \mid q_i(\mu) = q_i^*\}$ is nonempty for all i. Furthermore, at least one of the functions q_i has bounded level sets.

(ii) The sequence of subgradients $\{g^{i,k}\}$ is bounded, i.e., for some $C > 0$ we have

$$||g^{i,k}|| \le C, \quad \forall\ i,\ k.$$

(This is true in particular if each function q_i is polyhedral.)

Show that the set of optimal solutions M^* is nonempty. Furthermore, the following hold:

(a) If the stepsize sequence $\{\alpha^k\}$ is bounded, then a sequence $\{\mu^k\}$ generated by the method is bounded.

(b) If the stepsize sequence $\{\alpha^k\}$ satisfies

$$\lim_{k\to\infty} \alpha^k = 0, \qquad \sum_{k=0}^{\infty} \alpha^k = \infty,$$

then either $\mu^{\bar{k}} \in M^*$ for some $\bar{k}$, or else

$$\lim_{k\to\infty} \min_{\mu^*\in M^*} ||\mu^k - \mu^*|| = 0, \qquad \lim_{k\to\infty} q(\mu^k) = \max_{\mu\in M} q(\mu).$$

(c) If the stepsize sequence $\{\alpha^k\}$ satisfies

$$\sum_{k=0}^{\infty} \alpha^k = \infty, \qquad \sum_{k=0}^{\infty} (\alpha^k)^2 < \infty,$$

then $\{\mu^k\}$ converges to some optimal solution.

6.3.16 (Incremental Subgradient Method with Dynamically Changing Stepsize [GeB99]) (www)

Consider the incremental subgradient method of Exercise 6.3.15 and assume that each $q_i : \Re^r \mapsto \Re$ is a concave function, the optimal value $q^* = \max_{\mu\in M} q(\mu)$ is finite, and the optimal solution set M^* is nonempty. Assume also that for each i, C_i is a scalar satisfying

$$||g|| \le C_i, \qquad \forall\ g \in \partial q_i(\mu^k) \cup \partial q_i(\psi^{i-1,k}),\ k = 0, 1, \ldots,$$

and let $C = \sum_{i=1}^m C_i$. Let the stepsize be given by

$$\alpha^k = \frac{\gamma^k \big(q^* - q(\mu^k)\big)}{C^2},$$

where the sequence $\{\gamma^k\}$ satisfies for some scalars γ_l and γ_u

$$0 < \gamma_l \le \gamma^k \le \gamma_u < 2, \quad \forall\ k \ge 0.$$

Show that either $\mu^{\bar{k}} \in M^*$ for some $\bar{k}$, or else $\{\mu^k\}$ converges to some optimal solution.

6.3.17 (ϵ-Complementary Slackness and Approximate Subgradients) (www)

Consider the separable problem

$$\begin{aligned} &\text{minimize} \quad \sum_{i=1}^{n} f_i(x_i) \\ &\text{subject to} \quad \sum_{i=1}^{n} g_{ij}(x_i) \le 0, \quad j = 1, \ldots, r, \qquad \alpha_i \le x_i \le \beta_i, \quad i = 1, \ldots, n, \end{aligned}$$

where $f_i : \Re \mapsto \Re$, $g_{ij} : \Re \mapsto \Re$ are convex functions. For an $\epsilon > 0$, we say that a pair $(\bar{x}, \bar{\mu})$ satisfies *ϵ-complementary slackness* (see e.g., [BeT89], [Ber98] and the references given there for more on this notion) if $\bar{\mu} \ge 0$, $\bar{x}_i \in [\alpha_i, \beta_i]$ for all i, and

$$0 \le f_i^+(\bar{x}_i) + \sum_{j=1}^{r} \bar{\mu}_j g_{ij}^+(\bar{x}_i) + \epsilon, \ \forall\, i \in I^-, \quad f_i^-(\bar{x}_i) + \sum_{j=1}^{r} \bar{\mu}_j g_{ij}^-(\bar{x}_i) - \epsilon \le 0, \ \forall\, i \in I^+$$

where $I^- = \{i \mid x_i < \beta_i\}$, $I^+ = \{i \mid \alpha_i < x_i\}$, f_i^-, g_{ij}^- and f_i^+, g_{ij}^+ denote the left and right derivatives of f_i, g_{ij}, respectively. Show that the r-dimensional vector with jth component $\sum_{i=1}^{n} g_{ij}(\bar{x}_i)$ is an $\bar{\epsilon}$-subgradient of the dual function q at $\bar{\mu}$, where $\bar{\epsilon} = \epsilon \sum_{i=1}^{n} (\beta_i - \alpha_i)$. (Based on unpublished joint work with P. Tseng.)

6.4 DECOMPOSITION METHODS*

In this section we consider nondifferentiable optimization methods for solving the duals of separable problems of the form

$$\begin{aligned} &\text{minimize} \quad \sum_{j=1}^{J} f_j(x_j) \\ &\text{subject to} \quad x_j \in X_j, \quad j = 1, \ldots, J, \qquad \sum_{j=1}^{J} A_j x_j = b. \end{aligned} \tag{6.44}$$

Here $f_j : \Re^{n_j} \mapsto \Re$ and X_j are given subsets of $\Re^{n_j}$. The $r \times n_j$ matrices A_j and the vector $b \in \Re^r$ specify the constraint $\sum_{j=1}^{J} A_j x_j = b$, which couples the components x_j. Note that we are making no convexity assumptions on X_j or f_j.

The methodology of this section is based on converting the basic problem with coupling constraints (6.44) into a nondifferentiable optimization problem using two different approaches:

(a) Lagrangian relaxation of the coupling constraints.

(b) Decomposition by right-hand side allocation.

We will apply the subgradient and cutting plane methods of the previous section within the context of each of these two approaches.

6.4.1 Lagrangian Relaxation of the Coupling Constraints

The basic idea here is to solve the dual problem obtained by relaxing the coupling constraints. The dual function is

$$q(\lambda) = \sum_{j=1}^{J} \min_{x_j \in X_j} \big\{ f_j(x_j) + \lambda' A_j x_j \big\} - \lambda' b.$$

We will assume throughout this subsection that for all j and λ one can find a vector $x_j(\lambda)$ attaining the minimum above. Thus the dual value is

$$q(\lambda) = \sum_{j=1}^{J} \big\{ f_j\big(x_j(\lambda)\big) + \lambda' A_j x_j(\lambda) \big\} - \lambda' b \tag{6.45}$$

and a subgradient at λ is

$$g_\lambda = \sum_{j=1}^{J} A_j x_j(\lambda) - b, \tag{6.46}$$

(cf. the discussion of Section 6.1). Thus, both $q(\lambda)$ and g_λ can be calculated by solving a number of relatively small subproblems (one for each j).

We now discuss the computational solution of the dual problem

$$\begin{aligned} &\text{minimize} \quad q(\lambda) \\ &\text{subject to} \quad \lambda \in \Re^r. \end{aligned} \tag{6.47}$$

The application of the subgradient method is straightforward, so we concentrate on the cutting plane method.

Cutting Plane Method – Dantzig-Wolfe Decomposition

The cutting plane method consists of the iteration

$$\lambda^k = \arg\max_{\lambda \in \Re^r} Q^k(\lambda),$$

where $Q^k(\lambda)$ is the piecewise linear approximation of the dual function based on the preceding function values $q(\lambda^0), \ldots, q(\lambda^{k-1})$, and the corresponding subgradients $g^0, \ldots, g^{k-1}$, i.e.,

$$Q^k(\lambda) = \min\big\{ q(\lambda^0) + (\lambda - \lambda^0)' g^0, \ldots, q(\lambda^{k-1}) + (\lambda - \lambda^{k-1})' g^{k-1} \big\}.$$

Let us take a closer look at the subproblem of maximizing $Q^k(\lambda)$. By introducing an auxiliary variable v, we can write this problem as

$$\begin{aligned} &\text{maximize} \quad v \\ &\text{subject to} \quad v \le q(\lambda^i) + (\lambda - \lambda^i)' g^i, \qquad i = 0, \ldots, k-1. \end{aligned}$$

This is a linear program in the variables v and λ. Its dual can be verified to have the form

$$\begin{aligned} &\text{minimize} \quad \sum_{i=0}^{k-1} \xi^i \big(q(\lambda^i) - \lambda^{i\prime} g^i\big) \\ &\text{subject to} \quad \sum_{i=0}^{k-1} \xi^i = 1, \qquad \sum_{i=0}^{k-1} \xi^i g^i = 0, \\ &\qquad \xi^i \geq 0, \quad i = 0, \ldots, k-1, \end{aligned}$$

where $\xi^0, \ldots, \xi^{k-1}$ are the dual variables. Using Eqs. (6.45) and (6.46), this problem can be written as

$$\begin{aligned} &\text{minimize} \quad \sum_{j=1}^{J} \left(\sum_{i=0}^{k-1} \xi^i f_j\big(x_j(\lambda^i)\big) \right) \\ &\text{subject to} \quad \sum_{i=0}^{k-1} \xi^i = 1, \qquad \sum_{j=1}^{J} A_j \left(\sum_{i=0}^{k-1} \xi^i x_j(\lambda^i) \right) = b, \\ &\qquad \xi^i \geq 0, \quad i = 0, \ldots, k-1. \end{aligned} \tag{6.48}$$

The preceding problem is called the *master problem*. It is the dual of the linear subproblem $\max_{\lambda \in \Re^r} Q^k(\lambda)$, which in turn approximates the dual problem $\max_\lambda q(\lambda)$; in short, it is the *dual of the approximate dual*. We have seen in Chapter 5 that duality is often symmetric; that is, the dual problem by appropriate dualization yields the primal. Thus, it is not surprising that the master problem (6.48) and the original primal problem (6.44) are closely related. In fact, the master problem may be viewed as a piecewise linear approximation of the primal, as shown in Fig. 6.4.1. In particular, the primal cost function terms $f_j(x_j)$ are approximated by

$$\sum_{i=0}^{k-1} \xi^i f_j\big(x_j(\lambda^i)\big)$$

and each constraint set X_j of the primal is approximated by the convex hull of the points $x_j(\lambda^0), \ldots, x_j(\lambda^{k-1})$. This type of approximation of the primal problem is often called *inner linearization*. It may be viewed as dual to the cutting plane approximation of the dual problem, which is often called *outer linearization*.

Suppose now that we solve the master problem with a method that yields a Lagrange multiplier vector λ^k corresponding to the constraints

$$\sum_{j=1}^{J} A_j \left(\sum_{i=0}^{k-1} \xi^i x_j(\lambda^i) \right) = b;$$

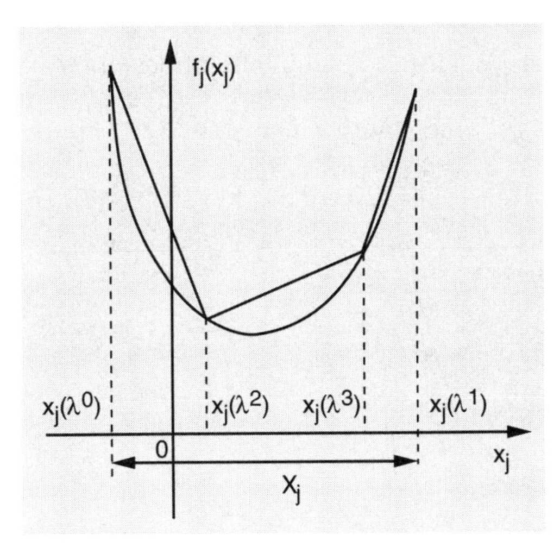

Figure 6.4.1. Viewing the master problem as a piecewise linear approximation of the primal problem. Each cost function term $f_j(x_j)$ of the primal is approximated by a piecewise linear function with break points at $x_j(\lambda^0), \ldots, x_j(\lambda^{k-1})$.

one possibility is to use a method that solves simultaneously both the primal and the dual of a linear program, such as the simplex method. Then, the dual of the master problem [which is the cutting plane subproblem $\max_\lambda Q^k(\lambda)$] is solved by the Lagrange multiplier λ^k. Therefore, λ^k is the next iterate of the cutting plane problem.

We can now piece together the typical cutting plane method.

Cutting Plane – Dantzig-Wolfe Decomposition Method

Start with any λ^0 and for all $j = 1, \ldots, J$, obtain a solution $x_j(\lambda^0)$ of the subproblem

$$\begin{aligned} &\text{minimize} \ \ f_j(x_j) + \lambda^{0\prime} A_j x_j \\ &\text{subject to} \ \ x_j \in X_j. \end{aligned}$$

Step 1: Given $\lambda^0, \ldots, \lambda^{k-1}$, and the vectors $x_j(\lambda^i)$ for $j = 1, \ldots, J$ and $i = 0, \ldots, k-1$, solve the master problem (6.48) and obtain λ^k, which is a Lagrange multiplier vector of the constraints

$$\sum_{j=1}^{J} A_j \left(\sum_{i=0}^{k-1} \xi^i x_j(\lambda^i) \right) = b.$$

Step 2: For all $j = 1, \ldots, J$, obtain a solution $x_j(\lambda^k)$ of the subproblem

$$\begin{aligned} &\text{minimize} \;\; f_j(x_j) + \lambda^{k\prime} A_j x_j \\ &\text{subject to} \;\; x_j \in X_j. \end{aligned}$$

Step 3: Modify the master problem by adding one more variable ξ^k with cost coefficient

$$\sum_{j=1}^{J} f_j\big(x_j(\lambda^k)\big)$$

and constraint column

$$\begin{pmatrix} 1 \\ \sum_{j=1}^{J} A_j x_j(\lambda^k) \end{pmatrix}$$

and go to Step 1.

Note that adding one more variable to the master problem (cf. Step 3), amounts to adding one more break point to the piecewise linear approximation of each primal cost term $f_j(x_j)$, as well as adding one more extreme point to the convex hull approximations of the sets X_j (cf. Fig. 6.4.1).

The method of this subsection is variously known as the *Dantzig-Wolfe decomposition* or *column generation* or *inner linearization* method. As we have seen, it is just the cutting plane method applied to a dual problem where only the coupling constraints $\sum_{j=1}^{J} A_j x_j = b$ of the original problem (6.44) are dualized.

6.4.2 Decomposition by Right-Hand Side Allocation

We now consider a different method for decomposing the original primal problem (6.44), which resembles the partitioning methodology of Section 6.2.2. By introducing auxiliary variables y_j, $j = 1, \ldots, J$, we can write this problem as

$$\begin{aligned} &\text{minimize} \;\; \sum_{j=1}^{J} f_j(x_j) \\ &\text{subject to} \;\; x_j \in X_j, \qquad j = 1, \ldots, J, \\ &\qquad\qquad \sum_{j=1}^{J} y_j = b, \qquad A_j x_j = y_j, \quad j = 1, \ldots, J, \end{aligned}$$

or equivalently,

$$\begin{aligned} &\text{minimize} \quad \sum_{j=1}^{J} \min_{A_j x_j = y_j,\ x_j \in X_j} f_j(x_j) \\ &\text{subject to} \quad \sum_{j=1}^{J} y_j = b, \qquad y_j \in Y_j, \quad j = 1, \ldots, J, \end{aligned} \tag{6.49}$$

where Y_j is the set of all vectors y_j for which the inner minimization problem

$$\begin{aligned} &\text{minimize} \quad f_j(x_j) \\ &\text{subject to} \quad A_j x_j = y_j, \qquad x_j \in X_j \end{aligned}$$

has at least one feasible solution.

Let us define

$$p_j(y_j) = \min_{A_j x_j = y_j,\, x_j \in X_j} f_j(x_j). \tag{6.50}$$

Then, problem (6.49) can be written as

$$\begin{aligned} &\text{minimize} \quad \sum_{j=1}^{J} p_j(y_j) \\ &\text{subject to} \quad \sum_{j=1}^{J} y_j = b, \qquad y_j \in Y_j, \quad j = 1, \ldots, J. \end{aligned} \tag{6.51}$$

This problem, called the *master problem*, may be solved by nondifferentiable optimization methods. Note from Eq. (6.50), that p_j may be viewed as a primal function of a certain minimization problem. Using this observation and the theory developed in Section 5.4.4, it can be seen that a subgradient of $p_j(y_j)$ at y_j is equal to $-\lambda_j$, where λ_j is the Lagrange multiplier vector corresponding to the constraint $A_j x_j = y_j$ in the definition (6.50) of p_j. Thus, we have

$$\text{Subgradient of } \sum_{j=1}^{J} p_j(y_j) = -(\lambda_1, \ldots, \lambda_J)'.$$

We can now write explicitly the subgradient and cutting plane methods for solving the master problem. In particular, the subgradient method is given by

$$y^{k+1} = \left[y^k + s^k \lambda^k\right]^+,$$

where s^k is a positive stepsize, $[\cdot]^+$ denotes projection on the constraint set

$$\left\{ y \;\middle|\; \sum_{j=1}^{J} y_j = b,\ y_j \in Y_j \right\}.$$

Note here that y^k is the $(n_1+\cdots+n_J)$-dimensional vector with components y_j^k, $j=1,\ldots,J$, and λ^k is the vector with components λ_j^k, $j=1,\ldots,J$, where λ_j^k is the Lagrange multiplier corresponding to the constraint $A_j x_j = y_j^k$ in the inner minimization subproblem

$$\begin{aligned} &\text{minimize} \quad f_j(x_j) \\ &\text{subject to} \quad A_j x_j = y_j^k, \qquad x_j \in X_j. \end{aligned}$$

Similarly, the cutting plane method is given by

$$y^k = \arg \min_{\substack{\sum_{j=1}^J y_j = b,\\ y_j \in Y_j,\, j=1,\ldots,J}} P^k(y),$$

where P^k is the piecewise linear function that approximates the cost function $\sum_{j=1}^J p_j(y_j)$ of the master problem (6.51), based on previous values of y^i and λ^i, i.e.,

$$P^k(y) = \max_{i=0,\ldots,k-1} \left\{ \sum_{j=1}^J \big(p_j(y_j^i) + (y_j - y_j^i)\lambda_j^i\big) \right\}.$$

EXERCISES

6.4.1 (Linear Problems with Coupling Variables)

Consider the problem with coupling variables of Example 6.2.2 in Section 6.2 for the case where the cost function and the constraints are linear. Show that its dual is a problem with coupling constraints of the type considered in this section.

6.4.2 (Outer Approximation Methods [GoP79], [PoH91])

Consider the problem

$$\begin{aligned} &\text{minimize} \quad \max_{y \in Y} \phi(x,y) \\ &\text{subject to} \quad x \in X, \end{aligned} \tag{6.52}$$

where X is a convex and compact subset of $\Re^n$, Y is a compact subset of $\Re^m$, and $\phi : \Re^n \times Y$ is a continuous function such that $\phi(\cdot, y)$ is convex as a function of x for each $y \in Y$. Let x^k be defined as

$$x^k = \arg\min_{x \in X} \max\big\{\phi(x,y^0), \phi(x,y^1), \ldots, \phi(x,y^{k-1})\big\},$$

where y^0 is any vector in Y and for each $i = 1, 2, \ldots$, $y^i = \arg\max_{y \in Y} \phi(x^i, y)$.

(a) What is the relation of this method with the cutting plane method?

(b) Show that every limit point of the sequence $\{x^k\}$ is an optimal solution of the original problem (6.52).

(c) Show that the conclusion of part (b) holds also in the case where $Y = [0, 1]$ and y^i is instead defined by

$$y^i = \arg \max_{y \in \{0, 1/i, 2/i, \ldots, 1\}} \phi(x^i, y).$$

6.5 NOTES AND SOURCES

Section 6.1: Extensive accounts of optimality conditions for nondifferentiable (possibly nonconvex) optimization are given by Clarke [Cla83], Demjanov and Vasilév [DeV85], Evtushenko [Evt85], Rockafellar [Roc93], and Rockafellar and Wets [RoW97].

Section 6.2: The material of this section has much in common with Section 3.4 of Bertsekas and Tsitsiklis [BeT89], which describes additional methods and gives many references. The dual coordinate ascent method for quadratic programming dates to Hildreth [Hil57]. See Cryer [Cry71], Censor and Herman [CeH87], Lin and Pang [LiP87], Mangasarian and De Leone [MaD87], [MaD88], and Censor and Zenios [1997] for subsequent work. For analysis and alternative proposals of dual ascent methods for quadratic and other convex problems, see Bertsekas, Hossein, and Tseng [BHT87], Tseng and Bertsekas [TsB87], Hearn and Lawphongpanich [HeL89], Tseng [Tse90], Tseng and Bertsekas [TsB91], Hager and Hearn [HaH93], Luo and Tseng [LuT93c], and Kiwiel [Kiw94].

Section 6.3: Subgradient methods were first introduced in the Soviet Union in the middle 60s by Shor; the works of Ermoliev and Poljak were also particularly influential. An extensive bibliography for the early period of the subject is given in the edited volume by Balinski and Wolfe [BaW75]. The convergence rate of subgradient methods is discussed by Goffin [Gof77]. Incremental subgradient methods were developed by the author in collaboration with A. Geary. Their analysis, as summarized in Exercises 6.3.15 and 6.3.16, is as yet unpublished. Another important class of methods, which we have not discussed, are the, so called, subgradient methods with space dilatation, which were proposed and developed extensively by Shor; see the textbooks [Sho85] and [Sho98].

Ascent methods for minimax problems and the steepest ascent method in particular, were proposed by Demjanov [Dem66], [Dem68], and Grinold [Gri72]. The general forms of the steepest ascent and ϵ-ascent methods, based on projection on the subdifferential and ϵ-subdifferential, respectively, were first proposed by Bertsekas and Mitter [BeM71], [BeM73]. Bundle methods, proposed by Lemarechal [Lem74], [Lem75], and Wolfe [Wol75], provided effective implementations of ϵ-ascent ideas, and stimulated a great deal of subsequent research on nondifferentiable optimization; see e.g. the book by Hiriart-Urruty and Lemarechal [HiL93].

The texts by Auslender [Aus76], Shapiro [Sha79], Evtushenko [Evt85], Shor [Sho85], Minoux [Min86], Poljak [Pol87], Hiriart-Urruty and Lemarechal [HiL93], and Shor [Sho98] give extensive accounts of subgradient methods that complement our treatment and give many references.

Cutting plane methods were introduced by Cheney and Goldstein [ChG59], and by Kelley [Kel60]. For analysis of proximal cutting plane and related methods, see Ruszczynski [Rus89], Lemaréchal and Sagastizábal [LeS93], Mifflin [Mif96], Bonnans et. al. [BGL95], Kiwiel [Kiw97b], Burke and Qian [BuQ98], and Mifflin, Sun, and Qi [MSQ98].

Central cutting plane methods were introduced by Elzinga and Moore [ElM75]. More recent proposals, some of which relate to interior point methods, are discussed in Goffin and Vial [GoV90], Goffin, Haurie, and Vial [GHV92], Ye [Ye92], Kortanek and No [KoN93], Goffin, Luo, and Ye [GLY94], Atkinson and Vaidya [AtV95], den Hertog et. al. [HKR95], Nesterov [Nes95], Goffin, Luo, and Ye [GLY96]. For a textbook treatment, see Ye [Ye97], and for a recent survey, see Goffin and Vial [GoV99].

Section 6.4: Three historically important references on decomposition methods are Dantzig and Wolfe [DaW60], Benders [Ben62], and Everett [Eve63]. The early text by Lasdon [Las70] on large-scale optimization was particularly influential; see also Geoffrion [Geo70], [Geo74].

The theoretical and applications literature on large-scale optimization and decomposition is quite voluminous. We provide a few references that complement the material we have covered in this chapter: Stephanopoulos and Westerberg [StW75], Kennington and Shalaby [KeS77], Bertsekas [Ber79a], Meyer [Mey79], Cohen [Coh80], Fortin and Glowinski [FoG83], Birge [Bir85], Golshtein [Gol85], Tanikawa and Mukai [TaM85], Spingarn [Spi85], Minoux [Min86], Ruszczynski [Rus86], Sen and Sherali [SeS86], Bertsekas and Tsitsiklis [BeT89], Hearn and Lawphongpanich [HeL89], Rockafellar [Roc90], Toint and Tuyttens [ToT90], Ferris and Mangasarian [FeM91], Kim and Nazareth [KiN91], Rockafellar and Wets [RoW91], Tseng [Tse91b], [Tse91c], Auslender [Aus92], Eckstein and Bertsekas [EcB92], Fukushima [Fuk92], Gaudioso and Monaco [GaM92], Mulvey and Ruszczynski [MuR92], Pinar and Zenios [PiZ92], Nagurney [Nag93], Patriksson [Pat93a], [Pat93b], Tseng [Tse93], Eckstein [Eck94b], Migdalas [Mig94], Pinar and Zenios [PiZ94], Mahey, Oualibouch, and Tao [MOT95], Mulvey and Ruszczynski [MuR95], Zhu [Zhu95], Censor and Zenios [1997], Kontogiorgis and Meyer [KoM98], Patriksson [Pat98], Zhao and Luh [ZhL98].

APPENDIX A: Mathematical Background

In this appendix, we collect definitions, notational conventions, and several results from linear algebra and analysis that are used extensively in nonlinear programming. Only a few proofs are given. Additional proofs can be found in Appendix A of the book by Bertsekas and Tsitsiklis [BeT89], which provides a similar but more extended summary of linear algebra and analysis. Related and additional material can be found in the books by Hager [Hag88], Hoffman and Kunze [HoK71], Lancaster and Tismenetsky [LaT85], and Strang [Str76] (linear algebra), and the books by Ash [Ash72], Ortega and Rheinboldt [OrR70], and Rudin [Rud76] (analysis).

Notation

If S is a set and x is an element of S, we write $x \in S$. A set can be specified in the form $S = \{x \mid x \text{ satisfies } P\}$, as the set of all elements satisfying property P. The union of two sets S and T is denoted by $S \cup T$ and their intersection by $S \cap T$. The symbols $\exists$ and $\forall$ have the meanings "there exists" and "for all," respectively. The set of real numbers (also referred to as scalars) is denoted by $\Re$.

If a and b are real numbers or $+\infty$, $-\infty$, we denote by $[a, b]$ the set of numbers x satisfying $a \leq x \leq b$ (including the possibility $x = +\infty$ or $x = -\infty$). A rounded, instead of square, bracket denotes strict inequality in the definition. Thus $(a, b]$, $[a, b)$, and (a, b) denote the set of all x satisfying $a < x \leq b$, $a \leq x < b$, and $a < x < b$, respectively.

If f is a function, we use the notation $f : A \mapsto B$ to indicate the fact that f is defined on a set A (its *domain*) and takes values in a set B (its *range*).

A.1 VECTORS AND MATRICES

We denote by $\Re$ the real line and by $\Re^n$ the set of n-dimensional real vectors. For any $x \in \Re^n$, we use x_i to indicate its ith *coordinate*, also called its ith *component*.

Vectors in $\Re^n$ will be viewed as column vectors, unless the contrary is explicitly stated. For any $x \in \Re^n$, x' denotes the transpose of x, which is an n-dimensional row vector. The *inner product* of two vectors $x, y \in \Re^n$ is defined by $x'y = \sum_{i=1}^n x_i y_i$. Any two vectors $x, y \in \Re^n$ satisfying $x'y = 0$ are called *orthogonal*.

If w is a vector in $\Re^n$, the notations $w > 0$ and $w \geq 0$ indicate that all coordinates of w are positive or nonnegative, respectively. For any two vectors w, v, the notation $w > v$ means that $w - v > 0$. The notations $w \geq v$, $w < v$, etc., are to be interpreted accordingly.

Subspaces and Linear Independence

A subset S of $\Re^n$ is called a *subspace* of $\Re^n$ if $ax + by \in S$ for every $x, y \in S$ and every $a, b \in \Re$. A *linear manifold* in $\Re^n$ is a translated subspace, that is, a set of the form

$$y + S = \{y + x \mid x \in S\},$$

where y is a vector in $\Re^n$ and S is a subspace of $\Re^n$. The *span* of a finite collection $\{x_1, \ldots, x_m\}$ of elements of $\Re^n$ is the subspace consisting of all vectors y of the form $y = \sum_{k=1}^m a_k x_k$, where each a_k is a scalar.

The vectors $x_1, \ldots, x_m \in \Re^n$ are called *linearly independent* if there exists no set of scalars $a_1, \ldots, a_m$ such that $\sum_{k=1}^m a_k x_k = 0$, unless $a_k = 0$ for each k. An equivalent definition is that $x_1 \neq 0$ and for every $k > 1$, the vector x_k does not belong to the span of $x_1, \ldots, x_{k-1}$.

Given a subspace S of $\Re^n$ containing at least one nonzero vector, a *basis* for S is a collection of vectors that are linearly independent and whose span is equal to S. Every basis of a given subspace has the same number of vectors. This number is called the *dimension* of S. By convention, the subspace $\{0\}$ is said to have dimension zero. The *dimension of a linear manifold* $y + S$ is the dimension of the corresponding subspace S. An important fact is that every subspace of nonzero dimension has an *orthogonal basis*, that is, a basis consisting of mutually orthogonal vectors.

Matrices

For any matrix A, we use A_{ij}, $[A]_{ij}$, or a_{ij} to denote its ijth entry. The *transpose* of A, denoted by A', is defined by $[A']_{ij} = a_{ji}$. For any two matrices A and B of compatible dimensions, we have $(AB)' = B'A'$.

Let A be a square matrix. We say that A is *symmetric* if $A' = A$. We say that A is *diagonal* if $[A]_{ij} = 0$ whenever $i \neq j$. It is *lower triangular* if $[A]_{ij} = 0$ whenever $i < j$. It is *upper triangular* if its transpose is lower triangular. We use I to denote the identity matrix. The *determinant* of A is denoted by $\det(A)$.

Let A be an $m \times n$ matrix. The *range space* of A is the set of all vectors $y \in \Re^m$ such that $y = Ax$ for some $x \in \Re^n$. The *null space* or *kernel* of A is the set of all vectors $x \in \Re^n$ such that $Ax = 0$. It is seen that the range space and the null space of A are subspaces. The *rank* of A is the minimum of the dimensions of the range space of A and the range space of the transpose A'. Clearly A and A' have the same rank. We say that A has *full rank*, if its rank is equal to $\min\{m, n\}$. It can be seen that A has full rank if and only if either the rows of A are linearly independent, or the columns of A are linearly independent.

A.2 NORMS, SEQUENCES, LIMITS, AND CONTINUITY

Definition A.1: A *norm* $\|\cdot\|$ on $\Re^n$ is a mapping that assigns a scalar $\|x\|$ to every $x \in \Re^n$ and that has the following properties:

(a) $\|x\| \geq 0$ for all $x \in \Re^n$.

(b) $\|cx\| = |c| \cdot \|x\|$ for every $c \in \Re$ and every $x \in \Re^n$.

(c) $\|x\| = 0$ if and only if $x = 0$.

(d) $\|x + y\| \leq \|x\| + \|y\|$ for all $x, y \in \Re^n$.

The *Euclidean norm* is defined by

$$\|x\| = (x'x)^{1/2} = \left(\sum_{i=1}^{n} |x_i|^2\right)^{1/2}.$$

The space $\Re^n$, equipped with this norm, is called a *Euclidean space*. We will use the Euclidean norm almost exclusively in this book. In particular, *in the absence of a clear indication to the contrary,* $\|\cdot\|$ *will denote the Euclidean norm.* Two important results for the Euclidean norm are:

Proposition A.1: (Pythagorean Theorem) If x and y are orthogonal then

$$\|x + y\|^2 = \|x\|^2 + \|y\|^2.$$

Proposition A.2: (Schwartz inequality) For any two vectors x and y, we have

$$|x'y| \leq \|x\| \cdot \|y\|,$$

with equality holding if and only if $x = \alpha y$ for some scalar α.

Two other important norms are the *maximum norm* $\|\cdot\|_\infty$ (also called *sup-norm* or ℓ_∞*-norm*), defined by

$$\|x\|_\infty = \max_i |x_i|,$$

and the ℓ_1*-norm* $\|\cdot\|_1$, defined by

$$\|x\|_1 = \sum_{i=1}^{n} |x_i|.$$

Sequences

We use both subscripts and superscripts in sequence notation. Generally, we use superscript notation for sequences of vectors generated by iterative algorithms whenever we need to reserve the subscript notation for indexing coordinates or components of vectors and functions.

A sequence $\{x_k \mid k = 1, 2, \ldots\}$ (or $\{x_k\}$ for short) of scalars is said to *converge* to a scalar x if for every $\epsilon > 0$ there exists some K (depending on ϵ) such that $|x_k - x| < \epsilon$ for every $k \geq K$. A sequence $\{x_k\}$ is said to converge to ∞ (respectively, $-\infty$) if for every b there exists some K (depending on b) such that $x_k \geq b$ (respectively, $x_k \leq b$) for all $k \geq K$. If a sequence $\{x_k\}$ converges to some x (possibly infinite), we say that x is the *limit* of $\{x_k\}$; symbolically, $x_k \to x$ or $\lim_{k\to\infty} x_k = x$. A sequence $\{x_k\}$ is called a *Cauchy sequence* if for every $\epsilon > 0$, there exists some K (depending on ϵ) such that $|x_k - x_m| < \epsilon$ for all $k \geq K$ and $m \geq K$.

A sequence $\{x_k\}$ is said to be *bounded above* (respectively, *below*) if there exists some scalar b such that $x_k \leq b$ (respectively, $x_k \geq b$) for all k. It is said to be *bounded* if it is bounded above and bounded below. The sequence $\{x_k\}$ is said to be *nonincreasing* (respectively, *nondecreasing*) if $x_{k+1} \leq x_k$ (respectively, $x_{k+1} \geq x_k$) for all k. If $\{x_k\}$ converges to x and is nonincreasing (nondecreasing) we also use the notation $x_k \downarrow x$ ($x_k \uparrow x$, respectively).

Proposition A.3: Every nonincreasing or nondecreasing scalar sequence converges to a possibly infinite number. If it is also bounded, then it converges to a finite real number.

The *supremum* of a nonempty set A of scalars, denoted by $\sup A$, is defined as the smallest scalar x such that $x \geq y$ for all $y \in A$. If no such scalar exists, we say that the supremum of A is ∞. Similarly, the *infimum* of A, denoted by $\inf A$, is defined as the largest scalar x such that $x \leq y$ for all $y \in A$, and is equal to $-\infty$ if no such scalar exists. Given a scalar sequence $\{x_k\}$, the supremum of the sequence, denoted by $\sup_k x_k$, is defined as $\sup\{x_k \mid k = 1, 2, \ldots\}$. The infimum of a sequence is similarly defined. Given a sequence $\{x_k\}$, let $y_m = \sup\{x_k \mid k \geq m\}$, $z_m = \inf\{x_k \mid k \geq m\}$. The sequences $\{y_m\}$ and $\{z_m\}$ are nonincreasing and nondecreasing, respectively, and therefore have a (possibly infinite) limit (Prop. A.3). The limit of y_m is denoted by $\limsup_{m\to\infty} x_m$ and the limit of z_m is denoted by $\liminf_{m\to\infty} x_m$.

Proposition A.4: Let $\{x_k\}$ be a scalar sequence.

(a) There holds

$$\inf_k x_k \leq \liminf_{k\to\infty} x_k \leq \limsup_{k\to\infty} x_k \leq \sup_k x_k.$$

(b) $\{x_k\}$ converges if and only if $\liminf_{k\to\infty} x_k = \limsup_{k\to\infty} x_k$ and, in that case, both of these quantities are equal to the limit of x_k.

(c) If $x_k \leq y_k$, then

$$\liminf_{k\to\infty} x_k \leq \liminf_{k\to\infty} y_k,$$

$$\limsup_{k\to\infty} x_k \leq \limsup_{k\to\infty} y_k.$$

A sequence $\{x_k\}$ of vectors in $\Re^n$ is said to converge to some $x \in \Re^n$ if the ith coordinate of x_k converges to the ith coordinate of x for every i. We use the notations $x_k \to x$ and $\lim_{k\to\infty} x_k = x$ to indicate convergence for vector sequences as well. The sequence $\{x_k\}$ is called bounded (or a Cauchy sequence) if each of its corresponding coordinate sequences is bounded (or a Cauchy sequence, respectively).

Definition A.2: We say that a vector $x \in \Re^n$ is a *limit point of a sequence* $\{x_k\}$ in $\Re^n$ if there exists a subsequence of $\{x_k\}$ that converges to x. Let A be a subset of $\Re^n$. We say that $x \in \Re^n$ is a *limit point of* A if there exists a sequence $\{x_k\}$, consisting of elements of A, that converges to x.

Proposition A.5:

(a) A bounded sequence of vectors in $\Re^n$ converges if and only if it has a unique limit point.

(b) A sequence in $\Re^n$ converges if and only if it is a Cauchy sequence.

(c) Every bounded sequence in $\Re^n$ has at least one limit point.

(d) Let $\{x_k\}$ be a scalar sequence. If $\limsup_{k\to\infty} x_k$ ($\liminf_{k\to\infty} x_k$) is finite, then it is the largest (respectively, smallest) limit point of $\{x_k\}$.

$o(\cdot)$ Notation

If p is a positive integer and $h : \Re^n \mapsto \Re^m$, then we write

$$h(x) = o\big(\|x\|^p\big)$$

if and only if

$$\lim_{x_k \to 0} \frac{h(x_k)}{\|x_k\|^p} = 0,$$

for all sequences $\{x_k\}$, with $x_k \neq 0$ for all k, that converge to 0.

Closed and Open Sets

Definition A.3: A set $A \subset \Re^n$ is called *closed* if it contains all of its limit points. It is called *open* if its complement (the set $\{x \mid x \notin A\}$) is closed. It is called *bounded* if there exists some $c \in \Re$ such that the magnitude of any coordinate of any element of A is less than c. The subset A is called *compact* if every sequence of elements of A has a subsequence that converges to an element of A. A *neighborhood* of a vector x is an open set containing x. If $A \subset \Re^n$ and $x \in A$, we say that x is an *interior* point of A if there exists a neighborhood of x that is contained in A. A vector $x \in A$ which is not an interior point of A is said to be a *boundary* point of A.

For any norm $\|\cdot\|$ in $\Re^n$, and any $\epsilon > 0$ and $x^* \in \Re^n$, consider the sets

$$\big\{x \mid \|x - x^*\| < \epsilon\big\}, \qquad \big\{x \mid \|x - x^*\| \leq \epsilon\big\}.$$

The first set is open and is called an *open sphere* centered at x^*, while the second set is closed and is called a *closed sphere* centered at x^*. Sometimes the terms *open ball* and *closed ball* are used, respectively.

Proposition A.6:

(a) The union of finitely many closed sets is closed.

(b) The intersection of closed sets is closed.

(c) The union of open sets is open.

(d) The intersection of finitely many open sets is open.

(e) A set is open if and only if all of its elements are interior points.

(f) Every subspace of $\Re^n$ is closed.

(g) A subset of $\Re^n$ is compact if and only if it is closed and bounded.

Continuity

Let A be a subset of $\Re^m$ and let $f : A \mapsto \Re^n$ be some function. Let x be a limit point of A. If the sequence $\{f(x_k)\}$ has a common limit z for every sequence $\{x_k\}$ of elements of A such that $\lim_{k\to\infty} x_k = x$, we write $\lim_{y\to x} f(y) = z$.

If A is a subset of $\Re$ and x is a limit point of A, the notation $\lim_{y\uparrow x} f(y)$ [respectively, $\lim_{y\downarrow x} f(y)$] will stand for the limit of $f(x_k)$, where $\{x_k\}$ is any sequence of elements of A converging to x and satisfying $x_k \leq x$ (respectively, $x_k \geq x$), assuming that the limit exists and is independent of the choice of the sequence $\{x_k\}$.

Definition A.4: Let A be a subset of $\Re^m$.

(a) A function $f : A \mapsto \Re^n$ is said to be *continuous* at a point $x \in A$ if $\lim_{y\to x} f(y) = f(x)$. It is said to be continuous on A (or over A) if it is continuous at every point $x \in A$.

(b) A real valued function $f : A \mapsto \Re$ is called *upper semicontinuous* (respectively, *lower semicontinuous*) at a vector $x \in A$ if $f(x) \geq \limsup_{k\to\infty} f(x_k)$ [respectively, $f(x) \leq \liminf_{k\to\infty} f(x_k)$] for every sequence $\{x_k\}$ of elements of A converging to x.

(c) A real valued function $f : A \mapsto \Re$ is called *coercive* if

$$\lim_{k\to\infty} f(x_k) = \infty$$

for every sequence $\{x_k\}$ of elements of A such that $\|x_k\| \to \infty$ for some norm $\|\cdot\|$.

(d) Let A be a subset of $\Re$. A function $f : A \mapsto \Re^n$ is called *right-continuous* (respectively, *left-continuous*) at a point $x \in A$ if $\lim_{y \downarrow x} f(y) = f(x)$ [respectively, $\lim_{y \uparrow x} f(y) = f(x)$].

It is easily seen that when A is a subset of $\Re$, a nondecreasing and right-continuous (respectively, left-continuous) function $f : A \mapsto \Re$ is upper (respectively, lower) semicontinuous.

Proposition A.7:

(a) The composition of two continuous functions is continuous.

(b) Any vector norm on $\Re^n$ is a continuous function.

(c) Let $f : \Re^m \mapsto \Re^n$ be continuous, and let $A \subset \Re^n$ be open (respectively, closed). Then the set $\{x \in \Re^m \mid f(x) \in A\}$ is open (respectively, closed).

An important property of compactness in connection with optimization problems is the following theorem.

Proposition A.8: (Weierstrass' Theorem) Let A be a nonempty subset of $\Re^n$ and let $f : A \mapsto \Re$ be lower semicontinuous at all points of A. Assume that one of the following three conditions holds:

(1) A is compact.

(2) A is closed and f is coercive.

(3) There exists a scalar γ such that the level set

$$\{x \in A \mid f(x) \leq \gamma\}$$

is nonempty and compact.

Then, there exists a vector $x \in A$ such that $f(x) = \inf_{z \in A} f(z)$.

Proof: Assume condition (1). Let $\{z_k\} \subset A$ be a sequence such that

$$\lim_{k \to \infty} f(z_k) = \inf_{z \in A} f(z).$$

Since A is bounded, this sequence has at least one limit point x [Prop.

A.5(c)]. Since A is closed, x belongs to A, while the lower semicontinuity of f implies that $f(x) \leq \lim_{k\to\infty} f(z_k) = \inf_{z\in A} f(z)$. Therefore, we must have $f(x) = \inf_{z\in A} f(z)$.

Assume condition (2). Consider a sequence $\{z_k\}$ as in the proof of part (a). Since f is coercive, $\{z_k\}$ must be bounded and the proof proceeds like the proof of part (a).

Assume condition (3). If the given γ is equal to $\inf_{z\in A} f(z)$, the set of minima of f over A is $\{x \in A \mid f(x) \leq \gamma\}$, and since by assumption this set is nonempty, we are done. If $\gamma > \inf_{z\in A} f(z)$, consider a sequence $\{z_k\}$ as in the proof of part (a). Then, for all k sufficiently large, z_k must belong to the set $\{x \in A \mid f(x) \leq \gamma\}$. Since this set is compact, $\{z_k\}$ must be bounded and the proof proceeds like the proof of part (a). **Q.E.D.**

Note that with appropriate adjustments, the above proposition applies to the existence of maxima of f over A. In particular, if f is upper semicontinuous at all points of A and A is compact, then there exists a vector $y \in A$ such that $f(y) = \sup_{z\in A} f(z)$. Vectors $x \in A$ or $y \in A$ that attain the minimum or the maximum of a function f over a set A, respectively, *even if they are not unique*, are denoted by

$$x = \arg\min_{z\in A} f(z), \qquad y = \arg\max_{z\in A} f(z).$$

Proposition A.9: For any two norms $\|\cdot\|$ and $\|\cdot\|'$ on $\Re^n$, there exists some positive constant $c \in \Re$ such that $\|x\| \leq c\|x\|'$ for all $x \in \Re^n$.

Proof: Let a be the minimum of $\|x\|'$ over the set of all $x \in \Re^n$ such that $\|x\| = 1$. The latter set is closed and bounded and, therefore, the minimum is attained at some $\tilde{x}$ (Prop. A.8) that must be nonzero since $\|\tilde{x}\| = 1$. For any $x \in \Re^n$, $x \neq 0$, the $\|\cdot\|$ norm of $x/\|x\|$ is equal to 1. Therefore,

$$0 < a = \|\tilde{x}\|' \leq \left\| \frac{x}{\|x\|} \right\|' = \frac{\|x\|'}{\|x\|}, \qquad \forall\, x \neq 0,$$

which proves the desired result with $c = 1/a$. **Q.E.D.**

The preceding proposition is referred to as the *norm equivalence property in* $\Re^n$. It shows that if a sequence converges with respect to one norm, it converges with respect to all other norms. From this we obtain the following.

Proposition A.10: If a subset of $\Re^n$ is open (respectively, closed, bounded, or compact) for some norm, it is open (respectively, closed, bounded, or compact), for all other norms.

Matrix Norms

A norm $\|\cdot\|$ on the set of $n \times n$ matrices is a real-valued mapping that has the same properties as vector norms do when the matrix is viewed as an element of $\Re^{n^2}$. The norm of an $n \times n$ matrix A is denoted by $\|A\|$.

We are mainly interested in *induced norms*, which are constructed as follows. Given any vector norm $\|\cdot\|$, the corresponding induced matrix norm, also denoted by $\|\cdot\|$, is defined by

$$\|A\| = \max_{\{x \in \Re^n \mid \|x\|=1\}} \|Ax\|. \tag{A.1}$$

The set over which the maximization takes place above is closed [Prop. A.7(c)] and bounded; the function being maximized is continuous [Prop. A.7(b)] and therefore the maximum is attained (Prop. A.8). It is easily verified that for any vector norm, Eq. (A.1) defines a bona fide matrix norm having all the required properties.

Note that by the Schwartz inequality (Prop. A.2), we have

$$\|A\| = \max_{\|x\|=1} \|Ax\| = \max_{\|y\|=\|x\|=1} |y'Ax|.$$

By reversing the roles of x and y in the above relation and by using the equality $y'Ax = x'A'y$, it follows that

$$\|A\| = \|A'\|. \tag{A.2}$$

A.3 SQUARE MATRICES AND EIGENVALUES

Definition A.5: A square matrix A is called *singular* if its determinant is zero. Otherwise it is called *nonsingular* or *invertible*.

Proposition A.11:

(a) Let A be an $n \times n$ matrix. The following are equivalent:

(i) The matrix A is nonsingular.

(ii) The matrix A' is nonsingular.

(iii) For every nonzero $x \in \Re^n$, we have $Ax \neq 0$.

(iv) For every $y \in \Re^n$, there exists a unique $x \in \Re^n$ such that $Ax = y$.

(v) There exists an $n \times n$ matrix B such that $AB = I = BA$.
(vi) The columns of A are linearly independent.
(vii) The rows of A are linearly independent.

(b) Assuming that A is nonsingular, the matrix B of statement (v) (called the *inverse* of A and denoted by A^{-1}) is unique.

(c) For any two square invertible matrices A and B of the same dimensions, we have $(AB)^{-1} = B^{-1}A^{-1}$.

Let A and B be square matrices and let C be a matrix of appropriate dimension. Then we have

$$(A + CBC')^{-1} = A^{-1} - A^{-1}C(B^{-1} + C'A^{-1}C)^{-1}C'A^{-1},$$

provided all the inverses appearing above exist. For a proof, multiply the right-hand side by $A + CBC'$ and show that the product is the identity.

Another useful formula provides the inverse of the partitioned matrix

$$M = \begin{bmatrix} A & B \\ C & D \end{bmatrix}.$$

There holds

$$M^{-1} = \begin{bmatrix} Q & -QBD^{-1} \\ -D^{-1}CQ & D^{-1} + D^{-1}CQBD^{-1} \end{bmatrix},$$

where

$$Q = (A - BD^{-1}C)^{-1},$$

provided all the inverses appearing above exist. For a proof, multiply M with the given expression for M^{-1} and verify that the product is the identity.

Definition A.6: The *characteristic polynomial* ϕ of an $n \times n$ matrix A is defined by $\phi(\lambda) = \det(\lambda I - A)$, where I is the identity matrix of the same size as A. The n (possibly repeated and complex) roots of ϕ are called the *eigenvalues* of A. A vector x (with possibly complex coordinates) such that $Ax = \lambda x$, where λ is an eigenvalue of A, is called an *eigenvector* of A associated with λ.

Proposition A.12: Let A be a square matrix.

(a) A complex number λ is an eigenvalue of A if and only if there exists a nonzero eigenvector associated with λ.

(b) A is singular if and only if it has an eigenvalue that is equal to zero.

Note that the only use of complex numbers in this book is in relation to eigenvalues and eigenvectors. All other matrices or vectors are implicitly assumed to have real components.

Proposition A.13: Let A be an $n \times n$ matrix.

(a) The eigenvalues of a triangular matrix are equal to its diagonal entries.

(b) If S is a nonsingular matrix and $B = SAS^{-1}$, then the eigenvalues of A and B coincide.

(c) The eigenvalues of $cI + A$ are equal to $c + \lambda_1, \ldots, c + \lambda_n$, where $\lambda_1, \ldots, \lambda_n$ are the eigenvalues of A.

(d) The eigenvalues of A^k are equal to $\lambda_1^k, \ldots, \lambda_n^k$, where $\lambda_1, \ldots, \lambda_n$ are the eigenvalues of A.

(e) If A is nonsingular, then the eigenvalues of A^{-1} are the reciprocals of the eigenvalues of A.

(f) The eigenvalues of A and A' coincide.

Definition A.7: The *spectral radius* $\rho(A)$ of a square matrix A is defined as the maximum of the magnitudes of the eigenvalues of A.

It can be shown that the roots of a polynomial depend continuously on the coefficients of the polynomial. For this reason, the eigenvalues of a square matrix A depend continuously on A, and we obtain the following.

Proposition A.14: The eigenvalues of a square matrix A depend continuously on the elements of A. In particular, $\rho(A)$ is a continuous function of A.

The next two propositions are fundamental for the convergence theory of linear iterative methods.

Proposition A.15: For any induced matrix norm $\|\cdot\|$ and any square matrix A we have

$$\lim_{k\to\infty} \|A^k\|^{1/k} = \rho(A) \leq \|A\|.$$

Furthermore, given any $\epsilon > 0$, there exists an induced matrix norm $\|\cdot\|$ such that

$$\|A\| = \rho(A) + \epsilon.$$

Proposition A.16: Let A be a square matrix. We have $\lim_{k\to\infty} A^k = 0$ if and only if $\rho(A) < 1$.

A corollary of the above proposition is that the iteration $x^{k+1} = Ax^k$ converges to 0 for every initial condition x^0 if and only if $\rho(A) < 1$.

A.4 SYMMETRIC AND POSITIVE DEFINITE MATRICES

Symmetric matrices have several special properties, particularly with respect to their eigenvalues and eigenvectors. In this section, $\|\cdot\|$ denotes the Euclidean norm throughout.

Proposition A.17: Let A be a symmetric $n \times n$ matrix. Then:

(a) The eigenvalues of A are real.

(b) The matrix A has a set of n mutually orthogonal, real, and nonzero eigenvectors $x_1, \ldots, x_n$.

(c) Suppose that the eigenvectors in part (b) have been normalized so that $\|x_i\| = 1$ for each i. Then

$$A = \sum_{i=1}^{n} \lambda_i x_i x_i',$$

where λ_i is the eigenvalue corresponding to x_i.

Proposition A.18: Let A be a symmetric $n \times n$ matrix, let $\lambda_1 \leq \cdots \leq \lambda_n$ be its (real) eigenvalues, and let $x_1, \ldots, x_n$ be associated orthogonal eigenvectors, normalized so that $\|x_i\| = 1$ for all i. Then:

(a) $\|A\| = \rho(A) = \max\{|\lambda_1|, |\lambda_n|\}$, where $\|\cdot\|$ is the matrix norm induced by the Euclidean norm.

(b) $\lambda_1 \|y\|^2 \leq y'Ay \leq \lambda_n \|y\|^2$ for all $y \in \Re^n$.

(c) (*Courant-Fisher Minimax Principle*) For all i, and for all i-dimensional subspaces $\overline{S}_i$ and all $(n-i+1)$-dimensional subspaces $\underline{S}_i$, there holds

$$\min_{\|y\|=1,\, y \in \underline{S}_i} y'Ay \leq \lambda_i \leq \max_{\|y\|=1,\, y \in \overline{S}_i} y'Ay.$$

Furthermore, equality on the left (right) side above is attained if $\underline{S}_i$ is the subspace spanned by $x_i, \ldots, x_n$ ($\overline{S}_i$ is the subspace spanned by $x_1, \ldots, x_i$, respectively).

(d) (*Interlocking Eigenvalues Lemma*) Let $\tilde{\lambda}_1 \leq \tilde{\lambda}_2 \leq \cdots \leq \tilde{\lambda}_n$ be the eigenvalues of $A + bb'$, where b is a vector in $\Re^n$. Then,

$$\lambda_1 \leq \tilde{\lambda}_1 \leq \lambda_2 \leq \tilde{\lambda}_2 \leq \cdots \leq \lambda_n \leq \tilde{\lambda}_n.$$

Proof: (a) We already know that $\|A\| \geq \rho(A)$ (Prop. A.15) and we need to show the reverse inequality. We express an arbitrary vector $y \in \Re^n$ in the form $y = \sum_{i=1}^n \xi_i x_i$, where each ξ_i is a suitable scalar. Using the orthogonality of the vectors x_i and the Pythagorean theorem (Prop. A.1), we obtain $\|y\|^2 = \sum_{i=1}^n |\xi_i|^2 \cdot \|x_i\|^2$. Using the Pythagorean theorem again, we obtain

$$\|Ay\|^2 = \left\| \sum_{i=1}^n \lambda_i \xi_i x_i \right\|^2 = \sum_{i=1}^n |\lambda_i|^2 \cdot |\xi_i|^2 \cdot \|x_i\|^2 \leq \rho^2(A) \|y\|^2.$$

Since this is true for every y, we obtain $\|A\| \leq \rho(A)$ and the desired result follows.

(b) As in part (a), we express the generic $y \in \Re^n$ as $y = \sum_{i=1}^n \xi_i x_i$. We have, using the orthogonality of the vectors x_i, $i = 1, \ldots, n$, and the fact $\|x_i\| = 1$,

$$y'Ay = \sum_{i=1}^n \lambda_i |\xi_i|^2 \|x_i\|^2 = \sum_{i=1}^n \lambda_i |\xi_i|^2$$

and

$$\|y\|^2 = \sum_{i=1}^{n} |\xi_i|^2 \|x_i\|^2 = \sum_{i=1}^{n} |\xi_i|^2.$$

These two relations prove the desired result.

(c) Let $\underline{X}_i$ be the subspace spanned by $x_1, \ldots, x_i$. The subspaces $\underline{X}_i$ and $\underline{S}_i$ must have a common vector x_0 with $\|x_0\| = 1$, since the sum of their dimensions is $n+1$ [if there was no common nonzero vector, we could take sets of basis vectors for $\underline{X}_i$ and S_i (a total of $n+1$ in number), which would have to be linearly independent, yielding a contradiction]. The vector x_0 can be expressed as a linear combination $x_0 = \sum_{j=1}^{i} \xi_j x_j$, and since $\|x_0\| = 1$ and $\|x_i\| = 1$ for all $i = 1, \ldots, n$, we must have

$$\sum_{j=1}^{i} \xi_j^2 = 1.$$

We also have using the expression $A = \sum_{j=1}^{n} \lambda_j x_j x_j'$ [cf. Prop. A.17(c)],

$$x_0' A x_0 = \sum_{j=1}^{i} \lambda_j \xi_j^2 \leq \lambda_i \left(\sum_{j=1}^{i} \xi_j^2 \right).$$

Combining the last two relations, we obtain $x_0' A x_0 \leq \lambda_i$, which proves the left-hand side of the desired inequality. The right-hand side is proved similarly. Furthermore, we have $x_i' A x_i = \lambda_i$, so equality is attained as in the final assertion.

(d) From part (c) we have

$$\lambda_i = \max_{\underline{S}_i} \min_{\|y\|=1,\, y \in \underline{S}_i} y' A y \leq \max_{\underline{S}_i} \min_{\|y\|=1,\, y \in \underline{S}_i} y'(A + bb')y \leq \tilde{\lambda}_i,$$

so that $\lambda_i \leq \tilde{\lambda}_i$ for all i. Furthermore, from part (c), for some $(n-i+1)$-dimensional subspace $\tilde{\underline{S}}_i$ we have

$$\tilde{\lambda}_i = \min_{\|y\|=1,\, y \in \tilde{\underline{S}}_i} y'(A + bb')y.$$

Using this relation and the left-hand side of the inequality of part (c), applied to the subspace $\{y \mid y \in \tilde{\underline{S}}_i,\, b'y = 0\}$, whose dimension is at least $(n-i)$, we obtain

$$\tilde{\lambda}_i \leq \min_{\|y\|=1,\, y \in \tilde{\underline{S}}_i,\, b'y=0} y'(A + bb')y = \min_{\|y\|=1,\, y \in \tilde{\underline{S}}_i,\, b'y=0} y' A y \leq \lambda_{i+1},$$

and the proof is complete. **Q.E.D.**

Proposition A.19: Let A be a square matrix, and let $\|\cdot\|$ be the matrix norm induced by the Euclidean norm. Then:

(a) If A is symmetric, then $\|A^k\| = \|A\|^k$ for any positive integer k.

(b) $\|A\|^2 = \|A'A\| = \|AA'\|$.

(c) If A is symmetric and nonsingular, then $\|A^{-1}\|$ is equal to the reciprocal of the smallest of the absolute values of the eigenvalues of A.

Proof: (a) If A is symmetric then A^k is symmetric. Using Prop. A.18(a), we have $\|A^k\| = \rho(A^k)$. Using Prop. A.13(d), we obtain $\rho(A^k) = \rho(A)^k$, which is equal to $\|A\|^k$ by Prop. A.18(a).

(b) For any vector x such that $\|x\| = 1$, we have, using the Schwartz inequality (Prop. A.2),

$$\|Ax\|^2 = x'A'Ax \leq \|x\| \cdot \|A'Ax\| \leq \|x\| \cdot \|A'A\| \cdot \|x\| = \|A'A\|.$$

Thus, $\|A\|^2 \leq \|A'A\|$. On the other hand,

$$\|A'A\| = \max_{\|y\|=\|x\|=1} |y'A'Ax| \leq \max_{\|y\|=\|x\|=1} \|Ay\| \cdot \|Ax\| = \|A\|^2.$$

Therefore, $\|A\|^2 = \|A'A\|$. The equality $\|A\|^2 = \|A'A\|$ is obtained by replacing A by A' and using Eq. (A.2).

(c) This follows by combining Prop. A.13(e) with Prop. A.18(a). **Q.E.D.**

Definition A.8: A symmetric $n \times n$ matrix A is called *positive definite* if $x'Ax > 0$ for all $x \in \Re^n$, $x \neq 0$. It is called *nonnegative definite* or *positive semidefinite* if $x'Ax \geq 0$ for all $x \in \Re^n$.

Throughout this book, the notion of positive and negative definiteness applies exclusively to symmetric matrices. Thus *whenever we say that a matrix is positive or negative (semi)definite, we implicitly assume that the matrix is symmetric.*

Proposition A.20:

(a) For any $m \times n$ matrix A, the matrix $A'A$ is symmetric and nonnegative definite. $A'A$ is positive definite if and only if A has rank n. In particular, if $m = n$, $A'A$ is positive definite if and only if A is nonsingular.

(b) A square symmetric matrix is nonnegative definite (respectively, positive definite) if and only if all of its eigenvalues are nonnegative (respectively, positive).

(c) The inverse of a symmetric positive definite matrix is symmetric and positive definite.

Proof: (a) Symmetry is obvious. For any vector $x \in \Re^n$, we have $x'A'Ax = \|Ax\|^2 \geq 0$, which establishes nonnegative definiteness. Positive definiteness is obtained if and only if the inequality is strict for every $x \neq 0$, which is the case if and only if $Ax \neq 0$ for every $x \neq 0$. This is equivalent to A having rank n.

(b) Let λ, $x \neq 0$, be an eigenvalue and a corresponding real eigenvector of a symmetric nonnegative definite matrix A. Then $0 \leq x'Ax = \lambda x'x = \lambda\|x\|^2$, which proves that $\lambda \geq 0$. For the converse result, let y be an arbitrary vector in $\Re^n$. Let $\lambda_1, \ldots, \lambda_n$ be the eigenvalues of A, assumed to be nonnegative, and let $x_1, \ldots, x_n$ be a corresponding set of nonzero, real, and orthogonal eigenvectors. Let us express y in the form $y = \sum_{i=1}^n \xi_i x_i$. Then $y'Ay = (\sum_{i=1}^n \xi_i x_i)'(\sum_{i=1}^n \xi_i \lambda_i x_i)$. From the orthogonality of the eigenvectors, the latter expression is equal to $\sum_{i=1}^n \xi_i^2 \lambda_i \|x_i\|^2 \geq 0$, which proves that A is nonnegative definite. The proof for the case of positive definite matrices is similar.

(c) The eigenvalues of A^{-1} are the reciprocal of the eigenvalues of A [Prop. A.13(e)], so the result follows using part (b). **Q.E.D.**

Proposition A.21: Let A be a square symmetric nonnegative definite matrix.

(a) There exists a symmetric matrix Q with the property $Q^2 = A$. Such a matrix is called a *symmetric square root* of A and is denoted by $A^{1/2}$.

(b) A symmetric square root $A^{1/2}$ is invertible if and only if A is invertible. Its inverse is denoted by $A^{-1/2}$.

(c) There holds $A^{-1/2}A^{-1/2} = A^{-1}$.

(d) There holds $AA^{1/2} = A^{1/2}A$.

Proof: (a) Let $\lambda_1, \ldots, \lambda_n$ be the eigenvalues of A and let $x_1, \ldots, x_n$ be corresponding nonzero, real, and orthogonal eigenvectors normalized so that $\|x_k\| = 1$ for each k. We let

$$A^{1/2} = \sum_{k=1}^{n} \lambda_k^{1/2} x_k x_k',$$

where $\lambda_k^{1/2}$ is the nonnegative square root of λ_k. We then have

$$A^{1/2}A^{1/2} = \sum_{i=1}^{n}\sum_{k=1}^{n} \lambda_i^{1/2}\lambda_k^{1/2} x_i x_i' x_k x_k' = \sum_{k=1}^{n} \lambda_k x_k x_k' = A.$$

Here the second equality follows from the orthogonality of distinct eigenvectors; the last equality follows from Prop. A.17(c). We now notice that each one of the matrices $x_k x_k'$ is symmetric, so $A^{1/2}$ is also symmetric.

(b) This follows from the fact that the eigenvalues of A are the squares of the eigenvalues of $A^{1/2}$ [Prop. A.13(d)].

(c) We have $(A^{-1/2}A^{-1/2})A = A^{-1/2}(A^{-1/2}A^{1/2})A^{1/2} = A^{-1/2}IA^{1/2} = I$.

(d) We have $AA^{1/2} = A^{1/2}A^{1/2}A^{1/2} = A^{1/2}A$. **Q.E.D.**

A symmetric square root of A is not unique. For example, let $A^{1/2}$ be as in the proof of Prop. A.21(a) and notice that the matrix $-A^{1/2}$ also has the property $(-A^{1/2})(-A^{1/2}) = A$. However, if A is positive definite, it can be shown that the matrix $A^{1/2}$ we have constructed is the only symmetric and positive definite square root of A.

A.5 DERIVATIVES

Let $f : \Re^n \mapsto \Re$ be some function, fix some $x \in \Re^n$, and consider the expression

$$\lim_{\alpha \to 0} \frac{f(x + \alpha e_i) - f(x)}{\alpha},$$

where e_i is the ith unit vector (all components are 0 except for the ith component which is 1). If the above limit exists, it is called the ith *partial derivative* of f at the point x and it is denoted by $(\partial f/\partial x_i)(x)$ or $\partial f(x)/\partial x_i$ (x_i in this section will denote the ith coordinate of the vector x). Assuming all of these partial derivatives exist, the *gradient* of f at x is defined as the column vector

$$\nabla f(x) = \begin{pmatrix} \frac{\partial f(x)}{\partial x_1} \\ \vdots \\ \frac{\partial f(x)}{\partial x_n} \end{pmatrix}.$$

For any $y \in \Re^n$, we define the one-sided *directional derivative* of f in the direction y, to be

$$f'(x; y) = \lim_{\alpha \downarrow 0} \frac{f(x + \alpha y) - f(x)}{\alpha},$$

provided that the limit exists. We note from the definitions that

$$f'(x; e_i) = -f'(x; -e_i) \qquad \Rightarrow \qquad f'(x; e_i) = (\partial f / \partial x_i)(x).$$

If the directional derivative of f at a vector x exists in all directions y and $f'(x; y)$ is a linear function of y, we say that f is *differentiable* at x. This type of differentiability is also called *Gateaux differentiability*. It is seen that f is differentiable at x if and only if the gradient $\nabla f(x)$ exists and satisfies $\nabla f(x)'y = f'(x; y)$ for every $y \in \Re^n$. The function f is called *differentiable over a given subset S of $\Re^n$* if it is differentiable at every $x \in S$. The function f is called differentiable (without qualification) if it is differentiable at all $x \in \Re^n$.

If f is differentiable over a set S and the gradient $\nabla f(x)$ is continuous at all $x \in S$, f is said to be *continuously differentiable over S*. Such a function is also continuous over S and has the property

$$\lim_{y \to 0} \frac{f(x + y) - f(x) - \nabla f(x)'y}{\|y\|} = 0, \qquad \forall\, x \in S, \tag{A.3}$$

where $\|\cdot\|$ is an arbitrary vector norm. The above equation can also be used as an alternative definition of differentiability. In particular, f is called *Frechet differentiable* at x if there exists a vector g satisfying Eq. (A.3) with $\nabla f(x)$ replaced by g. If such a vector g exists, it can be seen that all the partial derivatives $(\partial f / \partial x_i)(x)$ exist and that $g = \nabla f(x)$. Frechet differentiability implies (Gateaux) differentiability but not conversely (see for example [OrR70] for a detailed discussion). In this book, when dealing with a differentiable function f, we will always assume that f is continuously differentiable over a given set [$\nabla f(x)$ is a continuous function of x over that set], in which case f is both Gateaux and Frechet differentiable, and the distinctions made above are of no consequence.

Note that the definitions concerning differentiability of f at a point x only involve the values of f in a neighborhood of x. Thus, these definitions can be used for functions f that are not defined on all of $\Re^n$, but are defined instead in a neighborhood of the point at which the derivative is computed.

If $f : \Re^n \mapsto \Re^m$ is a vector-valued function, it is called differentiable (respectively, continuously differentiable) if each component f_i of f is differentiable (respectively, continuously differentiable). The *gradient matrix* of f, denoted $\nabla f(x)$, is the $n \times m$ matrix whose ith column is the gradient $\nabla f_i(x)$ of f_i. Thus,

$$\nabla f(x) = \Big[\nabla f_1(x) \cdots \nabla f_m(x) \Big].$$

The transpose of ∇f is called the *Jacobian* of f and is a matrix whose ijth entry is equal to the partial derivative $\partial f_i/\partial x_j$.

Now suppose that each one of the partial derivatives of a function $f : \Re^n \mapsto \Re$ is a continuously differentiable function of x. We use the notation $(\partial^2 f/\partial x_i \partial x_j)(x)$ to indicate the ith partial derivative of $\partial f/\partial x_j$ at a point $x \in \Re^n$. The *Hessian* of f is the matrix whose ijth entry is equal to $(\partial^2 f/\partial x_i \partial x_j)(x)$, and is denoted by $\nabla^2 f(x)$. We have $(\partial^2 f/\partial x_i \partial x_j)(x) = (\partial^2 f/\partial x_j \partial x_i)(x)$ for every x, which implies that $\nabla^2 f(x)$ is symmetric.

If $f : \Re^{m+n} \mapsto \Re$ is function of (x, y), where $x = (x_1, \ldots, x_m) \in \Re^m$ and $y = (y_1, \ldots, y_n) \in \Re^n$, we write

$$\nabla_x f(x,y) = \begin{pmatrix} \frac{\partial f(x,y)}{\partial x_1} \\ \vdots \\ \frac{\partial f(x,y)}{\partial x_m} \end{pmatrix}, \qquad \nabla_y f(x,y) = \begin{pmatrix} \frac{\partial f(x,y)}{\partial y_1} \\ \vdots \\ \frac{\partial f(x,y)}{\partial y_n} \end{pmatrix},$$

$$\nabla^2_{xx} f(x,y) = \left(\frac{\partial^2 f(x,y)}{\partial x_i \partial x_j} \right), \qquad \nabla^2_{xy} f(x,y) = \left(\frac{\partial^2 f(x,y)}{\partial x_i \partial y_j} \right),$$

$$\nabla^2_{yy} f(x,y) = \left(\frac{\partial^2 f(x,y)}{\partial y_i \partial y_j} \right).$$

If $f : \Re^{m+n} \mapsto \Re^r$, $f = (f_1, f_2, \ldots, f_r)$, we write

$$\nabla_x f(x,y) = \big(\nabla_x f_1(x,y) \cdots \nabla_x f_r(x,y)\big),$$

$$\nabla_y f(x,y) = \big(\nabla_y f_1(x,y) \cdots \nabla_y f_r(x,y)\big).$$

Let $f : \Re^k \mapsto \Re^m$ and $g : \Re^m \mapsto \Re^n$ be continuously differentiable functions, and let h be their composition, i.e.,

$$h(x) = g\big(f(x)\big).$$

Then, the *chain rule* for differentiation states that

$$\nabla h(x) = \nabla f(x) \nabla g\big(f(x)\big), \qquad \forall\, x \in \Re^k.$$

Some examples of useful relations that follow from the chain rule are:

$$\nabla\big(f(Ax)\big) = A' \nabla f(Ax), \qquad \nabla^2\big(f(Ax)\big) = A' \nabla^2 f(Ax) A,$$

where A is a matrix,

$$\nabla_x\Big(f\big(h(x), y\big)\Big) = \nabla h(x) \nabla_h f\big(h(x), y\big),$$

$$\nabla_x\Big(f\big(h(x), g(x)\big)\Big) = \nabla h(x) \nabla_h f\big(h(x), g(x)\big) + \nabla g(x) \nabla_g f\big(h(x), g(x)\big).$$

We now state the principal theorems relating to differentiable functions that will be useful for our purposes.

Proposition A.22: (Mean Value Theorem) If $f : \Re \mapsto \Re$ is continuously differentiable over an open interval I, then for every $x, y \in I$, there exists some $\xi \in [x, y]$ such that

$$f(y) - f(x) = \nabla f(\xi)(y - x).$$

Proposition A.23: (Second Order Expansions) Let $f : \Re^n \mapsto \Re$ be twice continuously differentiable over an open sphere S centered at a vector x.

(a) For all y such that $x + y \in S$,

$$f(x+y) = f(x) + y'\nabla f(x) + \tfrac{1}{2}y' \left(\int_0^1 \left(\int_0^t \nabla^2 f(x + \tau y) d\tau \right) dt \right) y.$$

(b) For all y such that $x + y \in S$, there exists an $\alpha \in [0, 1]$ such that

$$f(x+y) = f(x) + y'\nabla f(x) + \tfrac{1}{2}y'\nabla^2 f(x + \alpha y)y.$$

(c) For all y such that $x + y \in S$ there holds

$$f(x+y) = f(x) + y'\nabla f(x) + \tfrac{1}{2}y'\nabla^2 f(x)y + o\big(\|y\|^2\big).$$

Proposition A.24: (Descent Lemma) Let $f : \Re^n \mapsto \Re$ be continuously differentiable, and let x and y be two vectors in $\Re^n$. Suppose that

$$\|\nabla f(x + ty) - \nabla f(x)\| \leq Lt\|y\|, \qquad \forall\, t \in [0, 1],$$

where L is some scalar. Then

$$f(x+y) \leq f(x) + y'\nabla f(x) + \frac{L}{2}\|y\|^2.$$

Proof: Let t be a scalar parameter and let $g(t) = f(x + ty)$. The chain rule yields $(dg/dt)(t) = y'\nabla f(x + ty)$. Now

$$\begin{aligned} f(x+y) - f(x) = g(1) - g(0) &= \int_0^1 \frac{dg}{dt}(t)\,dt = \int_0^1 y'\nabla f(x+ty)\,dt \\ &\le \int_0^1 y'\nabla f(x)\,dt + \left| \int_0^1 y'\big(\nabla f(x+ty) - \nabla f(x)\big)\,dt \right| \\ &\le \int_0^1 y'\nabla f(x)\,dt + \int_0^1 \|y\| \cdot \|\nabla f(x+ty) - \nabla f(x)\| dt \\ &\le y'\nabla f(x) + \|y\| \int_0^1 Lt\|y\|\,dt \\ &= y'\nabla f(x) + \frac{L}{2}\|y\|^2. \end{aligned}$$

Q.E.D.

Proposition A.25: (Implicit Function Theorem) Let $f : \Re^{n+m} \mapsto \Re^m$ be a function of $x \in \Re^n$ and $y \in \Re^m$ such that:

(1) $f(\overline{x}, \overline{y}) = 0$.

(2) f is continuous, and has a continuous and nonsingular gradient matrix $\nabla_y f(x, y)$ in an open set containing $(\overline{x}, \overline{y})$.

Then there exist open sets $S_{\overline{x}} \subset \Re^n$ and $S_{\overline{y}} \subset \Re^m$ containing $\overline{x}$ and $\overline{y}$, respectively, and a continuous function $\phi : S_{\overline{x}} \mapsto S_{\overline{y}}$ such that $\overline{y} = \phi(\overline{x})$ and $f\big(x, \phi(x)\big) = 0$ for all $x \in S_{\overline{x}}$. The function ϕ is unique in the sense that if $x \in S_{\overline{x}}$, $y \in S_{\overline{y}}$, and $f(x, y) = 0$, then $y = \phi(x)$. Furthermore, if for some $p > 0$, f is p times continuously differentiable the same is true for ϕ, and we have

$$\nabla \phi(x) = -\nabla_x f\big(x, \phi(x)\big)\Big(\nabla_y f\big(x, \phi(x)\big)\Big)^{-1}, \qquad \forall\, x \in S_{\overline{x}}.$$

As a final word of caution to the reader, let us mention that one can easily get confused with gradient notation and its use in various formulas, such as for example the order of multiplication of various gradients in the chain rule and the implicit function theorem. Perhaps the safest guideline to minimize errors is to remember our conventions:

(a) A vector is viewed as a column vector (an $n \times 1$ matrix).

(b) The gradient ∇f of a scalar function $f : \Re^n \mapsto \Re$ is also viewed as a column vector.

(c) The gradient matrix ∇f of a vector function $f : \Re^n \mapsto \Re^m$ with components $f_1, \ldots, f_m$ is the $n \times m$ matrix whose columns are the (column) vectors $\nabla f_1, \ldots, \nabla f_m$.

With these rules in mind one can use "dimension matching" as an effective guide to writing correct formulas quickly.

A.6 CONTRACTION MAPPINGS

Many iterative algorithms can be written as

$$x^{k+1} = g(x^k), \qquad k = 0, 1, \ldots,$$

where g is a mapping from a subset X of $\Re^n$ into itself and has the property

$$\|g(x) - g(y)\| \leq \gamma \|x - y\|, \qquad \forall\, x, y \in X. \tag{A.4}$$

Here $\|\cdot\|$ is some norm, and γ is a scalar with $0 \leq \gamma < 1$. Such a mapping is called a *contraction mapping*, or simply a *contraction*. The scalar γ is called the *contraction modulus* of g. Note that a mapping g may be a contraction for some choice of the norm $\|\cdot\|$ and fail to be a contraction under a different choice of norm.

Let there be given a mapping $g : X \mapsto X$. Any vector $x^* \in X$ satisfying $g(x^*) = x^*$ is called a *fixed point* of g and the iteration $x^{k+1} = g(x^k)$ is an important algorithm for finding such a fixed point. The following is the central result regarding contraction mappings.

Proposition A.26: (Contraction Mapping Theorem) Suppose that $g : X \mapsto X$ is a contraction with modulus $\gamma \in [0, 1)$ and that X is a closed subset of $\Re^n$. Then:

(a) (*Existence and Uniqueness of Fixed Point*) The mapping g has a unique fixed point $x^* \in X$.

(b) (*Convergence*) For every initial vector $x^0 \in X$, the sequence $\{x^k\}$ generated by $x^{k+1} = g(x^k)$ converges to x^*. In particular,

$$\|x^k - x^*\| \leq \gamma^k \|x^0 - x^*\|, \qquad \forall\, k \geq 0.$$

Proof: (a) Fix some $x^0 \in X$ and consider the sequence $\{x^k\}$ generated by $x^{k+1} = g(x^k)$. We have, from inequality (A.4),

$$\|x^{k+1} - x^k\| \leq \gamma \|x^k - x^{k-1}\|,$$

for all $k \geq 1$, which implies

$$\|x^{k+1} - x^k\| \leq \gamma^k \|x^1 - x^0\|, \qquad \forall\, k \geq 0.$$

It follows that for every $k \geq 0$ and $m \geq 1$, we have

$$\begin{aligned} \|x^{k+m} - x^k\| &\leq \sum_{i=1}^{m} \|x^{k+i} - x^{k+i-1}\| \\ &\leq \gamma^k (1 + \gamma + \cdots + \gamma^{m-1}) \|x^1 - x^0\| \\ &\leq \frac{\gamma^k}{1-\gamma} \|x^1 - x^0\|. \end{aligned}$$

Therefore, $\{x^k\}$ is a Cauchy sequence and must converge to a limit x^* (Prop. A.5). Furthermore, since X is closed, x^* belongs to X. We have for all $k \geq 1$,

$$\|g(x^*) - x^*\| \leq \|g(x^*) - x^k\| + \|x^k - x^*\| \leq \gamma \|x^* - x^{k-1}\| + \|x^k - x^*\|$$

and since x^k converges to x^*, we obtain $g(x^*) = x^*$. Therefore, the limit x^* of x^k is a fixed point of g. It is a unique fixed point because if y^* were another fixed point, we would have

$$\|x^* - y^*\| = \|g(x^*) - g(y^*)\| \leq \gamma \|x^* - y^*\|,$$

which implies that $x^* = y^*$.

(b) We have

$$\|x^{k'} - x^*\| = \big\|g\big(x^{k'-1}\big) - g(x^*)\big\| \leq \gamma \|x^{k'-1} - x^*\|,$$

for all $k' \geq 1$, so by applying this relation successively for $k' = k,\ k-1,\ \ldots,1$, we obtain the desired result. **Q.E.D.**

APPENDIX B: *Convex Analysis*

Convexity is a central concept in nonlinear programming. In this appendix, we collect definitions, notational conventions, and several results from the theory of convex sets and functions. A classical and extensive reference on convex analysis is Rockafellar's book [Roc70]. Related and additional material can be found in Stoer and Witzgall [StW70], Ekeland and Teman [EkT76], Rockafellar [Roc84], Hiriart-Urruty and Lemarechal [HiL93], and Rockafellar and Wets [RoW97]. A discussion of generalized notions of convexity, including quasiconvexity and pseudoconvexity, and their applications in optimization can be found in the books by Avriel [Avr76], Bazaraa, Sherali, and Shetty [BSS93], Mangasarian [Man69], and the references quoted therein.

B.1 CONVEX SETS AND FUNCTIONS

The notions of a convex set and a convex function are defined below and are illustrated in Figs. B.1 and B.2, respectively.

Definition B.1: Let C be a subset of $\Re^n$. We say that C is *convex* if

$$\alpha x + (1 - \alpha) y \in C, \qquad \forall\, x, y \in C,\ \forall\, \alpha \in [0, 1]. \tag{B.1}$$

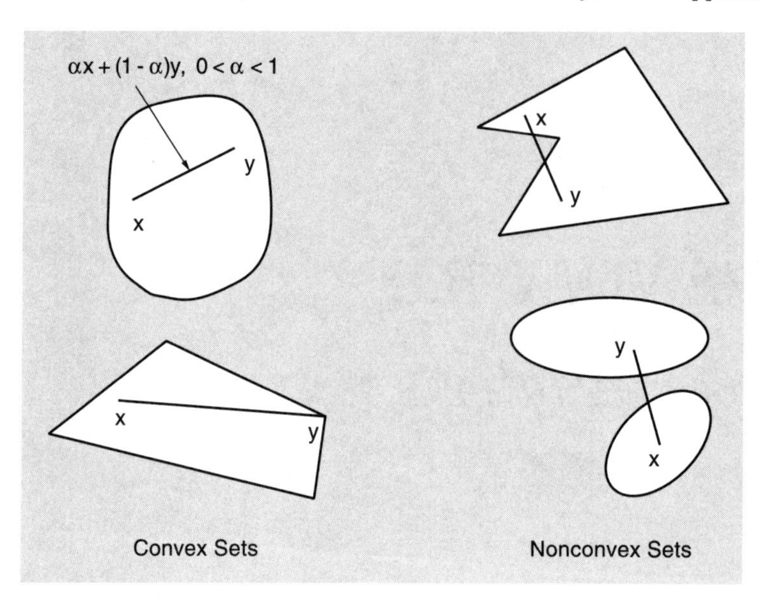

Figure B.1. Illustration of the definition of a convex set. For convexity, linear interpolation between two points in the set must yield a point within the set.

Definition B.2: Let C be a convex subset of $\Re^n$. A function $f : C \mapsto \Re$ is called *convex* if

$$f\big(\alpha x + (1-\alpha)y\big) \le \alpha f(x) + (1-\alpha)f(y), \qquad \forall\, x, y \in C,\ \forall\, \alpha \in [0,1]. \tag{B.2}$$

The function f is called *concave* if $-f$ is convex. The function f is called *strictly convex* if the above inequality is strict for all $x, y \in C$ with $x \neq y$, and all $\alpha \in (0,1)$. For a function $f : \Re^n \mapsto \Re$, we also say that f is *convex over the convex set* C if Eq. (B.2) holds.

The following proposition provides some means for verifying convexity of a set.

Proposition B.1:

(a) For any collection $\{C_i \mid i \in I\}$ of convex sets, the set intersection $\cap_{i \in I} C_i$ is convex.

(b) The vector sum $\{x_1 + x_2 \mid x_1 \in C_1,\ x_2 \in C_2\}$ of two convex sets C_1 and C_2 is convex.

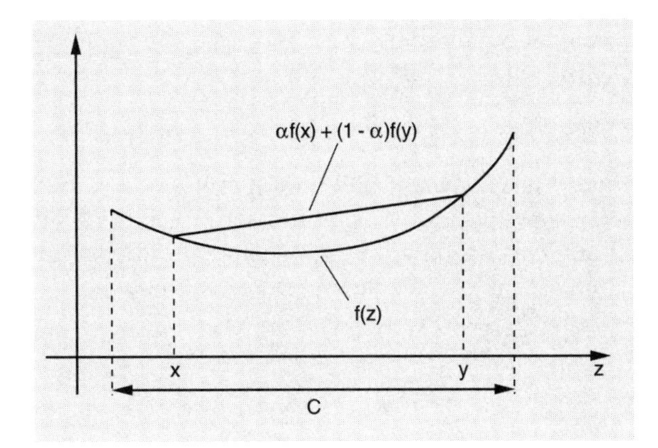

Figure B.2. Illustration of the definition of a convex function. The linear interpolation $\alpha f(x) + (1-\alpha)f(y)$ overestimates the function value $f(\alpha x + (1-\alpha)y)$. Note that the domain of the function must be a convex set.

(c) The image of a convex set under a linear transformation is convex.

(d) If C is a convex set and $f : C \mapsto \Re$ is a convex function, the level sets $\{x \in C \mid f(x) \leq \alpha\}$ and $\{x \in C \mid f(x) < \alpha\}$ are convex for all scalars α.

Proof: The proof is straightforward using the definitions (B.1) and (B.2). For example, to prove part (a), we take two points x and y from $\cap_{i \in I} C_i$, and we use the convexity of C_i to argue that the line segment connecting x and y belongs to all the sets C_i, and hence, to their intersection. The proofs of parts (b)-(d) are similar and are left as exercises for the reader. **Q.E.D.**

We occasionally deal with convex functions that can take the value of infinity. A function $f : C \mapsto (-\infty, \infty]$, where C is a convex subset of $\Re^n$, is also called convex if condition (B.2) holds. (Here the rules of arithmetic are extended to include $\infty + \infty = \infty$, $0 \cdot \infty = 0$, and $\alpha \cdot \infty = \infty$ if $\alpha > 0$.) The *effective domain* of f is the convex set

$$\text{dom}(f) = \{x \in C \mid f(x) < \infty\}.$$

By restricting the definition of a convex function to its effective domain we can avoid calculations with ∞, and we will often do this. However, in some analyses it is more economical to use convex functions that can take the value of infinity.

The *epigraph* of a function $f : C \mapsto (-\infty, \infty]$, where C is a convex subset of $\Re^n$, is the subset of $\Re^{n+1}$ given by

$$\text{epi}(f) = \{(x, w) \mid x \in C,\, w \in \Re,\, f(x) \leq w\}.$$

It can be seen that f is convex if and only if epi(f) is a convex set. This is a useful property, since it allows us to translate results about convex sets into results about convex functions. Another useful property, obtained by repeated application of inequality (B.2), is that if $x_1, \ldots, x_m \in C$, $\alpha_1, \ldots, \alpha_m \geq 0$, and $\sum_{i=1}^m \alpha_i = 1$, then

$$f\left(\sum_{i=1}^m \alpha_i x_i\right) \leq \sum_{i=1}^m \alpha_i f(x_i). \tag{B.3}$$

This is a special case of *Jensen's inequality* and can be used to prove a number of interesting inequalities in applied mathematics and probability theory.

The following proposition provides some means for recognizing convex functions.

Proposition B.2:

(a) A linear function is convex.

(b) Any vector norm is convex.

(c) The weighted sum of convex functions, with positive weights, is convex.

(d) If I is an index set, $C \subset \Re^n$ is a convex set, and $f_i : C \mapsto \Re$ is convex for each $i \in I$, then the function $h : C \mapsto (-\infty, \infty]$ defined by

$$h(x) = \sup_{i \in I} f_i(x)$$

is also convex.

Proof: Parts (a) and (c) are immediate consequences of the definition of convexity.

Let $\|\cdot\|$ be a vector norm. For any $x, y \in \Re^n$ and any $\alpha \in [0, 1]$, we have

$$\|\alpha x + (1-\alpha)y\| \leq \|\alpha x\| + \|(1-\alpha)y\| = \alpha\|x\| + (1-\alpha)\|y\|,$$

which proves part (b).

For part (d), let us fix some $x, y \in C$, $\alpha \in [0, 1]$, and let $z = \alpha x + (1-\alpha)y$. For every $i \in I$, we have

$$f_i(z) \leq \alpha f_i(x) + (1-\alpha) f_i(y) \leq \alpha h(x) + (1-\alpha) h(y).$$

Taking the supremum over all $i \in I$, we conclude that $h(z) \leq \alpha h(x) + (1 - \alpha)h(y)$, so h is convex. **Q.E.D.**

Characterizations of Differentiable Convex Functions

For differentiable functions, there is an alternative characterization of convexity, given in the following proposition and illustrated in Fig. B.3.

Proposition B.3: Let $C \subset \Re^n$ be a convex set and let $f : \Re^n \mapsto \Re$ be differentiable over C.

(a) The function f is convex over C if and only if

$$f(z) \geq f(x) + (z - x)'\nabla f(x), \qquad \forall\, x, z \in C. \tag{B.4}$$

(b) If the inequality (B.4) is strict whenever $x \neq z$, then f is strictly convex over C.

Proof: (a) Suppose that f is convex. Let $x \in C$ and $z \in C$. By the convexity of C, we obtain $x + \alpha(z - x) \in C$ for every $\alpha \in [0, 1]$. Furthermore,

$$\lim_{\alpha \downarrow 0} \frac{f\big(x + \alpha(z - x)\big) - f(x)}{\alpha} = (z - x)'\nabla f(x). \tag{B.5}$$

Using the convexity of f, we have

$$f\big(x + \alpha(z - x)\big) \leq \alpha f(z) + (1 - \alpha)f(x), \qquad \forall\, \alpha \in [0, 1],$$

from which

$$\frac{f\big(x + \alpha(z - x)\big) - f(x)}{\alpha} \leq f(z) - f(x), \qquad \forall\, \alpha \in [0, 1].$$

Taking the limit as $\alpha \downarrow 0$ and using Eq. (B.5), we obtain Eq. (B.4).

For the proof of the converse, suppose that inequality (B.4) is true. We fix some $x, y \in C$ and some $\alpha \in [0, 1]$. Let $z = \alpha x + (1 - \alpha)y$. Using inequality (B.4) twice, we obtain

$$f(x) \geq f(z) + (x - z)'\nabla f(z),$$

$$f(y) \geq f(z) + (y - z)'\nabla f(z).$$

We multiply the first inequality by α, the second by $(1 - \alpha)$, and add them to obtain

$$\alpha f(x) + (1 - \alpha)f(y) \geq f(z) + \big(\alpha x + (1 - \alpha)y - z\big)'\nabla f(z) = f(z),$$

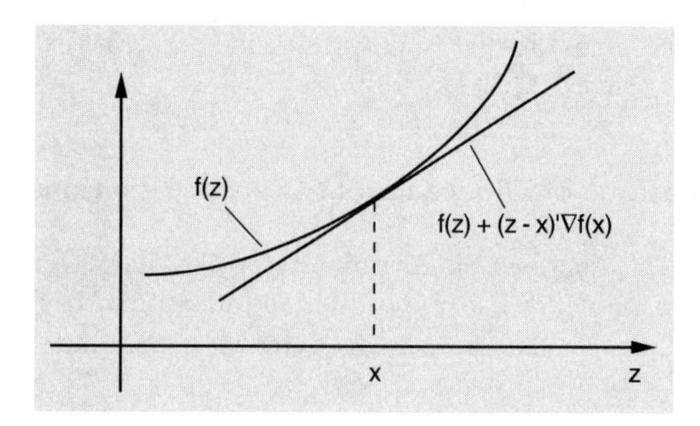

Figure B.3. Characterization of convexity in terms of first derivatives. The condition $f(z) \geq f(x) + (z-x)'\nabla f(x)$ states that a linear approximation, based on the first order Taylor series expansion, underestimates a convex function.

which proves that f is convex.

(b) The proof for the strictly convex case is almost identical to the proof of the corresponding statement of part (a) and is left for the reader. **Q.E.D.**

For twice differentiable convex functions, there is another characterization of convexity as shown by the following proposition.

Proposition B.4: Let $C \subset \Re^n$ be a convex set, let $f : \Re^n \mapsto \Re$ be twice continuously differentiable over C, and let Q be a real symmetric $n \times n$ matrix.

(a) If $\nabla^2 f(x)$ is positive semidefinite for all $x \in C$, then f is convex over C.

(b) If $\nabla^2 f(x)$ is positive definite for every $x \in C$, then f is strictly convex over C.

(c) If $C = \Re^n$ and f is convex, then $\nabla^2 f(x)$ is positive semidefinite for all $x \in C$.

(d) The quadratic function $f(x) = x'Qx$, where Q is a symmetric matrix, is convex if and only if Q is positive semidefinite. Furthermore, f is strictly convex if and only if Q is positive definite.

Proof: (a) By Prop. A.23(b) of Appendix A, for all $x, y \in C$ we have

$$f(y) = f(x) + (y-x)'\nabla f(x) + \tfrac{1}{2}(y-x)'\nabla^2 f\big(x + \alpha(y-x)\big)(y-x)$$

for some $\alpha \in [0,1]$. Therefore, using the positive semidefiniteness of $\nabla^2 f$, we obtain

$$f(y) \geq f(x) + (y-x)'\nabla f(x), \qquad \forall\, x, y \in C.$$

From Prop. B.3(a), we conclude that f is convex.

(b) Similar to the proof of part (a), we have $f(y) > f(x) + (y-x)'\nabla f(x)$ for all $x, y \in C$ with $x \neq y$, and the result follows from Prop. B.3(b).

(c) Suppose that $f : \Re^n \mapsto \Re$ is convex and suppose, to derive a contradiction, that there exist some $x \in \Re^n$ and some $z \in \Re^n$ such that $z'\nabla^2 f(x) z < 0$. Using the continuity of $\nabla^2 f$, we see that we can choose the magnitude of z to be small enough so that $z'\nabla^2 f(x+\alpha z)z < 0$ for every $\alpha \in [0,1]$. Then, using again Prop. A.23(b) of Appendix A, we obtain $f(x+z) < f(x) + z'\nabla f(x)$, which, in view of Prop. B.3(a), contradicts the convexity of f.

(d) An easy calculation shows that $\nabla^2 f(x) = 2Q$ for all $x \in \Re^n$. Hence, from parts (a) and (c), we obtain that f is convex if and only if Q is positive semidefinite.

If Q is positive definite, then strict convexity of f follows from part (b). For the converse, suppose that f is strictly convex. Then part (c) implies that Q is positive semidefinite and it remains to show that Q is actually positive definite. In view of Prop. A.20(b) of Appendix A, it suffices to show that zero is not an eigenvalue of Q. Suppose the contrary. Then there exists some $x \neq 0$ such that $Qx = 0$. It follows that

$$\tfrac{1}{2}\big(f(x) + f(-x)\big) = 0 = f(0),$$

which contradicts the strict convexity of f. **Q.E.D.**

The conclusion of Prop. B.4(c) also holds if C is assumed to have nonempty interior instead of being equal to $\Re^n$; see Exercise B.1.2. The following proposition considers a strengthened form of strict convexity characterized by the following equation:

$$\big(\nabla f(x) - \nabla f(y)\big)'(x-y) \geq \alpha \|x-y\|^2, \qquad \forall\, x, y \in \Re^n, \tag{B.6}$$

Convex functions with this property are called *strongly convex*.

Proposition B.5: (Strong Convexity) Let $f : \Re^n \mapsto \Re$ be continuously differentiable and let α be a positive scalar. If f is strongly convex then f is strictly convex. Furthermore, if f is twice continuously differentiable, then strong convexity of f is equivalent to the positive semidefiniteness of $\nabla^2 f(x) - \alpha I$ for every $x \in \Re^n$, where I is the identity matrix.

Proof: Fix some $x, y \in \Re^n$ such that $x \neq y$, and define the function $h : \Re \mapsto \Re$ by $h(t) = f\big(x + t(y-x)\big)$. Consider some $t, t' \in \Re$ such that

$t < t'$. Using the chain rule and Eq. (B.6), we have

$$
\begin{aligned}
\Big(\frac{dh}{dt}(t') - \frac{dh}{dt}(t)\Big)(t'-t) &\\
&= \Big(\nabla f\big(x + t'(y-x)\big) - \nabla f\big(x + t(y-x)\big)\Big)'(y-x)(t'-t) \\
&\geq \alpha (t'-t)^2 \|x-y\|^2 > 0.
\end{aligned}
$$

Thus, dh/dt is strictly increasing and for any $t \in (0,1)$, we have

$$
\frac{h(t)-h(0)}{t} = \frac{1}{t}\int_0^t \frac{dh}{d\tau}(\tau)\,d\tau < \frac{1}{1-t}\int_t^1 \frac{dh}{d\tau}(\tau)\,d\tau = \frac{h(1)-h(t)}{1-t}.
$$

Equivalently, $th(1) + (1-t)h(0) > h(t)$. The definition of h yields $tf(y) + (1-t)f(x) > f\big(ty + (1-t)x\big)$. Since this inequality has been proved for arbitrary $t \in (0,1)$ and $x \neq y$, we conclude that f is strictly convex.

Suppose now that f is twice continuously differentiable and Eq. (B.6) holds. Let c be a scalar. We use Prop. A.23(b) of Appendix A twice to obtain

$$
f(x+cy) = f(x) + cy'\nabla f(x) + \frac{c^2}{2} y'\nabla^2 f(x+tcy)y,
$$

and

$$
f(x) = f(x+cy) - cy'\nabla f(x+cy) + \frac{c^2}{2} y'\nabla^2 f(x+scy)y,
$$

for some t and s belonging to $[0,1]$. Adding these two equations and using Eq. (B.6), we obtain

$$
\frac{c^2}{2} y'\big(\nabla^2 f(x+scy) + \nabla^2 f(x+tcy)\big)y = \big(\nabla f(x+cy) - \nabla f(x)\big)'(cy) \geq \alpha c^2 \|y\|^2.
$$

We divide both sides by c^2 and then take the limit as $c \to 0$ to conclude that $y'\nabla^2 f(x)y \geq \alpha \|y\|^2$. Since this inequality is valid for every $y \in \Re^n$, it follows that $\nabla^2 f(x) - \alpha I$ is positive semidefinite.

For the converse, assume that $\nabla^2 f(x) - \alpha I$ is positive semidefinite for all $x \in \Re^n$. Consider the function $g : \Re \mapsto \Re$ defined by

$$
g(t) = \nabla f\big(tx + (1-t)y\big)'(x-y).
$$

Using the mean value theorem (Prop. A.22 in Appendix A), we have $\big(\nabla f(x) - \nabla f(y)\big)'(x-y) = g(1) - g(0) = (dg/dt)(t)$ for some $t \in [0,1]$. The result follows because

$$
\frac{dg}{dt}(t) = (x-y)'\nabla^2 f\big(tx + (1-t)y\big)(x-y) \geq \alpha \|x-y\|^2,
$$

where the last inequality is a consequence of the positive semidefiniteness of $\nabla^2 f\big(tx + (1-t)y\big) - \alpha I$. **Q.E.D.**

Convex and Affine Hulls

Let X be a subset of $\Re^n$. A *convex combination* of elements of X, is a vector of the form $\sum_{i=1}^m \alpha_i x_i$, where $x_1, \ldots, x_m$ belong to X and $\alpha_1, \ldots, \alpha_m$ are scalars such that

$$\alpha_i \geq 0, \quad i = 1, \ldots, m, \qquad \sum_{i=1}^m \alpha_i = 1.$$

The *convex hull* of X, denoted $\text{conv}(X)$, is the set of all convex combinations of elements of X. In particular, if X consists of a finite number of vectors $x_1, \ldots, x_m$, its convex hull is

$$\text{conv}\big(\{x_1, \ldots, x_m\}\big) = \left\{ \sum_{i=1}^m \alpha_i x_i \;\middle|\; \alpha_i \geq 0,\, i = 1, \ldots, m,\, \sum_{i=1}^m \alpha_i = 1 \right\}.$$

It is straightforward to verify that $\text{conv}(X)$ is a convex set, and using this, to assert that $\text{conv}(X)$ is the intersection of all convex sets containing X.

We recall that a linear manifold M is a set of the form $x + S = \{z \mid z - x \in S\}$, where S is a subspace, called the *subspace parallel to* M. If S is a subset of $\Re^n$, the *affine hull* of S, denoted $\text{aff}(S)$, is the intersection of all linear manifolds containing S. Note that $\text{aff}(S)$ is itself a linear manifold and that it contains $\text{conv}(S)$. It can be seen that the affine hull of S and the affine hull of $\text{conv}(S)$ coincide.

The following is a fundamental characterization of convex sets.

> **Proposition B.6: (Caratheodory's Theorem)** Let X be a subset of $\Re^n$. Every element of $\text{conv}(X)$ can be represented as a convex combination of no more than $n + 1$ elements of X.

Proof: Let $x \in \text{conv}(X)$. Then, we can represent x as $\sum_{i=1}^m \alpha_i x_i$ for some vectors $x_i \in X$ and scalars $\alpha_i \geq 0$ with $\sum_{i=1}^m \alpha_i = 1$. Let us assume that m is the minimal number of vectors for which such a representation of x is possible; in particular, this implies that $\alpha_i > 0$ for all i. Suppose, in order to arrive at a contradiction, that $m > n + 1$, and let S be the subspace parallel to $\text{aff}(X)$. The $m - 1$ vectors $x_2 - x_1, \ldots, x_m - x_1$ belong to S, and since $m - 1 > n$, they must be linearly dependent. Therefore, there exist scalars $\lambda_2, \ldots, \lambda_m$ at least one of which is positive, such that

$$\sum_{i=2}^m \lambda_i (x_i - x_1) = 0.$$

Letting $\mu_i = \lambda_i$ for $i = 2, \ldots, m$ and $\mu_1 = -\sum_{i=2}^{m} \lambda_i$, we see that

$$\sum_{i=1}^{m} \mu_i x_i = 0, \qquad \sum_{i=1}^{m} \mu_i = 0,$$

while at least one of the scalars $\mu_2, \ldots, \mu_m$ is positive. Define

$$\overline{\alpha}_i = \alpha_i - \overline{\gamma}\mu_i, \qquad i = 1, \ldots, m,$$

where $\overline{\gamma} > 0$ is the largest γ such that $\alpha_i - \gamma\mu_i \geq 0$ for all i. Then, since $\sum_{i=1}^{m} \mu_i x_i = 0$, we see that x is also represented as $\sum_{i=1}^{m} \overline{\alpha}_i x_i$. Furthermore, in view of the fact $\sum_{i=1}^{m} \mu_i = 0$ and the choice of $\overline{\gamma}$, the coefficients $\overline{\alpha}_i$ are nonnegative, sum to one, and at least one of them is zero. Thus, x can be represented as a convex combination of fewer that m vectors of X, contradicting our earlier assumption. **Q.E.D.**

Closure and Continuity Properties

We now explore some generic topological properties of convex sets and functions.

Let C be a convex subset of $\Re^n$. We say that x is a *relative interior point* of C, if $x \in C$ and there exists a neighborhood N of x such that $N \cap \text{aff}(C) \subset C$, that is, if x is an interior point of C relative to aff(C). The *relative interior of* C, denoted ri(C), is the set of all relative interior points of C. For example, if C is a line segment connecting two distinct points in the plane, then ri(C) consists of all points of C except for the end points.

Proposition B.7:

(a) (*Nonemptiness of Relative Interior*) If C is a nonempty convex set, ri(C) is nonempty and has the same affine hull as C.

(b) (*Line Segment Principle*) If C is a convex set, $x \in \text{ri}(C)$ and $\overline{x} \in C$, then all points on the line segment connecting x and $\overline{x}$, except possibly $\overline{x}$, belong to ri(C), i.e., $\alpha x + (1-\alpha)\overline{x} \in \text{ri}(C)$ for all $\alpha \in (0, 1]$.

Proof: (a) By using a transformation argument if necessary, we assume without loss of generality that $0 \in C$. Then, the affine hull of C, aff(C), is a subspace with dimension denoted by m. If $m = 0$, then C and aff(C) consist of a single point, which satisfies the definition of a relative interior point. If $m > 0$, we can find m linearly independent vectors $x_1, \ldots, x_m$ from C; otherwise there would exist a set of $r < m$ linearly independent vectors from C, whose span contains C, contradicting the fact that the dimension of $\mathrm{aff}(C)$ is m. Thus $x_1, \ldots, x_m$ form a basis for the subspace $\mathrm{aff}(C)$. It can be seen that the set

$$S = \left\{ x \;\middle|\; x = \sum_{i=1}^{m} \alpha_i x_i,\ \sum_{i=1}^{m} \alpha_i < 1,\ \alpha_i > 0,\ i = 1, \ldots, m \right\}$$

is open relative to $\mathrm{aff}(C)$; that is, if $x \in S$, there exists an open set N_x such that $x \in N_x$ and $N_x \cap \mathrm{aff}(C) \subset S$. [To see this, note that S is the image of the open subset of $\Re^m$

$$\left\{ (\alpha_1, \ldots, \alpha_m) \;\middle|\; \sum_{i=1}^{m} \alpha_i < 1,\ \alpha_i > 0,\ i = 1, \ldots, m \right\}$$

under the invertible linear transformation from $\Re^m$ onto $\mathrm{aff}(C)$ that maps $(\alpha_1, \ldots, \alpha_m)$ into $\sum_{i=1}^{m} \alpha_i x_i$; openness of sets is preserved by invertible linear transformations.] Since $S \subset C$, it follows that all points of S are relative interior points of C.

(b) See Fig. B.4. **Q.E.D.**

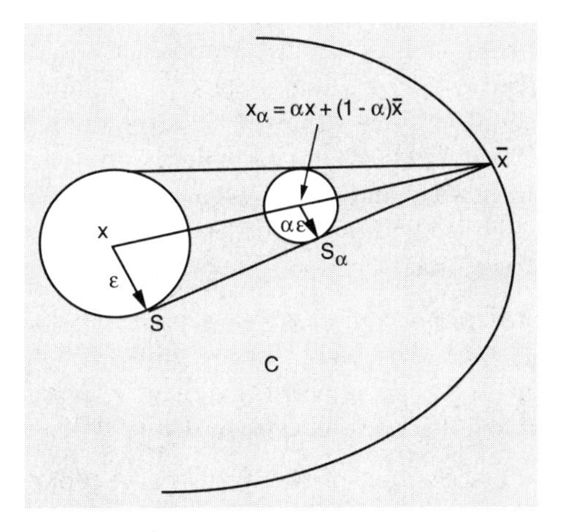

Figure B.4. Proof of the line segment principle. Since $x \in \mathrm{ri}(C)$, there exists a sphere $S = \{z \mid \|z - x\| < \epsilon\}$ such that $S \cap \mathrm{aff}(C) \subset C$. For all $\alpha \in (0, 1]$, let $x_\alpha = \alpha x + (1 - \alpha)\overline{x}$ and let $S_\alpha = \{z \mid \|z - x_\alpha\| < \alpha\epsilon\}$. It can be seen that each point of $S_\alpha \cap \mathrm{aff}(C)$ is a convex combination of $\overline{x}$ and some point of $S \cap \mathrm{aff}(C)$. Therefore, $S_\alpha \cap \mathrm{aff}(C) \subset C$, implying that $x_\alpha \in \mathrm{ri}(C)$.

The closure of a set $X \subset \Re^n$, denoted $\mathrm{cl}(X)$, is the set of all limit points of sequences from X. It is not generally true that the closedness of convex sets is preserved by taking vector sums, applying linear transformations, or forming convex hulls (for examples, see the subsequent Fig. B.8). We have, however, the following:

Proposition B.8:

(a) The closure $\mathrm{cl}(C)$ and the relative interior $\mathrm{ri}(C)$ of a convex set C are convex sets.

(b) The vector sum of two closed convex sets at least one of which is compact, is a closed convex set.

(c) The image of a convex and compact set under a linear transformation is a convex and compact set.

(d) The convex hull of a compact set is compact.

Proof: (a) Let $S_\epsilon = \{x + z \mid x \in C, \|x - z\| \le \epsilon\}$. Then $\mathrm{cl}(C) = \cap_{\epsilon>0} S_\epsilon$, and since each set S_ϵ can be seen to be convex, the same is true of $\mathrm{cl}(C)$. The convexity of $\mathrm{ri}(C)$ follows from the line segment principle [Prop. B.7(b)].

(b) Let C_1 and C_2 be closed convex sets and suppose that C_2 is compact. Their vector sum $C = \{x_1 + x_2 \mid x_1 \in C_1, x_2 \in C_2\}$ is convex by Prop. B.1(b). To show that C is also closed, consider a convergent sequence $\{x_1^k + x_2^k\} \subset C$ with $\{x_1^k\} \subset C_1$ and $\{x_2^k\} \subset C_2$. Then $\{x_2^k\}$ is bounded, since C_2 is compact, and since $\{x_1^k + x_2^k\}$ converges, it follows that $\{x_1^k\}$ is also bounded. Thus, $\{(x_1^k, x_2^k)\}$ is bounded, and $\{(x_1^k, x_2^k)\}$ has a limit point $(\tilde{x}_1, \tilde{x}_2)$ with $\tilde{x}_1 \in C_1$ and $\tilde{x}_2 \in C_2$, since C_1 and C_2 are closed. The vector $\tilde{x}_1 + \tilde{x}_2$, which is the limit of $\{x_1^k + x_2^k\}$, must therefore belong to C, proving that C is closed.

(c) Let C be a convex and compact set, A be a matrix, and $\{Ax_k\}$ be a sequence with $\{x_k\} \subset C$. Then, $\{x_k\}$ has a convergent subsequence $\{x_k\}_K$ and the subsequence $\{Ax_k\}_K$ is also convergent. Therefore, the image of C under A is compact. It is also convex by Prop. B.1(c).

(d) Let X be a compact subset of $\Re^n$. By Caratheodory's theorem (Prop. B.6), a sequence in $\mathrm{conv}(X)$ can be expressed as $\left\{\sum_{i=1}^{n+1} \alpha_i^k x_i^k\right\}$, where for all k and i, $\alpha_i^k \ge 0$, $x_i^k \in X$, and $\sum_{i=1}^{n+1} \alpha_i^k = 1$. Since the sequence

$$\{(\alpha_1^k, \ldots, \alpha_{n+1}^k, x_1^k, \ldots, x_{n+1}^k)\}$$

belongs to a compact set, it has a limit point $\{(\alpha_1, \ldots, \alpha_{n+1}, x_1, \ldots, x_{n+1})\}$ such that $\sum_{i=1}^{n+1} \alpha_i = 1$, and for all i, $\alpha_i \ge 0$, and $x_i \in X$. Thus, the vector $\sum_{i=1}^{n+1} \alpha_i x_i$, which belongs to $\mathrm{conv}(X)$, is a limit point of the sequence $\left\{\sum_{i=1}^{n+1} \alpha_i^k x_i^k\right\}$, showing that $\mathrm{conv}(X)$ is compact. **Q.E.D.**

The following result will also be very useful to us.

Proposition B.9:

(a) If $f : \Re^n \mapsto \Re$ is convex, then it is continuous. More generally, if $C \subset \Re^n$ is convex and $f : C \mapsto \Re$ is convex, then f is continuous in the relative interior of C.

(b) Let X^* be the set of minimizing points of a convex function $f : \Re^n \mapsto \Re$ over a closed convex set X, and assume that X^* is nonempty and bounded. Then the level set

$$L_a = \{x \in X \mid f(x) \leq a\}$$

is compact for each scalar a.

Proof: (a) Restricting attention to the affine hull of C and using a transformation argument if necessary, we assume without loss of generality, that the origin is an interior point of C and that the unit cube $S = \{z \mid \|x\|_\infty \leq 1\}$ is contained in C. Let e_i, $i = 1, \ldots, 2^n$, be the corners of S, that is, each e_i is a vector whose entries belong to $\{-1, 1\}$. It is not difficult to see that any $x \in S$ can be expressed in the form $x = \sum_{i=1}^{2^n} a_i e_i$, where each a_i is a nonnegative scalar and $\sum_{i=1}^{2^n} a_i = 1$. Let $A = \max_i f(e_i)$. From Jensen's inequality [Eq. (B.3)], it follows that $f(x) \leq A$ for every $x \in S$.

Let $\{x_k\}$ be a sequence in $\Re^n$ that converges to zero. For the purpose of proving continuity at zero, we can assume that $x_k \in S$ for all k. Using the definition of a convex function [Eq. (B.2)], we have

$$f(x_k) \leq \big(1 - \|x_k\|_\infty\big) f(0) + \|x_k\|_\infty f\left(\frac{x_k}{\|x_k\|_\infty}\right).$$

Letting k tend to infinity, $\|x_k\|_\infty$ goes to zero and we obtain

$$\limsup_{k\to\infty} f(x_k) \leq f(0) + A \limsup_{k\to\infty} \|x_k\|_\infty = f(0).$$

Inequality (B.2) also implies that

$$f(0) \leq \frac{\|x_k\|_\infty}{\|x_k\|_\infty + 1} f\left(\frac{-x_k}{\|x_k\|_\infty}\right) + \frac{1}{\|x_k\|_\infty + 1} f(x_k)$$

and letting k tend to infinity, we obtain $f(0) \leq \liminf_{k\to\infty} f(x_k)$. Thus, $\lim_{k\to\infty} f(x_k) = f(0)$ and f is continuous at zero.

(b) If X is bounded, then using also the continuity of f, which was proved in part (a), it follows that L_a is compact. We thus assume that X is unbounded. Fix some $x^* \in X^*$ and let $b \in \Re$ be such that $x \notin X^*$ for all x with $\|x - x^*\| = b$ (there exists such a b because X^* is bounded). Denote

$$S_b = \{x \in X \mid \|x - x^*\| = b\}, \qquad \tilde{f} = \inf_{x \in S_b} f(x).$$

Since X is unbounded, closed, and convex, it is seen that S_b is nonempty and compact, and since f is continuous, it follows from Weierstrass' theorem (Prop. A.8 in Appendix A) that the infimum above is attained at some point of S_b and we have

$$\tilde{f} > f(x^*).$$

For each $x \in X$ with $\|x - x^*\| > b$, let

$$\hat{\alpha} = \frac{b}{\|x - x^*\|}, \qquad \hat{x} = (1 - \hat{\alpha})x^* + \hat{\alpha}x.$$

By convexity of X, we have $\hat{x} \in X$, and by convexity of f, we have

$$(1 - \hat{\alpha})f(x^*) + \hat{\alpha}f(x) \geq f(\hat{x}).$$

Since $\|\hat{x} - x^*\| = \hat{\alpha}\|x - x^*\| = b$, we also have $\hat{x} \in S_b$, so that

$$f(\hat{x}) \geq \tilde{f}.$$

Combining these two relations and using the definition of $\hat{\alpha}$, we obtain

$$\begin{aligned} f(x) &\geq f(x^*) + \frac{f(\hat{x}) - f(x^*)}{\hat{\alpha}} \\ &\geq f(x^*) + \frac{\tilde{f} - f(x^*)}{\hat{\alpha}} \\ &= f(x^*) + \frac{\tilde{f} - f(x^*)}{b}\|x - x^*\|. \end{aligned}$$

Since $\tilde{f} > f(x^*)$, we see that if $x \in X$ and $f(x) \leq a$, then

$$\|x - x^*\| \leq \max\left\{b, \frac{b\big(a - f(x^*)\big)}{\tilde{f} - f(x^*)}\right\}.$$

Hence the level set L_a is bounded and it is also closed by continuity of f. **Q.E.D.**

Another way to phrase Prop. B.9(b) is that *if one level set of a convex function $f : \Re^n \mapsto \Re$ is compact, all level sets are compact.*

Local and Global Minima

Let $X \subset \Re^n$ and let $f : X \mapsto \Re$ be a function. A vector $x \in X$ is called a *local minimum* of f if there exists some $\epsilon > 0$ such that $f(x) \leq f(y)$ for every $y \in X$ satisfying $\|x - y\| \leq \epsilon$, where $\|\cdot\|$ is some vector norm. A vector $x \in X$ is called a *global minimum* of f if $f(x) \leq f(y)$ for every $y \in X$. A local or global maximum is defined similarly (compare also with Section 1.1).

Under convexity assumptions, the distinction between local and global minima is unnecessary as shown by the following proposition.

Proposition B.10: If $C \subset \Re^n$ is a convex set and $f : C \mapsto \Re$ is a convex function, then a local minimum of f is also a global minimum. If in addition f is strictly convex, then there exists at most one global minimum of f.

Proof: Suppose that x is a local minimum of f but not a global minimum. Then there exists some $y \neq x$ such that $f(y) < f(x)$. Using inequality (B.2), we conclude that $f\big(\alpha x + (1-\alpha)y\big) < f(x)$ for every $\alpha \in [0,1)$. This contradicts the assumption that x is a local minimum.

Suppose that f is strictly convex, and two distinct global minima x and y exist. Then their average $(x+y)/2$ must belong to C, since C is convex, and the value of f must be smaller at the average than at x and y by the strict convexity of f. Since x and y are global minima, we obtain a contradiction. **Q.E.D.**

The Projection Theorem

We close this section with a basic result of analysis and optimization, which will also be used later in this appendix.

Proposition B.11: (Projection Theorem) Let C be a closed convex set and let $\|\cdot\|$ be the Euclidean norm.

(a) For every $x \in \Re^n$, there exists a unique vector $z \in C$ that minimizes $\|z - x\|$ over all $z \in C$. This vector is called the *projection of x on C*, and is denoted by $[x]^+$, i.e.,

$$[x]^+ = \arg\min_{z \in C} \|z - x\|.$$

(b) Given some $x \in \Re^n$, a vector $z \in C$ is equal to $[x]^+$ if and only if

$$(y - z)'(x - z) \leq 0, \qquad \forall\ y \in C.$$

(c) The mapping $f : \Re^n \mapsto C$ defined by $f(x) = [x]^+$ is continuous and nonexpansive, i.e.,

$$\big\|[x]^+ - [y]^+\big\| \leq \|x - y\|, \qquad \forall\ x, y \in \Re^n.$$

Proof: (a) Fix x and let w be some element of C. Minimizing $\|x-z\|$ over all $z \in C$ is equivalent to minimizing the same function over all $z \in C$ such that $\|x - z\| \leq \|x - w\|$, which is a compact set. Furthermore, the function g defined by $g(z) = \|z - x\|^2$ is continuous. Existence of a minimizing vector follows by Weierstrass' theorem (Prop. A.8 in Appendix A).

To prove uniqueness, notice that the square of the Euclidean norm is a strictly convex function of its argument [Prop. B.4(d)]. Therefore, g is strictly convex and it follows that its minimum is attained at a unique point (Prop. B.10).

(b) For all y and z in C we have

$$\|y-x\|^2 = \|y-z\|^2 + \|z-x\|^2 - 2(y-z)'(x-z) \geq \|z-x\|^2 - 2(y-z)'(x-z).$$

Therefore, if z is such that $(y - z)'(x - z) \leq 0$ for all $y \in C$, we have $\|y - x\|^2 \geq \|z - x\|^2$ for all $y \in C$, implying that $z = [x]^+$.

Conversely, let $z = [x]^+$, consider any $y \in C$, and for $\alpha > 0$, define $y_\alpha = \alpha y + (1 - \alpha)z$. We have

$$\begin{aligned}\|x - y_\alpha\|^2 &= \|(1 - \alpha)(x - z) + \alpha(x - y)\|^2 \\ &= (1 - \alpha)^2\|x - z\|^2 + \alpha^2\|x - y\|^2 + 2(1 - \alpha)\alpha(x - z)'(x - y).\end{aligned}$$

Viewing $\|x - y_\alpha\|^2$ as a function of α, we have

$$\frac{\partial}{\partial\alpha}\big\{\|x - y_\alpha\|^2\big\}\Big|_{\alpha=0} = -2\|x - z\|^2 + 2(x - z)'(x - y) = -2(y - z)'(x - z).$$

Therefore, if $(y - z)'(x - z) > 0$ for some $y \in C$, then

$$\frac{\partial}{\partial\alpha}\big\{\|x - y_\alpha\|^2\big\}\Big|_{\alpha=0} < 0$$

and for positive but small enough α, we obtain $\|x - y_\alpha\| < \|x - z\|$. This contradicts the fact $z = [x]^+$ and shows that $(y - z)'(x - z) \leq 0$ for all $y \in C$.

(c) Let x and y be elements of $\Re^n$. From part (b), we have $\big(w-[x]^+\big)'\big(x-[x]^+\big) \leq 0$ for all $w \in C$. Since $[y]^+ \in C$, we obtain

$$\big([y]^+ - [x]^+\big)'\big(x-[x]^+\big) \leq 0.$$

Similarly,

$$\big([x]^+ - [y]^+\big)'\big(y-[y]^+\big) \leq 0.$$

Adding these two inequalities, we obtain

$$\big([y]^+ - [x]^+\big)'\big(x-[x]^+ - y + [y]^+\big) \leq 0.$$

By rearranging and by using the Schwartz inequality, we have

$$\big\|[y]^+ - [x]^+\big\|^2 \leq \big([y]^+ - [x]^+\big)'(y-x) \leq \big\|[y]^+ - [x]^+\big\| \cdot \|y-x\|,$$

showing that $[\cdot]^+$ is nonexpansive and *a fortiori* continuous. **Q.E.D.**

Figure B.5 illustrates the necessary and sufficient condition of part (b) of the projection theorem.

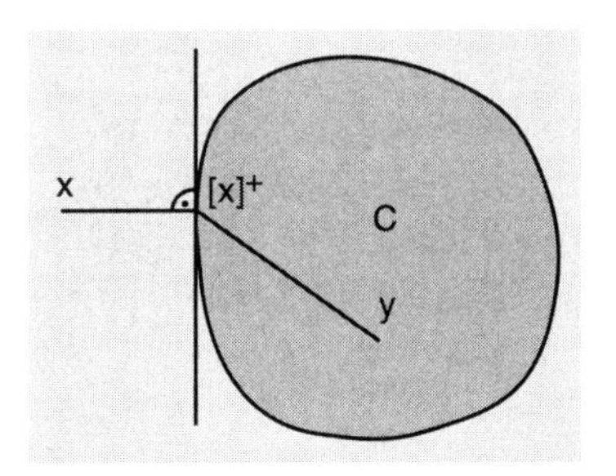

Figure B.5. Illustration of the condition satisfied by the projection $[x]^+$. For each vector $y \in C$, the vectors $x-[x]^+$ and $y-[x]^+$ form an angle larger than or equal to 90 degrees or, equivalently, $(y-[x]^+)'(x-[x]^+) \leq 0$.

EXERCISES

B.1.1

Let g be a convex, monotonically nondecreasing function of a single variable [i.e., $g(y) \leq g(\overline{y})$ for $y < \overline{y}$], and let f be a convex function defined on a convex set $C \subset \Re^n$. Show that the function h defined by

$$h(x) = g\big(f(x)\big)$$

is convex on C. Use this fact to show that the function $h(x) = e^{\beta x'Qx}$, where β is a positive scalar and Q is a positive semidefinite symmetric matrix, is convex over $\Re^n$.

B.1.2

Use the line segment principle and the method of proof of Prop. B.4(c) to show that if C is a convex set with nonempty interior, and $f : \Re^n \mapsto \Re$ is twice continuously differentiable over C with $\nabla^2 f(x)$ positive semidefinite for all $x \in C$, then f is convex over C.

B.1.3 (Arithmetic-Geometric Mean Inequality)

Show that if $\alpha_1, \ldots, \alpha_n$ are positive scalars with $\sum_{i=1}^n \alpha_i = 1$, then for every set of positive scalars $x_1, \ldots, x_n$, we have

$$x_1^{\alpha_1} x_2^{\alpha_2} \cdots x_n^{\alpha_n} \leq \alpha_1 x_1 + \alpha_2 x_2 + \cdots + \alpha_n x_n,$$

with equality if and only if $x_1 = x_2 = \cdots = x_n$. *Hint*: Show that $-\ln x$ is a strictly convex decreasing function on $(0, \infty)$.

B.1.4

Use the result of Exercise B.1.3 to verify Young's inequality

$$xy \leq \frac{x^p}{p} + \frac{y^q}{q},$$

where $p > 0$, $q > 0$, $1/p + 1/q = 1$, $x \geq 0$, and $y \geq 0$. Then, use Young's inequality to verify Holder's inequality

$$\sum_{i=1}^n |x_i y_i| \leq \left(\sum_{i=1}^n |x_i|^p\right)^{1/p} \left(\sum_{i=1}^n |y_i|^q\right)^{1/q}.$$

B.1.5

Let $f : \Re^{n+m} \mapsto \Re$ be a convex function. Consider the function $F : \Re^n \mapsto \Re$ given by

$$F(x) = \inf_{u \in U} f(x, u),$$

where U be any nonempty and convex subset of $\Re^m$ such that $F(x) > -\infty$ for all $x \in \Re^n$. Show that F is convex. *Hint*: There cannot exist $\alpha \in [0, 1]$, x_1, x_2, $u_1 \in U$, $u_2 \in U$ such that $F\big(\alpha x_1 + (1 - \alpha) x_2\big) > \alpha f(x_1, u_1) + (1 - \alpha) f(x_2, u_2)$.

B.1.6

Let $f : \Re^n \mapsto \Re$ be a differentiable function. Show that f is convex over a convex set C if and only if

$$\big(\nabla f(x) - \nabla f(y)\big)'(x - y) \geq 0, \qquad \forall\, x, y \in C.$$

Hint: The condition above says that the function f, restricted on the line segment connecting x and y, has monotonically nondecreasing gradient; see also the proof of Prop. B.5.

B.1.7 (Caratheodory's Theorem for Cones)

Let X be the cone generated by a subset of vectors $S \subset \Re^n$, i.e., the set of vectors x of the form

$$x = \sum_{i \in I} a_i x_i,$$

where I is a finite index set, and for all $i \in I$, $x_i \in S$ and a_i is a nonnegative scalar. Show that any nonzero vector from X can be represented as a positive combination of no more than n vectors from S. Furthermore, these vectors can be chosen to be linearly independent. *Hint*: Let x be a nonzero vector from X, and let m be the smallest integer such that x has the form $\sum_{i=1}^m a_i x_i$, where $a_i > 0$ and $x_i \in S$ for all $i = 1, \ldots, m$. If the vectors x_i were linearly dependent, there would exist scalars $\lambda_1, \ldots, \lambda_m$, at least one of which is positive, and such that $\sum_{i=1}^m \lambda_i x_i = 0$. Consider the linear combination $\sum_{i=1}^m (a_i - \overline{\gamma}\lambda_i) x_i$, where $\overline{\gamma}$ is the largest γ such that $a_i - \gamma\lambda_i \geq 0$ for all i, to arrive at a contradiction.

B.1.8 (Properties of Relative Interiors) [Roc70]

(a) If C is a convex set in $\Re^n$, then:

  (i) $\mathrm{cl}(C) = \mathrm{cl}\big(\mathrm{ri}(C)\big)$.

  (ii) $\mathrm{ri}(C) = \mathrm{ri}\big(\mathrm{cl}(C)\big)$.

  (iii) $\mathrm{ri}(A \cdot C) = A \cdot \mathrm{ri}(C)$ for all $m \times n$ matrices A.

(b) If C_1 and C_2 are convex sets in $\Re^n$, then:

  (i) $\mathrm{ri}(C_1 + C_2) = \mathrm{ri}(C_1) + \mathrm{ri}(C_2)$.

  (ii) $\mathrm{ri}(C_1 \cap C_2) = \mathrm{ri}(C_1) \cap \mathrm{ri}(C_2)$, provided the sets $\mathrm{ri}(C_1)$ and $\mathrm{ri}(C_1)$ have a nonempty intersection.

(c) If C_1 and C_2 are convex subsets of $\Re^n$ and $\Re^m$, respectively, then

$$\mathrm{ri}(C_1 \times C_2) = \mathrm{ri}(C_1) \times \mathrm{ri}(C_2).$$

B.2 SEPARATING HYPERPLANES

A *hyperplane* is a set of the form $\{x \mid a'x = b\}$, where $a \in \Re^n$, $a \neq 0$, and $b \in \Re$, as illustrated in Fig. B.6. An equivalent definition is that a hyperplane in $\Re^n$ is a linear manifold of dimension $n-1$. The vector a called the *normal* vector of the hyperplane (it is orthogonal to the difference $x - y$ of any two vectors x and y of the hyperplane). The two sets

$$\{x \mid a'x \geq b\}, \qquad \{x \mid a'x \leq b\},$$

are called the *halfspaces* associated with the hyperplane (also referred to as the *positive and negative halfspaces*, respectively). We have the following

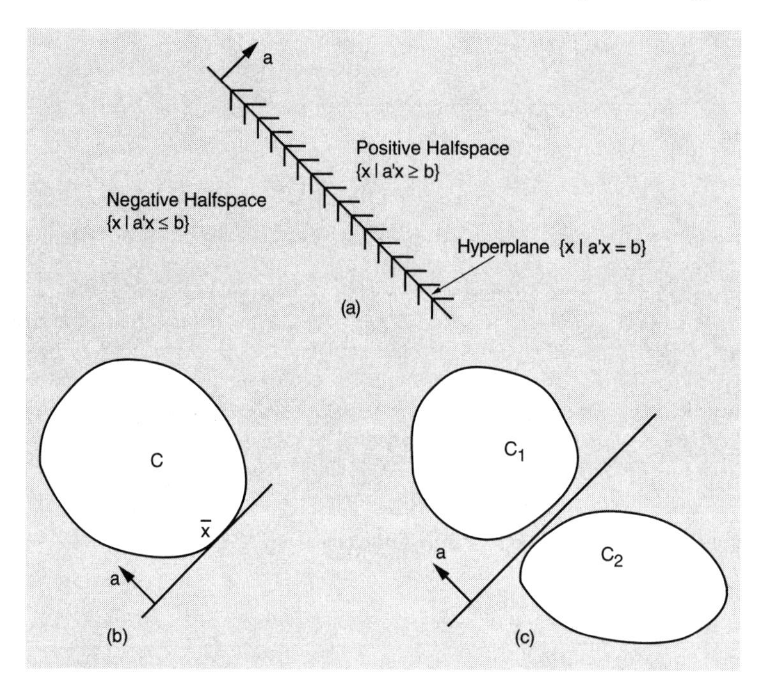

Figure B.6. (a) A hyperplane $\{x \mid a'x = b\}$ divides the space in two halfspaces as illustrated. (b) Geometric interpretation of the supporting hyperplane theorem. (c) Geometric interpretation of the separating hyperplane theorem.

result, which is also illustrated in Fig. B.6. The proof is based on the projection theorem and is illustrated in Fig. B.7.

Proposition B.12: (Supporting Hyperplane Theorem) If $C \subset \Re^n$ is a convex set and $\overline{x}$ is a point that does not belong to the interior of C, there exists a vector $a \neq 0$ such that

$$a'x \geq a'\overline{x}, \qquad \forall\, x \in C. \tag{B.7}$$

Proof: Denote by $\overline{C}$ the closure of C, which is a convex set by Prop. B.8. Let $\{x_k\}$ be a sequence of vectors not belonging to $\overline{C}$, which converges to $\overline{x}$; such a sequence exists because $\overline{x}$ does not belong to the interior of C. If $\hat{x}_k$ is the projection of x_k on $\overline{C}$, we have by part (b) of the projection theorem (Prop. B.11)

$$(\hat{x}_k - x_k)'(x - \hat{x}_k) \geq 0, \qquad \forall\, x \in \overline{C}.$$

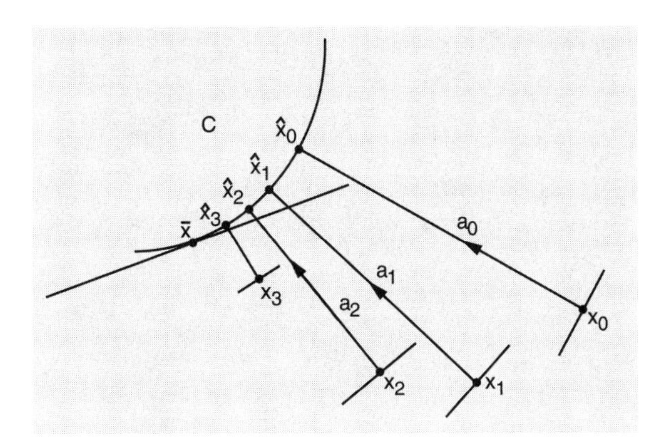

Figure B.7. Illustration of the proof of the supporting hyperplane theorem for the case where the vector $\overline{x}$ belongs to the closure of C. We choose a sequence $\{x_k\}$ of vectors not belonging to the closure of C which converges to $\overline{x}$, and we project x_k on the closure of C. We then consider, for each k, the hyperplane that is orthogonal to the line segment connecting x_k and its projection, and passes through x_k. These hyperplanes "converge" to a hyperplane that supports C at $\overline{x}$.

Hence we obtain for all k and $x \in \overline{C}$,

$$(\hat{x}_k - x_k)'x \geq (\hat{x}_k - x_k)'\hat{x}_k = (\hat{x}_k - x_k)'(\hat{x}_k - x_k) + (\hat{x}_k - x_k)'x_k \geq (\hat{x}_k - x_k)'x_k.$$

We can write this inequality as

$$a_k'x \geq a_k'x_k, \qquad \forall\, x \in \overline{C},\ k = 0, 1, \ldots, \tag{B.8}$$

where

$$a_k = \frac{\hat{x}_k - x_k}{\|\hat{x}_k - x_k\|}.$$

We have $\|a_k\| = 1$ for all k, and hence the sequence $\{a_k\}$ has a subsequence that converges to a nonzero limit a. By considering Eq. (B.8) for all a_k belonging to this subsequence and by taking the limit as $k \to \infty$, we obtain Eq. (B.7). **Q.E.D.**

Proposition B.13: (Separating Hyperplane Theorem) If C_1 and C_2 are two nonempty and disjoint convex subsets of $\Re^n$, there exists a hyperplane that separates them, i.e., a vector $a \neq 0$ such that

$$a'x_1 \leq a'x_2, \qquad \forall\, x_1 \in C_1,\ x_2 \in C_2. \tag{B.9}$$

Proof: Consider the convex set

$$C = \{x \mid x = x_2 - x_1,\, x_1 \in C_1,\, x_2 \in C_2\}.$$

Since C_1 and C_2 are disjoint, the origin does not belong to C, so by the supporting hyperplane theorem there exists a vector $a \neq 0$ such that

$$0 \leq a'x, \qquad \forall\, x \in C,$$

which is equivalent to Eq. (B.9). **Q.E.D.**

Proposition B.14: (Strict Separation Theorem) If C_1 and C_2 are two nonempty and disjoint convex sets such that C_1 is closed and C_2 is compact, there exists a hyperplane that strictly separates them, i.e., a vector $a \neq 0$ and a scalar b such that

$$a'x_1 < b < a'x_2, \qquad \forall\, x_1 \in C_1,\ x_2 \in C_2. \tag{B.10}$$

Proof: Consider the problem

$$\begin{aligned} &\text{minimize} \ \ \|x_1 - x_2\| \\ &\text{subject to} \ \ x_1 \in C_1,\ x_2 \in C_2. \end{aligned} \tag{B.11}$$

The set

$$C = \{x_1 - x_2 \mid x_1 \in C_1,\, x_2 \in C_2\}$$

is convex and closed by Prop. B.8(b). Since problem (B.11) is the problem of projecting the origin on C, we conclude using Prop. B.11(a), that problem (B.11) has at least one solution $(\overline{x}_1, \overline{x}_2)$. Let

$$a = \frac{\overline{x}_2 - \overline{x}_1}{2}, \qquad \overline{x} = \frac{\overline{x}_1 + \overline{x}_2}{2}, \qquad b = a'\overline{x}.$$

Then, $a \neq 0$, since $\overline{x}_1 \in C_1$, $\overline{x}_2 \in C_2$, and C_1 and C_2 are disjoint. The hyperplane

$$\{x \mid a'x = b\}$$

contains $\overline{x}$, and it can be seen from problem (B.11) that $\overline{x}_1$ is the projection of $\overline{x}$ on C_1, and $\overline{x}_2$ is the projection of $\overline{x}$ on C_2 (see Fig. B.8). By Prop. B.11(b), we have

$$(\overline{x} - \overline{x}_1)'(x_1 - \overline{x}_1) \leq 0, \qquad \forall\, x_1 \in C_1$$

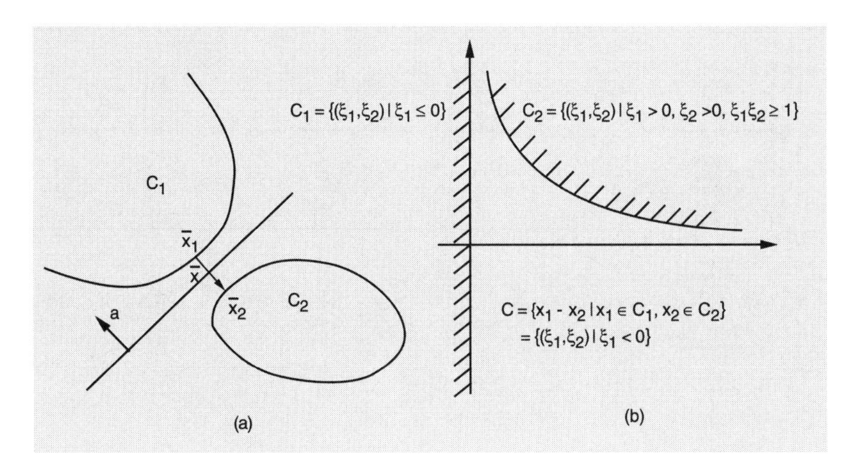

Figure B.8. (a) Illustration of the construction of a strictly separating hyperplane of two disjoint closed convex sets C_1 and C_2 one of which is also bounded (cf. Prop. B.14). (b) An example showing that if none of the two sets is compact, there may not exist a strictly separating hyperplane. This is due to the fact that the set $C = \{x_1 - x_2 \mid x_1 \in C_1,\, x_2 \in C_2\}$ is equal to $\{(\xi_1, \xi_2) \mid \xi_1 < 0\}$ and is not closed, even though C_1 and C_2 are closed. This is also an example where vector sum as well as linear transformation of closed convex sets does not preserve closure.

or equivalently, since $\overline{x} - \overline{x}_1 = a$,

$$a'x_1 \le a'\overline{x}_1 = a'\overline{x} + a'(\overline{x}_1 - \overline{x}) = b - \|a\|^2 < b, \qquad \forall\, x_1 \in C_1.$$

Thus, the left-hand side of Eq. (B.10) is proved. The right-hand side is proved similarly. **Q.E.D.**

The preceding proposition may be used to provide a fundamental characterization of closed convex sets.

Proposition B.15: Every closed convex subset of $\Re^n$ is the intersection of the halfspaces that contain it.

Proof: Let C be the set at issue. Clearly C is contained in the intersection of the halfspaces that contain C, so we focus on proving the reverse inclusion. Let $x \notin C$. Applying the strict separation theorem (Prop. B.14) to the sets C and $\{x\}$, we see that there exists a halfspace containing C but not containing x. Hence, if $x \notin C$, then x cannot belong to the intersection of the halfspaces containing C, proving the result. **Q.E.D.**

EXERCISES

B.2.1

Let C_1 and C_2 be two nonempty, convex sets, which are at positive Euclidean distance from each other, that is,

$$\inf_{x_1 \in C_1,\, x_2 \in C_2} \|x_1 - x_2\| > 0.$$

Show that there exists a hyperplane that strictly separates them. *Hint*: Adapt the proof of Prop. B.14.

B.3 CONES AND POLYHEDRAL CONVEXITY

We now develop some basic results regarding cones and also discuss the geometry of polyhedral sets. A set $C \subset \Re^n$ is said to be a *cone* if $ax \in C$ for all $a \geq 0$ and $x \in C$. We introduce three important types of cones.

Given a cone C, the cone given by

$$C^{\perp} = \{y \mid y'x \leq 0,\ \forall\ x \in C\},$$

is called the *polar cone* of C.

A cone C is said to be *finitely generated*, if it has the form

$$C = \left\{ x \;\middle|\; x = \sum_{j=1}^{r} \mu_j a_j,\ \mu_j \geq 0,\ j = 1, \ldots, r \right\},$$

where $a_1, \ldots, a_r$ are some vectors.

A cone C is said to be *polyhedral*, if it has the form

$$C = \{x \mid a_j'x \leq 0,\ j = 1, \ldots, r\},$$

where $a_1, \ldots, a_r$ are some vectors.

Figure B.9 illustrates the above definitions. It is straightforward to show that the polar cone of any cone, as well as all finitely generated and polyhedral cones are convex, by verifying the definition of convexity of Eq. (B.1). Furthermore, polar and polyhedral cones are closed, since they are intersections of closed halfspaces. Finitely generated cones are also closed

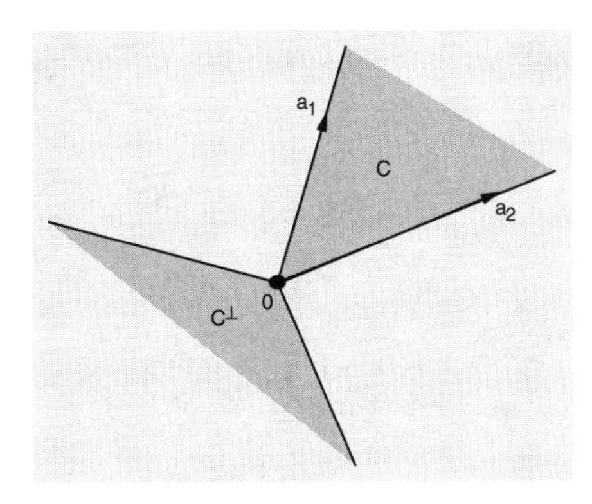

Figure B.9. Illustration of a cone and its polar in $\Re^2$. Here, a_1 and a_2 are given vectors, $C = \{x \mid x = \mu_1 a_1 + \mu_2 a_2,\ \mu_1 \geq 0,\ \mu_2 \geq 0\}$, which is a finitely generated cone, and $C^\perp = \{y \mid y'a_1 \leq 0,\ y'a_2 \leq 0\}$, which is a polyhedral cone.

as shown in part (b) of the following proposition, which also provides some additional important results.

Proposition B.16:

(a) (*Polar Cone Theorem*) For any nonempty closed convex cone C, we have $(C^\perp)^\perp = C$.

(b) Let $a_1, \ldots, a_r$ be vectors of $\Re^n$. Then the finitely generated cone

$$C = \left\{ x \;\middle|\; x = \sum_{j=1}^{r} \mu_j a_j,\ \mu_j \geq 0,\ j = 1, \ldots, r \right\} \tag{B.12}$$

is closed and its polar cone is the polyhedral cone given by

$$C^\perp = \{x \mid x'a_j \leq 0,\ j = 1, \ldots, r\}. \tag{B.13}$$

(c) (*Minkowski – Weyl Theorem*) A cone is polyhedral if and only if it is finitely generated.

(d) (*Farkas' Lemma*) Let x, $e_1, \ldots, e_m$, and $a_1, \ldots, a_r$ be vectors of $\Re^n$. We have $x'y \leq 0$ for all vectors $y \in \Re^n$ such that

$$y'e_i = 0, \quad \forall\, i = 1, \ldots, m, \qquad y'a_j \leq 0, \quad \forall\, j = 1, \ldots, r,$$

if and only if x can be expressed as

$$x = \sum_{i=1}^{m} \lambda_i e_i + \sum_{j=1}^{r} \mu_j a_j,$$

where λ_i and μ_j are some scalars with $\mu_j \geq 0$ for all j.

Proof: (a) See Fig. B.10.

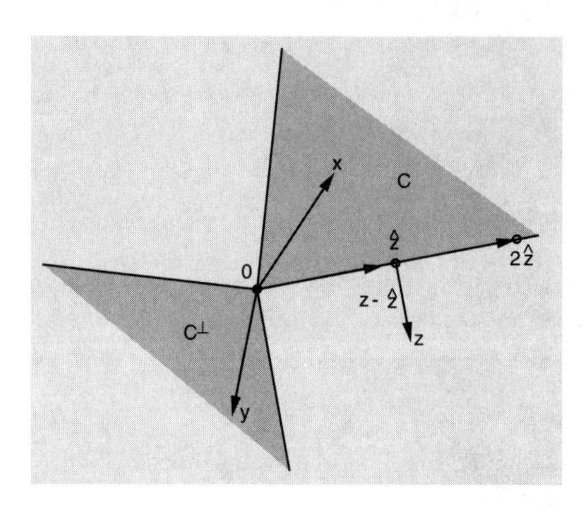

Figure B.10. Proof of the polar cone theorem. If $x \in C$, then for all $y \in C^\perp$, we have $x'y \leq 0$, which implies that $x \in (C^\perp)^\perp$. Hence, $C \subset (C^\perp)^\perp$. To prove the reverse inclusion, take $z \in (C^\perp)^\perp$, and let $\hat{z}$ be the unique projection of z on C, as shown in the figure. Since C is closed, the projection exists by the projection theorem (Prop. B.11), which also implies that

$$(z - \hat{z})'(x - \hat{z}) \leq 0, \qquad \forall\, x \in C.$$

By taking $x = 0$ and $x = 2\hat{z}$ in the preceding relation, it is seen that

$$(z - \hat{z})'\hat{z} = 0.$$

Combining the last two relations, we obtain $(z-\hat{z})'x \leq 0$ for all $x \in C$. Therefore, $(z - \hat{z}) \in C^\perp$, and since $z \in (C^\perp)^\perp$, we obtain $(z - \hat{z})'z \leq 0$, which when added to $(z - \hat{z})'\hat{z} = 0$ yields $\|z - \hat{z}\|^2 \leq 0$. Therefore, $z = \hat{z}$ and $z \in C$. It follows that $(C^\perp)^\perp \subset C$.

(b) We first show that the polar cone of C has the desired form (B.13). If y satisfies $y'a_j \leq 0$ for all j, then $y'x \leq 0$ for all $x \in C$, so the set in the right-hand side of Eq. (B.13) is a subset of $C^\perp$. Conversely, if $y \in C^\perp$, that is, if $y'x \leq 0$ for all $x \in C$, then (since a_j belong to C) we have $y'a_j \leq 0$, for all j. Thus, $C^\perp$ is a subset of the set in the right-hand side of Eq. (B.13).

To show that C is closed, it will suffice to show that C is polyhedral; this will also prove half of the Minkowski-Weyl theorem [part (c)]. Our proof, due to [Wet90], is constructive and uses induction on the number of vectors r. We will also give an alternative proof, which is simpler than the first but does not show simultaneously half of part (c).

To start the induction, we assume without loss of generality that $a_1 = 0$. Then, for $r = 1$, we have $C = \{0\}$, which is polyhedral, since it can be expressed as

$$\{x \mid u_i'x \leq 0,\, -u_i'x \leq 0,\, i = 1, \ldots, n\},$$

where u_i is the ith unit coordinate vector.

Assume that for some $r \geq 2$, the set

$$C_{r-1} = \left\{ x \;\middle|\; x = \sum_{j=1}^{r-1} \mu_j a_j,\, \mu_j \geq 0 \right\}$$

has a polyhedral representation

$$P_{r-1} = \big\{x \mid b_j'x \leq 0,\, j = 1, \ldots, m\big\}.$$

Let

$$\beta_j = a_r' b_j, \qquad j = 1, \ldots, m,$$

and define the index sets

$$J^- = \{j \mid \beta_j < 0\}, \qquad J^0 = \{j \mid \beta_j = 0\}, \qquad J^+ = \{j \mid \beta_j > 0\}.$$

Let also

$$b_{l,k} = b_l - \frac{\beta_l}{\beta_k} b_k, \qquad \forall\, l \in J^+,\ k \in J^-.$$

We will show that the set

$$C_r = \left\{ x \;\middle|\; x = \sum_{j=1}^{r} \mu_j a_j,\, \mu_j \geq 0 \right\}$$

has the polyhedral representation

$$P_r = \big\{x \mid b_j'x \leq 0,\, j \in J^- \cup J^0,\, b_{l,k}'x \leq 0,\, l \in J^+,\, k \in J^-\big\},$$

thus completing the induction.

We have $C_r \subset P_r$ because by construction, all the vectors $a_1, \ldots, a_r$ satisfy the inequalities defining P_r. To show the reverse inclusion, we fix a vector $x \in P_r$ and we verify that there exists $\mu_r \geq 0$ such that

$$x - \mu_r a_r \in P_{r-1},$$

which is equivalent to

$$\gamma \le \mu_r \le \delta,$$

where

$$\gamma = \max\left\{0, \max_{j\in J^+} \frac{b_j'x}{\beta_j}\right\}, \qquad \delta = \min_{j\in J^-} \frac{b_j'x}{\beta_j}.$$

Since $x \in P_r$, we have

$$0 \le \frac{b_k'x}{\beta_k}, \qquad \forall\, k \in J^-, \tag{B.14}$$

and also $b_{l,k}'x \le 0$ for all $l \in J^+$, $k \in J^-$, or equivalently

$$\frac{b_l'x}{\beta_l} \le \frac{b_k'x}{\beta_k}, \qquad \forall\, l \in J^+,\ k \in J^-. \tag{B.15}$$

Equations (B.14) and (B.15) imply that $\gamma \le \delta$, thereby completing the proof.

We now give an alternative proof that C is closed, which is based again on induction on the number of vectors r. When $r = 1$, C is either $\{0\}$ (if $a_1 = 0$) or a halfline, and is therefore closed. Suppose, for some $r \ge 1$, all cones of the form

$$\left\{x \;\middle|\; x = \sum_{j=1}^{r} \mu_j a_j,\ \mu_j \ge 0\right\}$$

are closed. Then, we will show that a cone of the form

$$C_{r+1} = \left\{x \;\middle|\; x = \sum_{j=1}^{r+1} \mu_j a_j,\ \mu_j \ge 0\right\}$$

is also closed. Without loss of generality, assume that $\|a_j\| = 1$ for all j. There are two cases: (i) The vectors $-a_1, \ldots, -a_{r+1}$ belong to C_{r+1}, in which case C_{r+1} is the subspace spanned by $a_1, \ldots, a_{r+1}$ and is therefore closed, and (ii) The negative of one of the vectors, say $-a_{r+1}$, does not belong to C_{r+1}. In this case, consider the cone

$$C_r = \left\{x \;\middle|\; x = \sum_{j=1}^{r} \mu_j a_j,\ \mu_j \ge 0\right\},$$

which is closed by the induction hypothesis. Let

$$m = \min_{x\in C_r,\ \|x\|=1} a_{r+1}'x.$$

Since, the set $\{x \in C_r \mid \|x\| = 1\}$ is nonempty and compact, the minimum above is attained at some x^* by Weierstrass' theorem. We have, using the Schwartz inequality,

$$m = a'_{r+1}x^* \geq -\|a_{r+1}\| \cdot \|x^*\| = -1,$$

with equality if and only if $x^* = -a_{r+1}$. It follows that

$$m > -1,$$

since otherwise we would have $x^* = -a_{r+1}$, which violates the hypothesis $(-a_{r+1}) \notin C_r$. Let $\{x_k\}$ be a convergent sequence in C_{r+1}. We will prove that its limit belongs to C_{r+1}, thereby showing that C_{r+1} is closed. Indeed, for all k, we have $x_k = \xi_k a_{r+1} + y_k$, where $\xi_k \geq 0$ and $y_k \in C_r$. Using the fact $\|a_{r+1}\| = 1$, we obtain

$$\begin{aligned} \|x_k\|^2 &= \xi_k^2 + \|y_k\|^2 + 2\xi_k a'_{r+1} y_k \\ &\geq \xi_k^2 + \|y_k\|^2 + 2m\xi_k \|y_k\| \\ &= (\xi_k - \|y_k\|)^2 + 2(1+m)\xi_k \|y_k\|. \end{aligned}$$

Since $\{x_k\}$ converges, $\xi_k \geq 0$, and $1 + m > 0$, it follows that the sequences $\{\xi_k\}$ and $\{y_k\}$ are bounded and hence, they have limit points denoted by ξ and y, respectively. The limit of $\{x_k\}$ is

$$\lim_{k \to \infty} (\xi_k a_{r+1} + y_k) = \xi a_{r+1} + y,$$

which belongs to C_{r+1}, since $\xi \geq 0$ and $y \in C_r$ (by the closure hypothesis on C_r). We conclude that C_{r+1} is closed, completing the proof.

(c) We have already shown in the proof of part (b) that a finitely generated cone is polyhedral. To show the reverse, we use parts (a) and (b) to conclude that the polar of any polyhedral cone [cf. Eq. (B.13)] is finitely generated [cf. Eq. (B.12)]. The finitely generated cone (B.12) has already been shown to be polyhedral, so its polar, which is the "typical" polyhedral cone (B.13), is finitely generated. This completes the proof.

(d) Define $a_{r+i} = e_i$ and $a_{r+m+i} = -e_i$, $i = 1, \ldots, m$. The result to be shown translates to

$$x \in C \qquad \Longleftrightarrow \qquad x \in P^\perp,$$

where

$$C = \left\{ x \;\middle|\; x = \sum_{j=1}^{r+2m} \mu_j a_j,\ \mu_j \geq 0 \right\},$$

$$P = \{ y \mid y' a_j \leq 0,\ j = 1, \ldots, r + 2m \}.$$

Since by part (b), $P = C^\perp$ and C is closed, we have by part (a), $P^\perp = (C^\perp)^\perp = C$. **Q.E.D.**

Polyhedral Sets

A nonempty subset of $\Re^n$ is said to be a *polyhedral set* (or *polyhedron*) if it is of the form

$$P = \{x \mid a_j'x \le b_j,\ j = 1, \ldots, r\},$$

where a_j are some vectors and b_j are some scalars.

The following is a fundamental result, showing that a polyhedral set can be represented as the sum of the convex hull of a finite set of points and a finitely generated cone. The proof is based on an interesting construction that can be used to translate results about polyhedral cones to results about polyhedral sets.

Proposition B.17: A set P is polyhedral if and only if there exist a nonempty and finite set of vectors $\{v_1, \ldots, v_m\}$, and a finitely generated cone C such that

$$P = \left\{ x \;\middle|\; x = y + \sum_{j=1}^{m} \mu_j v_j,\ y \in C,\ \sum_{j=1}^{m} \mu_j = 1,\ \mu_j \ge 0,\ j = 1, \ldots, m \right\}.$$

Proof: Assume that P is polyhedral. Then, it has the form

$$P = \{x \mid a_j'x \le b_j,\ j = 1, \ldots, r\},$$

for some vectors a_j and some scalars b_j. Consider the polyhedral cone of $\Re^{n+1}$

$$\hat{P} = \{(x, w) \mid 0 \le w,\ a_j'x \le b_j w,\ j = 1, \ldots, r\}$$

and note that

$$P = \{x \mid (x, 1) \in \hat{P}\}.$$

By the Minkowski – Weyl theorem [Prop. B.16(c)], $\hat{P}$ is finitely generated, so it has the form

$$\hat{P} = \left\{ (x, w) \;\middle|\; x = \sum_{j=1}^{m} \mu_j v_j,\ w = \sum_{j=1}^{m} \mu_j d_j,\ \mu_j \ge 0,\ j = 1, \ldots, m \right\},$$

for some vectors v_j and scalars d_j. Since $w \ge 0$ for all vectors $(x, w) \in \hat{P}$, we see that $d_j \ge 0$ for all j. Let

$$J^+ = \{j \mid d_j > 0\}, \qquad J^0 = \{j \mid d_j = 0\}.$$

By replacing μ_j by μ_j/d_j for all $j \in J^+$, we obtain the equivalent description

$$\hat{P} = \left\{ (x, w) \;\middle|\; x = \sum_{j=1}^{m} \mu_j v_j,\ w = \sum_{j \in J^+} \mu_j,\ \mu_j \geq 0,\ j = 1, \ldots, m \right\}.$$

Since $P = \{x \mid (x, 1) \in \hat{P}\}$, we obtain

$$P = \left\{ x \;\middle|\; x = \sum_{j \in J^+} \mu_j v_j + \sum_{j \in J^0} \mu_j v_j,\ \sum_{j \in J^+} \mu_j = 1,\ \mu_j \geq 0,\ j = 1, \ldots, m \right\}.$$

Thus, P is the vector sum of the convex hull of the vectors v_j, $j \in J^+$, plus the finitely generated cone

$$\Big\{ \sum_{j \in J^0} \mu_j v_j \mid \mu_j \geq 0,\ j \in J^0 \Big\}.$$

To prove that the vector sum of the convex hull of a finite set of points with a finitely generated cone is a polyhedral set, we use a reverse argument; we pass to a finitely generated cone description, we use the Minkowski – Weyl theorem to assert that this cone is polyhedral, and we finally construct a polyhedral set description. The details are left as an exercise for the reader. **Q.E.D.**

B.4 EXTREME POINTS

A vector x is said to be an *extreme point* of a convex set C if x belongs to C and there do not exist vectors $y \in C$ and $z \in C$, with $y \neq x$ and $z \neq x$, and a scalar $\alpha \in (0, 1)$ such that $x = \alpha y + (1 - \alpha) z$. An equivalent definition is that x cannot be expressed as a convex combination of some vectors of C, all of which are different from x.

An important fact that forms the basis for the simplex method of linear programming, is that if a linear function f attains a minimum over a polyhedral set C having at least one extreme point, then f attains a minimum at some extreme point of C (as well as possibly at some other nonextreme points). We will prove this fact after considering the more general case where f is concave and C is closed and convex. We first show a preliminary result.

Proposition B.18: Let C be a nonempty, closed, convex set in $\Re^n$.

(a) If H is a hyperplane that passes through a boundary point of C and contains C in one of its halfspaces, then every extreme point of $T = C \cap H$ is also an extreme point of C.

(b) C has at least one extreme point if and only if it does not contain a line, that is, a set L of the form $L = \{x + \alpha d \mid \alpha \in \Re\}$ with $d \neq 0$.

Proof: (a) Let $\overline{x}$ be an element of T which is not an extreme point of C. Then we have $\overline{x} = \alpha y + (1-\alpha)z$ for some $\alpha \in (0,1)$, and some $y \in C$ and $z \in C$, with $y \neq x$ and $z \neq x$. Since $\overline{x} \in H$, $\overline{x}$ is a boundary point of C, and the halfspace containing C is of the form $\{x \mid a'x \geq a'\overline{x}\}$, where $a \neq 0$. Then $a'y \geq a'\overline{x}$ and $a'z \geq a'\overline{x}$, which in view of $\overline{x} = \alpha y + (1-\alpha)z$, implies that $a'y = a'\overline{x}$ and $a'z = a'\overline{x}$. Therefore, $y \in T$ and $z \in T$, showing that $\overline{x}$ cannot be an extreme point of T.

(b) Assume that C has an extreme point x and contains a line $L = \{\overline{x}+\alpha d \mid \alpha \in \Re\}$, where $d \neq 0$. We will arrive at a contradiction. For each integer $n > 0$, the vector

$$x_n = \left(1 - \frac{1}{n}\right) x + \frac{1}{n}(\overline{x} + nd) = x + d + \frac{1}{n}(\overline{x} - x)$$

lies in the line segment connecting x and $\overline{x} + nd$, so it belongs to C. Since C is closed, $x + d = \lim_{n\to\infty} x_n$ must also belong to C. Similarly, we show that $x - d$ must belong to C. Thus $x - d$, x, and $x + d$ all belong to C, contradicting the hypothesis that x is an extreme point.

Conversely, we use induction on the dimension of the space to show that if C does not contain a line, it must have an extreme point. This is true in the real line $\Re^1$, so assume it is true in $\Re^{n-1}$. If a nonempty, closed, convex subset C of $\Re^n$ contains no line, it must have some boundary point $\overline{x}$. Take any hyperplane H passing through $\overline{x}$ and containing C in one of its halfspaces. Then, since H is an $(n-1)$-dimensional manifold, the set $C \cap H$ lies in an $(n-1)$-dimensional space and contains no line, so by the induction hypothesis, it must have an extreme point. By part (a), this extreme point must also be an extreme point of C. **Q.E.D.**

We say that a set $C \subset \Re^n$ is *bounded from below* if there exists a vector $b \in \Re^n$ such that $x \geq b$ for all $x \in C$.

Proposition B.19: Let C be a closed convex set which is bounded from below and let $f : C \mapsto \Re$ be a concave function. Then if f attains a minimum over C, it attains a minimum at some extreme point of C.

Proof: We first show that f attains a minimum at some boundary point of C. Let x^* be a vector where f attains a minimum over C. If x^* is a boundary point we are done, so assume that x^* is an interior point of C. Let

$$L = \{x \mid x = x^* + \lambda d, \ \lambda \in \Re\}$$

be a line passing through x^*, where d is a vector with strictly positive coordinates. Then, using the boundedness from below, convexity, and closure of C, we see that the set $C \cap L$ contains a set of the form

$$\{x^* + \lambda d \mid \lambda_1 \leq \lambda \leq \lambda_2\}$$

for some $\lambda_2 > 0$ and some $\lambda_1 < 0$ for which the vector

$$\overline{x} = x^* + \lambda_1 d$$

is a boundary point of C. If $f(\overline{x}) > f(x^*)$, we have by concavity of f,

$$\begin{aligned} f(x^*) &\geq \frac{\lambda_2}{\lambda_2 - \lambda_1} f(\overline{x}) + \left(1 - \frac{\lambda_2}{\lambda_2 - \lambda_1}\right) f(x^* + \lambda_2 d) \\ &> \frac{\lambda_2}{\lambda_2 - \lambda_1} f(x^*) + \left(1 - \frac{\lambda_2}{\lambda_2 - \lambda_1}\right) f(x^* + \lambda_2 d). \end{aligned}$$

It follows that $f(x^*) > f(x^* + \lambda_2 d)$. This contradicts the optimality of x^*, proving that $f(\overline{x}) = f(x^*)$.

We have shown that the minimum of f is attained at some boundary point $\overline{x}$ of C. If $\overline{x}$ is an extreme point of C, we are done. If it is not an extreme point, consider a hyperplane H passing through $\overline{x}$ and containing C in one of its halfspaces. The intersection $T_1 = C \cap H$ is closed, convex, bounded from below, and lies in a linear manifold M_1 of dimension $n-1$. Furthermore, f attains its minimum over T_1 at $\overline{x}$. Thus, by the preceding argument, it also attains its minimum at some boundary point x_1 of T_1. If x_1 is an extreme point of T_1, then by Prop. B.18, it is also an extreme point of C and the result follows. If x_1 is not an extreme point of T_1, then we view M_1 as a space of dimension $n-1$ and we form T_2, the intersection of T_1 with a hyperplane in M_1 that passes through x_1 and contains T_1 in one of its halfspaces. This hyperplane will be of dimension $n-2$. We can continue this process for at most n times, when a set T_n consisting of a single point is obtained. This point is an extreme point of T_n and, by repeated application of Prop. B.18, an extreme point of C. **Q.E.D.**

As a corollary we have the following:

Proposition B.20: Let C be a closed convex set and let $f : C \mapsto \Re$ be a concave function. Assume that for some invertible $n \times n$ matrix A and some $b \in \Re^n$ we have

$$Ax \geq b, \qquad \forall\, x \in C.$$

Then if f attains a minimum over C, it attains a minimum at some extreme point of C.

Proof: Consider the transformation $x = A^{-1}y$ and the problem of minimizing

$$h(y) = f\big(A^{-1}y\big)$$

over $Y = \{y \mid A^{-1}y \in C\}$. The function h is concave over the closed convex set Y. Furthermore, $y \geq b$ for al $y \in Y$ and hence Y is bounded from below. By Prop. B.19, h attains a minimum at some extreme point y^* of Y. Then f attains its minimum over C at $x^* = A^{-1}y^*$, while x^* is an extreme point of C, since it can be verified that invertible transformations of sets map extreme points to extreme points. **Q.E.D.**

Extreme Points of Polyhedral Sets

We now consider a polyhedral set P and we characterize the set of its extreme points (also called *vertices*). By Prop. B.17, P can be represented as

$$P = C + \hat{P},$$

where C is a finitely generated cone C and $\hat{P}$ is the convex hull of some vectors $v_1, \ldots, v_m$:

$$\hat{P} = \left\{ x \;\middle|\; x = \sum_{j=1}^{m} \mu_j v_j,\; \sum_{j=1}^{m} \mu_j = 1,\; \mu_j \geq 0,\; j = 1, \ldots, m \right\}.$$

We note that an extreme point $\overline{x}$ of P cannot be of the form $\overline{x} = c + \hat{x}$, where $c \neq 0$, $c \in C$, and $\hat{x} \in \hat{P}$, since in this case $\overline{x}$ would be the midpoint of the line segment connecting the distinct vectors $\hat{x}$ and $2c + \hat{x}$. Therefore, an extreme point of P must belong to $\hat{P}$, and since $\hat{P} \subset P$, it must also be an extreme point of $\hat{P}$. An extreme point of $\hat{P}$ must be one of the vectors $v_1, \ldots, v_m$, since otherwise this point would be expressible as a convex combination of $v_1, \ldots, v_m$. Thus the set of extreme points of P is either empty or finite. Using Prop. B.18(b), it follows that *the set of extreme points of P is nonempty and finite if and only if P contains no line.*

If P is bounded, then we must have $P = \hat{P}$, and it can be shown that *P is equal to the convex hull of its extreme points* (not just the convex hull of the vectors $v_1, \ldots, v_m$). The proof is sketched in Exercise B.4.1.

The following proposition gives another and more specific characterization of extreme points of polyhedral sets, and is central in the theory of linear programming.

Proposition B.21: Let P be a polyhedral set in $\Re^n$.

(a) If P has the form

$$P = \{x \mid a_j'x \leq b_j,\ j = 1, \ldots, r\},$$

where a_j and b_j are given vectors and scalars, respectively, then a vector $v \in P$ is an extreme point of P if and only if the set

$$A_v = \{a_j \mid a_j'v = b_j,\ j = 1, \ldots, r\}$$

contains n linearly independent vectors.

(b) If P has the form

$$P = \{x \mid Ax = b,\ x \geq 0\},$$

where A is a given $m \times n$ matrix and b is a given vector, then a vector $v \in P$ is an extreme point of P if and only if the columns of A corresponding to the nonzero coordinates of v are linearly independent.

(c) *(Fundamental Theorem of Linear Programming)* Assume that P has at least one extreme point. Then if a linear function attains a minimum over P, it attains a minimum at some extreme point of P.

Proof: (a) If the set A_v contains fewer than n linearly independent vectors, then the system of equations

$$a_j'w = 0, \qquad \forall\ a_j \in A_v$$

has a nonzero solution $\overline{w}$. For sufficiently small $\gamma > 0$, we have $v + \gamma\overline{w} \in P$ and $v - \gamma\overline{w} \in P$, thus showing that v is not an extreme point. Thus, if v is an extreme point, A_v must contain n linearly independent vectors.

Conversely, suppose that A_v contains a subset $\bar{A}_v$ consisting of n linearly independent vectors. Suppose that for some $y \in P$, $z \in P$, and $\alpha \in (0, 1)$, we have $v = \alpha y + (1 - \alpha)z$. Then for all $a_j \in \bar{A}_v$, we have

$$b_j = a_j'v = \alpha a_j'y + (1 - \alpha)a_j'z \leq \alpha b_j + (1 - \alpha)b_j = b_j.$$

Thus v, y, and z are all solutions of the system of n linearly independent equations

$$a_j' w = b_j, \qquad \forall \ a_j \in \bar{A}_v.$$

Hence $v = y = z$, implying that v is an extreme point.

(b) Let k be the number of zero coordinates of v, and consider the matrix $\bar{A}$, which is the same as A except that the columns corresponding to the zero coordinates of v are set to zero. We write P in the form

$$P = \{x \mid Ax \leq b, \ -Ax \leq -b, \ -x \leq 0\},$$

and apply the result of part (a). We obtain that v is an extreme point if and only if $\bar{A}$ contains $n-k$ linearly independent rows, which is equivalent to the $n-k$ nonzero columns of $\bar{A}$ (corresponding to the nonzero coordinates of v) being linearly independent.

(c) Since P is polyhedral, it has a representation

$$P = \{x \mid Ax \geq b\},$$

for some $m \times n$ matrix A and some $b \in \Re^m$. If A had rank less than n, then its nullspace would contain some nonzero vector $\overline{x}$, so P would contain a line parallel to $\overline{x}$, contradicting the existence of an extreme point [cf. Prop. B.18(b)]. Thus A has rank n and hence it must contain n linearly independent rows that constitute an $n \times n$ invertible submatrix $\hat{A}$. If $\hat{b}$ is the corresponding subvector of b, we see that every $x \in P$ satisfies $\hat{A}x \geq \hat{b}$. The result then follows using Prop. B.20. **Q.E.D.**

EXERCISES

B.4.1

Show that a polyhedron of the form

$$P = \left\{ x \;\middle|\; x = \sum_{j=1}^{m} \mu_j v_j, \ \sum_{j=1}^{m} \mu_j = 1, \ \mu_j \geq 0, \ j = 1, \ldots, m \right\}. \tag{B.16}$$

is the convex hull of its extreme points. *Hint*: Use induction on the dimension of the space. Suppose that all bounded polyhedra of $(n-1)$-dimensional spaces have a representation of the form (B.16), but there is a bounded polyhedron $P \subset \Re^n$

and a vector $x \in P$, which is not in the convex hull P_E of the extreme points of P. Let $\hat{x}$ be the projection of x on P_E and let $\overline{x}$ be a solution of the problem

$$\begin{aligned} &\text{maximize} \quad (x - \hat{x})'z \\ &\text{subject to} \quad z \in P. \end{aligned}$$

The polyhedron

$$\hat{P} = P \cap \{z \mid (x - \hat{x})'z = (x - \hat{x})'\overline{x}\}$$

is equal to the convex hull of its extreme points by the induction hypothesis. Show that $P_E \cap \hat{P} = \emptyset$, while, by Prop. B.18(a), each of the extreme points of $\hat{P}$ is also an extreme point of P, arriving at a contradiction.

B.5 DIFFERENTIABILITY ISSUES

Convex functions have interesting differentiability properties, which we discuss in this section. We first consider convex functions of a single variable.

Let I be an interval of real numbers, and let $f : I \mapsto \Re$ be convex. If $x, y, z \in I$ and $x < y < z$, then we can show the relation

$$\frac{f(y) - f(x)}{y - x} \leq \frac{f(z) - f(x)}{z - x} \leq \frac{f(z) - f(y)}{z - y}, \tag{B.17}$$

which is illustrated in Fig. B.11. For a formal proof, note that, using the definition of a convex function [cf. Eq. (B.2)], we obtain

$$f(y) \leq \left(\frac{y - x}{z - x}\right) f(z) + \left(\frac{z - y}{z - x}\right) f(x)$$

and either of the desired inequalities follows by appropriately rearranging terms.

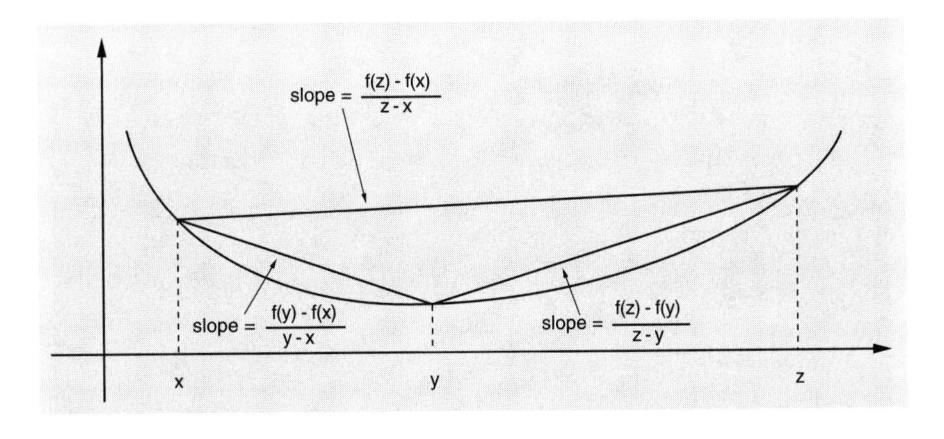

Figure B.11. Illustration of the inequalities (B.17). The rate of change of the function f is nondecreasing with its argument.

Let a and b be the infimum and the supremum, respectively, of I, also referred to as the *end points of* I. For any $x \in I$, $x \neq b$, and for any $\alpha > 0$ such that $x + \alpha \in I$, we define

$$s^+(x, \alpha) = \frac{f(x + \alpha) - f(x)}{\alpha}.$$

Let $0 < \alpha \leq \alpha'$. We use the first inequality in Eq. (B.17) with $y = x + \alpha$ and $z = x + \alpha'$ to obtain $s^+(x, \alpha) \leq s^+(x, \alpha')$. Therefore, $s^+(x, \alpha)$ is a nondecreasing function of α and, as α decreases to zero, it converges either to a finite number or to $-\infty$. Let $f^+(x)$ be the value of the limit, which we call the *right derivative* of f at the point x. Similarly, if $x \in I$, $x \neq a$, $\alpha > 0$, and $x - \alpha \in I$, we define

$$s^-(x, \alpha) = \frac{f(x) - f(x - \alpha)}{\alpha},$$

which is, by a symmetrical argument, a nonincreasing function of α. Its limit as α decreases to zero, denoted by $f^-(x)$, is called the *left derivative* of f at the point x, and is either finite or equal to ∞.

In the case where the end points a and b belong to the domain I of f, we define for completeness $f^-(a) = -\infty$ and $f^+(b) = \infty$.

Proposition B.22: Let $I \subset \Re$ be a convex interval and let $f : I \mapsto \Re$ be a convex function. Let a and b be the end points of I.

(a) We have $f^-(y) \leq f^+(y)$ for every $y \in I$.

(b) If x belongs to the interior of I, then $f^+(x)$ and $f^-(x)$ are finite.

(c) If $x, z \in I$ and $x < z$, then $f^+(x) \leq f^-(z)$.

(d) The functions $f^-, f^+ : I \mapsto [-\infty, +\infty]$ are nondecreasing.

(e) The function f^+ (respectively, f^-) is right– (respectively, left–) continuous at every interior point of I. Also, if $a \in I$ (respectively, $b \in I$) and f is continuous at a (respectively, b), then f^+ (respectively, f^-) is right– (respectively, left–) continuous at a (respectively, b).

(f) If f is differentiable at a point x belonging to the interior of I, then $f^+(x) = f^-(x) = (df/dx)(x)$.

(g) For any $x, z \in I$ and any d satisfying $f^-(x) \leq d \leq f^+(x)$, we have

$$f(z) \geq f(x) + d(z - x).$$

(h) The function $f^+ : I \mapsto (-\infty, \infty]$ [respectively, $f^- : I \mapsto [-\infty, \infty)$] is upper (respectively, lower) semicontinuous at every $x \in I$.

Proof: (a) If y is an end point of I, the result is trivial because $f^-(a) = -\infty$ and $f^+(b) = \infty$. We assume that y is an interior point, we let $\alpha > 0$, and use Eq. (B.17), with $x = y - \alpha$ and $z = y + \alpha$, to obtain $s^-(y, \alpha) \leq s^+(y, \alpha)$. Taking the limit as α decreases to zero, we obtain $f^-(y) \leq f^+(y)$.

(b) Let x belong to the interior of I and let $\alpha > 0$ be such that $x - \alpha \in I$. Then $f^-(x) \geq s^-(x, \alpha) > -\infty$. For similar reasons, we obtain $f^+(x) < \infty$. Part (a) then implies that $f^-(x) < \infty$ and $f^+(x) > -\infty$.

(c) We use Eq. (B.17), with $y = (z + x)/2$, to obtain $s^+\big(x, (z - x)/2\big) \leq s^-\big(z, (z - x)/2\big)$. The result then follows because $f^+(x) \leq s^+\big(x, (z - x)/2\big)$ and $s^-\big(z, (z - x)/2\big) \leq f^-(z)$.

(d) This follows by combining parts (a) and (c).

(e) Fix some $x \in I$, $x \neq b$, and some positive δ and α such that $x + \delta + \alpha < b$. We allow x to be equal to a, in which case f is assumed to be continuous at a. We have $f^+(x + \delta) \leq s^+(x + \delta, \alpha)$. We take the limit, as δ decreases to zero, to obtain $\lim_{\delta \downarrow 0} f^+(x + \delta) \leq s^+(x, \alpha)$. We have used here the fact that $s^+(x, \alpha)$ is a continuous function of x, which is a consequence of the continuity of f (Prop. B.9). We now let α decrease to zero to obtain $\lim_{\delta \downarrow 0} f^+(x + \delta) \leq f^+(x)$. The reverse inequality is also true because f^+ is nondecreasing and this proves the right–continuity of f^+. The proof for f^- is similar.

(f) This is immediate from the definition of f^+ and f^-.

(g) Fix some $x, z \in I$. The result is trivially true for $x = z$. We only consider the case $x < z$; the proof for the case $x > z$ is similar. Since $s^+(x, \alpha)$ is nondecreasing in α, we have $\big(f(z) - f(x)\big)/(z - x) \geq s^+(x, \alpha)$ for α belonging to $(0, z - x)$. Letting α decrease to zero, we obtain $\big(f(z) - f(x)\big)/(z - x) \geq f^+(x) \geq d$ and the result follows.

(h) This follows from parts (a), (d), (e), and the definition of semicontinuity (Definition A.4 in Appendix A). **Q.E.D.**

We now consider the *directional derivative* $f'(x; y)$ of a convex function $f : \Re^n \mapsto \Re$ at a vector $x \in \Re^n$ in the direction $y \in \Re^n$. This derivative is equal to the right derivative $F_y^+(0)$ of the convex scalar function $F_y(\alpha) = f(x + \alpha y)$ at $\alpha = 0$, i.e.,

$$f'(x; y) = \lim_{\alpha \downarrow 0} \frac{f(x + \alpha y) - f(x)}{\alpha} = \lim_{\alpha \downarrow 0} \frac{F_y(\alpha) - F_y(0)}{\alpha} = F_y^+(0), \quad \text{(B.18)}$$

and the limit in the above equation is guaranteed to exist. Similarly, the left derivative $F_y^-(0)$ of F_y is equal to $-f'(x; -y)$ and, by using Prop. B.22(a), we obtain $F_y^-(0) \leq F_y^+(0)$, or equivalently,

$$-f'(x; -y) \leq f'(x; y), \qquad \forall\ y \in \Re^n. \quad \text{(B.19)}$$

The directional derivative can be used to provide a necessary and sufficient condition for optimality in the problem of minimizing a convex function $f : \Re^n \mapsto \Re$ over a convex set $X \subset \Re^n$. In particular, x^* is a global minimum of f over X if and only if

$$f'(x^*; x - x^*) \geq 0, \qquad \forall\, x \in X.$$

This follows from the definition (B.18) of directional derivative, and from the fact that the difference quotient

$$\frac{f\big(x^* + \alpha(x - x^*)\big) - f(x^*)}{\alpha}$$

is a monotonically nondecreasing function of α.

The following proposition generalizes the upper semicontinuity property of right derivatives of scalar convex functions [Prop. B.22(h)], and shows that if f is differentiable, then its gradient is continuous.

Proposition B.23: Let $f : \Re^n \mapsto \Re$ be convex, and let $\{f_k\}$ be a sequence of convex functions $f_k : \Re^n \mapsto \Re$ with the property that $\lim_{k\to\infty} f_k(x_k) = f(x)$ for every $x \in \Re^n$ and every sequence $\{x_k\}$ that converges to x. Then for any $x \in \Re^n$ and $y \in \Re^n$, and any sequences $\{x_k\}$ and $\{y_k\}$ converging to x and y, respectively, we have

$$\limsup_{k\to\infty} f_k'(x_k; y_k) \leq f'(x; y). \tag{B.20}$$

Furthermore, if f is differentiable at all $x \in \Re^n$, then its gradient $\nabla f(x)$ is a continuous function of x.

Proof: For any $\mu > f'(x; y)$, there exists an $\overline{\alpha} > 0$ such that

$$\frac{f(x + \alpha y) - f(x)}{\alpha} < \mu, \qquad \forall\, \alpha \leq \overline{\alpha}.$$

Hence, for $\alpha \leq \overline{\alpha}$, we have

$$\frac{f_k(x_k + \alpha y_k) - f_k(x_k)}{\alpha} < \mu$$

for all sufficiently large k, and using Eq. (B.18), we obtain

$$\limsup_{k\to\infty} f_k'(x_k; y_k) < \mu.$$

Since this is true for all $\mu > f'(x; y)$, inequality (B.20) follows.

If f is differentiable at all $x \in \Re^n$, then using the continuity of f and the part of the proposition just proved, we have for every sequence $\{x_k\}$ converging to x and every $y \in \Re^n$,

$$\limsup_{k\to\infty} \nabla f(x_k)'y = \limsup_{k\to\infty} f'(x_k; y) \leq f'(x; y) = \nabla f(x)'y.$$

By replacing y by $-y$ in the preceding argument, we obtain

$$-\liminf_{k\to\infty} \nabla f(x_k)'y = \limsup_{k\to\infty} \big(-\nabla f(x_k)'y\big) \leq -\nabla f(x)'y.$$

Therefore, we have $\nabla f(x_k)'y \to \nabla f(x)'y$ for every y, which implies that $\nabla f(x_k) \to \nabla f(x)$. Hence, the gradient is continuous. **Q.E.D.**

Subgradients and Subdifferentials

Given a convex function $f : \Re^n \mapsto \Re$, we say that a vector $d \in \Re^n$ is a *subgradient* of f at a point $x \in \Re^n$ if

$$f(z) \geq f(x) + (z - x)'d, \qquad \forall\, z \in \Re^n. \tag{B.21}$$

If instead f is a concave function, we say that d is a subgradient of f at x if $-d$ is a subgradient of the convex function $-f$ at x. The set of all subgradients of a convex (or concave) function f at $x \in \Re^n$ is called the *subdifferential* of f at x, and is denoted by $\partial f(x)$. Figure B.12 provides some examples of subdifferentials.

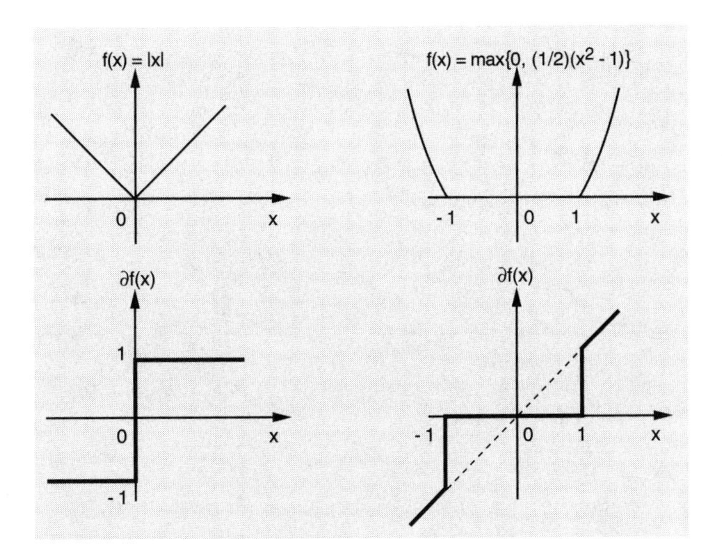

Figure B.12. The subdifferential of some scalar convex functions as a function of the argument x.

We next provide the relationship between the directional derivative and the subdifferential, and prove some basic properties of subgradients.

Proposition B.24: Let $f : \Re^n \mapsto \Re$ be convex. For every $x \in \Re^n$, the following hold:

(a) A vector d is a subgradient of f at x if and only if

$$f'(x; y) \geq y'd, \qquad \forall\, y \in \Re^n.$$

(b) The subdifferential $\partial f(x)$ is a nonempty, convex, and compact set, and there holds

$$f'(x; y) = \max_{d \in \partial f(x)} y'd, \qquad \forall\, y \in \Re^n. \tag{B.22}$$

In particular, f is differentiable at x with gradient $\nabla f(x)$, if and only if it has $\nabla f(x)$ as its unique subgradient at x. Furthermore, if X is a bounded set, the set $\cup_{x \in X} \partial f(x)$ is bounded.

(c) If a sequence $\{x_k\}$ converges to x and $d_k \in \partial f(x_k)$ for all k, the sequence $\{d_k\}$ is bounded and each of its limit points is a subgradient of f at x.

(d) If f is equal to the sum $f_1 + \cdots + f_m$ of convex functions $f_j : \Re^n \mapsto \Re$, $j = 1, \ldots, m$, then $\partial f(x)$ is equal to the vector sum $\partial f_1(x) + \cdots + \partial f_m(x)$.

(e) If f is equal to the composition of a convex function $h : \Re^m \mapsto \Re$ and an $m \times n$ matrix A $[f(x) = h(Ax)]$, then $\partial f(x)$ is equal to $A'\partial h(Ax) = \big\{A'g \mid g \in \partial h(Ax)\big\}$.

(f) x minimizes f over a convex set $X \subset \Re^n$ if and only if there exists a subgradient $d \in \partial f(x)$ such that

$$d'(z - x) \geq 0, \qquad \forall\, z \in X.$$

Proof: (a) The subgradient inequality (B.21) is equivalent to

$$\frac{f(x + \alpha y) - f(x)}{\alpha} \geq y'd, \qquad \forall\, y \in \Re^n,\ \alpha > 0.$$

Since the quotient on the left above decreases monotonically to $f'(x; y)$ as $\alpha \downarrow 0$ [Eq. (B.17)], we conclude that the subgradient inequality (B.21) is equivalent to $f'(x; y) \geq y'd$ for all $y \in \Re^n$. Therefore we obtain

$$d \in \partial f(x) \qquad \Longleftrightarrow \qquad f'(x; y) \geq y'd, \qquad \forall\, y \in \Re^n. \tag{B.23}$$

(b) From Eq. (B.23), we see that $\partial f(x)$ is the intersection of the closed halfspaces $\{d \mid y'd \leq f'(x;y)\}$, where y ranges over the nonzero vectors of $\Re^n$. It follows that $\partial f(x)$ is closed and convex. It is also bounded, since otherwise, for some $y \in \Re^n$, $y'd$ could be made unbounded by proper choice of $d \in \partial f(x)$, contradicting Eq. (B.23). Since $\partial f(x)$ is both closed and bounded, it is compact.

To show that $\partial f(x)$ is nonempty and that Eq. (B.22) holds, we first observe that Eq. (B.23) implies that $f'(x;y) \geq \max_{d\in\partial f(x)} y'd$ [where the maximum is $-\infty$ if $\partial f(x)$ is empty]. To show the reverse inequality, take any x and y in $\Re^n$, and consider the subset of $\Re^{n+1}$

$$C_1 = \{(\mu, z) \mid \mu > f(z)\},$$

and the half-line

$$C_2 = \{(\mu, z) \mid \mu = f(x) + \alpha f'(x;y),\ z = x + \alpha y,\ \alpha \geq 0\};$$

see Fig. B.13. Using the definition of directional derivative and the convexity of f, it follows that these two sets are nonempty, convex, and disjoint. By applying the separating hyperplane theorem (Prop. B.13), we see that there exists a nonzero vector $(\gamma, w) \in \Re^{n+1}$ such that

$$\gamma\mu + w'z \leq \gamma\big(f(x) + \alpha f'(x;y)\big) + w'(x + \alpha y), \qquad \forall\, \alpha \geq 0,\ z \in \Re^n,\ \mu > f(z). \tag{B.24}$$

We cannot have $\gamma > 0$ since then the left-hand side above could be made arbitrarily large by choosing μ sufficiently large. Also if $\gamma = 0$, then Eq. (B.24) implies that $w = 0$, which is a contradiction. Therefore, $\gamma < 0$ and by dividing with γ in Eq. (B.24), we obtain

$$\mu + (z - x)'(w/\gamma) \geq f(x) + \alpha f'(x;y) + \alpha y'(w/\gamma),\ \ \forall\, \alpha \geq 0,\ z \in \Re^n,\ \mu > f(z). \tag{B.25}$$

By taking the limit in the above relation as $\alpha \downarrow 0$ and $\mu \downarrow f(z)$, we obtain $f(z) \geq f(x) + (z-x)'(-w/\gamma)$ for all $z \in \Re^n$, implying that $(-w/\gamma) \in \partial f(x)$. By taking $z = x$ and $\alpha = 1$ in Eq. (B.25), and by taking the limit as $\mu \downarrow f(x)$, we obtain $y'(-w/\gamma) \geq f'(x;y)$, which implies that $\max_{d\in\partial f(x)} y'd \geq f'(x;y)$. The proof of Eq. (B.22) is complete.

From the definition of directional derivative, we see that f is differentiable at x with gradient $\nabla f(x)$ if and only if the directional derivative $f'(x;y)$ is a linear function of the form $f'(x;y) = \nabla f(x)'y$. Thus, from Eq. (B.22), f is differentiable at x with gradient $\nabla f(x)$, if and only if it has $\nabla f(x)$ as its unique subgradient at x.

Finally, let X be a bounded set. To show that $\cup_{x\in X}\partial f(x)$ is bounded, we assume the contrary, i.e. that there exists a sequence $\{x_k\} \subset X$, and a sequence $\{d_k\}$ with $d_k \in \partial f(x_k)$ for all k and $\|d_k\| \to \infty$. Without loss of generality, we assume that $d_k \neq 0$ for all k, and we denote $y_k = d_k/\|d_k\|$.

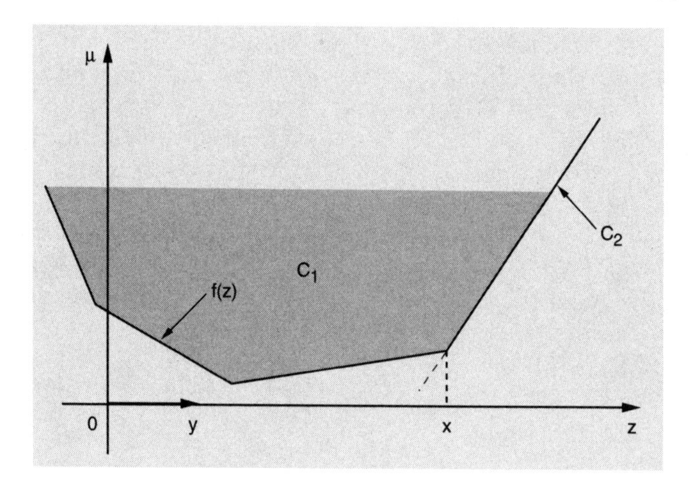

Figure B.13. Illustration of the sets C_1 and C_2 used in the hyperplane separation argument of the proof of Prop. B.24.

Since both $\{x_k\}$ and $\{y_k\}$ are bounded, they must contain convergent subsequences. We assume without loss of generality that x_k converges to some x and y_k converges to some y with $\|y\| = 1$. By Eq. (B.22), we have

$$f'(x_k; y_k) \geq d_k' y_k = \|d_k\|,$$

so it follows that $f'(x_k; y_k) \to \infty$. This contradicts, however, Eq. (B.20), which implies that $\limsup_{k\to\infty} f'(x_k; y_k) \leq f'(x; y)$.

(c) By part (b), the sequence $\{d_k\}$ is bounded, and by part (a), we have

$$y' d_k \leq f'(x_k; y), \qquad \forall\, y \in \Re^n.$$

If d is a limit point of $\{d_k\}$, we have by taking limit in the above relation and by using Prop. B.23

$$y'd \leq \limsup_{k\to\infty} f'(x_k; y) \leq f'(x; y), \qquad \forall\, y \in \Re^n.$$

Therefore, by part (a), we have $d \in \partial f(x)$.

(d) It will suffice to prove the result for the case where $f = f_1 + f_2$. If $d_1 \in \partial f_1(x)$ and $d_2 \in \partial f_2(x)$, then from the subgradient inequality (B.21), we have

$$f_1(z) \geq f_1(x) + (z - x)' d_1, \qquad \forall\, z \in \Re^n,$$

$$f_2(z) \geq f_2(x) + (z - x)' d_2, \qquad \forall\, z \in \Re^n,$$

so by adding, we obtain

$$f(z) \geq f(x) + (z - x)'(d_1 + d_2), \qquad \forall\, z \in \Re^n.$$

Hence $d_1 + d_2 \in \partial f(x)$, implying that $\partial f_1(x) + \partial f_2(x) \subset \partial f(x)$.

To prove the reverse inclusion, suppose to come to a contradiction, that there exists a $d \in \partial f(x)$ such that $d \notin \partial f_1(x) + \partial f_2(x)$. Since by part (b), the sets $\partial f_1(x)$ and $\partial f_2(x)$ are compact, the set $\partial f_1(x) + \partial f_2(x)$ is compact (cf. Prop. B.8), and by Prop. B.14, there exists a hyperplane strictly separating $\{d\}$ from $\partial f_1(x) + \partial f_2(x)$, i.e., a vector y and a scalar b such that

$$y'(d_1 + d_2) < b < y'd, \qquad \forall\ d_1 \in \partial f_1(x),\ d_2 \in \partial f_2(x).$$

From this we obtain

$$\max_{d_1 \in \partial f_1(x)} y'd_1 + \max_{d_2 \in \partial f_2(x)} y'd_2 < y'd,$$

or using part (b),

$$f_1'(x; y) + f_2'(x; y) < y'd.$$

By using the definition of directional derivative, $f_1'(x; y) + f_2'(x; y) = f'(x; y)$, so we have

$$f'(x; y) < y'd,$$

which is a contradiction in view of part (a).

(e) It is seen using the definition of directional derivative that

$$f'(x; y) = h'(Ax; Ay), \qquad \forall\ y \in \Re^n.$$

Let $g \in \partial h(Ax)$ and $d = A'g$. Then by part (a), we have

$$g'z \leq h'(Ax; z) \qquad \forall\ z \in \Re^m,$$

and in particular,

$$g'Ay \leq h'(Ax; Ay) \qquad \forall\ y \in \Re^n,$$

or

$$(A'g)'y \leq f(x; y), \qquad \forall\ y \in \Re^n.$$

Hence, by part (a), we have $A'g \in \partial f(x)$, so that $A'\partial h(Ax) \subset \partial f(x)$.

To prove the reverse inclusion, suppose to come to a contradiction, that there exists a $d \in \partial f(x)$ such that $d \notin A'\partial h(Ax)$. Since by part (b), the set $\partial h(Ax)$ is compact, the set $A'\partial h(Ax)$ is also compact (cf. Prop. B.8), and by Prop. B.14, there exists a hyperplane strictly separating $\{d\}$ from $A'\partial h(Ax)$, i.e., a vector y and a scalar b such that

$$y'(A'g) < b < y'd, \qquad \forall\ g \in \partial h(Ax).$$

From this we obtain

$$\max_{g \in \partial h(Ax)} (Ay)'g < y'd,$$

or using part (b),

$$h'(Ax; Ay) < y'd.$$

Since $h'(Ax; Ay) = f'(x; y)$, it follows that

$$f'(x; y) < y'd,$$

which is a contradiction in view of part (a).

(f) Suppose that for some $d \in \partial f(x)$ and all $z \in X$, we have $d'(z - x) \geq 0$. Then, since from the definition of a subgradient we have $f(z) - f(x) \geq d'(z - x)$ for all $z \in X$, we obtain $f(z) - f(x) \geq 0$ for all $z \in X$, so x minimizes f over X.

Conversely, suppose that x minimizes f over X. Consider the set of feasible directions of X at x

$$W = \{w \neq 0 \mid x + \alpha w \in X \text{ for some } \alpha > 0\},$$

and the cone

$$\hat{W} = \{d \mid d'w \geq 0, \ \forall\, w \in W\}$$

(this is equal to $-W^{\perp}$, the set of all d such that $-d$ belongs to the polar cone $W^{\perp}$). If $\partial f(x)$ and $\hat{W}$ have a point in common, we are done, so to arrive at a contradiction, assume the opposite, i.e., $\partial f(x) \cap \hat{W} = \emptyset$. Since $\partial f(x)$ is compact and $\hat{W}$ is closed, by Prop. B.14 there exists a hyperplane strictly separating $\partial f(x)$ and $\hat{W}$, i.e., a vector y and a scalar c such that

$$g'y < c < d'y, \qquad \forall\, g \in \partial f(x),\ \forall\, d \in \hat{W}.$$

Using the fact that $\hat{W}$ is a closed cone, it follows that

$$c < 0 \leq d'y, \qquad \forall\, d \in \hat{W}, \tag{B.26}$$

which when combined with the preceding inequality, also yields

$$\max_{g \in \partial f(x)} g'y < c < 0.$$

Thus, using part (b), we have $f'(x; y) < 0$, while from Eq. (B.26), we see that y belongs to the polar cone of $W^{\perp}$, which by the polar cone theorem [Prop. B.16(a)], implies that y is in the closure of the set of feasible directions W. Hence for a sequence y^k of feasible directions converging to y we have $f'(x; y^k) < 0$, and this contradicts the optimality of x. **Q.E.D.**

Note that Prop. B.24(f) generalizes the optimality condition of Chapter 2 for the case where f is differentiable: $\nabla f(x)'(z - x) \geq 0$ for all $z \in X$. In the special case where $X = \Re^n$, we obtain a basic necessary and sufficient condition for unconstrained optimality of x:

$$0 \in \partial f(x).$$

This optimality condition is also evident from the subgradient inequality (B.21).

Danskin's Min-Max Theorem

We next consider the directional derivative and the subdifferential of the function $f(x) = \max_{z \in Z} \phi(x, z)$.

Proposition B.25: (Danskin's Theorem) Let $Z \subset \Re^m$ be a compact set, and let $\phi : \Re^n \times Z \mapsto \Re$ be continuous and such that $\phi(\cdot, z) : \Re^n \mapsto \Re$ is convex for each $z \in Z$.

(a) The function $f : \Re^n \mapsto \Re$ given by

$$f(x) = \max_{z \in Z} \phi(x, z) \tag{B.27}$$

is convex and has directional derivative given by

$$f'(x; y) = \max_{z \in Z(x)} \phi'(x, z; y), \tag{B.28}$$

where $\phi'(x, z; y)$ is the directional derivative of the function $\phi(\cdot, z)$ at x in the direction y, and $Z(x)$ is the set of maximizing points in Eq. (B.27)

$$Z(x) = \left\{ \overline{z} \;\middle|\; \phi(x, \overline{z}) = \max_{z \in Z} \phi(x, z) \right\}.$$

In particular, if $Z(x)$ consists of a unique point $\overline{z}$ and $\phi(\cdot, \overline{z})$ is differentiable at x, then f is differentiable at x, and $\nabla f(x) = \nabla_x \phi(x, \overline{z})$, where $\nabla_x \phi(x, \overline{z})$ is the vector with coordinates

$$\frac{\partial \phi(x, \overline{z})}{\partial x_i}, \qquad i = 1, \ldots, n.$$

(b) If $\phi(\cdot, z)$ is differentiable for all $z \in Z$ and $\nabla_x \phi(x, \cdot)$ is continuous on Z for each x, then

$$\partial f(x) = \operatorname{conv}\big\{ \nabla_x \phi(x, z) \mid z \in Z(x) \big\}, \qquad \forall\, x \in \Re^n.$$

In particular, if ϕ is linear in x for all $z \in Z$, i.e.,

$$\phi(x, z) = a_z' x + b_z, \qquad \forall\, z \in Z,$$

then

$$\partial f(x) = \operatorname{conv}\big\{ a_z \mid z \in Z(x) \big\}.$$

Proof: (a) The convexity of f has been established in Prop. B.2(d). We note that since ϕ is continuous and Z is compact, the set $Z(x)$ is nonempty by Weierstrass' theorem (Prop. A.8 in Appendix A) and f is finite. For any $z \in Z(x)$, $y \in \Re^n$, and $\alpha > 0$, we use the definition of f to obtain

$$\frac{f(x+\alpha y) - f(x)}{\alpha} \geq \frac{\phi(x+\alpha y, z) - \phi(x, z)}{\alpha}.$$

Taking the limit as α decreases to zero, we obtain $f'(x; y) \geq \phi'(x, z; y)$. Since this is true for every $z \in Z(x)$, we conclude that

$$f'(x; y) \geq \sup_{z \in Z(x)} \phi'(x, z; y), \qquad \forall\, y \in \Re^n. \tag{B.29}$$

To prove the reverse inequality and that the supremum in the right-hand side of the above inequality is attained, consider a sequence $\{\alpha_k\}$ of positive scalars that converges to zero and let $x_k = x + \alpha_k y$. For each k, let z_k be a vector in $Z(x_k)$. Since $\{z_k\}$ belongs to the compact set Z, it has a subsequence converging to some $\overline{z} \in Z$. Without loss of generality, we assume that the entire sequence $\{z_k\}$ converges to $\overline{z}$. We have

$$\phi(x_k, z_k) \geq \phi(x_k, z), \qquad \forall\, z \in Z,$$

so by taking the limit as $k \to \infty$ and by using the continuity of ϕ, we obtain

$$\phi(x, \overline{z}) \geq \phi(x, z), \qquad \forall\, z \in Z.$$

Therefore, $\overline{z} \in Z(x)$. We now have

$$\begin{aligned} f'(x; y) &\leq \frac{f(x+\alpha_k y) - f(x)}{\alpha_k} \\ &= \frac{\phi(x+\alpha_k y, z_k) - \phi(x, \overline{z})}{\alpha_k} \\ &\leq \frac{\phi(x+\alpha_k y, z_k) - \phi(x, z_k)}{\alpha_k} \\ &\leq -\phi'(x+\alpha_k y, z_k; -y) \\ &\leq \phi'(x+\alpha_k y, z_k; y), \end{aligned} \tag{B.30}$$

where the last inequality follows from inequality (B.19). We apply Prop. B.23 to the functions f_k defined by $f_k(\cdot) = \phi(\cdot, z_k)$, and with $x_k = x + \alpha_k y$, to obtain

$$\limsup_{k \to \infty} \phi'(x + \alpha_k y, z_k; y) \leq \phi'(x, \overline{z}; y). \tag{B.31}$$

We take the limit in inequality (B.30) as $k \to \infty$, and we use inequality (B.31) to conclude that

$$f'(x; y) \leq \phi'(x, \overline{z}; y).$$

This relation together with inequality (B.29) proves Eq. (B.28).

For the last statement of part (a), if $Z(x)$ consists of the unique point $\overline{z}$, Eq. (B.28) and the differentiability assumption on ϕ yield

$$f'(x;y) = \phi'(x,\overline{z};y) = y'\nabla_x\phi(x,\overline{z}), \qquad \forall\, y \in \Re^n,$$

which implies that $\nabla f(x) = \nabla_x\phi(x,\overline{z})$.

(b) By part (a), we have

$$f'(x;y) = \max_{z\in Z(x)} \nabla_x\phi(x,z)'y,$$

while by Prop. B.24, we have

$$f'(x;y) = \max_{z\in\partial f(x)} d'y.$$

For all $\overline{z} \in Z(x)$ and $y \in \Re^n$, we have

$$\begin{aligned} f(y) &= \max_{z\in Z} \phi(y,z) \\ &\geq \phi(y,\overline{z}) \\ &\geq \phi(x,\overline{z}) + \nabla_x\phi(x,\overline{z})'(y-x) \\ &= f(x) + \nabla_x\phi(x,\overline{z})'(y-x). \end{aligned}$$

Therefore, $\nabla_x\phi(x,\overline{z})$ is a subgradient of f at x, implying that

$$\mathrm{conv}\big\{\nabla_x\phi(x,z) \mid z \in Z(x)\big\} \subset \partial f(x).$$

To prove the reverse inclusion, we use a hyperplane separation argument. By the continuity of $\nabla_x\phi(x,\cdot)$ and the compactness of Z, we see that $Z(x)$ is compact, and therefore also the set $\big\{\nabla_x\phi(x,z) \mid z \in Z(x)\big\}$ is compact. By Prop. B.8(d), it follows that $\mathrm{conv}\big\{\nabla_x\phi(x,z) \mid z \in Z(x)\big\}$ is compact. If $d \in \partial f(x)$ while $d \notin \mathrm{conv}\big\{\nabla_x\phi(x,z) \mid z \in Z(x)\big\}$, by the strict separation theorem (Prop. B.14), there exists $y \neq 0$, and $\gamma \in \Re$, such that

$$d'y > \gamma > \nabla_x\phi(x,z)'y, \qquad \forall\, z \in Z(x).$$

Therefore, we have

$$d'y > \max_{z\in Z(x)} \nabla_x\phi(x,z)'y = f'(x;y),$$

contradicting Prop. B.24. Therefore, $\partial f(x) \subset \mathrm{conv}\big\{\nabla_x\phi(x,z) \mid z \in Z(x)\big\}$ and the proof is complete. **Q.E.D.**

Subgradients of Extended-Real Valued Convex Functions

In this book the major emphasis is on real-valued convex functions $f : \Re^n \mapsto \Re$, which are defined over the entire space $\Re^n$ and are convex over $\Re^n$. There are, however, important cases, prominently arising in the context of duality, where we must deal with functions $g : D \mapsto \Re$ that are defined over a convex subset D of $\Re^n$, and are convex over D. This type of function may also be specified as the extended real-valued function $f : \Re^n \mapsto (-\infty, \infty]$ given by

$$f(x) = \begin{cases} g(x) & \text{if } x \in D, \\ \infty & \text{otherwise,} \end{cases}$$

with D referred to as the *effective domain* of f.

The notion of a subdifferential and a subgradient of such a function can be developed along the lines of the present section. In particular, given a convex function $f : \Re^n \mapsto (-\infty, \infty]$, a vector d is a subgradient of f at a vector x such that $f(x) < \infty$ if the subgradient inequality holds, i.e.,

$$f(z) \geq f(x) + (z - x)'d, \qquad \forall\ z \in \Re^n.$$

If $g : D \mapsto \Re$ is a concave function (that is, $-g$ is a convex function over the convex set D), it can also be represented as the extended real-valued function $f : \Re^n \mapsto [-\infty, \infty)$, where

$$f(x) = \begin{cases} g(x) & \text{if } x \in D, \\ -\infty & \text{otherwise.} \end{cases}$$

As earlier, we say that d is a subgradient of f at an $x \in D$ if $-d$ is a subgradient of the convex function $-g$ at x.

The subdifferential $\partial f(x)$ is the set of all subgradients of the convex (or concave) function f. By convention, $\partial f(x)$ is considered empty for all x with $f(x) = \infty$. Note that contrary to the case of real-valued functions, $\partial f(x)$ may be empty, or closed but unbounded. For example, the extended real-valued convex function given by

$$f(x) = \begin{cases} -\sqrt{x} & \text{if } 0 \leq x \leq 1, \\ \infty & \text{otherwise,} \end{cases}$$

has the subdifferential

$$\partial f(x) = \begin{cases} -\frac{1}{2\sqrt{x}} & \text{if } 0 < x < 1, \\ [-1/2, \infty) & \text{if } x = 1, \\ \varnothing & \text{if } x \leq 0 \text{ or } 1 < x. \end{cases}$$

Thus, $\partial f(x)$ can be empty and can be unbounded at points x that belong to the effective domain of f (as in the cases $x = 0$ and $x = 1$, respectively, of the above example). However, it can be shown that $\partial f(x)$ is nonempty and compact at points x that are *interior* points of the effective domain of f, as also illustrated by the above example.

One can provide generalized versions of the results of Props. B.24 and B.25 within the context of extended real-valued convex functions, but with appropriate adjustments and additional assumptions to deal with cases where $\partial f(x)$ may be empty or noncompact. The reader will find a detailed account of the corresponding theory in the book by Rockafellar [Roc70].

APPENDIX C:

Line Search Methods

In this appendix we describe algorithms for one-dimensional minimization. These are iterative algorithms, used to implement (approximately) the line minimization stepsize rules.

We briefly present three practical methods. The first two use polynomial interpolation, one requiring derivatives, the second only function values. The third, the Golden Section method, also requires just function values. By contrast with the interpolation methods, it does not depend on the existence of derivatives of the minimized function and may be applied even to discontinuous functions. Its validity depends, however, on a certain unimodality assumption.

In our presentation of the interpolation methods, we consider minimization of the function

$$g(\alpha) = f(x + \alpha d),$$

where f is continuously differentiable. By the chain rule, we have

$$g'(\alpha) = \frac{dg(\alpha)}{d\alpha} = \nabla f(x + \alpha d)'d.$$

We assume that $g'(0) = \nabla f(x)'d < 0$, that is, d is a descent direction at x. We give no convergence or rate of convergence results, but under some fairly natural assumptions, it can be shown that the interpolation methods converge superlinearly.

C.1 CUBIC INTERPOLATION

The cubic interpolation method successively determines at each iteration an appropriate interval $[a, b]$ within which a local minimum of g is guaranteed to exist. It then fits a cubic polynomial to the values $g(a)$, $g(b)$, $g'(a)$,

$g'(b)$. The minimizing point $\bar{\alpha}$ of this cubic polynomial lies within $[a, b]$ and replaces one of the two points a or b for the next iteration.

Cubic Interpolation

Step 1: (Determination of the Initial Interval) Let $s > 0$ be some scalar. (Note: If d "approximates well" the Newton direction, then we take $s = 1$.) Evaluate $g(\alpha)$ and $g'(\alpha)$ at the points $\alpha = 0$, s, $2s$, $4s$, $8s, \ldots$, until two successive points a and b are found such that either $g'(b) \geq 0$ or $g(b) \geq g(a)$. Then, it can be seen that a local minimum of g exists within the interval $(a, b]$. [Note: If $g(s)$ is "much larger" than $g(0)$, it is advisable to replace s by βs, where $\beta \in (0, 1)$, for example $\beta = \frac{1}{2}$ or $\beta = \frac{1}{5}$, and repeat this step.] One can show that this step can be carried out if $\lim_{\alpha \to \infty} g(\alpha) > g(0)$.

Step 2: (Updating of the Current Interval) Given the current interval $[a, b]$, a cubic polynomial is fitted to the four values $g(a)$, $g'(a)$, $g(b)$, $g'(b)$. The cubic can be shown to have a unique minimum $\bar{\alpha}$ in the interval $(a, b]$ given by

$$\bar{\alpha} = b - \frac{g'(b) + w - z}{g'(b) - g'(a) + 2w}(b - a),$$

where

$$z = \frac{3\big(g(a) - g(b)\big)}{b - a} + g'(a) + g'(b),$$

$$w = \sqrt{z^2 - g'(a)g'(b)}.$$

If $g'(\bar{\alpha}) \geq 0$ or $g(\bar{\alpha}) \geq g(a)$ replace b by $\bar{\alpha}$. If $g'(\bar{\alpha}) < 0$ and $g(\bar{\alpha}) < g(a)$ replace a by $\bar{\alpha}$. (Note: In practice the computation is terminated once the length of the current interval becomes smaller than a prespecified tolerance or else we obtain $\bar{\alpha} = b$.)

C.2 QUADRATIC INTERPOLATION

This method uses three points a, b, and c such that $a < b < c$, and $g(a) > g(b)$ and $g(b) < g(c)$. Such a set of points is referred to as a *three-point pattern*. It can be seen that a local minimum of g must lie between the extreme points a and c of a three-point pattern a, b, c. At each iteration, the method fits a quadratic polynomial to the three values $g(a)$, $g(b)$, and $g(c)$, and replaces one of the points a, b, and c by the minimizing point of this quadratic polynomial (see Fig. C.1).

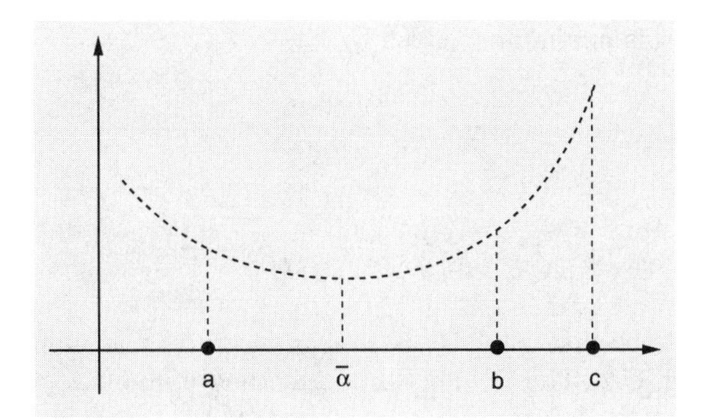

Figure C.1. A three-point pattern and the associated quadratic polynomial. If $\bar{\alpha}$ minimizes the quadratic, a new three point pattern is obtained using $\bar{\alpha}$ and two of the three points a, b, and c (a, b, and $\bar{\alpha}$ in the example of the figure).

Quadratic Interpolation

Step 1: (Determination of Initial Three-Point Pattern) We search along the line as in the cubic interpolation method until we find three successive points a, b, and c with $a < b < c$ such that $g(a) > g(b)$ and $g(b) < g(c)$. As for the cubic interpolation method, we assume that this stage can be carried out, and we can show that this is guaranteed if $\lim_{\alpha \to \infty} g(\alpha) > g(0)$.

Step 2: (Updating the Current Three-Point Pattern) Given the current three-point pattern a, b, c, we fit a quadratic polynomial to the values $g(a)$, $g(b)$, and $g(c)$, and we determine its unique minimum $\bar{\alpha}$. It can be shown that $\bar{\alpha} \in (a, c)$ and that

$$\bar{\alpha} = \frac{1}{2} \frac{g(a)(c^2-b^2)+g(b)(a^2-c^2)+g(c)(b^2-a^2)}{g(a)(c-b)+g(b)(a-c)+g(c)(b-a)}.$$

Then, we form a new three-point pattern as follows. If $\bar{\alpha} > b$, we replace a or c by $\bar{\alpha}$ depending on whether $g(\bar{\alpha} < g(b)$ or $g(\bar{\alpha}) > g(b)$, respectively. If $\bar{\alpha} < b$, we replace c or a by $\bar{\alpha}$ depending on whether $g(\bar{\alpha}) < g(b)$ or $g(\bar{\alpha}) > g(b)$, respectively. [Note: If $g(\bar{\alpha}) = g(b)$ then a special local search near $\bar{\alpha}$ should be conducted to replace $\bar{\alpha}$ by a point $\bar{\alpha}'$ with $g(\bar{\alpha}') \neq g(b)$. The computation is terminated when the length of the three-point pattern is smaller than a certain tolerance.]

An alternative possibility for quadratic interpolation is to determine the minimum $\bar{a}$ of the quadratic polynomial that has the same value as g at the points 0 and a, and the same first derivative as g at 0. It can be verified that this minimum is given by

$$\bar{a} = \frac{g'(0)a^2}{2\big(g'(0)a + g(0) - g(a)\big)}.$$

C.3 THE GOLDEN SECTION METHOD

Here, we assume that $g(\alpha)$ is *strictly unimodal* in the interval $[0, s]$, as defined in Fig. C.2. The Golden Section method minimizes g over $[0, s]$ by determining at the kth iteration an interval $[\alpha_k, \bar{\alpha}_k]$ containing α^*. These intervals are obtained using the number

$$\tau = \frac{3 - \sqrt{5}}{2},$$

which satisfies $\tau = (1-\tau)^2$ and is related to the Fibonacci number sequence. The significance of this number will be seen shortly.

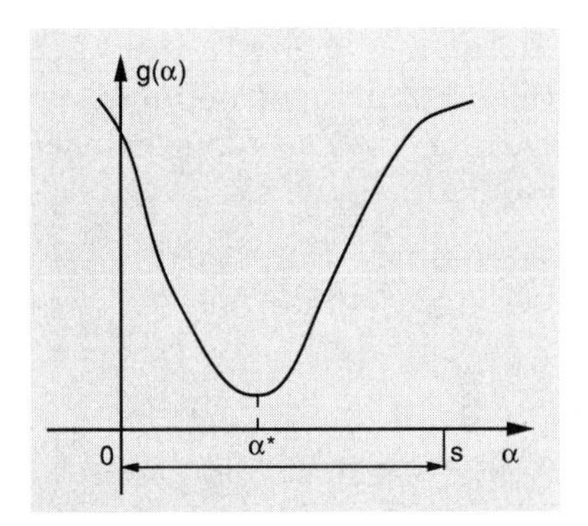

Figure C.2. A strictly unimodal function g over an interval $[0, s]$ is defined as a function that has a unique global minimum α^* in $[0, s]$ and if α_1, α_2 are two points in $[0, s]$ such that $\alpha_1 < \alpha_2 < \alpha^*$ or $\alpha^* < \alpha_1 < \alpha_2$, then $g(\alpha_1) > g(\alpha_2) > g(\alpha^*)$ or $g(\alpha^*) < g(\alpha_1) < g(\alpha_2)$, respectively. An example of a strictly unimodal function, is a function which is strictly convex over $[0, s]$.

Initially, we take

$$[\alpha_0, \bar{\alpha}_0] = [0, s].$$

Given $[\alpha_k, \bar{\alpha}_k]$, we determine $[\alpha_{k+1}, \bar{\alpha}_{k+1}]$ so that $\alpha^* \in [\alpha_{k+1}, \bar{\alpha}_{k+1}]$ as follows. We calculate

$$b_k = \alpha_k + \tau(\bar{\alpha}_k - \alpha_k)$$

$$\bar{b}_k = \bar{\alpha}_k - \tau(\bar{\alpha}_k - \alpha_k)$$

and $g(b_k)$, $g(\bar{b}_k)$. Then:

(1) If $g(b_k) < g(\bar{b}_k)$ we set

$$\alpha_{k+1} = \alpha_k, \quad \bar{\alpha}_{k+1} = b_k \qquad \text{if} \qquad g(\alpha_k) \leq g(b_k)$$

$$\alpha_{k+1} = \alpha_k, \quad \bar{\alpha}_{k+1} = \bar{b}_k \qquad \text{if} \qquad g(\alpha_k) > g(b_k).$$

(2) If $g(b_k) > g(\bar{b}_k)$ we set

$$\alpha_{k+1} = \bar{b}_k, \quad \bar{\alpha}_{k+1} = \bar{\alpha}_k \qquad \text{if} \qquad g(\bar{b}_k) \geq g(\bar{\alpha}_k)$$

$$\alpha_{k+1} = b_k, \quad \bar{\alpha}_{k+1} = \bar{a}_k \qquad \text{if} \qquad g(\bar{b}_k) < g(\alpha_k).$$

(3) If $g(b_k) = g(\bar{b}_k)$ we set

$$\alpha_{k+1} = b_k, \qquad \bar{\alpha}_{k+1} = \bar{b}_k.$$

Based on the definition of a strictly unimodal function it can be shown (see Fig. C.3) that the intervals $[\alpha_k, \bar{\alpha}_k]$ contain α^* and their lengths converge to zero. In practice, the computation is terminated once $(\bar{\alpha}_k - \alpha_k)$ becomes smaller than a prespecified tolerance.

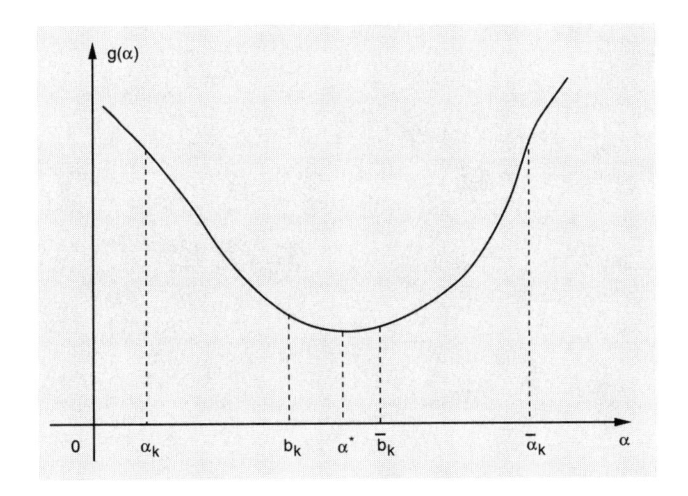

Figure C.3. Golden Section search. Given the interval $[\alpha_k, \bar{\alpha}_k]$ containing the minimum α^*, we calculate

$$b_k = \alpha_k + \tau(\bar{\alpha}_k - \alpha_k)$$

and

$$\bar{b}_k = \bar{\alpha}_k - \tau(\bar{\alpha}_k - \alpha_k).$$

The new interval $[\alpha_{k+1}, \bar{\alpha}_{k+1}]$ has either b_k or $\bar{b}_k$ as one of its endpoints.

An important fact, which rests on the choice of the particular number τ is that

$$[\alpha_{k+1}, \bar{\alpha}_{k+1}] = [\alpha_k, \bar{b}_k] \qquad \Longrightarrow \qquad \bar{b}_{k+1} = b_k,$$

$$[\alpha_{k+1}, \bar{\alpha}_{k+1}] = [b_k, \bar{\alpha}_k] \qquad \Longrightarrow \qquad b_{k+1} = \bar{b}_k.$$

In other words, a trial point b_k or $\bar{b}_k$ that is not used as the end point of the next interval continues to be a trial point for the next iteration. The reader can verify this, using the property

$$\tau = (1 - \tau)^2.$$

Thus, in either of the above situations, the values $\bar{b}_{k+1}$, $g(\bar{b}_{k+1})$ or b_{k+1}, $g(b_{k+1})$ are available and need not be recomputed at the next iteration, requiring a single function evaluation instead of two.

APPENDIX D: Implementation of Newton's Method

In this appendix we describe a globally convergent version of Newton's method based on the modified Cholesky factorization approach discussed in Section 1.4. A computer code implementing the method is available from the author on request.

D.1 CHOLESKY FACTORIZATION

We will give an algorithm for factoring a positive definite symmetric matrix A as

$$A = LL',$$

where L is lower triangular. This is the *Cholesky factorization*. Let a_{ij} be the elements of A and let A_i be the ith leading principal submatrix of A, that is, the submatrix

$$A_i = \begin{bmatrix} a_{11} & a_{12} & \cdots & a_{1i} \\ a_{21} & a_{22} & \cdots & a_{2i} \\ \vdots & \vdots & \ddots & \vdots \\ a_{i1} & a_{i2} & \cdots & a_{ii} \end{bmatrix}.$$

It is seen that this submatrix is positive definite, since for any $y \in \Re_i$, $y \neq 0$, we have by the positive definiteness of A

$$y'A_i y = [\,y' \quad 0\,]\, A \begin{bmatrix} y \\ 0 \end{bmatrix} > 0.$$

The factorization of A is obtained by successive factorization of $A_1, A_2, \ldots$. Indeed we have $A_1 = L_1 L_1'$, where $L_1 = [\sqrt{a_{11}}]$. Suppose we have the Cholesky factorization of A_{i-1},

$$A_{i-1} = L_{i-1} L_{i-1}'. \tag{D.1}$$

Let us write

$$A_i = \begin{bmatrix} A_{i-1} & \beta_i \\ \beta_i' & a_{ii} \end{bmatrix}, \tag{D.2}$$

where β_i is the column vector

$$\beta_i = \begin{bmatrix} a_{1i} \\ \vdots \\ a_{i-1,i} \end{bmatrix}. \tag{D.3}$$

Based on Eqs. (D.1)-(D.3), it can be verified that

$$A_i = L_i L_i',$$

where

$$L_i = \begin{bmatrix} L_{i-1} & 0 \\ l_i' & \lambda_{ii} \end{bmatrix}, \tag{D.4}$$

and

$$l_i = L_{i-1}^{-1} \beta_i, \qquad \lambda_{ii} = \sqrt{a_{ii} - l_i' l_i}. \tag{D.5}$$

The scalar λ_{ii} is well defined because it can be shown that $a_{ii} - l_i' l_i > 0$. This is seen by defining $b = A_{i-1}^{-1} \beta_i$, and by using the positive definiteness of A_i to write

$$\begin{aligned} 0 < [\,b' \quad -1\,]\, A_i \begin{bmatrix} b \\ -1 \end{bmatrix} &= b' A_{i-1} b - 2b' \beta_i + a_{ii} \\ &= b' \beta_i - 2b' \beta_i + a_{ii} = a_{ii} - b' \beta_i \\ &= a_{ii} - \beta_i' A_{i-1}^{-1} \beta_i = a_{ii} - \beta_i' (L_{i-1} L_{i-1}')^{-1} \beta_i \\ &= a_{ii} - (L_{i-1}^{-1} \beta_i)' (L_{i-1}^{-1} \beta_i) = a_{ii} - l_i' l_i. \end{aligned}$$

The preceding construction can also be used to show that the Cholesky factorization is unique among factorizations involving lower triangular matrices with positive elements along the diagonal. Indeed, A_1 has a unique such factorization, and if A_{i-1} has a unique factorization $A_{i-1} = L_{i-1} L_{i-1}'$, then L_i is uniquely determined from the requirement $A_i = L_i L_i'$ with the diagonal elements of L_i positive, and Eqs. (D.4) and (D.5).

Cholesky Factorization by Columns

In the preceding algorithm, we calculate L by rows, that is, we first calculate the first row of L, then the second row, etc. An alternative and equivalent method is to calculate L by columns, that is, first calculate the first column of L, then the second column, etc. To see how this can be done, we note that the first column of A is equal to the first column of L multiplied with l_{11}, that is,

$$a_{i1} = l_{11} l_{i1}, \qquad i = 1, \ldots, n,$$

from which we obtain

$$l_{11} = \sqrt{a_{11}},$$

$$l_{i1} = \frac{a_{i1}}{l_{11}}, \qquad i = 2, \ldots, n.$$

Similarly, given columns $1, 2, \ldots, j-1$ of L, we equate the elements of the jth column of A with the corresponding elements of LL' and we obtain the elements of the jth column of L as follows:

$$l_{jj} = \sqrt{a_{jj} - \sum_{m=1}^{j-1} l_{jm}^2},$$

$$l_{ij} = \frac{a_{ij} - \sum_{m=1}^{j-1} l_{jm} l_{im}}{l_{jj}}, \qquad i = j+1, \ldots, n.$$

D.2 APPLICATION TO A MODIFIED NEWTON METHOD

Consider now adding to A a diagonal correction E and simultaneously factoring the matrix

$$F = A + E,$$

where E is such that F is positive definite. The elements of E are introduced sequentially during the factorization process as some diagonal elements of the triangular factor are discovered, which are either negative or are close to zero, indicating that A is either not positive definite or is nearly singular. As discussed in Section 1.4, this is a principal method by which Newton's method is modified to enhance its global convergence properties. The precise mechanization is as follows:

We first fix positive scalars μ_1 and μ_2, where $\mu_1 < \mu_2$. We calculate the first column of the triangular factor L of F by

$$l_{11} = \begin{cases} \sqrt{a_{11}} & \text{if } \mu_1 < a_{11}, \\ \sqrt{\mu_2} & \text{otherwise}, \end{cases}$$

$$l_{i1} = \frac{a_{i1}}{l_{11}}, \qquad i = 2, \ldots, n.$$

Similarly, given columns $1, 2, \ldots, j-1$ of L, we obtain the elements of the jth column from the equations

$$l_{jj} = \begin{cases} \sqrt{a_{jj} - \sum_{m=1}^{j-1} l_{jm}^2} & \text{if } \mu_1 < a_{11} - \sum_{m=1}^{j-1} l_{jm}^2, \\ \sqrt{\mu_2} & \text{otherwise,} \end{cases}$$

$$l_{ij} = \frac{a_{ij} - \sum_{m=1}^{j-1} l_{jm} l_{im}}{l_{jj}}, \qquad i = j+1, \ldots, n.$$

In words, if the diagonal element of LL' comes out less than μ_1, we bring it up to μ_2.

Note that the jth diagonal element of the correction matrix E is equal to zero if $\mu_1 < a_{jj} - \sum_{m=1}^{j-1} l_{jm}^2$ and is equal to

$$\mu_2 - \left(a_{jj} - \sum_{m=1}^{j-1} l_{jm}^2 \right)$$

otherwise.

The preceding scheme can be used to modify Newton's method, where at the kth iteration, we add a diagonal correction Δ^k to the Hessian $\nabla^2 f(x^k)$ and simultaneously obtain the Cholesky factorization $L^k L^{k\prime}$ of $\nabla^2 f(x^k) + \Delta^k$ as described above. A modified Newton direction d^k is then obtained by first solving the triangular system

$$L^k y = -\nabla f(x^k),$$

and then solving the triangular system

$$L^{k\prime} d^k = y.$$

Solving the first system is called *forward elimination* and is accomplished in $O(n^2)$ arithmetic operations using the equations

$$y_1 = -\frac{\partial f(x^k)/\partial x_1}{l_{11}},$$

$$y_i = -\frac{\partial f(x^k)/\partial x_i + \sum_{m=1}^{i-1} l_{im} y^m}{l_{ii}}, \qquad i = 2, \ldots, n,$$

where l_{im} is the imth element of L^k. Solving the second system is called *back substitution* and is accomplished again in $O(n^2)$ arithmetic operations using the equations

$$d^n = \frac{y^n}{l_{nn}},$$

$$d_i = \frac{y_i - \sum_{m=i+1}^{n} l_{mi} d^m}{l_{ii}}, \qquad i = 1, \ldots, n-1.$$

The next point x^{k+1} is obtained from

$$x^{k+1} = x^k + \alpha^k d^k,$$

where α^k is chosen by the Armijo rule with unity initial step whenever the Hessian is not modified ($\Delta^k = 0$) and by means of a line minimization otherwise.

Assuming fixed values of μ_1 and μ_2, the following may be verified for the modified Newton's method just described:

(a) The algorithm is globally convergent in the sense that every limit point of $\{x^k\}$ is a stationary point of f. This can be shown using Prop. 1.2.1 in Section 1.2.

(b) For each local minimum x^* with positive definite Hessian, there exist scalars $\mu > 0$ and $\epsilon > 0$ such that if $\mu_1 \leq \mu$ and $\|x^0 - x^*\| \leq \epsilon$, then $x^k \to x^*$, $\Delta^k = 0$, and $\alpha^k = 1$ for all k. In other words if μ_1 is not chosen too large, the Hessian will never be modified near x^*, the method will be reduced to the pure form of Newton's method, and the convergence to x^* will be superlinear. The theoretical requirement that μ_1 be sufficiently small can be eliminated by making μ_1 dependent on the norm of the gradient (e.g. $\mu_1 = c\|\nabla f(x^k)\|$, where c is some positive scalar).

Practical Choice of Parameters and Stepsize Selection

We now address some practical issues. As discussed earlier, one should try to choose μ_1 small in order to avoid detrimental modification of the Hessian. Some trial and error with one's particular problem may be required here. As a practical matter, we recommend choosing initially $\mu_1 = 0$ and increasing μ_1 only if difficulties arise due to roundoff error or extremely large norm of calculated direction. (Choosing $\mu_1 = 0$, runs counter to our convergence theory because the generated directions are not guaranteed to be gradient related, but the practical consequences of this are typically insignificant.)

The parameter μ_2 should generally be chosen considerably larger than μ_1. It can be seen that choosing μ_2 very small can make the modified Hessian matrix $L^k L^{k'}$ nearly singular. On the other hand, choosing μ_2 very large has the effect of making nearly zero the coordinates of d^k that correspond to nonzero diagonal elements of the correction matrix Δ^k. Generally, some trial and error is necessary to determine a proper value of μ_2. A good guideline is to try a relatively small value of μ_2 and to increase μ_2 if the stepsize generated by the line minimization algorithm is substantially smaller than unity. The idea here is that small values of μ_2 tend to produce directions d^k with large value of norm and hence small values of stepsize. Thus a small value of stepsize indicates that μ_2 is chosen smaller than appropriate, and suggests that an increase of μ_2 is desirable. It is also possible to construct along these lines an adaptive scheme that changes the values of μ_1 and μ_2 in the course of the algorithm.

The following scheme to set and adjust μ_1 and μ_2 has worked well for the author. At each iteration k, we determine the maximal absolute diagonal element of the Hessian, that is,

$$w^k = \max\left\{\left|\frac{\partial^2 f(x^k)}{(x_1)^2}\right|, \ldots, \left|\frac{\partial^2 f(x^k)}{(x_n)^2}\right|\right\},$$

and we set μ_1 and μ_2 to

$$\mu_1 = r_1 w^k, \qquad \mu_2 = r_2 w^k.$$

The scalar r_1 is set at some "small" (or zero) value. The scalar r_2 is changed each time the Hessian is modified; it is multiplied by 5 if the stepsize obtained by the minimization rule is less than 0.2, and it is divided by 5 each time the stepsize is larger than 0.9.

Finally, regarding stepsize selection, any of a large number of possible line minimization algorithms can be used for those iterations where the Hessian is modified (in other iterations the Armijo rule with unity initial stepsize is used). One possibility is to use quadratic interpolation based on function values; see Section C.2 in Appendix C.

It is worth noting that if the cost function is quadratic, then it can be shown that a unity stepsize results in cost reduction for any values of μ_1 and μ_2. In other words if f is quadratic (not necessarily positive definite), we have

$$f\big(x^k - (F^k)^{-1}\nabla f(x^k)\big) \leq f(x^k),$$

where $F^k = \nabla^2 f(x^k) + \Delta^k$ and Δ^k is any positive definite matrix such that F^k is positive definite. As a result, a stepsize near unity is appropriate for initiating the line minimization algorithm. This fact can be used to guide the implementation of the line minimization routine.

References

[AHR97] Auslender, A., Cominetti, R., and Haddou, M., 1997. "Asymptotic Analysis for Penalty and Barrier Methods in Convex and Linear Programming, Math. Operations Res., Vol. 22, pp. 43-62.

[AHR93] Anstreicher, K. M., den Hertog, D., Roos, C., and Terlaky, T., 1993. "A Long Step Barrier Method for Convex Quadratic Programming," Algorithmica, Vol. 10, pp. 365-382.

[AHU58] Arrow, K. J., Hurwicz, L., and Uzawa, H., (Eds.), 1958. Studies in Linear and Nonlinear Programming, Stanford Univ. Press, Stanford, CA.

[AHU61] Arrow, K. J., Hurwicz, L., and Uzawa, H., 1961. "Constraint Qualifications in Maximization Problems," Naval Research Logistics Quarterly, Vol. 8, pp. 175-191.

[AaL97] Aarts, E., and Lenstra, J. K., 1997. Local Search in Combinatorial Optimization, Wiley, N. Y.

[Aba67] Abadie, J., 1967. "On the Kuhn-Tucker Theorem," in Nonlinear Programming, Abadie, J., (Ed.), North Holland, Amsterdam.

[AlK90] Al-Khayyal, F., and Kyparisis, J., 1990. "Finite Convergence of Algorithms for Nonlinear Programs and Variational Inequalities," J. Opt. Theory and Appl., Vol. 70, pp. 319-332.

[Ali92] Alizadeh, F., 1992. "Optimization over the Positive-Definite Cone: Interior Point Methods and Combinatorial Applications," in Pardalos, P., (Ed.), Advances in Optimization and Parallel Computing, North Holland, Amsterdam.

[Ali95] Alizadeh, F., 1995. "Interior-Point Methods in Semidefinite Programming with Applications in Combinatorial Applications," SIAM J. on Optimization, Vol. 5, pp. 13-51.

[AnV94] Anstreicher, K. M., and Vial, J.-P., 1994. "On the Convergence of an Infeasible Primal-Dual Interior-Point Method for Convex Programming," Optimization Methods and Software, Vol. 3, pp. 273-283.

[Arm66] Armijo, L., 1966. "Minimization of Functions Having Continuous Partial Derivatives," Pacific J. Math., Vol. 16, pp. 1-3.

[Ash72] Ash, R. B., 1972. Real Analysis and Probability, Academic Press, N. Y.

[AtV95] Atkinson, D. S., and Vaidya, P. M., 1995. "A Cutting Plane Algorithm for Convex Programming that Uses Analytic Centers," Math. Programming, Vol. 69, pp. 1-44.

[AuC90] Auslender, A., and Cominetti, R., 1990. "First and Second Order Sensitivity Conditions," Optimization, Vol. 21, pp. 1-13.

[AuE76] Aubin, J. P., and Ekeland, I., 1976. "Estimates of the Duality Gap in Nonconvex Optimization," Math. Operations Res., Vol. 1, pp. 225-245.

[Aus76] Auslender, A., 1976. Optimization: Methodes Numeriques, Mason, Paris.

[Aus92] Auslender, A., 1992. "Asymptotic Properties of the Fenchel Dual Functional and Applications to Decomposition Properties," Vol. 73, pp. 427-449.

[Avr76] Avriel, M., 1976. Nonlinear Programming: Analysis and Methods, Prentice-Hall, Englewood Cliffs, N. J.

[BCS99] Bonnans, J. F., Cominetti, R., and Shapiro, A., 1999. "Second Order Optimality Conditions Based on Parabolic Second Order Tangent Sets," SIAM J. on Optimization, Vol. 9, pp. 466-492.

[BJS90] Bazaraa, M. S., Jarvis, J. J., and Sherali, H. D., 1990. Linear Programming and Network Flows, (2nd Ed.), Wiley, N. Y.

[BGG84] Bertsekas, D. P., Gafni, E. M., and Gallager, R. G., 1984. "Second Derivative Algorithms for Minimum Delay Distributed Routing in Networks," IEEE Trans. on Communications, Vol. 32, pp. 911-919.

[BGL95] Bonnans, J. F., Gilbert, J. C., Lemaréchal, C., Sagastizábal, S. C., 1995. "A Family of Variable Metric Proximal Methods," Math. Programming, Vol. 68, pp. 15-47.

[BGS72] Bazaraa, M. S., Goode, J. J., and Shetty, C. M., 1972. "Constraint Qualifications Revisited," Management Science, Vol. 18, pp. 567-573.

[BGT81] Bland, R. G., Goldfarb, D., and Todd, M. J., 1981. "The Ellipsoid Method: A Survey," Operations Research, Vol. 29, pp. 1039-91.

[BHT87] Bertsekas, D. P., Hossein, P., and Tseng, P., 1987. "Relaxation Methods for Network Flow Problems with Convex Arc Costs," SIAM J. on Control and Optimization, Vol. 25, pp. 1219-1243.

[BLS83] Bertsekas, D. P., Lauer, G. S., Sandell, N. R., Jr., and Posberg, T. A., 1983. "Optimal Short Term Scheduling of Large-Scale Power Systems," IEEE Trans. Automat. Control, Vol. AC-28, pp. 1-11.

[BMM95a] Ball, M. O., Magnanti, T. L., Monma, C. L., and Nemhauser,

G. L., 1995. Network Models, Handbooks in OR and MS, Vol. 7, North-Holland, Amsterdam.

[BMM95b] Ball, M. O., Magnanti, T. L., Monma, C. L., and Nemhauser, G. L., 1995. Network Routing, Handbooks in OR and MS, Vol. 8, North-Holland, Amsterdam.

[BMS99] Boltyanski, V., Martini, H., and Soltan, V., 1999. Geometric Methods and Optimization Problems, Kluwer, Boston.

[BMT90] Burke, J. V., Moré, J. J., and Toraldo, G., 1990. "Convergence Properties of Trust Region Methods for Linear and Convex Constraints," Math. Programming, Vol. 47, pp. 305-336.

[BPT92] Bonnans, J. F., Panier, E. R., Tits, A. L., and Zhou, J. L., 1992. "Avoiding the Maratos Effect by Means of a Nonmonotone Line Search II. Inequality Constrained Problems – Feasible Iterates," SIAM J. Numer. Anal., Vol. 29, pp. 1187-1202.

[BPT97a] Bertsekas, D. P., Polymenakos, L. C., and Tseng, P., 1997. "An ϵ-Relaxation Method for Separable Convex Cost Network Flow Problems," SIAM J. on Optimization, Vol. 7, pp. 853-870.

[BPT97b] Bertsekas, D. P., Polymenakos, L. C., and Tseng, P., 1997. "Epsilon-Relaxation and Auction Methods for Separable Convex Cost Network Flow Problems," in Network Optimization, Pardalos, P. M., Hearn, D. W., and Hager, W. W., (Eds.), Lecture Notes in Economics and Mathematical Systems, Springer-Verlag, N. Y., pp. 103-126.

[BSS93] Bazaraa, M. S., Sherali, H. D., and Shetty, C. M., 1993. Nonlinear Programming Theory and Algorithms, (2nd Ed.), Wiley, N. Y.

[BTW82] Boggs, P. T., Tolle, J. W., and Wang, P., 1982. "On the Local Convergence of Quasi-Newton Methods for Constrained Optimization," SIAM J. on Control and Optimization, Vol. 20, pp. 161-171.

[BaB88] Barzilai, J., and Borwein, J. M., 1988. "Two-Point Step Size Gradient Methods," IMA J. Numerical Analysis, Vol. 8, pp. 141-148.

[BaD93] Barzilai, J., and Dempster, M. A. H., 1993. "Measuring Rates of Convergence of Numerical Algorithms," J. Opt. Theory and Appl., Vol. 78, pp. 109-125.

[BaL89] Bayer, D. A., and Lagarias, J. C., 1989. "The Nonlinear Geometry of Linear Programming. I. Affine and Projective Scaling Trajectories. II. Legendre Transform Coordinates and Central Trajectories. III. Projective Legendre Transform Coordinates and Hilbert Geometry," Trans. Amer. Math. Soc., Vol. 314, pp. 499-581.

[BaT85] Balas, E., and Toth, P., 1985. "Branch and Bound Methods," in The Traveling Salesman Problem, Lawler, E., Lenstra, J. K., Rinnoy Kan, A. H. G., and Shmoys, D. B., (Eds.), Wiley, N. Y., pp. 361-401.

[BaW75] Balinski, M., and Wolfe, P., (Eds.), 1975. Nondifferentiable Optimization, Math. Programming Study 3, North-Holland, Amsterdam.

[Bar86] Barnes, E. R., 1986. "A Variation on Karmarkar's Algorithm for Solving Linear Programming Problems," Math. Programming, Vol. 36, pp. 174-182.

[BeE88] Bertsekas, D. P., and Eckstein, J., 1988. "Dual Coordinate Step Methods for Linear Network Flow Problems," Math. Programming, Vol. 42, pp. 203-243.

[BeG82] Bertsekas, D. P., and Gafni, E., 1982. "Projection Methods for Variational Inequalities with Application to the Traffic Assignment Problem," Math. Programming Studies, Vol. 17, pp. 139-159.

[BeG83] Bertsekas, D. P., and Gafni, E., 1983. "Projected Newton Methods and Optimization of Multicommodity Flows," IEEE Trans. Automat. Control, Vol. AC-28, pp. 1090-1096.

[BeG92] Bertsekas, D. P., and Gallager, R. G., 1992. Data Networks, (2nd Ed.), Prentice-Hall, Englewood Cliffs, N. J.

[BeM71] Bertsekas, D. P, and Mitter, S. K., 1971. "Steepest Descent for Optimization Problems with Nondifferentiable Cost Functionals," Proc. 5th Annual Princeton Confer. Inform. Sci. Systems, Princeton, N. J., pp. 347-351.

[BeM73] Bertsekas, D. P., and Mitter, S. K., 1973. "A Descent Numerical Method for Optimization Problems with Nondifferentiable Cost Functionals," SIAM J. on Control, Vol. 11, pp. 637-652.

[BeT88] Bertsekas, D. P., and Tseng, P., 1988. "Relaxation Methods for Minimum Cost Ordinary and Generalized Network Flow Problems," Operations Research, Vol. 36, pp. 93-114.

[BeT89] Bertsekas, D. P., and Tsitsiklis, J. N., 1989. Parallel and Distributed Computation: Numerical Methods, Prentice-Hall, Englewood Cliffs, N. J; republished by Athena Scientific, Belmont, MA, 1997.

[BeT94] Bertsekas, D. P., and Tseng, P., 1994. "Partial Proximal Minimization Algorithms for Convex Programming," SIAM J. on Optimization, Vol. 4, pp. 551-572.

[BeT97] Bertsimas, D., and Tsitsiklis, J. N., 1997. Introduction to Linear Optimization, Athena Scientific, Belmont, MA.

[BeT99] Bertsekas, D. P., and Tsitsiklis, J. N., 1999. "Gradient Convergence of Gradient Methods with Errors," Report LIDS-P-2404, Lab. for Info. and Dec. Systems, M.I.T.; to appear in SIAM J. on Optimization.

[BeZ82] Ben-Tal, A., and Zowe, J., 1982. "A Unified Theory of First and Second-Order Conditions for Extremum Problems in Topological Vector

Spaces," Math. Programming Studies, Vol. 19, pp. 39-76.

[BeZ97] Ben-Tal, A., and Zibulevsky, M., 1997. "Penalty/Barrier Multiplier Methods for Convex Programming Problems," SIAM J. on Optimization, Vol. 7, pp. 347-366.

[Ben62] Benders, J. F., 1962. "Partitioning Procedures for Solving Mixed Variables Programming Problems," Numer. Math., Vol. 4, pp. 238-252.

[Ben80] Ben-Tal, A., 1980. "Second-Order and Related Extremality Conditions in Nonlinear Programming," J. Opt. Th. and Appl., Vol. 31, pp. 143-165.

[Ber74] Bertsekas, D. P., 1974. "Partial Conjugate Gradient Methods for a Class of Optimal Control Problems," IEEE Trans. Automat. Control, Vol. 19, pp. 209-217.

[Ber75a] Bertsekas, D. P., 1975. "Necessary and Sufficient Conditions for a Penalty Method to be Exact," Math. Programming, Vol. 9, pp. 87-99.

[Ber75b] Bertsekas, D. P., 1975. "Combined Primal-Dual and Penalty Methods for Constrained Optimization, SIAM J. on Control, Vol. 13, pp. 521-544.

[Ber75c] Bertsekas, D. P., 1975. "Nondifferentiable Optimization Via Approximation," Math. Programming Study 3, Balinski, M., and Wolfe, P., (Eds.), North-Holland, Amsterdam, pp. 1-25.

[Ber76a] Bertsekas, D. P., 1976. "On Penalty and Multiplier Methods for Constrained Optimization," SIAM J. on Control and Optimization, Vol. 14, pp. 216-235.

[Ber76b] Bertsekas, D. P., 1976. "Multiplier Methods: A Survey," Automatica, Vol. 12, pp. 133-145.

[Ber76c] Bertsekas, D. P., 1976. "On the Goldstein-Levitin-Poljak Gradient Projection Method," IEEE Trans. Automat. Control, Vol. 21, pp. 174-184.

[Ber77] Bertsekas, D. P., 1977. "Approximation Procedures Based on the Method of Multipliers," J. Opt. Th. and Appl., Vol. 23, pp. 487-510.

[Ber79a] Bertsekas, D. P., 1979. "Convexification Procedures and Decomposition Algorithms for Large-Scale Nonconvex Optimization Problems, J. Opt. Th. and Appl., Vol. 29, pp. 169-197.

[Ber79b] Bertsekas, D. P., 1979. "A Distributed Algorithm for the Assignment Problem," Lab. for Information and Decision Systems Working Paper, M.I.T.

[Ber80] Bertsekas, D. P., 1980. "Variable Metric Methods for Constrained Optimization Based on Differentiable Exact Penalty Functions," Proc. Allerton Conference on Communication, Control, and Computation, Allerton Park, Ill., pp. 584-593.

[Ber81] Bertsekas, D. P., 1981. "A New Algorithm for the Assignment Problem," Math. Programming, Vol. 21, pp. 152-171.

[Ber82a] Bertsekas, D. P., 1982. Constrained Optimization and Lagrange Multiplier Methods, Academic Press, N. Y; republished by Athena Scientific, Belmont, MA, 1997.

[Ber82b] Bertsekas, D. P., 1982. "Projected Newton Methods for Optimization Problems with Simple Constraints," SIAM J. on Control and Optimization, Vol. 20, pp. 221-246.

[Ber82c] Bertsekas, D. P., 1982. "Enlarging the Region of Convergence of Newton's Method for Constrained Optimization," J. Opt. Th. and Appl., Vol. 36, pp. 221-252.

[Ber85] Bertsekas, D. P., 1985. "A Unified Framework for Minimum Cost Network Flow Problems," Math. Programming, Vol. 32, pp. 125-145.

[Ber86] Bertsekas, D. P., 1986. "Distributed Relaxation Methods for Linear Network Flow Problems," Proceedings of 25th IEEE Conference on Decision and Control, pp. 2101-2106.

[Ber91] Bertsekas, D. P., 1991. Linear Network Optimization: Algorithms and Codes, M.I.T. Press, Cambridge, MA.

[Ber92] Bertsekas, D. P., 1992. "Auction Algorithms for Network Problems: A Tutorial Introduction," Computational Optimization and Applications, Vol. 1, pp. 7-66.

[Ber93] D. P. Bertsekas, 1993. "Thevenin Decomposition and Network Optimization," Report LIDS-P-2204, Lab. for Info. and Dec. Systems, M.I.T.; J. Opt. Theory and Appl., Vol. 89, 1996, pp. 1-15.

[Ber94] Bertsekas, D. P., 1994. "Incremental Least Squares Methods and the Extended Kalman Filter," Report LIDS-P-2237, Lab. for Info. and Dec. Systems, M.I.T.; SIAM J. on Optimization, Vol. 6, 1996, pp. 807-822.

[Ber95a] Bertsekas, D. P., 1995. Dynamic Programming and Optimal Control, Vols. I and II, Athena Scientific, Belmont, MA.

[Ber95b] Bertsekas, D. P., 1995. "A New Class of Incremental Gradient Methods for Least Squares Problems," Report LIDS-P-2257, Lab. for Info. and Dec. Systems, M.I.T.; SIAM J. on Optimization, Vol. 7, 1997, pp. 913-926.

[Ber97] Bertsekas, D. P., 1997. "A Note on Error Bounds for Convex and Nonconvex Programs," Report LIDS-P-2393, Lab. for Info. and Dec. Systems, M.I.T.; to appear in Computational Optimization and Applications.

[Ber98] Bertsekas, D. P., 1998. Network Optimization: Continuous and Discrete Models, Athena Scientific, Belmont, MA.

[BiL97] Birge, J. R., and Louveaux, 1997. Introduction to Stochastic Programming, Springer-Verlag, New York, N. Y.

[Bis95] Bishop, C. M, 1995. Neural Networks for Pattern Recognition, Oxford University Press, N. Y.

[BoS98] Bonnans, J. F., and Shapiro, A., 1998. "Optimization Problems with Perturbations: A Guided Tour," SIAM Review, Vol. 40, pp. 228-264.

[BoT80] Boggs, P. T., and Tolle, J. W., 1980. "Augmented Lagrangians which are Quadratic in the Multiplier," J. Opt. Th. and Appl., Vol. 31, pp. 17-26.

[Bog90] Bogart, K. P., 1990. Introductory Combinatorics, Harcourt Brace Jovanovich, Inc., New York, N. Y.

[Bon89a] Bonnans, J. F., 1989. "A Variant of a Projected Variable Metric Method for Bound Constrained Optimization Problems," Report, INRIA, France.

[Bon89b] Bonnans, J. F., 1989. "Asymptotic Admissibility of the Unit Stepsize in Exact Penalty Methods," SIAM J. on Control and Optimization, Vol. 27, pp. 631-641.

[Bon92] Bonnans, J. F., 1992. "Directional Derivatives of Optimal Solutions in Smooth Nonlinear Programming," J. Opt. Theory and Appl., Vol. 73, pp. 27-45.

[Bon94] Bonnans, J. F., 1994. "Local Analysis of Newton Type Methods for Variational Inequalities and Nonlinear Programming," J. Applied Math. Optimization, Vol. 29, pp. 161-186

[Bro70] Broyden, C. G., 1970. "The Convergence of a Class of Double Rank Minimization Algorithms," J. Inst. Math. Appl., Vol. 6, pp. 76-90.

[BuM88] Burke, J. V., and Moré, J. J., 1988. "On the Identification of Active Constraints," SIAM J. Numer. Anal., Vol. 25, pp. 1197-1211.

[BuQ98] Burke, J. V., and Qian, M., 1998. "A Variable Metric Proximal Point Algorithm for Monotone Operators," SIAM J. on Control and Optimization, Vol. 37, pp. 353-375.

[CCP70] Canon, M. D., Cullum, C. D., and Polak, E., 1970. Theory of Optimal Control and Mathematical Programming, McGraw-Hill, N. Y.

[CCP98] Cook, W., Cunningham, W., Pulleyblank, W., and Schrijver, A., 1998. Combinatorial Optimization, Wiley, N. Y.

[CFM75] Camerini, P. M., Fratta, L., and Maffioli, F., 1975. "On Improving Relaxation Methods by Modified Gradient Techniques," Math. Programming Studies, Vol. 3, pp. 26-34.

[CGT91] Conn, A. R., Gould, N. I. M., and Toint, P. L., 1991. "A Globally Convergent Augmented Lagrangian Algorithm for Optimization with General Constraints and Simple Bounds, SIAM J. Numer. Anal., Vol. 28, pp. 545-572.

[CPS92] Cottle, R., Pang, J. S., and Stone, R. E., 1992. The Linear Complementarity Problem, Academic Press, Boston.

[CaC68] Canon, M. D., and Cullum, C. D., 1968. "A Tight Upper Bound on the Rate of Convergence of the Frank-Wolfe Algorithm," SIAM J. on Control, Vol. 6, pp. 509-516.

[CaF97] Caprara, A., and Fischetti, M., 1997. "Branch and Cut Algorithms," in Annotated Bibliographies in Combinatorial Optimization, Dell'-Amico, M., Maffioli, F., and Martello, S., (Eds.), Wiley, Chisester, Ch. 4.

[CaM87] Calamai, P. H., and Moré, J. J., 1987. "Projected Gradient Methods for Linearly Constrained Problems," Math. Programming, Vol. 39, pp. 98-116.

[Car61] Carroll, C. W., 1961. "The Created Response Surface Technique for Optimizing Nonlinear Restrained Systems," Operations Research, Vol. 9, pp. 169-184.

[Cau47] Cauchy, M. A., 1847. "Analyse Mathématique–Méthode Générale Pour La Résolution des Systémes d' Eq́uations Simultanées," Comptes Vendus Acad. Sc., Paris.

[CeH87] Censor, Y., and Herman, G. T., 1987. "On Some Optimization Techniques in Image Reconstruction from Projections," Applied Numer. Math., Vol. 3, pp. 365-391.

[CeZ92] Censor, Y., and Zenios, S. A., 1992. "The Proximal Minimization Algorithm with D-Functions," J. Opt. Theory and Appl., Vol. 73, pp. 451-464.

[CeZ97] Censor, Y., and Zenios, S. A., 1997. Parallel Optimization: Theory, Algorithms, and Applications, Oxford University Press, N. Y.

[ChG59] Cheney, E. W., and Goldstein, A. A., 1959. "Newton's Method for Convex Programming and Tchebycheff Approximation," Numer. Math., Vol. I, pp. 253-268.

[ChT93] Chen, G., and Teboulle, M., 1993. "Convergence Analysis of a Proximal-Like Minimization Algorithm Using Bregman Functions," SIAM J. on Optimization, Vol. 3, pp. 538-543.

[ChT94] Chen, G., and Teboulle, M., 1994. "A Proximal-Based Decomposition Method for Convex Minimization Problems," Math. Programming, Vol. 64, pp. 81-101.

[Cla83] Clarke, F. H., 1983. Nonsmooth Analysis and Optimization, Wiley-Interscience, N. Y.

[CoC82a] Coleman, T. F., and Conn, A. R., 1982. "Nonlinear Programming Via an Exact Penalty Function: Asymptotic Analysis," Math. Progamming, Vol. 24, pp. 123-136.

[CoC82b] Coleman, T. F., and Conn, A. R., 1982. "Nonlinear Programming Via an Exact Penalty Function: Global Analysis," Math. Programming, Vol. 24, pp. 137-161.

[CoL94] Correa, R., and Lemarechal, C., 1994. "Convergence of Some Algorithms for Convex Minimization," Math. Programming, Vol. 62, pp. 261-276.

[Coh80] Cohen, G., 1980. "Auxiliary Problem Principle and Decomposition of Optimization Problems," J. Opt. Theory and Appl., Vol. 32, pp. 277-305.

[Cro58] Croes, G. A., 1958. "A Method for Solving Traveling Salesman Problems," Operations Research, Vol. 6, pp. 791-812.

[Cry71] Cryer, C. W., 1971. "The Solution of a Quadratic Programming Problem Using Systematic Overrelaxation," SIAM J. on Control, Vol. 9, pp. 385-392.

[Cul71] Cullum, J., 1971. "An Explicit Procedure for Discretizing Continuous Optimal Control Problems," J. Opt. Theory and Appl., Vol. 8, pp. 15-34.

[Cyb89] Cybenko, 1989. "Approximation by Superpositions of a Sigmoidal Function," Math. of Control, Signals, and Systems, Vol. 2, pp. 303-314.

[DFJ54] Dantzig, G. B., Fulkerson, D. R., and Johnson, S. M., 1954. "Solution of a Large-Scale Traveling-Salesman Problem," Operations Research, Vol. 2, pp. 393-410.

[DES82] Dembo, R. S., Eisenstadt, S. C., and Steihaug, T., 1982. "Inexact Newton Methods," SIAM J. Numer. Anal., Vol. 19, pp. 400-408.

[DKK83] Decker, D. W., Keller, H. B., and Kelley, C. T., 1983. "Convergence Rates for Newton's Method at Singular Points," SIAM J. Numer. Anal., Vol. 20, pp. 296-314.

[DaW60] Dantzig, G. B., and Wolfe, P., 1960. "Decomposition Principle for Linear Programs," Operations Research, Vol. 8, pp. 101-111.

[Dan71] Daniel, J. W., 1971. The Aproximate Minimization of Functionals, Prentice-Hall, Englewood Cliffs, N. J.

[Dav59] Davidon, W. C., 1959. "Variable Metric Method for Minimization," Argonne National Lab., Report ANL-5990 (Rev.), Argonne, Ill. Reprinted with a new preface in SIAM J. on Optimization, Vol. 1, 1991, pp. 1-17.

[Dav76] Davidon, W. C., 1976. "New Least Squares Algorithms," J. Opt. Theory and Appl., Vol. 18, pp. 187-197.

[DeK80] Decker, D. W., and Kelley, C. T., 1980. "Newton's Method at Singular Points, Parts I and II," SIAM J. Numer. Anal., Vol. 17, pp. 66-70, 465-471.

[DeM77] Dennis, J. E., and Moré, J. J., 1977. "Quasi-Newton Methods: Motivation and Theory," SIAM Review, Vol. 19, pp. 46-89.

[DeR70] Demjanov, V. F., and Rubinov, A. M., 1970. Approximate Methods in Optimization Problems, American Elsevier, N. Y.

[DeS83] Dennis, J. E., and Schnabel, R. E., 1983. Numerical Methods for Unconstrained Optimization and Nonlinear Equations, Prentice-Hall, Englewood Cliffs, N. J.

[DeT83] Dembo, R. S., and Tulowitzki, U., 1983. "On the Minimization of Quadratic Functions Subject to Box Constraints," Working Paper Series B No. 71, School of Organization and Management, Yale Univ., New Haven, Conn.

[DeT91] Dennis, J. E., and Torczon, V., 1991. "Direct Search Methods on Parallel Machines," SIAM J. on Optimization, Vol. 1, pp. 448-474.

[DeT93] De Angelis, P. L., and Toraldo, G., 1993. "On the Identification Property of a Projected Gradient Method," SIAM J. Numer. Anal., Vol. 30, pp. 1483-1497.

[DeV85] Demjanov, V. F., and Vasilév, L. V., 1985. Nondifferentiable Optimization, Optimization Software, N. Y.

[Dem66] Demjanov, V. F., 1966. "The Solution of Several Minimax Problems," Kibernetika, Vol. 2, pp. 58-66.

[Dem68] Demjanov, V. F., 1968. "Algorithms for Some Minimax Problems," J. of Computer and Systems Science, Vol. 2, pp. 342-380.

[DiG79] DiPillo, G., and Grippo, L., 1979. "A New Class of Augmented Lagrangians in Nonlinear Programming," SIAM J. on Control and Optimization, Vol. 17, pp. 618-628.

[DiG89] DiPillo, G., and Grippo, L., 1989. "Exact Penalty Functions in Constrained Optimization," SIAM J. on Control and Optimization, Vol. 27, pp. 1333-1360.

[Dik67] Dikin, I. I, 1967. "Iterative Solution of Problems of Linear and Quadratic Programming," Soviet Math. Doklady, Vol. 8, pp. 674-675.

[Dix72a] Dixon, L. C. W., 1972. "Quasi-Newton Algorithms Generate Identical Points," Math. Programming, Vol. 2, pp. 383-387.

[Dix72b] Dixon, L. C. W., 1972. "Quasi-Newton Algorithms Generate Identical Points. II. The Proofs of Four New Theorems," Math. Programming, Vol. 3, pp. 345-358.

[DoT86] Dontchev, A. L., and Tongen, H. Th., 1986. "On the Regularity of the Kuhn-Tucker Curve," SIAM J. on Control and Optimization, Vol. 24, pp. 169-176.

[DoT98] Doljansky, M., and Teboulle, M., 1998. "An Interior Proximal Algorithm and the Exponential Multiplier Method for Semidefinite Programming," SIAM J. on Optimization, Vol. 9, pp. 1-13.

[DuB89] Dunn, J. C., and Bertsekas, D. P., 1989. "Efficient Dynamic Programming Implementations of Newton's Method for Unconstrained Optimal Control Problems," J. Opt. Theory and Appl., Vol. 63, pp. 23-38.

[DuM65] Dubovitskii, M. D., and Milyutin, A. A., 1965. "Extremum Problems in the Presence of Restriction," USSR Comp. Math. and Math. Phys., Vol. 5, pp. 1-80.

[DuS83] Dunn, J. C., and Sachs, E., 1983. "The Effect of Perturbations on the Convergence Rates of Optimization Algorithms," Appl. Math. Optim., Vol. 10, pp. 143-157.

[DuZ89] Du, D.-Z., and Zhang, X.-S., 1989. "Global Convergence of Rosen's Gradient Projection Method," Math. Programming, Vol. 44, pp. 357-366.

[Dun79] Dunn, J. C., 1979. "Rates of Convergence for Conditional Gradient Algorithms Near Singular and Nonsingular Extremals," SIAM J. on Control and Optimization, Vol. 17, pp. 187-211.

[Dun80a] Dunn, J. C., 1980. "Convergence Rates for Conditional Gradient Sequences Generated by Implicit Step Length Rules," SIAM J. on Control and Optimization, Vol. 18, pp. 473-487.

[Dun80b] Dunn, J. C., 1980. "Newton's Method and the Goldstein Step Length Rule for Constrained Minimization Problems," SIAM J. on Control and Optimization, Vol. 18, pp. 659-674.

[Dun81] Dunn, J. C., 1981. "Global and Asymptotic Convergence Rate Estimates for a Class of Projected Gradient Processes," SIAM J. on Control and Optimization, Vol. 19, pp. 368-400.

[Dun87] Dunn, J. C., 1987. "On the Convergence of Projected Gradient Processes to Singular Critical Points," J. Opt. Theory and Appl., Vol. 55, pp. 203-216.

[Dun88a] Dunn, J. C., 1988. "Gradient Projection Methods for Systems Optimization Problems," Control and Dynamic Systems, Vol. 29, pp. 135-195.

[Dun88b] Dunn, J. C., 1988. "A Projected Newton Method for Minimization Problems with Nonlinear Inequality Constraints," Numer. Math., Vol. 53, pp. 377-409.

[Dun91a] Dunn, J. C., 1991. "Scaled Gradient Projection Methods for Optimal Control Problems and Other Structured Nonlinear Programs," in New Trends in Systems Theory, Conte, G., et al, (Eds.), Birkhöuser, Boston, MA.

[Dun91b] Dunn, J. C., 1991. "A Subspace Decomposition Principle for Scaled Gradient Projection Methods: Global Theory," SIAM J. on Control and Optimization, Vol. 29, pp. 219-246.

[Dun93a] Dunn, J. C., 1993. "A Subspace Decomposition Principle for Scaled Gradient Projection Methods: Local Theory," SIAM J. on Control and Optimization, Vol. 31, pp. 219-246.

[Dun93b] Dunn, J. C., 1993. "Second-Order Multiplier Update Calculations for Optimal Control Problems and Related Large Scale Nonlinear Programs," SIAM J. on Optimization, Vol. 3, pp. 489-502.

[Dun93c] Dunn, J. C., 1993. Private Communication.

[Dun94] Dunn, J. C., 1994. "Gradient-Related Constrained Minimization Algorithms in Function Spaces: Convergence Properties and Computational Implications," in Large Scale Optimization: State of the Art, Hager, W. W., Hearn, D. W., and Pardalos, P. M., (Eds.), Kluwer, Boston.

[Eas58] Eastman, W. L., 1958. Linear Programming with Pattern Constraints, Ph.D. Thesis, Harvard University, Cambridge, MA.

[EcB92] Eckstein, J., and Bertsekas, D. P., 1992. "On the Douglas-Rachford Splitting Method and the Proximal Point Algorithm for Maximal Monotone Operators," Math. Programming, Vol. 55, pp. 293-318.

[Eck94a] Eckstein, J., 1994. "Nonlinear Proximal Point Algorithms Using Bregman Functions, with Applications to Convex Programming," Math. Operations Res., Vol. 18, pp. 202-226.

[Eck94b] Eckstein, J., 1994. "Parallel Alternating Direction Multiplier Decomposition of Convex Programs," J. Opt. Theory and Appl., Vol. 80, pp. 39-62.

[ElM75] Elzinga, J., and Moore, T. G., 1975. "A Central Cutting Plane Algorithm for the Convex Programming Problem," Math. Programming, Vol. 8, pp. 134-145.

[EkT76] Ekeland, I., and Teman, R., 1976. Convex Analysis and Variational Problems, North-Holland Publ., Amsterdam.

[Eve63] Everett, H., 1963. "Generalized Lagrange Multiplier Method for Solving Problems of Optimal Allocation of Resources," Operations Research, Vol. 11, pp. 399-417.

[Evt85] Evtushenko, Y. G., 1985. Numerical Optimization Techniques, Optimization Software, N. Y.

[FaF63] Fadeev, D. K., and Fadeeva, V. N., 1963. Computational Methods of Linear Algebra, Freeman, San Francisco, CA.

[FeM91] Ferris, M. C., and Mangasarian, O. L., 1991. "Parallel Constraint Distribution," SIAM J. on Optimization, Vol. 1, pp. 487-500.

[Fen49] Fenchel, W., 1949. "On Conjugate Convex Functions," Canad. J. Math., Vol. 1, pp. 73-77.

[Fen51] Fenchel, W., 1951. "Convex Cones, Sets, and Functions," Mimeographed Notes, Princeton Univ.

[FiM68] Fiacco, A. V., and McCormick, G. P., 1968. Nonlinear Programming: Sequential Unconstrained Minimization Techniques, Wiley, N. Y.

[Fia83] Fiacco, A. V., 1983. Introduction to Sensitivity and Stability Analysis in Nonlinear Programming, Academic Press, N. Y.

[FlH95] Florian, M. S., and Hearn, D., 1995. "Network Equilibrium Models and Algorithms," Handbooks in OR and MS, Ball, M. O., Magnanti, T. L., Monma, C. L., and Nemhauser, G. L., (Eds.), Vol. 8, North-Holland, Amsterdam, pp. 485-550.

[FlP63] Fletcher, R., and Powell, M. J. D., 1963. "A Rapidly Convergent Descent Algorithm for Minimization," Comput. J., Vol. 6, pp. 163-168.

[FlP95] Floudas, C., and Pardalos, P. M., (Eds.), 1995. State of the Art in Global Optimization : Computational Methods and Applications, Kluwer, Boston.

[Fla92] Flam, S. D., 1992. "On Finite Convergence and Constraint Identification of Subgradient Projection Methods," Math. Programming, Vol. 57, pp 427-437.

[Fle70a] Fletcher, R., 1970. "A New Approach to Variable Metric Algorithms," Computer J., Vol. 13, pp. 317-322.

[Fle70b] Fletcher, R., 1970. "A Class of Methods for Nonlinear Programming with Termination and Convergence Properties," in Integer and Nonlinear Programming, Abadie, J., (Ed.), pp. 157-173, North-Holland Publ., Amsterdam.

[Flo95] Floudas, C. A., 1995. Nonlinear and Mixed-Integer Optimization: Fundamentals and Applications, Oxford University Press, N. Y.

[FoG83] Fortin, M., and Glowinski, R., (Eds.), 1983. Augmented Lagrangian Methods: Applications to the Numerical Solution of Boundary-Value Problems, North-Holland, Amsterdam.

[FrW56] Frank, M., and Wolfe, P., 1956. "An Algorithm for Quadratic Programming," Naval Research Logistics Quarterly, Vol. 3, pp. 95-110.

[Fre91] Freund, R. M., 1991. "Theoretical Efficiency of a Shifted Barrier

Function Algorithm in Linear Programming," Linear Algebra and Appl., Vol. 152, pp. 19-41.

[Fri56] Frisch, M. R., 1956. "La Resolution des Problemes de Programme Lineaire par la Methode du Potential Logarithmique," Cahiers du Seminaire D'Econometrie, Vol. 4, pp. 7-20.

[Fuk92] Fukushima, M., 1992. "Application of the Alternating Direction Method of Multipliers to Separable Convex Programming," Comp. Opt. and Appl., Vol. 1, pp. 93-111.

[GHV92] Goffin, J. L., Haurie, A., and Vial, J. P., 1992. "Decomposition and Nondifferentiable Optimization with the Projective Algorithm," Management Science, Vol. 38, pp. 284-302.

[GKT51] Gale, D., Kuhn, H. W., and Tucker, A. W., 1951. "Linear Programming and the Theory of Games," in Activity Analysis of Production and Allocation, Koopmans, T. C., (Ed.), Wiley, N. Y.

[GLL91] Grippo, L., Lampariello, F., and Lucidi, S., 1991. "A Class of Nonmonotone Stabilization Methods in Unconstrained Minimization," Numer. Math., Vol. 59, pp. 779-805.

[GLY94] Goffin, J. L., Luo, Z.-Q., and Ye, Y., 1994. "On the Complexity of a Column Generation Algorithm for Convex or Quasiconvex Feasibility Problems," in Large Scale Optimization: State of the Art, Hager, W. W., Hearn, D. W., and Pardalos, P. M., (Eds.), Kluwer, Boston.

[GLY96] Goffin, J. L., Luo, Z.-Q., and Ye, Y., 1996. "Complexity Analysis of an Interior Cutting Plane Method for Convex Feasibility Problems," SIAM J. on Optimization, Vol. 6, pp. 638-652.

[GMW81] Gill, P. E., Murray, W., and Wright, M. H., 1981. Practical Optimization, Academic Press, N. Y.

[GMW91] Gill, P. E., Murray, W., and Wright, M. H., 1991. Numerical Linear Algebra and Optimization, Vol. I, Addison-Wesley, Redwood City, CA.

[GaB82] Gafni, E. M., and Bertsekas, D. P., 1982. "Convergence of a Gradient Projection Method," Report LIDS-P-1201, Lab. for Info. and Dec. Systems, M.I.T.

[GaB84] Gafni, E. M., and Bertsekas, D. P., 1984. "Two-Metric Projection Methods for Constrained Optimization," SIAM J. on Control and Optimization, Vol. 22, pp. 936-964.

[GaD88] Gawande, M., and Dunn, J. C., 1988. "Variable Metric Gradient Projection Processes in Convex Feasible Sets Defined by Nonlinear Inequalities," Appl. Math. Optim., Vol. 17, pp. 103-119.

[GaJ88] Gauvin, J., and Janin, R., 1988. "Directional Behavior of Optimal

Solutions in Nonlinear Mathematical Programming," Math. of Operations Res., Vol. 13, pp. 629-649.

[GaM92] Gaudioso, M., and Monaco, M. F, 1992. "Variants to the Cutting Plane Approach for Convex Nondifferentiable Optimization," Optimization, Vol. 25, pp. 65-75.

[Gab82] Gabay, D., 1982. "Reduced Quasi-Newton Methods with Feasibility Improvement for Nonlinearly Constrained Optimization," Math. Programming Studies, Vol. 16, pp. 18-44.

[Gai94] Gaivoronski, A. A., 1994. "Convergence Analysis of Parallel Backpropagation Algorithm for Neural Networks," Optimization Methods and Software, Vol. 4, pp. 117-134.

[Gau77] Gauvin, J., 1977. "A Necessary and Sufficient Condition to Have Bounded Multipliers in Convex Programming," Math. Programming., Vol. 12, pp. 136-138.

[GeB99] Geary, A., and Bertsekas, D. P., 1999. "Incremental Subgradient Methods," to appear in Proc. of 1999 IEEE Conf. on Decision and Control.

[Geo70] Geoffrion, A. M., 1970. "Elements of Large-Scale Mathematical Programming, I, II," Management Science, Vol. 16, pp. 652-675, 676-691.

[Geo74] Geoffrion, A. M., 1974. "Lagrangian Relaxation for Integer Programming," Math. Programming Studies, Vol. 2, pp. 82-114.

[Geo77] Geoffrion, A. M., 1977. "Objective Function Approximations in Mathematical Programming," Math. Programming, Vol. 13, pp. 23-27.

[GiK95] Gilmore, P., and Kelley, C. T., 1995. "An Implicit Filtering Algorithm for Optimization of Functions with Many Local Minima," SIAM J. on Optimization, Vol. 5, pp. 269-285.

[GiM74] Gill, P. E., and Murray, W., (Eds.), 1974. Numerical Methods for Constrained Optimization, Academic Press, N. Y.

[Gil92] Gilbert, J. C., 1992. "Automatic Differentiation and Iterative Processes," Optimization Methods and Software, Vol. 1, pp. 13-21.

[GlL97] Glover, F., and Laguna, M., 1997. Tabu Search, Kluwer, Boston.

[GlP79] Glad, T., and Polak, E., 1979. "A Multiplier Method with Automatic Limitation of Penalty Growth," Math. Programming, Vol. 17, pp. 140-155.

[Gla79] Glad, T., 1979. "Properties of Updating Methods for the Multipliers in Augmented Lagrangians," J. Opt. Th. and Appl., Vol. 28, pp. 135-156.

[Gom58] Gomory, R. E., 1958. "Outline of an Algorithm for Integer Solutions to Linear Programs," Bulletin of the American Mathematical Society, Vol. 64, pp. 275-278.

[GoP79] Gonzaga, C., and Polak, E., 1979. "On Constraint Dropping Schemes and Optimality Functions for a Class of Outer Approximations Algorithms," SIAM J. on Control and Optimization, Vol. 17, pp.477-493.

[GoT71] Gould, F. J., and Tolle, J., 1971. "A Necessary and Sufficient Condition for Constrained Optimization," SIAM J. Applied Math., Vol. 20, pp. 164-172.

[GoT72] Gould, F. J., and Tolle, J., 1972. "Geometry of Optimality Conditions and Constraint Qualifications," Math. Programming, Vol. 2, pp. 1-18.

[GoK99] Goffin, J. L., and Kiwiel, K. C, 1999. "Convergence of a Simple Subgradient Level Method," Math. Programming, Vol. 85, pp. 207-211.

[GoV90] Goffin, J. L., and Vial, J. P., 1990. "Cutting Planes and Column Generation Techniques with the Projective Algorithm," J. Opt. Th. and Appl., Vol. 65, pp. 409-429.

[GoV99] Goffin, J. L., and Vial, J. P., 1999. "Convex Nondifferentiable Optimization: A Survey Focussed on the Analytic Center Cutting Plane Method," Logilab Technical Report, Department of Management Studies, University of Geneva, Switzerland; also GERAD Tech. Report G-99-17, McGill Univ., Montreal, Canada.

[GoI82] Goldfarb, D., and Idnani, A., 1982. "A Numerically Stable Dual Method for Solving Strictly Convex Quadratic Programs," Math. Programming, Vol. 27, pp. 1-33.

[Gol89] Goldberg, D. E., 1989. Genetic Algorithms in Search, Optimization, and Machine Learning, Addison Wesley, Reading, MA.

[Gof77] Goffin, J. L., 1977. "On Convergence Rates of Subgradient Optimization Methods," Math. Programming, Vol. 13, pp. 329-347.

[Gol62] Goldstein, A. A., 1962. "Cauchy's Method of Minimization," Numer. Math., Vol. 4, pp. 146-150.

[Gol64] Goldstein, A. A., 1964. "Convex Programming in Hibert Space," Bull. Amer. Math. Soc., Vol. 70, pp. 709-710.

[Gol67] Goldstein, A. A., 1967. Constructive Real Analysis, Harper and Row, N. Y.

[Gol70] Goldfarb, D., 1970. "A Family of Variable-Metric Methods Derived by Variational Means," Math. Comp., Vol. 24, pp. 23-26.

[Gol85] Golshtein, E. G., 1985. "A Decomposition Method for Linear and Convex Programming Problems," Matecon, Vol. 21, pp. 1077-1091.

[Gon91] Gonzaga, C. C., 1991. "Large Step Path-Following Methods for Linear Programming, Part I: Barrier Function Method," SIAM J. on Optimization, Vol. 1, pp. 268-279.

[Gon92] Gonzaga, C. C., 1992. "Path Following Methods for Linear Programming," SIAM Review, Vol. 34, pp. 167-227.

[Gri72] Grinold, R. C., 1972. "Steepest Ascent for Large-Scale Linear Programs," SIAM Review, Vol. 14, pp. 447-464.

[GrS98] Grippo, L., and Sciandrone, M., 1998. "On the Convergence of the Block Nonlinear Gauss-Seidel Method Under Convex Constraints," Report R. 467, Istituto di Analisi dei Sistemi ed Informatica, Universita di Roma "La Sapienza," Italy.

[Gri89] Griewank, A., 1989. "On Automatic Differentiation," in Mathematical Programming: Recent Developments, Iri, M., and Tanabe, K., (Eds.), Kluwer, Dordrecht, pp. 83-108.

[Gri94] Grippo, L., 1994. "A Class of Unconstrained Minimization Methods for Neural Network Training," Optimization Methods and Software, Vol. 4, pp. 135-150.

[Gui69] Guignard, M., 1969. "Generalized Kuhn-Tucker Conditions for Mathematical Programming Problems in a Banach Space," SIAM J. on Control, Vol. 7, pp. 232-241.

[Gul92] Guler, O., 1992. "New Proximal Point Algorithms for Convex Minimization," SIAM J. on Optimization, Vol. 2, pp. 649-664.

[Gul94] Guler, O., 1994. "Limiting Behavior of Weighted Central Paths in Linear Programming," Math. Programming, Vol. 65, pp. 347-363.

[HKR95] den Hertog, D., Kaliski, J., Roos, C., and Terlaky, T., 1993. "A Path-Following Cutting Plane Method for Convex Programming," Annals of Operations Research, Vol. 58, pp. 69-98.

[HLV87] Hearn, D. W., Lawphongpanich, S., and Ventura, J. A., 1987. "Restricted Simplicial Decomposition: Computation and Extensions," Math. Programming Studies, Vol. 31, pp. 119-136.

[HPT95] Horst, R., Pardalos, P. M., and Thoai, N. V., 1995. Introduction to Global Optimization, Kluwer, Boston.

[HaB70] Haarhoff, P. C., and Buys, J. D, 1970. "A New Method for the Optimization of a Nonlinear Function Subject to Nonlinear Constraints," Computer J., Vol. 13, pp. 178-184.

[HaH93] Hager, W. W., and Hearn, D. W., 1993. "Application of the Dual Active Set Algorithm to Quadratic Network Optimization," Computational Optimization and Applications, Vol. 1, pp. 349-373.

[HaM79] Han, S. P., and Mangasarian, O. L., 1979. "Exact Penalty Functions in Nonlinear Programming," Math. Programming, Vol. 17, pp. 251-269.

[Hag88] Hager, W. W., 1988. Applied Numerical Linear Algebra, Prentice-Hall, Englewood Cliffs, N. J.

[Hag99] Hager, W. W., 1999. "Stabilized Sequential Quadratic Programming," Computational Optimization and Applications, Vol. 12.

[Han77] Han, S. P., 1977. "A Globally Convergent Method for Nonlinear Programming," J. Opt. Th. and Appl., Vol. 22, pp. 297-309.

[Hay98] Haykin, S., 1998. Neural Networks: A Comprehensive Foundation, (2nd Ed.), McMillan, N. Y.

[HeK70] Held, M., and Karp, R. M., 1970. "The Traveling Salesman Problem and Minimum Spanning Trees," Operations Research, Vol. 18, pp. 1138-1162.

[HeK71] Held, M., and Karp, R. M., 1971. "The Traveling Salesman Problem and Minimum Spanning Trees: Part II," Math. Programming, Vol. 1, pp. 6-25.

[HeL89] Hearn, D. W., and Lawphongpanich, S., 1989. "Lagrangian Dual Ascent by Generalized Linear Programming," Operations Res. Letters, Vol. 8, pp. 189-196.

[HeS52] Hestenes, M. R., and Stiefel, E. L., 1952. "Methods of Conjugate Gradients for Solving Linear Systems," J. Res. Nat. Bur. Standards Sect. B, Vol. 49, pp. 409-436.

[Hei96] Heinkenschloss, M., 1996. "Projected Sequential Quadratic Programming Methods," SIAM J. on Optimization, Vol. 6, pp. 373-417.

[Her94] den Hertog, D., 1994. Interior Point Approach to Linear, Quadratic, and Convex Programming, Kluwer, Dordrecht, The Netherlands.

[Hes69] Hestenes, M. R., 1969. "Multiplier and Gradient Methods," J. Opt. Th. and Appl., Vol. 4, pp. 303-320.

[Hes75] Hestenes, M. R., 1975. Optimization Theory: The Finite Dimensional Case, Wiley, N. Y.

[Hes80] Hestenes, M. R., 1980. Conjugate Direction Methods in Optimization, Springer-Verlag, Berlin and N. Y.

[HiL93] Hiriart-Urruty, J.-B., and Lemarechal, C., 1993. Convex Analysis and Minimization Algorithms, Vols. I and II, Springer-Verlag, Berlin and N. Y.

[Hil57] Hildreth, C., 1957. "A Quadratic Programming Procedure," Naval Res. Logist. Quart., Vol. 4, pp. 79-85. See also "Erratum," Naval Res. Logist. Quart., Vol. 4, p. 361.

[HoJ61] Hooke, R., and Jeeves, T. A., 1961. "Direct Search Solution of Numerical and Statistical Problems," J. Assoc. Comp. Mach., Vol. 8, pp.

212-221.

[HoK71] Hoffman, K., and Kunze, R., 1971. Linear Algebra, Prentice-Hall, Englewood Cliffs, N. J.

[Hoh77] Hohenbalken, B. von, 1977. "Simplicial Decomposition in Nonlinear Programming," Math. Programming, Vol. 13, pp. 49-68.

[Hol74] Holloway, C. A., 1974. "An Extension of the Frank and Wolfe Method of Feasible Directions," Math. Programming, Vol. 6, pp. 14-27.

[HuD84] Hughes, G. C., and Dunn, J. C., 1984. "Newton-Goldstein Convergence Rates for Convex Constrained Minimization Problems with Singular Solutions," Appl. Math. Optim., Vol. 12, pp. 203-230, 1984.

[HuL88] Huang, C., and Litzenberger, R. H., 1988. Foundations of Financial Economics, Prentice-Hall, Englewood Cliffs, N. J.

[IST94] Iusem, A. N., Svaiter, B., and Teboulle, M., 1994. "Entropy-Like Proximal Methods in Convex Programming," Math. Operations Res., Vol. 19, pp. 790-814.

[IbK88] Ibaraki, T., and Katoh, N., 1988. Resource Allocation Problems: Algorithmic Approaches, M.I.T. Press, Cambridge, MA.

[Iof94] Ioffe, A., 1994. "On Sensitivity Analysis of Nonlinear Programs in Banach Spaces: The Approach via Composite Unconstrained Minimization," SIAM J. on Optimization, Vol. 4, pp. 1-43.

[IuT95] Iusem, A. N., and Teboulle, M., 1995. "Convergence Rate Analysis of Nonquadratic Proximal Methods for Convex and Linear Programming," Math. Operations Res., Vol. 20.

[Ius98] Iusem, A. N., 1998. "Augmented Lagrangian Methods and Proximal Point Methods for Convex Minimization," unpublished report; to appear in Investigacion Operativa.

[JRT95] Junger, M., Reinelt, G., and Rinaldi, G., 1995. "Practical Problem Solving with Cutting Plane Algorithms in Combinatorial Oprimization," in Combinatorial Optimization, Cook, W. J., Lovasz, L., and Seymour, P., (Eds.), DIMACS Series in Discrete Mathematics and Computer Science, AMS, pp. 11-152.

[JaS95] Jarre, F., and Saunders, M. A., 1995. "A Practical Interior-Point Method for Convex Programming," SIAM J. Optimization, Vol. 5, pp. 149-171.

[JeW92] Jeyakumar, V., and Wolkowicz, H., 1992. "Generalizations of Slater's Constraint Qualification for Infinite Convex Programs," Math. Programming, Vol. 57, pp. 85-101.

[Joh48] John, F., 1948. "Extremum Problems with Inequalities as Subsidiary Conditions," in Studies and Essays: Courant Anniversary Volume,

K. O. Friedrichs, Neugebauer, O. E., and Stoker, J. J., (Eds.), Wiley-Interscience, N. Y., pp. 187-204.

[KAK89] Korst, J., Aarts, E. H., and Korst, A., 1989. Simulated Annealing and Boltzmann Machines: A Stochastic Approach to Combinatorial Optimization and Neural Computing, Wiley, N. Y.

[KMN91] Kojima, M., Meggido, N., Noma, T., and Yoshise, A., 1991. A Unified Approach to Interior Point Algorithms for Linear Complementarity Problems, Vol. 538, Lecture Notes in Computer Science, Springer-Verlag, Berlin.

[KMY89] Kojima, M., Mizuno, S., and Yoshise, A., 1989. "A Primal-Dual Interior Point Algorithm for Linear Programming," in Progress in Mathematical Programming, Interior Point and Related Methods, Meggido, N., (Ed.), Springer-Verlag, N. Y., pp. 29-47.

[KaW94] Kall, P., and Wallace, S. W., 1994. Stochastic Programming, Wiley, Chichester, UK.

[Kan45] Kantorovich, L. V., 1945. "On an Effective Method of Solution of Extremal Problems for a Quadratic Functional," Dokl. Akad. Nauk SSSR, Vol. 48, pp. 483-487.

[Kan49] Kantorovich, L. V., 1945. "On Newton's Method," Trudy Mat. Inst. Steklov, Vol. 28, pp. 104-144. Translation in Selected Articles in Numerical Analysis by C. D. Benster, 104.1-144.2.

[Kar39] Karush, W., 1939. "Minima of Functions of Several Variables with Inequalities as Side Conditions," M.S. Thesis, Department of Math., University of Chicago.

[Kar84] Karmarkar, N., 1984. "A New Polynomial-Time Algorithm for Linear Programming," Combinatorica, Vol. 4, pp. 373-395.

[KeS77] Kennington, J., and Shalaby, M., 1977. "An Effective Subgradient Procedure for Minimal Cost Multicommodity Flow Problems," Management Science, Vol. 23, pp. 994-1004.

[Kel60] Kelley, J. E., 1960. "The Cutting-Plane Method for Solving Convex Programs," J. Soc. Indust. Appl. Math., Vol. 8, pp. 703-712.

[Kha79] Khachiyan, L. G., 1979. "A Polynomial Algorithm for Linear Programming," Soviet Math. Doklady, Vol. 20, pp. 191-194.

[KiN91] Kim, S., and Nazareth, J. L., 1991. "The Decomposition Principle and Algorithms for Linear Programming," Linear Algebra and its Applications, Vol. 152, pp. 119-133.

[KiR92] King, A., and Rockafellar, R. T., 1992. "Sensitivity Analysis for Nonsmooth Generalized Equations," Math. Programming, Vol. 55, pp. 193-212.

[Kiw94] Kiwiel, K. C., 1994. "Free-Steering Relaxation Methods for Problems with Strictly Convex Costs and Linear Constraints," Systems Research Institute, Newelska 6, 01-447 Warsaw, Poland.

[Kiw97a] Kiwiel, K. C., 1997. "Proximal Minimization Methods with Generalized Bregman Functions," SIAM J. on Control and Optimization, Vol. 35, pp. 1142-1168.

[Kiw97b] Kiwiel, K. C., 1997. "Efficiency of the Analytic Center Cutting Plane Method for Convex Minimization," SIAM J. on Optimization, Vol. 7, pp. 336-346.

[KlH98] Klatte, D., and Henrion, R., 1998. "Regularity and Stability in Nonlinear Semi-Infinite Optimization," in Semi-Infinite Programming, Reemtsen, R., and Ruckman, J. J., (Eds.), Kluwer, Boston, pp. 69-102.

[KoB72] Kort, B. W., and Bertsekas, D. P., 1972. "A New Penalty Function Method for Constrained Minimization," Proc. 1972 IEEE Confer. Decision Control, New Orleans, LA, pp. 162-166.

[KoB76] Kort, B. W., and Bertsekas, D. P., 1976. "Combined Primal-Dual and Penalty Methods for Convex Programming," SIAM J. on Control and Optimization, Vol. 14, pp. 268-294.

[KoM98] Kontogiorgis, S., and Meyer, R. R., 1998. "A Variable-Penalty Alternating Directions Method for Convex Optimization," Math. Programming, Vol. 83, pp. 29-53.

[KoN93] Kortanek, K. O., and No, H., 1993. "A Central Cutting Plane Algorithm for Convex Semi-Infinite Programming Problems," SIAM J. on Optimization, Vol. 3, pp. 901-918.

[KoZ93] Kortanek, K. O., and Zhu, J., 1993. "A Polynomial Barrier Algorithm for Linearly Constrained Convex Programming Problems," Math. Operations Res., Vol. 18, pp. 116-127.

[KoZ95] Kortanek, K. O., and Zhu, J., 1995. "On Controlling the Parameter in the Logarithmic Barrier Term for Convex Programming Problems," J. Opt. Th. and Appl., Vol. 84, pp. 117-143.

[Koh74] Kohonen, T., 1974. "An Adaptive Associative Memory Principle," IEEE Trans. on Computers, Vol. C-23, pp. 444-445.

[Kor75] Kort, B. W., 1975. "Combined Primal-Dual and Penalty Function Algorithms for Nonlinear Programming," Ph.D. Thesis, Dept. of Engineering-Economic Systems, Stanford Univ., Stanford, Ca.

[KuC78] Kushner, H. J., and Clark, D. S., 1978. Stochastic Approximation Methods for Constrained and Unconstrained Systems, Springer-Verlag, N. Y.

[KuY97] Kushner, H. J., and Yin, G., 1997. Stochastic Approximation Methods, Springer-Verlag, N. Y.

[KuT51] Kuhn, H. W., and Tucker, A. W., 1951. "Nonlinear Programming," in Proc. of the Second Berkeley Symposium on Math. Statistics and Probability, Neyman, J., (Ed.), Univ. of California Press, Berkeley, CA, pp. 481-492.

[Kuh76] Kuhn, H. W., 1976. "Nonlinear Programming: A Historical View," in Nonlinear Programming, Cottle, R. W., and Lemke, C. E., (Eds.), SIAM-AMS Proc., Vol. IX, American Math. Soc., Providence, RI, pp. 1-26.

[Kyp90] Kyparisis, J., 1990. "Solution Differentiability for Variational Inequalities," Math. Programming, Vol. 48, pp. 285-301.

[LBV98] Lobo, M. S., Vandenberghe, L., Boyd, S., and Lebret, H., 1998. "Applications of Second-Order Cone Programming," Linear Algebra and Applications, Vol. 284, pp. 193-228.

[LPW92] Ljung, L., Pflug, G., and Walk, H., 1992. Stochastic Approximation and Optimization of Random Systems, Birkhauser, Boston.

[LMS63] Little, J. D. C., Murty, K. G., Sweeney, D. W., and Karel, C., 1963. "An Algorithm for the Traveling Salesman Problem," Operations Research, Vol. 11, pp. 972-989.

[LMS92] Lustig, I. J., Marsten, R. E., and Shanno, D. F., 1992. "On Implementing Mehrotra's Predictor-Corrector Interior-Point Method for Linear Programming," SIAM J. on Optimization, Vol. 2, pp. 435-449.

[LaD60] Land, A. H., and Doig, A. G., 1960. "An Automatic Method for Solving Discrete Programming Problems," Econometrica, Vol. 28, pp. 497-520.

[LaT85] Lancaster, P., and Tismenetsky, M., 1985. The Theory of Matrices, Academic Press, N. Y.

[LaW78] Lasdon, L. S., and Waren, A. D., 1978. "Generalized Reduced Gradient Software for Linearly and Nonlinearly Constrained Problems," in Design and Implementation of Optimization Software, Greenberg, H. J., (Ed.), Sijthoff and Noordhoff, Holland, pp. 335-362.

[Las70] Lasdon, L. S., 1970. Optimization Theory for Large Systems, Macmillian, N. Y.

[Law76] Lawler, E., 1976. Combinatorial Optimization: Networks and Matroids, Holt, Reinhart, and Winston, N. Y.

[LeP65] Levitin, E. S., and Poljak, B. T., 1965. "Constrained Minimization Methods," Ž. Vyčisl. Mat. i Mat. Fiz., Vol. 6, pp. 787-823.

[LeS93] Lemaréchal, C., and Sagastizábal, C., 1993. "An Approach to Variable Metric Bundle Methods," in Systems Modelling and Optimization,

Proc. of the 16th IFIP-TC7 Conference, Compiègne, Henry, J., and Yvon, J.-P., (Eds.), Lecture Notes in Control and Information Sciences 197, pp. 144-162.

[Lem74] Lemarechal, C., 1974. "An Algorithm for Minimizing Convex Functions," in Information Processing '74, Rosenfeld, J. L., (Ed.), pp. 552-556, North-Holland, Amsterdam.

[Lem75] Lemarechal, C., 1975. "An Extension of Davidon Methods to Nondifferentiable Problems," Math. Programming Study 3, Balinski, M., and Wolfe, P., (Eds.), North-Holland, Amsterdam, pp. 95-109.

[Lem89] Lemaire, B., 1989. "The Proximal Algorithm," in International Series of Numerical Mathematics, Vol. 87, Penot, J. P., (Ed.), Birkhöuser-Verlag, Basel, Switzerland, pp. 73-87.

[LiP87] Lin, Y. Y., and Pang, J.-S., 1987. "Iterative Methods for Large Convex Quadratic Programs: A Survey," SIAM J. on Control and Optimization, Vol. 18, pp. 383-411.

[LuB96] Lucena, A., and Beasley, J. E., 1996. "Branch and Cut Algorithms," in Advances in Linear and Integer Programming, Beasley, J. E., (Ed.), Oxford University Press, N. Y., Chapter 5.

[LuT91] Luo, Z. Q., and Tseng, P., 1991. "On the Convergence of a Matrix-Splitting Algorithm for the Symmetric Monotone Linear Complementarity Problem," SIAM J. on Control and Optimization, Vol. 29, pp. 1037-1060.

[LuT92a] Luo, Z. Q., and Tseng, P., 1992. "On the Convergence of the Coordinate Descent Method for Convex Differentiable Minimization," J. Opt. Th. and Appl., Vol. 72, pp. 7-35.

[LuT92b] Luo, Z. Q., and Tseng, P., 1992. "On the Linear Convergence of Descent Methods for Convex Essentially Smooth Minimization," SIAM J. on Control and Optimization, Vol. 30, pp. 408-425.

[LuT93a] Luo, Z. Q., and Tseng, P., 1993. "Error Bounds and Convergence Analysis of Feasible Descent Methods: A General Approach," Annals of Operations Res., Vol. 46, pp. 157-178.

[LuT93b] Luo, Z. Q., and Tseng, P., 1993. "Error Bound and Reduced Gradient Projection Algorithms for Convex Minimization over a Polyhedral Set," SIAM J. on Optimization, Vol. 3, pp. 43-59.

[LuT93c] Luo, Z. Q., and Tseng, P., 1993. "On the Convergence Rate of Dual Ascent Methods for Linearly Constrained Convex Minimization," Math. of Operations Res., Vol. 18, pp. 846-867.

[LuT94a] Luo, Z. Q., and Tseng, P., 1994. "Analysis of an Approximate Gradient Projection Method with Applications to the Backpropagation Algorithm," Optimization Methods and Software, Vol. 4, pp. 85-101.

[LuT94b] Luo, Z. Q., and Tseng, P., 1994. "On the Rate of Convergence of a Distributed Asynchronous Routing Algorithm," IEEE Transactions on Automatic Control, Vol. 39, pp. 1123-1129.

[Lue69] Luenberger, D. G., 1969. Optimization by Vector Space Methods, Wiley, N. Y.

[Lue84] Luenberger, D. G., 1984. Introduction to Linear and Nonlinear Programming, (2nd Ed.), Addison-Wesley, Reading, MA.

[Lue98] Luenberger, D. G., 1998. Investment Science, Oxford University Press, N. Y.

[Luo91] Luo, Z. Q., 1991. "On the Convergence of the LMS Algorithm with Adaptive Learning Rate for Linear Feedforward Networks," Neural Computation, Vol. 3, pp. 226-245.

[Luq84] Luque, F.J., 1984. "Asymptotic Convergence Analysis of the Proximal Point Algorithm," SIAM J. on Control and Optimization, Vol. 22, pp. 277-293.

[MMS91] McShane, K. A., Monma, C. L., and Shanno, D., 1991. "An Implementation of a Primal-Dual Interior Point Method for Linear Programming," ORSA J. on Computing, Vol. 1, pp. 70-83.

[MMZ95] McKenna, M. P., Mesirov, J. P., and Zenios, S. A., 1995. "Data Parallel Quadratic Programming on Box-Constrained Problems," SIAM J. on Optimization, Vol. 5, pp. 570-589.

[MOT95] Mahey, P., Oualibouch, S., and Tao, P. D., 1995. "Proximal Decomposition on the Graph of a Maximal Monotone Operator," SIAM J. on Optimization, Vol. 5, pp. 454-466.

[MSQ98] Mifflin, R., Sun, D., and Qi, L., 1998. "Quasi-Newton Bundle-Type Methods for Nondifferentiable Convex Optimization, SIAM J. on Optimization, Vol. 8, pp. 583-603.

[MTW93] Monteiro, R. D. C., Tsuchiya, T., and Wang, Y., 1993. "A Simplified Global Convergence Proof of the Affine Scaling Algorithm," Annals of Operations Res., Vol. 47, pp. 443-482.

[MaD87] Mangasarian, O. L., and De Leone, R., 1987. "Parallel Successive Overelaxation Methods for Symmetric Linear Complementarity Problems and Linear Programs," J. of Optimization Th. and Appl., Vol. 54, pp. 437-446.

[MaD88] Mangasarian, O. L., and De Leone, R., 1988. "Parallel Gradient Projection Successive Overrelaxation for Symmetric Linear Complementarity Problems," Annals of Operations Res., Vol. 14, pp. 41-59.

[MaF67] Mangasarian, O. L., and Fromovitz, S., 1967. "The Fritz John Necessary Optimality Conditions in the Presence of Equality and Inequality

Constraints," J. Math. Anal. and Appl., Vol. 17, pp. 37-47.

[MaP82] Mayne, D. Q., and Polak, E., 1982. "A Superlinearly Convergent Algorithm for Constrained Optimization Problems," Math. Programming Studies, Vol. 16, pp. 45-61.

[MaS94] Mangasarian, O. L., and Solodov, M. V., 1994. "Serial and Parallel Backpropagation Convergence Via Nonmonotone Perturbed Minimization," Optimization Methods and Software, Vol. 4, pp. 103-116.

[Man69] Mangasarian, O. L., 1969. Nonlinear Programming, Prentice-Hall, Englewood Cliffs, N. J; also SIAM, Classics in Applied Mathematics 10, Phila., PA., 1994.

[Man93] Mangasarian, O. L., 1993. "Mathematical Programming in Neural Networks," ORSA Journal on Computing, Vol. 5, pp. 349-360.

[Mar70] Martinet, B., 1970. "Regularisation d'Inequations Variationnelles par Approximations Successives," Rev. Francaise Inf. Rech. Oper., pp. 154-159.

[Mar72] Martinet, B., 1972. "Determination Approchee d'un Point Fixe d'une Application Pseudo-Contractante," C. R. Acad. Sci. Paris, 274A, pp. 163-165.

[Mar78] Maratos, N., 1978. "Exact Penalty Function Algorithms for Finite Dimensional and Control Optimization Problems," Ph.D. Thesis, Imperial College Sci. Tech, Univ. of London.

[McK98] McKinnon, K. I. M., 1998. "Convergence of the Nelder-Mead Simplex Method to a Non-Stationary Point," SIAM J. on Optimization, Vol. 9, pp. 148-158.

[McL80] McLinden, L., 1980. "The Complementarity Problem for Maximal Monotone Multifunctions," in Variational Inequalities and Complementarity Problems, Cottle, R., Giannessi, F., and Lions, J.-L., (Eds.), Wiley, N. Y., pp. 251-270.

[McS73] McShane, E. J., 1973. "The Lagrange Multiplier Rule," American Mathematical Monthly, Vol. 80, pp. 922-925.

[Meg88] Megiddo, N., 1988. "Pathways to the Optimal Set in Linear Programming," in Progress in Mathematical Programming, Megiddo, N., (Ed.), Springer-Verlag, N. Y., pp. 131-158.

[Meh92] Mehrotra, S., 1992. "On the Implementation of a Primal-Dual Interior Point Method," SIAM J. on Optimization, Vol. 2, pp. 575-601.

[Mey79] Meyer, R. R., 1979. "Two-Segment Separable Programming," Management Science, Vol. 25, pp. 385-395.

[Mif96] Mifflin, R., 1996. "A Quasi-Second-Order Proximal Bundle Algorithm," Math. Programming, Vol. 73, pp. 51-72.

[Mig94] Migdalas, A., 1994. "A Regularization of the Frank-Wolfe Method and Unification of Certain Nonlinear Programming Methods," Math. Programming, Vol. 65, pp. 331-345.

[Min60] Minty, G. J., 1960. "Monotone Networks," Proc. Roy. Soc. London, A, Vol. 257, pp. 194-212.

[Min86] Minoux, M., 1986. Mathematical Programming: Theory and Algorithms, Wiley, N. Y.

[Mit66] Mitter, S. K., 1966. "Successive Approximation Methods for the Solution of Optimal Control Problems," Automatica, Vol. 3, pp. 135-149.

[MoA89a] Monteiro, R. D. C., and Adler, I., 1989. "Interior Path Following Primal-Dual Algorithms, Part I: Linear Programming," Math. Programming, Vol. 44, pp. 27-41.

[MoA89b] Monteiro, R. D. C., and Adler, I., 1989. "Interior Path Following Primal-Dual Algorithms, Part I: Convex Quadratic Programming," Math. Programming, Vol. 44, pp. 43-66.

[MoS83] Moré, J. J., and Sorensen, D. C., 1983. "Computing a Trust Region Step," SIAM J. on Scientific and Statistical Computing, Vol. 4, pp. 553-572.

[MoT89] Moré, J. J., and Toraldo, G., 1989. "Algorithms for Bound Constrained Quadratic Programming Problems," Numer. Math., Vol. 55, pp. 377-400.

[MoW93] Moré, J. J., and Wright, S. J., 1993. Optimization Software Guide, SIAM, Frontiers in Applied Mathematics 14, Phila., PA.

[Mor88] Mordukhovich, B., 1988. Approximation Methods in Problems of Optimization and Control, Nauka, Moscow.

[MuP75] Mukai, H., and Polak, E., 1975. "A Quadratically Convergent Primal-Dual Algorithm With Global Convergence Properties for Solving Optimization Problems With Equality Constraints," Math. Programming, Vol. 9, pp. 336-349.

[MuR92] Mulvey, J. M., and Ruszcynski, A., 1992. "A Diagonal Quadratic Approximation Method for Large Scale Linear Programs," Operations Res. Letters, Vol. 12, pp. 205-215.

[MuR95] Mulvey, J. M., and Ruszcynski, A., 1995. "A New Scenario Decomposition Method for Large Scale Stochastic Optimization," Operations Research, Vol. 43, pp. 477-490.

[Mur92] Murty, K. G., 1992. Network Programming, Prentice-Hall, Englewood Cliffs, N. J.

[NaQ96] Nazareth, J. L., and Qi, L., 1996. "Globalization of Newton's Method for Solving Nonlinear Equations," Numerical Linear Algebra with Applications, Vol. 3, pp. 239-249.

[NaS89] Nash, S. G., and Sofer, 1989. "Block Truncated-Newton Methods for Parallel Optimization," Math. Programming, Vol. 45, pp. 529-546.

[Nag93] Nagurney, A., 1993. Network Economics: A Variational Inequality Approach, Kluwer, Dordrecht, The Netherlands.

[Nas85] Nash, S. G., 1985. "Preconditioning of Truncated-Newton Methods," SIAM J. on Scientific and Statistical Computing, Vol. 6, pp. 599-616.

[Naz94] Nazareth, J. L., 1994. The Newton-Cauchy Framework: A Unified Approach to Unconstrained Nonlinear Minimization, Lecture Notes in Computer Science No. 769, Springer-Verlag, Berlin and New York.

[Naz96] Nazareth, J. L., 1996. "Lagrangian Globalization: Solving Nonlinear Equations via Constrained Optimization," in Mathematics of Numerical Analysis, Renegar, J., Shub, M., and Smale, S., (Eds.), Lectures in Applied Mathematics, Vol. 32, The American Mathematical Society, Providence, RI, pp. 533-542.

[NeM65] Nelder, J. A., and Mead, R., 1965. "A Simplex Method for Function Minimization," Computer J., Vol. 7, pp. 308-313.

[NeN94] Nesterov, Y., and Nemirovskii, A., 1994. Interior Point Polynomial Algorithms in Convex Programming, SIAM, Studies in Applied Mathematics 13, Phila., PA.

[NeW88] Nemhauser, G. L., and Wolsey, L. A., 1988. Integer and Combinatorial Optimization, Wiley, N. Y.

[NeY83] Nemirovsky, A., and Yudin, D. B., 1983. Problem Complexity and Method Efficiency, Wiley, N. Y.

[Nes95] Nesterov, Y., 1995. "Complexity Estimates of Some Cutting Plane Methods Based on Analytic Barrier," Math. Programming, Vol. 69, pp. 149-176.

[NgS79] Nguyen, V. H., and Strodiot, J. J., 1979. "On the Convergence Rate of a Penalty Function Method of Exponential Type," J. Opt. Th. and Appl., Vol. 27, pp. 495-508.

[Noc80] Nocedal, J., 1980. "Updating Quasi-Newton Matrices with Limited Storage," Math. of Computation, Vol. 35, pp. 773-782.

[OLR85] O'hEigeartaigh, M., Lenstra, S. K., and Rinnoy Kan, A. H. G., (Eds.), 1985. Combinatorial Optimization: Annotated Bibliographies, Wiley, N. Y.

[OrL74] Oren, S. S., and Luenberger, D. G., 1974. "Self-Scaling Variable Metric Algorithm, Part I," Management Science, Vol. 20, pp. 845-862.

[OrR70] Ortega, J. M., and Rheinboldt, W. C., 1970. Iterative Solution of Nonlinear Equations in Several Variables, Academic Press, N. Y.

[Ore73] Oren, S. S., 1973. "Self-Scaling Variable Metric Algorithm, Part II," Management Science, Vol. 20, pp. 863-874.

[PCC93] Pulleyblank, W., Cook, W., Cunningham, W., and Schrijver, A., 1993. An Introduction to Combinatorial Optimization, Wiley, N. Y.

[PaM89] Pantoja, J. F. A. D., and Mayne, D. Q., 1989. "Sequential Quadratic Programming Algorithm for Discrete Optimal Control Problems with Control Inequality Constraints," Intern. J. on Control, Vol. 53, pp. 823-836.

[PaR87] Pardalos, P. M., and Rosen, J. B., 1987. Constrained Global Optimization: Algorithms and Applications, Springer-Verlag, N. Y.

[PaS82] Papadimitriou, C. H., and Steiglitz, K., 1982. Combinatorial Optimization: Algorithms and Complexity, Prentice-Hall, Englewood Cliffs, N. J.

[PaT91] Panier, E. R., and Tits, A. L., 1991. "Avoiding the Maratos Effect by Means of a Nonmonotone Line Search. I.," SIAM J. on Numer. Anal., Vol. 28, pp. 1183-1195.

[Pan84] Pang, J.-S., 1984. "On the Convergence of Dual Ascent Methods for Large-Scale Linearly Constrained Optimization Problems," Unpublished Manuscript, School of Management, Univ. of Texas, Dallas, Texas.

[Pap81] Papavassilopoulos, G., 1981. "Algorithms for a Class of Nondifferentiable Problems," J. Opt. Th. and Appl., Vol. 34, pp. 41-82.

[Pap82] Pappas, T. N., 1982. "Solution of Nonlinear Equations by Davidon's Least Squares Method," M.S. Thesis, Dept. of Electrical Engineering and Computer Science, M.I.T., Cambridge, MA.

[Pat93a] Patriksson, M., 1993. "A Unified Framework of Descent Algorithms for Nonlinear Programs and Variational Inequalities," Ph.D. Thesis, Dept. of Math., Linkoping Inst. of Technology, Linkoping, Sweden.

[Pat93b] Patriksson, M., 1993. "Partial Linearization Methods for Nonlinear Programming," J. Opt. Th. and Appl., Vol. 78, pp. 227-246.

[Pat98] Patriksson, M., 1998. Nonlinear Programming and Variational Inequalities: A Unified Approach, Kluwer, Dordtrecht, The Netherlands.

[Per78] Perry, A., 1978. "A Modified Conjugate Gradient Algorithm," Operations Research, Vol. 26, pp. 1073-1078.

[Pfl96] Pflug, G. C., 1996. Optimization of Stochastic Models, Kluwer, Boston.

[PiP73] Pironneau, O., and Polak, E., 1973. "Rate of Convergence of a Class of Methods of Feasible Directions," SIAM J. Numer. Anal., Vol. 10, pp. 161-173.

[PiZ92] Pinar, M. C., and Zenios, S. A., 1992. "Parallel Decomposition of

Multicommodity Network Flows Using a Linear-Quadratic Penalty Algorithm," ORSA J. on Computing, Vol. 4, pp. 235-249.

[PiZ94] Pinar, M. C., and Zenios, S. A., 1994. "On Smoothing Exact Penalty Functions for Convex Constrained Problems," SIAM J. on Optimization, Vol. 4, pp. 486-511.

[PoH91] Polak, E., and He, L., 1991. "Finite-Termination Schemes for Solving Semi-infinite Satisfycing Problems," J. Opt. Theory and Appl., Vol. 70, pp. 429-442.

[PoR69] Polak, E., and Ribiere, G., 1969. "Note sur la Convergence de Methodes de Directions Conjugees." Rev. Fr. Inform. Rech. Oper., Vol. 16-R1, pp. 35-43.

[PoT73a] Poljak, B. T, and Tsypkin, Y. Z., 1973. "Pseudogradient Adaptation and Training Algorithms," Automation and Remote Control, pp. 45-68.

[PoT73b] Poljak, B. T, and Tretjakov, N. V., 1973. "The Method of Penalty Estimates for Conditional Extremum Problems," Z. VyČisl. Mat. i Mat. Fiz., Vol. 13, pp. 34-46.

[PoT74] Poljak, B. T, and Tretjakov, N. V., 1974. "An Iterative Method for Linear Programming and its Economic Interpretation," Matecon, Vol. 10, pp. 81-100.

[PoT80] Polak, E., and Tits, A. L., 1980. "A Globally Convergent, Implementable Multiplier Method with Automatic Penalty Limitation," Applied Math. and Optimization, Vol. 6, pp. 335-360.

[PoT97] Polyak, R., and Teboulle, M., 1997. "Nonlinear Rescaling and Proximal-Like Methods in Convex Optimization," Math. Programming, Vol. 76, pp. 265-284.

[Pol64] Poljak, B. T., 1964. "Some Methods of Speeding up the Convergence of Iteration Methods," Z. VyČisl. Mat. i Mat. Fiz., Vol. 4, pp. 1-17.

[Pol69a] Poljak, B. T., 1969. "The Conjugate Gradient Method in Extremal Problems," Z. Vyčisl. Mat. i Mat. Fiz., Vol. 9, pp. 94-112.

[Pol69b] Poljak, B. T., 1969. "Minimization of Unsmooth Functionals," Z. Vyčisl. Mat. i Mat. Fiz., Vol. 9, pp. 14-29.

[Pol70] Poljak, B. T., 1970. "Iterative Methods Using Lagrange Multipliers for Solving Extremal Problems with Constraints of the Equation Type," Z. VyČisl. Mat. i Mat. Fiz., Vol. 10, pp. 1098-1106.

[Pol71] Polak, E., 1971. Computational Methods in Optimization: A Unified Approach, Academic Press, N. Y.

[Pol73] Polak, E., 1973. "A Historical Survey of Computational Methods in Optimal Control," SIAM Review, Vol. 15, pp. 553-584.

[Pol79] Poljak, B. T., 1979. "On Bertsekas' Method for Minimization of Composite Functions," Internat. Symp. Systems Opt. Analysis, Benoussan, A., and Lions, J. L., (Eds.), pp. 179-186, Springer-Verlag, Berlin and N. Y.

[Pol87] Poljak, B. T., 1987. Introduction to Optimization, Optimization Software Inc., N. Y.

[Pol92] Polyak, R., 1992. "Modified Barrier Functions (Theory and Methods)," Math. Programming, Vol. 54, pp. 177-222.

[Pol97] Polak, E., 1997. Optimization: Algorithms and Consistent Approximations, Springer-Verlag, N. Y.

[Pot94] Potra, F. A., 1994. "A Quadratically Convergent Predictor-Corrector Method for Solving Linear Programs from Infeasible Starting Points," Math. Programming, Vol. 67, pp. 383–406.

[Pow64] Powell, M. J. D., 1964. "An Efficient Method for Finding the Minimum of a Function of Several Variables without Calculating Derivatives," The Computer Journal, Vol. VII, pp. 155-162.

[Pow69] Powell, M. J. D., 1969. "A Method for Nonlinear Constraints in Minimizing Problems," in Optimization, Fletcher, R., (Ed.), Academic Press, N. Y, pp. 283-298.

[Pow73] Powell, M. J. D., 1973. "On Search Directions for Minimization Algorithms," Math. Programming, Vol. 4, pp. 193-201.

[Pre95] Prekopa, A., 1995. Stochastic Programming, Kluwer, Boston.

[PsD75] Pschenichny, B. N., and Danilin, Y. M., 1975. "Numerical Methods in Extremal Problems," MIR, Moscow, (Eng. trans., 1978).

[Psc70] Pschenichny, B. N., 1970. "Algorithms for the General Problem of Mathematical Proramming," Kibernetika (Kiev), Vol. 6, pp. 120-125.

[Pyt98] Pytlak, R., 1998. "An Efficient Algorithm for Large-Scale Nonlinear Programming Problems with Simple Bounds on the Variables," SIAM J. on Optimization, Vol. 8, pp. 532-560.

[Ray93] Raydan, M., 1993. "On the Barzilai and Borwein Choice of Steplength for the Gradient Method," IMA J. Num. Anal., Vol. 13, pp. 321-326.

[ReR98] Reemtsen, R., and Ruckman, J. J., (Eds.), 1998. Semi-Infinite Programming, Kluwer, Boston.

[RoW91] Rockafellar, R. T., and Wets, R. J.-B., 1991. "Scenarios and Policy Aggergation in Optimization under Uncertainty," Math. of Operations Res., Vol. 16, pp. 119-147.

[RoW97] Rockafellar, R. T., and Wets, R. J.-B., 1997. Variational Analysis, Springer-Verlag, Berlin.

[Rob74] Robinson, S. M., 1974. "Perturbed Kuhn-Tucker Points and Rates of Convergence for a Class of Nonlinear Programming Algorithms," Math. Programming, Vol. 7, pp. 1-16.

[Rob87] Robinson, S. M., 1987. "Local Structure of Feasible Sets in Nonlinear Programming, Part III. Stability and Sensitivity," Math. Programming Studies, Vol. 30, pp. 45-66.

[Roc70] Rockafellar, R. T., 1970. Convex Analysis, Princeton Univ. Press, Princeton, N. J.

[Roc73] Rockafellar, R. T., 1973. "A Dual Approach to Solving Nonlinear Programming Problems by Unconstrained Optimization," Math. Programming, pp. 354-373.

[Roc76a] Rockafellar, R. T., 1976. "Monotone Operators and the Proximal Point Algorithm," SIAM J. on Control and Optimization, Vol. 14, pp. 877-898.

[Roc76b] Rockafellar, R. T., 1976. "Augmented Lagrangians and Applications of the Proximal Point Algorithm in Convex Programming," Math. Operations Res., Vol. 1, pp. 97-116.

[Roc76c] Rockafellar, R. T., 1976. "Solving a Nonlinear Programming Problem by Way of a Dual Problem," Symp. Matematica, Vol. 27, pp. 135-160.

[Roc81] Rockafellar, R. T., 1981. "Monotropic Programming: Descent Algorithms and Duality," in Nonlinear Programming 4, by Mangasarian, O. L., Meyer, R. R., and Robinson, S. M., (Eds.), Academic Press, N. Y., pp. 327-366.

[Roc84] Rockafellar, R. T., 1984. Network Flows and Monotropic Optimization, Wiley, N. Y.; republished by Athena Scientific, Belmont, MA, 1998.

[Roc90] Rockafellar, R. T., 1990. "Computational Schemes for Solving Large-Scale Problems in Extended Linear-Quadratic Programming," Math. Programming, Vol. 48, pp. 447-474.

[Roc93] Rockafellar, R. T., 1993. "Lagrange Multipliers and Optimality," SIAM Review, Vol. 35, pp. 183-238.

[Ros60a] Rosenbrock, H. H., 1960. "An Automatic Method for Finding the Greatest or Least Value of a Function," Computer J., Vol. 3, pp. 175-184.

[Ros60b] Rosen, J. B., 1960. "The Gradient Projection Method for Nonlinear Programming, Part I, Linear Constraints," SIAM J. Applied Math., Vol. 8, pp. 514-553.

[Rud76] Rudin, W., 1976. Real Analysis, McGraw-Hill, N. Y.

[Rus86] Ruszczynski, A., 1986. "A Regularized Decomposition Method for Minimizing a Sum of Polyhedral Functions," Math. Programming, Vol. 35, pp. 309-333.

[Rus89] Ruszczynski, A., 1989. "An Augmented Lagrangian Decomposition Method for Block Diagonal Linear Programming Problems," Operations Res. Letters, Vol. 8, pp. 287-294.

[Rus95] Ruszczynski, A., 1995. "On Convergence of an Augmented Lagrangian Decomposition Method for Sparse Convex Optimization," Math. of Operations Res., Vol. 20, pp. 634-656.

[Rus97] Ruszczynski, A., 1997. "Decomposition Methods in Stochastic Programming," Math. Programming, Vol. 79, pp. 333-353.

[SBC93] Saarinen, S., Bramley, R., and Cybenko, G., 1993. "Ill-Conditioning in Neural Network Training Problems," SIAM J. Sci. Comput., Vol. 14, pp. 693-714.

[SHH62] Spendley, W. G., Hext, G. R., and Himsworth, F. R., 1962. "Sequential Application of Simplex Designs in Optimisation and Evolutionary Operation," Technometrics, Vol. 4, pp. 441-461.

[SaS86] Saad, Y., and Schultz, M. H., 1986. "GMRES: A Generalized Minimal Residual Algorithm for Solving Nonsymmetric Linear Systems," SIAM J. Sci. Statist. Comput., Vol. 7, pp. 856-869.

[SBK64] Shah, B., Buehler, R., and Kempthorne, O., 1964. "Some Algorithms for Minimizing a Function of Several Variables," J. Soc. Indust. Appl. Math., Vol. 12, pp. 74-92.

[Sch82] Schnabel, R. B., 1982. "Determining Feasibility of a Set of Nonlinear Inequality Constraints," Math. Programming Studies, Vol. 16, pp. 137-148.

[Sch86] Schrijver, A., 1986. Theory of Linear and Integer Programming, Wiley, N. Y.

[SeS86] Sen, S., and Sherali, H. D., 1986. "A Class of Convergent Primal-Dual Subgradient Algorithms for Decomposable Convex Programs," Math. Programming, Vol. 35, pp. 279-297.

[ShA99] Sherali, H. D., and Adams, W. P., 1999. A Reformulation-Linearization Technique for Solving Discrete and Continuous Nonconvex Problems, Kluwer, Boston.

[Sha70] Shanno, D. F., 1970. "Conditioning of Quasi-Newton Methods for Function Minimization," Math. Comput., Vol. 27, pp. 647-656.

[Sha78] Shanno, D. F., 1978. "Conjugate Gradient Methods with Inexact Line Searches," Math. of Operations Res., Vol. 3, pp. 244-256.

[Sha79] Shapiro, J. E., 1979. Mathematical Programming Structures and Algorithms, Wiley, N. Y.

[Sha88] Shapiro, A., 1988. "Sensitivity Analysis of Nonlinear Programs and Differentiability Properties of Metric Projections," SIAM J. on Control and

Optimization, Vol. 26, pp. 628-645.

[Sch86] Schrijver, A., 1986. Theory of Linear and Integer Programming, Wiley, N. Y.

[Sho85] Shor, N. Z., 1985. Minimization Methods for Nondifferentiable Functions, Springer-Verlag, Berlin.

[Sho98] Shor, N. Z., 1998. Nondifferentiable Optimization and Polynomial Problems, Kluwer, Dordrecht, the Netherlands.

[Sla50] Slater, M., 1950. "Lagrange Multipliers Revisited: A Contribution to Non-Linear Programming," Cowles Commission Discussion paper, Math. 403.

[Sol98] Solodov, M. V., 1998. "Incremental Gradient Algorithms with Stepsizes Bounded Away from Zero," Computational Optimization and Applications, Vol. 11, pp. 23-35.

[Son86] Sonnevend, G., 1986. "An "Analytical Centre" for Polyhedrons and New Classes of Global Algorithms for Linear (Smooth, Convex) Programming," Lecture Notes in Control and Information Sciences, Vol. 84, pp. 866-878.

[Spi85] Spingarn, J. E., 1985. "Applications of the Method of Partial Inverses to Convex Programming: Decomposition," Math. Programming, Vol. 32, pp. 199-223.

[StW70] Stoer, J., and Witzgall, C., 1970. Convexity and Optimization in Finite Dimensions, Springer-Verlag, Berlin.

[StW75] Stephanopoulos, G., and Westerberg, A. W., 1975. "The Use of Hestenes' Method of Multipliers to Resolve Dual Gaps in Engineering System Optimization," J. Opt. Th. and Applications, Vol. 15, pp. 285-309.

[Str76] Strang, G., 1976. Linear Algebra and Its Applications, Academic Press, N. Y.

[TBA86] Tsitsiklis, J. N., Bertsekas, D. P., and Athans, M., 1986. "Distributed Asynchronous Deterministic and Stochastic Gradient Optimization Algorithms," IEEE Trans. on Aut. Control, Vol. AC-31, pp. 803-812.

[TZY95] Tapia, R. A., Zhang, Y., and Ye, Y., 1995. "On the Convergence of the Iteration Sequence in Primal-Dual Interior-Point Methods," Math. Programming, Vol. 68, pp. 141-154.

[TaM85] Tanikawa, A., and Mukai, H., 1985. "A New Technique for Nonconvex Primal-Dual Decomposition," IEEE Trans. on Aut. Control, Vol. AC-30, pp. 133-143.

[Tap77] Tapia, R. A., 1977. "Diagonalized Multiplier Methods and Quasi-Newton Methods for Constrained Minimization," J. Opt. Th. and Applications, Vol. 22, pp. 135-194.

[Teb92] Teboulle, M., 1992. "Entropic Proximal Mappings with Applications to Nonlinear Programming," Math. Operations Res., Vol. 17, pp. 1-21.

[ToT90] Toint, P. L., and Tuyttens, D., 1990. "On Large Scale Nonlinear Network Optimization," Math. Programming, Vol. 48, pp. 125-159.

[ToV67] Topkis, D. M., and Veinott, A. F., 1967. "On the Convergence of Some Feasible Directions Algorithms for Nonlinear Programming," SIAM J. on Control, Vol. 5, pp. 268-279.

[Tor91] Torczon, V., 1991. "On the Convergence of the Multidimensional Search Algorithm," SIAM J. on Optimization, Vol. 1, pp. 123-145.

[TrW80] Traub, J. F., and Wozniakowski, H., 1980. A General Theory of Optimal Algorithms, Academic Press, N. Y.

[TsB86] Tsitsiklis, J. N., and Bertsekas, D. P., 1986. "Distributed Asynchronous Optimal Routing in Data Networks," IEEE Trans. on Automatic Control, Vol. 31, pp. 325-331.

[TsB87] Tseng, P., and Bertsekas, D. P., 1987. "Relaxation Methods for Problems with Strictly Convex Separable Costs and Linear Constraints," Math. Programming, Vol. 38, pp. 303-321.

[TsB90] Tseng, P., and Bertsekas, D. P., 1990. "Relaxation Methods for Monotropic Programs," Math. Programming, Vol. 46, 1990, pp. 127-151.

[TsB91] Tseng, P., and Bertsekas, D. P., 1991. "Relaxation Methods for Problems with Strictly Convex Costs and Linear Constraints," Math. Operations Res., Vol. 16, pp. 462-481.

[TsB93] Tseng, P., and Bertsekas, D. P., 1993. "On the Convergence of the Exponential Multiplier Method for Convex Programming," Math. Programming, Vol. 60, pp. 1-19.

[TsL92] Tseng, P., and Luo, Z.-Q., 1992. "On the Convergence of the Affine-Scaling Algorithm," Math. Programming, Vol. 56, pp. 301-319.

[TsM95] Tsuchiya, T., and Muramatsu, M., 1995. "Global Convergence of a Long-Step Affine Scaling Algorithm for Degenerate Linear Programming Problems." SIAM J. on Optimization, Vol. 5, pp. 525-551.

[Tse89] Tseng, P., 1989. "A Simple Complexity Proof for a Polynomial-Time Linear Programming Algorithm," Operations Res. Letters, Vol. 8, pp. 155-159.

[Tse90] Tseng, P., 1990. "Dual Ascent Methods for Problems with Strictly Convex Costs and Linear Constraints: A Unified Approach," SIAM J. on Control and Optimization, Vol. 28, pp. 214-242.

[Tse91a] Tseng, P., 1991. "On the Rate of Convergence of a Partially Asynchronous Gradient Projection Algorithm," SIAM J. on Optimization, Vol.

4, pp. 603-619.

[Tse91b] Tseng, P., 1991. "Relaxation Method for Large Scale Linear Programming using Decomposition," Math. of Operations Res., Vol. 17, pp. 859-880.

[Tse91c] Tseng, P., 1991. "Decomposition Algorithm for Convex Differentiable Minimization," J. Opt. Theory and Appl., Vol. 70, pp. 109-135.

[Tse92] Tseng, P., 1992. "Complexity Analysis of a Linear Complementarity Algorithm Based on a Lyapunov Function," Math. Programming, Vol. 53, pp. 297-306.

[Tse93] Tseng, P., 1993. "Dual Coordinate Ascent Methods for Non-Strictly Convex Minimization," Math. Programming, Vol. 59, pp. 231-247.

[Tse95a] Tseng, P., 1995. "Fortified-Descent Simplicial Search Method," Report, Dept. of Math., University of Washington, Seattle, Wash.; to appear in SIAM J. on Optimization.

[Tse95b] Tseng, P., 1995. "Simplified Analysis of an $O(nL)$-Iteration Infeasible Predictor-Corrector Path Following Method for Monotone LCP,"in Recent Trends in Optimization Theory and Applications, Agarwal, R. P., (Ed.), World Scientific, pp. 423-434.

[Tse98] Tseng, P., 1998. "Incremental Gradient(-Projection) Method with Momentum Term and Adaptive Stepsize Rule," SIAM J. on Optimization, Vol. 8, pp. 506-531.

[Tsu91] Tsuchiya, T., 1991. "Global Convergence of the Affine-Scaling Methods for Degenerate Linear Programming Problems," Math. Programming, Vol. 52, pp. 377-404.

[VMF86] Vanderbei, R. J., Meketon, M. S., and Freedman, B. A., 1986. "A Modification of Karmarkar's Linear Programming Algorithm," Algorithmica, Vol. 1, pp. 395-407.

[VaB95] Vandenberghe, L., and Boyd, S., 1995. "A Primal-Dual Potential Reduction Method for Problems Involving Matrix Inequalities," Math. Programming, Vol. 69, pp. 205-236.

[VeH93] Ventura, J. A., and Hearn, D. W., 1993. "Restricted Simplicial Decomposition for Convex Constrained Problems," Math. Programming, Vol. 59, pp. 71-85.

[WQB98] Wei, Z., Qi, L., and Birge, J. R., 1998. "New Method for Nonsmooth Convex Optimization," J. of Inequalities and Applications, Vol. 2, pp. 157-179.

[Web29] Weber, A., 1929. Theory of Location of Industries, (Engl. Transl. by C. J. Friedrich), Univ. of Chicago Press, Chicago, Ill.

[Wet90] Wets, R. J., 1990. "Elementary Constructive Proofs of the Theorems of Farkas, Minkowski and Weyl," in Economic Decision Making: Games, Econometrics and Optimization. Contributions in Honour of Jacques Dreze, Gabszewicz, J. J., Richard, J.-F., and Wolsey, L. A., (Eds.), North-Holland, Elsevier-Science, Amsterdam, pp. 427-432.

[WiH60] Widrow, B., and Hoff, M. E., 1960. "Adaptive Switching Circuits," Institute of Radio Engineers, Western Electronic Show and Convention, Convention Record, part 4, pp. 96-104.

[Wil63] Wilson, R. B., 1963. "A Simplicial Algorithm for Concave Programming," Ph.D. Thesis, Grad. Sch. Business Admin., Harvard Univ., Cambridge, MA.

[Wol75] Wolfe, P., 1975. "A Method of Conjugate Subgradients for Minimizing Nondifferentiable Functions," Math. Programming Study 3, Balinski, M., and Wolfe, P., (Eds.), North-Holland, Amsterdam, pp. 145-173.

[Wol98] Wolsey, L. A., 1998. Integer Programming, Wiley, N. Y.

[Wri92] Wright, S. J., 1992. "An Interior Point Algorithm for Linearly Constrained Optimization," SIAM J. on Optimization, Vol. 2, pp. 450-473.

[Wri93a] Wright, S. J., 1993. "Identifiable Surfaces in Constrained Optimization," SIAM J. on Control and Optimization, Vol. 31, pp. 1063-1079.

[Wri93b] Wright, S. J., 1993. "Interior Point Methods for Optimal Control of Discrete Time Systems," J. Opt. Theory and Appl., Vol. 77, pp. 161-187.

[Wri93c] Wright, S. J., 1993. 'A Path-Following Infeasible-Interior-Point Algorithm for Linear Complementarity Problems," Optimization Methods and Software, Vol. 2, pp. 79-106.

[Wri94] Wright, S. J., 1994. "An Infeasible-Interior-Point Algorithm for Linear Complementarity Problems," Math. Programming, Vol. 67, pp. 29-52.

[Wri96] Wright, S. J., 1996. "A Path-Following Interior-Point Algorithm for Linear and Quadratic Problems," Annals of Operations Res., Vol. 62, pp. 103-130.

[Wri97] Wright, S. J., 1997. Primal-Dual Interior Point Methods, SIAM, Phila., PA.

[Wri98] Wright, S. J., 1998. "Superlinear Convergence of a Stabilized SQP Method to a Degenerate Solution," Computational Optimization and Applications, Vol. 11, pp. 253-275.

[WuB99] Wu, C., and Bertsekas, D. P., 1999. "Distributed Power Control Algorithms for Wireless Networks," to appear in Proc. of 1999 IEEE Conf. on Decision and Control.

[Ye92] Ye, Y., 1992. "A Potential Reduction Algorithm Allowing Column Generation," SIAM J. on Optimization, Vol. 2, pp. 7-20.

[Ye97] Ye, Y., 1997. Interior Point Algorithms: Theory and Analysis, Wiley Interscience, N. Y.

[Yok92] Yokohama, K., 1992. "ϵ-Optimality Criteria for Convex Programming Problems via Exact Penalty Functions," Math. Programming, Vol. 56, pp. 233-243.

[ZLW99] Zhao, X., Luh, P. B., Wang, J., 1999. "Surrogate Gradient Algorithm for Lagrangian Relaxation," J. Opt. Theory and Appl., Vol. 100, pp. 699-712.

[ZTP93] Zhang, Y., Tapia, R. A., and Potra, F., 1993. "On the Superlinear Convergence of Interior-Point Algorithms for a General Class of Problems," SIAM J. on Optimization, Vol. 3 pp. 413-422.

[Zan67a] Zangwill, W. I., 1967. "Minimizing a Function Without Calculating Derivatives," The Computer Journal, Vol. X, pp. 293-296.

[Zan67b] Zangwill, W. I., 1967. "Nonlinear Programming Via Penalty Functions," Management Science, Vol. 13, pp. 344-358.

[Zan69] Zangwill, W. I., 1969. Nonlinear Programming, Prentice-Hall, Englewood Cliffs, N. J.

[ZhT92] Zhang, Y., and Tapia, R. A., 1992. "Superlinear and Quadratic Convergence of Primal-Dual Interior-Point Algorithms for Linear Programming Revisited," J. Opt. Theory and Appl., Vol. 73, pp. 229-242.

[ZhT93] Zhang, Y., and Tapia, R. A., 1993. "A Superlinearly Convergent Polynomial Primal-Dual Interior-Point Algorithm for Linear Programming," SIAM J. on Optimization, Vol. 3, pp. 118-133.

[Zhu95] Zhu, C., 1995. "On the Primal-Dual Steepest Descent Algorithm for Extended Linear-Quadratic Programming," SIAM J. on Optimization, Vol. 5, pp. 114-128.

[Zou60] Zoutendijk, G., 1960. Methods of Feasible Directions, Elsevier Publ. Co., Amsterdam.

[Zou76] Zoutendijk, G., 1976. Mathematical Programming Methods, North Holland, Amsterdam.

INDEX

H

I

J

K

L

M

ATHENA SCIENTIFIC

OPTIMIZATION AND COMPUTATION SERIES

1. Dynamic Programming and Optimal Control, Vols. I and II, by Dimitri P. Bertsekas, 1995, ISBN 1-886529-11-6, 704 pages
2. Nonlinear Programming, Second Edition, by Dimitri P. Bertsekas, 1999, ISBN 1-886529-00-0, 791 pages
3. Neuro-Dynamic Programming, by Dimitri P. Bertsekas and John N. Tsitsiklis, 1996, ISBN 1-886529-10-8, 512 pages
4. Constrained Optimization and Lagrange Multiplier Methods, by Dimitri P. Bertsekas, 1996, ISBN 1-886529-04-3, 410 pages
5. Stochastic Optimal Control: The Discrete-Time Case by Dimitri P. Bertsekas and Steven E. Shreve, 1996, ISBN 1-886529-03-5, 330 pages
6. Introduction to Linear Optimization by Dimitris Bertsimas and John N. Tsitsiklis, 1997, ISBN 1-886529-19-1, 608 pages
7. Parallel and Distributed Computation: Numerical Methods by Dimitri P. Bertsekas and John N. Tsitsiklis, 1997, ISBN 1-886529-01-9, 718 pages
8. Network Flows and Monotropic Optimization by R. Tyrrell Rockafellar, 1998, ISBN 1-886529-06-X, 634 pages
9. Network Optimization: Continuous and Discrete Models by Dimitri P. Bertsekas, 1998, ISBN 1-886529-02-7, 608 pages

Network Optimization: Continuous and Discrete Models

Dimitri P. Bertsekas
Massachusetts Institute of Technology

A comprehensive, insightful, and up-to-date treatment of one of the most elegant and applications-rich classes of optimization problems.

Linear, nonlinear, and discrete network optimization problems are discussed extensively, including their theory and algorithms, and their applications in fields such as communication, transportation, manufacturing, logistics, and production planning. A unified approach bridges the gap between continuous (linear/nonlinear) problems, and discrete (integer/combinatorial) problems.

From the review by Panos Pardalos (J. of Opt. and Software, 1998):

"This beautifully written book provides an introductory treatment of linear, nonlinear, and discrete network optimization problems... The textbook is addressed not only to students of optimization but to all scientists in numerous disciplines who need network optimization methods to model and solve problems. This book is an engaging read and it is highly recommended either as a textbook or as a reference on network optimization."

Among its special features, the book:

- provides a comprehensive account of the theory and the practical application of the principal algorithms for linear and nonlinear network flow problems, including simplex, dual ascent, and auction
- describes the main models for discrete network optimization, such as traveling salesman, constrained shortest path, vehicle routing, multidimensional assignment, facility location, network design, etc
- describes the main methods for integer-constrained network problems, such as branch-and-bound, Lagrangian relaxation, genetic algorithms, tabu search, simulated annealing, and rollout algorithms
- discusses extensively auction algorithms, based on the author's extensive research on the subject
- contains many examples, practical applications, and exercises

Contents: 1. Introduction. 2. Shortest Path Problems. 3. The Max-Flow Problem. 4. The Min-Cost Flow Problem. 5. Simplex Methods for Min-Cost Flow. 6. Dual Ascent Methods for Min-Cost Flow. 7.Auction Algorithms for Min-Cost Flow. 8. Nonlinear Network Optimization. 9. Convex Separable Network Problems. 10. Network Problems with Integer Constraints. Appendix: Mathematical Background.

ISBN: 1-886529-02-7, 608 pp., hardcover, 1998

Introduction to Linear Optimization

Dimitris Bertsimas and John N. Tsitsiklis
Massachusetts Institute of Technology

This book provides a unified, insightful, and modern treatment of linear optimization, that is, linear programming, network flow problems, and discrete optimization. It includes classical topics as well as the state of the art, in both theory and practice.

From the review by T. V. Marana (Optima, June 1997):

"The true merit of the book, however, lies in its pedagogical qualities which are so impressive that I have decided to adopt it for a course in linear programming ..."

"... the overall writing style is pleasant and to-the-point."

"One reading of this book is sufficient to appreciate the tremendous amount of the quality effort that the authors have put into the writing, and I strongly recommend it ..."

Among its special features, the book:

- develops the major algorithms and duality theory through a geometric perspective
- provides a thorough treatment of the geometry, convergence, and complexity of interior point methods
- covers the main methods for network flow problems
- contains a detailed treatment of integer programming formulations and algorithms
- discusses the art of formulating and solving large scale problems through practical case studies
- includes a large number of examples and exercises. Has been developed through extensive classroom use in graduate courses.

Contents: 1. Introduction. 2. The geometry of linear programming. 3. The simplex method. 4. Duality theory. 5. Sensitivity analysis. 6. Large scale optimization. 7. Network flow problems. 8. Complexity of linear programming and the ellipsoid method. 9. Interior point methods. 10. Integer programming formulations. 11. Integer programming methods. 12. The art in linear optimization.

ISBN 1-886529-19-1, 608 pp., hardcover, 1997

Dynamic Programming and Optimal Control

Dimitri P. Bertsekas
Massachusetts Institute of Technology

This two-volume textbook develops in depth dynamic programming, a central algorithmic method for optimal control, sequential decision making under uncertainty, and combinatorial optimization.

The first volume is oriented towards modeling, conceptualization, and finite-horizon problems, but also includes a substantive introduction to infinite horizon problems that is suitable for classroom use. The second volume is oriented towards mathematical analysis and computation, and treats infinite horizon problems extensively. The text contains many illustrations, worked-out examples, and exercises.

Reviews:

"Here is a tour-de-force in its field." David K. Smith, in the Journal of Operational Research Society. "In conclusion, this book is an excellent source of reference ... The main strengths of the book are the clarity of the exposition, the quality and variety of the examples, and its coverage of the most recent advances." Thomas W. Archibald, in IMA Jnl. of Math. Appl. in Business and Industry.

Among its special features, the book:

- provides a unifying framework for sequential decision making
- treats simultaneously deterministic and stochastic control problems popular in control theory and operations research
- develops the theory of deterministic optimal control problems including the Pontryagin Minimum Principle
- describes the recent simulation-based approximation techniques of neuro-dynamic programming/reinforcement learning
- provides a comprehensive treatment of infinite horizon problems

Contents of Volume 1: 1. The Dynamic Programming Algorithm. 2. Deterministic Systems and the Shortest Path Problem. 3. Deterministic Continuous-Time Optimal Control. 4. Problems with Perfect State Information. 5. Problems with Imperfect State Information. 6. Suboptimal and Adaptive Control. 7. Introduction to Infinite Horizon Problems.

Contents of Volume 2: 1. Infinite Horizon - Discounted Problems. 2. Stochastic Shortest Path Problems. 3. Undiscounted Problems. 4. Average Cost per Stage Problems. 5. Continuous-Time Problems.

ISBN: 1-886529-11-6, 704 pp., hardcover, 1995

Neuro-Dynamic Programming

Dimitri P. Bertsekas and John N. Tsitsiklis
Massachusetts Institute of Technology

This is the first textbook that fully explains the neuro-dynamic programming/reinforcement learning methodology, which is a recent breakthrough in the practical application of neural networks and dynamic programming to complex problems of planning, optimal decision making, and intelligent control.

From the review by George Cybenko for IEEE Computational Science and Engineering, May 1998:

" Neurodynamic Programming is a remarkable monograph that integrates a sweeping mathematical and computational landscape into a coherent body of rigorous knowledge. The topics are current, the writing is clear and to the point, the examples are comprehensive and the historical notes and comments are scholarly."

"In this monograph, Bertsekas and Tsitsiklis have performed a Herculean task that will be studied and appreciated by generations to come. I strongly recommend it to scientists and engineers eager to seriously understand the mathematics and computations behind modern behavioral machine learning."

Among its special features, the book:

- Describes and unifies a large number of NDP methods, including several that are new
- Describes new approaches to formulation and solution of important problems in stochastic optimal control, sequential decision making, and discrete optimization
- Rigorously explains the mathematical principles behind NDP
- Illustrates through examples and case studies the practical application of NDP to complex problems from optimal resource allocation, optimal feedback control, data communications, game playing, and combinatorial optimization
- Presents extensive background and new research material on dynamic programming and neural network training

Neuro-Dynamic Programming is the winner of the 1997 INFORMS CSTS prize for research excellence in the interface between Operations Research and Computer Science

ISBN 1-886529-10-8, 512 pp., hardcover, 1996

Parallel and Distributed Computation: Numerical Methods

Dimitri P. Bertsekas and John N. Tsitsiklis
Massachusetts Institute of Technology

This highly acclaimed work, first published by Prentice Hall in 1989, is a comprehensive and theoretically sound treatment of parallel and distributed numerical methods. It focuses on algorithms that are naturally suited for massive parallelization, and it explores the fundamental convergence, rate of convergence, communication, and synchronization issues associated with such algorithms. This is an extensive book, which aside from its focus on parallel and distributed algorithms, contains a wealth of material on a broad variety of computation and optimization topics.

Reviews:

"This major contribution to the literature belongs on the bookshelf of every scientist with an interest in computational science, directly beside Knuth's three volumes and Numerical Recipes..." Anna Nagurney, University of Massachusetts, in the Intern. J. of Supercomputer Applications

"This major work of exceptional scholarship summarizes more than three decades of research into general-purpose algorithms for solving systems of equations and optimization problems." W. Smyth, in Comp. Rev.

Among its special features, the book:

- quantifies the performance of parallel algorithms, including the limitations imposed by the communication and synchronization penalties
- provides a comprehensive convergence analysis of asynchronous methods and a comparison with their asynchronous counterparts
- covers extensively parallel and distributed algorithms for systems of equations, variational inequalities, nonlinear programming, shortest paths, dynamic programming, network flows, and large-scale decomposition
- includes extensive research material on optimization methods, asynchronous algorithm convergence, rollback synchronization, asynchronous communication network protocols, and others
- contains many exercises with solutions posted on the internet
- contains much in depth research not found in any other textbook

ISBN 1-886529-01-9, 718 pp., softcover, 1997

Network Flows and Monotropic Optimization

R. Tyrrell Rockafellar
University of Washington

A rigorous and comprehensive treatment of network flow theory and monotropic optimization by one of the world's most renowned applied mathematicians.

This classic textbook, first published by J. Wiley in 1984, covers extensively the duality theory and the algorithms of linear and nonlinear network optimization optimization, and their significant extensions to monotropic programming (separable convex constrained optimization problems, including linear programs).

Monotropic programming problems are characterized by a rich interplay between combinatorial structure and convexity properties. Rockafellar develops, for the first time, algorithms and a remarkably complete duality theory for these problems.

From the review by Dimitri P. Bertsekas in SIAM Review:

"By creating an elegant unifying framework for a broad range of subjects, Rockafellar's book represents an important event in the evolution of optimization theory. Besides its creative aspect, the book is thoughtful, well written, and packed with a wealth of material. The large number of exercises (a total of 479!), most of them extensions and elaborations of the theory, enhance its value as a class textbook, and provide fertile grounds for self-study and inspiration. Every student and practitioner of optimization should take a careful look at this book."

Among its special features, the book:

- treats in-depth the duality theory for linear and nonlinear network optimization
- uses a rigorous step-by-step approach to develop the principal network optimization algorithms
- covers the main algorithms for specialized network problems, such as max-flow, feasibility, assignment, and shortest path
- develops in detail the theory of monotropic programming, based on the author's highly acclaimed research
- contains many examples, illustrations, and exercises
- contains much new material not found in any other textbook

ISBN 1-886529-06-X, 634 pp., hardcover, 1998

Constrained Optimization and Lagrange Multiplier Methods

Dimitri P. Bertsekas
Massachusetts Institute of Technology

This reference textbook, first published in 1982 by Academic Press, remains the authoritative and comprehensive treatment of some of the most widely used constrained optimization methods, including the augmented Lagrangian/multiplier and sequential quadratic programming methods. It is an excellent supplement to the present Nonlinear Programming book.

From the review by S. Zlobec in SIAM Review:

"This is an excellent reference book. The author has done a great job in at least three directions. First, he expertly, systematically and with ever-present authority guides the reader through complicated areas of numerical optimization. This is achieved by carefully explaining and illustrating (by figures, if necessary) the underlying principles and theory. Second, he provides extensive guidance on the merits of various types of methods. This is expremely useful to practitioners. Finally, this is truly a state of the art book on numerical optimization."

Among its special features, the book:

- treats extensively augmented Lagrangian methods, including an exhaustive analysis of the associated convergence and rate of convergence properties
- develops comprehensively sequential quadratic programming and other Lagrangian methods
- provides a detailed analysis of differentiable and nondifferentiable exact penalty methods
- presents nondifferentiable and minimax optimization methods based on smoothing
- contains much in depth research not found in any other textbook

Contents: 1. Introduction. 2. The method of multipliers for equality constrained problems. 3. The method of multipliers for inequality constrained problems and nondifferentiable optimization. 4. Exact penalty methods and Lagrangian methods. 5. Nonquadratic penalty functions – convex programming .

ISBN 1-886529-04-3, 410 pp., softcover, 1996